U.S. Customary Units and Their SI Equivalents

Quantity	U.S. Customary Unit	SI Equivalent
Acceleration	ft/s^2	0.3048 m/s^2
	in./s^2	0.0254 m/s^2
Area	ft^2	0.0929 m^2
	in^2	645.2 mm^2
Energy	ft · lb	1.356 J
Force	kip	4.448 kN
	lb	4.448 N
	oz	0.2780 N
Impulse	lb · s	4.448 N · s
Length	ft	0.3048 m
	in.	25.40 mm
	mi	1.609 km
Mass	oz mass	28.35 g
	lb mass	0.4536 kg
	slug	14.59 kg
	ton	907.2 kg
Moment of a force	lb · ft	1.356 N · m
	lb · in.	0.1130 N · m
Moment of inertia		
Of an area	in^4	0.4162×10^6 mm^4
Of a mass	lb · ft · s^2	1.356 kg · m^2
Momentum	lb · s	4.448 kg · m/s
Power	ft · lb/s	1.356 W
	hp	745.7 W
Pressure or stress	lb/ft^2	47.88 Pa
	lb/in^2 (psi)	6.895 kPa
Velocity	ft/s	0.3048 m/s
	in./s	0.0254 m/s
	mi/h (mph)	0.4470 m/s
	mi/h (mph)	1.609 km/h
Volume, solids	ft^3	0.02832 m^3
	in^3	16.39 cm^3
Liquids	gal	3.785 l
	qt	0.9464 l
Work	ft · lb	1.356 J

Vector Mechanics for Engineers DYNAMICS

Third Edition

Ferdinand P. Beer

Professor and Chairman
Department of Mechanical Engineering and Mechanics
Lehigh University

E. Russell Johnston, Jr.

Professor and Head
Department of Civil Engineering
University of Connecticut

McGraw-Hill Book Company

New York
St. Louis
San Francisco
Auckland
Bogotá
Düsseldorf
Johannesburg
London
Madrid
Mexico
Montreal
New Delhi
Panama
Paris
São Paulo
Singapore
Sydney
Tokyo
Toronto

Vector Mechanics for Engineers
DYNAMICS

0 DODO 8 3 2 1

This book was set in Laurel by York Graphic Services, Inc. The editors were B. J. Clark and J. W. Maisel; the designer was Merrill Haber; the production supervisor was Thomas J. LoPinto. The drawings were done by Felix Cooper.
R. R. Donnelley & Sons Company was printer and binder.

Library of Congress Cataloging in Publication Data

Beer, Ferdinand Pierre, date
Vector mechanics for engineers.

Includes index.
CONTENTS: [1] Statics.—[2] Dynamics.
1. Mechanics, Applied. 2. Vector analysis.
3. Mechanics, Applied—Problems, exercises, etc.
I. Johnston, Elwood Russell, date joint author.
II. Title.
TA350.B3552 1977 531'.01'51563 76-54914
ISBN 0-07-004281-0

Contents

17
PLANE MOTION OF RIGID BODIES: ENERGY AND MOMENTUM METHODS 779

18
KINETICS OF RIGID BODIES IN THREE DIMENSIONS 828

19
MECHANICAL VIBRATIONS 875

Preface

The main objective of a first course in mechanics should be to develop in the engineering student the ability to analyze any problem in a simple and logical manner and to apply to its solution a few, well-understood, basic principles. It is hoped that this text, as well as the preceding volume, *Vector Mechanics for Engineers: Statics*, will help the instructor achieve this goal.†

Vector algebra was introduced at the beginning of the first volume and used in the presentation of the basic principles of statics, as well as in the solution of many problems, particularly three-dimensional problems. Similarly, the concept of vector differentiation will be introduced early in this volume, and vector analysis will be used throughout the presentation of dynamics. This approach results in a more concise derivation of the fundamental principles. It also makes it possible to analyze many problems in kinematics and kinetics which could not be solved by the standard scalar methods. The emphasis in this text, however, remains on the correct understanding of the principles of mechanics and on their application to the solution of engineering problems, and vector analysis is presented chiefly as a convenient tool.‡

One of the characteristics of the approach used in these volumes is that the mechanics of *particles* has been clearly separated from the mechanics of *rigid bodies*. This approach makes it possible to consider simple practical applications at an early stage and to postpone the introduction of more difficult concepts. In the volume on statics, the statics of particles was treated first, and the principle of equilibrium was immediately applied to practical situations involving only concurrent forces. The statics of rigid bodies was considered later, at which time the vector and scalar products of two vectors were introduced and used to define the moment of a force about a point and about an axis. In this volume, the same division is observed. The basic

† Both texts are also available in a single volume, *Vector Mechanics for Engineers: Statics and Dynamics*, third edition.

‡ In a parallel text, *Mechanics for Engineers: Dynamics*, third edition, the use of vector algebra is limited to the addition and subtraction of vectors, and vector differentiation is omitted.

concepts of force, mass, and acceleration, of work and energy, and of impulse and momentum are introduced and first applied to problems involving only particles. Thus the student may familiarize himself with the three basic methods used in dynamics and learn their respective advantages before facing the difficulties associated with the motion of rigid bodies.

Since this text is designed for a first course in dynamics, new concepts have been presented in simple terms and every step explained in detail. On the other hand, by discussing the broader aspects of the problems considered and by stressing methods of general applicability, a definite maturity of approach has been achieved. For example, the concept of potential energy is discussed in the general case of a conservative force. Also, the study of the plane motion of rigid bodies had been designed to lead naturally to the study of their general motion in space. This is true in kinematics as well as in kinetics, where the principle of equivalence of external and effective forces is applied directly to the analysis of plane motion, thus facilitating the transition to the study of three-dimensional motion.

The fact that mechanics is essentially a *deductive* science based on a few fundamental principles has been stressed. Derivations have been presented in their logical sequence and with all the rigor warranted at this level. However, the learning process being largely *inductive*, simple applications have been considered first. Thus the dynamics of particles precedes the dynamics of rigid bodies; and, in the latter, the fundamental principles of kinetics are first applied to the solution of two-dimensional problems, which can be more easily visualized by the student (Chaps. 16 and 17), while three-dimensional problems are postponed until Chap. 18.

The third edition of *Vector Mechanics for Engineers* retains the unified presentation of the principles of kinetics which characterized the second edition. The concepts of linear and angular momentum are introduced in Chap. 12 so that Newton's second law of motion may be presented, not only in its conventional form $\mathbf{F} = m\mathbf{a}$, but also as a law relating, respectively, the sum of the forces acting on a particle and the sum of their moments to the rates of change of the linear and angular momentum of the particle. This makes possible an earlier introduction of the principle of conservation of angular momentum and a more meaningful discussion of the motion of a particle under a central force (Sec. 12.8). More importantly, this approach may be readily extended to the study of the motion of a system of particles (Chap. 14) and leads to a more concise and unified treatment of the kinetics of rigid bodies in two and three dimensions (Chaps. 16 through 18).

Free-body diagrams were introduced early in statics. They were used not only to solve equilibrium problems but also to express the equivalence of two systems of forces or, more generally, of two systems of vectors. The advantage of this approach becomes apparent in the study of the dynamics of rigid bodies, where it is used to solve three-dimensional as well as two-dimensional problems. By placing the emphasis on "free-body-diagram equations" rather than on the standard algebraic equations of motion, a more intuitive and more complete understanding of the fundamental principles of dynamics may be achieved. This approach, which was first introduced in 1962 in the first edition of *Vector Mechanics for Engineers*, has now gained wide acceptance among mechanics teachers in this country. It is, therefore, used in preference to the method of dynamic equilibrium and to the equations of motion in the solution of all sample problems in this new edition.

Color has again been used in this edition to distinguish forces from other elements of the free-body diagrams. This makes it easier for the students to identify the forces acting on a given particle or rigid body and to follow the discussion of sample problems and other examples given in the text.

Because of the current trend among American engineers to adopt the international system of units (SI metric units), the SI units most frequently used in mechanics were introduced in Chap. 1 of *Statics*. They are discussed again in Chap. 12 of this volume and used throughout the text. Half the sample problems and problems to be assigned have been stated in these units, while the other half retain U.S. customary units. The authors believe that this approach will best serve the needs of the students, who will be entering the engineering profession during the period of transition from one system of units to the other. It also should be recognized that the passage from one system to the other entails more than the use of conversion factors. Since the SI system of units is an absolute system based on the units of time, length, and mass, whereas the U.S. customary system is a gravitational system based on the units of time, length, and force, different approaches are required for the solution of many problems. For example, when SI units are used, a body is generally specified by its mass expressed in kilograms; in most problems of statics it was necessary to determine the weight of the body in newtons, and an additional calculation was required for this purpose. On the other hand, when U.S. customary units are used, a body is specified by its weight in pounds and, in dynamics problems, an additional calculation will be required to determine its mass in slugs (or $\text{lb} \cdot \text{sec}^2/\text{ft}$). The authors, therefore, believe that problems assignments should include both types of units. A sufficient number of problems, however, have been provided so that, if so desired, two complete sets of assignments may be selected from problems stated in SI units only and two others from problems stated in U.S. customary units. Since the answers to all even-numbered problems stated in U.S. customary units have been given in both systems of units, teachers who wish to give special instruction to their students in the conversion of units may assign these problems and ask their students to use SI units in their solutions. This has been illustrated in two sample problems involving, respectively, the kinetics of particles (Sample Prob. 12.2) and the computation of mass moments of inertia (Sample Prob. 9.13 in Appendix B).

A number of optional sections have been included. These sections are indicated by asterisks and may thus easily be distinguished from those which form the core of the basic dynamics course. They may be omitted without prejudice to the understanding of the rest of the text. The topics covered in these additional sections include graphical methods for the solution of rectilinear-motion problems, the trajectory of a particle under a central force, the deflection of fluid streams, problems involving jet and rocket propulsion, the kinematics and kinetics of rigid bodies in three dimensions, damped mechanical vibrations, and electrical analogues. These topics will be found of particular interest when dynamics is taught in the junior year.

The material presented in this volume and most of the problems require no previous mathematical knowledge beyond algebra, trigonometry, elementary calculus, and the elements of vector algebra presented in Chaps. 2 and 3 of the volume on statics.† However, special problems have been included, which make use of a more advanced knowledge of calculus, and certain sections, such as Secs. 19.8 and 19.9 on damped vibrations, should be assigned only if the students possess the proper mathematical background.

The text has been divided into units, each consisting of one or several theory sections, one or several sample problems, and a large number of problems to be assigned. Each unit corresponds to a well-defined topic and generally may be covered in one lesson. In a number of cases, however, the instructor will find it desirable to devote more than one lesson to a given topic. The sample problems have been set up in much the same form that a student will use in solving the assigned problems. They thus serve the double purpose of amplifying the text and demonstrat-

†Some useful definitions and properties of vector algebra have been summarized in Appendix A at the end of this volume for the convenience of the reader. Also, Secs. 9.10 through 9.16 of the volume on statics, which deal with the moments of inertia of masses, have been reproduced in Appendix B.

ing the type of neat and orderly work that the student should cultivate in his own solutions. Most of the problems to be assigned are of a practical nature and should appeal to the engineering student. They are primarily designed, however, to illustrate the material presented in the text and to help the student understand the basic principles of mechanics. The problems have been grouped according to the portions of material they illustrate and have been arranged in order of increasing difficulty. Problems requiring special attention have been indicated by asterisks. Answers to all even-numbered problems are given at the end of the book.

The authors wish to acknowledge gratefully the many helpful comments and suggestions offered by the users of the previous editions of *Mechanics for Engineers* and of *Vector Mechanics for Engineers*.

FERDINAND P. BEER
E. RUSSELL JOHNSTON, JR.

List of Symbols

$\mathbf{a}, a$	Acceleration
a	Constant; radius; distance; semimajor axis of ellipse
$\bar{\mathbf{a}}, \bar{a}$	Acceleration of mass center
$\mathbf{a}_{B/A}$	Acceleration of B relative to frame in translation with A
$\mathbf{a}_c$	Coriolis acceleration
$\mathbf{A}, \mathbf{B}, \mathbf{C}, \ldots$	Reactions at supports and connections
$A, B, C, \ldots$	Points
A	Area
b	Width; distance; semiminor axis of ellipse
c	Constant; coefficient of viscous damping
C	Centroid; instantaneous center of rotation; capacitance
d	Distance
e	Coefficient of restitution; base of natural logarithms
E	Total mechanical energy; voltage
f	Frequency; scalar function
$\mathbf{F}$	Force; friction force
g	Acceleration of gravity
G	Center of gravity; mass center; constant of gravitation
h	Angular momentum per unit mass
$\mathbf{H}_O$	Angular momentum about point O
$\dot{\mathbf{H}}_G$	Rate of change of angular momentum $\mathbf{H}_G$ with respect to frame of fixed orientation
$(\dot{\mathbf{H}}_G)_{Gxyz}$	Rate of change of angular momentum $\mathbf{H}_G$ with respect to rotating frame $Gxyz$
$\mathbf{i}, \mathbf{j}, \mathbf{k}$	Unit vectors along coordinate axes
$\mathbf{i}_n, \mathbf{i}_t$	Unit vectors along normal and tangent
$\mathbf{i}_r, \mathbf{i}_\theta$	Unit vectors in radial and transverse directions
i	Current
$I, I_x, \ldots$	Moment of inertia
$\bar{I}$	Centroidal moment of inertia
J	Polar moment of inertia
k	Spring constant
k_x, k_y, k_o	Radius of gyration

$\bar{k}$	Centroidal radius of gyration
l	Length
$\mathbf{L}$	Linear momentum
L	Length; inductance
m	Mass; mass per unit length
$\mathbf{M}$	Couple; moment
$\mathbf{M}_O$	Moment about point O
$\mathbf{M}_O^R$	Moment resultant about point O
M	Magnitude of couple or moment; mass of earth
M_{OL}	Moment about axis OL
n	Normal direction
$\mathbf{N}$	Normal component of reaction
O	Origin of coordinates
p	Circular frequency
$\mathbf{P}$	Force; vector
$\dot{\mathbf{P}}$	Rate of change of vector **P** with respect to frame of fixed orientation
$P_{xy}, \ldots$	Product of inertia
q	Mass rate of flow; electric charge
$\mathbf{Q}$	Force; vector
$\dot{\mathbf{Q}}$	Rate of change of vector **Q** with respect to frame of fixed orientation
$(\dot{\mathbf{Q}})_{Oxyz}$	Rate of change of vector **Q** with respect to frame $Oxyz$
$\mathbf{r}$	Position vector
r	Radius; distance; polar coordinate
$\mathbf{R}$	Resultant force; resultant vector; reaction
R	Radius of earth; resistance
$\mathbf{s}$	Position vector
s	Length of arc
t	Time; thickness; tangential direction
$\mathbf{T}$	Force
T	Tension; kinetic energy
$\mathbf{u}$	Velocity
u	Rectangular coordinate; variable
U	Work
$\mathbf{v}, v$	Velocity
v	Speed; rectangular coordinate
$\bar{\mathbf{v}}, \bar{v}$	Velocity of mass center
$\mathbf{v}_{B/A}$	Velocity of B relative to frame in translation with A.
$\mathbf{V}$	Vector product
V	Volume; potential energy
w	Load per unit length
$\mathbf{W}, W$	Weight; load
x, y, z	Rectangular coordinates; distances

$\dot{x}, \dot{y}, \dot{z}$	Time derivatives of coordinates x, y, z
$\bar{x}, \bar{y}, \bar{z}$	Rectangular coordinates of centroid, center of gravity, or mass center
$\boldsymbol{\alpha}, \alpha$	Angular acceleration
α, β, γ	Angles
γ	Specific weight
δ	Elongation
ε	Eccentricity of conic section or of orbit
$\boldsymbol{\lambda}$	Unit vector along a line
η	Efficiency
θ	Angular coordinate; Eulerian angle; angle; polar coordinate
μ	Coefficient of friction
ρ	Density; radius of curvature
τ	Period; periodic time
ϕ	Angle of friction; Eulerian angle; phase angle; angle
φ	Phase difference
ψ	Eulerian angle
$\boldsymbol{\omega}, \omega$	Angular velocity
ω	Circular frequency of forced vibration
$\boldsymbol{\Omega}$	Angular velocity of frame of reference

Kinematics of Particles

CHAPTER 11

11.1. Introduction to Dynamics. Chapters 1 to 10 were devoted to *statics*, i.e., to the analysis of bodies at rest. We shall now begin the study of *dynamics*, which is the part of mechanics dealing with the analysis of bodies in motion.

While the study of statics goes back to the time of the Greek philosophers, the first significant contribution to dynamics was made by Galileo (1564–1642). His experiments on uniformly accelerated bodies led Newton (1642–1727) to formulate his fundamental laws of motion.

Dynamics is divided into two parts: (1) *Kinematics*, which is the study of the geometry of motion; kinematics is used to relate displacement, velocity, acceleration, and time, without reference to the cause of the motion. (2) *Kinetics*, which is the study of the relation existing between the forces acting on a body, the mass of the body, and the motion of the body; kinetics is used to predict the motion caused by given forces or to determine the forces required to produce a given motion.

Chapters 11 to 14 are devoted to the *dynamics of particles*, and Chap. 11 more particularly to the *kinematics of particles*. The use of the word particles does not imply that we shall restrict our study to that of small corpuscles; it rather indicates that in these first chapters we shall study the motion of bodies—possibly as large as cars, rockets, or airplanes—without regard to their size. By saying that the bodies are analyzed as particles, we mean that only their motion as an entire unit will be considered; any rotation about their own mass center will be neglected. There are cases, however, when such a rotation is not negligible; the bodies, then, may not be considered as particles. The analysis of such motions will be carried out in later chapters dealing with the *dynamics of rigid bodies*.

RECTILINEAR MOTION OF PARTICLES

11.2. Position, Velocity, and Acceleration. A particle moving along a straight line is said to be in *rectilinear motion*. At any given instant t, the particle will occupy a certain position on the straight line. To define the position P of the particle, we choose a fixed origin O on the straight line and a positive direction along the line. We measure the distance x from O to P and record it with a plus or minus sign, according to whether P is reached from O by moving along the line in the positive or the negative direction. The distance x, with the appropriate sign, completely defines the position of the particle; it is called the *position coordinate* of the particle considered. For example, the position coordinate corresponding to P in Fig. 11.1a is $x = +5$ m, while the coordinate corresponding to P' in Fig. 11.1b is $x' = -2$ m.

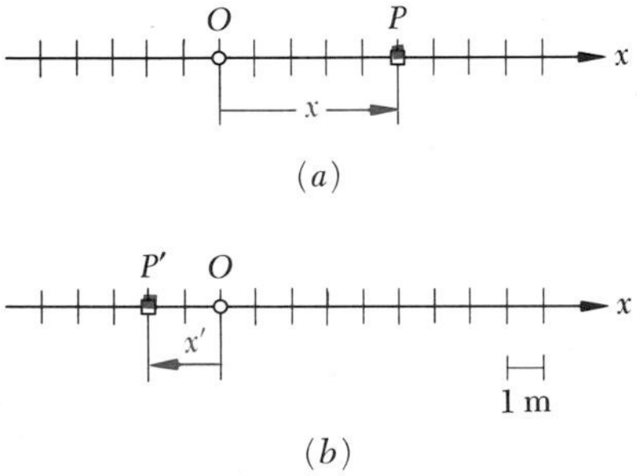

Fig. 11.1

When the position coordinate x of a particle is known for every value of time t, we say that the motion of the particle is known. The "timetable" of the motion may be given in the form of an equation in x and t, such as $x = 6t^2 - t^3$, or in the form of a graph of x vs. t as shown in Fig. 11.6. The units most generally used to measure the position coordinate x are the meter (m) in the SI system of units,† and the foot (ft) in the U.S. customary system of units. Time t will generally be measured in seconds (s).

† Cf. Sec. 1.3.

Consider the position P occupied by the particle at time t and the corresponding coordinate x (Fig. 11.2). Consider also the position P' occupied by the particle at a later time $t + \Delta t$; the position coordinate of P' may be obtained by adding to the coordinate x of P the small displacement Δx, which will be positive or negative according to whether P' is to the right or to the left of P. The *average velocity* of the particle over the time interval Δt is defined as the quotient of the displacement Δx and the time interval Δt,

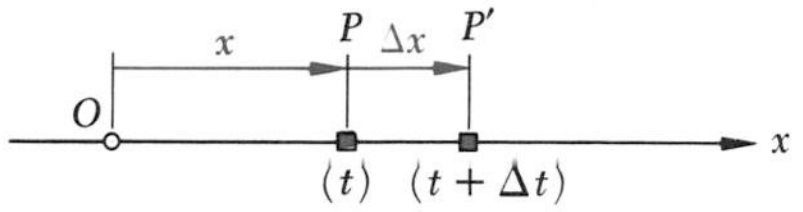

Fig. 11.2

$$\text{Average velocity} = \frac{\Delta x}{\Delta t}$$

If SI units are used, Δx is expressed in meters and Δt in seconds; the average velocity will thus be expressed in meters per second (m/s). If U.S. customary units are used, Δx is expressed in feet and Δt in seconds; the average velocity will then be expressed in feet per second (ft/s).

The *instantaneous velocity* v of the particle at the instant t is obtained from the average velocity by choosing shorter and shorter time intervals Δt and displacements Δx,

$$\text{Instantaneous velocity} = v = \lim_{\Delta t \to 0} \frac{\Delta x}{\Delta t}$$

The instantaneous velocity will also be expressed in m/s or ft/s. Observing that the limit of the quotient is equal, by definition, to the derivative of x with respect to t, we write

$$v = \frac{dx}{dt} \tag{11.1}$$

The velocity v is represented by an algebraic number which may be positive or negative.† A positive value of v indicates that x increases, i.e., that the particle moves in the positive direction (Fig. 11.3*a*); a negative value of v indicates that x decreases, i.e., that the particle moves in the negative direction (Fig. 11.3*b*). The magnitude of v is known as the *speed* of the particle.

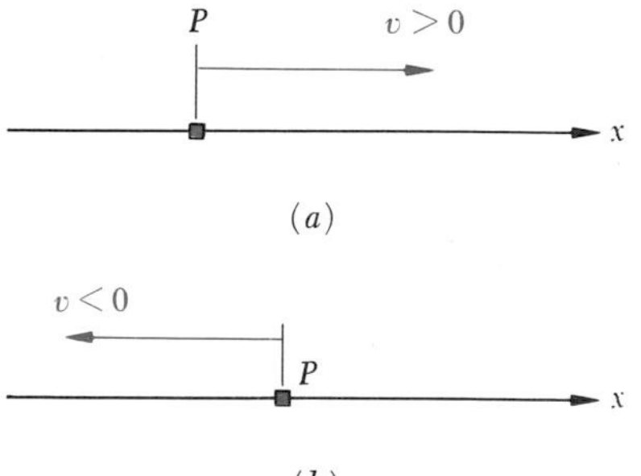

Fig. 11.3

† As we shall see in Sec. 11.9, the velocity is actually a vector quantity. However, since we are considering here the rectilinear motion of a particle, where the velocity of the particle has a known and fixed direction, we need only specify the sense and magnitude of the velocity; this may be conveniently done by using a scalar quantity with a plus or minus sign. The same remark will apply to the acceleration of a particle in rectilinear motion.

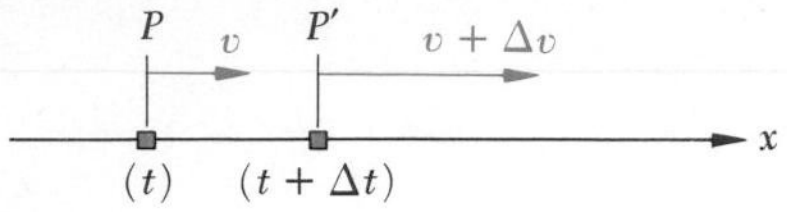

Fig. 11.4

Consider the velocity v of the particle at time t and also its velocity $v + \Delta v$ at a later time $t + \Delta t$ (Fig. 11.4). The *average acceleration* of the particle over the time interval Δt is defined as the quotient of Δv and Δt,

$$\text{Average acceleration} = \frac{\Delta v}{\Delta t}$$

If SI units are used, Δv is expressed in m/s and Δt in seconds; the average acceleration will thus be expressed in m/s^2. If U.S. customary units are used, Δv is expressed in ft/s and Δt in seconds; the average acceleration will then be expressed in ft/s^2.

The *instantaneous acceleration* a of the particle at the instant t is obtained from the average acceleration by choosing smaller and smaller values for Δt and Δv,

$$\text{Instantaneous acceleration} = a = \lim_{\Delta t \to 0} \frac{\Delta v}{\Delta t}$$

The instantaneous acceleration will also be expressed in m/s^2 or ft/s^2. The limit of the quotient is by definition the derivative of v with respect to t and measures the rate of change of the velocity. We write

$$a = \frac{dv}{dt} \tag{11.2}$$

or, substituting for v from (11.1),

$$a = \frac{d^2x}{dt^2} \tag{11.3}$$

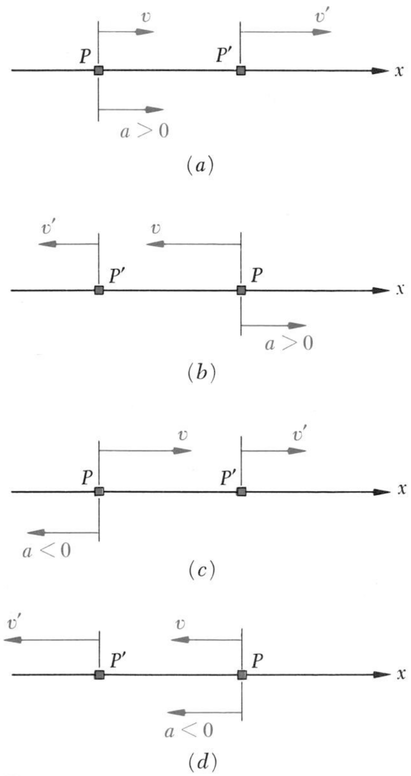

Fig. 11.5

The acceleration a is represented by an algebraic number which may be positive or negative.† A positive value of a indicates that the velocity (i.e., the algebraic number v) increases. This may mean that the particle is moving faster in the positive direction (Fig. 11.5*a*) or that it is moving more slowly in the negative direction (Fig. 11.5*b*); in both cases, Δv is positive. A negative value of a indicates that the velocity decreases; either the particle is moving more slowly in the positive direction (Fig. 11.5*c*), or it is moving faster in the negative direction (Fig. 11.5*d*).

The term *deceleration* is sometimes used to refer to a when the speed of the particle (i.e., the magnitude of v) decreases; the particle is then moving more slowly. For example, the particle of Fig. 11.5 is decelerated in parts b and c, while it is truly accelerated (i.e., moves faster) in parts a and d.

Another expression may be obtained for the acceleration by eliminating the differential dt in Eqs. (11.1) and (11.2). Solving

†See footnote, page 437.

(11.1) for dt, we obtain $dt = dx/v$; carrying into (11.2), we write

$$a = v\frac{dv}{dx} \tag{11.4}$$

Example. Consider a particle moving in a straight line, and assume that its position is defined by the equation

$$x = 6t^2 - t^3$$

where t is expressed in seconds and x in meters. The velocity v at any time t is obtained by differentiating x with respect to t,

$$v = \frac{dx}{dt} = 12t - 3t^2$$

The acceleration a is obtained by differentiating again with respect to t,

$$a = \frac{dv}{dt} = 12 - 6t$$

The position coordinate, the velocity, and the acceleration have been plotted against t in Fig. 11.6. The curves obtained are known as *motion curves*. It should be kept in mind, however, that the particle does not move along any of these curves; the particle moves in a straight line. Since the derivative of a function measures the slope of the corresponding curve, the slope of the x–t curve at any given time is equal to the value of v at that time and the slope of the v–t curve is equal to the value of a. Since $a = 0$ at $t = 2$ s, the slope of the v–t curve must be zero at $t = 2$ s; the velocity reaches a maximum at this instant. Also, since $v = 0$ at $t = 0$ and at $t = 4$ s, the tangent to the x–t curve must be horizontal for both of these values of t.

A study of the three motion curves of Fig. 11.6 shows that the motion of the particle from $t = 0$ to $t = \infty$ may be divided into four phases:

1. The particle starts from the origin, $x = 0$, with no velocity but with a positive acceleration. Under this acceleration, the particle gains a positive velocity and moves in the positive direction. From $t = 0$ to $t = 2$ s, x, v, and a are all positive.
2. At $t = 2$ s, the acceleration is zero; the velocity has reached its maximum value. From $t = 2$ s to $t = 4$ s, v is positive, but a is negative; the particle still moves in the positive direction but more and more slowly; the particle is decelerated.
3. At $t = 4$ s, the velocity is zero; the position coordinate x has reached its maximum value. From then on, both v and a are negative; the particle is accelerated and moves in the negative direction with increasing speed.
4. At $t = 6$ s, the particle passes through the origin; its coordinate x is then zero, while the total distance traveled since the beginning of the motion is 64 m. For values of t larger than 6 s, x, v, and a will all be negative. The particle keeps moving in the negative direction, away from O, faster and faster.

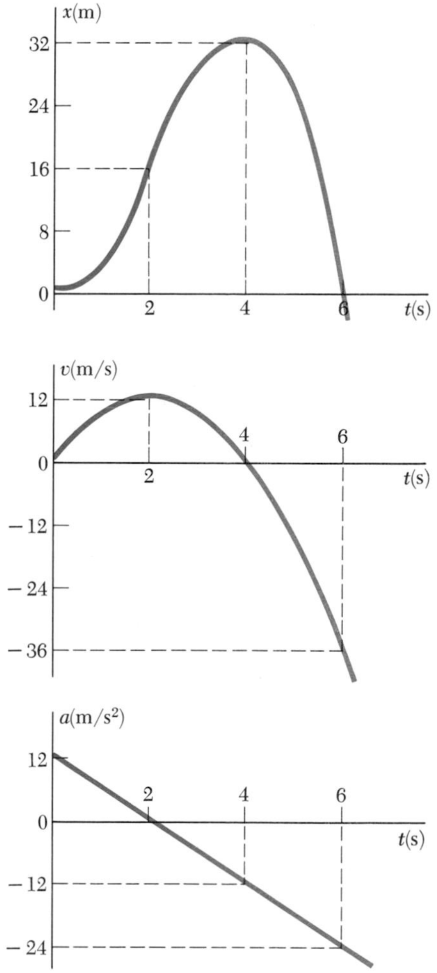

Fig. 11.6

11.3. Determination of the Motion of a Particle. We saw in the preceding section that the motion of a particle is said to be known if the position of the particle is known for every value of the time t. In practice, however, a motion is seldom defined by a relation between x and t. More often, the conditions of the motion will be specified by the type of acceleration that the particle possesses. For example, a freely falling body will have a constant acceleration, directed downward and equal to $9.81\ m/s^2$ or $32.2\ ft/s^2$; a mass attached to a spring which has been stretched will have an acceleration proportional to the instantaneous elongation of the spring measured from the equilibrium position; etc. In general, the acceleration of the particle may be expressed as a function of one or more of the variables x, v, and t. In order to determine the position coordinate x in terms of t, it will thus be necessary to perform two successive integrations.

We shall consider three common classes of motion:

1. $a = f(t)$. *The Acceleration Is a Given Function of* t. Solving (11.2) for dv and substituting $f(t)$ for a, we write

$$dv = a\,dt$$
$$dv = f(t)\,dt$$

Integrating both members, we obtain the equation

$$\int dv = \int f(t)\,dt$$

which defines v in terms of t. It should be noted, however, that an arbitrary constant will be introduced as a result of the integration. This is due to the fact that there are many motions which correspond to the given acceleration $a = f(t)$. In order to uniquely define the motion of the particle, it is necessary to specify the *initial conditions* of the motion, i.e., the value v_0 of the velocity and the value x_0 of the position coordinate at $t = 0$. Replacing the indefinite integrals by *definite integrals* with lower limits corresponding to the initial conditions $t = 0$ and $v = v_0$ and upper limits corresponding to $t = t$ and $v = v$, we write

$$\int_{v_0}^{v} dv = \int_0^t f(t)\,dt$$
$$v - v_0 = \int_0^t f(t)\,dt$$

which yields v in terms of t.

We shall now solve (11.1) for dx,

$$dx = v\,dt$$

and substitute for v the expression just obtained. Both members are then integrated, the left-hand member with respect to x from $x = x_0$ to $x = x$, and the right-hand member with respect to t from $t = 0$ to $t = t$. The position coordinate x is thus obtained in terms of t; the motion is completely determined.

Two important particular cases will be studied in greater detail in Secs. 11.4 and 11.5: the case when $a = 0$, corresponding to a *uniform motion*, and the case when $a =$ constant, corresponding to a *uniformly accelerated motion*.

2. $a = f(x)$. *The Acceleration Is a Given Function of x.* Rearranging Eq. (11.4) and substituting $f(x)$ for a, we write

$$v\,dv = a\,dx$$
$$v\,dv = f(x)\,dx$$

Since each member contains only one variable, we may integrate the equation. Denoting again by v_0 and x_0, respectively, the initial values of the velocity and of the position coordinate, we obtain

$$\int_{v_0}^{v} v\,dv = \int_{x_0}^{x} f(x)\,dx$$
$$\tfrac{1}{2}v^2 - \tfrac{1}{2}v_0^2 = \int_{x_0}^{x} f(x)\,dx$$

which yields v in terms of x. We now solve (11.1) for dt,

$$dt = \frac{dx}{v}$$

and substitute for v the expression just obtained. Both members may be integrated, and the desired relation between x and t is obtained.

3. $a = f(v)$. *The Acceleration Is a Given Function of v.* We may then substitute $f(v)$ for a either in (11.2) or in (11.4) to obtain either of the following relations:

$$f(v) = \frac{dv}{dt} \qquad\qquad f(v) = v\frac{dv}{dx}$$
$$dt = \frac{dv}{f(v)} \qquad\qquad dx = \frac{v\,dv}{f(v)}$$

Integration of the first equation will yield a relation between v and t; integration of the second equation will yield a relation between v and x. Either of these relations may be used in conjunction with Eq. (11.1) to obtain the relation between x and t which characterizes the motion of the particle.

SAMPLE PROBLEM 11.1

The position of a particle which moves along a straight line is defined by the relation $x = t^3 - 6t^2 - 15t + 40$, where x is expressed in feet and t in seconds. Determine (*a*) the time at which the velocity will be zero, (*b*) the position and distance traveled by the particle at that time, (*c*) the acceleration of the particle at that time, (*d*) the distance traveled by the particle from $t = 4$ s to $t = 6$ s.

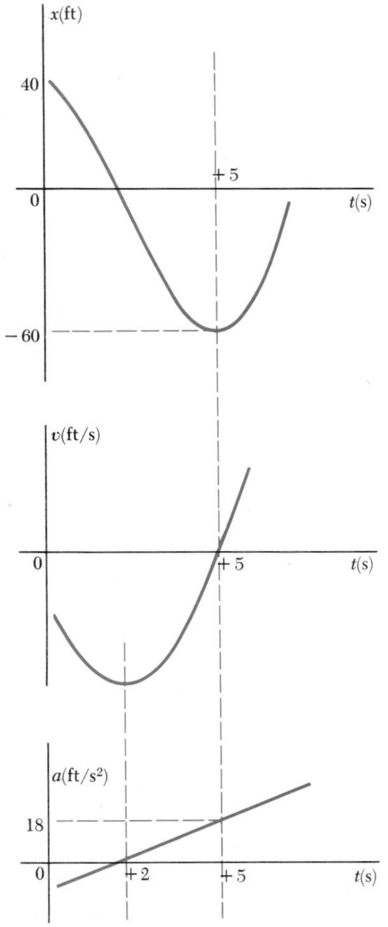

Solution. The equations of motion are

$$x = t^3 - 6t^2 - 15t + 40 \tag{1}$$

$$v = \frac{dx}{dt} = 3t^2 - 12t - 15 \tag{2}$$

$$a = \frac{dv}{dt} = 6t - 12 \tag{3}$$

a. Time at Which $v = 0$. We make $v = 0$ in (2),

$$3t^2 - 12t - 15 = 0 \qquad t = -1\text{ s} \qquad \text{and} \qquad t = +5\text{ s} \blacktriangleleft$$

Only the root $t = +5$ s corresponds to a time after the motion has begun: for $t < 5$ s, $v < 0$, the particle moves in the negative direction; for $t > 5$ s, $v > 0$, the particle moves in the positive direction.

b. Position and Distance Traveled When $v = 0$. Carrying $t = +5$ s into (1), we have

$$x_5 = (5)^3 - 6(5)^2 - 15(5) + 40 \qquad x_5 = -60\text{ ft} \blacktriangleleft$$

The initial position at $t = 0$ was $x_0 = +40$ ft. Since $v \neq 0$ during the interval $t = 0$ to $t = 5$ s, we have

$$\text{Distance traveled} = x_5 - x_0 = -60\text{ ft} - 40\text{ ft} = -100\text{ ft}$$

Distance traveled = 100 ft in the negative direction ◄

c. Acceleration When $v = 0$. We carry $t = +5$ s into (3):

$$a_5 = 6(5) - 12 \qquad a_5 = +18\text{ ft/s}^2 \blacktriangleleft$$

d. Distance Traveled from $t = 4$ s to $t = 6$ s. Since the particle moves in the negative direction from $t = 4$ s to $t = 5$ s and in the positive direction from $t = 5$ s to $t = 6$ s, we shall compute separately the distance traveled during each of these time intervals.

From $t = 4$ s to $t = 5$ s: $\quad x_5 = -60$ ft

$$x_4 = (4)^3 - 6(4)^2 - 15(4) + 40 = -52\text{ ft}$$

$$\text{Distance traveled} = x_5 - x_4 = -60\text{ ft} - (-52\text{ ft}) = -8\text{ ft}$$
$$= 8\text{ ft in the negative direction}$$

From $t = 5$ s to $t = 6$ s: $\quad x_5 = -60$ ft

$$x_6 = (6)^3 - 6(6)^2 - 15(6) + 40 = -50\text{ ft}$$

$$\text{Distance traveled} = x_6 - x_5 = -50\text{ ft} - (-60\text{ ft}) = +10\text{ ft}$$
$$= 10\text{ ft in the positive direction}$$

Total distance traveled from $t = 4$ s to $t = 6$ s is

$$8\text{ ft} + 10\text{ ft} = 18\text{ ft} \blacktriangleleft$$

SAMPLE PROBLEM 11.2

A ball is thrown from the top of a tower 18 m high, with a velocity of 12 m/s directed vertically upward. Knowing that the acceleration of the ball is constant and equal to 9.81 m/s^2 downward, determine (*a*) the velocity v and elevation y of the ball above the ground at any time t, (*b*) the highest elevation reached by the ball and the corresponding value of t, (*c*) the time when the ball will hit the ground and the corresponding velocity. Draw the v–t and y–t curves.

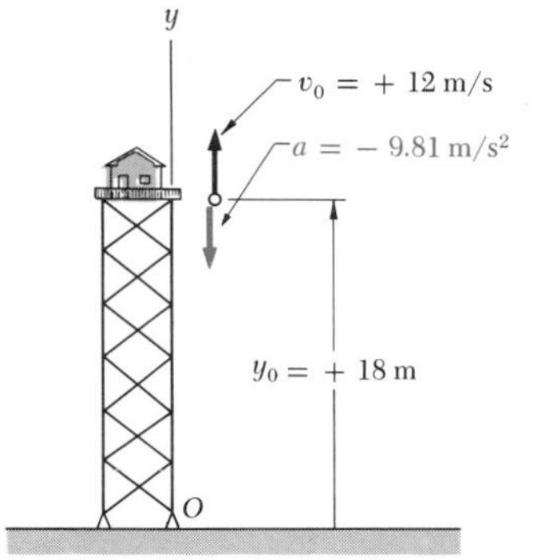

a. Velocity and Elevation. The y axis measuring the position coordinate (or elevation) is chosen with its origin O on the ground and its positive sense upward. The value of the acceleration and the initial values of v and y are as indicated. Substituting for a in $a = dv/dt$ and noting that, at $t = 0$, $v_0 = +12$ m/s, we have

$$\frac{dv}{dt} = a = -9.81 \text{ m/s}^2$$

$$\int_{v_0=12}^{v} dv = -\int_0^t 9.81\, dt$$

$$[v]_{12}^{v} = -[9.81t]_0^t$$

$$v - 12 = -9.81t$$

$$v = 12 - 9.81t \quad (1)$$

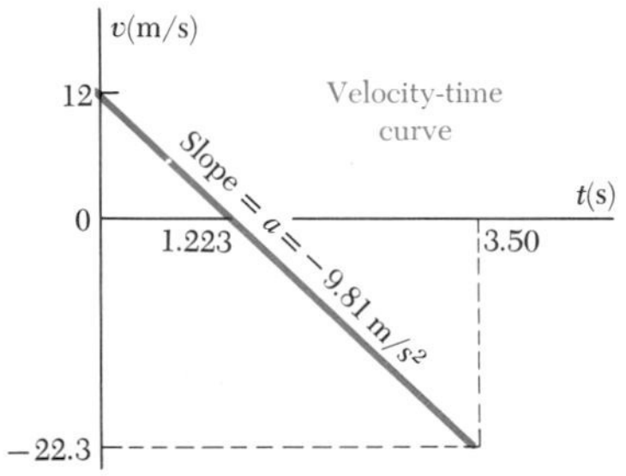

Substituting for v in $v = dy/dt$ and noting that, at $t = 0$, $y_0 = 18$ m, we have

$$\frac{dy}{dt} = v = 12 - 9.81t$$

$$\int_{y_0=18}^{y} dy = \int_0^t (12 - 9.81t)\, dt$$

$$[y]_{18}^{y} = [12t - 4.90t^2]_0^t$$

$$y - 18 = 12t - 4.90t^2$$

$$y = 18 + 12t - 4.90t^2 \quad (2)$$

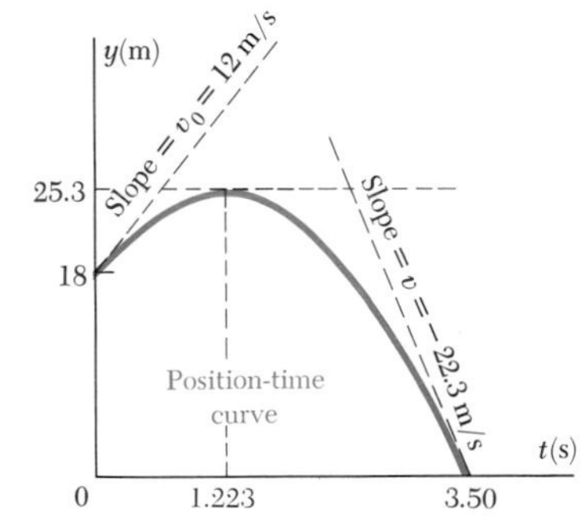

b. Highest Elevation. When the ball reaches its highest elevation, we have $v = 0$. Substituting into (1), we obtain

$$12 - 9.81t = 0 \qquad t = 1.223 \text{ s}$$

Carrying $t = 1.223$ s into (2), we have

$$y = 18 + 12(1.223) - 4.90(1.223)^2 \qquad y = 25.3 \text{ m}$$

c. Ball Hits the Ground. When the ball hits the ground, we have $y = 0$. Substituting into (2), we obtain

$$18 + 12t - 4.90t^2 = 0 \qquad t = -1.05 \text{ s} \quad \text{and} \quad t = +3.50 \text{ s}$$

Only the root $t = +3.50$ s corresponds to a time after the motion has begun. Carrying this value of t into (1), we have

$$v = 12 - 9.81(3.50) = -22.3 \text{ m/s} \qquad v = 22.3 \text{ m/s} \downarrow$$

SAMPLE PROBLEM 11.3

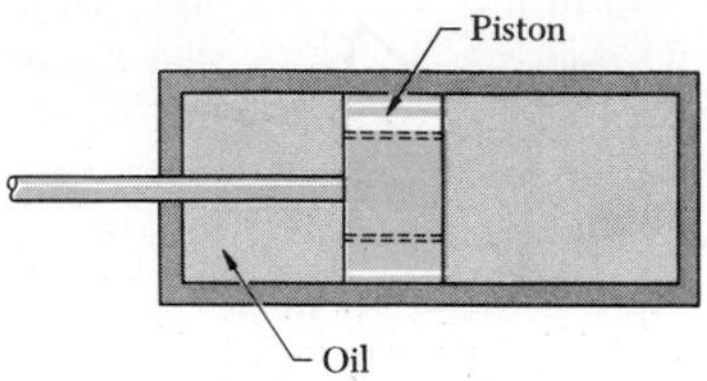

The brake mechanism used to reduce recoil in certain types of guns consists essentially of a piston which is attached to the barrel and may move in a fixed cylinder filled with oil. As the barrel recoils with an initial velocity v_0, the piston moves and oil is forced through orifices in the piston, causing the piston and the barrel to decelerate at a rate proportional to their velocity, i.e., $a = -kv$. Express (*a*) v in terms of t, (*b*) x in terms of t, (*c*) v in terms of x. Draw the corresponding motion curves.

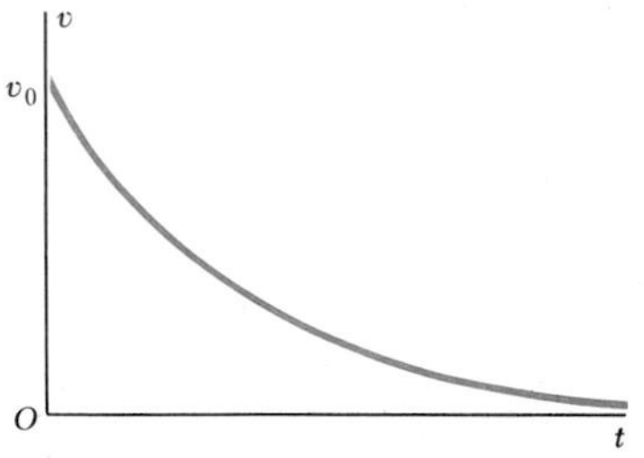

a. v in Terms of t. Substituting $-kv$ for a in the fundamental formula defining acceleration, $a = dv/dt$, we write

$$-kv = \frac{dv}{dt} \qquad \frac{dv}{v} = -k\,dt \qquad \int_{v_0}^{v} \frac{dv}{v} = -k \int_0^t dt$$

$$\ln \frac{v}{v_0} = -kt \qquad\qquad v = v_0 e^{-kt}$$ ◄

b. x in Terms of t. Substituting the expression just obtained for v into $v = dx/dt$, we write

$$v_0 e^{-kt} = \frac{dx}{dt}$$

$$\int_0^x dx = v_0 \int_0^t e^{-kt}\,dt$$

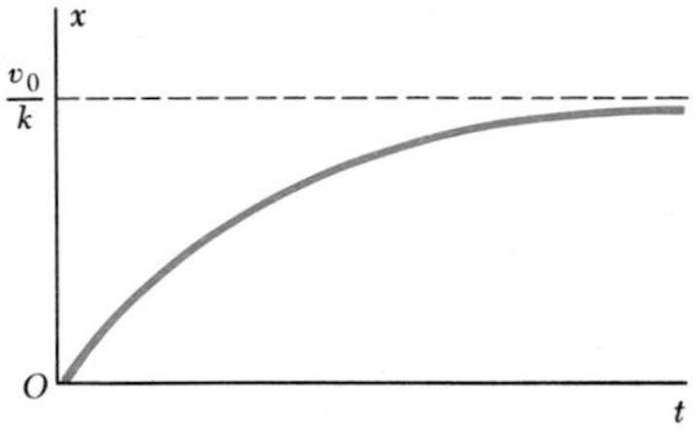

$$x = -\frac{v_0}{k}[e^{-kt}]_0^t = -\frac{v_0}{k}(e^{-kt} - 1)$$

$$x = \frac{v_0}{k}(1 - e^{-kt})$$ ◄

c. v in Terms of x. Substituting $-kv$ for a in $a = v\,dv/dx$, we write

$$-kv = v\frac{dv}{dx}$$

$$dv = -k\,dx$$

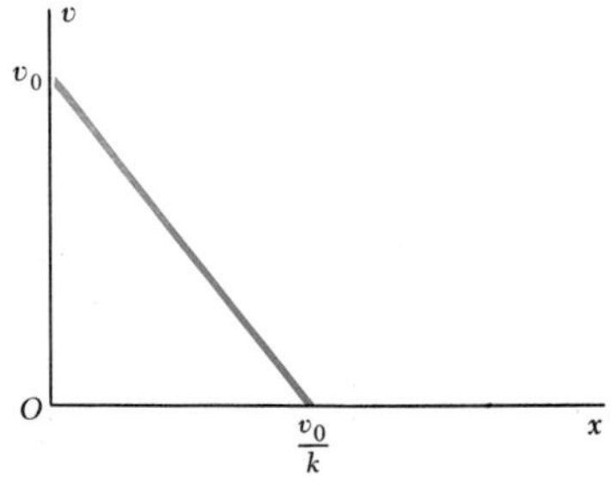

$$\int_{v_0}^{v} dv = -k \int_0^x dx$$

$$v - v_0 = -kx \qquad\qquad v = v_0 - kx$$ ◄

Check. Part *c* could have been solved by eliminating t from the answers obtained for parts *a* and *b*. This alternate method may be used as a check. From part *a* we obtain $e^{-kt} = v/v_0$; substituting in the answer of part *b*, we obtain

$$x = \frac{v_0}{k}(1 - e^{-kt}) = \frac{v_0}{k}\left(1 - \frac{v}{v_0}\right) \qquad v = v_0 - kx \qquad \text{(checks)}$$

PROBLEMS

11.1 The motion of a particle is defined by the relation $x = 2t^3 - 8t^2 + 5t + 9$, where x is expressed in inches and t in seconds. Determine the position, velocity, and acceleration when $t = 3$ s.

11.2 The motion of a particle is defined by the relation $x = 2t^3 - 9t^2 + 12$, where x is expressed in inches and t in seconds. Determine the time, position, and acceleration when $v = 0$.

11.3 The motion of a particle is defined by the relation $x = t^2 - 10t + 30$, where x is expressed in meters and t in seconds. Determine (*a*) when the velocity is zero, (*b*) the position and the total distance traveled when $t = 8$ s.

11.4 The motion of a particle is defined by the relation $x = \frac{1}{3}t^3 - 3t^2 + 8t + 2$, where x is expressed in meters and t in seconds. Determine (*a*) when the velocity is zero, (*b*) the position and the total distance traveled when the acceleration is zero.

11.5 The acceleration of a particle is directly proportional to the time t. At $t = 0$, the velocity of the particle is $v = -16$ m/s. Knowing that both the velocity and the position coordinate are zero when $t = 4$ s, write the equations of motion for the particle.

11.6 The acceleration of a particle is defined by the relation $a = -2\ \text{m/s}^2$. If $v = +8$ m/s and $x = 0$ when $t = 0$, determine the velocity, position, and total distance traveled when $t = 6$ s.

11.7 The acceleration of a particle is defined by the relation $a = kt^2$. (*a*) Knowing that $v = -250$ in./s when $t = 0$ and that $v = +250$ in./s when $t = 5$ s, determine the constant k. (*b*) Write the equations of motion knowing also that $x = 0$ when $t = 2$ s.

11.8 The acceleration of a particle is defined by the relation $a = 18 - 6t^2$. The particle starts at $t = 0$ with $v = 0$ and $x = 100$ in. Determine (*a*) the time when the velocity is again zero, (*b*) the position and velocity when $t = 4$ s, (*c*) the total distance traveled by the particle from $t = 0$ to $t = 4$ s.

11.9 The acceleration of a particle is defined by the relation $a = 21 - 12x^2$, where a is expressed in m/s^2 and x in meters. The particle starts with no initial velocity at the position $x = 0$. Determine (*a*) the velocity when $x = 1.5$ m, (*b*) the position where the velocity is again zero, (*c*) the position where the velocity is maximum.

11.10 The acceleration of an oscillating particle is defined by the relation $a = -kx$. Find the value of k such that $v = 10$ m/s when $x = 0$ and $x = 2$ m when $v = 0$.

11.11 The acceleration of a particle moving in a straight line is directed toward a fixed point O and is inversely proportional to the distance of the particle from O. At $t = 0$, the particle is 8 in. to the right of O, has a velocity of 16 in./s to the right, and has an acceleration of 12 in./s^2 to the left. Determine (*a*) the velocity of the particle when it is 12 in. away from O, (*b*) the position of the particle at which its velocity is zero.

11.12 The acceleration of a particle is defined by the relation $a = -kx^{-2}$. The particle starts with no initial velocity at $x = 12$ in., and it is observed that its velocity is 8 in./s when $x = 6$ in. Determine (*a*) the value of k, (*b*) the velocity of the particle when $x = 3$ in.

11.13 The acceleration of a particle is defined by the relation $a = -10v$, where a is expressed in m/s^2 and v in m/s. Knowing that at $t = 0$ the velocity is 30 m/s, determine (*a*) the distance the particle will travel before coming to rest, (*b*) the time required for the particle to come to rest, (*c*) the time required for the velocity of the particle to be reduced to 1 percent of its initial value.

11.14 The acceleration of a particle is defined by the relation $a = -0.0125v^2$, where a is the acceleration in m/s^2 and v is the velocity in m/s. If the particle is given an initial velocity v_0, find the distance it will travel (*a*) before its velocity drops to half the initial value, (*b*) before it comes to rest.

11.15 The acceleration of a particle falling through the atmosphere is defined by the relation $a = g(1 - k^2v^2)$. Knowing that the particle starts at $t = 0$ and $x = 0$ with no initial velocity, (*a*) show that the velocity at any time t is $v = (1/k) \tanh kgt$, (*b*) write an equation defining the velocity for any value of x. (*c*) Why is $v_t = 1/k$ called the terminal velocity?

11.16 It has been determined experimentally that the magnitude in ft/s^2 of the deceleration due to air resistance of a projectile is $0.001v^2$, where v is expressed in ft/s. If the projectile is released from rest and keeps pointing downward, determine its velocity after it has fallen 500 ft. (*Hint.* The total acceleration is $g - 0.001v^2$, where $g = 32.2$ ft/s^2.)

11.17 The acceleration of a particle is defined by the relation $a = -kv^{1.5}$. The particle starts at $t = 0$ and $x = 0$ with an initial velocity v_0. (*a*) Show that the velocity and position coordinate at any time t are related by the equation $x/t = \sqrt{v_0 v}$. (*b*) Determine the value of k, knowing that for $v_0 = 100$ ft/s the particle comes to rest after traveling 5 ft.

11.18 The acceleration of a particle is defined by the relation $a = k \sin(\pi t/T)$. Knowing that both the velocity and the position coordinate of the particle are zero when $t = 0$, determine (*a*) the equations of motion, (*b*) the maximum velocity, (*c*) the position at $t = 2T$, (*d*) the average velocity during the interval $t = 0$ to $t = 2T$.

11.19 The position of an oscillating particle is defined by the relation $x = A \sin(pt + \phi)$. Denoting the velocity and position coordinate when $t = 0$ by v_0 and x_0, respectively, show (*a*) that $\tan \phi = x_0 p/v_0$, (*b*) that the maximum value of the position coordinate is

$$A = \sqrt{x_0^2 + \left(\frac{v_0}{p}\right)^2}$$

11.20 The acceleration due to gravity at an altitude y above the surface of the earth may be expressed as

$$a = \frac{-32.2}{\left(1 + \dfrac{y}{20.9 \times 10^6}\right)^2}$$

where a is measured in ft/s² and y in feet. Using this expression, compute the height reached by a bullet fired vertically upward from the surface of the earth with the following initial velocities: (*a*) 1000 ft/s, (*b*) 10,000 ft/s, (*c*) 36,700 ft/s.

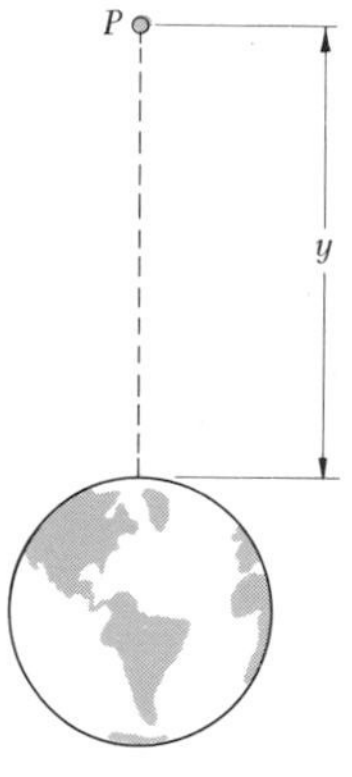

Fig. P11.20

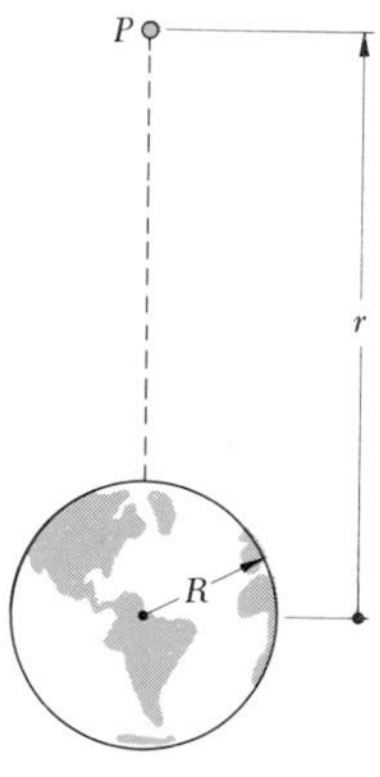

Fig. P11.21

11.21 The acceleration due to gravity of a particle falling toward the earth is $a = -gR^2/r^2$, where r is the distance from the *center* of the earth to the particle, R is the radius of the earth, and g is the acceleration due to gravity at the surface of the earth. Derive an expression for the *escape velocity,* i.e., for the minimum velocity with which a particle should be projected vertically upward from the surface of the earth if it is not to return to the earth. (*Hint.* $v = 0$ for $r = \infty$.)

***11.22** When a package is dropped on a rigid surface, the acceleration of its cushioned contents may be defined by the relation $a = -k \tan(\pi x/2L)$, where L is the distance through which the cushioning material can be compressed. Denoting by v_0 the velocity when $x = 0$, show that $v^2 = v_0^2 + (4kL/\pi) \ln \cos(\pi x/2L)$. If $k = 300$ m/s² and $L = 0.36$ m, compute the initial velocity v_0 for which the maximum value of the position coordinate x is (*a*) 0.18 m, (*b*) 0.36 m.

***11.23** Using the expression for the acceleration due to gravity given in Prob. 11.21, derive an expression for the time required for a particle to reach the surface of the earth if it is released with no velocity at a distance r_0 from the center of the earth.

11.4. Uniform Rectilinear Motion. This is a type of straight-line motion which is frequently encountered in practical applications. In this motion, the acceleration a of the particle is zero for every value of t. The velocity v is therefore constant, and Eq. (11.1) becomes

$$\frac{dx}{dt} = v = \text{constant}$$

The position coordinate x is obtained by integrating this equation. Denoting by x_0 the initial value of x, we write

$$\int_{x_0}^{x} dx = v \int_0^t dt$$

$$x - x_0 = vt$$

$$x = x_0 + vt \tag{11.5}$$

This equation may be used *only if the velocity of the particle is known to be constant.*

11.5. Uniformly Accelerated Rectilinear Motion. This is another common type of motion. In this motion, the acceleration a of the particle is constant, and Eq. (11.2) becomes

$$\frac{dv}{dt} = a = \text{constant}$$

The velocity v of the particle is obtained by integrating this equation,

$$\int_{v_0}^{v} dv = a \int_0^t dt$$

$$v - v_0 = at$$

$$v = v_0 + at \tag{11.6}$$

where v_0 is the initial velocity. Substituting for v into (11.1), we write

$$\frac{dx}{dt} = v_0 + at$$

Denoting by x_0 the initial value of x and integrating, we have

$$\int_{x_0}^{x} dx = \int_0^t (v_0 + at)\, dt$$

$$x - x_0 = v_0 t + \tfrac{1}{2}at^2$$

$$x = x_0 + v_0 t + \tfrac{1}{2}at^2 \qquad (11.7)$$

We may also use Eq. (11.4) and write

$$v\frac{dv}{dx} = a = \text{constant}$$

$$v\, dv = a\, dx$$

Integrating both sides, we obtain

$$\int_{v_0}^{v} v\, dv = a\int_{x_0}^{x} dx$$

$$\tfrac{1}{2}(v^2 - v_0^2) = a(x - x_0)$$

$$v^2 = v_0^2 + 2a(x - x_0) \qquad (11.8)$$

The three equations we have derived provide useful relations among position coordinate, velocity, and time in the case of a uniformly accelerated motion, as soon as appropriate values have been substituted for a, v_0, and x_0. The origin O of the x axis should first be defined and a positive direction chosen along the axis; this direction will be used to determine the signs of a, v_0, and x_0. Equation (11.6) relates v and t and should be used when the value of v corresponding to a given value of t is desired, or inversely. Equation (11.7) relates x and t; Eq. (11.8) relates v and x. An important application of uniformly accelerated motion is the motion of a *freely falling body*. The acceleration of a freely falling body (usually denoted by g) is equal to 9.81 m/s^2 or 32.2 ft/s^2.

It is important to keep in mind that the three equations above may be used *only when the acceleration of the particle is known to be constant*. If the acceleration of the particle is variable, its motion should be determined from the fundamental equations (11.1) to (11.4), according to the methods outlined in Sec. 11.3.

11.6. Motion of Several Particles. When several particles move independently along the same line, independent equations of motion may be written for each particle. Whenever possible, time should be recorded from the same initial instant for all particles, and displacements should be measured from the same origin and in the same direction. In other words, a single clock and a single measuring tape should be used.

Relative Motion of Two Particles. Consider two particles A and B moving along the same straight line (Fig. 11.7). If the position coordinates x_A and x_B are measured from the same origin, the difference $x_B - x_A$ defines the *relative position coordinate of B with respect to A* and is denoted by $x_{B/A}$. We write

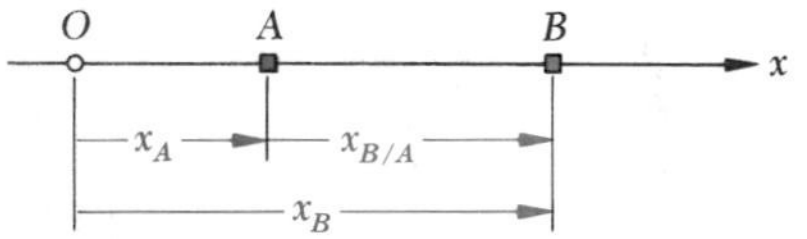

Fig. 11.7

$$x_{B/A} = x_B - x_A \qquad \text{or} \qquad x_B = x_A + x_{B/A} \tag{11.9}$$

A positive sign for $x_{B/A}$ means that B is to the right of A, a negative sign that B is to the left of A, regardless of the position of A and B with respect to the origin.

The rate of change of $x_{B/A}$ is known as the *relative velocity of B with respect to A* and is denoted by $v_{B/A}$. Differentiating (11.9), we write

$$v_{B/A} = v_B - v_A \qquad \text{or} \qquad v_B = v_A + v_{B/A} \tag{11.10}$$

A positive sign for $v_{B/A}$ means that B is *observed from* A to move in the positive direction; a negative sign, that it is observed to move in the negative direction.

The rate of change of $v_{B/A}$ is known as the *relative acceleration of B with respect to A* and is denoted by $a_{B/A}$. Differentiating (11.10), we obtain

$$a_{B/A} = a_B - a_A \qquad \text{or} \qquad a_B = a_A + a_{B/A} \tag{11.11}$$

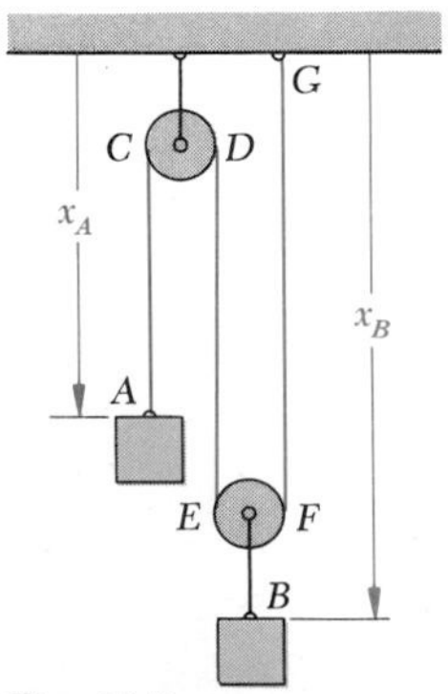

Fig. 11.8

Dependent Motions. Sometimes, the position of a particle will depend upon the position of another or of several other particles. The motions are then said to be dependent. For example, the position of block B in Fig. 11.8 depends upon the position of block A. Since the rope ACDEFG is of constant length, and since the lengths of the portions of rope CD and EF wrapped around the pulleys remain constant, it follows that the sum of the lengths of the segments AC, DE, and FG is constant. Observing that the length of the segment AC differs from x_A only by a constant, and that, similarly, the lengths of the segments DE and FG differ from x_B only by a constant, we write

$$x_A + 2x_B = \text{constant}$$

Since only one of the two coordinates x_A and x_B may be chosen arbitrarily, we say that the system shown in Fig. 11.8 has *one degree of freedom.* From the relation between the position coordinates x_A and x_B, it follows that if x_A is given an increment Δx_A, i.e., if block A is lowered by an amount Δx_A, the coordinate x_B will receive an increment $\Delta x_B = -\frac{1}{2}\Delta x_A$, that is, block B will rise by half the same amount; this may easily be checked directly from Fig. 11.8.

In the case of the three blocks of Fig. 11.9, we may again observe that the length of the rope which passes over the pulleys is constant, and thus that the following relation must be satisfied by the position coordinates of the three blocks:

$$2x_A + 2x_B + x_C = \text{constant}$$

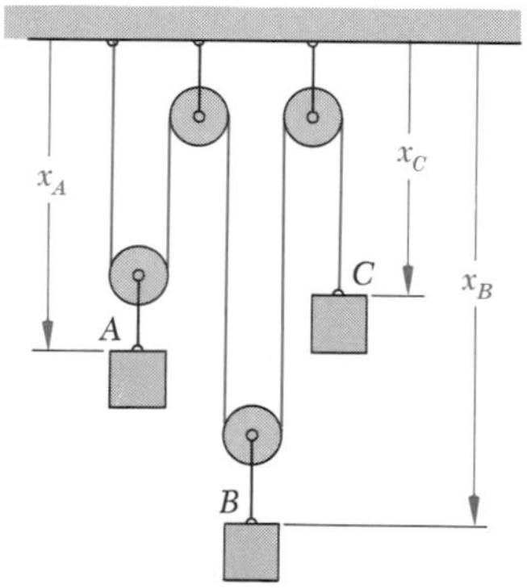

Fig. 11.9

Since two of the coordinates may be chosen arbitrarily, we say that the system shown in Fig. 11.9 has *two degrees of freedom.*

When the relation existing between the position coordinates of several particles is *linear,* a similar relation holds between the velocities and between the accelerations of the particles. In the case of the blocks of Fig. 11.9, for instance, we differentiate twice the equation obtained and write

$$2\frac{dx_A}{dt} + 2\frac{dx_B}{dt} + \frac{dx_C}{dt} = 0 \qquad \text{or} \qquad 2v_A + 2v_B + v_C = 0$$

$$2\frac{dv_A}{dt} + 2\frac{dv_B}{dt} + \frac{dv_C}{dt} = 0 \qquad \text{or} \qquad 2a_A + 2a_B + a_C = 0$$

SAMPLE PROBLEM 11.4

A ball is thrown vertically upward from the 40-ft level in an elevator shaft, with an initial velocity of 50 ft/s. At the same instant an open-platform elevator passes the 10-ft level, moving upward with a constant velocity of 5 ft/s. Determine (*a*) when and where the ball will hit the elevator, (*b*) the relative velocity of the ball with respect to the elevator when the ball hits the elevator.

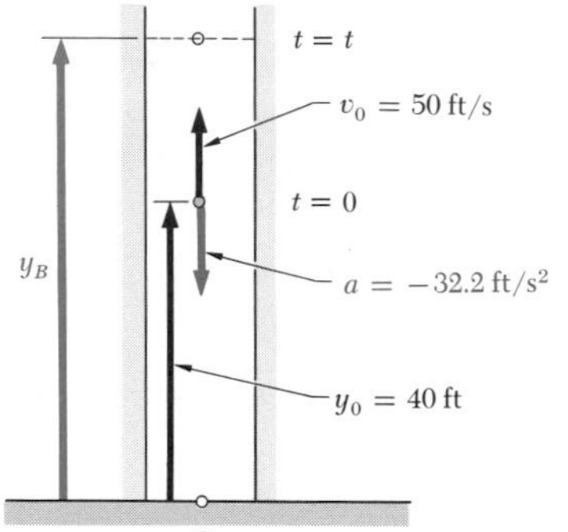

Motion of Ball. Since the ball has a constant acceleration, its motion is *uniformly accelerated*. Placing the origin O of the y axis at ground level and choosing its positive direction upward, we find that the initial position is $y_0 = +40$ ft, the initial velocity is $v_0 = +50$ ft/s, and the acceleration is $a = -32.2$ ft/s^2. Substituting these values in the equations for uniformly accelerated motion, we write

$$v_B = v_0 + at \qquad v_B = 50 - 32.2t \tag{1}$$

$$y_B = y_0 + v_0 t + \tfrac{1}{2}at^2 \qquad y_B = 40 + 50t - 16.1t^2 \tag{2}$$

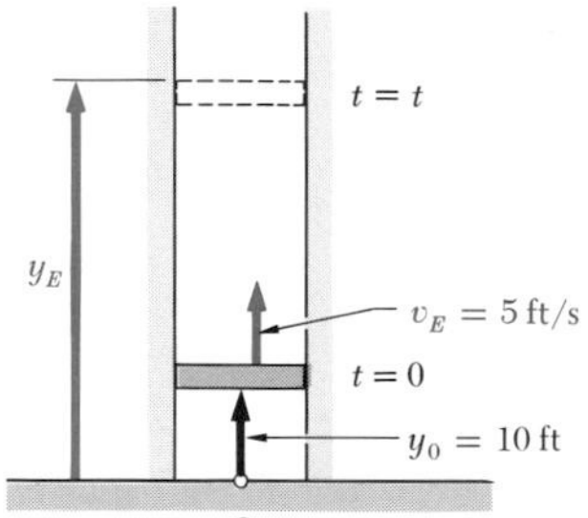

Motion of Elevator. Since the elevator has a constant velocity, its motion is *uniform*. Again placing the origin O at the ground level and choosing the positive direction upward, we note that $y_0 = +10$ ft and write

$$v_E = +5 \text{ ft/s} \tag{3}$$

$$y_E = y_0 + v_E t \qquad y_E = 10 + 5t \tag{4}$$

Ball Hits Elevator. We first note that the same time t and the same origin O were used in writing the equations of motion of both the ball and the elevator. We see from the figure that, when the ball hits the elevator,

$$y_E = y_B \tag{5}$$

Substituting for y_E and y_B from (2) and (4) into (5), we have

$$10 + 5t = 40 + 50t - 16.1t^2$$

$$t = -0.56 \text{ s} \qquad \text{and} \qquad t = +3.35 \text{ s} \quad ◄$$

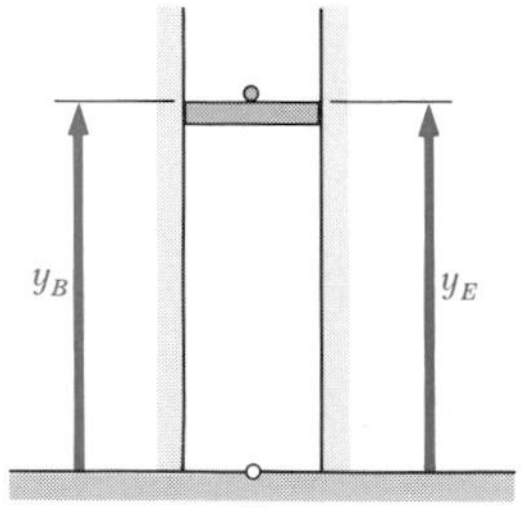

Only the root $t = 3.35$ s corresponds to a time after the motion has begun. Substituting this value into (4), we have

$$y_E = 10 + 5(3.35) = 26.7 \text{ ft}$$

Elevation from ground = 26.7 ft ◄

The relative velocity of the ball with respect to the elevator is

$$v_{B/E} = v_B - v_E = (50 - 32.2t) - 5 = 45 - 32.2t$$

When the ball hits the elevator at time $t = 3.35$ s, we have

$$v_{B/E} = 45 - 32.2(3.35) \qquad v_{B/E} = -62.9 \text{ ft/s} \quad ◄$$

The negative sign means that the ball is observed from the elevator to be moving in the negative sense (downward).

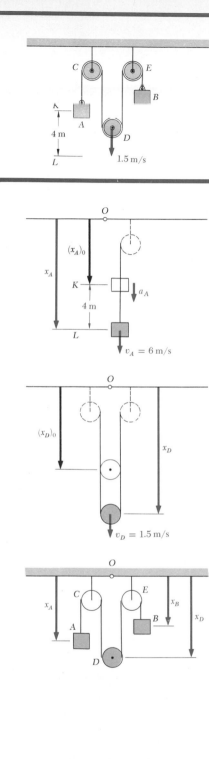

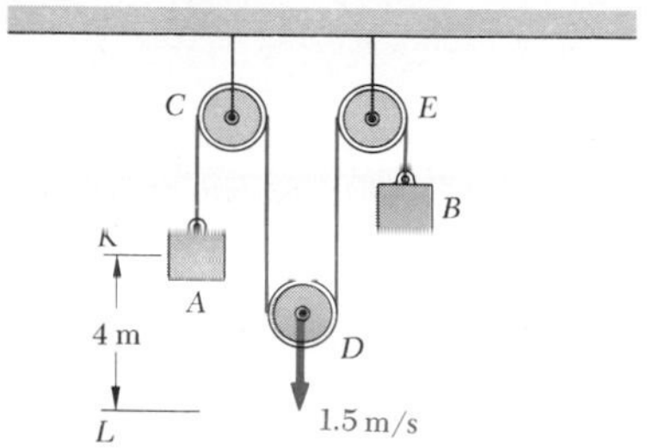

SAMPLE PROBLEM 11.5

Two blocks A and B are connected by a cord passing over three pulleys C, D, and E as shown. Pulleys C and E are fixed, while D is pulled downward with a constant velocity of 1.5 m/s. At $t = 0$, block A starts moving downward from the position K with a constant acceleration and no initial velocity. Knowing that the velocity of block A is 6 m/s as it passes through point L, determine the change in elevation, the velocity, and the acceleration of block B when A passes through L.

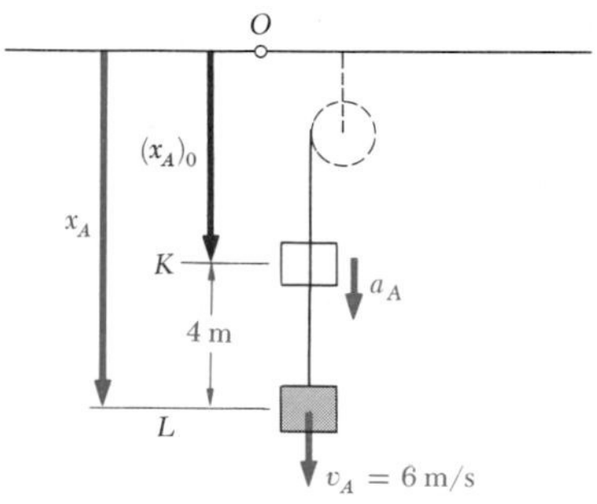

Motion of Block A. We place the origin O at the horizontal surface and choose the positive direction downward. We observe that when $t = 0$, block A is at position K and $(v_A)_0 = 0$. Since $v_A = 6$ m/s and $x_A - (x_A)_0 = 4$ m when the block passes through L, we write

$$v_A^2 = (v_A)_0^2 + 2a_A[x_A - (x_A)_0] \qquad (6)^2 = 0 + 2a_A(4)$$
$$a_A = 4.50 \text{ m/s}^2$$

The time at which block A reaches point L is obtained by writing

$$v_A = (v_A)_0 + a_A t \qquad 6 = 0 + 4.50t \qquad t = 1.333 \text{ s}$$

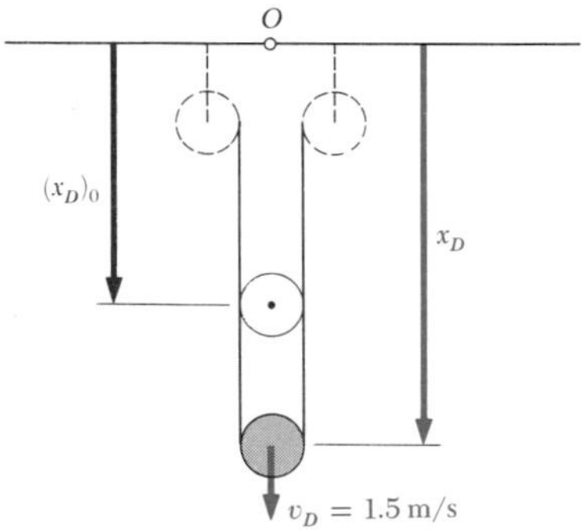

Motion of Pulley D. Recalling that the positive direction is downward, we write

$$a_D = 0 \qquad v_D = 1.5 \text{ m/s} \qquad x_D = (x_D)_0 + v_D t = (x_D)_0 + 1.5t$$

When block A reaches L, at $t = 1.333$ s, we have

$$x_D = (x_D)_0 + 1.5(1.333) = (x_D)_0 + 2$$

Thus,
$$x_D - (x_D)_0 = 2 \text{ m}$$

Motion of Block B. We note that the total length of cord $ACDEB$ differs from the quantity $(x_A + 2x_D + x_B)$ only by a constant. Since the cord length is constant during the motion, this quantity must also remain constant. Thus considering the times $t = 0$ and $t = 1.333$ s, we write

$$x_A + 2x_D + x_B = (x_A)_0 + 2(x_D)_0 + (x_B)_0 \qquad (1)$$
$$[x_A - (x_A)_0] + 2[x_D - (x_D)_0] + [x_B - (x_B)_0] = 0 \qquad (2)$$

But we know that $x_A - (x_A)_0 = 4$ m and $x_D - (x_D)_0 = 2$ m; substituting these values in (2), we find

$$4 + 2(2) + [x_B - (x_B)_0] = 0 \qquad x_B - (x_B)_0 = -8 \text{ m}$$

Thus: Change in elevation of $B = 8 \text{ m} \uparrow$ ◄

Differentiating (1) twice, we obtain equations relating the velocities and the accelerations of A, B, and D. Substituting for the velocities and accelerations of A and D at $t = 1.333$ s, we have

$$v_A + 2v_D + v_B = 0: \qquad 6 + 2(1.5) + v_B = 0$$
$$v_B = -9 \text{ m/s} \qquad v_B = 9 \text{ m/s} \uparrow$$ ◄

$$a_A + 2a_D + a_B = 0: \qquad 4.50 + 2(0) + a_B = 0$$
$$a_B = -4.50 \text{ m/s}^2 \qquad a_B = 4.50 \text{ m/s}^2 \uparrow$$ ◄

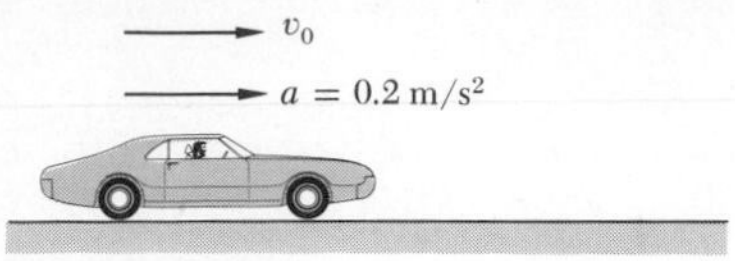

Fig. P11.24

PROBLEMS

11.24 An automobile travels 240 m in 30 s while being accelerated at a constant rate of $0.2\ \text{m/s}^2$. Determine (*a*) its initial velocity, (*b*) its final velocity, (*c*) the distance traveled during the first 10 s.

11.25 A stone is released from an elevator moving up at a speed of 5 m/s and reaches the bottom of the shaft in 3 s. (*a*) How high was the elevator when the stone was released? (*b*) With what speed does the stone strike the bottom of the shaft?

11.26 A stone is thrown vertically upward from a point on a bridge located 135 ft above the water. Knowing that it strikes the water 4 s after release, determine (*a*) the speed with which the stone was thrown upward, (*b*) the speed with which the stone strikes the water.

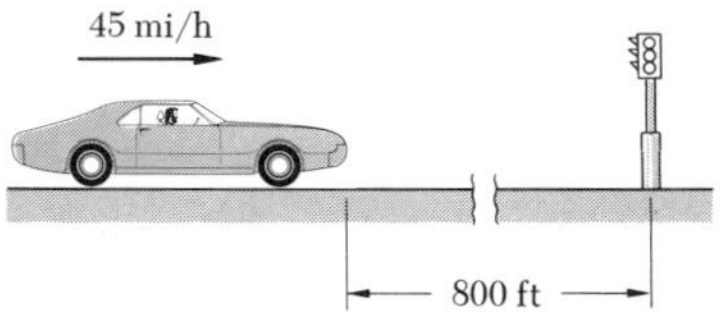

Fig. P11.27

11.27 A motorist is traveling at 45 mi/h when he observes that a traffic light 800 ft ahead of him turns red. The traffic light is timed to stay red for 15 s. If the motorist wishes to pass the light without stopping just as it turns green again, determine (*a*) the required uniform deceleration of the car, (*b*) the speed of the car as it passes the light.

11.28 Automobile A starts from O and accelerates at the constant rate of $4\ \text{ft/s}^2$. A short time later it is passed by truck B which is traveling in the opposite direction at a constant speed of 45 ft/s. Knowing that truck B passes point O, 25 s after automobile A started from there, determine when and where the vehicles passed each other.

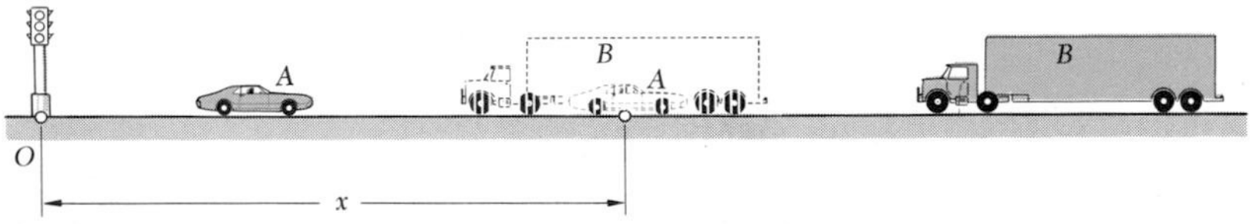

Fig. P11.28

11.29 An open-platform elevator is moving down a mine shaft at a constant velocity v_e when the elevator platform hits and dislodges a stone. Assuming that the stone starts falling with no initial velocity, (*a*) show that the stone will hit the platform with a relative velocity of magnitude v_e. (*b*) If $v_e = 16\ \text{ft/s}$, determine when and where the stone will hit the elevator platform.

11.30 Two automobiles A and B are traveling in the same direction in adjacent highway lanes. Automobile B is stopped when it is passed by A, which travels at a constant speed of 36 km/h. Two seconds later automobile B starts and accelerates at a constant rate of $1.5\ \text{m/s}^2$. Determine (a) when and where B will overtake A, (b) the speed of B at that time.

11.31 Drops of water are observed to drip from a faucet at uniform intervals of time. As any drop B begins to fall freely, the preceding drop A has already fallen 0.3 m. Determine the distance drop A will have fallen by the time the distance between A and B will have increased to 0.9 m.

11.32 The elevator shown in the figure moves upward at the constant velocity of 18 ft/s. Determine (a) the velocity of the cable C, (b) the velocity of the counterweight W, (c) the relative velocity of the cable C with respect to the elevator, (d) the relative velocity of the counterweight W with respect to the elevator.

11.33 The elevator shown starts from rest and moves upward with a constant acceleration. If the counterweight W moves through 24 ft in 4 s, determine (a) the accelerations of the elevator and the cable C, (b) the velocity of the elevator after 4 s.

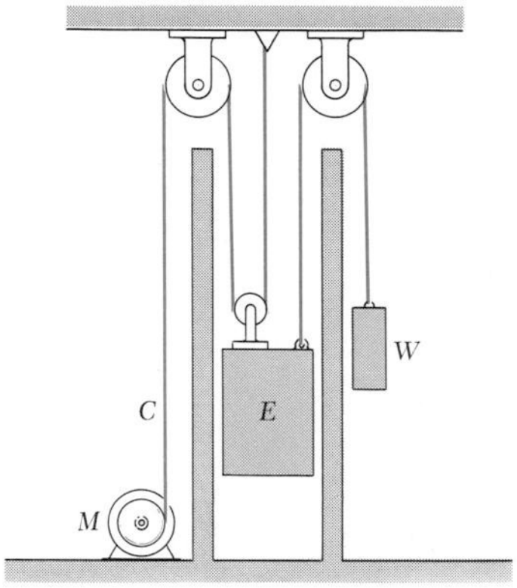

Fig. P11.32 and P11.33

11.34 The slider block A moves to the left at a constant velocity of 300 mm/s. Determine (a) the velocity of block B, (b) the velocities of portions C and D of the cable, (c) the relative velocity of A with respect to B, (d) the relative velocity of portion C of the cable with respect to portion D.

11.35 The slider block B starts from rest and moves to the right with a constant acceleration. After 4 s the relative velocity of A with respect to B is 60 mm/s. Determine (a) the accelerations of A and B, (b) the velocity and position of B after 3 s.

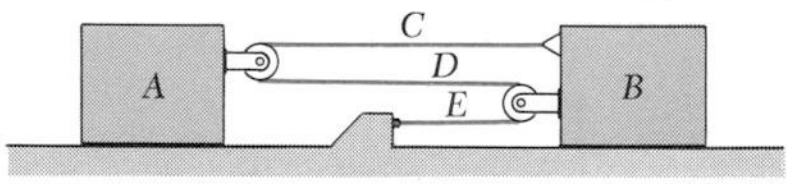

Fig. P11.34 and P11.35

11.36 Collars A and B start from rest and move with the following accelerations: $a_A = 3\ \text{in./s}^2$ upward and $a_B = 6t\ \text{in./s}^2$ downward. Determine (a) the time at which the velocity of block C is again zero, (b) the distance through which block C will have moved at that time.

11.37 (a) Choosing the positive sense upward, express the velocity of block C in terms of the velocities of collars A and B. (b) Knowing that both collars start from rest and move upward with the accelerations $a_A = 4\ \text{in./s}^2$ and $a_B = 3\ \text{in./s}^2$, determine the velocity of block C at $t = 4$ s and the distance through which it will have moved at that time.

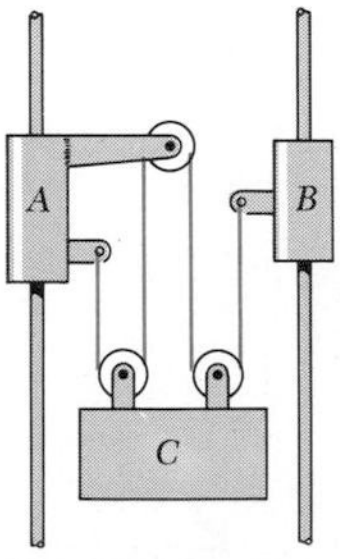

Fig. P11.36 and P11.37

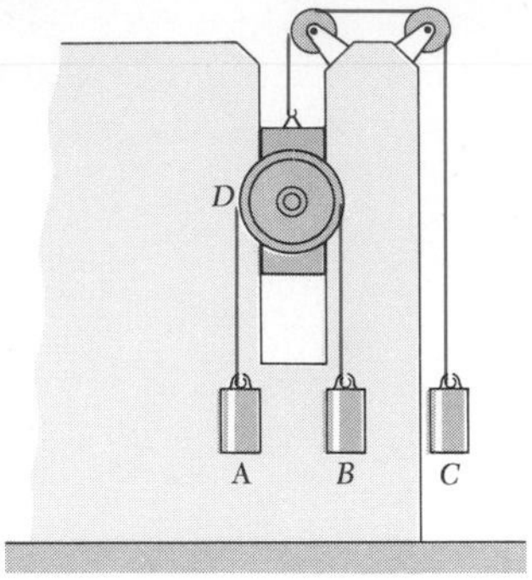

Fig. P11.38 and P11.40

11.38 The three blocks shown move with constant velocities. Find the velocity of each block, knowing that the relative velocity of C with respect to A is 200 mm/s upward and that the relative velocity of B with respect to C is 120 mm/s downward.

11.39 The three blocks of Fig. 11.9 move with constant velocities. Find the velocity of each block, knowing that C is observed from B to move downward with a relative velocity of 180 mm/s and A is observed from B to move downward with a relative velocity of 160 mm/s.

***11.40** The three blocks shown are equally spaced horizontally and move vertically with constant velocities. Knowing that initially they are at the same level and that the relative velocity of A with respect to B is 160 mm/s downward, determine the velocity of each block so that the three blocks will remain aligned during their motion.

***11.7. Graphical Solution of Rectilinear-Motion Problems.** It was observed in Sec. 11.2 that the fundamental formulas

$$v = \frac{dx}{dt} \qquad \text{and} \qquad a = \frac{dv}{dt}$$

have a geometrical significance. The first formula expresses that the velocity at any instant is equal to the slope of the x–t curve at the same instant (Fig. 11.10). The second formula expresses that the acceleration is equal to the slope of the v–t curve. These

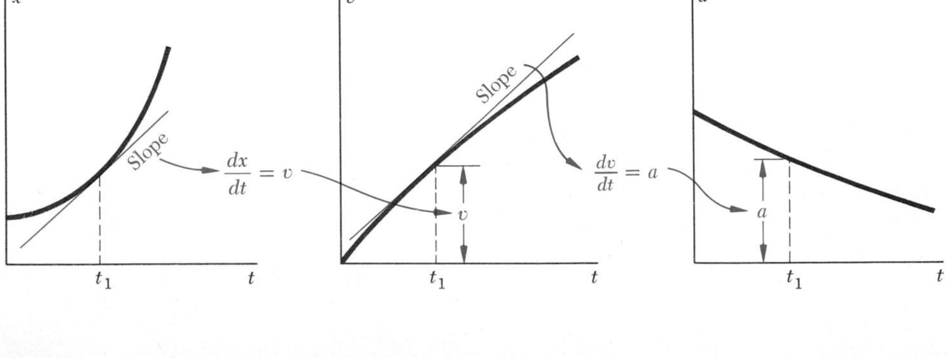

Fig. 11.10

two properties may be used to derive graphically the v–t and a–t curves of a motion when the x–t curve is known.

Integrating the two fundamental formulas from a time t_1 to a time t_2, we write

$$x_2 - x_1 = \int_{t_1}^{t_2} v\, dt \qquad \text{and} \qquad v_2 - v_1 = \int_{t_1}^{t_2} a\, dt \quad (11.12)$$

The first formula expresses that the area measured under the v–t curve from t_1 to t_2 is equal to the change in x during that time interval (Fig. 11.11). The second formula expresses similarly that the area measured under the a–t curve from t_1 to t_2 is equal to the change in v during that time interval. These two properties may be used to determine graphically the x–t curve of a motion when its v–t curve or its a–t curve is known (see Sample Prob. 11.6).

Graphical solutions are particularly useful when the motion considered is defined from experimental data and when x, v, and a are not analytical functions of t. They may also be used to advantage when the motion consists of distinct parts and when its analysis requires writing a different equation for each of its parts. When using a graphical solution, however, one should be careful to note (1) that the area under the v–t curve measures the *change in* x, not x itself, and, similarly, that the area under the a–t curve measures the change in v; (2) that, while an area above the t axis corresponds to an increase in x or v, an area located below the t axis measures a decrease in x or v.

It will be useful to remember, in drawing motion curves, that, if the velocity is constant, it will be represented by a horizontal straight line; the position coordinate x will then be a linear function of t and will be represented by an oblique straight line. If the acceleration is constant and different from zero, it will be represented by a horizontal straight line; v will then be a linear function of t, represented by an oblique straight line; and x will be expressed as a second-degree polynomial in t, represented by a parabola. If the acceleration is a linear function of t, the velocity and the position coordinate will be equal, respectively, to second-degree and third-degree polynomials; a is then represented by an oblique straight line, v by a parabola, and x by a cubic. In general, if the acceleration is a polynomial of degree n in t, the velocity will be a polynomial of degree $n + 1$ and the position coordinate a polynomial of degree $n + 2$; these polynomials are represented by motion curves of a corresponding degree.

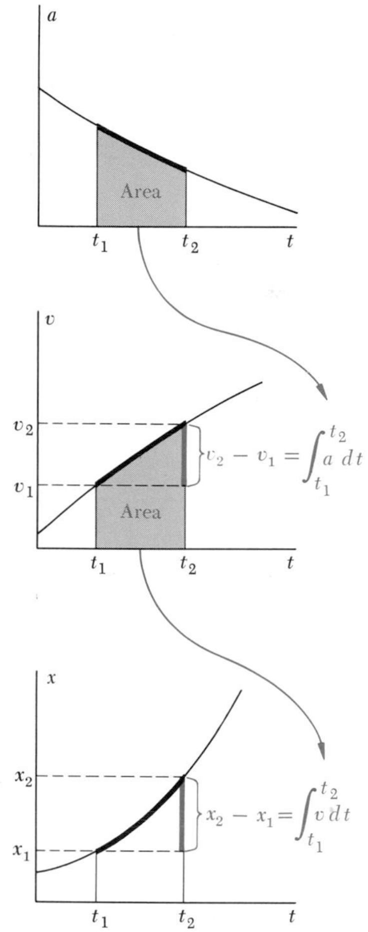

Fig. 11.11

*11.8. Other Graphical Methods.

An alternate graphical solution may be used to determine directly from the a–t curve the position of a particle at a given instant. Denoting respectively by x_0 and v_0 the values of x and v at $t = 0$, by x_1 and v_1 their values at $t = t_1$, and observing that the area under the v–t curve may be divided into a rectangle of area $v_0 t_1$ and horizontal differential elements of area $(t_1 - t)\,dv$ (Fig. 11.12*a*), we write

$$x_1 - x_0 = \text{area under } v\text{–}t \text{ curve} = v_0 t_1 + \int_{v_0}^{v_1} (t_1 - t)\,dv$$

Substituting $dv = a\,dt$ in the integral, we obtain

$$x_1 - x_0 = v_0 t_1 + \int_0^{t_1} (t_1 - t) a\,dt$$

Referring to Fig. 11.12*b*, we note that the integral represents the first moment of the area under the a–t curve with respect to the line $t = t_1$ bounding the area on the right. This method of solution is known, therefore, as the *moment-area method.* If the abscissa $\bar{t}$ of the centroid C of the area is known, the position coordinate x_1 may be obtained by writing

$$x_1 = x_0 + v_0 t_1 + (\text{area under } a\text{–}t \text{ curve})(t_1 - \bar{t}) \qquad (11.13)$$

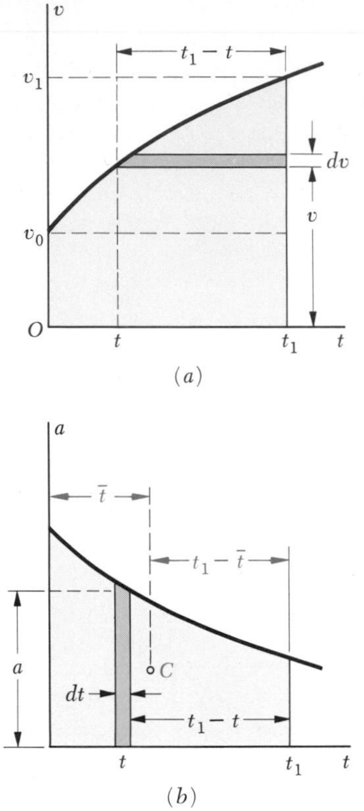

Fig. 11.12

If the area under the a–t curve is a composite area, the last term in (11.13) may be obtained by multiplying each component area by the distance from its centroid to the line $t = t_1$. Areas above the t axis should be considered as positive and areas below the t axis as negative.

Another type of motion curve, the v–x curve, is sometimes used. If such a curve has been plotted (Fig. 11.13), the acceleration a may be obtained at any time by drawing the normal to the curve and *measuring the subnormal BC.* Indeed, observing that the angle between AC and AB is equal to the angle θ between the horizontal and the tangent at A (the slope of which is $\tan\theta = dv/dx$), we write

$$BC = AB \tan\theta = v\frac{dv}{dx}$$

and thus, recalling formula (11.4),

$$BC = a$$

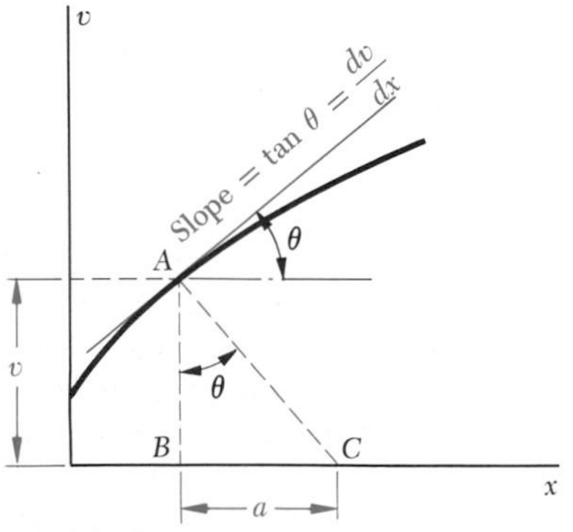

Fig. 11.13

SAMPLE PROBLEM 11.6

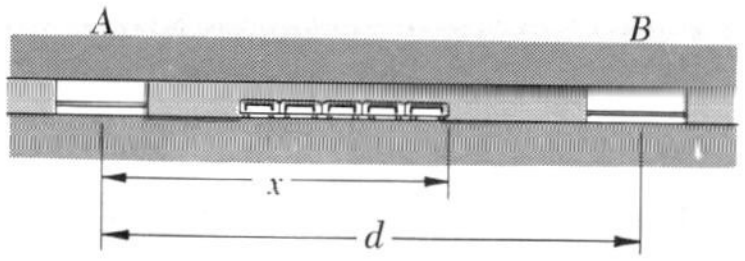

A subway train leaves station A; it gains speed at the rate of 4 ft/s^2 for 6 s, and then at the rate of 6 ft/s^2 until it has reached the speed of 48 ft/s. The train maintains the same speed until it approaches station B; brakes are then applied, giving the train a constant deceleration and bringing it to a stop in 6 s. The total running time from A to B is 40 s. Draw the a–t, v–t, and x–t curves, and determine the distance between stations A and B.

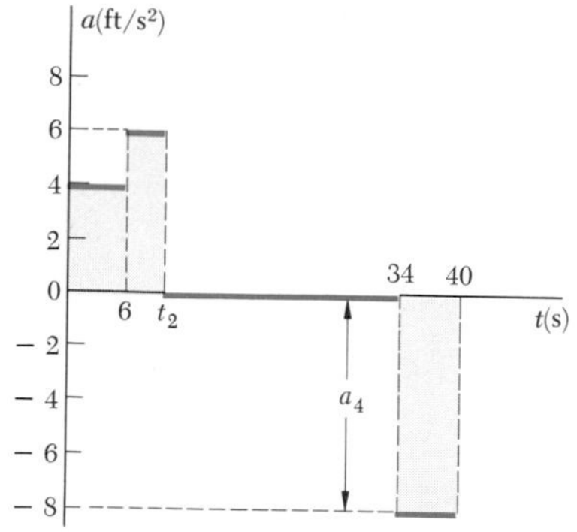

Acceleration-Time Curve. Since the acceleration is either constant or zero, the a–t curve is made of horizontal straight-line segments. The values of t_2 and a_4 are determined as follows:

$0 < t < 6$: Change in v = area under a–t curve

$$v_6 - 0 = (6 \text{ s})(4 \text{ ft/s}^2) = 24 \text{ ft/s}$$

$6 < t < t_2$: Since the velocity increases from 24 to 48 ft/s,

Change in v = area under a–t curve

$$48 - 24 = (t_2 - 6)(6 \text{ ft/s}^2) \qquad t_2 = 10 \text{ s}$$

$t_2 < t < 34$: Since the velocity is constant, the acceleration is zero.

$34 < t < 40$: Change in v = area under a–t curve

$$0 - 48 = (6 \text{ s})a_4 \qquad a_4 = -8 \text{ ft/s}^2$$

The acceleration being negative, the corresponding area is below the t axis; this area represents a decrease in velocity.

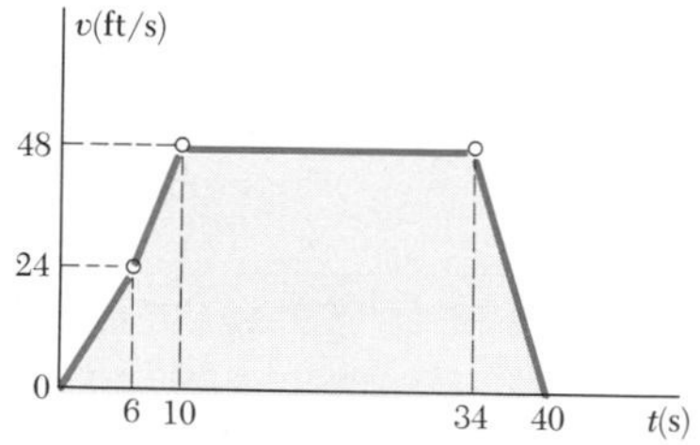

Velocity-Time Curve. Since the acceleration is either constant or zero, the v–t curve is made of segments of straight line connecting the points determined above.

Change in x = area under v–t curve

$0 < t < 6$: $\quad x_6 - 0 = \frac{1}{2}(6)(24) = 72 \text{ ft}$

$6 < t < 10$: $\quad x_{10} - x_6 = \frac{1}{2}(4)(24 + 48) = 144 \text{ ft}$

$10 < t < 34$: $\quad x_{34} - x_{10} = (24)(48) = 1152 \text{ ft}$

$34 < t < 40$: $\quad x_{40} - x_{34} = \frac{1}{2}(6)(48) = 144 \text{ ft}$

Adding the changes in x, we obtain the distance from A to B:

$$d = x_{40} - 0 = 1512 \text{ ft}$$

$d = 1512$ ft ◄

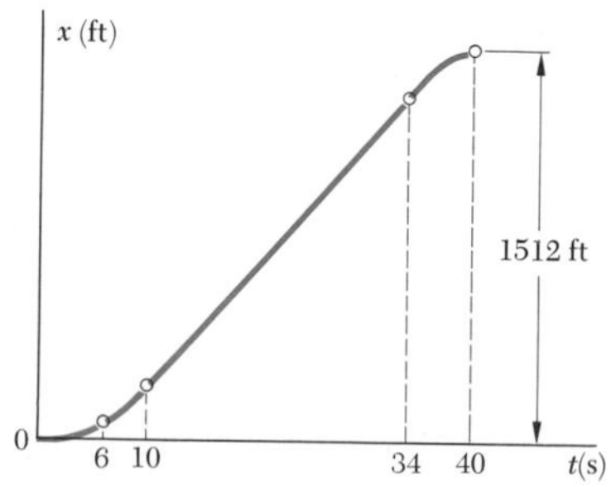

Position-Time Curve. The points determined above should be joined by three arcs of parabola and one segment of straight line. The construction of the x–t curve will be performed more easily and more accurately if we keep in mind that for any value of t the slope of the tangent to the x–t curve is equal to the value of v at that instant.

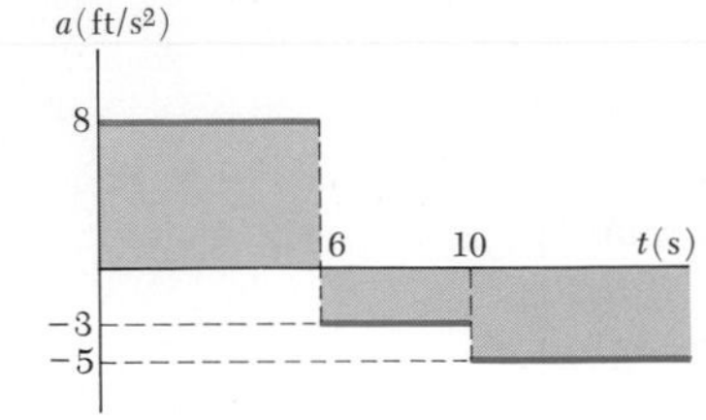

Fig. P11.41

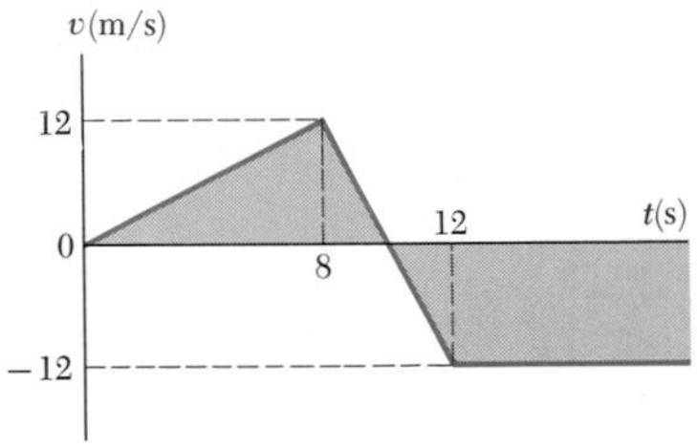

Fig. P11.43

PROBLEMS

11.41 A particle moves in a straight line with the acceleration shown in the figure. Knowing that it starts from the origin with $v_0 = -16$ ft/s, (*a*) plot the v–t and x–t curves for $0 < t < 16$ s, (*b*) determine its velocity, its position, and the total distance traveled after 12 s.

11.42 For the particle and motion of Prob. 11.41, plot the v–t and x–t curves for $0 < t < 16$ s and determine (*a*) the maximum value of the velocity of the particle, (*b*) the maximum value of its position coordinate.

11.43 A particle moves in a straight line with the velocity shown in the figure. Knowing that $x = -12$ m at $t = 0$, draw the a–t and x–t curves for $0 < t < 16$ s and determine (*a*) the total distance traveled by the particle after 12 s, (*b*) the two values of t for which the particle passes through the origin.

11.44 For the particle and motion of Prob. 11.43, plot the a–t and x–t curves for $0 < t < 16$ s and determine (*a*) the maximum value of the position coordinate of the particle, (*b*) the values of t for which the particle is at a distance of 15 m from the origin.

11.45 A series of city traffic signals is timed so that an automobile traveling at a constant speed of 25 mi/h will reach each signal just as it turns green. A motorist misses a signal and is stopped at signal *A*. Knowing that the next signal *B* is 750 ft ahead and that the maximum acceleration of his automobile is 6 ft/s^2, determine what the motorist should do to keep his maximum speed as small as possible, yet reach signal *B* just as it turns green. What is the maximum speed reached?

11.46 A bus starts from rest at point *A* and accelerates at the rate of 0.9 m/s^2 until it reaches a speed of 7.2 m/s. It then proceeds at 7.2 m/s until the brakes are applied; it comes to rest at point *B*, 18 m beyond the point where the brakes were applied. Assuming uniform deceleration and knowing that the distance between *A* and *B* is 90 m, determine the time required for the bus to travel from *A* to *B*.

11.47 The firing of a howitzer causes the barrel to recoil 800 mm before a braking mechanism brings it to rest. From a high-speed photographic record, it is found that the maximum value of the recoil velocity is 5.4 m/s and that this is reached 0.02 s after firing. Assuming that the recoil period consists of two phases during which the acceleration has, respectively, a constant positive value a_1 and a constant negative value a_2, determine (*a*) the values of a_1 and a_2, (*b*) the position of the barrel 0.02 s after firing, (*c*) the time at which the velocity of the barrel is zero.

11.48 A motorist is traveling at 45 mi/h when he observes that a traffic signal 1200 ft ahead of him turns red. He knows that the signal is timed to stay red for 24 s. What should he do to pass the signal at 15 mi/h just as it turns green again? Draw the v–t curve, selecting the solution which calls for the smallest possible deceleration and acceleration, and determine (*a*) the common value of the deceleration and acceleration in ft/s^2, (*b*) the minimum speed reached in mi/h.

11.49 A policeman on a motorcycle is escorting a motorcade which is traveling at 54 km/h. The policeman suddenly decides to take a new position in the motorcade, 70 m ahead. Assuming that he accelerates and decelerates at the rate of $2.5 m/s^2$ and that he does not exceed at any time a speed of 72 km/h, draw the a–t and v–t curves for his motion and determine (*a*) the shortest time in which he can occupy his new position in the motorcade, (*b*) the distance he will travel in that time.

11.50 A freight elevator moving upward with a constant velocity of 5 m/s passes a passenger elevator which is stopped. Three seconds later, the passenger elevator starts upward with an acceleration of $1.25 m/s^2$. When the passenger elevator has reached a velocity of 10 m/s, it proceeds at constant speed. Draw the v–t and y–t curves, and from them determine the time and distance required by the passenger elevator to overtake the freight elevator.

11.51 A car and a truck are both traveling at the constant speed of 35 mi/h; the car is 40 ft behind the truck. The driver of the car wants to pass the truck, i.e., he wishes to place his car at *B*, 40 ft in front of the truck, and then resume the speed of 35 mi/h. The maximum acceleration of the car is $5 ft/s^2$ and the maximum deceleration obtained by applying the brakes is $20 ft/s^2$. What is the shortest time in which the driver of the car can complete the passing operation if he does not at any time exceed a speed of 50 mi/h? Draw the v–t curve.

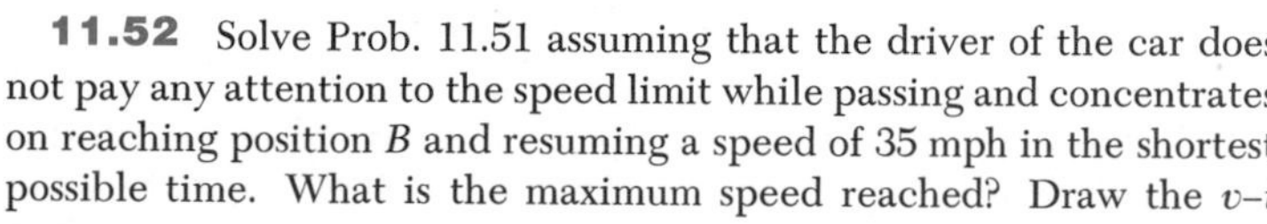

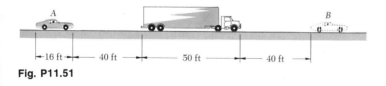

Fig. P11.51

11.52 Solve Prob. 11.51 assuming that the driver of the car does not pay any attention to the speed limit while passing and concentrates on reaching position *B* and resuming a speed of 35 mph in the shortest possible time. What is the maximum speed reached? Draw the v–t curve.

11.53 A car and a truck are both traveling at the constant speed of 60 mi/h; the car is 30 ft behind the truck. The truck driver suddenly applies his brakes, causing the truck to decelerate at the constant rate of 9 ft/s^2. Two seconds later the driver of the car applies his brakes and just manages to avoid a rear-end collision. Determine the constant rate at which the car decelerated.

11.54 Two cars are traveling toward each other on a single-lane road at 16 and 12 m/s, respectively. When 120 m apart, both drivers realize the situation and apply their brakes. They succeed in stopping simultaneously, and just short of colliding. Assuming a constant deceleration for each car, determine (*a*) the time required for the cars to stop, (*b*) the deceleration of each car, and (*c*) the distance traveled by each car while slowing down.

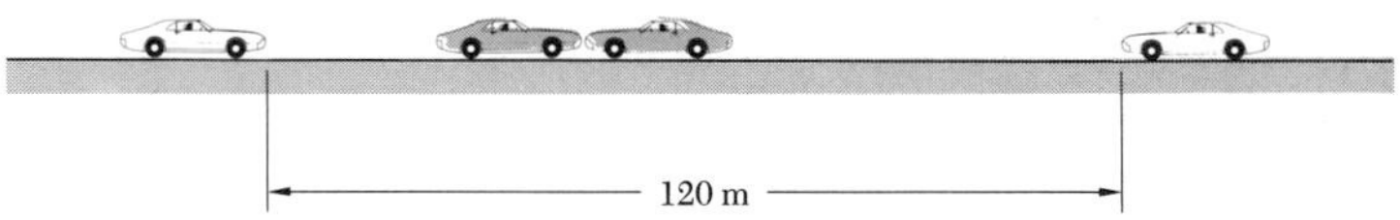

Fig. P11.54

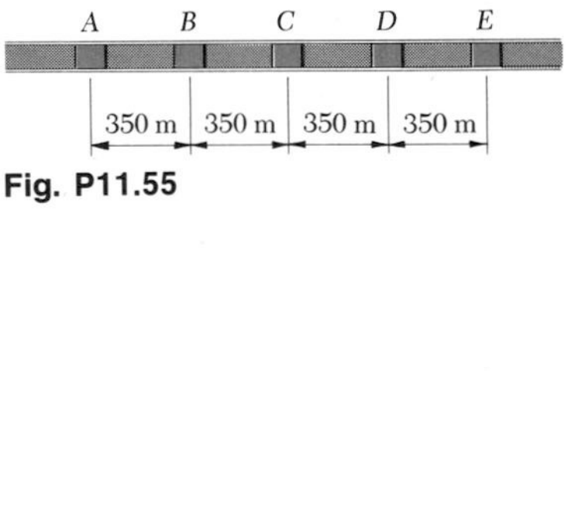

Fig. P11.55

11.55 An express subway train and a train making local stops run on parallel tracks between stations *A* and *E*, which are 1400 m apart. The local train makes stops of 30-s duration at each of the stations *B*, *C*, and *D*; the express train proceeds to station *E* without any intermediate stop. Each train accelerates at a rate of 1.25 m/s^2 until it reaches a speed of 12.5 m/s; it then proceeds at that constant speed. As the train approaches its next stop, the brakes are applied, providing a constant deceleration of 1.5 m/s^2. If the express train leaves station *A* 4 min after the local train has left *A*, determine (*a*) which of the two trains will arrive at station *E* first, (*b*) how much later the other train will arrive at station *E*.

11.56 The acceleration of a particle varies uniformly from $a = 75$ in./s^2 at $t = 0$, to $a = -75$ in./s^2 at $t = 8$ s. Knowing that $x = 0$ and $v = 0$ when $t = 0$, determine (*a*) the maximum velocity of the particle, (*b*) its position at $t = 8$ s, (*c*) its *average* velocity over the interval $0 < t < 8$ s. Draw the a–t, v–t, and x–t curves for the motion.

11.57 The rate of change of acceleration is known as the *jerk;* large or abrupt rates of change of acceleration cause discomfort to elevator passengers. If the jerk, or rate of change of the acceleration, of an elevator is limited to ± 0.5 m/s^2 per second, determine the shortest time required for an elevator, starting from rest, to rise 8 m and stop.

11.58 The acceleration record shown was obtained for a truck traveling on a straight highway. Knowing that the initial velocity of the truck was 18 km/h, determine the velocity and distance traveled when (*a*) $t = 4$ s, (*b*) $t = 6$ s.

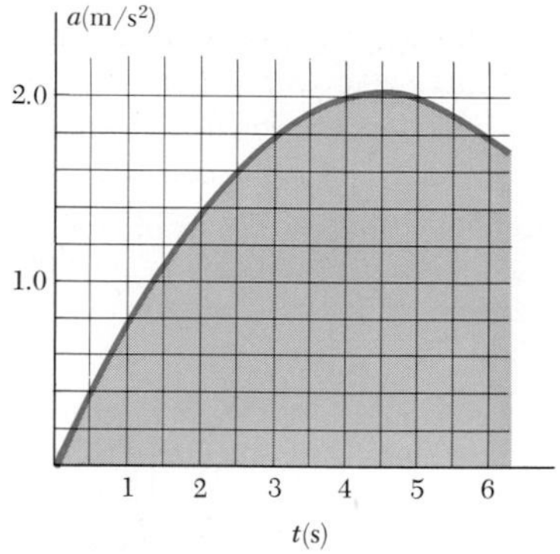

Fig. P11.58

11.59 A training airplane lands on an aircraft carrier and is brought to rest in 4 s by the arresting gear of the carrier. An accelerometer attached to the airplane provides the acceleration record shown. Determine by approximate means (*a*) the initial velocity of the airplane relative to the deck, (*b*) the distance the airplane travels along the deck before coming to rest.

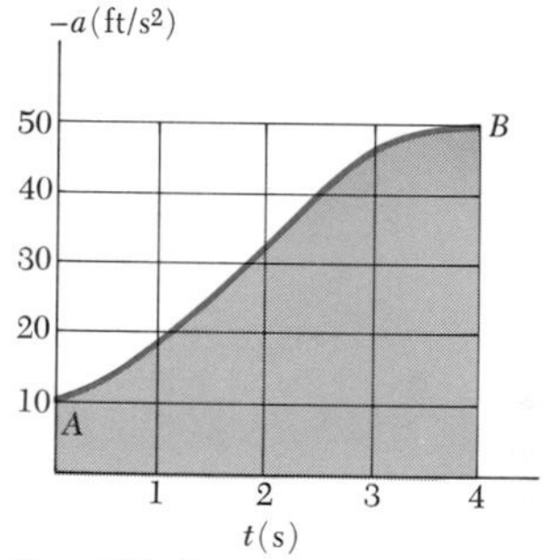

Fig. P11.59

11.60 The v–x curve shown was obtained experimentally during the motion of the bed of an industrial planer. Determine by approximate means the acceleration (*a*) when $x = 3$ in., (*b*) when $v = 40$ in./s.

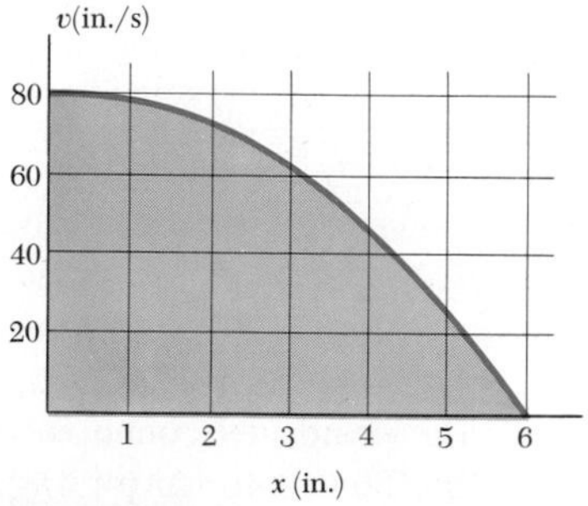

Fig. P11.60

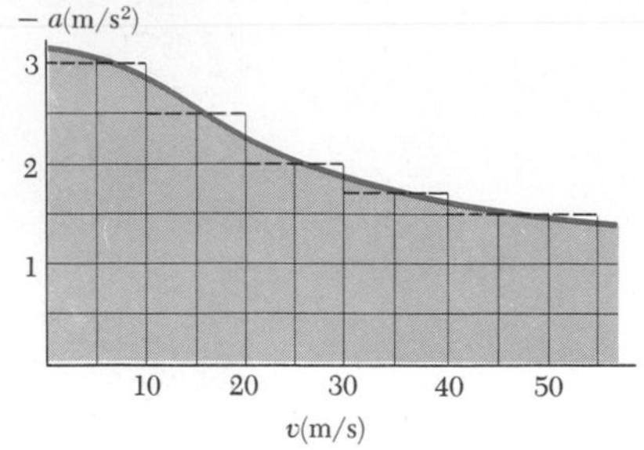

Fig. P11.61

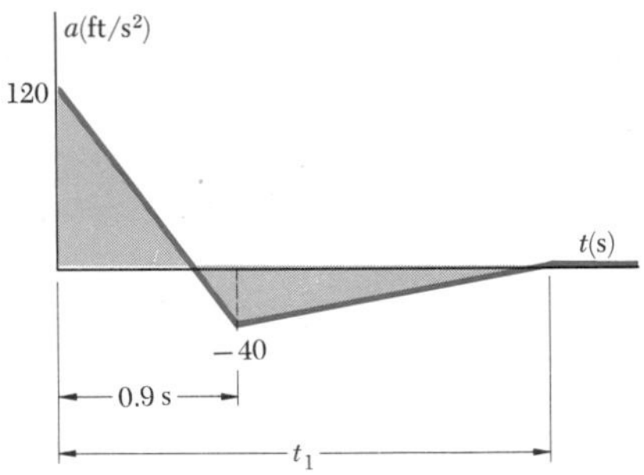

Fig. P11.64

11.61 The maximum possible acceleration of a passenger train under emergency conditions was determined experimentally; the results are shown (solid curve) in the figure. If the brakes are applied when the train is traveling at 90 km/h, determine by approximate means (*a*) the time required for the train to come to rest, (*b*) the distance traveled in that time.

11.62 Using the method of Sec. 11.8, derive the formula $x = x_0 + v_0 t + \frac{1}{2}at^2$, for the position coordinate of a particle in uniformly accelerated rectilinear motion.

11.63 Using the method of Sec. 11.8, obtain an approximate solution for Prob. 11.59, assuming that the a–t curve is a straight line from point A to point B.

11.64 The acceleration of an object subjected to the pressure wave of a large explosion is defined approximately by the curve shown. The object is initially at rest and is again at rest at time t_1. Using the method of Sec. 11.8, determine (*a*) the time t_1, (*b*) the distance through which the object is moved by the pressure wave.

11.65 Using the method of Sec. 11.8, determine the position of the particle of Prob. 11.41 when $t = 12$ s.

11.66 For the particle of Prob. 11.43, draw the a–t curve and, using the method of Sec. 11.8, determine (*a*) the position of the particle when $t = 14$ s, (*b*) the maximum value of its position coordinate.

CURVILINEAR MOTION OF PARTICLES

11.9. Position Vector, Velocity, and Acceleration. When a particle moves along a curve other than a straight line, we say that the particle is in *curvilinear motion*. To define the position P occupied by the particle at a given time t, we select a fixed reference system, such as the x, y, z axes shown in Fig. 11.14*a*, and draw the vector $\mathbf{r}$ joining the origin O and point P. Since the vector $\mathbf{r}$ is characterized by its magnitude r and its direction with respect to the reference axes, it completely defines the position of the particle with respect to those axes; the vector $\mathbf{r}$ is referred to as the *position vector* of the particle at time t.

Consider now the vector $\mathbf{r}'$ defining the position P' occupied by the same particle at a later time $t + \Delta t$. The vector $\Delta\mathbf{r}$ joining P and P' represents the change in the position vector during the time interval Δt since, as we may easily check from Fig. 11.14*a*, the vector $\mathbf{r}'$ is obtained by adding the vectors $\mathbf{r}$ and $\Delta\mathbf{r}$ according to the triangle rule. We note that $\Delta\mathbf{r}$ represents a change in *direction* as well as a change in *magnitude* of the position vector $\mathbf{r}$. The *average velocity* of the particle over the time interval Δt is defined as the quotient of $\Delta\mathbf{r}$ and Δt. Since $\Delta\mathbf{r}$ is a vector and Δt a scalar, the quotient $\Delta\mathbf{r}/\Delta t$ is a vector attached at P, of the same direction as $\Delta\mathbf{r}$, and of magnitude equal to the magnitude of $\Delta\mathbf{r}$ divided by Δt (Fig. 11.14*b*).

The *instantaneous velocity* of the particle at time t is obtained by choosing shorter and shorter time intervals Δt and, correspondingly, shorter and shorter vector increments $\Delta\mathbf{r}$. The instantaneous velocity is thus represented by the vector

$$\mathbf{v} = \lim_{\Delta t \to 0} \frac{\Delta\mathbf{r}}{\Delta t} \tag{11.14}$$

As Δt and $\Delta\mathbf{r}$ become shorter, the points P and P' get closer; the vector $\mathbf{v}$ obtained at the limit must therefore be tangent to the path of the particle (Fig. 11.14*c*).

Since the position vector $\mathbf{r}$ depends upon the time t, we may refer to it as a *vector function* of the scalar variable t and denote it by $\mathbf{r}(t)$. Extending the concept of derivative of a scalar function introduced in elementary calculus, we shall refer to the limit of the quotient $\Delta\mathbf{r}/\Delta t$ as the *derivative* of the vector function $\mathbf{r}(t)$. We write

$$\mathbf{v} = \frac{d\mathbf{r}}{dt} \tag{11.15}$$

The magnitude v of the vector $\mathbf{v}$ is called the *speed* of the particle. It may be obtained by substituting for the vector $\Delta\mathbf{r}$ in formula (11.14) its magnitude represented by the straight-line segment PP'. But the length of the segment PP' approaches the length Δs of the arc PP' as Δt decreases (Fig. 11.14*a*), and we may write

$$v = \lim_{\Delta t \to 0} \frac{PP'}{\Delta t} = \lim_{\Delta t \to 0} \frac{\Delta s}{\Delta t}$$

$$v = \frac{ds}{dt} \tag{11.16}$$

The speed v may thus be obtained by differentiating with respect to t the length s of the arc described by the particle.

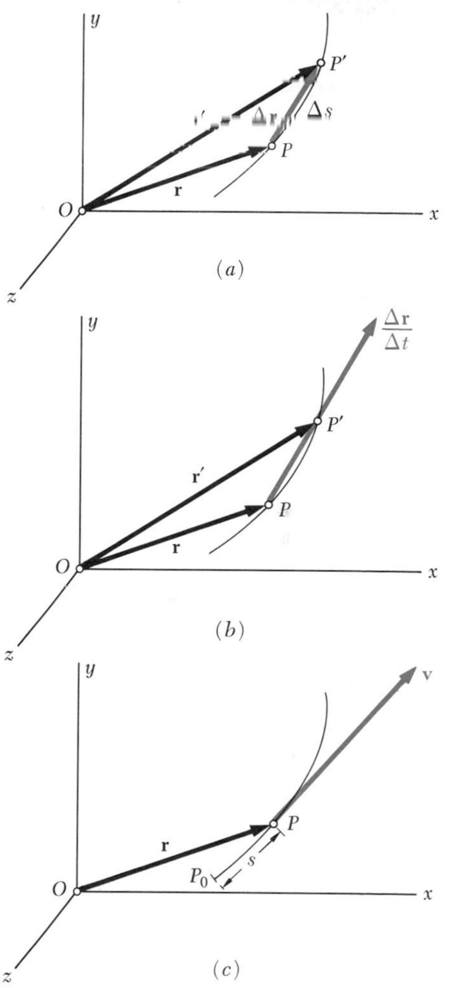

Fig. 11.14

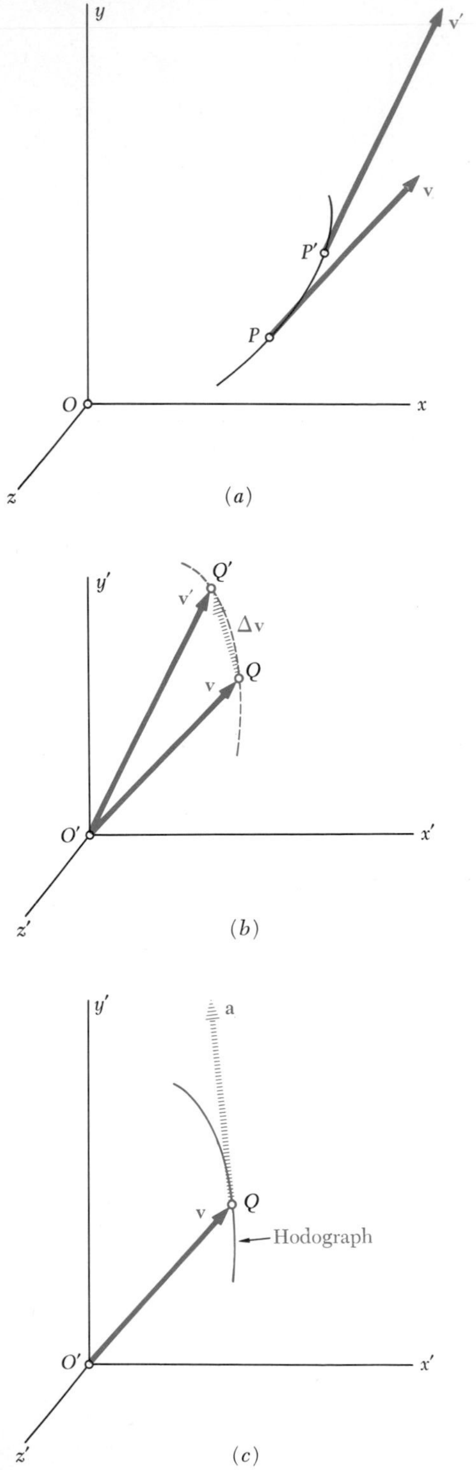

Consider the velocity $\mathbf{v}$ of the particle at time t and also its velocity $\mathbf{v}'$ at a later time $t + \Delta t$ (Fig. 11.15*a*). Let us draw both vectors $\mathbf{v}$ and $\mathbf{v}'$ from the same origin O' (Fig. 11.15*b*). The vector $\Delta\mathbf{v}$ joining Q and Q' represents the change in the velocity of the particle during the time interval Δt, since the vector $\mathbf{v}'$ may be obtained by adding the vectors $\mathbf{v}$ and $\Delta\mathbf{v}$. We should note that $\Delta\mathbf{v}$ represents a change in the *direction* of the velocity as well as a change in *speed*. The *average acceleration* of the particle over the time interval Δt is defined as the quotient of $\Delta\mathbf{v}$ and Δt. Since $\Delta\mathbf{v}$ is a vector and Δt a scalar, the quotient $\Delta\mathbf{v}/\Delta t$ is a vector of the same direction as $\Delta\mathbf{v}$.

The *instantaneous acceleration* of the particle at time t is obtained by choosing smaller and smaller values for Δt and $\Delta\mathbf{v}$. The instantaneous acceleration is thus represented by the vector

$$\mathbf{a} = \lim_{\Delta t \to 0} \frac{\Delta\mathbf{v}}{\Delta t} \tag{11.17}$$

Noting that the velocity $\mathbf{v}$ is a vector function $\mathbf{v}(t)$ of the time t, we may refer to the limit of the quotient $\Delta\mathbf{v}/\Delta t$ as the derivative of $\mathbf{v}$ with respect to t. We write

$$\mathbf{a} = \frac{d\mathbf{v}}{dt} \tag{11.18}$$

We observe that the acceleration $\mathbf{a}$ is tangent to the curve described by the tip Q of the vector $\mathbf{v}$ when the latter is drawn from a fixed origin O' (Fig. 11.15*c*) and that, in general, the acceleration is *not* tangent to the path of the particle (Fig. 11.15*d*). The curve described by the tip of $\mathbf{v}$ and shown in Fig. 11.15*c* is called the *hodograph* of the motion.

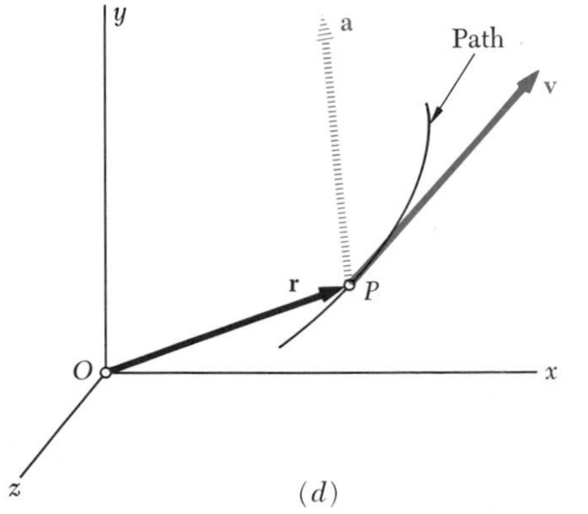

Fig. 11.15

11.10. Derivatives of Vector Functions. We saw in the preceding section that the velocity **v** of a particle in curvilinear motion may be represented by the derivative of the vector function **r**(t) characterizing the position of the particle. Similarly, the acceleration **a** of the particle may be represented by the derivative of the vector function **v**(t). In this section, we shall give a formal definition of the derivative of a vector function and establish a few rules governing the differentiation of sums and products of vector functions.

Let **P**(u) be a vector function of the scalar variable u. By that we mean that the scalar u completely defines the magnitude and direction of the vector **P**. If the vector **P** is drawn from a fixed origin O and the scalar u is allowed to vary, the tip of **P** will describe a given curve in space. Consider the vectors **P** corresponding respectively to the values u and $u + \Delta u$ of the scalar variable (Fig. 11.16a). Let $\Delta\mathbf{P}$ be the vector joining the tips of the two given vectors; we write

$$\Delta\mathbf{P} = \mathbf{P}(u + \Delta u) - \mathbf{P}(u)$$

Dividing through by Δu and letting Δu approach zero, *we define the derivative of the vector function* **P**(u):

$$\frac{d\mathbf{P}}{du} = \lim_{\Delta u \to 0} \frac{\Delta\mathbf{P}}{\Delta u} = \lim_{\Delta u \to 0} \frac{\mathbf{P}(u + \Delta u) - \mathbf{P}(u)}{\Delta u} \qquad (11.19)$$

As Δu approaches zero, the line of action of $\Delta\mathbf{P}$ becomes tangent to the curve of Fig. 11.16a. Thus, the derivative $d\mathbf{P}/du$ of the vector function **P**(u) *is tangent to the curve described by the tip of* **P**(u) (Fig. 11.16b).

We shall now show that the standard rules for the differentiation of the sums and products of scalar functions may be extended to vector functions. Consider first the *sum of two vector functions* **P**(u) and **Q**(u) of the same scalar variable u. According to the definition given in (11.19), the derivative of the vector $\mathbf{P} + \mathbf{Q}$ is

$$\frac{d(\mathbf{P} + \mathbf{Q})}{du} = \lim_{\Delta u \to 0} \frac{\Delta(\mathbf{P} + \mathbf{Q})}{\Delta u} = \lim_{\Delta u \to 0} \left(\frac{\Delta\mathbf{P}}{\Delta u} + \frac{\Delta\mathbf{Q}}{\Delta u}\right)$$

or, since the limit of a sum is equal to the sum of the limits of its terms,

$$\frac{d(\mathbf{P} + \mathbf{Q})}{du} = \lim_{\Delta u \to 0} \frac{\Delta\mathbf{P}}{\Delta u} + \lim_{\Delta u \to 0} \frac{\Delta\mathbf{Q}}{\Delta u}$$

$$\frac{d(\mathbf{P} + \mathbf{Q})}{du} = \frac{d\mathbf{P}}{du} + \frac{d\mathbf{Q}}{du} \qquad (11.20)$$

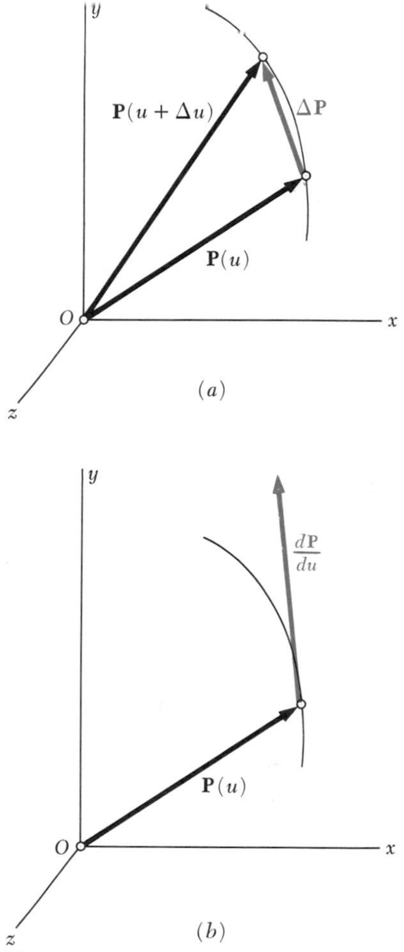

Fig. 11.16

Next, we shall consider the *product of a scalar function* $f(u)$ *and of a vector function* $\mathbf{P}(u)$ of the same scalar variable u. The derivative of the vector $f\mathbf{P}$ is

$$\frac{d(f\mathbf{P})}{du} = \lim_{\Delta u \to 0} \frac{(f + \Delta f)(\mathbf{P} + \Delta \mathbf{P}) - f\mathbf{P}}{\Delta u}$$

$$= \lim_{\Delta u \to 0} \left(\frac{\Delta f}{\Delta u}\mathbf{P} + f\frac{\Delta \mathbf{P}}{\Delta u} \right)$$

or, recalling the properties of the limits of sums and products,

$$\frac{d(f\mathbf{P})}{du} = \frac{df}{du}\mathbf{P} + f\frac{d\mathbf{P}}{du} \qquad (11.21)$$

The derivatives of the *scalar product* and of the *vector product* of two vector functions $\mathbf{P}(u)$ and $\mathbf{Q}(u)$ may be obtained in a similar way. We have

$$\frac{d(\mathbf{P} \cdot \mathbf{Q})}{du} = \frac{d\mathbf{P}}{du} \cdot \mathbf{Q} + \mathbf{P} \cdot \frac{d\mathbf{Q}}{du} \qquad (11.22)$$

$$\frac{d(\mathbf{P} \times \mathbf{Q})}{du} = \frac{d\mathbf{P}}{du} \times \mathbf{Q} + \mathbf{P} \times \frac{d\mathbf{Q}}{du} \qquad (11.23)\dagger$$

We shall use the properties established above to determine the *rectangular components of the derivative of a vector function* $\mathbf{P}(u)$. Resolving $\mathbf{P}$ into components along fixed rectangular axes x, y, z, we write

$$\mathbf{P} = P_x\mathbf{i} + P_y\mathbf{j} + P_z\mathbf{k} \qquad (11.24)$$

where P_x, P_y, P_z are the rectangular scalar components of the vector $\mathbf{P}$, and $\mathbf{i}, \mathbf{j}, \mathbf{k}$ the unit vectors corresponding respectively to the x, y, and z axes (Sec. 2.11). By (11.20), the derivative of $\mathbf{P}$ is equal to the sum of the derivatives of the terms in the right-hand member. Since each of these terms is the product of a scalar and a vector function, we should use (11.21). But the unit vectors $\mathbf{i}, \mathbf{j}, \mathbf{k}$ have a constant magnitude (equal to 1) and fixed directions. Their derivatives are therefore zero, and we write

$$\frac{d\mathbf{P}}{du} = \frac{dP_x}{du}\mathbf{i} + \frac{dP_y}{du}\mathbf{j} + \frac{dP_z}{du}\mathbf{k} \qquad (11.25)$$

† Since the vector product is not commutative (Sec. 3.3), the order of the factors in (11.23) must be maintained.

Noting that the coefficients of the unit vectors are, by definition, the scalar components of the vector $d\mathbf{P}/du$, we conclude that *the rectangular scalar components of the derivative* $d\mathbf{P}/du$ *of the vector function* $\mathbf{P}(u)$ *are obtained by differentiating the corresponding scalar components of* **P**.

Rate of Change of a Vector. When the vector **P** is a function of the time t, its derivative $d\mathbf{P}/dt$ represents the *rate of change* of **P** with respect to the frame $Oxyz$. Resolving **P** into rectangular components, we have, by (11.25),

$$\frac{d\mathbf{P}}{dt} = \frac{dP_x}{dt}\mathbf{i} + \frac{dP_y}{dt}\mathbf{j} + \frac{dP_z}{dt}\mathbf{k}$$

or, using dots to indicate differentiation with respect to t,

$$\dot{\mathbf{P}} = \dot{P}_x\mathbf{i} + \dot{P}_y\mathbf{j} + \dot{P}_z\mathbf{k} \qquad (11.25')$$

As we shall see in Sec. 15.10, the rate of change of a vector, as observed from a *moving frame of reference*, is, in general, different from its rate of change as observed from a fixed frame of reference. However, if the moving frame $O'x'y'z'$ is in *translation*, i.e., if its axes remain parallel to the corresponding axes of the fixed frame $Oxyz$ (Fig. 11.17), the same unit vectors **i**, **j**, **k** are

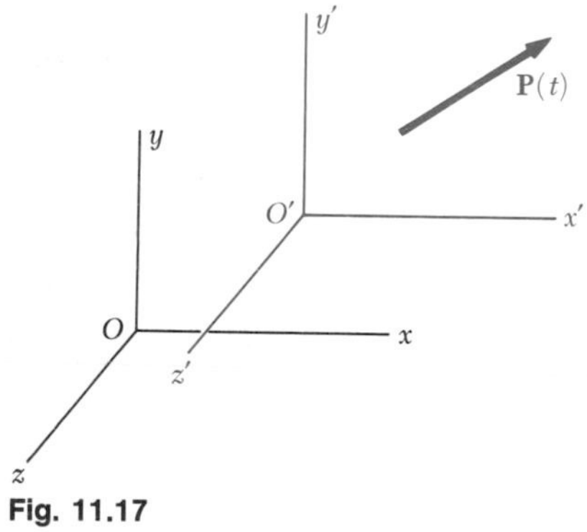

Fig. 11.17

used in both frames, and the vector **P** has, at any given instant, the same components P_x, P_y, P_z in both frames. It follows from (11.25′) that the rate of change $\dot{\mathbf{P}}$ is the same with respect to the frames $Oxyz$ and $O'x'y'z'$. We state, therefore: *The rate of change of a vector is the same with respect to a fixed frame and with respect to a frame in translation.* This property will greatly simplify our work, since we shall deal mainly with frames in translation.

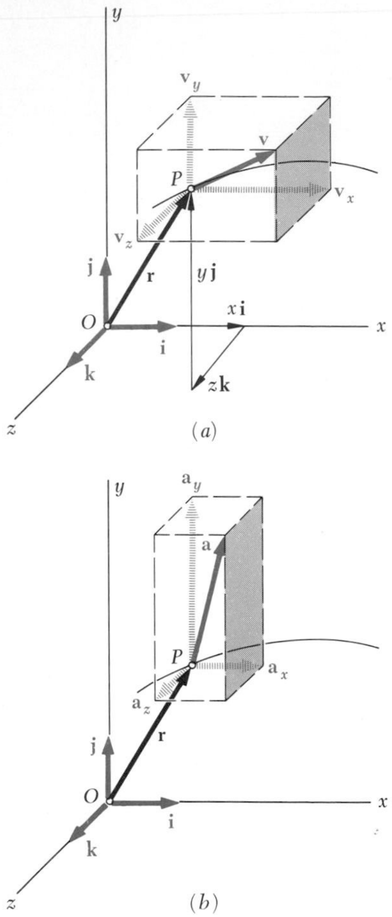

Fig. 11.18

11.11. Rectangular Components of Velocity and Acceleration. When the position of a particle P is defined at any instant by its rectangular coordinates x, y, and z, it is convenient to resolve the velocity $\mathbf{v}$ and the acceleration $\mathbf{a}$ of the particle into rectangular components (Fig. 11.18).

Resolving the position vector $\mathbf{r}$ of the particle into rectangular components, we write

$$\mathbf{r} = x\mathbf{i} + y\mathbf{j} + z\mathbf{k} \tag{11.26}$$

where the coordinates x, y, z are functions of t. Differentiating twice, we obtain

$$\mathbf{v} = \frac{d\mathbf{r}}{dt} = \dot{x}\mathbf{i} + \dot{y}\mathbf{j} + \dot{z}\mathbf{k} \tag{11.27}$$

$$\mathbf{a} = \frac{d\mathbf{v}}{dt} = \ddot{x}\mathbf{i} + \ddot{y}\mathbf{j} + \ddot{z}\mathbf{k} \tag{11.28}$$

where $\dot{x}$, $\dot{y}$, $\dot{z}$ and $\ddot{x}$, $\ddot{y}$, $\ddot{z}$ represent, respectively, the first and second derivatives of x, y, and z with respect to t. It follows from (11.27) and (11.28) that the scalar components of the velocity and acceleration are

$$v_x = \dot{x} \qquad v_y = \dot{y} \qquad v_z = \dot{z} \tag{11.29}$$

$$a_x = \ddot{x} \qquad a_y = \ddot{y} \qquad a_z = \ddot{z} \tag{11.30}$$

A positive value for v_x indicates that the vector component $\mathbf{v}_x$ is directed to the right, a negative value that it is directed to the left; the sense of each of the other vector components may be determined in a similar way from the sign of the corresponding scalar component. If desired, the magnitudes and directions of the velocity and acceleration may be obtained from their scalar components by the methods of Secs. 2.6 and 2.11.

The use of rectangular components to describe the position, the velocity, and the acceleration of a particle is particularly effective when the component a_x of the acceleration depends only upon t, x, and/or v_x, and when, similarly, a_y depends only upon t, y, and/or v_y, and a_z upon t, z, and/or v_z. Equations (11.30) may then be integrated independently, and so may Eqs. (11.29). In other words, the motion of the particle in the x direction, its motion in the y direction, and its motion in the z direction may be considered separately.

In the case of the *motion of a projectile*, for example, it may be shown (see Sec. 12.4) that the components of the acceleration are

$$a_x = \ddot{x} = 0 \qquad a_y = \ddot{y} = -g \qquad a_z = \ddot{z} = 0$$

if the resistance of the air is neglected. Denoting by x_0, y_0, z_0 the coordinates of the gun, and by $(v_x)_0$, $(v_y)_0$, $(v_z)_0$ the components of the initial velocity $\mathbf{v}_0$ of the projectile, we integrate twice in t and obtain

$$v_x = \dot{x} = (v_x)_0 \qquad v_y = \dot{y} = (v_y)_0 - gt \qquad v_z = \dot{z} = (v_z)_0$$
$$x = x_0 + (v_x)_0 t \qquad y = y_0 + (v_y)_0 t - \tfrac{1}{2}gt^2 \qquad z = z_0 + (v_z)_0 t$$

If the projectile is fired in the xy plane from the origin O, we have $x_0 = y_0 = z_0 = 0$ and $(v_z)_0 = 0$, and the equations of motion reduce to

$$v_x = (v_x)_0 \qquad v_y = (v_y)_0 - gt \qquad v_z = 0$$
$$x = (v_x)_0 t \qquad y = (v_y)_0 t - \tfrac{1}{2}gt^2 \qquad z = 0$$

These equations show that the projectile remains in the xy plane and that its motion in the horizontal direction is uniform, while its motion in the vertical direction is uniformly accelerated. The motion of a projectile may thus be replaced by two independent rectilinear motions, which are easily visualized if we assume that the projectile is fired vertically with an initial velocity $(\mathbf{v}_y)_0$ from a platform moving with a constant horizontal velocity $(\mathbf{v}_x)_0$ (Fig. 11.19). The coordinate x of the projectile is equal at any instant to the distance traveled by the platform, while its coordinate y may be computed as if the projectile were moving along a vertical line.

It may be observed that the equations defining the coordinates x and y of a projectile at any instant are the parametric equations of a parabola. Thus, the trajectory of a projectile is *parabolic*. This result, however, ceases to be valid when the resistance of the air or the variation with altitude of the acceleration of gravity is taken into account.

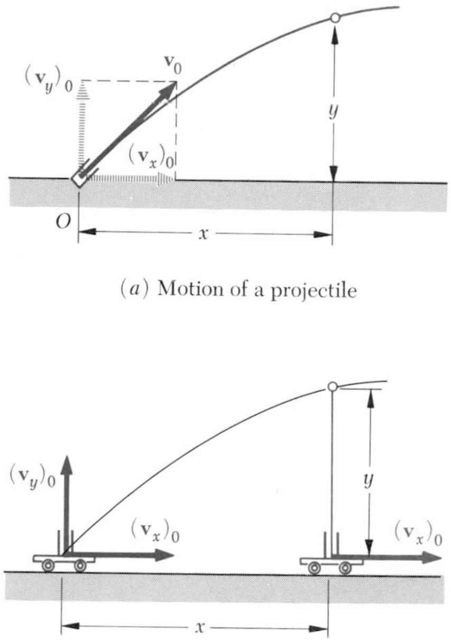

(*a*) Motion of a projectile

(*b*) Equivalent rectilinear motions

Fig. 11.19

11.12. Motion Relative to a Frame in Translation. In the preceding section, a single frame of reference was used to describe the motion of a particle. In most cases this frame was attached to the earth and was considered as fixed. We shall now analyze situations in which it is convenient to use simultaneously several frames of reference. If one of the frames is attached to the earth, we shall call it a *fixed frame of reference* and the other frames will be referred to as *moving frames of reference*. It should be understood, however, that the selection of a fixed frame of reference is purely arbitrary. Any frame may be designated as "fixed"; all other frames not rigidly attached to this frame will then be described as "moving."

Consider two particles A and B moving in space (Fig. 11.20); the vectors $\mathbf{r}_A$ and $\mathbf{r}_B$ define their positions at any given instant with respect to the fixed frame of reference $Oxyz$. Consider now

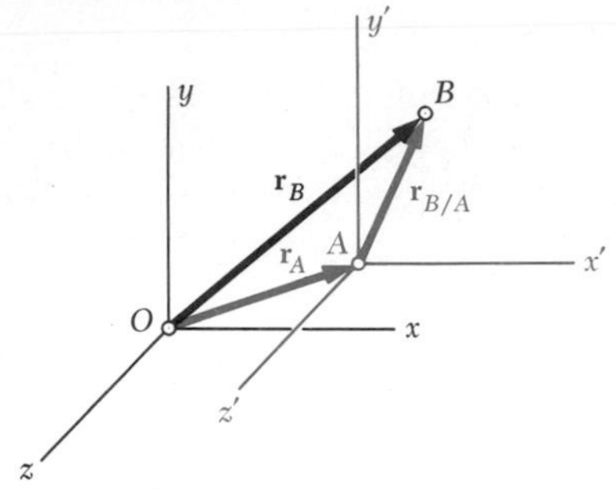

Fig. 11.20

a system of axes x', y', z' centered at A and parallel to the x, y, z axes. While the origin of these axes moves, their orientation remains the same; the frame of reference $Ax'y'z'$ is in *translation* with respect to $Oxyz$. The vector $\mathbf{r}_{B/A}$ joining A and B defines *the position of B relative to the moving frame Ax'y'z'* (or, for short, *the position of B relative to A*).

We note from Fig. 11.20 that the position vector $\mathbf{r}_B$ of particle B is the sum of the position vector $\mathbf{r}_A$ of particle A and of the position vector $\mathbf{r}_{B/A}$ of B relative to A; we write

$$\mathbf{r}_B = \mathbf{r}_A + \mathbf{r}_{B/A} \tag{11.31}$$

Differentiating (11.31) with respect to t within the fixed frame of reference, and using dots to indicate time derivatives, we have

$$\dot{\mathbf{r}}_B = \dot{\mathbf{r}}_A + \dot{\mathbf{r}}_{B/A} \tag{11.32}$$

The derivatives $\dot{\mathbf{r}}_A$ and $\dot{\mathbf{r}}_B$ represent, respectively, the velocities $\mathbf{v}_A$ and $\mathbf{v}_B$ of the particles A and B. The derivative $\dot{\mathbf{r}}_{B/A}$ represents the rate of change of $\mathbf{r}_{B/A}$ with respect to the frame $Ax'y'z'$, as well as with respect to the fixed frame, since $Ax'y'z'$ is in translation (Sec. 11.10). This derivative, therefore, defines *the velocity $\mathbf{v}_{B/A}$ of B relative to the frame Ax'y'z'* (or, for short, *the velocity $\mathbf{v}_{B/A}$ of B relative to A*). We write

$$\mathbf{v}_B = \mathbf{v}_A + \mathbf{v}_{B/A} \tag{11.33}$$

Differentiating Eq. (11.33) with respect to t, and using the derivative $\dot{\mathbf{v}}_{B/A}$ to define *the acceleration $\mathbf{a}_{B/A}$ of B relative to the frame Ax'y'z'* (or, for short, *the acceleration of $\mathbf{a}_{B/A}$ of B relative to A*), we write

$$\mathbf{a}_B = \mathbf{a}_A + \mathbf{a}_{B/A} \tag{11.34}$$

The motion of B with respect to the fixed frame $Oxyz$ is referred to as the *absolute motion of B*. The equations derived in this section show that *the absolute motion of B may be obtained by combining the motion of A and the relative motion of B with respect to the moving frame attached to A*. Equation (11.33), for example, expresses that the absolute velocity $\mathbf{v}_B$ of particle B may be obtained by adding vectorially the velocity of A and the velocity of B relative to the frame $Ax'y'z'$. Equation (11.34) expresses a similar property in terms of the accelerations. We should keep in mind, however, that *the frame Ax'y'z' is in translation;* i.e., while it moves with A, it maintains the same orientation. As we shall see later (Sec. 15.14), different relations must be used in the case of a rotating frame of reference.

SAMPLE PROBLEM 11.7

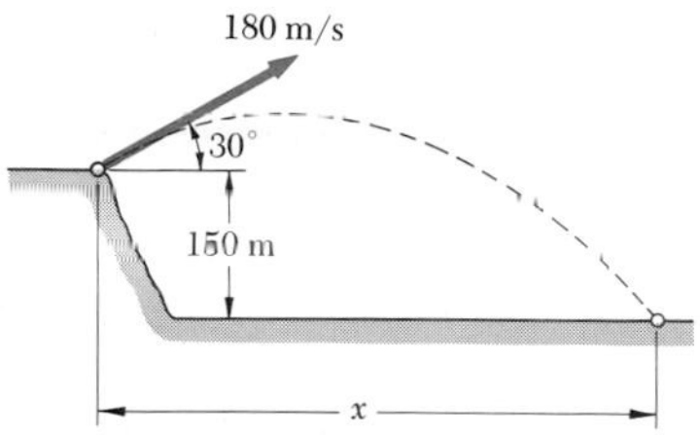

A projectile is fired from the edge of a 150-m cliff with an initial velocity of 180 m/s, at an angle of 30° with the horizontal. Neglecting air resistance, find (*a*) the horizontal distance from the gun to the point where the projectile strikes the ground, (*b*) the greatest elevation above the ground reached by the projectile.

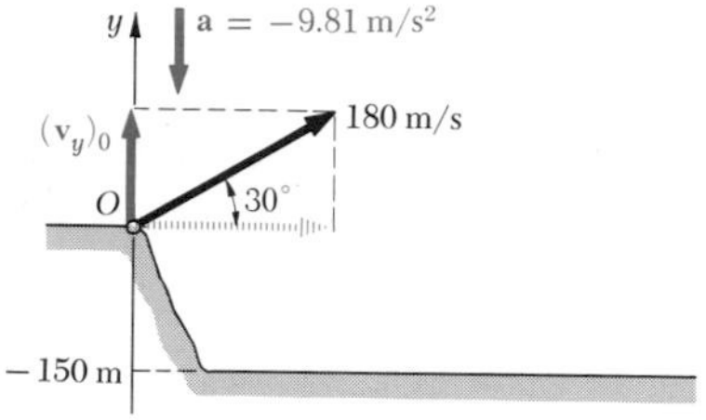

Solution. We shall consider separately the vertical and the horizontal motion.

Vertical Motion. Uniformly accelerated motion. Choosing the positive sense of the y axis upward and placing the origin O at the gun, we have

$$(v_y)_0 = (180 \text{ m/s}) \sin 30° = +90 \text{ m/s}$$
$$a = -9.81 \text{ m/s}^2$$

Substituting into the equations of uniformly accelerated motion, we have

$$v_y = (v_y)_0 + at \qquad v_y = 90 - 9.81t \qquad (1)$$
$$y = (v_y)_0 t + \tfrac{1}{2}at^2 \qquad y = 90t - 4.90t^2 \qquad (2)$$
$$v_y^2 = (v_y)_0^2 + 2ay \qquad v_y^2 = 8100 - 19.62y \qquad (3)$$

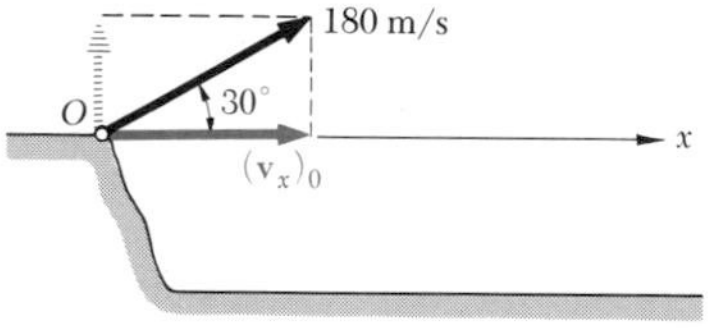

Horizontal Motion. Uniform motion. Choosing the positive sense of the x axis to the right, we have

$$(v_x)_0 = (180 \text{ m/s}) \cos 30° = +155.9 \text{ m/s}$$

Substituting into the equation of uniform motion, we obtain

$$x = (v_x)_0 t \qquad x = 155.9t \qquad (4)$$

***a.* Horizontal Distance.** When the projectile strikes the ground, we have

$$y = -150 \text{ m}$$

Carrying this value into Eq. (2) for the vertical motion, we write

$$-150 = 90t - 4.90t^2 \qquad t^2 - 18.37t - 30.6 = 0 \qquad t = 19.91 \text{ s}$$

Carrying $t = 19.91$ s into Eq. (4) for the horizontal motion, we obtain

$$x = 155.9(19.91) \qquad x = 3100 \text{ m} \quad ◄$$

***b.* Greatest Elevation.** When the projectile reaches its greatest elevation, we have $v_y = 0$; carrying this value into Eq. (3) for the vertical motion, we write

$$0 = 8100 - 19.62y \qquad y = 413 \text{ m}$$

Greatest elevation above ground = 150 m + 413 m = 563 m ◄

SAMPLE PROBLEM 11.8

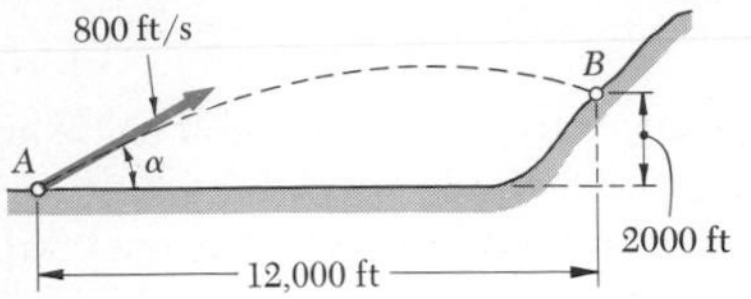

A projectile is fired with an initial velocity of 800 ft/s at a target B located 2000 ft above the gun A and at a horizontal distance of 12,000 ft. Neglecting air resistance, determine the value of the firing angle α.

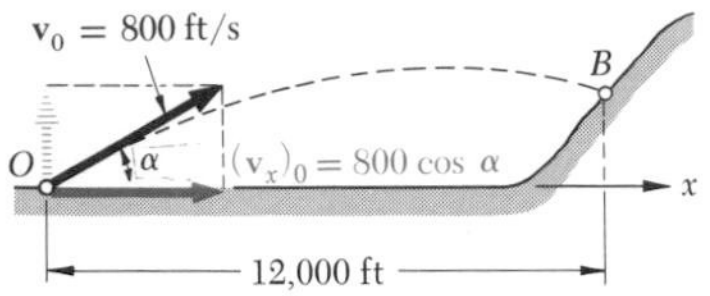

Solution. We shall consider separately the horizontal and the vertical motion.

Horizontal Motion. Placing the origin of coordinates at the gun, we have

$$(v_x)_0 = 800 \cos \alpha$$

Substituting into the equation of uniform horizontal motion, we obtain

$$x = (v_x)_0 t \qquad x = (800 \cos \alpha)t$$

The time required for the projectile to move through a horizontal distance of 12,000 ft is obtained by making x equal to 12,000 ft.

$$12{,}000 = (800 \cos \alpha)t$$

$$t = \frac{12{,}000}{800 \cos \alpha} = \frac{15}{\cos \alpha}$$

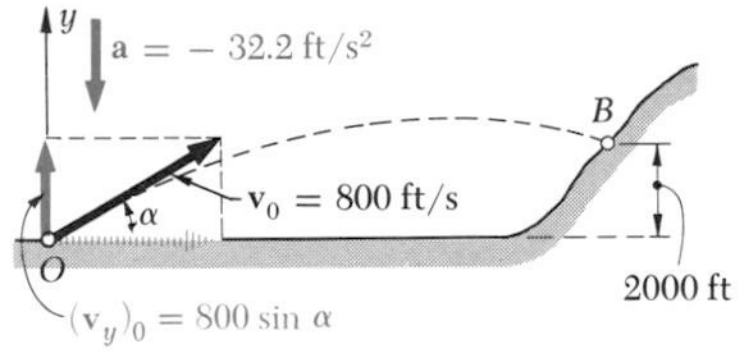

Vertical Motion

$$(v_y)_0 = 800 \sin \alpha \qquad a = -32.2 \text{ ft/s}^2$$

Substituting into the equation of uniformly accelerated vertical motion, we obtain

$$y = (v_y)_0 t + \tfrac{1}{2}at^2 \qquad y = (800 \sin \alpha)t - 16.1t^2$$

Projectile Hits Target. When $x = 12{,}000$ ft, we must have $y = 2000$ ft. Substituting for y and making t equal to the value found above, we write

$$2000 = 800 \sin \alpha \frac{15}{\cos \alpha} - 16.1\left(\frac{15}{\cos \alpha}\right)^2$$

Since $1/\cos^2 \alpha = \sec^2 \alpha = 1 + \tan^2 \alpha$, we have

$$2000 = 800(15) \tan \alpha - 16.1(15^2)(1 + \tan^2 \alpha)$$
$$3622 \tan^2 \alpha - 12{,}000 \tan \alpha + 5622 = 0$$

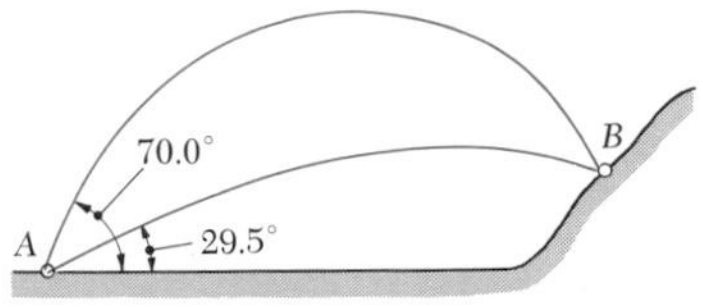

Solving this quadratic equation for $\tan \alpha$, we have

$$\tan \alpha = 0.565 \qquad \text{and} \qquad \tan \alpha = 2.75$$

$$\alpha = 29.5° \qquad \text{and} \qquad \alpha = 70.0°$$ ◄

The target will be hit if either of these two firing angles is used (see figure).

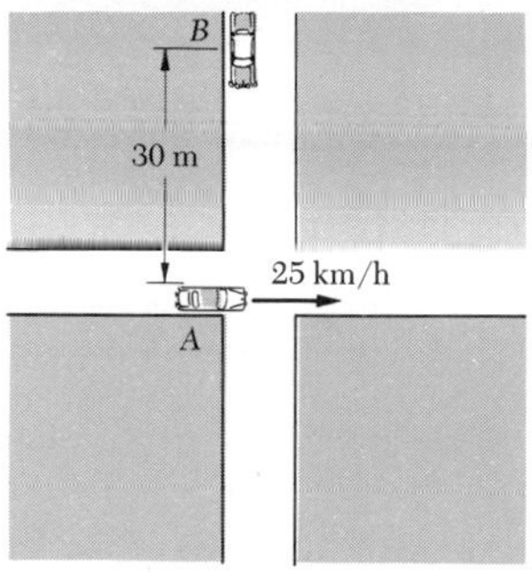

SAMPLE PROBLEM 11.9

Automobile A is traveling east at the constant speed of 25 km/h. As automobile A crosses the intersection shown, automobile B starts from rest 30 m north of the intersection and moves south with a constant acceleration of $1.2\ \text{m/s}^2$. Determine the position, velocity, and acceleration of B relative to A five seconds after A crosses the intersection.

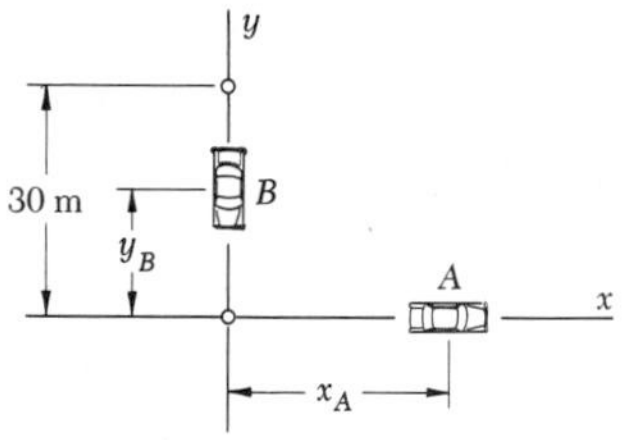

Solution. We choose x and y axes with origin at the intersection of the two streets and with positive senses directed respectively east and north.

Motion of Automobile A. First the speed is expressed in m/s:

$$25\ \text{km/h} = \frac{25\ \text{km}}{1\ \text{h}} = \frac{25\,000\ \text{m}}{3600\ \text{s}} = 6.94\ \text{m/s}$$

Noting that the motion of A is uniform, we write, for any time t,

$$a_A = 0$$
$$v_A = +6.94\ \text{m/s}$$
$$x_A = (x_A)_0 + v_A t = 0 + 6.94t$$

For $t = 5$ s, we have

$$a_A = 0 \qquad\qquad \mathbf{a}_A = 0$$
$$v_A = +6.94\ \text{m/s} \qquad\qquad \mathbf{v}_A = 6.94\ \text{m/s} \rightarrow$$
$$x_A = +(6.94\ \text{m/s})(5\ \text{s}) = +34.7\ \text{m} \qquad\qquad \mathbf{r}_A = 34.7\ \text{m} \rightarrow$$

Motion of Automobile B. We note that the motion of B is uniformly accelerated, and write

$$a_B = -1.2\ \text{m/s}^2$$
$$v_B = (v_B)_0 + at = 0 - 1.2t$$
$$y_B = (y_B)_0 + (v_B)_0 t + \tfrac{1}{2}a_B t^2 = 30 + 0 - \tfrac{1}{2}(1.2)t^2$$

For $t = 5$ s, we have

$$a_B = -1.2\ \text{m/s}^2 \qquad\qquad \mathbf{a}_B = 1.2\ \text{m/s}^2 \downarrow$$
$$v_B = -(1.2\ \text{m/s})(5\ \text{s}) = -6\ \text{m/s} \qquad\qquad \mathbf{v}_B = 6\ \text{m/s} \downarrow$$
$$y_B = 30 - \tfrac{1}{2}(1.2\ \text{m/s})(5\ \text{s})^2 = +15\ \text{m} \qquad\qquad \mathbf{r}_B = 15\ \text{m} \uparrow$$

Motion of B Relative to A. We draw the triangle corresponding to the vector equation $\mathbf{r}_B = \mathbf{r}_A + \mathbf{r}_{B/A}$ and obtain the magnitude and direction of the position vector of B relative to A.

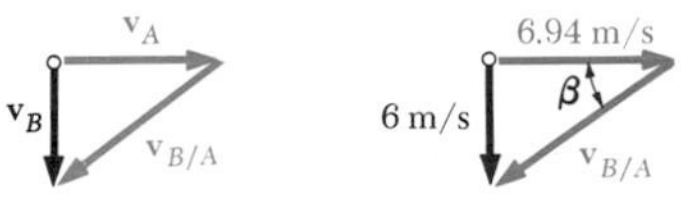

$$r_{B/A} = 37.8\ \text{m} \qquad \alpha = 23.4° \qquad \mathbf{r}_{B/A} = 37.8\ \text{m} \nwarrow 23.4° \quad ◀$$

Proceeding in a similar fashion, we find the velocity and acceleration of B relative to A.

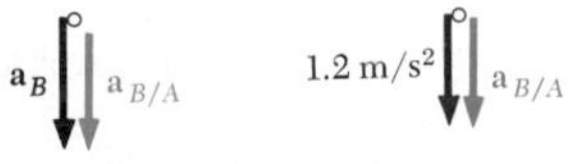

$$\mathbf{v}_B = \mathbf{v}_A + \mathbf{v}_{B/A}$$
$$v_{B/A} = 9.17\ \text{m/s} \qquad \beta = 40.8° \qquad \mathbf{v}_{B/A} = 9.17\ \text{m/s} \swarrow 40.8° \quad ◀$$
$$\mathbf{a}_B = \mathbf{a}_A + \mathbf{a}_{B/A} \qquad\qquad \mathbf{a}_{B/A} = 1.2\ \text{m/s}^2 \downarrow \quad ◀$$

PROBLEMS

Note. Neglect air resistance in problems concerning projectiles.

11.67 The motion of a particle is defined by the equations $x = \frac{1}{3}t^3 - 2t^2$ and $y = \frac{1}{2}t^2 - 2t$, where x and y are expressed in meters and t in seconds. Determine the velocity and acceleration when (*a*) $t = 1$ s, (*b*) $t = 3$ s.

11.68 In Prob. 11.67, determine (*a*) the time at which the value of the y coordinate is minimum, (*b*) the corresponding velocity and acceleration of the particle.

11.69 The motion of a particle is defined by the equations $x = e^{t/2}$ and $y = e^{-t/2}$, where x and y are expressed in feet and t in seconds. Show that the path of the particle is a rectangular hyperbola and determine the velocity and acceleration when (*a*) $t = 0$, (*b*) $t = 1$ s.

11.70 The motion of a particle is defined by the equations $x = 5(1 - e^{-t})$ and $y = 5t/(t + 1)$, where x and y are expressed in feet and t in seconds. Determine the velocity and acceleration when $t = 1$ s.

11.71 The motion of a vibrating particle is defined by the position vector $\mathbf{r} = (100 \sin \pi t)\mathbf{i} + (25 \cos 2\pi t)\mathbf{j}$, where r is expressed in millimeters and t in seconds. (*a*) Determine the velocity and acceleration when $t = 1$ s. (*b*) Show that the path of the particle is parabolic.

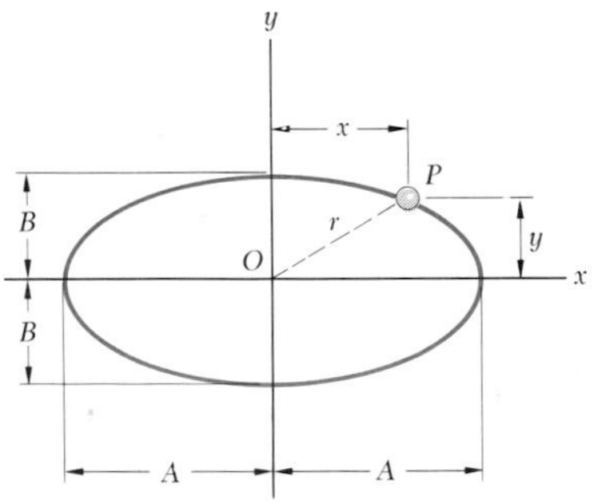

Fig. P11.72

11.72 A particle moves in an elliptic path defined by the position vector $\mathbf{r} = (A \cos pt)\mathbf{i} + (B \sin pt)\mathbf{j}$. Show that the acceleration (*a*) is directed toward the origin, (*b*) is proportional to the distance from the origin to the particle.

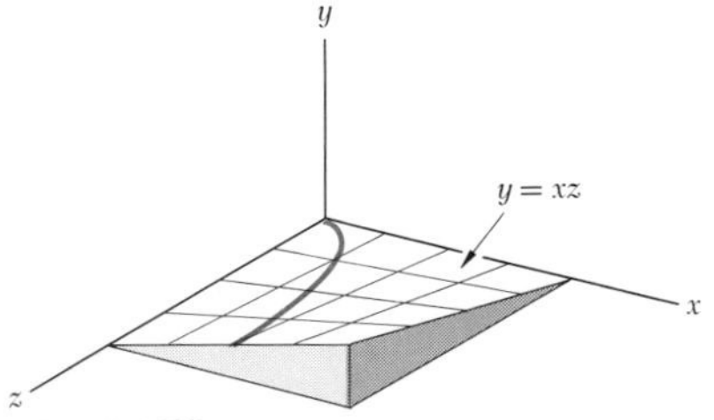

Fig. P11.73

11.73 The three-dimensional motion of a particle is defined by the position vector $\mathbf{r} = At\mathbf{i} + ABt^3\mathbf{j} + Bt^2\mathbf{k}$, where r is expressed in feet and t in seconds. Show that the space curve described by the particle lies on the hyperbolic paraboloid $y = xz$. For $A = B = 1$, determine the magnitudes of the velocity and acceleration when (*a*) $t = 0$, (*b*) $t = 2$ s.

11.74 The three-dimensional motion of a particle is defined by the position vector $\mathbf{r} = (R \sin pt)\mathbf{i} + ct\mathbf{j} + (R \cos pt)\mathbf{k}$. Determine the magnitudes of the velocity and acceleration of the particle. (The space curve described by the particle is a helix.)

11.75 A man standing on a bridge 20 m above the water throws a stone in a horizontal direction. Knowing that the stone hits the water 30 m from a point on the water directly below the man, determine (*a*) the initial velocity of the stone, (*b*) the distance at which the stone would hit the water if it were thrown with the same velocity from a bridge 5 m lower.

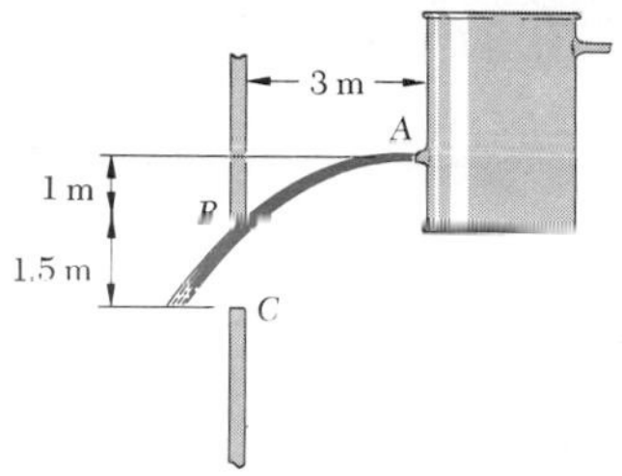

Fig. P11.76

11.76 Water issues at *A* from a pressure tank with a horizontal velocity $\mathbf{v}_0$. For what range of values v_0 will the water enter the opening *BC*?

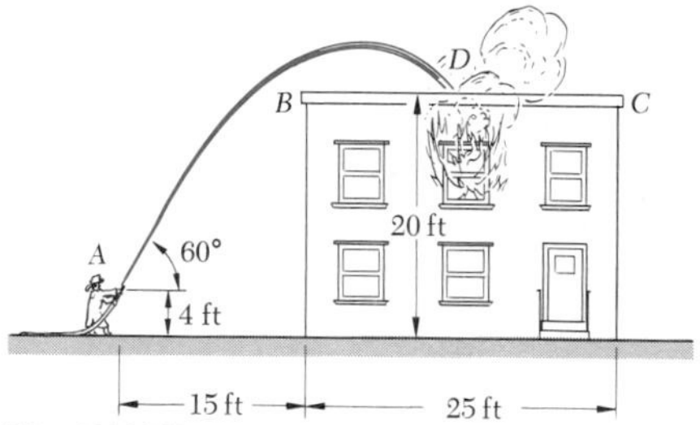

Fig. P11.77

11.77 A nozzle at *A* discharges water with an initial velocity of 40 ft/s at an angle of 60° with the horizontal. Determine where the stream of water strikes the roof. Check that the stream will clear the edge of the roof.

11.78 In Prob. 11.77, determine the largest and smallest initial velocity for which the water will fall on the roof.

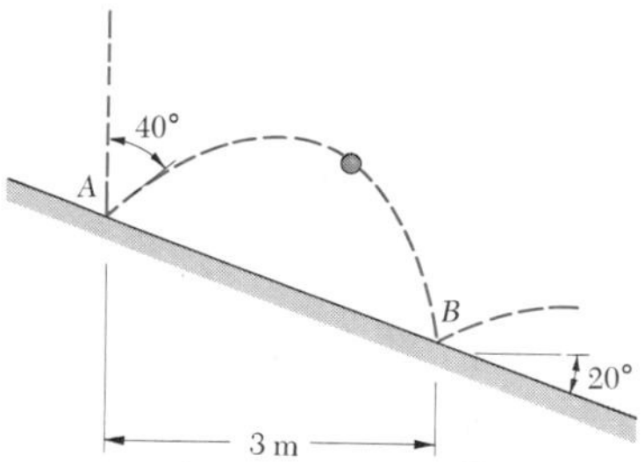

Fig. P11.79

11.79 A ball is dropped vertically onto a 20° incline at *A*; the direction of rebound forms an angle of 40° with the vertical. Knowing that the ball next strikes the incline at *B*, determine (*a*) the velocity of rebound at *A*, (*b*) the time required for the ball to travel from *A* to *B*.

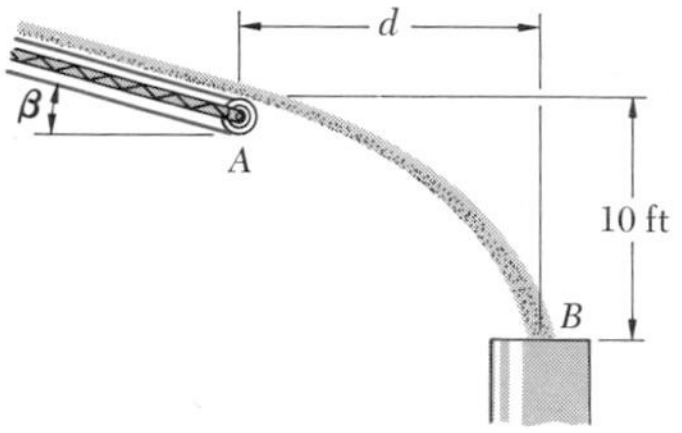

Fig. P11.80

11.80 Sand is discharged at *A* from a conveyor belt and falls into a collection pipe at *B*. Knowing that the conveyor belt forms an angle $\beta = 15°$ with the horizontal and moves at a constant speed of 20 ft/s, determine what the distance *d* should be so that the sand will hit the center of the pipe.

11.81 The conveyor belt moves at a constant speed of 12 ft/s. Knowing that $d = 8$ ft, determine the angle β for which the sand reaches the center of the pipe *B*.

11.82 A projectile is fired with an initial velocity of 210 m/s. Find the angle at which it should be fired if it is to hit a target located at a distance of 3600 m on the same level.

11.83 A boy can throw a baseball a maximum distance of 30 m in New York, where $g = 9.81\ \text{m/s}^2$. How far could he throw the baseball (*a*) in Singapore, where $g = 9.78\ \text{m/s}^2$? (*b*) On the moon, where $g = 1.618\ \text{m/s}^2$?

11.84 If the maximum horizontal range of a given gun is R, determine the firing angle which should be used to hit a target located at a distance $\frac{1}{2}R$ on the same level.

11.85 A projectile is fired with an initial velocity $\mathbf{v}_0$ at an angle α with the horizontal. Determine (*a*) the maximum height h reached by the projectile, (*b*) the horizontal range R of the projectile, (*c*) the maximum horizontal range R and the corresponding firing angle α.

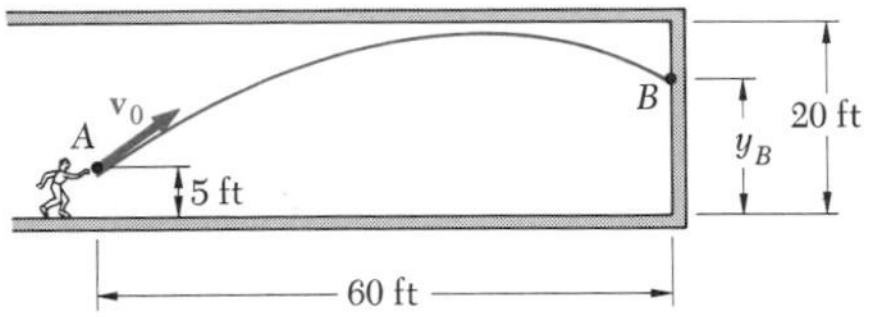

Fig. P11.86

11.86 A player throws a ball with an initial velocity $\mathbf{v}_0$ of 50 ft/s from a point A located 5 ft above the floor. Knowing that the ceiling of the gymnasium is 20 ft high, determine the highest point B at which the ball can strike the wall 60 ft away.

11.87 A fire nozzle discharges water with an initial velocity $\mathbf{v}_0$ of 80 ft/s. Knowing that the nozzle is located 100 ft from a building, determine (*a*) the maximum height h that can be reached by the water, (*b*) the corresponding angle α.

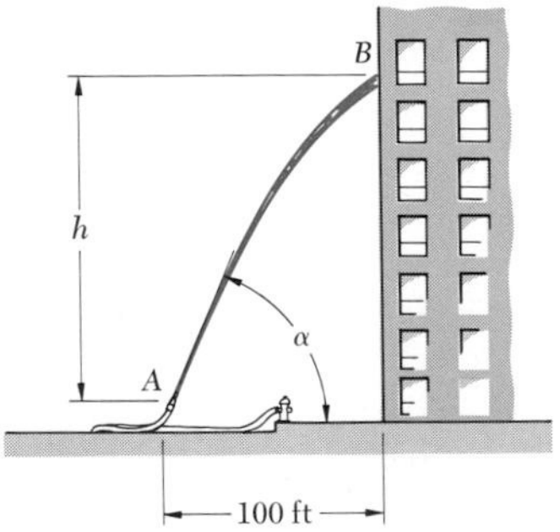

Fig. P11.87

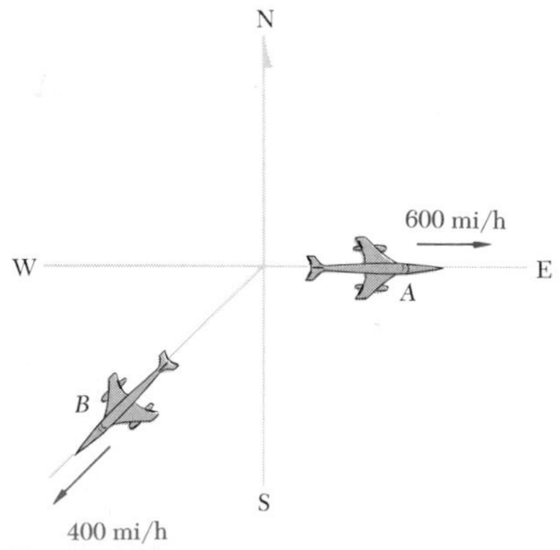

Fig. P11.88

11.88 Two airplanes A and B are each flying at a constant altitude; plane A is flying due east at a constant speed of 600 mi/h while plane B is flying southwest at a constant speed of 400 mi/h. Determine the change in position of plane B relative to plane A which takes place during a 1.5-min interval.

11.89 Instruments in an airplane indicate that, with respect to the air, the plane is moving east at a speed of 350 mi/h. At the same time ground-based radar indicates the plane to be moving at a speed of 325 mi/h in a direction 8° north of east. Determine the magnitude and direction of the velocity of the air.

11.90 As he passes a pole, a man riding in a truck tries to hit the pole by throwing a stone with a horizontal velocity of 20 m/s relative to the truck. Knowing that the speed of the truck is 40 km/h, determine (*a*) the direction in which he must throw the stone, (*b*) the horizontal velocity of the stone with respect to the ground.

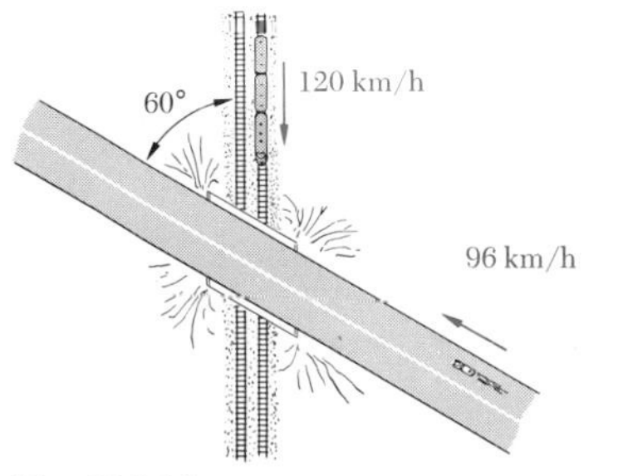

Fig. P11.91

11.91 An automobile and a train travel at the constant speeds shown. Three seconds after the train passes under the highway bridge the automobile crosses the bridge. Determine (*a*) the velocity of the train relative to the automobile, (*b*) the change in position of the train relative to the automobile during a 4-s interval, (*c*) the distance between the train and the automobile 5 s after the automobile has crossed the bridge.

11.92 During a rainstorm the paths of the raindrops appear to form an angle of 30° with the vertical when observed from a side window of a train moving at a speed of 15 km/h. A short time later, after the speed of the train has increased to 30 km/h, the angle between the vertical and the paths of the drops appears to be 45°. If the train were stopped, at what angle and with what velocity would the drops be observed to fall?

11.93 As the speed of the train of Prob. 11.92 increases, the angle between the vertical and the paths of the drops becomes equal to 60°. Determine the speed of the train at that time.

11.94 As observed from a ship moving due south at 10 mi/h, the wind appears to blow from the east. After the ship has changed course, and as it is moving due west at 10 mi/h, the wind appears to blow from the northeast. Assuming that the wind velocity is constant during the period of observation, determine the magnitude and direction of the true wind velocity.

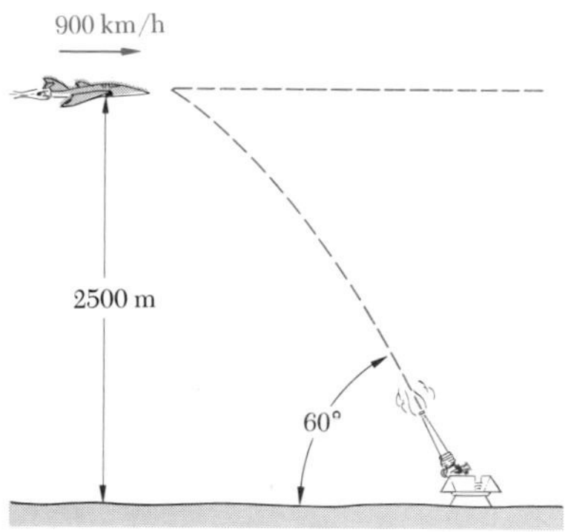

Fig. P11.95

11.95 An airplane is flying horizontally at an altitude of 2500 m and at a constant speed of 900 km/h on a path which passes directly over an antiaircraft gun. The gun fires a shell with a muzzle velocity of 500 m/s and hits the airplane. Knowing that the firing angle of the gun is 60°, determine the velocity and acceleration of the shell relative to the airplane at the time of impact.

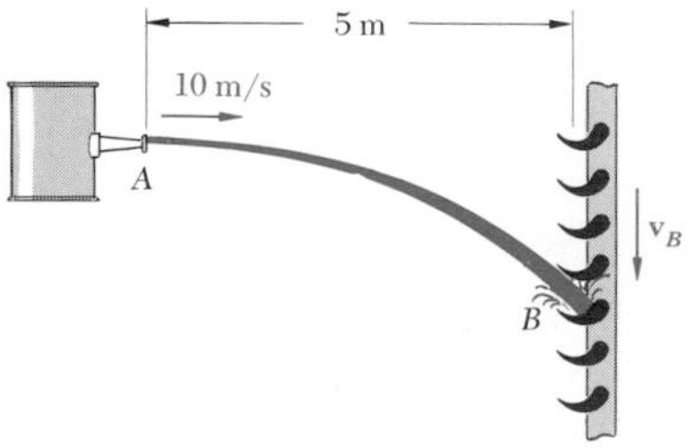

Fig. 11.96

11.96 Water is discharged at *A* with an initial velocity of 10 m/s and strikes a series of vanes at *B*. Knowing that the vanes move downward with a constant speed of 3 m/s, determine the velocity and acceleration of the water relative to the vane at *B*.

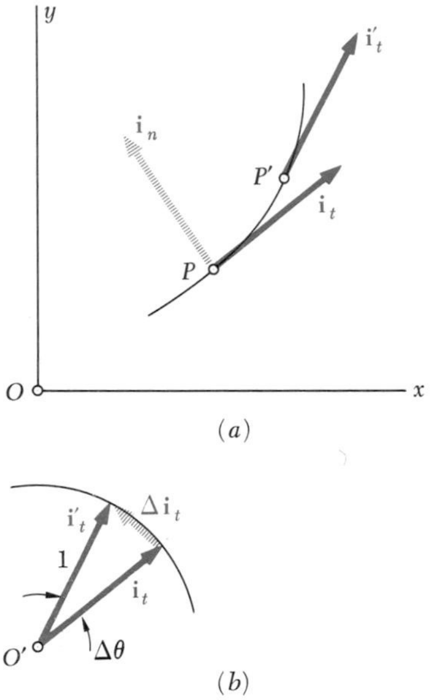

Fig. 11.21

11.13. Tangential and Normal Components. We saw in Sec. 11.9 that the velocity of a particle is a vector tangent to the path of the particle but that, in general, the acceleration is not tangent to the path. It is sometimes convenient to resolve the acceleration into components directed, respectively, along the tangent and the normal to the path of the particle.

Plane Motion of a Particle. We shall first consider a particle which moves along a curve contained in the plane of the figure. Let P be the position of the particle at a given instant. We attach at P a unit vector $\mathbf{i}_t$ tangent to the path of the particle and pointing toward the direction of motion (Fig. 11.21*a*). Let $\mathbf{i}'_t$ be the unit vector corresponding to the position P' of the particle at a later instant. Drawing both vectors from the same origin O', we define the vector $\Delta\mathbf{i}_t = \mathbf{i}'_t - \mathbf{i}_t$ (Fig. 11.21*b*). Since $\mathbf{i}_t$ and $\mathbf{i}'_t$ are of unit length, their tips lie on a circle of radius 1. Denoting by $\Delta\theta$ the angle formed by $\mathbf{i}_t$ and $\mathbf{i}'_t$, we find that the magnitude of $\Delta\mathbf{i}_t$ is $2 \sin (\Delta\theta/2)$. Considering now the vector $\Delta\mathbf{i}_t/\Delta\theta$, we note that, as $\Delta\theta$ approaches zero, this vector becomes tangent to the unit circle of Fig. 11.21*b*, i.e., perpendicular to $\mathbf{i}_t$, and that its magnitude approaches

$$\lim_{\Delta\theta\to 0} \frac{2 \sin (\Delta\theta/2)}{\Delta\theta} = \lim_{\Delta\theta\to 0} \frac{\sin (\Delta\theta/2)}{\Delta\theta/2} = 1$$

Thus, the vector obtained at the limit is a unit vector along the normal to the path of the particle, in the direction toward which $\mathbf{i}_t$ turns. Denoting this vector by $\mathbf{i}_n$, we write

$$\mathbf{i}_n = \lim_{\Delta\theta\to 0} \frac{\Delta\mathbf{i}_t}{\Delta\theta} \qquad \mathbf{i}_n = \frac{d\mathbf{i}_t}{d\theta} \tag{11.35}$$

Since the velocity $\mathbf{v}$ of the particle is tangent to the path, we may express it as the product of the scalar v and the unit vector $\mathbf{i}_t$. We have

$$\mathbf{v} = v\mathbf{i}_t \tag{11.36}$$

To obtain the acceleration of the particle, we shall differentiate (11.36) with respect to t. Applying the rule for the differentiation of the product of a scalar and a vector function (Sec. 11.10), we write

$$\mathbf{a} = \frac{d\mathbf{v}}{dt} = \frac{dv}{dt}\mathbf{i}_t + v\frac{d\mathbf{i}_t}{dt} \tag{11.37}$$

But

$$\frac{d\mathbf{i}_t}{dt} = \frac{d\mathbf{i}_t}{d\theta}\frac{d\theta}{ds}\frac{ds}{dt}$$

Recalling from (11.16) that $ds/dt = v$, from (11.35) that $d\mathbf{i}_t/d\theta = \mathbf{i}_n$, and from elementary calculus that $d\theta/ds$ is equal to $1/\rho$, where ρ is the radius of curvature of the path at *P* (Fig. 11.22), we have

$$\frac{d\mathbf{i}_t}{dt} = \frac{v}{\rho}\mathbf{i}_n \tag{11.38}$$

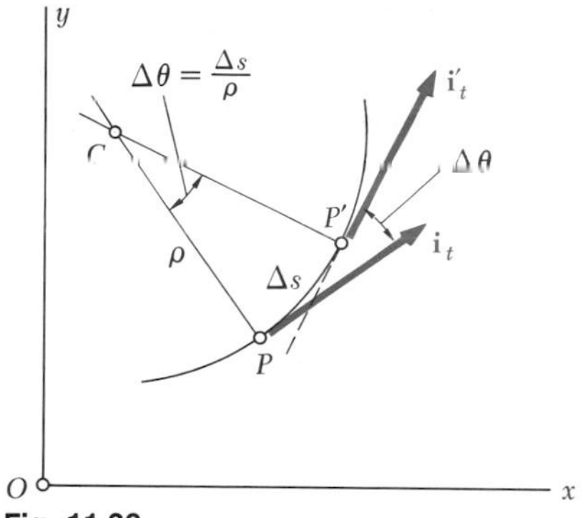

Fig. 11.22

Substituting into (11.37), we obtain

$$\mathbf{a} = \frac{dv}{dt}\mathbf{i}_t + \frac{v^2}{\rho}\mathbf{i}_n \tag{11.39}$$

Thus, the scalar components of the acceleration are

$$a_t = \frac{dv}{dt} \qquad a_n = \frac{v^2}{\rho} \tag{11.40}$$

The relations obtained express that the *tangential component* of the acceleration is equal to the *rate of change of the speed of the particle*, while the *normal component* is equal to the *square of the speed divided by the radius of curvature of the path at P.* Depending upon whether the speed of the particle increases or decreases, a_t is positive or negative, and the vector component $\mathbf{a}_t$ points in the direction of motion or against the direction of motion. The vector component $\mathbf{a}_n$, on the other hand, *is always directed toward the center of curvature C of the path* (Fig. 11.23).

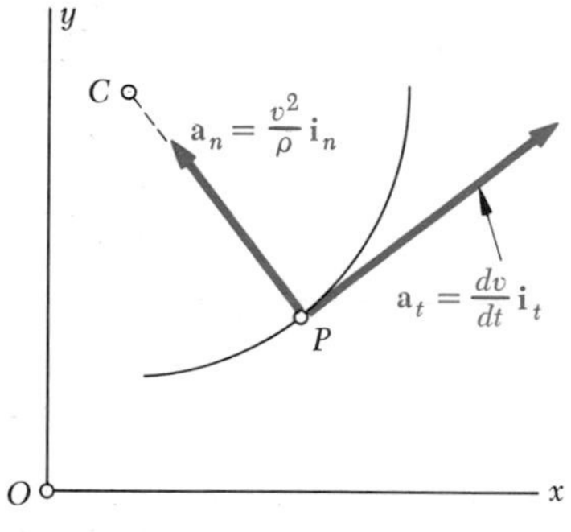

Fig. 11.23

It appears from the above that the tangential component of the acceleration reflects a change in the speed of the particle, while its normal component reflects a change in the direction of motion of the particle. The acceleration of a particle will be zero only if both its components are zero. Thus, the acceleration of a particle moving with constant speed along a curve will not be zero, unless the particle happens to pass through a point of inflection of the curve (where the radius of curvature is infinite) or unless the curve is a straight line.

The fact that the normal component of the acceleration depends upon the radius of curvature of the path followed by the particle is taken into account in the design of structures or mechanisms as widely different as airplane wings, railroad tracks, and cams. In order to avoid sudden changes in the acceleration of the air particles flowing past a wing, wing profiles are designed without any sudden change in curvature. Similar care is taken in designing railroad curves, to avoid sudden changes in the acceleration of the cars (which would be hard on the equipment and unpleasant for the passengers). A straight section of track, for instance, is never directly followed by a circular section. Special transition sections are used, to help pass smoothly from the infinite radius of curvature of the straight section to the finite radius of the circular track. Likewise, in the design of high-speed cams, abrupt changes in acceleration are avoided by using transition curves which produce a continuous change in acceleration.

Motion of a Particle in Space. The relations (11.39) and (11.40) still hold in the case of a particle moving along a space curve. However, since there is an infinite number of straight lines which are perpendicular to the tangent at a given point P of a space curve, it is then necessary to define more precisely the direction of the unit vector $\mathbf{i}_n$.

Let us consider again the unit vectors $\mathbf{i}_t$ and $\mathbf{i}'_t$ tangent to the path of the particle at two neighboring points P and P' (Fig. 11.24a) and the vector $\Delta\mathbf{i}_t$ representing the difference between $\mathbf{i}_t$ and $\mathbf{i}'_t$ (Fig. 11.24b). Let us now imagine a plane through P (Fig. 11.24a), parallel to the plane defined by the vectors $\mathbf{i}_t$, $\mathbf{i}'_t$, and $\Delta\mathbf{i}_t$ (Fig. 11.24b). This plane contains the tangent to the curve at P and is parallel to the tangent at P'. If we let P' approach P, we obtain at the limit the plane which fits the curve most closely in the neighborhood of P. This plane is called the *osculating plane* at P.† It follows from this definition that the osculating plane contains the unit vector $\mathbf{i}_n$, since this vector represents the limit of the vector $\Delta\mathbf{i}_t/\Delta\theta$. The normal defined by $\mathbf{i}_n$ is thus contained in the osculating plane; it is called the *principal normal* at P. The unit vector $\mathbf{i}_b = \mathbf{i}_t \times \mathbf{i}_n$ which completes the right-handed triad $\mathbf{i}_t$, $\mathbf{i}_n$, $\mathbf{i}_b$ (Fig. 11.24c) defines the *binormal* at P. The binormal is thus perpendicular to the osculating plane. We conclude that, as stated in (11.39), the acceleration of the particle at P may be resolved into two components, one along the tangent, the other along the principal normal at P. The acceleration has no component along the binormal.

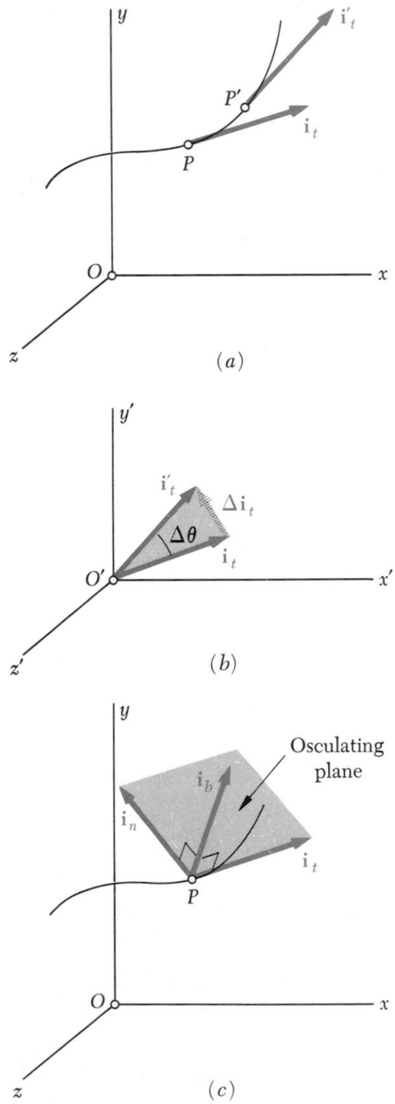

Fig. 11.24

†From the Latin *osculari,* to embrace.

11.14. Radial and Transverse Components. In certain problems of plane motion, the position of the particle P is defined by its polar coordinates r and θ (Fig. 11.25*a*). It is then convenient to resolve the velocity and acceleration of the particle into components parallel and perpendicular, respectively, to the line OP. These components are called *radial* and *transverse components*.

We attach at P two unit vectors, $\mathbf{i}_r$ and $\mathbf{i}_\theta$ (Fig. 11.25*b*). The vector $\mathbf{i}_r$ is directed along OP and the vector $\mathbf{i}_\theta$ is obtained by rotating $\mathbf{i}_r$ through 90° counterclockwise. The unit vector $\mathbf{i}_r$ defines the *radial* direction, i.e., the direction in which P would move if r were increased and θ kept constant; the unit vector $\mathbf{i}_\theta$ defines the *transverse* direction, i.e., the direction in which P would move if θ were increased and r kept constant. A derivation similar to the one we used in Sec. 11.13 to determine the derivative of the unit vector $\mathbf{i}_t$ leads to the relations

$$\frac{d\mathbf{i}_r}{d\theta} = \mathbf{i}_\theta \qquad \frac{d\mathbf{i}_\theta}{d\theta} = -\mathbf{i}_r \tag{11.41}$$

where $-\mathbf{i}_r$ denotes a unit vector of sense opposite to that of $\mathbf{i}_r$ (Fig. 11.25*c*).

Expressing the position vector $\mathbf{r}$ of the particle P as the product of the scalar r and the unit vector $\mathbf{i}_r$, and differentiating with respect to t, we write

$$\mathbf{r} = r\mathbf{i}_r \tag{11.42}$$

$$\mathbf{v} = \frac{d\mathbf{r}}{dt} = \frac{dr}{dt}\mathbf{i}_r + r\frac{d\mathbf{i}_r}{dt}$$

$$= \frac{dr}{dt}\mathbf{i}_r + r\frac{d\theta}{dt}\frac{d\mathbf{i}_r}{d\theta}$$

Recalling the first of the relations (11.41), and using dots to indicate time derivatives, we have

$$\mathbf{v} = \dot{r}\mathbf{i}_r + r\dot{\theta}\mathbf{i}_\theta \tag{11.43}$$

Differentiating again with respect to t, we write

$$\mathbf{a} = \frac{d\mathbf{v}}{dt} = \ddot{r}\mathbf{i}_r + \dot{r}\frac{d\mathbf{i}_r}{dt} + \dot{r}\dot{\theta}\mathbf{i}_\theta + r\ddot{\theta}\mathbf{i}_\theta + r\dot{\theta}\frac{d\mathbf{i}_\theta}{dt}$$

or, since $d\mathbf{i}_r/dt = \dot{\theta}\mathbf{i}_\theta$ and $d\mathbf{i}_\theta/dt = -\dot{\theta}\mathbf{i}_r$,

$$\mathbf{a} = (\ddot{r} - r\dot{\theta}^2)\mathbf{i}_r + (r\ddot{\theta} + 2\dot{r}\dot{\theta})\mathbf{i}_\theta \tag{11.44}$$

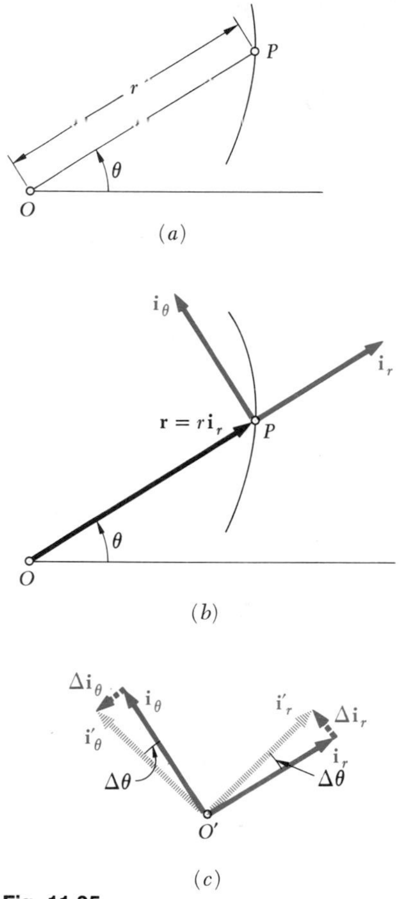

Fig. 11.25

The scalar components of the velocity and acceleration in the radial and transverse directions are therefore

$$v_r = \dot{r} \qquad v_\theta = r\dot{\theta} \tag{11.45}$$
$$a_r = \ddot{r} - r\dot{\theta}^2 \qquad a_\theta = r\ddot{\theta} + 2\dot{r}\dot{\theta} \tag{11.46}$$

It is important to note that a_r is *not* equal to the time derivative of v_r, and that a_θ is *not* equal to the time derivative of v_θ.

In the case of a particle moving along a circle of center O, we have $r =$ constant, $\dot{r} = \ddot{r} = 0$, and the formulas (11.43) and (11.44) reduce, respectively, to

$$\mathbf{v} = r\dot{\theta}\mathbf{i}_\theta \qquad \mathbf{a} = -r\dot{\theta}^2\mathbf{i}_r + r\ddot{\theta}\mathbf{i}_\theta \tag{11.47}$$

Extension to the Motion of a Particle in Space: Cylindrical Coordinates. The position of a particle P in space is sometimes defined by its cylindrical coordinates R, θ, and z (Fig. 11.26*a*). It is then convenient to use the unit vectors $\mathbf{i}_R$, $\mathbf{i}_\theta$, and $\mathbf{k}$

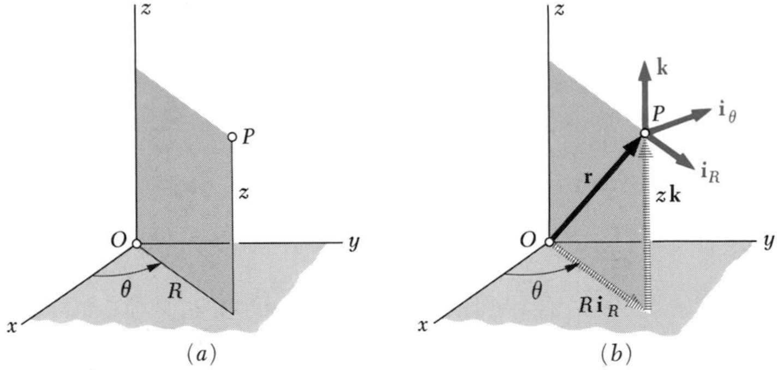

Fig. 11.26

shown in Fig. 11.26*b*. Resolving the position vector $\mathbf{r}$ of the particle P into components along the unit vectors, we write

$$\mathbf{r} = R\mathbf{i}_R + z\mathbf{k} \tag{11.48}$$

Observing that $\mathbf{i}_R$ and $\mathbf{i}_\theta$ define, respectively, the radial and transverse direction in the horizontal xy plane, and that the vector $\mathbf{k}$, which defines the *axial* direction, is constant in direction as well as in magnitude, we easily verify that

$$\mathbf{v} = \frac{d\mathbf{r}}{dt} = \dot{R}\mathbf{i}_R + R\dot{\theta}\mathbf{i}_\theta + \dot{z}\mathbf{k} \tag{11.49}$$

$$\mathbf{a} = \frac{d\mathbf{v}}{dt} = (\ddot{R} - R\dot{\theta}^2)\mathbf{i}_R + (R\ddot{\theta} + 2\dot{R}\dot{\theta})\mathbf{i}_\theta + \ddot{z}\mathbf{k} \tag{11.50}$$

SAMPLE PROBLEM 11.10

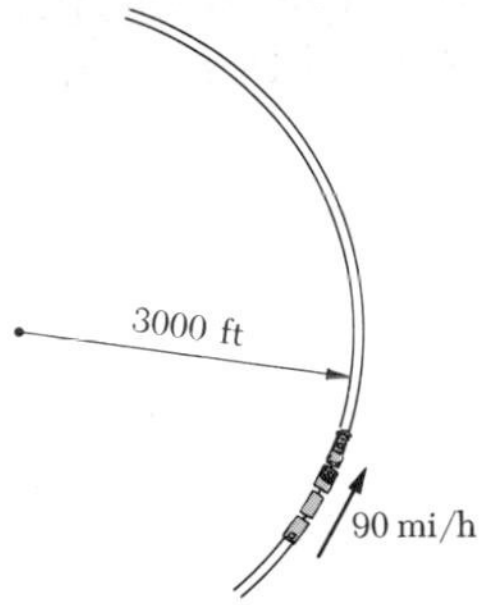

A train is traveling on a curved section of track of radius 3000 ft at the speed of 90 mi/h. The brakes are suddenly applied, causing the train to slow down at a constant rate; after 6 s, the speed has been reduced to 60 mi/h. Determine the acceleration of a car immediately after the brakes have been applied.

Tangential Component of Acceleration. First the speeds are expressed in ft/s.

$$90 \text{ mi/h} = \left(90 \frac{\text{mi}}{\text{h}}\right)\left(\frac{5280 \text{ ft}}{1 \text{ mi}}\right)\left(\frac{1 \text{ h}}{3600 \text{ s}}\right) = 132 \text{ ft/s}$$

$$60 \text{ mi/h} = 88 \text{ ft/s}$$

Since the train slows down at a constant rate, we have

$$a_t = \text{average } a_t = \frac{\Delta v}{\Delta t} = \frac{88 \text{ ft/s} - 132 \text{ ft/s}}{6 \text{ s}} = -7.33 \text{ ft/s}^2$$

Normal Component of Acceleration. Immediately after the brakes have been applied, the speed is still 132 ft/s, and we have

$$a_n = \frac{v^2}{\rho} = \frac{(132 \text{ ft/s})^2}{3000 \text{ ft}} = 5.81 \text{ ft/s}^2$$

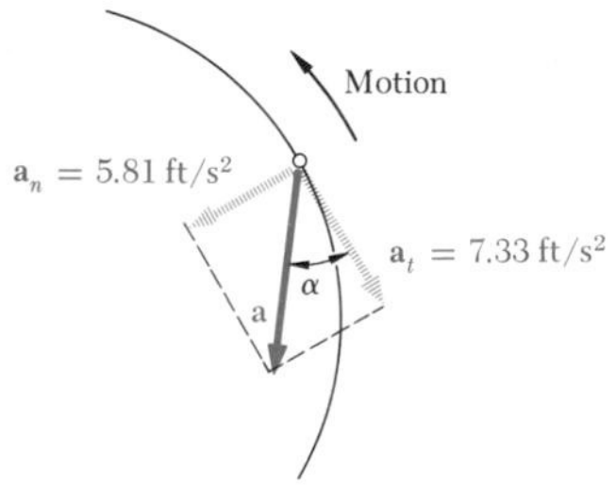

Magnitude and Direction of Acceleration. The magnitude and direction of the resultant $\mathbf{a}$ of the components $\mathbf{a}_n$ and $\mathbf{a}_t$ are

$$\tan \alpha = \frac{a_n}{a_t} = \frac{5.81 \text{ ft/s}^2}{7.33 \text{ ft/s}^2} \qquad \alpha = 38.4° \blacktriangleleft$$

$$a = \frac{a_n}{\sin \alpha} = \frac{5.81 \text{ ft/s}^2}{\sin 38.4°} \qquad a = 9.35 \text{ ft/s}^2 \blacktriangleleft$$

SAMPLE PROBLEM 11.11

Determine the minimum radius of curvature of the trajectory described by the projectile considered in Sample Prob. 11.7.

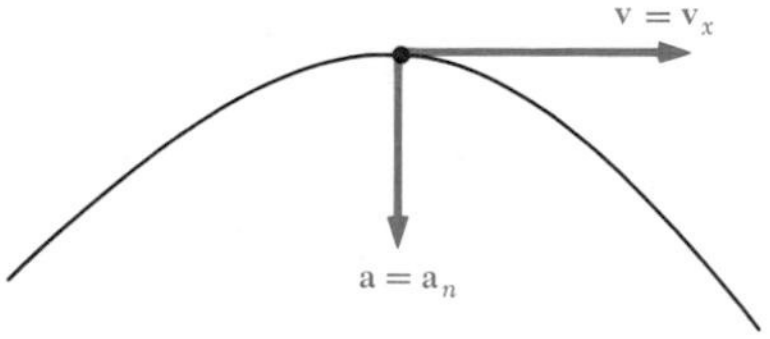

Solution. Since $a_n = v^2/\rho$, we have $\rho = v^2/a_n$. The radius will be small when v is small or when a_n is large. The speed v is minimum at the top of the trajectory since $v_y = 0$ at that point; a_n is maximum at that same point, since the direction of the vertical coincides with the direction of the normal. Therefore, the minimum radius of curvature occurs at the top of the trajectory. At this point, we have

$$v = v_x = 155.9 \text{ m/s} \qquad a_n = a = 9.81 \text{ m/s}^2$$

$$\rho = \frac{v^2}{a_n} = \frac{(155.9 \text{ m/s})^2}{9.81 \text{ m/s}^2} \qquad \rho = 2480 \text{ m} \blacktriangleleft$$

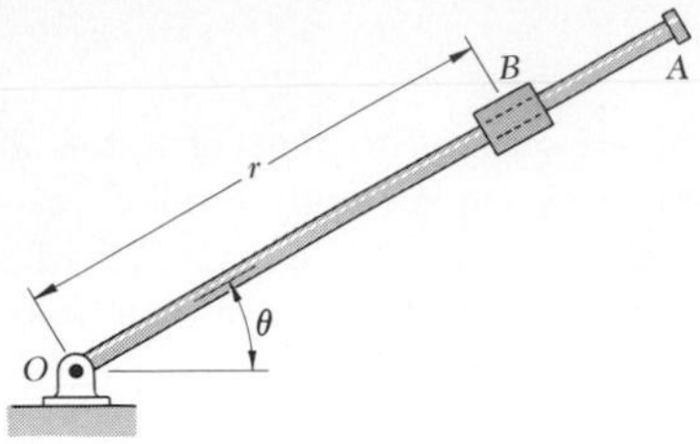

SAMPLE PROBLEM 11.12

The rotation of the 3-ft arm OA about O is defined by the relation $\theta = 0.15t^2$, where θ is expressed in radians and t in seconds. Block B slides along the arm in such a way that its distance from O is $r = 3 - 0.40t^2$, where r is expressed in feet and t in seconds. Determine the total velocity and the total acceleration of block B after the arm OA has rotated through 30°.

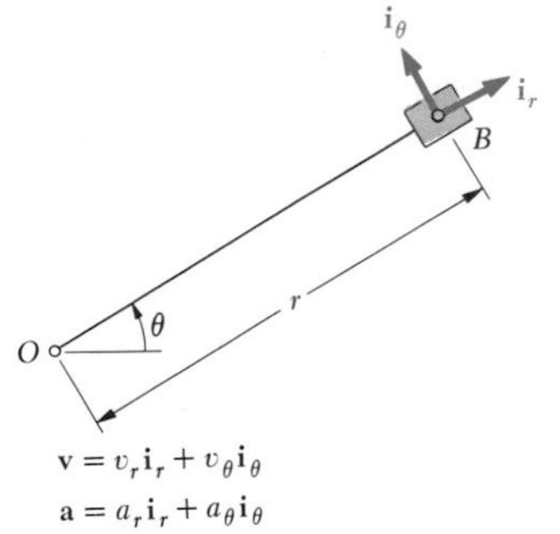

Solution. We first find the time t at which $\theta = 30°$. Substituting $\theta = 30° = 0.524$ rad into the expression for θ, we obtain

$$\theta = 0.15t^2 \qquad 0.524 = 0.15t^2 \qquad t = 1.869 \text{ s}$$

Equations of Motion. Substituting $t = 1.869$ s in the expressions for r, θ, and their first and second derivatives, we have

$$\begin{aligned} r &= 3 - 0.40t^2 = 1.603 \text{ ft} & \theta &= 0.15t^2 = 0.524 \text{ rad} \\ \dot{r} &= -0.80t = -1.495 \text{ ft/s} & \dot{\theta} &= 0.30t = 0.561 \text{ rad/s} \\ \ddot{r} &= -0.80 = -0.800 \text{ ft/s}^2 & \ddot{\theta} &= 0.30 = 0.300 \text{ rad/s}^2 \end{aligned}$$

Velocity of B. Using Eqs. (11.45), we obtain the values of v_r and v_θ when $t = 1.869$ s.

$$\begin{aligned} v_r &= \dot{r} = -1.495 \text{ ft/s} \\ v_\theta &= r\dot{\theta} = 1.603(0.561) = 0.899 \text{ ft/s} \end{aligned}$$

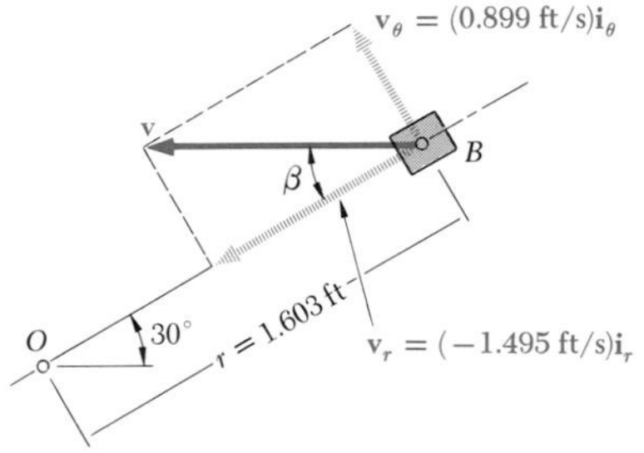

Solving the right triangle shown, we obtain the magnitude and direction of the velocity,

$$v = 1.744 \text{ ft/s} \qquad \beta = 31.0° \quad ◀$$

Acceleration of B. Using Eqs. (11.46), we obtain

$$\begin{aligned} a_r &= \ddot{r} - r\dot{\theta}^2 \\ &= -0.800 - 1.603(0.561)^2 = -1.304 \text{ ft/s}^2 \\ a_\theta &= r\ddot{\theta} + 2\dot{r}\dot{\theta} \\ &= 1.603(0.300) + 2(-1.495)(0.561) = -1.196 \text{ ft/s}^2 \end{aligned}$$

$$a = 1.770 \text{ ft/s} \qquad \gamma = 42.5° \quad ◀$$

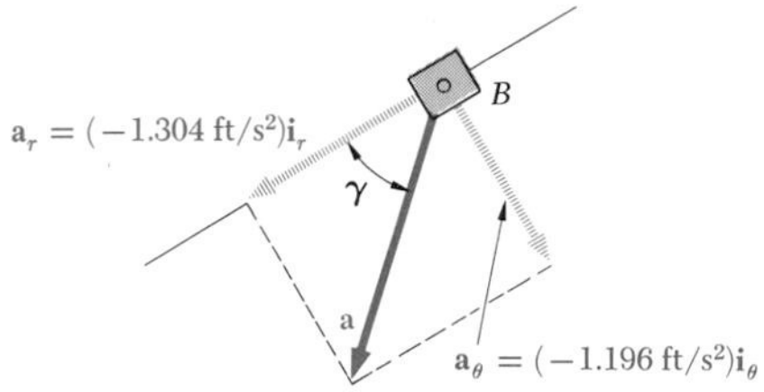

PROBLEMS

11.97 An automobile travels at a constant speed on a highway curve of 1000-m radius. If the normal component of the acceleration is not to exceed 1.2 m/s^2, determine the maximum allowable speed.

11.98 A car goes around a highway curve of 300-m radius at a speed of 90 km/h. (*a*) What is the normal component of its acceleration? (*b*) At what speed is the normal component of the acceleration one-half as large as that found in part *a*?

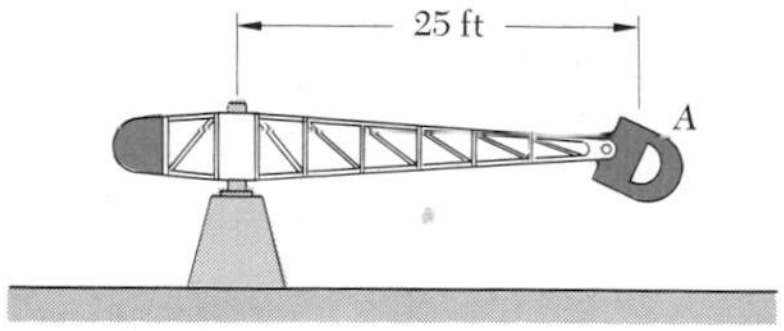

Fig. P11.99

11.99 Determine the peripheral speed of the centrifuge test cab *A* for which the normal component of the acceleration is 10*g*.

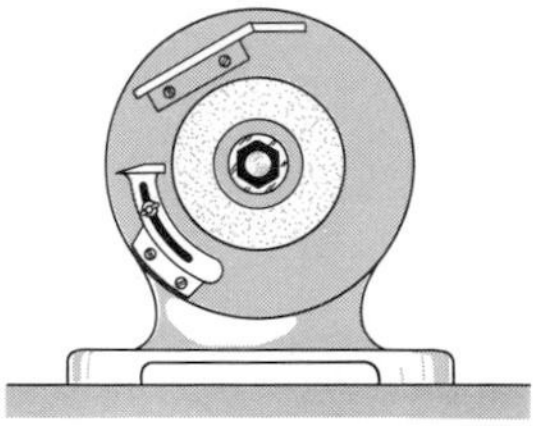
Fig. P11.100

11.100 A small grinding wheel has a 5-in. diameter and is attached to the shaft of an electric motor which has a rated speed of 3600 rpm. Determine the normal component of the acceleration of a point on the circumference of the wheel when the wheel is rotating at the rated speed.

11.101 A motorist starts from rest on a curve of 400-ft radius and accelerates at the uniform rate of 3 ft/s^2. Determine the distance that his automobile will travel before the magnitude of its total acceleration is 6 ft/s^2.

11.102 A motorist enters a curve of 500-ft radius at a speed of 45 mi/h. As he applies his brakes, he decreases his speed at a constant rate of 5 ft/s^2. Determine the magnitude of the total acceleration of the automobile when its speed is 40 mi/h.

11.103 The speed of a racing car is increased at a constant rate from 90 km/h to 126 km/h over a distance of 150 m along a curve of 250-m radius. Determine the magnitude of the total acceleration of the car after it has traveled 100 m along the curve.

11.104 A monorail train is traveling at a speed of 144 km/h along a curve of 1000-m radius. Determine the maximum rate at which the speed may be decreased if the total acceleration of the train is not to exceed 2 m/s^2.

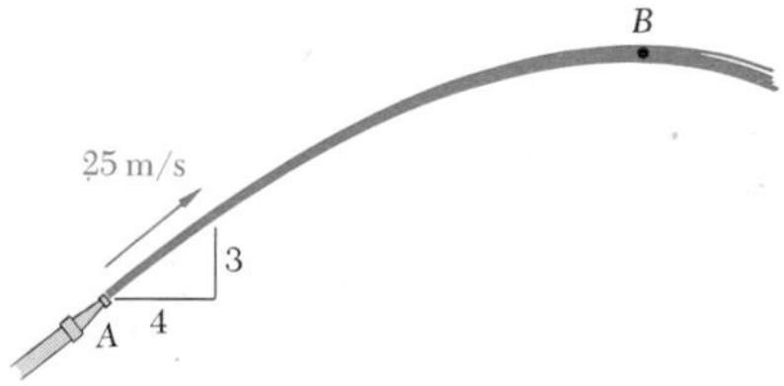

Fig. P11.105

11.105 A nozzle discharges a stream of water in the direction shown with an initial velocity of 25 m/s. Determine the radius of curvature of the stream (*a*) as it leaves the nozzle, (*b*) at the maximum height of the stream.

11.106 Determine the radius of curvature of the trajectory described by the projectile of Sample Prob. 11.7 as the projectile leaves the gun.

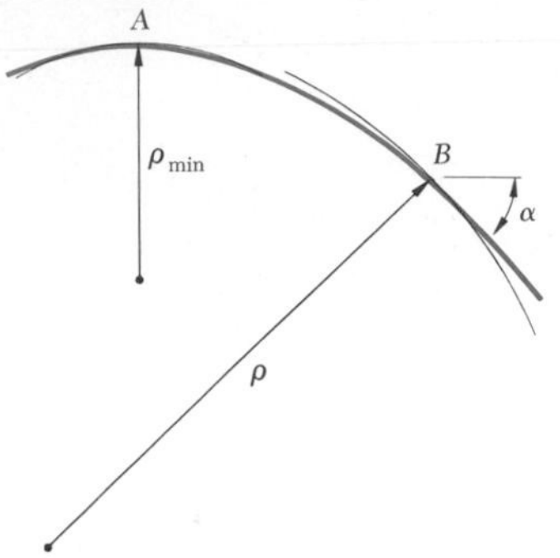

Fig. P11.107

11.107 (*a*) Show that the radius of curvature of the trajectory of a projectile reaches its minimum value at the highest point A of the trajectory. (*b*) Denoting by α the angle formed by the trajectory and the horizontal at a given point B, show that the radius of curvature of the trajectory at B is $\rho = \rho_{min}/\cos^3 \alpha$.

11.108 For each of the two firing angles obtained in Sample Prob. 11.8, determine the radius of curvature of the trajectory described by the projectile as it leaves the gun.

***11.109** Determine the radius of curvature of the path described by the particle of Prob. 11.73 when (*a*) $t = 0$, (*b*) $t = 2$ s.

***11.110** Determine the radius of curvature of the helix of Prob. 11.74.

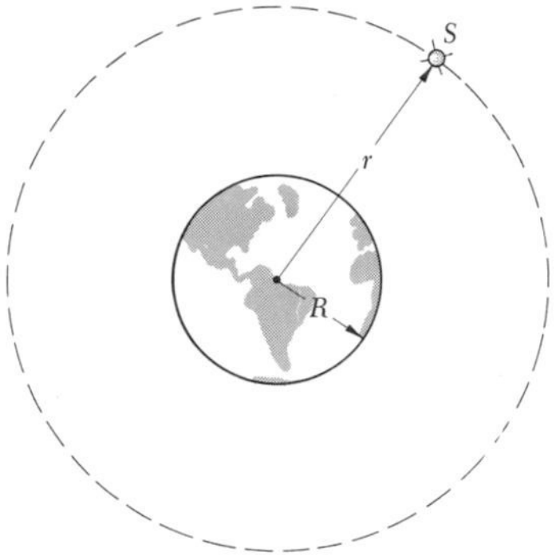

Fig. P11.111

11.111 A satellite will travel indefinitely in a circular orbit around the earth if the normal component of its acceleration is equal to $g(R/r)^2$, where $g = 32.2$ ft/s^2, R = radius of the earth = 3960 mi, and r = distance from the center of the earth to the satellite. Determine the height above the surface of the earth at which a satellite will travel indefinitely around the earth at a speed of 15,000 mi/h.

11.112 Determine the speed of an earth satellite traveling in a circular orbit 300 mi above the surface of the earth. (See information given in Prob. 11.111.)

11.113 Assuming the orbit of the moon to be a circle of radius 239,000 mi, determine the speed of the moon relative to the earth. (See information given in Prob. 11.111.)

11.114 Show that the speed of an earth satellite traveling in a circular orbit is inversely proportional to the square root of the radius of its orbit. Also, determine the minimum time in which a satellite can circle the earth. (See information given in Prob. 11.111.)

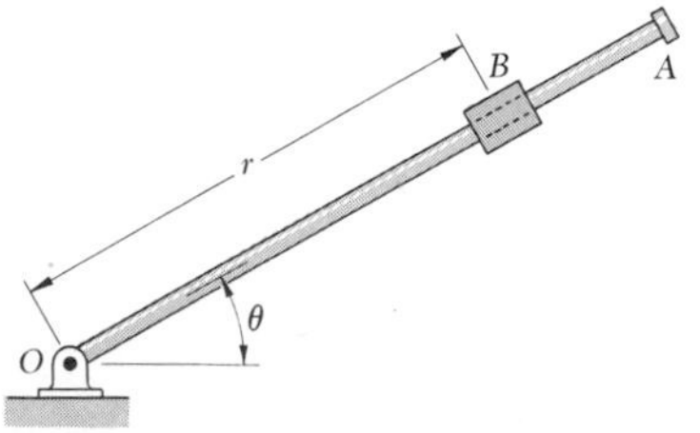

Fig. P11.115

11.115 The two-dimensional motion of a particle is defined by the relations $r = 60t^2 - 20t^3$ and $\theta = 2t^2$, where r is expressed in millimeters, t in seconds, and θ in radians. Determine the velocity and acceleration of the particle when (*a*) $t = 0$, (*b*) $t = 1$ s.

11.116 The particle of Prob. 11.115 is at the origin at $t = 0$. Determine its velocity and acceleration as it returns to the origin.

11.117 The two-dimensional motion of a particle is defined by the relations $r = 2b \sin \omega t$ and $\theta = \omega t$, where b and ω are constants. Determine (*a*) the velocity and acceleration of the particle at any instant, (*b*) the radius of curvature of its path. What conclusion can you draw regarding the path of the particle?

11.118 As circle B rolls on the fixed circle A, point P describes a cardioid defined by the relations $r = 2b(1 + \cos 2\pi t)$ and $\theta = 2\pi t$. Determine the velocity and acceleration of P when (*a*) $t = 0.25$, (*b*) $t = 0.50$.

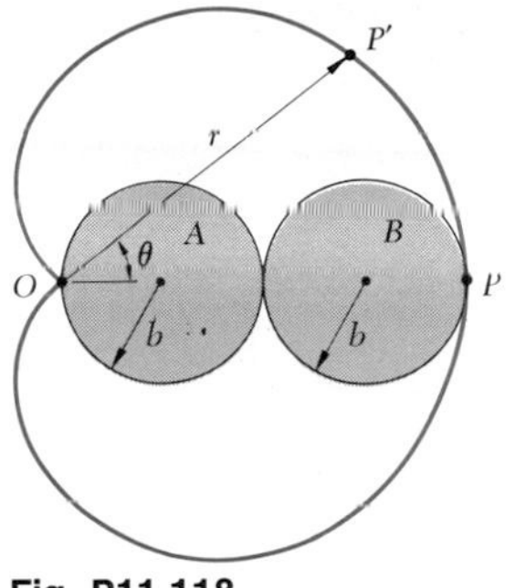

Fig. P11.118

11.119 A wire OA connects the collar A and a reel located at O. Knowing that the collar moves to the right with a constant speed v_0, determine $d\theta/dt$ in terms of v_0, b, and θ.

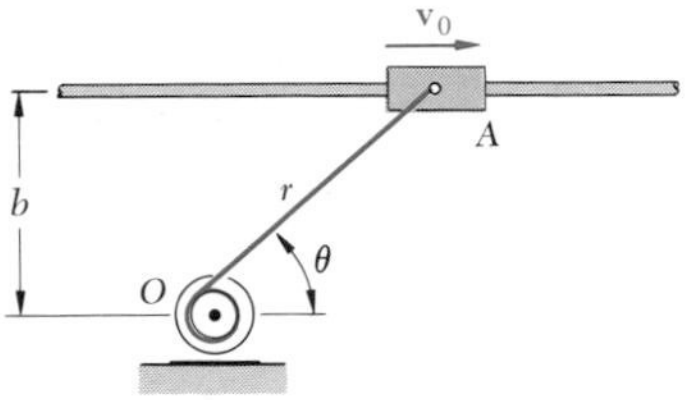

Fig. P11.119

11.120 A rocket is fired vertically from a launching pad at B. Its flight is tracked by radar from point A. Determine the velocity of the rocket in terms of b, θ, and $\dot{\theta}$.

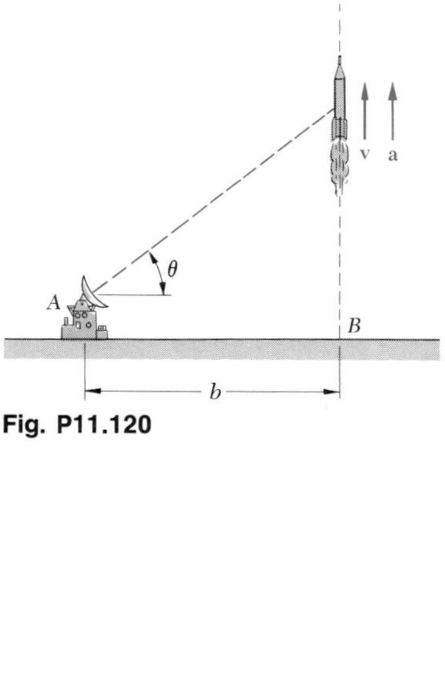

Fig. P11.120

11.121 Determine the acceleration of the rocket of Prob. 11.120 in terms of b, θ, $\dot{\theta}$, and $\ddot{\theta}$.

11.122 As the rod OA rotates, the pin P moves along the parabola BCD. Knowing that the equation of the parabola is $r = 2b/(1 + \cos\theta)$ and that $\theta = kt$, determine the velocity and acceleration of P when (*a*) $\theta = 0$, (*b*) $\theta = 90°$.

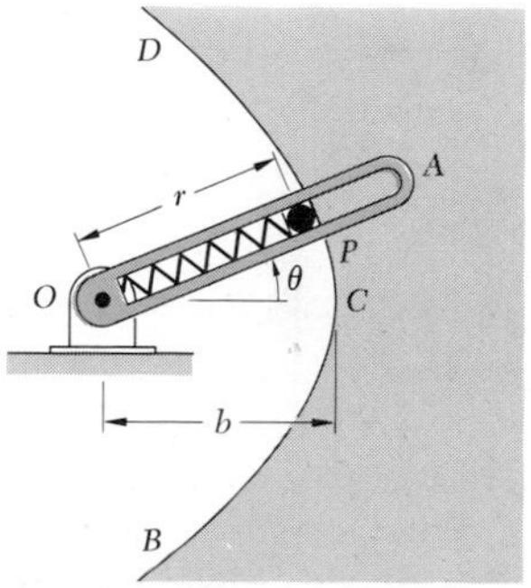

Fig. P11.122

11.123 The pin at B is free to slide along the circular slot and along the rotating rod OC. If pin B slides counterclockwise around the circular slot at a constant speed v_0, determine the rate $d\theta/dt$ at which rod OC rotates and the radial component v_r of the velocity of the pin B (*a*) when $\phi = 0°$, (*b*) when $\phi = 90°$.

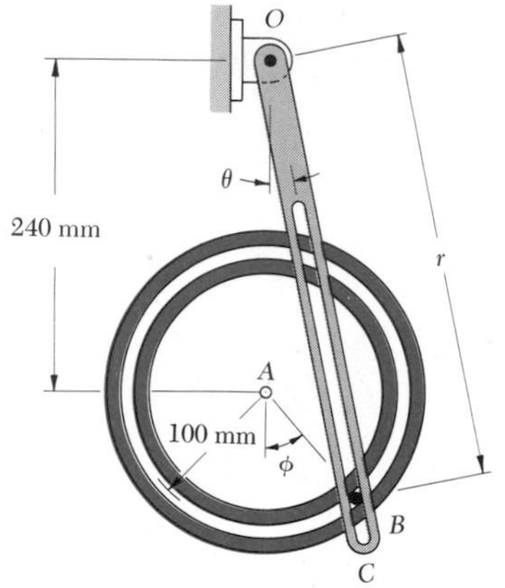

Fig. P11.123

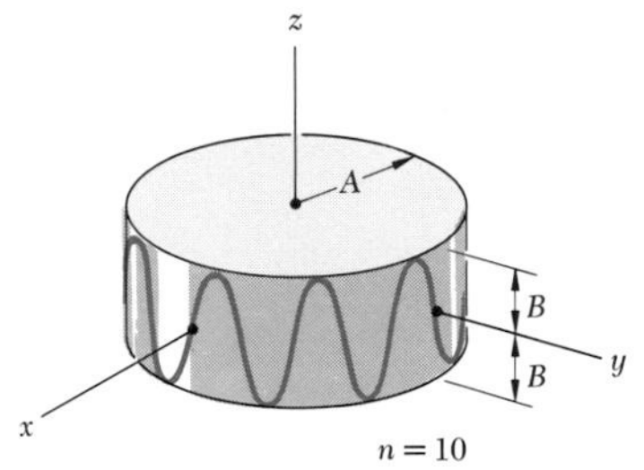

Fig. P11.124

11.124 The motion of a particle on the surface of a right circular cylinder is defined by the relations $R = A$, $\theta = 2\pi t$, and $z = B \sin 2\pi nt$, where A and B are constants and n is an integer. Determine the magnitudes of the velocity and acceleration of the particle at any time t.

11.125 For the case when $n = 1$ in Prob. 11.124, (*a*) show that the path of the particle is contained in a plane, (*b*) determine the maximum and minimum radii of curvature of the path.

11.126 The motion of a particle on the surface of a right circular cone is defined by the relations $R = ht \tan \beta$, $\theta = 2\pi t$, and $z = ht$, where β is the apex angle of the cone and h is the distance the particle rises in one passage around the cone. Determine the magnitudes of the velocity and acceleration at any time t.

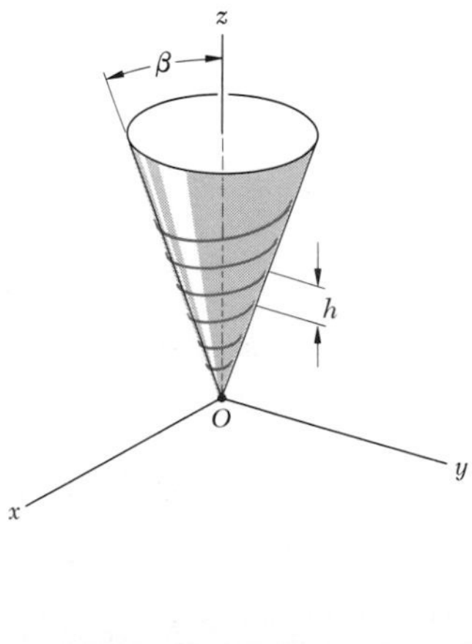

Fig. P11.126

11.127 The three-dimensional motion of a particle is defined by the relations $R = A$, $\theta = 2\pi t$, and $z = A \sin^2 2\pi t$. Determine the magnitudes of the velocity and acceleration at any time t.

***11.128** For the helix of Prob. 11.74, determine the angle that the osculating plane forms with the y axis.

***11.129** Determine the direction of the binormal of the path described by the particle of Prob. 11.73 when (*a*) $t = 0$, (*b*) $t = 2$ s.

***11.130** The position vector of a particle is defined by the relation

$$\mathbf{r} = x\mathbf{i} + y\mathbf{j} + z\mathbf{k}$$

where x, y, z are known functions of the time t, and $\mathbf{i}$, $\mathbf{j}$, $\mathbf{k}$ are unit vectors along fixed rectangular axes. Express in terms of the functions x, y, z and their first and second derivatives (*a*) the tangential component of the acceleration of the particle, (*b*) the normal component of its acceleration, (*c*) the radius of curvature of the path described by the particle.

***11.131** For the particle of Prob. 11.130, express the direction cosines of (*a*) the tangent, (*b*) the binormal, (*c*) the principal normal of the path described by the particle, in terms of the functions x, y, z and their first and second derivatives.

REVIEW PROBLEMS

11.132 The a–t curve shown was obtained during the motion of a test sled. Knowing that the sled started from rest at $t = 0$, determine the velocity and position of the sled at $t = 0.08$ s.

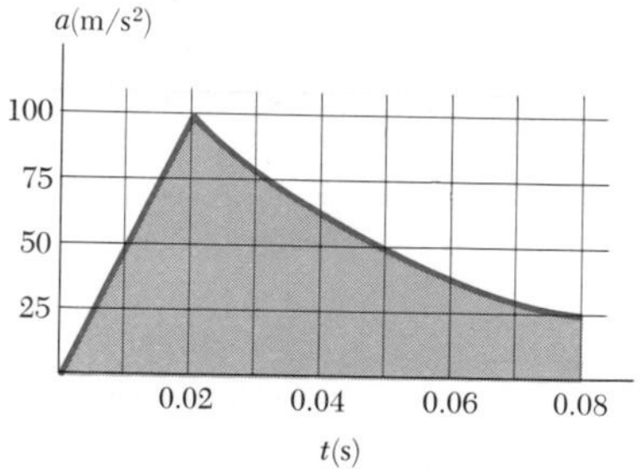

Fig. P11.132

11.133 An experimental ion-propulsion engine is capable of giving a space vehicle a constant acceleration of 0.01 ft/s². If the engine is placed in operation when the speed of the vehicle is 21,000 mi/h, determine the time required to bring the speed of the vehicle to 22,000 mi/h. Assume that the vehicle is moving in a straight line, far from the sun or any planet.

11.134 The velocity of a particle is given by the relation $v = 100 - 10x$, where v is expressed in meters per second and x in meters. Knowing that $x = 0$ at $t = 0$, determine (*a*) the distance traveled before the particle comes to rest, (*b*) the time t when $x = 5$ m, (*c*) the acceleration at $t = 0$.

11.135 A nozzle discharges a stream of water with an initial velocity $\mathbf{v}_0$ of 50 ft/s into the end of a horizontal pipe of inside diameter $d = 5$ ft. Determine the largest distance x that the stream can reach.

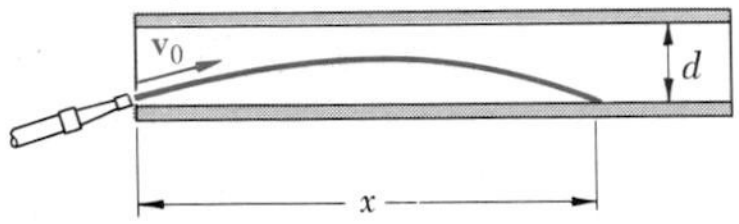

Fig. P11.135

11.136 The magnitude in m/s² of the deceleration due to air resistance of the nose cone of a small experimental rocket is known to be $6 \times 10^{-4}\, v^2$, where v is expressed in m/s. If the nose cone is projected vertically from the ground with an initial velocity of 100 m/s, determine the maximum height that it will reach.

11.137 Determine the velocity of the nose cone of Prob. 11.136 when it returns to the ground.

11.138 Standing on the side of a hill, an archer shoots an arrow with an initial velocity of 250 ft/s at an angle $\alpha = 15°$ with the horizontal. Determine the horizontal distance d traveled by the arrow before it strikes the ground at B.

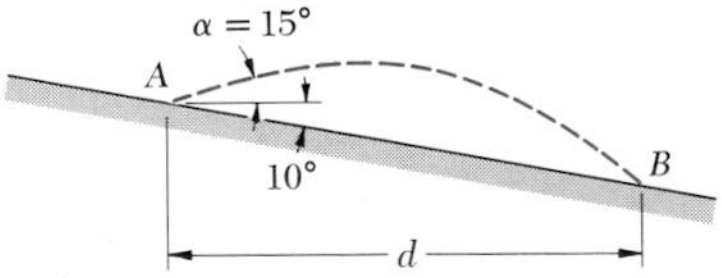

Fig. P11.138

11.139 In Prob. 11.138, determine the radius of curvature of the trajectory (*a*) immediately after the arrow has been shot, (*b*) as the arrow passes through its point of maximum elevation.

11.140 A train starts at a station and accelerates uniformly at a rate of $0.6\ \mathrm{m/s^2}$ until it reaches a speed of 24 m/s; it then proceeds at the constant speed of 24 m/s. Determine the time and the distance traveled if its average velocity is (*a*) 16 m/s, (*b*) 22 m/s.

11.141 A man jumps from a 20-ft cliff with no initial velocity. (*a*) How long does it take him to reach the ground, and with what velocity does he hit the ground? (*b*) If this takes place on the moon, where $g = 5.31\ \mathrm{ft/s^2}$, what are the values obtained for the time and velocity? (*c*) If a motion picture is taken on the earth, but if the scene is supposed to take place on the moon, how many frames per second should be used so that the scene would appear realistic when projected at the standard speed of 24 frames per second?

11.142 Drops of water fall down a mine shaft at the uniform rate of one drop per second. A mine elevator moving up the shaft at 30 ft/s is struck by a drop of water when it is 300 ft below ground level. When and where will the next drop of water strike the elevator?

11.143 Knowing that block *B* moves downward with a constant velocity of 180 mm/s, determine (*a*) the velocity of block *A*, (*b*) the velocity of pulley *D*.

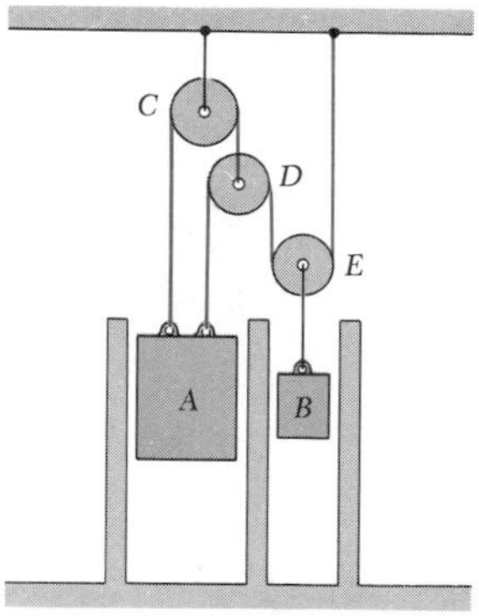

Fig. P11.143

CHAPTER 12 Kinetics of Particles: Newton's Second Law

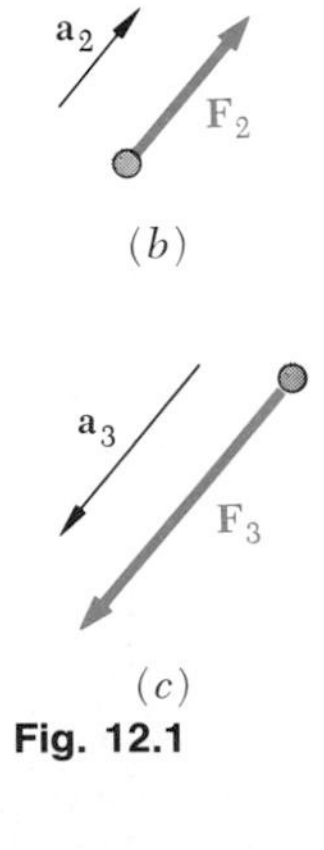

Fig. 12.1

12.1. Newton's Second Law of Motion. Newton's first and third laws of motion were used extensively in statics to study bodies at rest and the forces acting upon them. These two laws are also used in dynamics; in fact, they are sufficient for the study of the motion of bodies which have no acceleration. However, when bodies are accelerated, i.e., when the magnitude or the direction of their velocity changes, it is necessary to use the second law of motion in order to relate the motion of the body with the forces acting on it. This law may be stated as follows:

If the resultant force acting on a particle is not zero, the particle will have an acceleration proportional to the magnitude of the resultant and in the direction of this resultant force.

Newton's second law of motion may best be understood if we imagine the following experiment: A particle is subjected to a force $\mathbf{F}_1$ of constant direction and constant magnitude F_1. Under the action of that force, the particle will be observed to move in a straight line and *in the direction of the force* (Fig. 12.1*a*). By determining the position of the particle at various instants, we find that its acceleration has a constant magnitude a_1. If the experiment is repeated with forces $\mathbf{F}_2$, $\mathbf{F}_3$, etc., of different magnitude or direction (Fig. 12.1*b* and *c*), we find each time that the particle moves in the direction of the force acting on it and that the magnitudes a_1, a_2, a_3, etc., of the accelerations are proportional to the magnitudes F_1, F_2, F_3, etc., of the corresponding forces,

$$\frac{F_1}{a_1} = \frac{F_2}{a_2} = \frac{F_3}{a_3} = \cdots = \text{constant}$$

The constant value obtained for the ratio of the magnitudes of the forces and accelerations is a characteristic of the particle under consideration. It is called the *mass* of the particle and is denoted by m. When a particle of mass m is acted upon by a force $\mathbf{F}$, the force $\mathbf{F}$ and the acceleration $\mathbf{a}$ of the particle must therefore satisfy the relation

$$\mathbf{F} = m\mathbf{a} \tag{12.1}$$

This relation provides a complete formulation of Newton's second law; it expresses not only that the magnitudes of $\mathbf{F}$ and $\mathbf{a}$ are proportional, but also (since m is a positive scalar) that the vectors $\mathbf{F}$ and $\mathbf{a}$ have the same direction (Fig. 12.2). We should note that Eq. (12.1) still holds when $\mathbf{F}$ is not constant but varies with t in magnitude or direction. The magnitudes of $\mathbf{F}$ and $\mathbf{a}$ remain proportional, and the two vectors have the same direction at any given instant. However, they will not, in general, be tangent to the path of the particle.

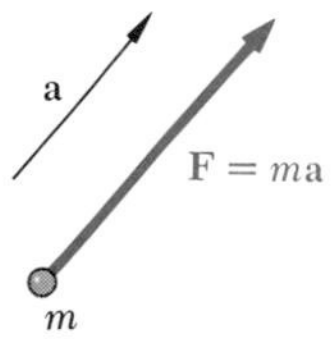

Fig. 12.2

When a particle is subjected simultaneously to several forces, Eq. (12.1) should be replaced by

$$\Sigma\mathbf{F} = m\mathbf{a} \tag{12.2}$$

where $\Sigma\mathbf{F}$ represents the sum, or resultant, of all the forces acting on the particle.

It should be noted that the system of axes with respect to which the acceleration $\mathbf{a}$ is determined is not arbitrary. These axes must have a constant orientation with respect to the stars, and their origin must either be attached to the sun† or move with a constant velocity with respect to the sun. Such a system of axes is called a *newtonian frame of reference.*‡ A system of axes attached to the earth does *not* constitute a newtonian frame of reference, since the earth rotates with respect to the stars and is accelerated with respect to the sun. However, in most

† More accurately, to the mass center of the solar system.

‡ Since the stars are not actually fixed, a more rigorous definition of a newtonian frame of reference (also called *inertial system*) is *one with respect to which Eq.* (12.2) *holds.*

engineering applications, the acceleration **a** may be determined with respect to axes attached to the earth and Eqs. (12.1) and (12.2) used without any appreciable error. On the other hand, these equations do not hold if **a** represents a relative acceleration measured with respect to moving axes, such as axes attached to an accelerated car or to a rotating piece of machinery.

We may observe that, if the resultant $\Sigma\mathbf{F}$ of the forces acting on the particle is zero, it follows from Eq. (12.2) that the acceleration **a** of the particle is also zero. If the particle is initially at rest ($\mathbf{v}_0 = 0$) with respect to the newtonian frame of reference used, it will thus remain at rest ($\mathbf{v} = 0$). If originally moving with a velocity $\mathbf{v}_0$, the particle will maintain a constant velocity $\mathbf{v} = \mathbf{v}_0$; that is, it will move with the constant speed v_0 in a straight line. This, we recall, is the statement of Newton's first law (Sec. 2.9). Thus, Newton's first law is a particular case of Newton's second law and may be omitted from the fundamental principles of mechanics.

12.2. Linear Momentum of a Particle. Rate of Change of Linear Momentum. Replacing the acceleration **a** by the derivative $d\mathbf{v}/dt$ in Eq. (12.2), we write

$$\Sigma\mathbf{F} = m\frac{d\mathbf{v}}{dt}$$

or, since the mass m of the particle is constant,

$$\Sigma\mathbf{F} = \frac{d}{dt}(m\mathbf{v}) \qquad (12.3)$$

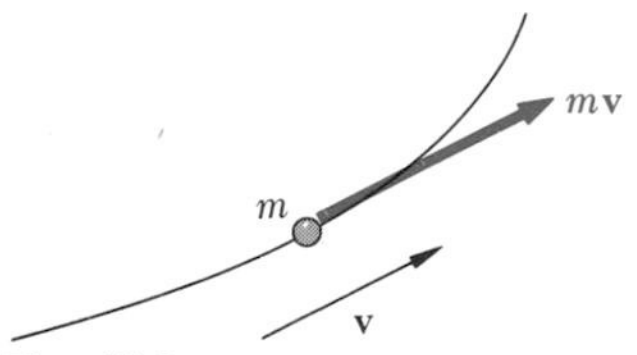

Fig. 12.3

The vector $m\mathbf{v}$ is called the *linear momentum*, or simply the *momentum*, of the particle. It has the same direction as the velocity of the particle and its magnitude is equal to the product of the mass m and the speed v of the particle (Fig. 12.3). Equation (12.3) expresses that *the resultant of the forces acting on the particle is equal to the rate of change of the linear momentum of the particle.* It is in this form that the second law of motion was originally stated by Newton. Denoting by **L** the linear momentum of the particle,

$$\mathbf{L} = m\mathbf{v} \qquad (12.4)$$

and by $\dot{\mathbf{L}}$ its derivative with respect to t, we may write Eq. (12.3) in the alternate form

$$\Sigma\mathbf{F} = \dot{\mathbf{L}} \qquad (12.5)$$

It should be noted that the mass m of the particle is assumed constant in Eqs. (12.3), (12.4), and (12.5). Equations (12.3) or (12.5), therefore, should not be used to solve problems involving the motion of bodies which gain or lose mass, such as rockets. Problems of that type will be considered in Sec. 14.11.†

It follows from Eq. (12.3) that the rate of change of the linear momentum $m\mathbf{v}$ is zero when $\Sigma\mathbf{F} = 0$. Thus, *if the resultant force acting on a particle is zero, the linear momentum of the particle remains constant, both in magnitude and direction.* This is the principle of *conservation of linear momentum* for a particle, which we may recognize as just an alternate statement of Newton's first law (Sec. 2.9).

12.3. Systems of Units. In using the fundamental equation $\mathbf{F} = m\mathbf{a}$, the units of force, mass, length, and time cannot be chosen arbitrarily. If they are, the magnitude of the force $\mathbf{F}$ required to give an acceleration $\mathbf{a}$ to the mass m will *not* be numerically equal to the product ma; it will only be proportional to this product. Thus, we may choose three of the four units arbitrarily but must choose the fourth unit so that the equation $\mathbf{F} = m\mathbf{a}$ is satisfied. The units are then said to form a system of consistent kinetic units.

Two systems of consistent kinetic units are currently used by American engineers, the International System of Units (SI units‡), and the U.S. customary units. Since both systems have been discussed in detail in Sec. 1.3, we shall describe them only briefly in this section.

International System of Units (SI Units). In this system, the base units are the units of length, mass, and time, and are called, respectively, the *meter* (m), the *kilogram* (kg), and the *second* (s). All three are arbitrarily defined (Sec. 1.3). The unit of force is a derived unit. It is called the *newton* (N) and is defined as the force which gives an acceleration of 1 m/s^2 to a mass of 1 kg (Fig. 12.4). From Eq. (12.1) we write

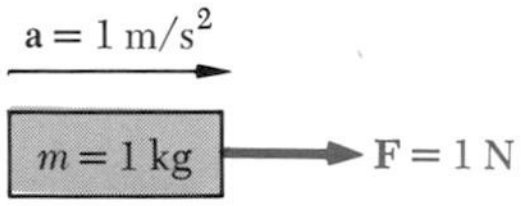

Fig. 12.4

$$1\text{ N} = (1\text{ kg})(1\text{ m/s}^2) = 1\text{ kg}\cdot\text{m/s}^2$$

The SI units are said to form an *absolute* system of units. This means that the three base units chosen are independent of the location where measurements are made. The meter, the kilogram, and the second may be used anywhere on the earth; they

†On the other hand, Eqs. (12.3) and (12.5) do hold in *relativistic mechanics*, where the mass m of the particle is assumed to vary with the speed of the particle.

‡SI stands for *Système International d'Unités* (French).

may even be used on another planet. They will always have the same significance.

Like any other force, the *weight* **W** of a body should be expressed in newtons. Since a body subjected to its own weight acquires an acceleration equal to the acceleration of gravity g, it follows from Newton's second law that the magnitude W of the weight of a body of mass m is

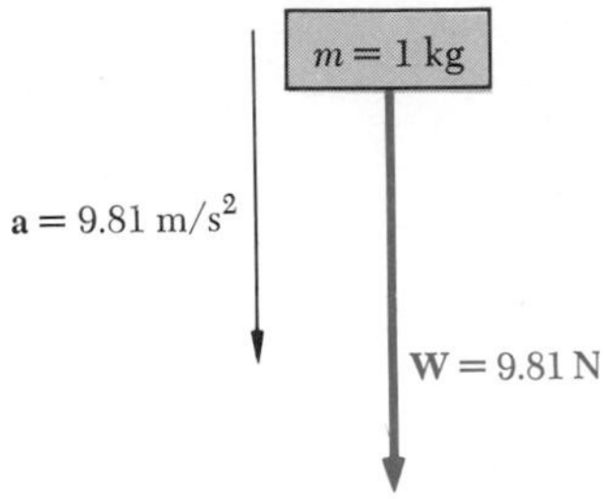

Fig. 12.5

$$W = mg \tag{12.6}$$

Recalling that $g = 9.81 \text{ m/s}^2$, we find that the weight of a body of mass 1 kg (Fig. 12.5) is

$$W = (1 \text{ kg})(9.81 \text{ m/s}^2) = 9.81 \text{ N}$$

Multiples and submultiples of the units of length, mass, and force are frequently used in engineering practice. They are, respectively, the *kilometer* (km) and the *millimeter* (mm); the *megagram*† (Mg) and the *gram* (g); and the *kilonewton* (kN). By definition

$$1 \text{ km} = 1000 \text{ m} \qquad 1 \text{ mm} = 0.001 \text{ m}$$
$$1 \text{ Mg} = 1000 \text{ kg} \qquad 1 \text{ g} = 0.001 \text{ kg}$$
$$1 \text{ kN} = 1000 \text{ N}$$

The conversion of these units to meters, kilograms, and newtons, respectively, can be effected by simply moving the decimal point three places to the right or to the left.

Units other than the units of mass, length, and time may all be expressed in terms of these three base units. For example, the unit of linear momentum may be obtained by recalling the definition of linear momentum and writing

$$mv = (\text{kg})(\text{m/s}) = \text{kg} \cdot \text{m/s}$$

U.S. Customary Units. Most practicing American engineers still commonly use a system in which the base units are the units of length, force, and time. These units are, respectively, the *foot* (ft), the *pound* (lb), and the *second* (s). The second is the same as the corresponding SI unit. The foot is defined as 0.3048 m. The pound is defined as the *weight* of a platinum standard, called the *standard pound* and kept at the National Bureau of Standards in Washington, the mass of which is 0.453 592 43 kg. Since the weight of a body depends upon the gravitational attraction of the earth, which varies with location, it is specified that the standard pound should be placed at sea level and at the latitude of 45° to properly define a force of 1 lb. Clearly the U.S. cus-

† Also known as a *metric ton*.

tomary units do not form an absolute system of units. Because of their dependence upon the gravitational attraction of the earth, they are said to form a *gravitational* system of units.

While the standard pound also serves as the unit of mass in commercial transactions in the United States, it cannot be so used in engineering computations since such a unit would not be consistent with the base units defined in the preceding paragraph. Indeed, when acted upon by a force of 1 lb, that is, when subjected to its own weight, the standard pound receives the acceleration of gravity, $g = 32.2 \text{ ft/s}^2$ (Fig. 12.6), not the unit acceleration required by Eq. (12.1). The unit of mass consistent with the foot, the pound, and the second is the mass which receives an acceleration of 1 ft/s^2 when a force of 1 lb is applied to it (Fig. 12.7). This unit, sometimes called a *slug*, can be derived from the equation $F = ma$ after substituting 1 lb and 1 ft/s^2 for F and a, respectively. We write

$$F = ma \qquad 1 \text{ lb} = (1 \text{ slug})(1 \text{ ft/s}^2)$$

and obtain

$$1 \text{ slug} = \frac{1 \text{ lb}}{1 \text{ ft/s}^2} = 1 \text{ lb} \cdot \text{s}^2/\text{ft}$$

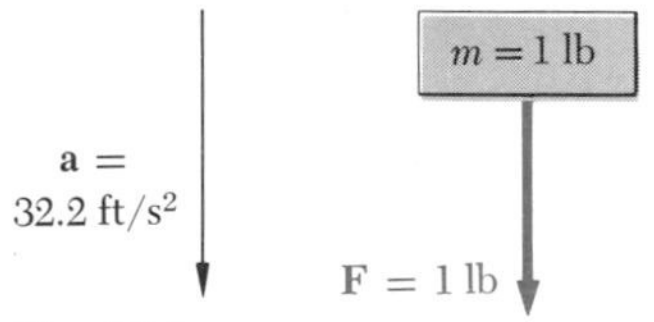

Fig. 12.6

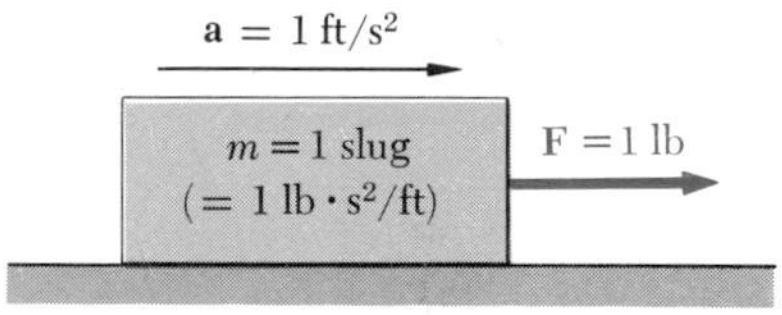

Fig. 12.7

Comparing Figs. 12.6 and 12.7, we conclude that the slug is a mass 32.2 times larger than the mass of the standard pound.

The fact that bodies are characterized in the U.S. customary system of units by their weight in pounds, rather than by their mass in slugs, was a convenience in the study of statics, where we were dealing constantly with weights and other forces and only seldom with masses. However, in the study of kinetics, where forces, masses, and accelerations are involved, we repeatedly shall have to express the mass m in slugs of a body, the weight W of which has been given in pounds. Recalling Eq. (12.6), we shall write

$$m = \frac{W}{g} \tag{12.7}$$

where g is the acceleration of gravity ($g = 32.2 \text{ ft/s}^2$).

Units other than the units of force, length, and time may all be expressed in terms of these three base units. For example, the unit of linear momentum may be obtained by recalling the definition of linear momentum and writing

$$mv = (\text{lb} \cdot \text{s}^2/\text{ft})(\text{ft/s}) = \text{lb} \cdot \text{s}$$

The conversion from U.S. customary units to SI units, and vice versa, has been discussed in Sec. 1.4. We shall recall the conversion factors obtained respectively for the units of length, force, and mass:

Length: $1 \text{ ft} = 0.3048 \text{ m}$
Force: $1 \text{ lb} = 4.448 \text{ N}$
Mass: $1 \text{ slug} = 1 \text{ lb} \cdot \text{s}^2/\text{ft} = 14.59 \text{ kg}$

Although it cannot be used as a consistent unit of mass, we also recall that the mass of the standard pound is, by definition,

$$1 \text{ pound-mass} = 0.4536 \text{ kg}$$

This constant may be used to determine the *mass* in SI units (kilograms) of a body which has been characterized by its *weight* in U.S. customary units (pounds).

12.4. Equations of Motion. Consider a particle of mass m acted upon by several forces. We recall from Sec. 12.1 that Newton's second law may be expressed by writing the equation

$$\Sigma \mathbf{F} = m\mathbf{a} \qquad (12.2)$$

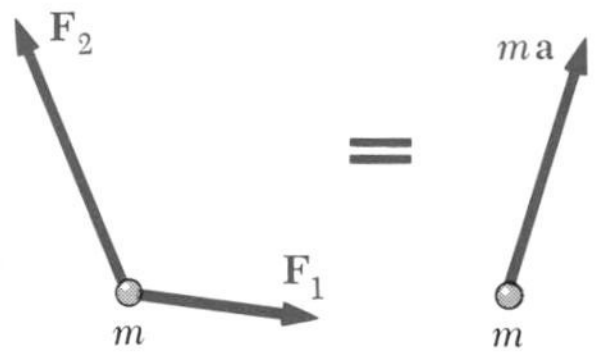

Fig. 12.8

which relates the forces acting on the particle and the vector $m\mathbf{a}$ (Fig. 12.8). In order to solve problems involving the motion of a particle, however, it will be found more convenient to replace Eq. (12.2) by equivalent equations involving scalar quantities.

Rectangular Components. Resolving each force $\mathbf{F}$ and the acceleration $\mathbf{a}$ into rectangular components, we write

$$\Sigma(F_x\mathbf{i} + F_y\mathbf{j} + F_z\mathbf{k}) = m(a_x\mathbf{i} + a_y\mathbf{j} + a_z\mathbf{k})$$

from which it follows that

$$\Sigma F_x = ma_x \qquad \Sigma F_y = ma_y \qquad \Sigma F_z = ma_z \qquad (12.8)$$

Recalling from Sec. 11.11 that the components of the acceleration are equal to the second derivatives of the coordinates of the particle, we have

$$\Sigma F_x = m\ddot{x} \qquad \Sigma F_y = m\ddot{y} \qquad \Sigma F_z = m\ddot{z} \qquad (12.8')$$

Consider, as an example, the motion of a projectile. If the resistance of the air is neglected, the only force acting on the projectile after it has been fired is its weight $\mathbf{W} = -W\mathbf{j}$. The equations defining the motion of the projectile are therefore

$$m\ddot{x} = 0 \qquad m\ddot{y} = -W \qquad m\ddot{z} = 0$$

and the components of the acceleration of the projectile are

$$\ddot{x} = 0 \qquad \ddot{y} = -\frac{W}{m} = -g \qquad \ddot{z} = 0$$

where g is 9.81 m/s² or 32.2 ft/s². The equations obtained may be integrated independently, as was shown in Sec. 11.11, to obtain the velocity and displacement of the projectile at any instant.

Tangential and Normal Components. Resolving the forces and the acceleration of the particle into components along the tangent to the path (in the direction of motion) and the normal (toward the inside of the path) (Fig. 12.9), and substituting into Eq. (12.2), we obtain the two scalar equations

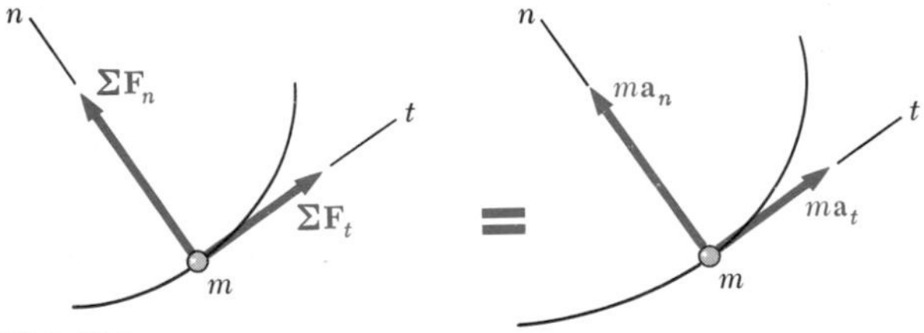

Fig. 12.9

$$\Sigma F_t = ma_t \qquad \Sigma F_n = ma_n \tag{12.9}$$

Substituting for a_t and a_n from Eqs. (11.40), we have

$$\Sigma F_t = m\frac{dv}{dt} \qquad \Sigma F_n = m\frac{v^2}{\rho} \tag{12.9'}$$

The equations obtained may be solved for two unknowns.

12.5. Dynamic Equilibrium. Returning to Eq. (12.2) and transposing the right-hand member, we write Newton's second law in the alternate form

$$\Sigma\mathbf{F} - m\mathbf{a} = 0 \tag{12.10}$$

which expresses that, if we add the vector $-m\mathbf{a}$ to the forces acting on the particle, *we obtain a system of vectors equivalent to zero* (Fig. 12.10). The vector $-m\mathbf{a}$, of magnitude ma and of *direction opposite* to that of the acceleration, is called an *inertia vector*. The particle may thus be considered to be in equilibrium

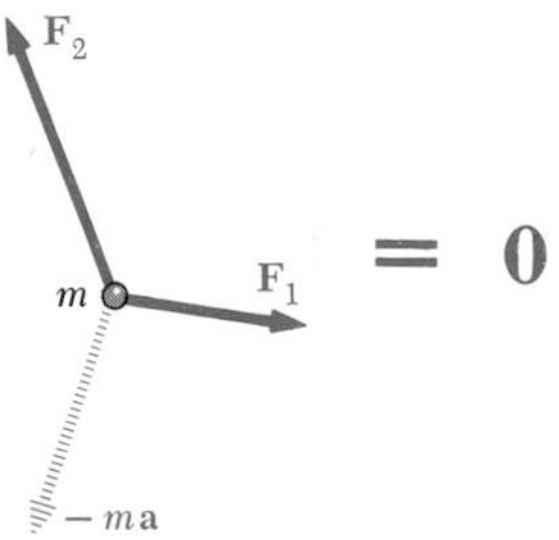

Fig. 12.10

under the given forces and the inertia vector. The particle is said to be in *dynamic equilibrium*, and the problem under consideration may be solved by the methods developed earlier in statics.

In the case of coplanar forces, we may draw in tip-to-tail fashion all the vectors shown in Fig. 12.10, *including the inertia vector*, to form a closed-vector polygon. Or we may write that the sums of the components of all the vectors in Fig. 12.10, including again the inertia vector, are zero. Using rectangular components, we therefore write

$$\Sigma F_x = 0 \qquad \Sigma F_y = 0 \qquad \textbf{\textit{including inertia vector}} \quad (12.11)$$

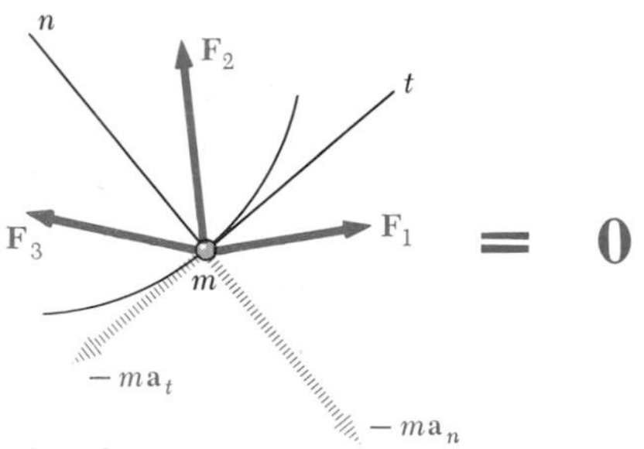

Fig. 12.11

When tangential and normal components are used, it is more convenient to represent the inertia vector by its two components $-m\mathbf{a}_t$ and $-m\mathbf{a}_n$ in the sketch itself (Fig. 12.11). The tangential component of the inertia vector provides a measure of the resistance the particle offers to a change in speed, while its normal component (also called *centrifugal force*) represents the tendency of the particle to leave its curved path. We should note that either of these two components may be zero under special conditions: (1) if the particle starts from rest, its initial velocity is zero and the normal component of the inertia vector is zero at $t = 0$; (2) if the particle moves at constant speed along its path, the tangential component of the inertia vector is zero and only its normal component needs to be considered.

Because they measure the resistance that particles offer when we try to set them in motion or when we try to change the conditions of their motion, inertia vectors are often called *inertia forces*. The inertia forces, however, are not forces like the forces found in statics, which are either contact forces or gravitational forces (weights). Many people, therefore, object to the use of the word "force" when referring to the vector $-m\mathbf{a}$ or even avoid altogether the concept of dynamic equilibrium. Others point out that inertia forces and actual forces, such as gravitational forces, affect our senses in the same way and cannot be distinguished by physical measurements. A man riding in an elevator which is accelerated upward will have the feeling that his weight has suddenly increased; and no measurement made within the elevator could establish whether the elevator is truly accelerated or whether the force of attraction exerted by the earth has suddenly increased.

Sample problems have been solved in this text by the direct application of Newton's second law, as illustrated in Figs. 12.8 and 12.9, rather than by the method of dynamic equilibrium.

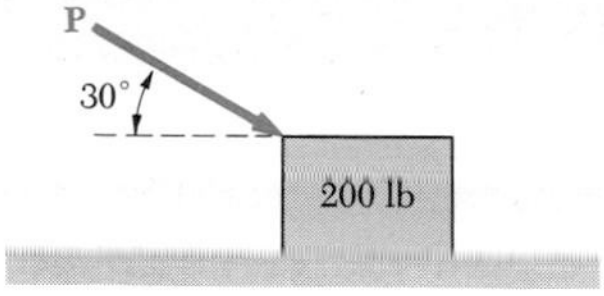

SAMPLE PROBLEM 12.1

A 200-lb block rests on a horizontal plane. Find the magnitude of the force **P** required to give the block an acceleration of 10 ft/s^2 to the right. The coefficient of friction between the block and the plane is $\mu = 0.25$.

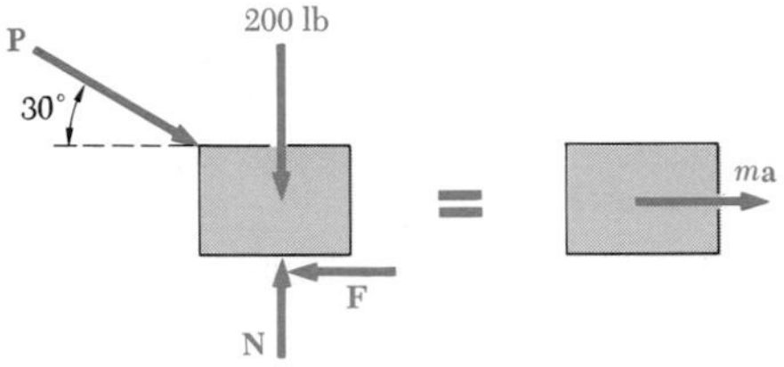

Solution. The mass of the block is

$$m = \frac{W}{g} = \frac{200 \text{ lb}}{32.2 \text{ ft/s}^2} = 6.21 \text{ lb} \cdot \text{s}^2/\text{ft}$$

We note that $F = \mu N = 0.25N$ and that $a = 10 \text{ ft/s}^2$. Expressing that the forces acting on the block are equivalent to the vector $m\mathbf{a}$, we write

$$\overset{+}{\rightarrow} \Sigma F_x = ma: \qquad P \cos 30° - 0.25N = (6.21 \text{ lb} \cdot \text{s}^2/\text{ft})(10 \text{ ft/s}^2)$$
$$P \cos 30° - 0.25N = 62.1 \text{ lb} \qquad (1)$$

$$+\uparrow \Sigma F_y = 0: \qquad N - P \sin 30° - 200 \text{ lb} = 0 \qquad (2)$$

Solving (2) for N and carrying the result into (1), we obtain

$$N = P \sin 30° + 200 \text{ lb}$$
$$P \cos 30° - 0.25(P \sin 30° + 200 \text{ lb}) = 62.1 \text{ lb} \qquad P = 151 \text{ lb} \blacktriangleleft$$

SAMPLE PROBLEM 12.2

Solve Sample Prob. 12.1 using SI units.

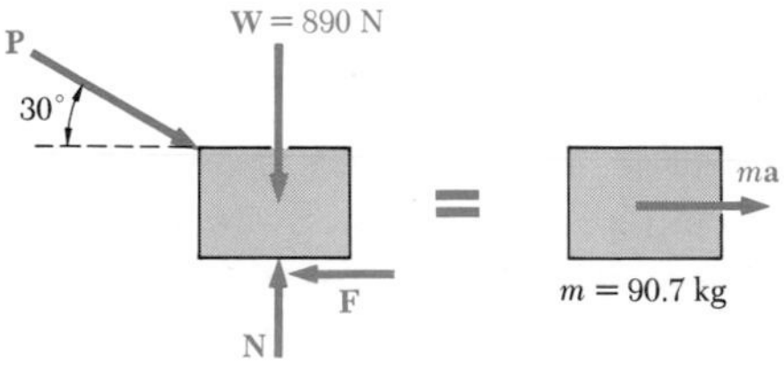

Solution. Using the conversion factors given in Sec. 12.3, we write

$$a = (10 \text{ ft/s}^2)(0.3048 \text{ m/ft}) = 3.05 \text{ m/s}^2$$
$$W = (200 \text{ lb})(4.448 \text{ N/lb}) = 890 \text{ N}$$

Recalling that, by definition, 1 lb is the weight of a mass of 0.4536 kg, we find that the mass of the 200-lb block is

$$m = 200(0.4536 \text{ kg}) = 90.7 \text{ kg}$$

Noting that $F = \mu N = 0.25N$ and expressing that the forces acting on the block are equivalent to the vector $m\mathbf{a}$, we write

$$\overset{+}{\rightarrow} \Sigma F_x = ma: \qquad P \cos 30° - 0.25N = (90.7 \text{ kg})(3.05 \text{ m/s}^2)$$
$$P \cos 30° - 0.25N = 277 \text{ N} \qquad (1)$$

$$+\uparrow \Sigma F_y = 0: \qquad N - P \sin 30° - 890 \text{ N} = 0 \qquad (2)$$

Solving (2) for N and carrying the result into (1), we obtain

$$N = P \sin 30° + 890 \text{ N}$$
$$P \cos 30° - 0.25(P \sin 30° + 890 \text{ N}) = 277 \text{ N} \qquad P = 674 \text{ N} \blacktriangleleft$$

or, in U.S. customary units,

$$P = (674 \text{ N}) \div (4.448 \text{ N/lb}) \qquad P = 151 \text{ lb} \blacktriangleleft$$

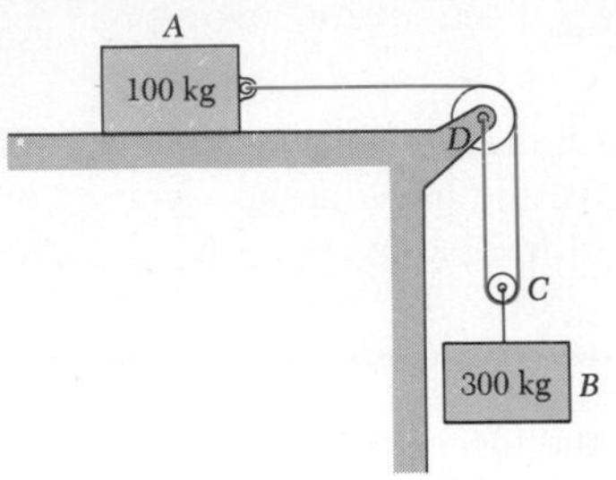

SAMPLE PROBLEM 12.3

The two blocks shown start from rest. The horizontal plane and the pulley are frictionless, and the pulley is assumed to be of negligible mass. Determine the acceleration of each block and the tension in each cord.

Solution. We denote by T_1 the tension in cord ACD and by T_2 the tension in cord BC. We note that if block A moves through s_A, block B moves through

$$s_B = \tfrac{1}{2}s_A$$

Differentiating twice with respect to t, we have

$$a_B = \tfrac{1}{2}a_A \tag{1}$$

We shall apply Newton's second law successively to block A, block B, and pulley C.

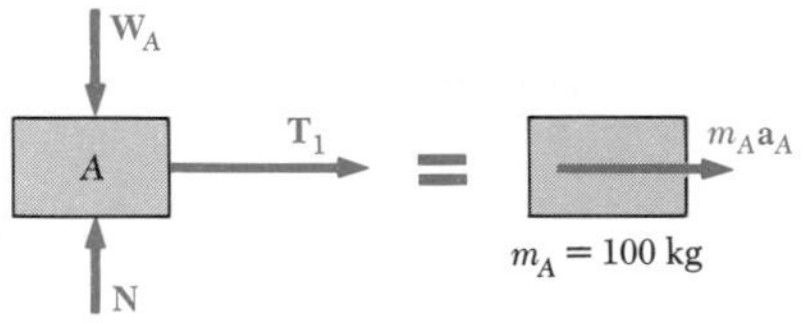

Block A

$$\overset{+}{\rightarrow} \Sigma F_x = m_A a_A: \qquad T_1 = 100a_A \tag{2}$$

Block B. Observing that the weight of block B is

$$W_B = m_B g = (300 \text{ kg})(9.81 \text{ m/s}^2) = 2940 \text{ N}$$

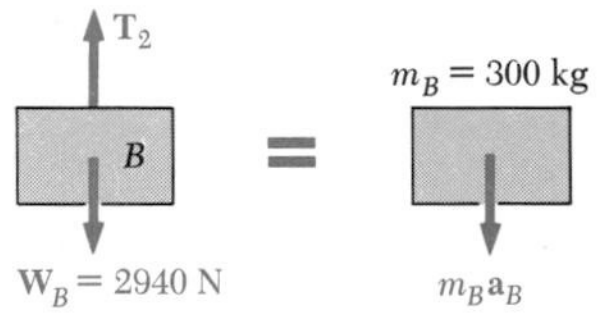

we write

$$+\downarrow \Sigma F_y = m_B a_B: \qquad 2940 - T_2 = 300a_B$$

or, substituting for a_B from (1),

$$2940 - T_2 = 300(\tfrac{1}{2}a_A)$$
$$T_2 = 2940 - 150a_A \tag{3}$$

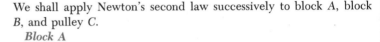

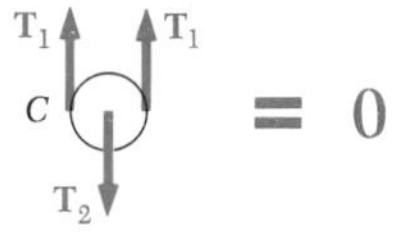

Pulley C. Since m_C is assumed to be zero, we have

$$+\downarrow \Sigma F_y = m_C a_C = 0: \qquad T_2 - 2T_1 = 0 \tag{4}$$

Substituting for T_1 and T_2 from (2) and (3), respectively, into (4), we write

$$2940 - 150a_A - 2(100a_A) = 0$$
$$2940 - 350a_A = 0 \qquad a_A = 8.40 \text{ m/s}^2 \blacktriangleleft$$

Substituting the value obtained for a_A into (1) and (2), we have

$$a_B = \tfrac{1}{2}a_A = \tfrac{1}{2}(8.40 \text{ m/s}^2) \qquad a_B = 4.20 \text{ m/s}^2 \blacktriangleleft$$
$$T_1 = 100a_A = (100 \text{ kg})(8.40 \text{ m/s}^2) \qquad T_1 = 840 \text{ N} \blacktriangleleft$$

Recalling (4), we write

$$T_2 = 2T_1 \qquad T_2 = 2(840 \text{ N}) \qquad T_2 = 1680 \text{ N} \blacktriangleleft$$

We note that the value obtained for T_2 is *not* equal to the weight of block B.

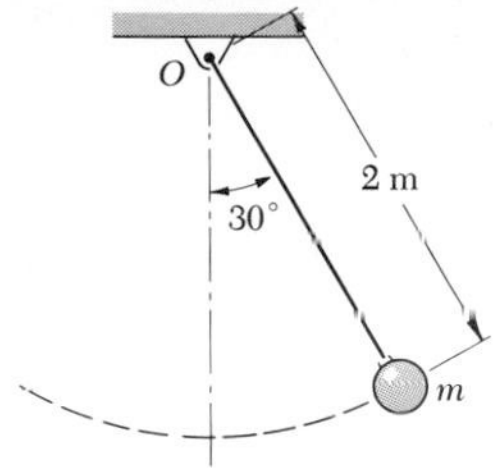

SAMPLE PROBLEM 12.4

The bob of a 2-m pendulum describes an arc of circle in a vertical plane. If the tension in the cord is 2.5 times the weight of the bob for the position shown, find the velocity and acceleration of the bob in that position.

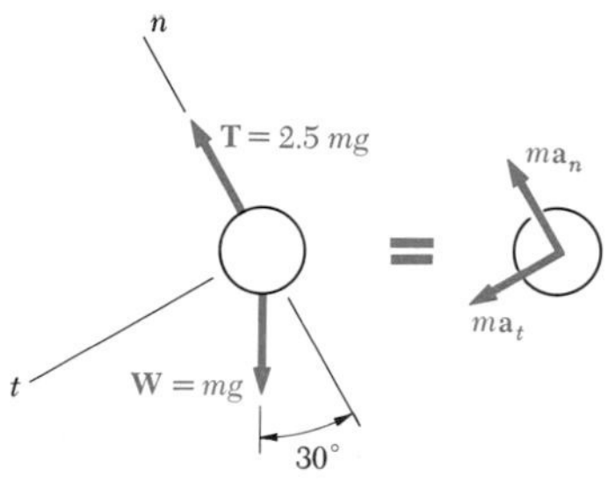

Solution. The weight of the bob is $W = mg$; the tension in the cord is thus 2.5 mg. Recalling that $\mathbf{a}_n$ is directed toward O and assuming $\mathbf{a}_t$ as shown, we apply Newton's second law and obtain

$+\swarrow \Sigma F_t = ma_t$: $\quad mg \sin 30° = ma_t$

$$a_t = g \sin 30° = +4.90 \text{ m/s}^2 \qquad \mathbf{a}_t = 4.90 \text{ m/s}^2 \swarrow \blacktriangleleft$$

$+\nwarrow \Sigma F_n = ma_n$: $\quad 2.5\, mg - mg \cos 30° = ma_n$

$$a_n = 1.634\, g = +16.03 \text{ m/s}^2 \qquad \mathbf{a}_n = 16.03 \text{ m/s}^2 \nwarrow \blacktriangleleft$$

Since $a_n = v^2/\rho$, we have $v^2 = \rho a_n = (2 \text{ m})(16.03 \text{ m/s}^2)$

$$v = \pm 5.66 \text{ m/s} \qquad \mathbf{v} = 5.66 \text{ m/s} \text{ (up or down)} \blacktriangleleft$$

SAMPLE PROBLEM 12.5

Determine the rated speed of a highway curve of radius $\rho = 400$ ft banked through an angle $\theta = 18°$. The rated speed of a banked curved road is the speed at which a car should travel if no lateral friction force is to be exerted on its wheels.

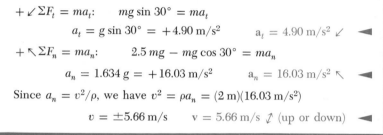

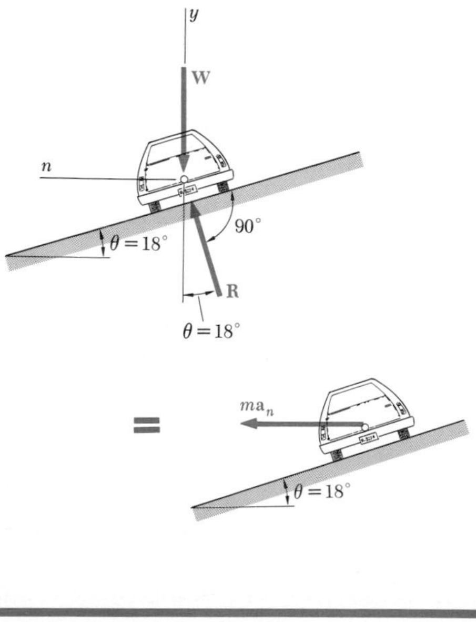

Solution. The car travels in a *horizontal* circular path of radius ρ. The normal component $\mathbf{a}_n$ of the acceleration is directed toward the center of the path; its magnitude is $a_n = v^2/\rho$, where v is the speed of the car in ft/s. The mass m of the car is W/g, where W is the weight of the car. Since no lateral friction force is to be exerted on the car, the reaction $\mathbf{R}$ of the road is shown perpendicular to the roadway. Applying Newton's second law, we write

$+\uparrow \Sigma F_y = 0$: $\quad R \cos\theta - W = 0 \qquad R = \dfrac{W}{\cos\theta} \qquad (1)$

$\overset{+}{\leftarrow} \Sigma F_n = ma_n$: $\quad R \sin\theta = \dfrac{W}{g} a_n \qquad (2)$

Substituting for R from (1) into (2), and recalling that $a_n = v^2/\rho$:

$$\frac{W}{\cos\theta} \sin\theta = \frac{W}{g}\frac{v^2}{\rho} \qquad v^2 = g\rho \tan\theta$$

Substituting the given data, $\rho = 400$ ft and $\theta = 18°$, into this equation, we obtain

$$v^2 = (32.2 \text{ ft/s}^2)(400 \text{ ft}) \tan 18°$$
$$v = 64.7 \text{ ft/s} \qquad v = 44.1 \text{ mi/h} \blacktriangleleft$$

PROBLEMS

12.1 The value of g at any latitude ϕ may be obtained from the formula

$$g = 9.7807(1 + 0.0053 \sin^2 \phi) \qquad \text{m/s}^2$$

Determine to four significant figures the weight in newtons and the mass in kilograms, at the latitudes of 0°, 45°, and 90°, of a silver bar whose mass is officially defined as 10 kg.

12.2 The acceleration due to gravity on the moon is 5.31 ft/s^2. Determine the weight in pounds, the mass in pounds, and the mass in lb · s^2/ft, on the moon, of a silver bar whose mass is officially defined as 100.00 lb.

12.3 A 100-kg satellite has been placed in a circular orbit 2000 km above the surface of the earth. The acceleration of gravity at this elevation is 5.68 m/s^2. Determine the linear momentum of the satellite, knowing that its orbital speed is 24 800 km/h.

12.4 Two boxes are weighed on the scales shown: scale a is a lever scale; scale b is a spring scale. The scales are attached to the roof of an elevator. When the elevator is at rest, each scale indicates a load of 20 lb. If the spring scale indicates a load of 18 lb, determine the acceleration of the elevator and the load indicated by the lever scale.

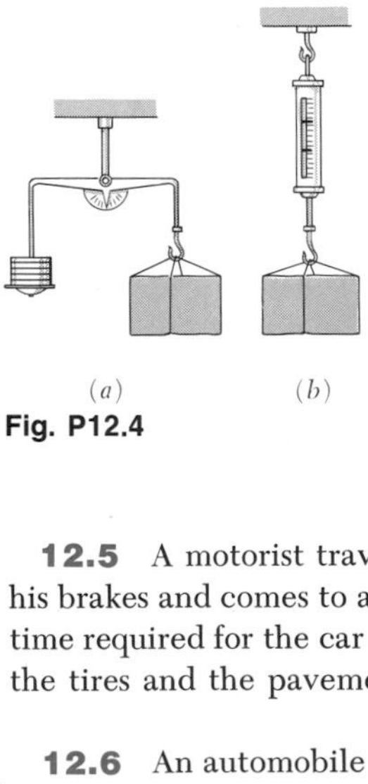

Fig. P12.4

12.5 A motorist traveling at a speed of 45 mi/h suddenly applies his brakes and comes to a stop after skidding 150 ft. Determine (a) the time required for the car to stop, (b) the coefficient of friction between the tires and the pavement.

12.6 An automobile skids 90 ft on a level road before coming to a stop. If the coefficient of friction between the tires and the pavement is 0.75, determine (a) the speed of the automobile before the brakes were applied, (b) the time required for the automobile to come to a stop.

12.7 A truck is proceeding up a long 3-percent grade at a constant speed of 60 km/h. If the driver does not change the setting of his throttle or shift gears, what will be the acceleration of the truck as it starts moving on the level section of the road?

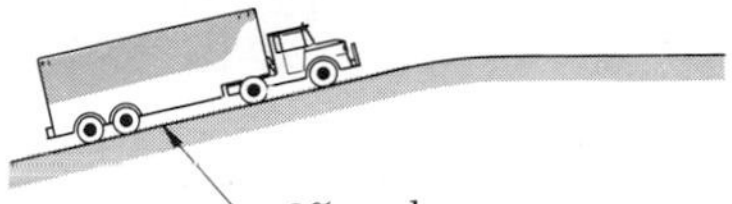

Fig. P12.7

12.8 A 5-kg package is projected down the incline with an initial velocity of 4 m/s. Knowing that the coefficient of friction between the package and the incline is 0.35, determine (*a*) the velocity of the package after 3 m of motion, (*b*) the distance *d* at which the package comes to rest.

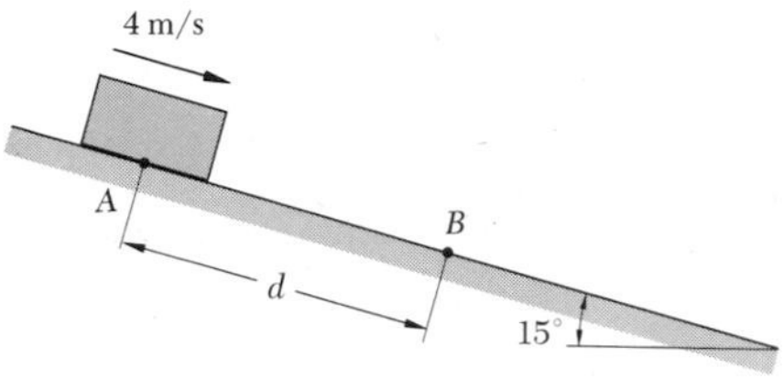

Fig. P12.8

12.9 The 3-kg collar was moving down the rod with a velocity of 3 m/s when a force **P** was applied to the horizontal cable. Assuming negligible friction between the collar and the rod, determine the magnitude of the force **P** if the collar stopped after moving 1 m more down the rod.

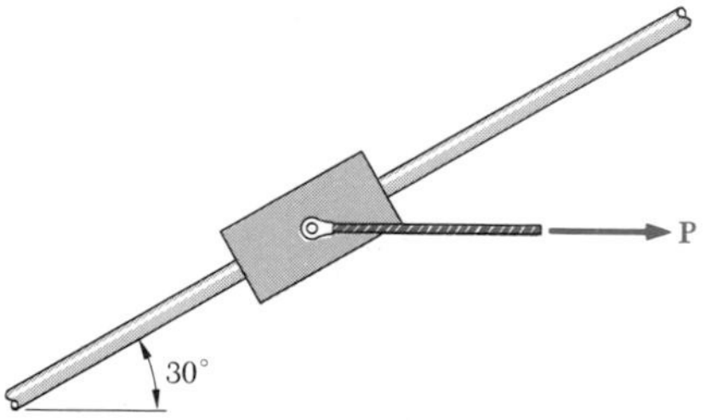

Fig. P12.9

12.10 Solve Prob. 12.9, assuming a coefficient of friction of 0.20 between the collar and the rod.

12.11 The subway train shown travels at a speed of 30 mi/h. Determine the force in each coupling when the brakes are applied, knowing that the braking force is 5000 lb on each car.

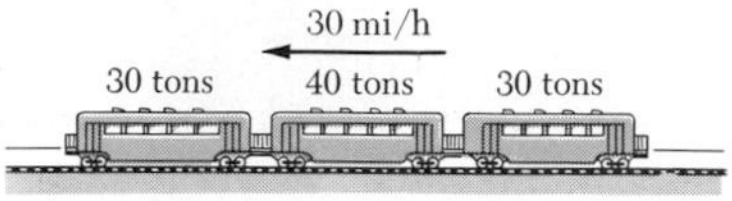

Fig. P12.11

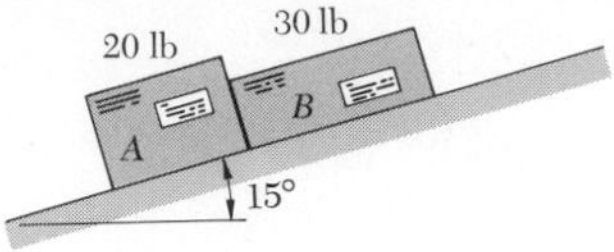

Fig. P12.12

12.12 Two packages are placed on an incline as shown. The coefficient of friction is 0.25 between the incline and package *A*, and 0.15 between the incline and package *B*. Knowing that the packages are in contact when released, determine (*a*) the acceleration of each package, (*b*) the force exerted by package *A* on package *B*.

12.13 Solve Prob. 12.12, assuming the positions of the packages are reversed so that package *A* is to the right of package *B*.

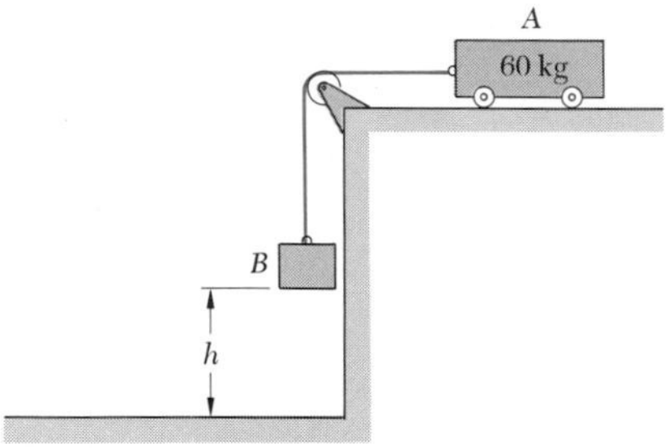

Fig. P12.14, P12.15, and P12.16

12.14 When the system shown is released from rest, the acceleration of block *B* is observed to be $3\ \text{m/s}^2$ downward. Neglecting the effect of friction, determine (*a*) the tension in the cable, (*b*) the mass of block *B*.

12.15 The system shown is released from rest when $h = 1.4$ m. (*a*) Determine the mass of block *B*, knowing that it strikes the ground with a speed of 3 m/s. (*b*) Attempt to solve part *a*, assuming the final speed to be 6 m/s; explain the difficulty encountered.

12.16 The system shown is released from rest. Knowing that the mass of block *B* is 30 kg, determine how far the cart will move before it reaches a speed of 2.5 m/s, (*a*) if the pulley may be considered as weightless and frictionless, (*b*) if the pulley "freezes" on its shaft and the cable must slip, with $\mu = 0.10$, over the pulley.

12.17 Each of the systems shown is initially at rest. Assuming the pulleys to be weightless and neglecting axle friction, determine for each system (*a*) the acceleration of block *A*, (*b*) the velocity of block *A* after 4 s, (*c*) the velocity of block *A* after it has moved 10 ft.

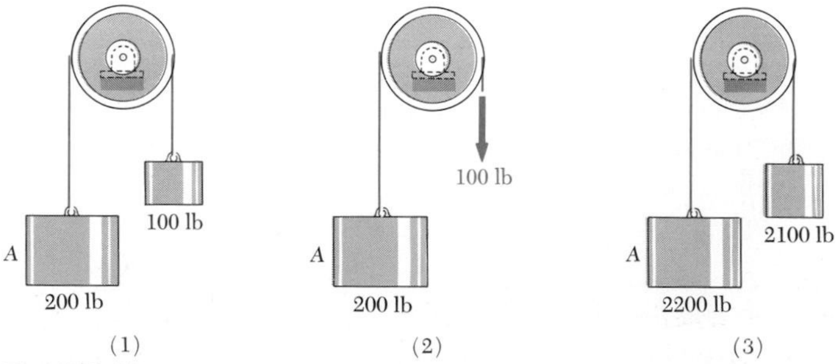

Fig. P12.17

12.18 The 100-kg block A is connected to a 25-kg counterweight B by the cable arrangement shown. If the system is released from rest, determine (*a*) the tension in the cable, (*b*) the velocity of B after 3 s, (*c*) the velocity of A after it has moved 1.2 m.

12.19 Block A is observed to move with an acceleration of 0.9 m/s² directed upward. Determine (*a*) the mass of block B, (*b*) the corresponding tension in the cable.

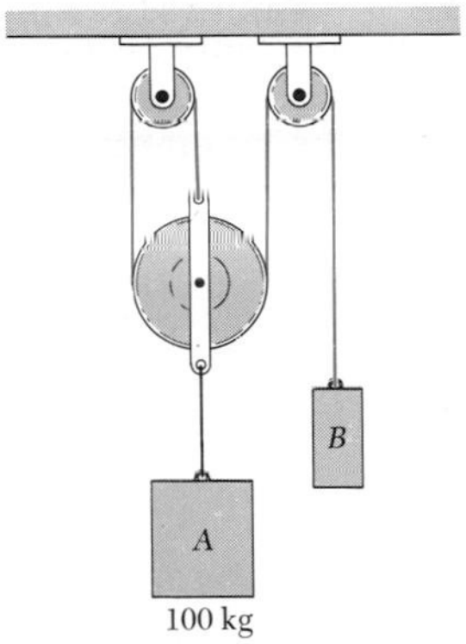

Fig. P12.18 and P12.19

12.20 The system shown is initially at rest. Neglecting the effect of friction, determine (*a*) the force **P** required if the velocity of collar B is to be 12 ft/s after it has moved 18 in. to the right, (*b*) the corresponding tension in the cable.

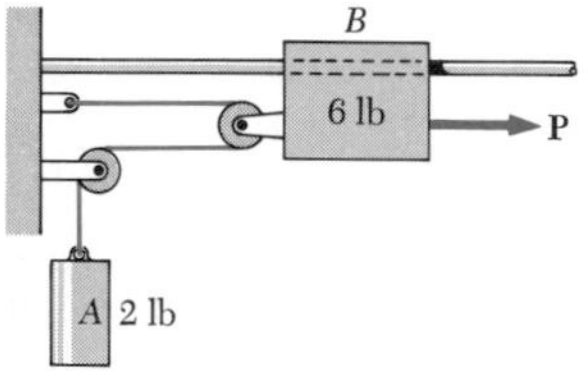

Fig. P12.20 and P12.21

12.21 A force **P** of magnitude 15 lb is applied to collar B, which is observed to move 3 ft in 0.5 s after starting from rest. Neglecting the effect of friction in the pulleys, determine the friction force that the rod exerts on collar B.

12.22 Neglecting the effect of friction, determine (*a*) the acceleration of each block, (*b*) the tension in the cable.

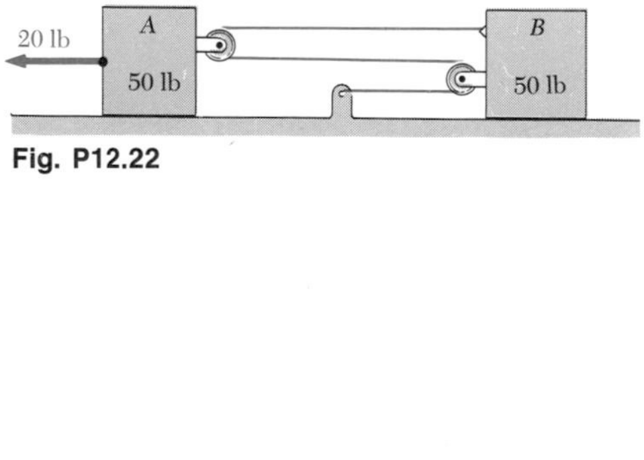

Fig. P12.22

12.23 The *rimpull* of a truck is defined as the tractive force between the rubber tires of the driving wheels and the ground. For a truck used to haul earth at a construction site, the rimpull actually utilized by the average driver in each of the first five forward gears and the maximum speed attained in each gear are as follows:

Gear	Max v (mi/h)	Average rimpull (lb)
1st	3	6000
2d	6	3800
3d	9	2800
4th	15	2000
5th	27	1500

Knowing that a truck (and load) weighs 44,000 lb and has a rolling resistance of 60 lb/ton for the unpaved surface encountered, determine the time required for the truck to attain a speed of 27 mi/h. Neglect the time needed to shift gears.

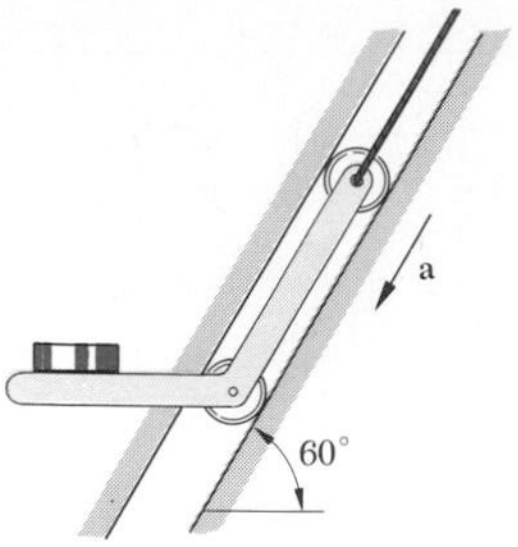

Fig. P12.24

12.24 In a manufacturing process, disks are moved from one elevation to another by the lifting arm shown; the coefficient of friction between a disk and the arm is 0.20. Determine the magnitude of the acceleration for which the disks slide on the arm, assuming the acceleration is directed (*a*) downward as shown, (*b*) upward.

12.25 The coefficient of friction between the load and the flat-bed trailer shown is 0.40. Knowing that the forward speed of the truck is 50 km/h, determine the shortest distance in which the truck can be brought to a stop if the load is not to shift.

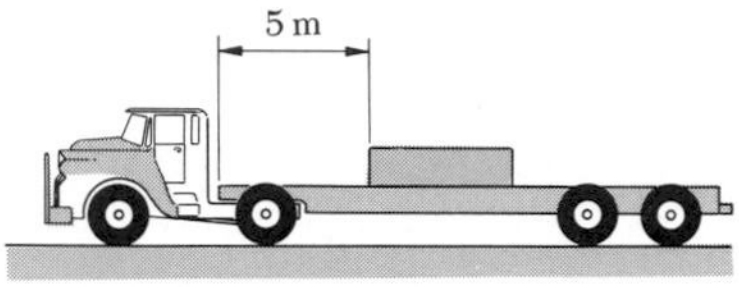

Fig. P12.25 and P12.26

12.26 The coefficient of friction between the load and the flat-bed trailer is 0.40. While traveling at 100 km/h, the driver makes an emergency stop and the truck skids to rest in 90 m. Determine the velocity of the load relative to the trailer as it reaches the forward edge of the trailer.

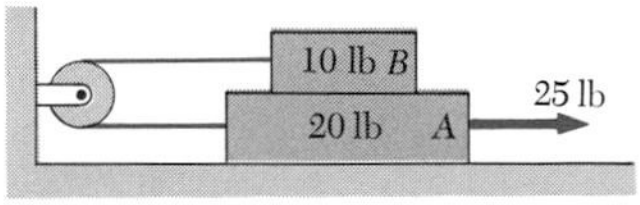

Fig. P12.27

12.27 Knowing that the coefficient of friction is 0.30 at all surfaces of contact, determine (*a*) the acceleration of plate *A*, (*b*) the tension in the cable. (Neglect bearing friction in the pulley.)

12.28 Solve Prob. 12.27, assuming that the 25-lb force is applied to plate *B*.

12.29 A 30-kg crate rests on a 20-kg cart; the coefficient of static friction between the crate and the cart is 0.25. If the crate is not to slip with respect to the cart, determine (*a*) the maximum allowable magnitude of **P**, (*b*) the corresponding acceleration of the cart.

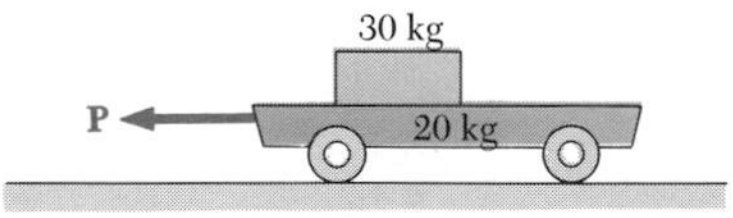

Fig. P12.29 and P12.30

12.30 The coefficients of friction between the 30-kg crate and the 20-kg cart are $\mu_s = 0.25$ and $\mu_k = 0.20$. If a force **P** of magnitude 150 N is applied to the cart, determine the acceleration (*a*) of the cart, (*b*) of the crate, (*c*) of the crate with respect to the cart.

12.31 The force exerted by a magnet on a small steel block varies inversely as the square of the distance between the block and the magnet. When the block is 250 mm from the magnet, the magnetic force is 1.5 N. The coefficient of friction between the steel block and the horizontal surface is 0.50. If the block is released from the position shown, determine its velocity when it is 100 mm from the magnet.

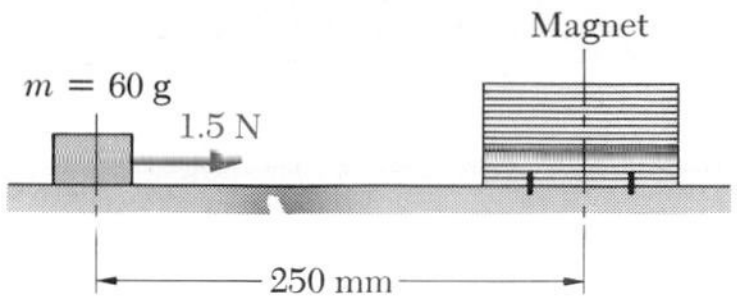

Fig. P12.31

12.32 A constant force **P** is applied to a piston and rod of total mass m in order to make them move in a cylinder filled with oil. As the piston moves, the oil is forced through orifices in the piston and exerts on the piston an additional force of magnitude kv, proportional to the speed v of the piston and in a direction opposite to its motion. Express the acceleration and velocity of the piston as a function of the time t, assuming that the piston starts from rest at time $t = 0$.

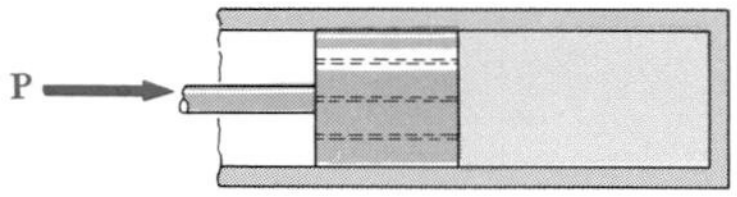

Fig. P12.32

12.33 A ship of total mass m is anchored in the middle of a river which is flowing with a constant velocity $\mathbf{v}_0$. The horizontal component of the force exerted on the ship by the anchor chain is $\mathbf{T}_0$. If the anchor chain suddenly breaks, determine the time required for the ship to attain a velocity equal to $\frac{1}{2}\mathbf{v}_0$. Assume that the frictional resistance of the water is proportional to the velocity of the ship relative to the water.

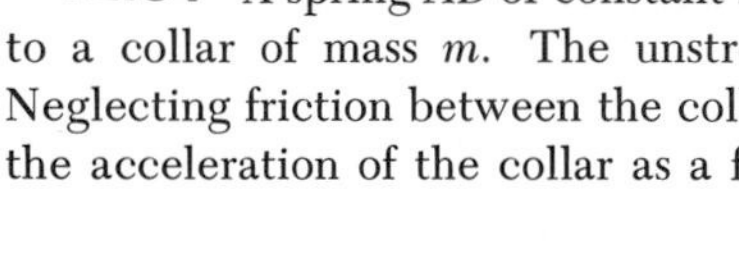

Fig. P12.33

12.34 A spring AB of constant k is attached to a support at A and to a collar of mass m. The unstretched length of the spring is l. Neglecting friction between the collar and the horizontal rod, express the acceleration of the collar as a function of the distance x.

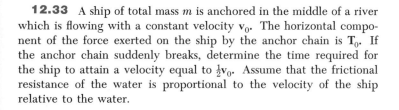

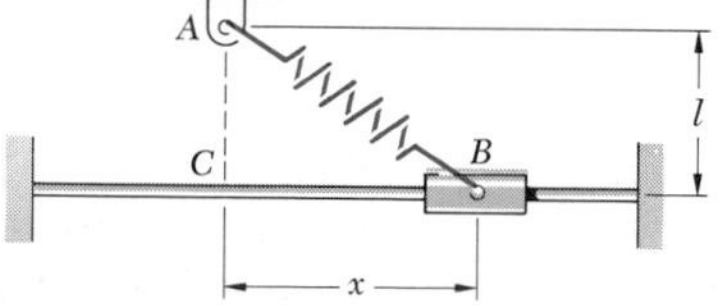

Fig. P12.34

12.35 Knowing that blocks B and C strike the ground simultaneously and exactly 1 s after the system is released from rest, determine W_B and W_C in terms of W_A.

12.36 Determine the acceleration of each block when W_A = 10 lb, W_B = 30 lb, and W_C = 20 lb. Which block strikes the ground first?

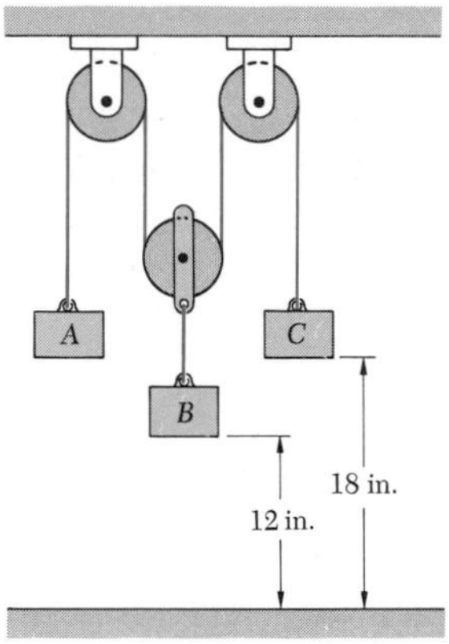

Fig. P12.35, P12.36, and P12.37

12.37 In the system shown, W_A = 10 lb and W_C = 20 lb. Determine the required weight W_B if block B is not to move when the system is released from rest.

12.38 Determine the acceleration of each block when m_A = 15 kg, m_B = 10 kg, and m_C = 5 kg.

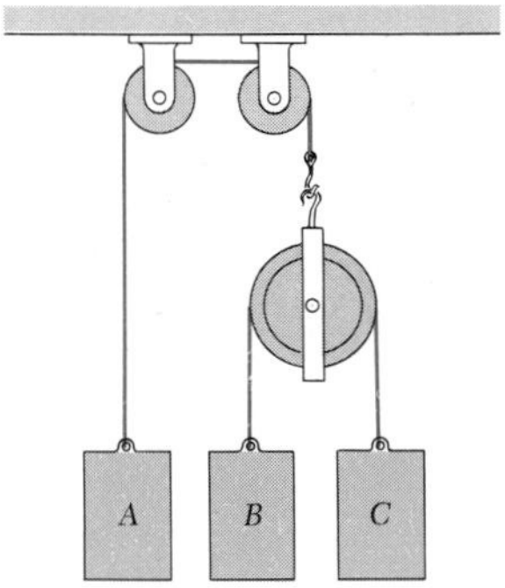

Fig. P12.38

12.39 Knowing that μ = 0.30, determine the acceleration of each block when $m_A = m_B = m_C$.

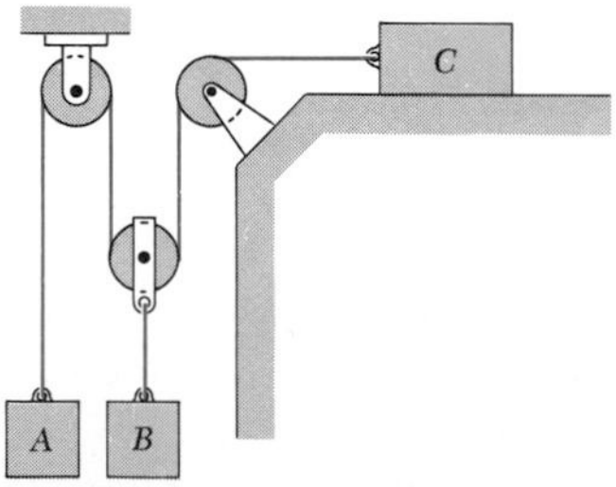

Fig. P12.39 and P12.40

12.40 Knowing that μ = 0.50, determine the acceleration of each block when m_A = 5 kg, m_B = 20 kg, and m_C = 15 kg.

12.41 A small ball of mass $m = 5$ kg is attached to a cord of length $L = 2$ m and is made to revolve in a horizontal circle at a constant speed v_0. Knowing that the cord forms an angle $\theta = 40°$ with the vertical, determine (*a*) the tension in the cord, (*b*) the speed v_0 of the ball.

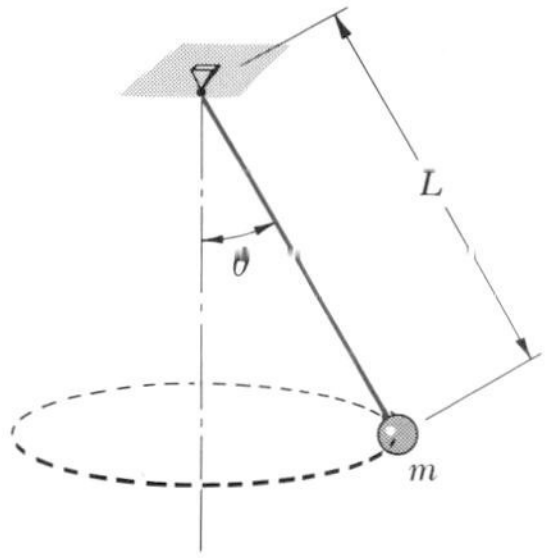

Fig. P12.41 and P12.42

12.42 A small ball of mass $m = 5$ kg is made to revolve in a horizontal circle as shown. Knowing that the maximum allowable tension in the cord is 100 N, determine (*a*) the maximum allowable velocity if $L = 2$ m, (*b*) the corresponding value of the angle θ.

12.43 Two wires *AC* and *BC* are each tied to a sphere at *C*. The sphere is made to revolve in a horizontal circle at a constant speed v. Determine the range of values of the speed v for which *both* wires are taut.

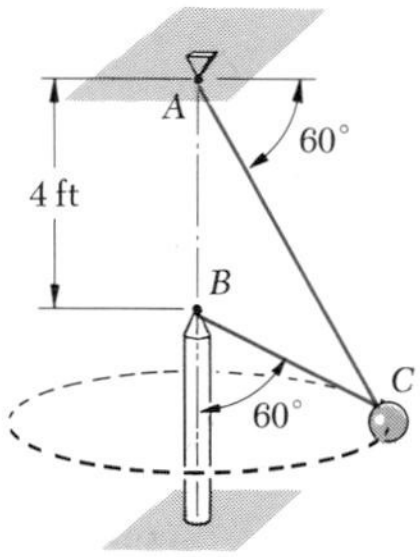

Fig. P12.43 and P12.44

12.44 Two wires *AC* and *BC* are each tied to a 10-lb sphere. The sphere is made to revolve in a horizontal circle at a constant speed v. Determine (*a*) the speed for which the tension is the same in both wires, (*b*) the corresponding tension.

12.45 A 3-kg ball is swung in a vertical circle at the end of a cord of length $l = 0.8$ m. Knowing that when $\theta = 60°$ the tension in the cord is 25 N, determine the instantaneous velocity and acceleration of the ball.

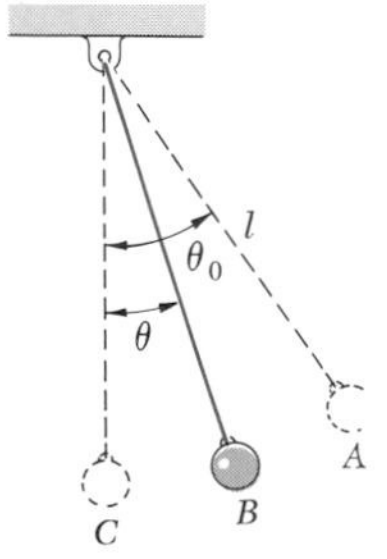

Fig. P12.45 and P12.46

12.46 A ball of weight W is released with no velocity from position *A* and oscillates in a vertical plane at the end of a cord of length l. Determine (*a*) the tangential component of the acceleration in position *B* in terms of the angle θ, (*b*) the velocity in position *B* in terms of θ, θ_0, and l, (*c*) the tension in the cord in terms of W and θ_0 when the ball passes through its lowest position *C*, (*d*) the value of θ_0 if the tension in the cord is $T = 2W$ when the ball passes through position *C*.

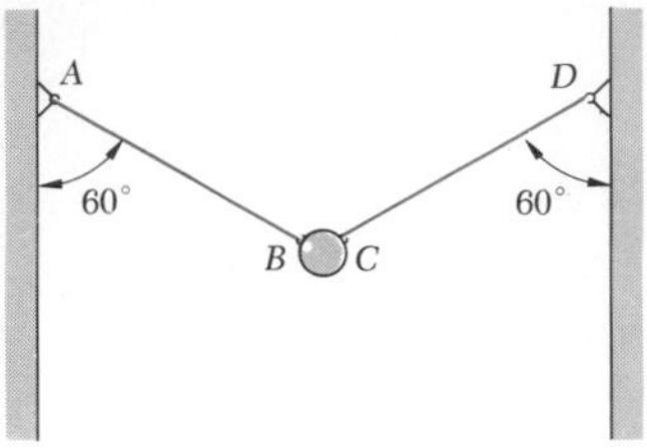

Fig. P12.47

12.47 A small sphere of weight W is held as shown by two wires AB and CD. Wire AB is then cut. Determine (a) the tension in wire CD before AB was cut, (b) the tension in wire CD and the acceleration of the sphere just after AB has been cut.

12.48 A man swings a bucket full of water in a vertical plane in a circle of radius 0.75 m. What is the smallest velocity that the bucket should have at the top of the circle if no water is to be spilled?

12.49 A 175-lb pilot flies a small plane in a vertical loop of 400-ft radius. Determine the speed of the plane at points A and B, knowing that at point A the pilot experiences weightlessness and that at point B the pilot's apparent weight is 600 lb.

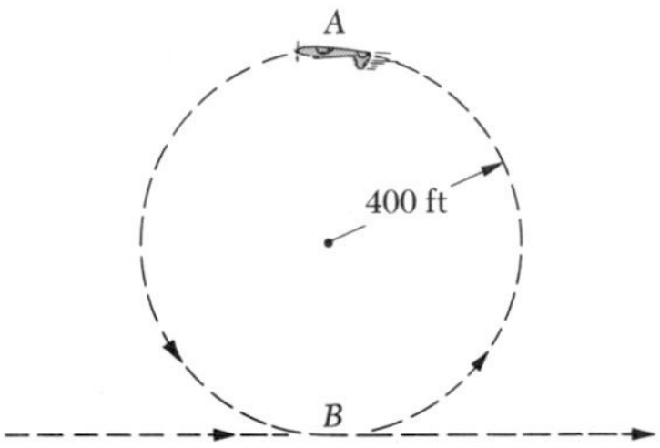

Fig. P12.49

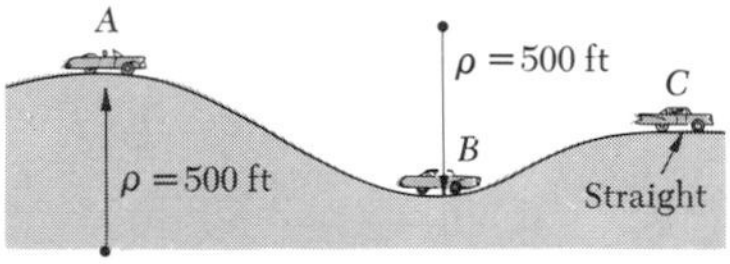

Fig. P12.50

12.50 Three automobiles are proceeding at a speed of 50 mi/h along the road shown. Knowing that the coefficient of friction between the tires and the road is 0.60, determine the tangential deceleration of each automobile if its brakes are suddenly applied and the wheels skid.

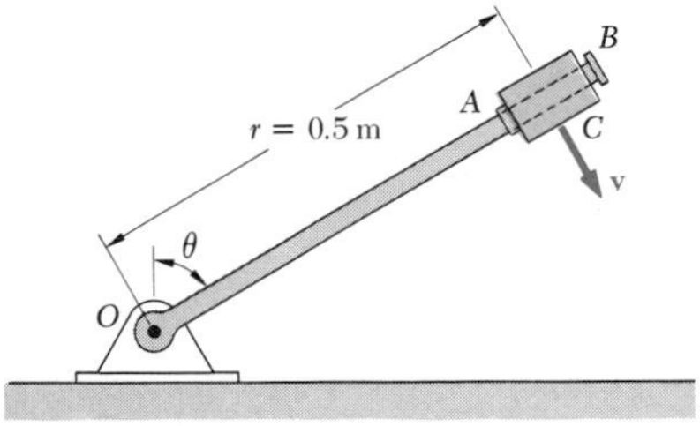

Fig. P12.51

12.51 The rod OAB rotates in a vertical plane at a constant rate such that the speed of collar C is 1.5 m/s. The collar is free to slide on the rod between two stops A and B. Knowing that the distance between the stops is only slightly larger than the collar and neglecting the effect of friction, determine the range of values of θ for which the collar is in contact with stop A.

12.52 Express the minimum and maximum safe speeds, with respect to skidding, of a car traveling on a banked road, in terms of the radius r of the curve, the banking angle θ, and the friction angle ϕ between the tires and the pavement.

12.53 A man on a motorcycle takes a turn on a flat unbanked road at 72 km/h. If the radius of the turn is 50 m, determine the minimum value of the coefficient of friction between the tires and the road which will ensure no skidding.

12.54 What angle of banking should be given to the road in Prob. 12.53 if the man on the motorcycle is to be able to take the turn at 72 km/h with a coefficient of friction $\mu = 0.30$?

12.55 A stunt driver proposes to drive a small automobile on the vertical wall of a circular pit of radius 40 ft. Knowing that the coefficient of friction between the tires and the wall is 0.65, determine the minimum speed at which the stunt can be performed.

12.56 The assembly shown rotates about a vertical axis at a constant rate. Knowing that the coefficient of friction between the small block A and the cylindrical wall is 0.20, determine the lowest speed v for which the block will remain in contact with the wall.

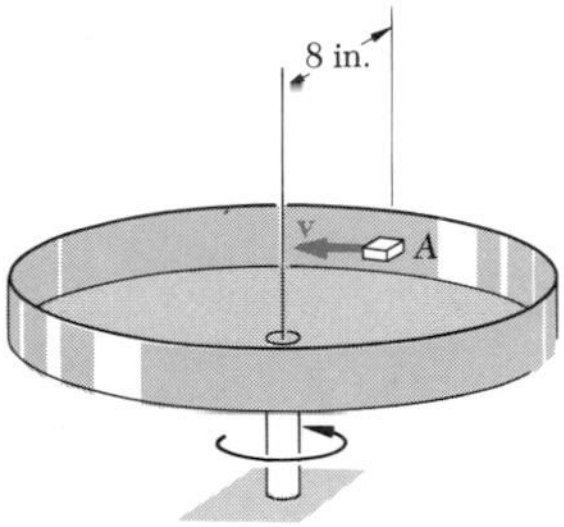

Fig. P12.56

12.57 A small ball rolls at a speed v_0 along a horizontal circle inside the circular cone shown. Express the speed v_0 in terms of the height y of the path above the apex of the cone.

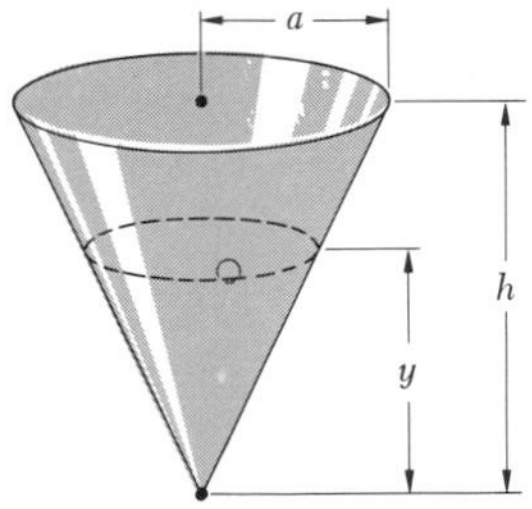

Fig. P12.57

12.58 A small ball rolls at a speed v_0 along a horizontal circle inside a bowl as shown. The inside surface of the bowl is a surface of revolution obtained by rotating the curve OA about the y axis. Determine the required equation of the curve OA if the speed v_0 of the ball is to be proportional to the distance x from the y axis to the ball.

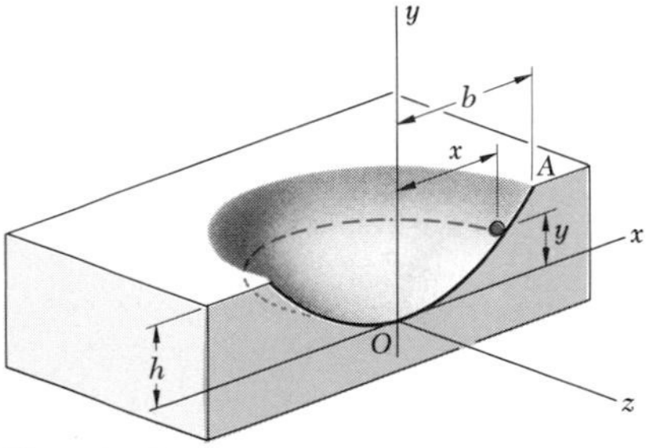

Fig. P12.58

12.59 Assuming that the equation of the curve OA in Prob. 12.58 is $y = kx^n$, where n is an arbitrary positive number, express the speed v_0 in terms of the height y of the path above the origin.

12.60 In the cathode-ray tube shown, electrons emitted by the cathode and attracted by the anode pass through a small hole in the anode and keep traveling in a straight line with a speed v_0 until they strike the screen at A. However, if a difference of potential V is established between the two parallel plates, each electron will be subjected to a force $\mathbf{F}$ perpendicular to the plates while it travels between the plates and will strike the screen at point B at a distance δ from A. The magnitude of the force $\mathbf{F}$ is $F = eV/d$, where $-e$ is the charge of the electron and d is the distance between the plates. Derive an expression for the deflection δ in terms of V, v_0, the charge $-e$ of the electron, its mass m, and the dimensions d, l, and L.

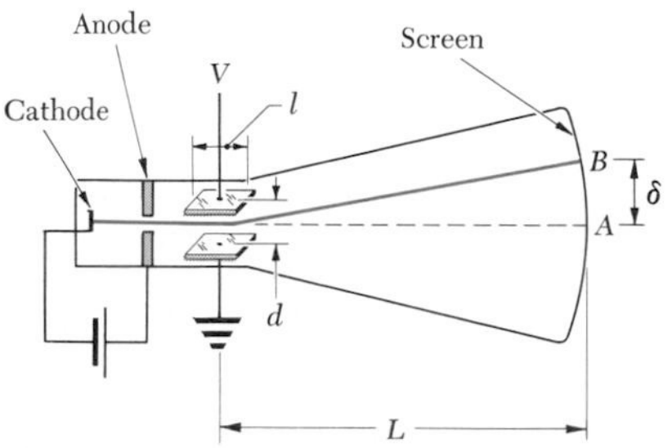

Fig. P12.60

12.61 A manufacturer wishes to design a new cathode-ray tube which will be only half as long as his current model. If the size of the screen is to remain the same, how should the length l of the plates be modified if all the other characteristics of the circuit are to remain unchanged? (See Prob. 12.60 for description of cathode-ray tube.)

12.62 In Prob. 12.60, determine the smallest allowable value of the ratio d/l in terms of e, m, v_0, and V if the electrons are not to strike the positive plate.

12.63 A cathode-ray tube emitting electrons with a velocity $\mathbf{v}_0$ is placed as shown between the poles of a large electromagnet which creates a uniform magnetic field of strength **B**. Determine the coordinates of the point where the electron beam strikes the tube screen when no difference of potential exists between the plates. It is known that an electron (mass m and charge $-e$) traveling with a velocity **v** at a right angle to the lines of force of a magnetic field of strength **B** is subjected to a force $\mathbf{F} = e\mathbf{B} \times \mathbf{v}$.

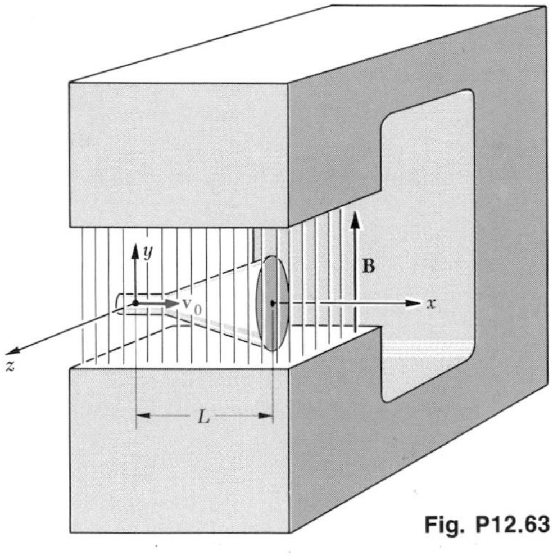

Fig. P12.63

12.6. Angular Momentum of a Particle. Rate of Change of Angular Momentum. Consider a particle P of mass m moving with respect to a newtonian frame of reference $Oxyz$. As we saw in Sec. 12.2, the linear momentum of the particle at a given instant is defined as the vector $m\mathbf{v}$ obtained by multiplying the velocity **v** of the particle by its mass m. The moment about O of the vector $m\mathbf{v}$ is called the *moment of momentum*, or the *angular momentum*, of the particle about O at that instant and is denoted by $\mathbf{H}_O$. Recalling the definition of the moment of a vector (Sec. 3.5), and denoting by **r** the position vector of P, we write

$$\mathbf{H}_O = \mathbf{r} \times m\mathbf{v} \tag{12.12}$$

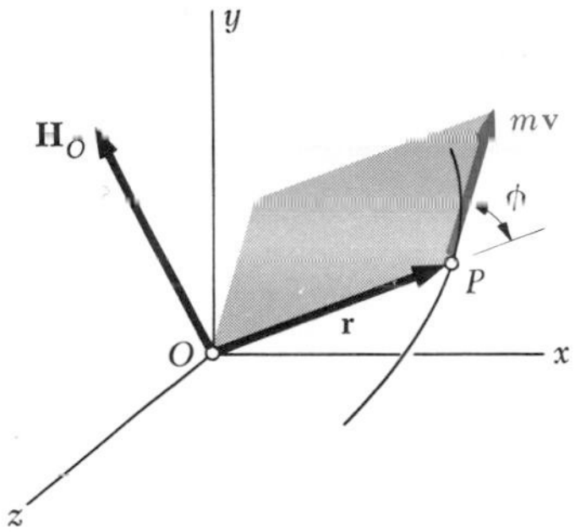

Fig. 12.12

and note that $\mathbf{H}_O$ is a vector perpendicular to the plane containing $\mathbf{r}$ and $m\mathbf{v}$, and of magnitude

$$H_O = rmv \sin \phi \tag{12.13}$$

where ϕ is the angle between $\mathbf{r}$ and $m\mathbf{v}$ (Fig. 12.12). The sense of $\mathbf{H}_O$ may be determined from the sense of $m\mathbf{v}$ by applying the right-hand rule. The unit of angular momentum is obtained by multiplying the units of length and of linear momentum (Sec. 12.3). With SI units we have

$$(\text{m})(\text{kg} \cdot \text{m/s}) = \text{kg} \cdot \text{m}^2/\text{s}$$

while, with U.S. customary units, we write

$$(\text{ft})(\text{lb} \cdot \text{s}) = \text{ft} \cdot \text{lb} \cdot \text{s}$$

Resolving the vectors $\mathbf{r}$ and $m\mathbf{v}$ into components, and applying formula (3.10), we write

$$\mathbf{H}_O = \begin{vmatrix} \mathbf{i} & \mathbf{j} & \mathbf{k} \\ x & y & z \\ mv_x & mv_y & mv_z \end{vmatrix} \tag{12.14}$$

The components of $\mathbf{H}_O$, which also represent the moments of the linear momentum $m\mathbf{v}$ about the coordinate axes, may be obtained by expanding the determinant in (12.14). We have

$$\begin{aligned} H_x &= m(yv_z - zv_y) \\ H_y &= m(zv_x - xv_z) \\ H_z &= m(xv_y - yv_x) \end{aligned} \tag{12.15}$$

In the case of a particle moving in the xy plane, we have $z = v_z = 0$ and the components H_x and H_y reduce to zero. The angular momentum is thus perpendicular to the xy plane; it is then completely defined by the scalar

$$H_O = H_z = m(xv_y - yv_x) \tag{12.16}$$

which will be positive or negative, according to the sense in which the particle is observed to move from O. If polar coordinates are used, we resolve the linear momentum of the particle into radial and transverse components (Fig. 12.13) and write

$$H_O = rmv \sin \phi = rmv_\theta \tag{12.17}$$

or, recalling from (11.45) that $v_\theta = r\dot{\theta}$,

$$H_O = mr^2\dot{\theta} \tag{12.18}$$

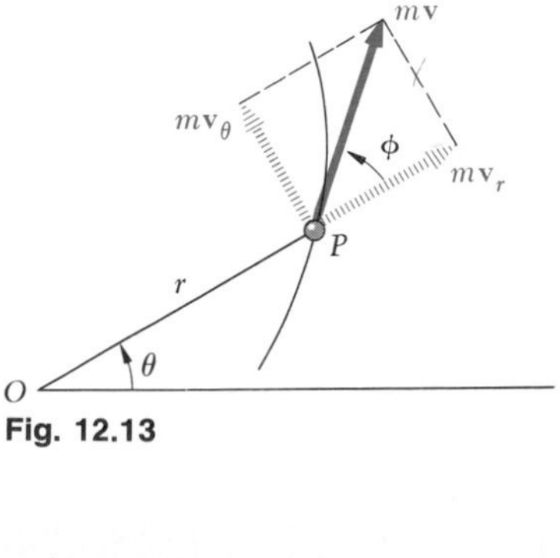

Fig. 12.13

We shall now compute the derivative with respect to t of the angular momentum $\mathbf{H}_O$ of a particle P moving in space. Differentiating both members of Eq. (12.12), and recalling the rule for the differentiation of a vector product (Sec. 11.10), we write

$$\dot{\mathbf{H}}_O = \dot{\mathbf{r}} \times m\mathbf{v} + \mathbf{r} \times m\dot{\mathbf{v}} = \mathbf{v} \times m\mathbf{v} + \mathbf{r} \times m\mathbf{a}$$

Since the vectors $\mathbf{v}$ and $m\mathbf{v}$ are collinear, the first term of the expression obtained is zero; and, by Newton's second law, $m\mathbf{a}$ is equal to the sum $\Sigma\mathbf{F}$ of the forces acting on P. Noting that $\mathbf{r} \times \Sigma\mathbf{F}$ represents the sum $\Sigma\mathbf{M}_O$ of the moments about O of these forces, we write

$$\Sigma\mathbf{M}_O = \dot{\mathbf{H}}_O \qquad (12.19)$$

Equation (12.19), which results directly from Newton's second law, expresses that *the sum of the moments about O of the forces acting on the particle is equal to the rate of change of the moment of momentum, or angular momentum, of the particle about O.*

12.7. Equations of Motion in Terms of Radial and Transverse Components. Consider a particle P, of polar coordinates r and θ, which moves in a plane under the action of several forces. Resolving the forces and the acceleration of the particle into radial and transverse components (Fig. 12.14), and substituting into Eq. (12.2), we obtain the two scalar equations

$$\Sigma F_r = ma_r \qquad \Sigma F_\theta = ma_\theta \qquad (12.20)$$

Substituting for a_r and a_θ from Eqs. (11.46), we have

$$\Sigma F_r = m(\ddot{r} - r\dot{\theta}^2) \qquad (12.21)$$

$$\Sigma F_\theta = m(r\ddot{\theta} + 2\dot{r}\dot{\theta}) \qquad (12.22)$$

The equations obtained may be solved for two unknowns.

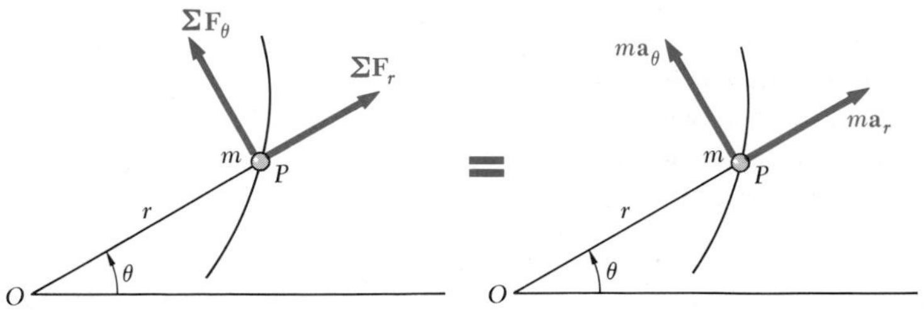

Fig. 12.14

Equation (12.22) could have been derived from Eq. (12.19). Recalling (12.18) and noting that $\Sigma M_O = r\Sigma F_\theta$, Eq. (12.19) yields

$$r\Sigma F_\theta = \frac{d}{dt}(mr^2\dot{\theta})$$
$$= m(r^2\ddot{\theta} + 2r\dot{r}\dot{\theta})$$

and, after dividing both members by r,

$$\Sigma F_\theta = m(r\ddot{\theta} + 2\dot{r}\dot{\theta}) \qquad (12.22)$$

12.8. Motion under a Central Force. Conservation of Angular Momentum. When the only force acting on a particle P is a force $\mathbf{F}$ directed toward or away from a fixed point O, the particle is said to be moving *under a central force,* and the point O is referred to as the *center of force* (Fig. 12.15). Since the line of action of $\mathbf{F}$ passes through O, we must have $\Sigma\mathbf{M}_O = 0$ at any given instant. Substituting into Eq. (12.19), we therefore obtain

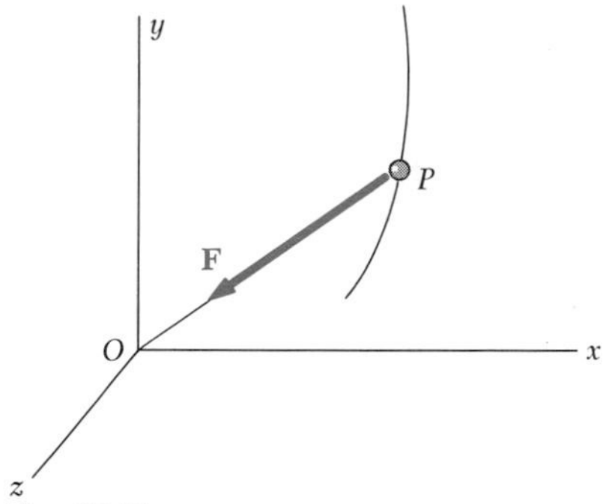

Fig. 12.15

$$\dot{\mathbf{H}}_O = 0$$

for all values of t or, integrating in t,

$$\mathbf{H}_O = \text{constant} \qquad (12.23)$$

We thus conclude that *the angular momentum of a particle moving under a central force is constant, both in magnitude and direction.*

Recalling the definition of the angular momentum of a particle (Sec. 12.6), we write

$$\mathbf{r} \times m\mathbf{v} = \mathbf{H}_O = \text{constant} \qquad (12.24)$$

from which it follows that the position vector $\mathbf{r}$ of the particle P must be perpendicular to the constant vector $\mathbf{H}_O$. Thus, a particle under a central force moves in a fixed plane perpendicular to $\mathbf{H}_O$. The vector $\mathbf{H}_O$ and the fixed plane are defined by the initial position vector $\mathbf{r}_0$ and the initial velocity $\mathbf{v}_0$ of the particle. For convenience, we shall assume that the plane of the figure coincides with the fixed plane of motion (Fig. 12.16).

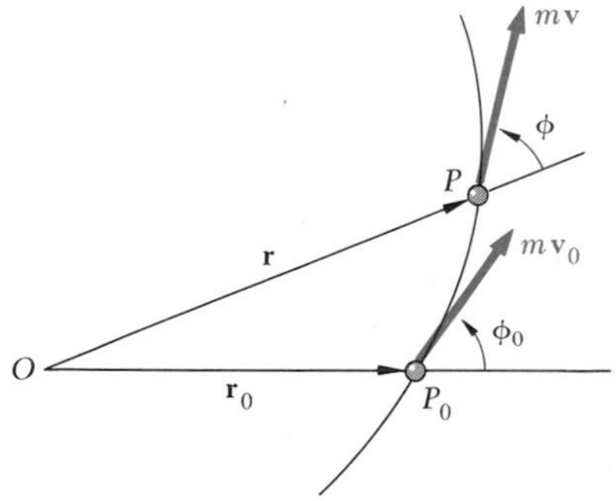

Fig. 12.16

Since the magnitude H_O of the angular momentum of the particle P is constant, the right-hand member in Eq. (12.13) must be constant. We therefore write

$$rmv \sin\phi = r_0 m v_0 \sin\phi_0 \qquad (12.25)$$

This relation applies to the motion of any particle under a

central force. Since the gravitational force exerted by the sun on a planet is a central force directed toward the center of the sun, Eq. (12.25) is fundamental to the study of planetary motion. For a similar reason, it is also fundamental to the study of the motion of space vehicles in orbit about the earth.

Recalling Eq. (12.18), we may alternatively express the fact that the magnitude H_O of the angular momentum of the particle P is constant by writing

$$mr^2\dot{\theta} = H_O = \text{constant} \qquad (12.26)$$

or, dividing by m and denoting by h the angular momentum per unit mass H_O/m,

$$r^2\dot{\theta} = h \qquad (12.27)$$

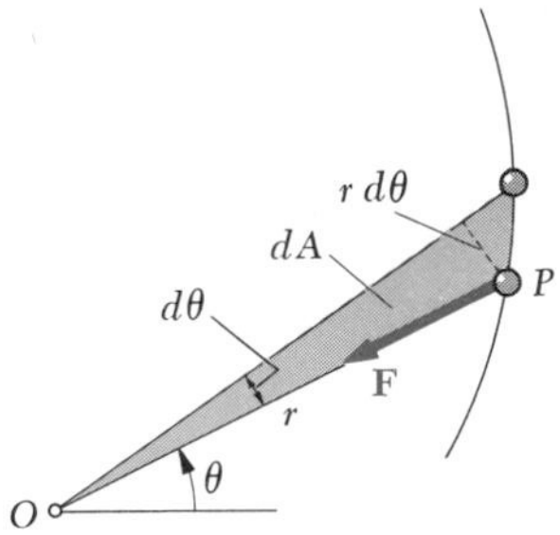

Fig. 12.17

Equation (12.27) may be given an interesting geometric interpretation. Observing from Fig. 12.17 that the radius vector OP sweeps an infinitesimal area $dA = \frac{1}{2}r^2\,d\theta$ as it rotates through an angle $d\theta$, and defining the *areal velocity* of the particle as the quotient dA/dt, we note that the left-hand member of Eq. (12.27) represents twice the areal velocity of the particle. We thus conclude that, *when a particle moves under a central force, its areal velocity is constant.*

12.9. Newton's Law of Gravitation. As we saw in the preceding section, the gravitational force exerted by the sun on a planet, or by the earth on an orbiting satellite, is an important example of a central force. In this section we shall learn how to determine the magnitude of a gravitational force.

In his *law of universal gravitation,* Newton states that two particles at a distance r from each other and, respectively, of mass M and m attract each other with equal and opposite forces $\mathbf{F}$ and $-\mathbf{F}$ directed along the line joining the particles (Fig. 12.18). The common magnitude F of the two forces is

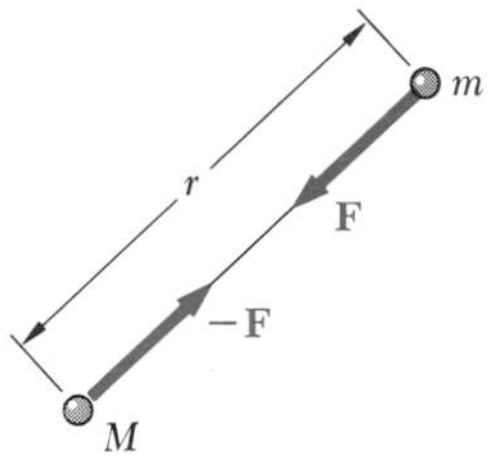

Fig. 12.18

$$F = G\frac{Mm}{r^2} \qquad (12.28)$$

where G is a universal constant, called the *constant of gravitation.* Experiments show that the value of G is $(6.673 \pm 0.003) \times 10^{-11}\ \text{m}^3/\text{kg}\cdot\text{s}^2$ in SI units, or approximately $3.44 \times 10^{-8}\ \text{ft}^4/\text{lb}\cdot\text{s}^4$ in U.S. customary units. While gravitational forces exist between any pair of bodies, their effect is appreciable only when one of the bodies has a very large mass. The effect of gravitational forces is apparent in the case of the motion of a planet about the sun, of satellites orbiting about the earth, or of bodies falling on the surface of the earth.

Since the force exerted by the earth on a body of mass m located on or near its surface is defined as the weight $\mathbf{W}$ of the body, we may substitute the magnitude $W = mg$ of the weight for F, and the radius R of the earth for r, in Eq. (12.28). We obtain

$$W = mg = \frac{GM}{R^2}m \qquad \text{or} \qquad g = \frac{GM}{R^2} \tag{12.29}$$

where M is the mass of the earth. Since the earth is not truly spherical, the distance R from the center of the earth depends upon the point selected on its surface, and the values of W and g will thus vary with the altitude and latitude of the point considered. Another reason for the variation of W and g with the latitude is that a system of axes attached to the earth does not constitute a newtonian frame of reference (see Sec. 12.1). A more accurate definition of the weight of a body should therefore include a component representing the centrifugal force due to the rotation of the earth. Values of g at sea level vary from 9.781 m/s^2 or 32.09 ft/s^2 at the equator to 9.833 m/s^2 or 32.26 ft/s^2 at the poles.†

The force exerted by the earth on a body of mass m located in space at a distance r from its center may be found from Eq. (12.28). The computations will be somewhat simplified if we note that, according to Eq. (12.29), the product of the constant of gravitation G and of the mass M of the earth may be expressed as

$$GM = gR^2 \tag{12.30}$$

where g and the radius R of the earth will be given their average values $g = 9.81 \text{ m/s}^2$ and $R = 6.37 \times 10^6 \text{ m}$ in SI units,‡ or $g = 32.2 \text{ ft/s}^2$ and $R = (3960 \text{ mi})(5280 \text{ ft/mi})$ in U.S. customary units.

The discovery of the law of universal gravitation has often been attributed to the fact that Newton, after observing an apple falling from a tree, had reflected that the earth must attract an apple and the moon in much the same way. While it is doubtful that this incident actually took place, it may be said that Newton would not have formulated his law if he had not first perceived that the acceleration of a falling body must have the same cause as the acceleration which keeps the moon in its orbit. This basic concept of continuity of the gravitational attraction is more easily understood now, when the gap between the apple and the moon is being filled with long-range ballistic missiles and artificial earth satellites.

† A formula expressing g in terms of the latitude ϕ was given in Prob. 12.1.

‡ The value of R is easily found if one recalls that the circumference of the earth is $2\pi R = 40 \times 10^6$ m.

SAMPLE PROBLEM 12.6

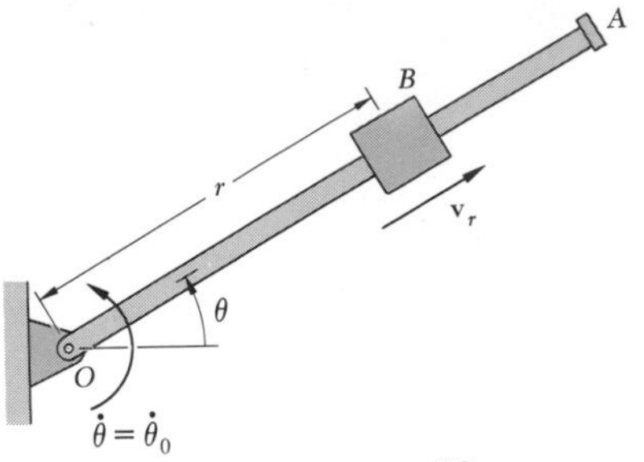

A block B of mass m may slide freely on a frictionless arm OA which rotates in a horizontal plane at a constant rate $\dot{\theta}_0$. Knowing that B is released at a distance r_0 from O, express as a function of r, (*a*) the component v_r of the velocity of B along OA, (*b*) the magnitude of the horizontal force $\mathbf{F}$ exerted on B by the arm OA.

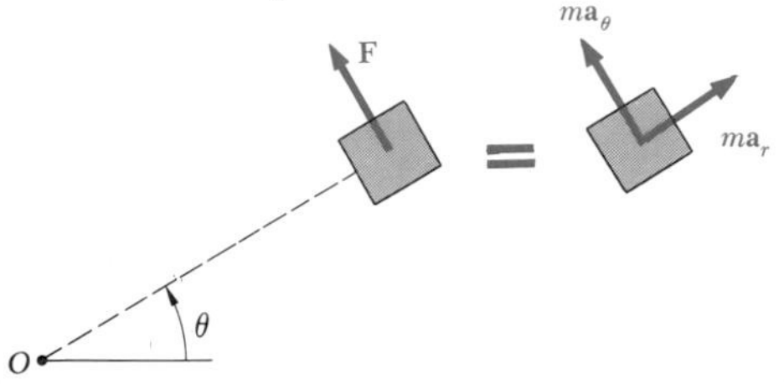

Solution. Since all other forces are perpendicular to the plane of the figure, the only force shown acting on B is the force $\mathbf{F}$ perpendicular to OA.

Equations of Motion. Using radial and transverse components:

$$+\nearrow \Sigma F_r = ma_r: \qquad 0 = m(\ddot{r} - r\dot{\theta}^2) \qquad (1)$$

$$+\nwarrow \Sigma F_\theta = ma_\theta: \qquad F = m(r\ddot{\theta} + 2\dot{r}\dot{\theta}) \qquad (2)$$

***a.* Component v_r of Velocity.** Since $v_r = \dot{r}$, we have

$$\ddot{r} = \dot{v}_r = \frac{dv_r}{dt} = \frac{dv_r}{dr}\frac{dr}{dt} = v_r \frac{dv_r}{dr}$$

Substituting for $\ddot{r}$ into (1), recalling that $\dot{\theta} = \dot{\theta}_0$, and separating the variables:

$$v_r \, dv_r = \dot{\theta}_0^2 r \, dr$$

Multiplying by 2, and integrating from 0 to v_r and from r_0 to r:

$$v_r^2 = \dot{\theta}_0^2(r^2 - r_0^2) \qquad v_r = \dot{\theta}_0(r^2 - r_0^2)^{1/2} \quad ◀$$

***b.* Horizontal Force F.** Making $\dot{\theta} = \dot{\theta}_0$, $\ddot{\theta} = 0$, $\dot{r} = v_r$ in Eq. (2), and substituting for v_r the expression obtained in part *a*:

$$F = 2m\dot{\theta}_0(r^2 - r_0^2)^{1/2}\dot{\theta}_0 \qquad F = 2m\dot{\theta}_0^2(r^2 - r_0^2)^{1/2} \quad ◀$$

SAMPLE PROBLEM 12.7

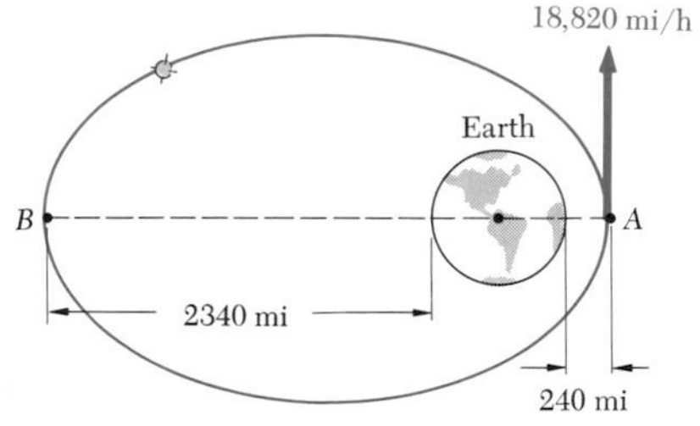

A satellite is launched in a direction parallel to the surface of the earth with a velocity of 18,820 mi/h from an altitude of 240 mi. Determine the velocity of the satellite as it reaches its maximum altitude of 2340 mi. It is recalled that the radius of the earth is 3960 mi.

Solution. Since the satellite is moving under a central force directed toward the center O of the earth, its angular momentum $\mathbf{H}_O$ is constant. From Eq. (12.13) we have

$$rmv \sin\phi = H_O = \text{constant}$$

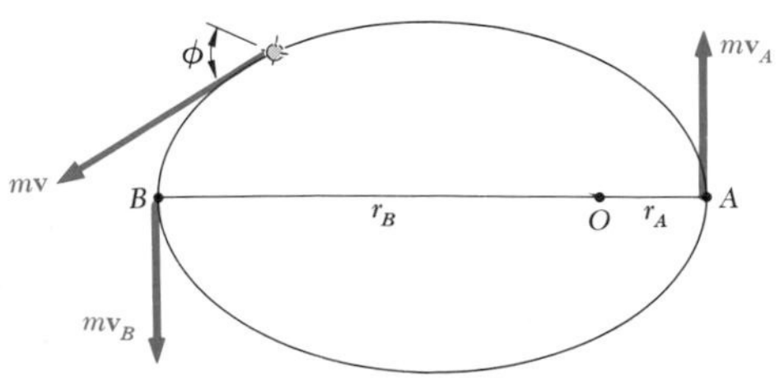

which shows that v is minimum at B, where both r and $\sin\phi$ are maximum. Expressing conservation of angular momentum between A and B:

$$r_A m v_A = r_B m v_B$$

$$v_B = v_A \frac{r_A}{r_B} = (18{,}820 \text{ mi/h})\frac{3960 \text{ mi} + 240 \text{ mi}}{3960 \text{ mi} + 2340 \text{ mi}}$$

$$v_B = 12{,}550 \text{ mi/h} \quad ◀$$

PROBLEMS

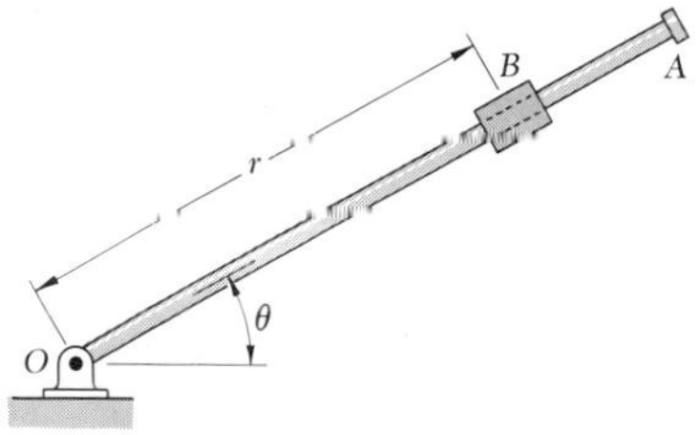

Fig. P12.64 and P12.66

12.64 The two-dimensional motion of particle B is defined by the relations $r = t^2 - \frac{1}{3}t^3$ and $\theta = 2t^2$, where r is expressed in meters, t in seconds, and θ in radians. If the particle has a mass of 2 kg and moves in a horizontal plane, determine the radial and transverse components of the force acting on the particle when (*a*) $t = 0$, (*b*) $t = 1$ s.

12.65 For the motion defined in Prob. 12.64, determine the radial and transverse components of the force acting on the 2-kg particle as it returns to the origin at $t = 3$ s.

12.66 The two-dimensional motion of a particle B is defined by the relations $r = 10(1 + \cos 2\pi t)$ and $\theta = 2\pi t$, where r is expressed in feet, t in seconds, and θ in radians. If the particle weighs 2 lb and moves in a horizontal plane, determine the radial and transverse components of the force acting on the particle when (*a*) $t = 0$, (*b*) $t = 0.25$ s.

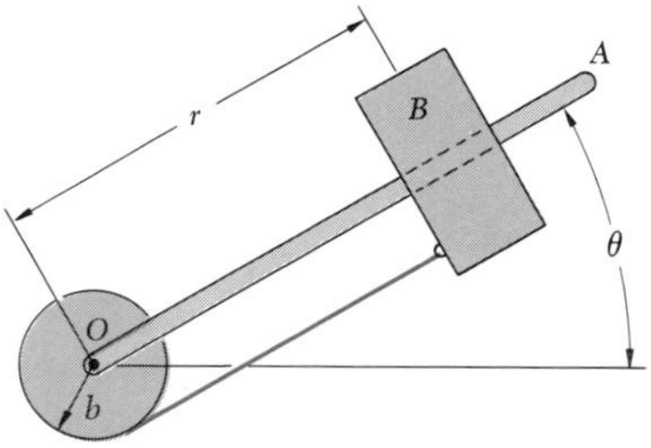

Fig. P12.67

12.67 A block B of mass m may slide on the frictionless arm OA which rotates in a *horizontal* plane at a constant rate $\dot{\theta}_0$. As the arm rotates, the cord wraps around a *fixed* drum of radius b and pulls the block toward O with a speed $b\dot{\theta}_0$. Express as a function of m, r, b, and $\dot{\theta}_0$, (*a*) the tension T in the cord, (*b*) the magnitude of the horizontal force $\mathbf{Q}$ exerted on B by the arm OA.

12.68 Solve Prob. 12.67, knowing that the weight of the block is 3 lb and that $r = 2$ ft, $b = 3$ in., and $\dot{\theta}_0 = 8$ rad/s.

12.69 Slider C has a mass of 250 g and oscillates in the radial slot in arm AB as the arm rotates in a horizontal plane at a constant rate $\dot{\theta} = 12$ rad/s. In the position shown, it is known that the slider is moving outward along the slot at the speed of 1.5 m/s and that the spring is compressed and exerts a force of 10 N on the slider. Neglecting the effect of friction, determine (*a*) the components of the acceleration of the slider, (*b*) the horizontal force exerted on the slider by the arm AB.

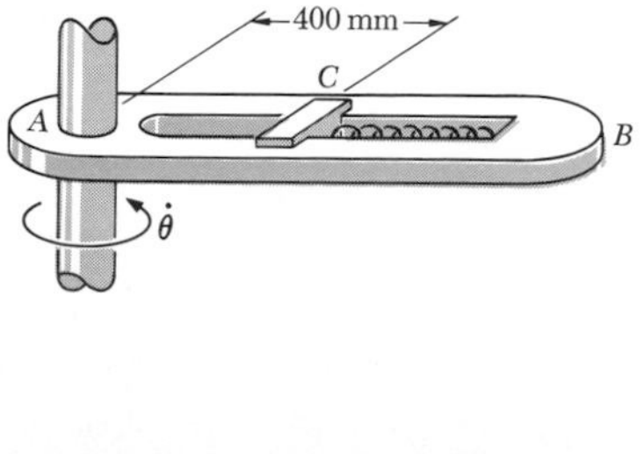

Fig. P12.69

12.70 While aiming at a moving target, a man rotates his rifle clockwise in a horizontal plane at the rate of 15° per second. Assuming that he can maintain the motion as the rifle is fired, determine the horizontal force exerted by the barrel on a 45-g bullet just before it leaves the barrel with a muzzle velocity of 550 m/s.

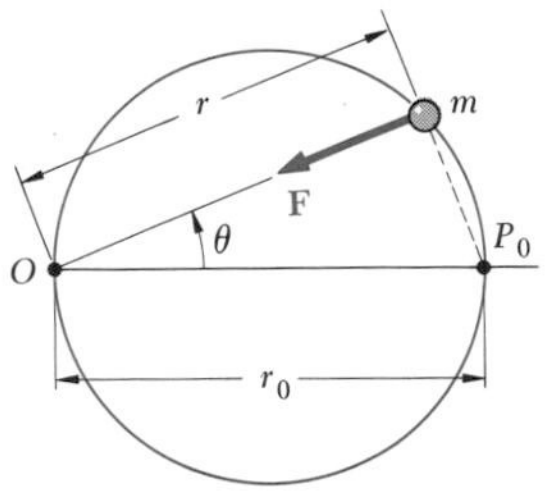

Fig. P12.71

12.71 A particle moves under a central force in a circular path of diameter r_0 which passes through the center of force O. Show that its speed is $v = v_0/\cos^2\theta$, where v_0 is the speed of the particle at point P_0 directly across the circle from O. [*Hint.* Use Eq. (12.27) with $r = r_0 \cos\theta$].

12.72 A particle moves under a central force in a path defined by the equation $r = r_0/\cos n\theta$, where n is a positive constant. Using Eq. (12.27) show that the radial and transverse components of the velocity are $v_r = nv_0 \sin n\theta$ and $v_\theta = v_0 \cos n\theta$, where v_0 is the velocity of the particle for $\theta = 0$. What is the motion of the particle when $n = 0$ and when $n = 1$?

12.73 For the particle and motion of Prob. 12.72, show that the radial and transverse components of the acceleration are $a_r = (n^2 - 1)(v_0^2/r_0)\cos^3 n\theta$ and $a_\theta = 0$.

12.74 If a particle of mass m is attached to the end of a very light circular rod as shown in (*1*), the rod exerts on the mass a force **F** of magnitude $F = kr$ directed toward the origin O, as shown in (2). The path of the particle is observed to be an ellipse with semiaxes $a = 6$ in. and $b = 2$ in. (*a*) Knowing that the speed of the particle at A is 8 in./s, determine the speed at B. (*b*) Further knowing that the constant k/m is equal to 16 s^{-2}, determine the radius of curvature of the path at A and at B.

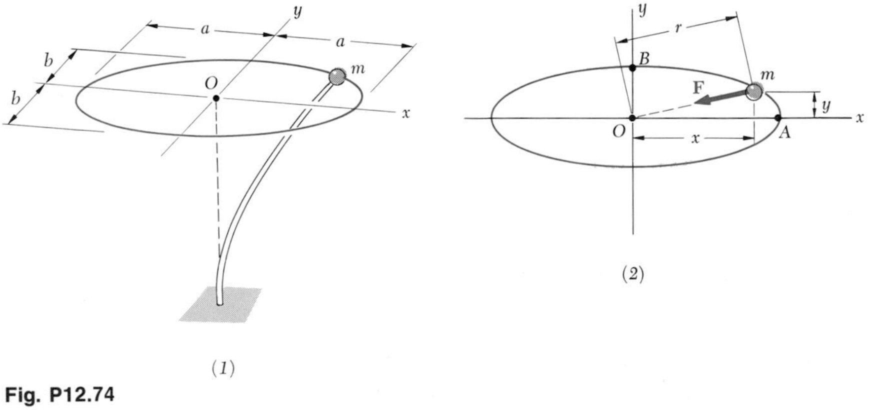

Fig. P12.74

12.75 Show that the radius r of the moon's orbit may be determined from the radius R of the earth, the acceleration of gravity g at the surface of the earth, and the time τ required by the moon to revolve once around the earth. Compute r knowing that $\tau = 27.3$ days.

12.76 Determine the mass of the earth from Newton's law of gravitation, knowing that it takes 94.14 min for a satellite to describe a circular orbit 300 mi above the surface of the earth.

12.77 Two solid steel spheres, each of radius 100 mm, are placed so that their surfaces are in contact. (*a*) Determine the force of gravitational attraction between the spheres, knowing that the density of steel is 7850 kg/m^3. (*b*) If the spheres are moved 2 mm apart and released with zero velocity, determine the approximate time required for their gravitational attraction to bring them back into contact. (*Hint.* Assume the gravitational forces to remain constant.)

12.78 Communication satellites have been placed in a geosynchronous orbit, i.e., in a circular orbit such that they complete one full revolution about the earth in one sidereal day (23 h 56 min), and thus appear stationary with respect to the ground. Determine (*a*) the altitude of the satellites above the surface of the earth, (*b*) the velocity with which they describe their orbit. Give the answers in both SI and U.S. customary units.

12.79 Collar B may slide freely on rod OA, which in turn may rotate freely in the horizontal plane. The collar is describing a circle of radius 0.5 m with a speed $v_1 = 0.28$ m/s when a spring located between A and B is released, projecting the collar along the rod with an initial relative speed $v_2 = 0.96$ m/s. Neglecting the mass of the rod, determine the minimum distance between the collar and point O in the ensuing motion.

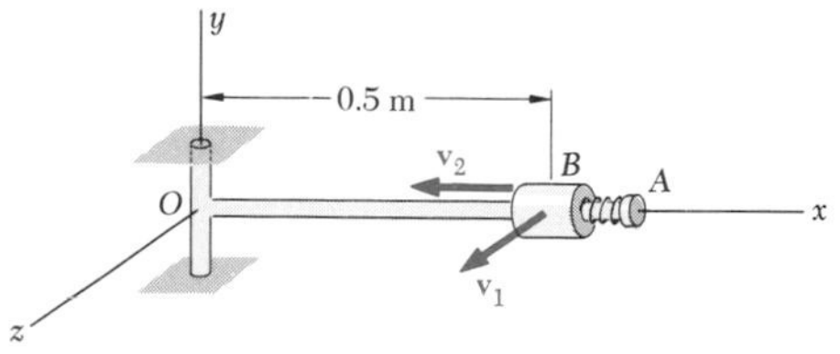

Fig. P12.79

12.80 A heavy ball is mounted on a horizontal rod which rotates freely about a vertical shaft. In the position shown, the speed of the ball is $v_1 = 30$ in./s and the ball is held by a cord attached to the shaft. The cord is suddenly cut and the ball moves to position A' as the rod rotates. Neglecting the mass of the rod, determine (*a*) the speed of the ball in position A', (*b*) the path (on the xz plane) of the ball as it moves from A to A'.

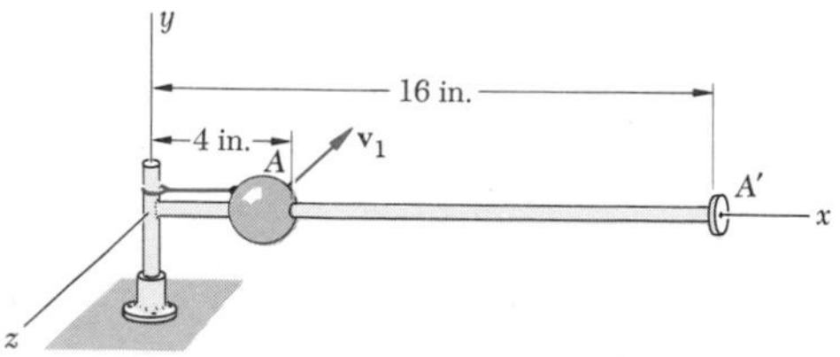

Fig. P12.80

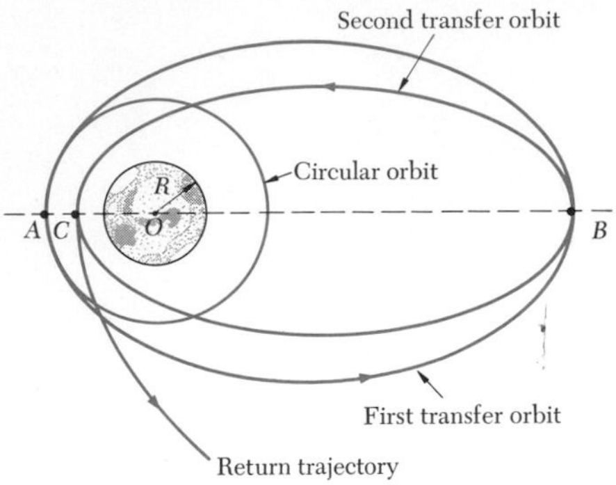

Fig. P12.81

12.81 Plans for an unmanned landing mission on the planet Mars call for the earth-return vehicle to first describe a circular orbit about the planet at an altitude $d_A = 2200$ km with a velocity of 2771 m/s. As it passes through point A, the vehicle will be inserted into an elliptic transfer orbit by firing its engine and increasing its speed by $\Delta v_A = 1046$ m/s. As it passes through point B, at an altitude $d_B = 100\,000$ km, the vehicle will be inserted into a second transfer orbit located in a slightly different plane, by changing the direction of its velocity and reducing its speed by $\Delta v_B = -22$ m/s. Finally, as the vehicle passes through point C, at an altitude $d_C = 1000$ km, its speed will be increased by $\Delta v_C = 660$ m/s to insert it into its return trajectory. Knowing that the radius of the planet Mars is $R = 3400$ km, determine the velocity of the vehicle after the last maneuver has been completed.

12.82 A space tug describes a circular orbit of 6000-mi radius around the earth. In order to transfer it to a larger circular orbit of 24,000-mi radius, the tug is first placed on an elliptic path AB by firing its engine as it passes through A, thus increasing its velocity by 3810 mi/h. By how much should the tug's velocity be increased as it reaches B to insert it into the larger circular orbit?

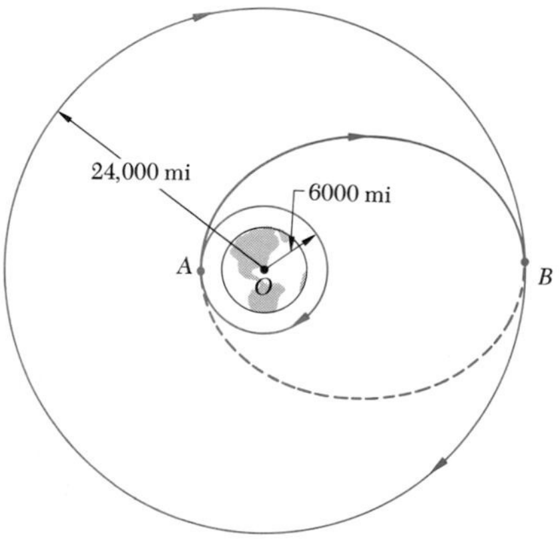

Fig. P12.82

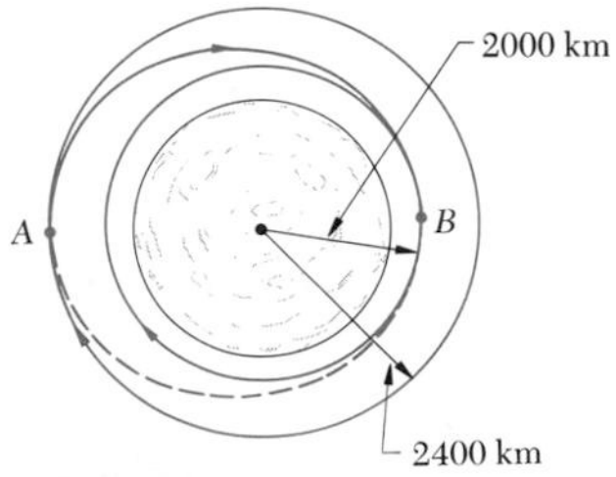

Fig. P12.83

12.83 An Apollo spacecraft describes a circular orbit of 2400-km radius around the moon with a velocity of 5140 km/h. In order to transfer it to a smaller circular orbit of 2000-km radius, the spacecraft is first placed on an elliptic path AB by reducing its velocity to 4900 km/h as it passes through A. Determine (a) the velocity of the spacecraft as it approaches B on the elliptic path, (b) the value to which its velocity must be reduced at B to insert it into the smaller circular orbit.

12.84 Solve Prob. 12.83, assuming that the Apollo spacecraft is to be transferred from the orbit of 2400-km radius to a circular orbit of 1800-km radius and that its velocity is reduced to 4760 km/h as it passes through A.

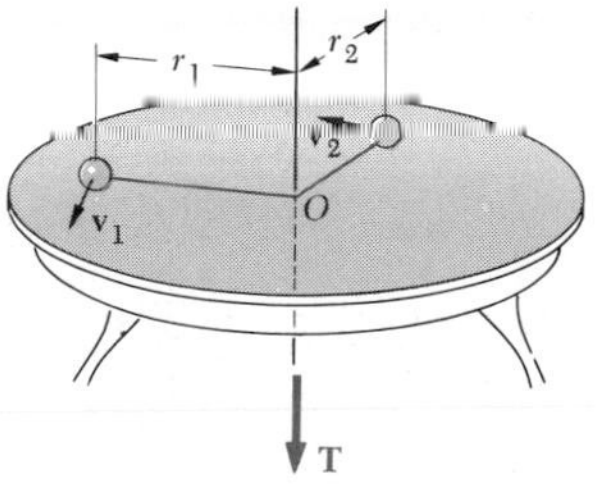

Fig. P12.85

12.85 A 3-oz ball slides on a smooth horizontal table at the end of a string which passes through a small hole in the table at O. When the length of string above the table is $r_1 = 15$ in., the speed of the ball is $v_1 = 8$ ft/s. Knowing that the breaking strength of the string is 3.00 lb, determine (a) the smallest distance r_2 which can be achieved by slowly drawing the string through the hole, (b) the corresponding speed v_2.

12.86 A small ball swings in a horizontal circle at the end of a cord of length l_1 which forms an angle θ_1 with the vertical. The cord is then slowly drawn through the support at O until the free end is l_2. (a) Derive a relation between l_1, l_2, θ_1, and θ_2. (b) If the ball is set in motion so that, initially, $l_1 = 600$ mm and $\theta_1 = 30°$, determine the length l_2 for which $\theta_2 = 60°$.

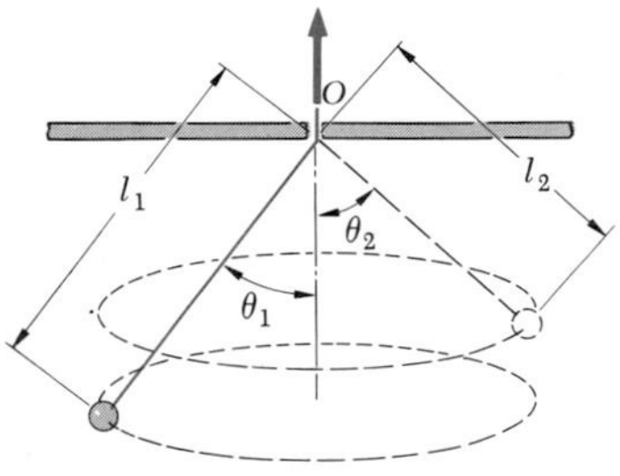

Fig. P12.86

*12.10. Trajectory of a Particle under a Central Force.

Consider a particle P moving under a central force $\mathbf{F}$. We propose to obtain the differential equation which defines its trajectory.

Assuming that the force $\mathbf{F}$ is directed toward the center of force O, we note that ΣF_r and ΣF_θ reduce respectively to $-F$ and zero in Eqs. (12.21) and (12.22). We therefore write

$$m(\ddot{r} - r\dot{\theta}^2) = -F \tag{12.31}$$

$$m(r\ddot{\theta} + 2\dot{r}\dot{\theta}) = 0 \tag{12.32}$$

These equations define the motion of P. We shall, however, replace Eq. (12.32) by Eq. (12.27), which is more convenient to use and which is equivalent to Eq. (12.32), as we may easily check by differentiating it with respect to t. We write

$$r^2\dot{\theta} = h \qquad \text{or} \qquad r^2\frac{d\theta}{dt} = h \tag{12.33}$$

Equation (12.33) may be used to eliminate the independent variable t from Eq. (12.31). Solving Eq. (12.33) for $\dot{\theta}$ or $d\theta/dt$, we have

$$\dot{\theta} = \frac{d\theta}{dt} = \frac{h}{r^2} \tag{12.34}$$

from which it follows that

$$\dot{r} = \frac{dr}{dt} = \frac{dr}{d\theta}\frac{d\theta}{dt} = \frac{h}{r^2}\frac{dr}{d\theta} = -h\frac{d}{d\theta}\left(\frac{1}{r}\right) \tag{12.35}$$

$$\ddot{r} = \frac{d\dot{r}}{dt} = \frac{d\dot{r}}{d\theta}\frac{d\theta}{dt} = \frac{h}{r^2}\frac{d\dot{r}}{d\theta}$$

or, substituting for $\dot{r}$ from (12.35),

$$\ddot{r} = \frac{h}{r^2}\frac{d}{d\theta}\left[-h\frac{d}{d\theta}\left(\frac{1}{r}\right)\right]$$

$$\ddot{r} = -\frac{h^2}{r^2}\frac{d^2}{d\theta^2}\left(\frac{1}{r}\right) \tag{12.36}$$

Substituting for $\dot{\theta}$ and $\ddot{r}$ from (12.34) and (12.36), respectively, into Eq. (12.31), and introducing the function $u = 1/r$, we obtain after reductions

$$\frac{d^2u}{d\theta^2} + u = \frac{F}{mh^2u^2} \tag{12.37}$$

In deriving Eq. (12.37), the force **F** was assumed directed toward O. The magnitude F should therefore be positive if **F** is actually directed toward O (attractive force) and negative if **F** is directed away from O (repulsive force). If F is a known function of r and thus of u, Eq. (12.37) is a differential equation in u and θ. This differential equation defines the trajectory followed by the particle under the central force **F**. The equation of the trajectory will be obtained by solving the differential equation (12.37) for u as a function of θ and determining the constants of integration from the initial conditions.

***12.11. Application to Space Mechanics.** After the last stage of their launching rockets has burned out, earth satellites and other space vehicles are subjected only to the gravitational pull of the earth. Their motion may therefore be determined from Eqs. (12.33) and (12.37), which govern the motion of a particle under a central force, after F has been replaced by the expression obtained for the force of gravitational attraction.† Setting in Eq. (12.37)

$$F = \frac{GMm}{r^2} = GMmu^2$$

where M = mass of earth
m = mass of space vehicle
r = distance from center of earth to vehicle
$u = 1/r$

we obtain the differential equation

$$\frac{d^2u}{d\theta^2} + u = \frac{GM}{h^2} \tag{12.38}$$

where the right-hand member is observed to be a constant.

The solution of the differential equation (12.38) is obtained by adding the particular solution $u = GM/h^2$ to the general solution $u = C\cos(\theta - \theta_0)$ of the corresponding homogeneous equation (i.e., the equation obtained by setting the right-hand member equal to zero). Choosing the polar axis so that $\theta_0 = 0$, we write

$$\frac{1}{r} = u = \frac{GM}{h^2} + C\cos\theta \tag{12.39}$$

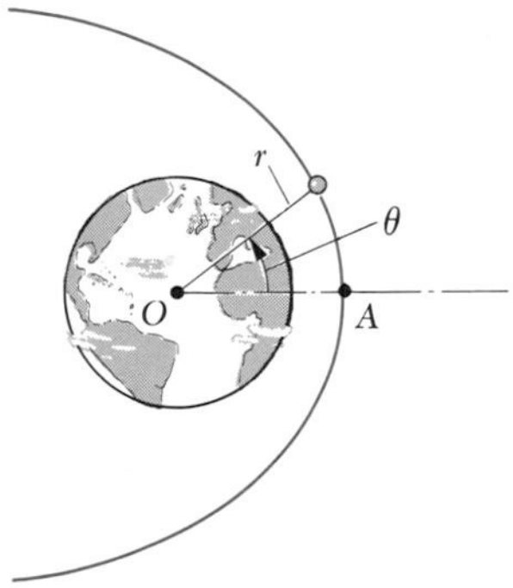

Fig. 12.19

Equation (12.39) is the equation of a *conic section* (ellipse, parabola, or hyperbola) in the polar coordinates r and θ. The origin O of the coordinates, which is located at the center of the earth, is a *focus* of this conic section, and the polar axis is one of its axes of symmetry (Fig. 12.19).

The ratio of the constants C and GM/h^2 defines the *eccentricity* ε of the conic section; setting

$$\varepsilon = \frac{C}{GM/h^2} = \frac{Ch^2}{GM} \tag{12.40}$$

we may write Eq. (12.39) in the form

$$\frac{1}{r} = \frac{GM}{h^2}(1 + \varepsilon\cos\theta) \tag{12.39'}$$

†It is assumed that the space vehicles considered here are attracted only by the earth and that their mass is negligible compared to the mass of the earth. If a vehicle moves very far from the earth, its path may be affected by the attraction of the sun, the moon, or another planet.

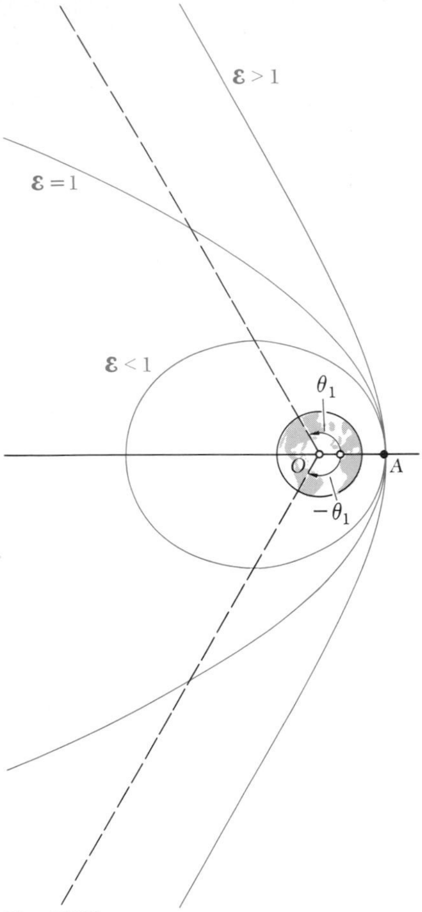

Fig. 12.20

Three cases may be distinguished:

1. $\varepsilon > 1$, or $C > GM/h^2$: There are two values θ_1 and $-\theta_1$ of the polar angle, defined by $\cos\theta_1 = -GM/Ch^2$, for which the right-hand member of Eq. (12.39) becomes zero. For both of these values, the radius vector r becomes infinite; the conic section is a *hyperbola* (Fig. 12.20).
2. $\varepsilon = 1$, or $C = GM/h^2$: The radius vector becomes infinite for $\theta = 180°$; the conic section is a *parabola*.
3. $\varepsilon < 1$, or $C < GM/h^2$: The radius vector remains finite for every value of θ; the conic section is an *ellipse*. In the particular case when $\varepsilon = C = 0$, the length of the radius vector is constant; the conic section is a circle.

We shall see now how the constants C and GM/h^2 which characterize the trajectory of a space vehicle may be determined from the position and the velocity of the space vehicle at the beginning of its free flight. We shall assume, as it is generally the case, that the powered phase of its flight has been programmed in such a way that, as the last stage of the launching rocket burns out, the vehicle has a velocity parallel to the surface of the earth (Fig. 12.21). In other words, we shall assume that the space vehicle begins its free flight at the vertex A of its trajectory.†

Denoting respectively by r_0 and v_0 the radius vector and speed of the vehicle at the beginning of its free flight, we observe, since the velocity reduces to its transverse component, that $v_0 = r_0\dot{\theta}_0$. Recalling Eq. (12.27), we express the angular momentum per unit mass h as

$$h = r_0^2\dot{\theta}_0 = r_0 v_0 \tag{12.41}$$

The value obtained for h may be used to determine the constant GM/h^2. We also note that the computation of this constant will be simplified if we use the relation indicated in Sec. 12.9,

$$GM = gR^2 \tag{12.30}$$

where R is the radius of the earth ($R = 6.37 \times 10^6$ m or 3960 mi) and g the acceleration of gravity at the surface of the earth.

The constant C will be determined by setting $\theta = 0$, $r = r_0$ in Eq. (12.39); we obtain

$$C = \frac{1}{r_0} - \frac{GM}{h^2} \tag{12.42}$$

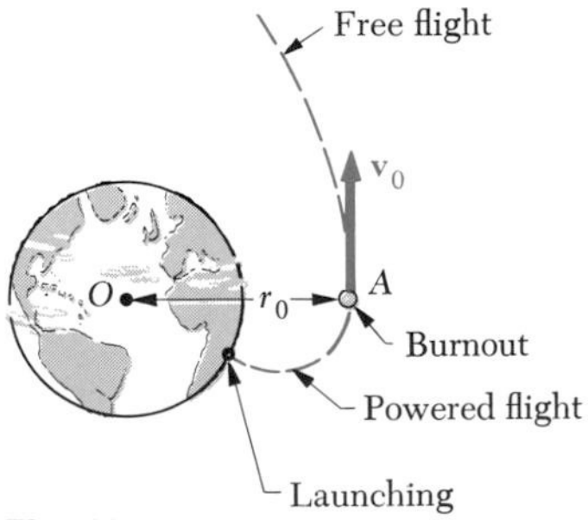

Fig. 12.21

† Problems involving oblique launchings will be considered in Sec. 13.9.

Substituting for h from (12.41), we may then easily express C in terms of r_0 and v_0.

Let us now determine the initial conditions corresponding to each of the three fundamental trajectories indicated above. Considering first the parabolic trajectory, we set C equal to GM/h^2 in Eq. (12.42) and eliminate h between Eqs. (12.41) and (12.42). Solving for v_0, we obtain

$$v_0 = \sqrt{\frac{2GM}{r_0}}$$

We may easily check that a larger value of the initial velocity corresponds to a hyperbolic trajectory, and a smaller value to an elliptic orbit. Since the value of v_0 obtained for the parabolic trajectory is the smallest value for which the space vehicle does not return to its starting point, it is called the *escape velocity*. We write therefore

$$v_{esc} = \sqrt{\frac{2GM}{r_0}} \qquad \text{or} \qquad v_{esc} = \sqrt{\frac{2gR^2}{r_0}} \tag{12.43}$$

if we make use of Eq. (12.30). We note that the trajectory will be (1) hyperbolic if $v_0 > v_{esc}$; (2) parabolic if $v_0 = v_{esc}$; (3) elliptic if $v_0 < v_{esc}$.

Among the various possible elliptic orbits, one is of special interest, the *circular orbit*, which is obtained when $C = 0$. The value of the initial velocity corresponding to a circular orbit is easily found to be

$$v_{circ} = \sqrt{\frac{GM}{r_0}} \qquad \text{or} \qquad v_{circ} = \sqrt{\frac{gR^2}{r_0}} \tag{12.44}$$

if Eq. (12.30) is taken into account. We may note from Fig. 12.22 that, for values of v_0 comprised between v_{circ} and v_{esc}, point A where free flight begins is the point of the orbit closest to the earth; this point is called the *perigee*, while point A', which is farthest away from the earth, is known as the *apogee*. For values of v_0 smaller than v_{circ}, point A becomes the apogee, while point A'', on the other side of the orbit, becomes the perigee. For values of v_0 much smaller than v_{circ}, the trajectory of the space vehicle intersects the surface of the earth; in such a case, the vehicle does not go into orbit.

Ballistic missiles, which are designed to hit the surface of the earth, also travel along elliptic trajectories. In fact, we should now realize that any object projected in vacuum with an initial

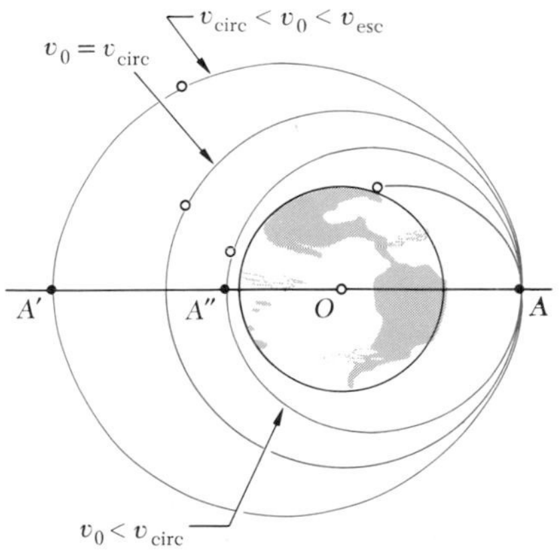

Fig. 12.22

velocity v_0 smaller than v_{esc} will move along an elliptic path. It is only when the distances involved are small that the gravitational field of the earth may be assumed uniform, and that the elliptic path may be approximated by a parabolic path, as was done earlier (Sec. 11.11) in the case of conventional projectiles.

Periodic Time. An important characteristic of the motion of an earth satellite is the time required by the satellite to describe its orbit. This time is known as the *periodic time* of the satellite and is denoted by τ. We first observe, in view of the definition of the areal velocity (Sec. 12.8), that τ may be obtained by dividing the area inside the orbit by the areal velocity. Since the area of an ellipse is equal to πab, where a and b denote, respectively, the semimajor and semiminor axes, and since the areal velocity is equal to $h/2$, we write

$$\tau = \frac{2\pi ab}{h} \qquad (12.45)$$

While h may be readily determined from r_0 and v_0 in the case of a satellite launched in a direction parallel to the surface of the earth, the semiaxes a and b are not directly related to the initial conditions. Since, on the other hand, the values r_0 and r_1 of r corresponding to the perigee and apogee of the orbit may easily be determined from Eq. (12.39), we shall express the semiaxes a and b in terms of r_0 and r_1.

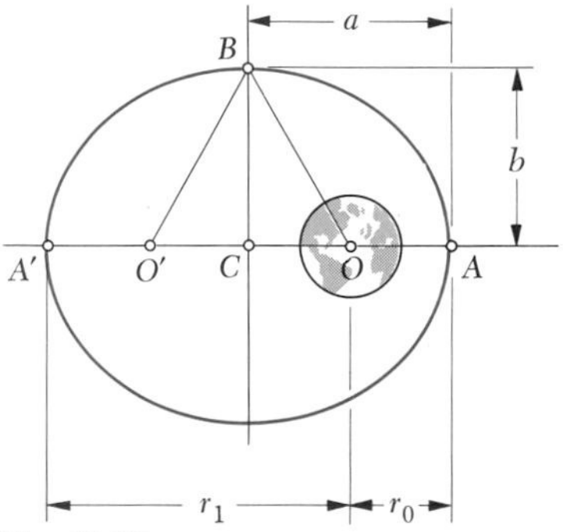

Fig. 12.23

Consider the elliptic orbit shown in Fig. 12.23. The earth's center is located at O and coincides with one of the two foci of the ellipse, while the points A and A' represent, respectively, the perigee and apogee of the orbit. We easily check that

$$r_0 + r_1 = 2a$$

and thus

$$a = \tfrac{1}{2}(r_0 + r_1) \qquad (12.46)$$

Recalling that the sum of the distances from each of the foci to any point of the ellipse is constant, we write

$$O'B + BO = O'A + OA = 2a \qquad \text{or} \qquad BO = a$$

On the other hand, we have $CO = a - r_0$. We may therefore write

$$b^2 = (BC)^2 = (BO)^2 - (CO)^2 = a^2 - (a - r_0)^2$$
$$b^2 = r_0(2a - r_0) = r_0 r_1$$

and thus

$$b = \sqrt{r_0 r_1} \tag{12.47}$$

Formulas (12.46) and (12.47) indicate that the semimajor and semiminor axes of the orbit are respectively equal to the arithmetic and geometric means of the maximum and minimum values of the radius vector. Once r_0 and r_1 have been determined, the lengths of the semiaxes may thus be easily computed and substituted for a and b in formula (12.45).

***12.12. Kepler's Laws of Planetary Motion.** The equations governing the motion of an earth satellite may be used to describe the motion of the moon around the earth. In that case, however, the mass of the moon is not negligible compared to the mass of the earth, and the results obtained are not entirely accurate.

The theory developed in the preceding sections may also be applied to the study of the motion of the planets around the sun. While another error is introduced by neglecting the forces exerted by the planets on each other, the approximation obtained is excellent. Indeed, the properties expressed by Eq. (12.39), where M now represents the mass of the sun, and by Eq. (12.33) had been discovered by the German astronomer Johann Kepler (1571–1630) from astronomical observations of the motion of the planets, even before Newton had formulated his fundamental theory.

Kepler's three *laws of planetary motion* may be stated as follows:

1. Each planet describes an ellipse, with the sun located at one of its foci.
2. The radius vector drawn from the sun to a planet sweeps equal areas in equal times.
3. The squares of the periodic times of the planets are proportional to the cubes of the semimajor axes of their orbits.

The first law states a particular case of the result established in Sec. 12.11, while the second law expresses that the areal velocity of each planet is constant (see Sec. 12.8). Kepler's third law may also be derived from the results obtained in Sec. 12.11.†

† See Prob. 12.104

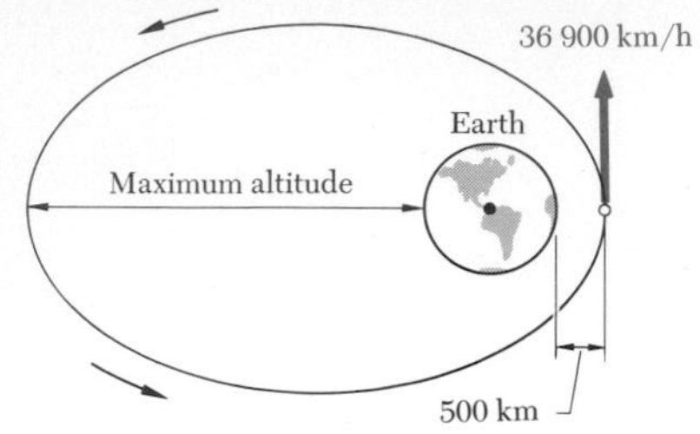

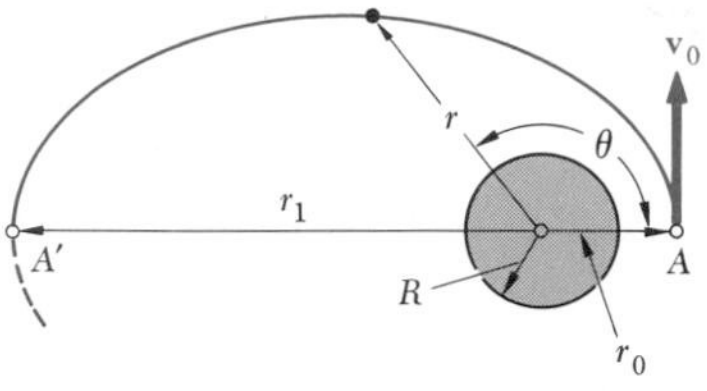

SAMPLE PROBLEM 12.8

A satellite is launched in a direction parallel to the surface of the earth with a velocity of 36 900 km/h from an altitude of 500 km. Determine (*a*) the maximum altitude reached by the satellite, (*b*) the periodic time of the satellite.

***a.* Maximum Altitude.** After launching, the satellite is subjected only to the gravitational attraction of the earth; its motion is thus governed by Eq. (12.39).

$$\frac{1}{r} = \frac{GM}{h^2} + C\cos\theta \qquad (1)$$

Since the radial component of the velocity is zero at the point of launching A, we have $h = r_0 v_0$. Recalling that the radius of the earth is $R = 6370$ km, we compute

$$r_0 = 6370 \text{ km} + 500 \text{ km} = 6870 \text{ km} = 6.87 \times 10^6 \text{ m}$$

$$v_0 = 36\,900 \text{ km/h} = \frac{3.69 \times 10^7 \text{ m}}{3.6 \times 10^3 \text{ s}} = 1.025 \times 10^4 \text{ m/s}$$

$$h = r_0 v_0 = (6.87 \times 10^6 \text{ m})(1.025 \times 10^4 \text{ m/s}) = 7.04 \times 10^{10} \text{ m}^2\text{/s}$$

$$h^2 = 4.96 \times 10^{21} \text{ m}^4\text{/s}^2$$

Since $GM = gR^2$, where R is the radius of the earth, we have

$$GM = gR^2 = (9.81 \text{ m/s}^2)(6.37 \times 10^6 \text{ m})^2 = 3.98 \times 10^{14} \text{ m}^3\text{/s}^2$$

$$\frac{GM}{h^2} = \frac{3.98 \times 10^{14} \text{ m}^3\text{/s}^2}{4.96 \times 10^{21} \text{ m}^4\text{/s}^2} = 8.03 \times 10^{-8} \text{ m}^{-1}$$

Substituting this value into (1), we obtain

$$\frac{1}{r} = 8.03 \times 10^{-8} + C\cos\theta \qquad (2)$$

Noting that at point A we have $\theta = 0$ and $r = r_0 = 6.87 \times 10^6$ m, we compute the constant C.

$$\frac{1}{6.87 \times 10^6 \text{ m}} = 8.03 \times 10^{-8} + C\cos 0^\circ \qquad C = 6.53 \times 10^{-8} \text{ m}^{-1}$$

At A', the point on the orbit farthest from the earth, we have $\theta = 180^\circ$. Using (2), we compute the corresponding distance r_1.

$$\frac{1}{r_1} = 8.03 \times 10^{-8} + (6.53 \times 10^{-8})\cos 180^\circ$$

$$r_1 = 0.667 \times 10^8 \text{ m} = 66\,700 \text{ km}$$

Maximum altitude = 66 700 km − 6370 km = 60 300 km ◄

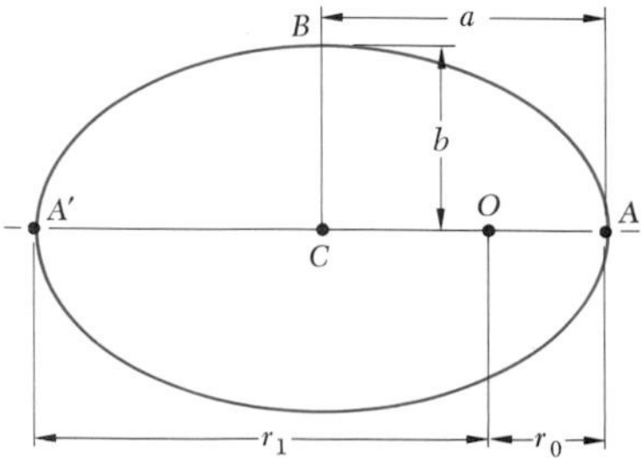

***b.* Periodic Time.** Since A and A' are the perigee and apogee, respectively, of the elliptic orbit, we use Eqs. (12.46) and (12.47) and compute the semimajor and semiminor axes of the orbit.

$$a = \tfrac{1}{2}(r_0 + r_1) = \tfrac{1}{2}(6.87 + 66.7)(10^6) \text{ m} = 36.8 \times 10^6 \text{ m}$$

$$b = \sqrt{r_0 r_1} = \sqrt{(6.87)(66.7)} \times 10^6 \text{ m} = 21.4 \times 10^6 \text{ m}$$

$$\tau = \frac{2\pi ab}{h} = \frac{2\pi(36.8 \times 10^6 \text{ m})(21.4 \times 10^6 \text{ m})}{7.04 \times 10^{10} \text{ m}^2\text{/s}}$$

$$\tau = 7.03 \times 10^4 \text{ s} = 1171 \text{ min} = 19 \text{ h } 31 \text{ min}$$ ◄

PROBLEMS

12.87 A spacecraft is describing a circular orbit at an altitude of 240 mi above the surface of the earth when its engine is fired and its speed increased by 4000 ft/s. Determine the maximum altitude reached by the spacecraft.

12.88 A space tug is used to place communication satellites into a geosynchronous orbit (see Prob. 12.78) at an altitude of 22,230 mi above the surface of the earth. Knowing that the tug initially describes a circular orbit at an altitude of 220 mi, determine (*a*) the increase in speed required at A to insert the tug into an elliptic transfer orbit, (*b*) the increase in speed required at B to insert the tug into the geosynchronous orbit.

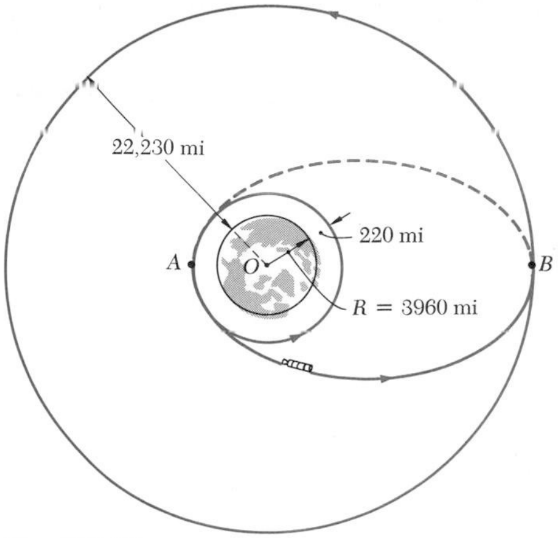

Fig. P12.88

12.89 Plans for an unmanned landing mission on the planet Mars call for the earth-return vehicle to first describe a circular orbit about the planet. As it passes through point A, the vehicle will be inserted into an elliptic transfer orbit by firing its engine and increasing its speed by Δv_A. As it passes through point B, the vehicle will be inserted into a second transfer orbit located in a slightly different plane, by changing the direction of its velocity and by reducing its speed by Δv_B. Finally, as the vehicle passes through point C, its speed will be increased by Δv_C to insert it into its return trajectory. Knowing that the radius of the planet Mars is $R = 3400$ km, that its mass is 0.108 times the mass of the earth, and that the altitudes of points A and B are, respectively, $d_A = 2500$ km and $d_B = 90\,000$ km, determine the increase in speed Δv_A required at point A to insert the vehicle into its first transfer orbit.

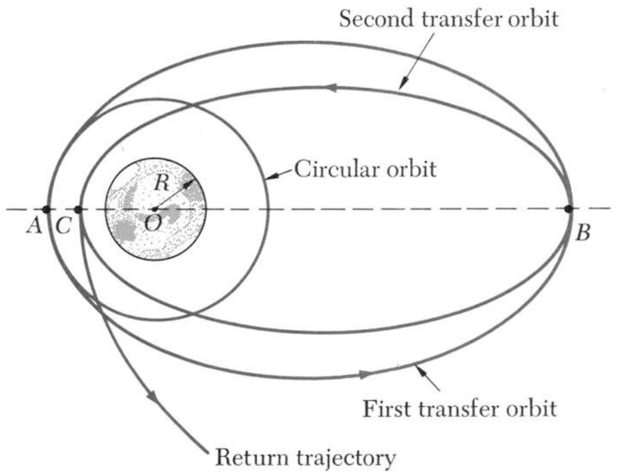

Fig. P12.89

12.90 For the vehicle of Prob. 12.89, it is known that the altitudes of points A, B, and C are, respectively, $d_A = 2500$ km, $d_B = 90\,000$ km, and $d_C = 1000$ km. Determine the change in speed Δv_B required at point B to insert the vehicle into its second transfer orbit.

12.91 For the vehicle of Prob. 12.89, it is known that the altitudes of points B and C are, respectively, $d_B = 90\,000$ km and $d_C = 1000$ km. Determine the minimum increase in speed Δv_C required at point C to insert the vehicle into an escape trajectory.

12.92 For the vehicle of Prob. 12.89, it is known that the altitude of point B is $d_B = 90\,000$ km. If, for a given mission, the speed of the vehicle is 215 m/s immediately after its insertion into the second transfer orbit, determine (*a*) the altitude of point C, (*b*) the speed of the vehicle as it approaches point C, (*c*) the eccentricity of the return trajectory if the speed of the vehicle is increased at C by $\Delta v_C = 630$ m/s.

12.93 After completing their moon-exploration mission, the two astronauts forming the crew of an Apollo lunar excursion module (LEM) prepare to rejoin the command module which is orbiting the moon at an altitude of 85 mi. They fire the LEM's engine, bring it along a curved path to a point A, 5 mi above the moon's surface, and shut off the engine. Knowing that the LEM is moving at that time in a direction parallel to the moon's surface and that it will coast along an elliptic path to a rendezvous at B with the command module, determine (*a*) the speed of the LEM at engine shutoff, (*b*) the relative velocity with which the command module will approach the LEM at B. (The radius of the moon is 1080 miles and its mass is 0.01230 times the mass of the earth.)

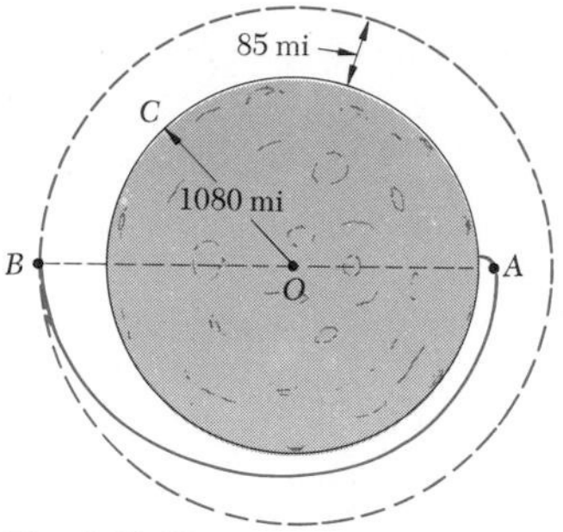

Fig. P12.93

12.94 Solve Prob. 12.93, assuming that the Apollo command module is orbiting the moon at an altitude of 55 mi.

12.95 Referring to Prob. 12.89, determine the time required for the vehicle to describe its first transfer orbit form A to B.

12.96 Referring to Probs. 12.89 and 12.90, determine the time required for the vehicle to describe its second transfer orbit from B to C.

12.97 Determine the time required for the LEM of Prob. 12.93 to travel from A to B.

12.98 Determine the time required for the space tug of Prob. 12.88 to travel from A to B.

12.99 Determine the approximate time required for an object to fall to the surface of the earth after being released with no velocity from a distance equal to the radius of the orbit of the moon, namely, 239,000 mi. (*Hint.* Assume that the object is given a very small initial velocity in a transverse direction, say, $v_\theta = 1$ ft/s, and determine the periodic time τ of the object on the resulting orbit. An examination of the orbit will show that the time of fall must be approximately equal to $\frac{1}{2}\tau$.)

12.100 A spacecraft describes a circular orbit at an altitude of 3200 km above the earth's surface. Preparatory to reentry it reduces its speed to a value $v_0 = 5400$ m/s, thus placing itself on an elliptic trajectory. Determine the value of θ defining the point B where splashdown will occur. (*Hint.* Point A is the apogee of the elliptic trajectory.)

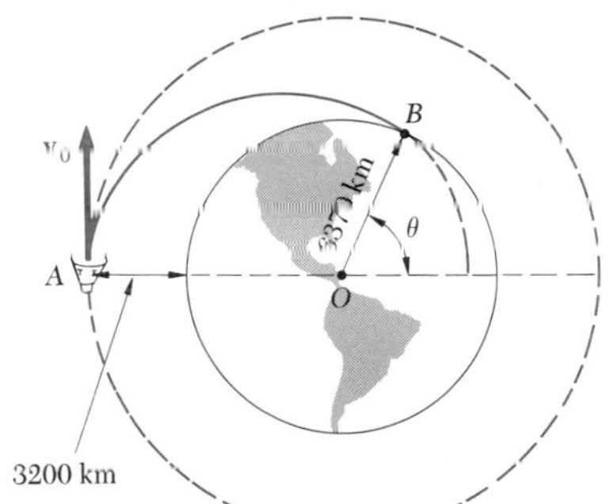

Fig. P12.100 and P12.101

12.101 A spacecraft describes a circular orbit at an altitude of 3200 km above the earth's surface. Preparatory to reentry it places itself on an elliptic trajectory by reducing its speed to a value v_0. Determine v_0 so that splashdown will occur at a point B corresponding to $\theta = 120°$. (See hint of Prob. 12.100.)

12.102 Upon the LEM's return to the command module, the Apollo spacecraft of Prob. 12.93 is turned around so that the LEM faces to the rear. After completing a full orbit, i.e., as the craft passes again through B, the LEM is cast adrift and crashes on the moon's surface at point C. Determine the velocity of the LEM relative to the command module as it is cast adrift, knowing that the angle BOC is 90°. (*Hint.* Point B is the apogee of the elliptic crash trajectory.)

12.103 Upon the LEM's return to the command module, the Apollo spacecraft of Prob. 12.93 is turned around so that the LEM faces to the rear. After completing a full orbit, i.e., as the craft passes again through B, the LEM is cast adrift with a velocity of 600 ft/s relative to the command module. Determine the point C where the LEM will crash on the moon's surface. (See hint of Prob. 12.102.)

12.104 Derive Kepler's third law of planetary motion from Eqs. (12.39) and (12.45).

12.105 (*a*) Express the eccentricity ε of the elliptic orbit described by a satellite about the earth (or any other planet) in terms of the distances r_0 and r_1 corresponding, respectively, to the perigee and apogee of the orbit. (*b*) Use the result obtained in part *a* to determine the eccentricities of the two transfer orbits described in Probs. 12.89 and 12.90.

12.106 Two space stations S_1 and S_2 are describing coplanar circular counterclockwise orbits of radius r_0 and $8r_0$, respectively, around the earth. It is desired to send a vehicle from S_1 to S_2. The vehicle is to be launched in a direction tangent to the orbit of S_1 and is to reach S_2 with a velocity tangent to the orbit of S_2. After a short powered phase, the vehicle will travel in free flight from S_1 to S_2. (*a*) Determine the launching velocity (velocity of the vehicle relative to S_1) in terms of the velocity $\mathbf{v}_0$ of S_1. (*b*) Determine the angle θ defining the required position of S_2 relative to S_1 at the time of launching.

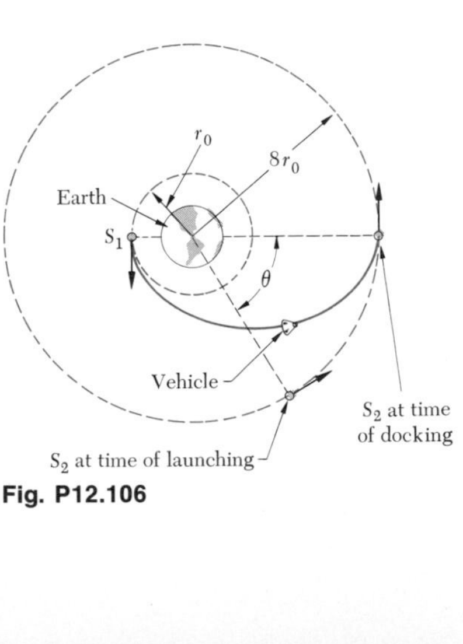

Fig. P12.106

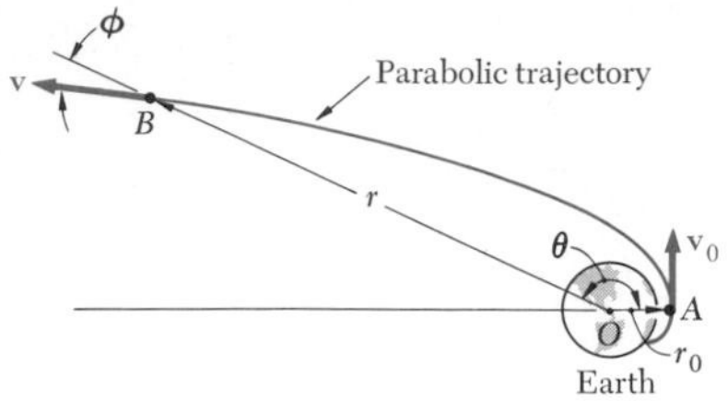

Fig. P12.107 and P12.108

$*$**12.107** A space vehicle is inserted at point A, at a distance r_0 from the center O of the earth, into a parabolic trajectory. (*a*) For any position B of the vehicle on its trajectory, express the distance r from O to B and the time t elapsed since the insertion of the vehicle into its trajectory in terms of θ, r_0, g, and the radius R of the earth. (*b*) Use the result obtained in part *a*, assuming $r_0 = 4300$ mi, to determine the time required for the space vehicle to reach a distance r equal to the radius of the orbit of the moon (239,000 mi).

$*$**12.108** A space vehicle is inserted at point A, at a distance r_0 from the center O of the earth, into a parabolic trajectory. (*a*) For any position B of the vehicle on its trajectory, express in terms of θ, r_0, g, and the radius R of the earth (*a*) the magnitude of the velocity **v** of the vehicle, (*b*) the angle ϕ that **v** forms with the line OB.

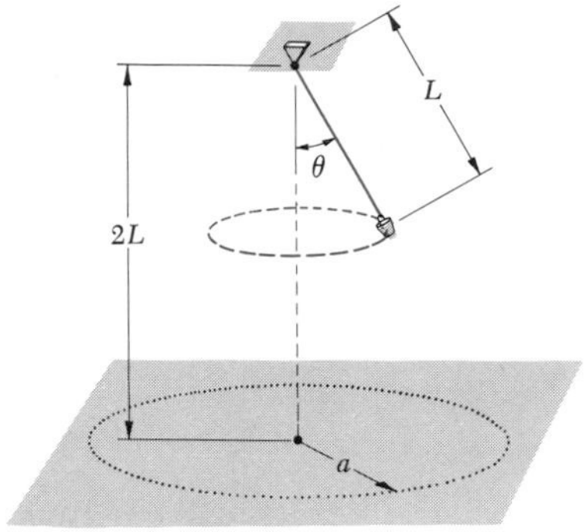

Fig. P12.109

REVIEW PROBLEMS

12.109 A bucket is attached to a rope of length $L = 1.2$ m and is made to revolve in a horizontal circle. Drops of water leaking from the bucket fall and strike the floor along the perimeter of a circle of radius a. Determine the radius a when $\theta = 30°$.

12.110 Determine the radius a in Prob. 12.109, assuming that the speed of the bucket is 5 m/s. (The angle θ is not 30° in this case.)

12.111 Determine the required tension T if the acceleration of the 500-lb cylinder is to be (*a*) 6 ft/s^2 upward, (*b*) 6 ft/s^2 downward.

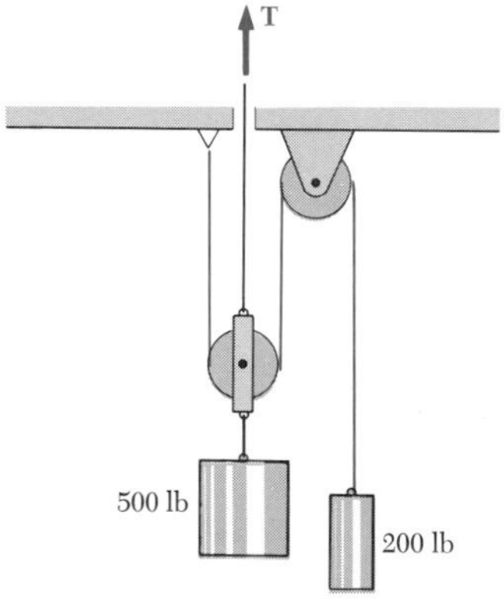

Fig. P12.111 and P12.112

12.112 Determine the acceleration of the 200-lb cylinder if (*a*) $T = 300$ lb, (*b*) $T = 800$ lb.

12.113 A series of small packages, being moved by a conveyor belt at a constant speed v, passes over an idler roller as shown. Knowing that the coefficient of friction between the packages and the belt is 0.75, determine the maximum value of v for which the packages do not slip with respect to the belt.

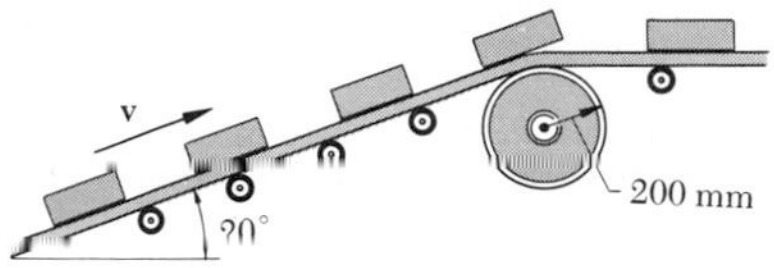

Fig. P12.113

12.114 A 4-kg collar slides without friction along a rod which forms an angle of 30° with the vertical. The spring, of constant $k = 150$ N/m, is unstretched when the collar is at A. Determine the initial acceleration of the collar if it is released from rest at point B.

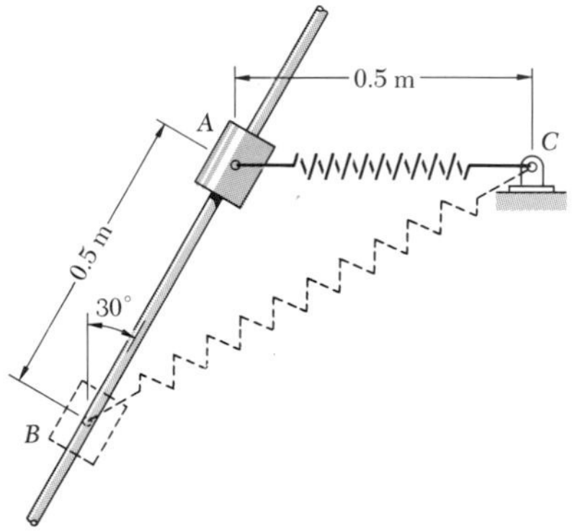

Fig. P12.114

12.115 (*a*) Express the rated speed of a banked road in terms of the radius r of the curve and the banking angle θ. (*b*) What is the apparent weight of an automobile traveling at the rated speed? (See Sample Prob. 12.5 for the definition of rated speed.)

12.116 Denoting by v_t the terminal speed of an object dropped from a great height, determine the distance the object will fall before its speed reaches the value $\frac{1}{2}v_t$. Assume that the frictional resistance of the air is proportional to the square of the speed of the object.

12.117 A spacecraft is describing a circular orbit of radius r_0 with a speed v_0 around an unspecified celestial body of center O, when its engine is suddenly fired, increasing the speed of the spacecraft from v_0 to αv_0, where $1 < \alpha^2 < 2$. Show that the maximum distance r_{max} from O reached by the spacecraft depends only upon r_0 and α, and express the ratio r_{max}/r_0 as a function of α.

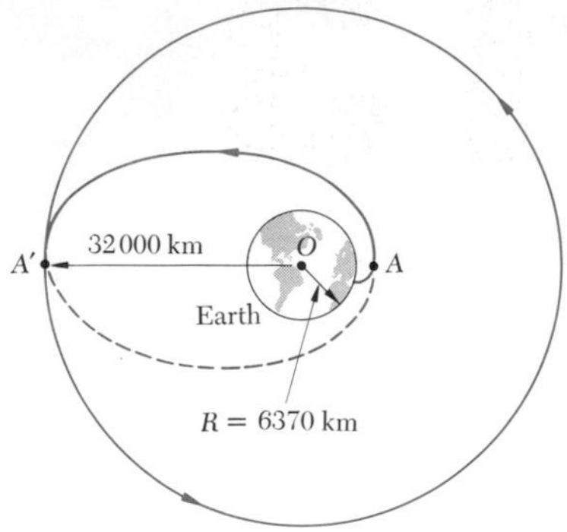

Fig. P12.118

12.118 In order to place a satellite in a circular orbit of radius 32×10^3 km around the earth, the satellite is first projected horizontally from A at an altitude of 500 km into an elliptic path whose apogee A' is at a distance of 32×10^3 km from the center of the earth. Auxiliary rockets are fired as the satellite reaches A' in order to place it in its final orbit. Determine (*a*) the initial velocity of the satellite at A, (*b*) the increase in velocity resulting from the firing of the rockets at A'.

12.119 Two packages are placed on a conveyor belt which is at rest. The coefficient of friction is 0.20 between the belt and package A, and 0.10 between the belt and package B. If the belt is suddenly started to the right and slipping occurs between the belt and the packages, determine (*a*) the acceleration of the packages, (*b*) the force exerted by package A on package B.

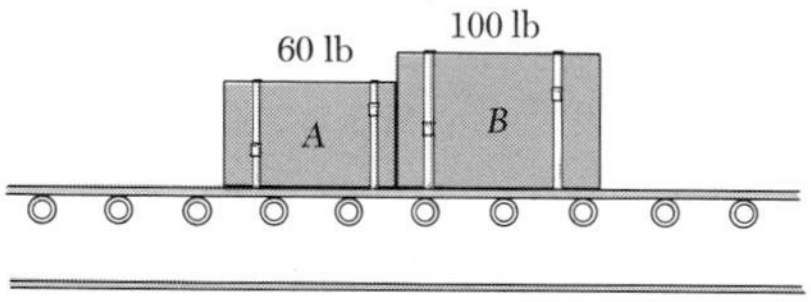

Fig. P12.119

12.120 Two plates A and B, each of mass 50 kg, are placed as shown on a 15° incline. The coefficient of friction between A and B is 0.10; the coefficient of friction between A and the incline is 0.20. (*a*) If the plates are released from rest, determine the acceleration of each plate. (*b*) Solve part *a* assuming that plates A and B are welded together and act as a single rigid body.

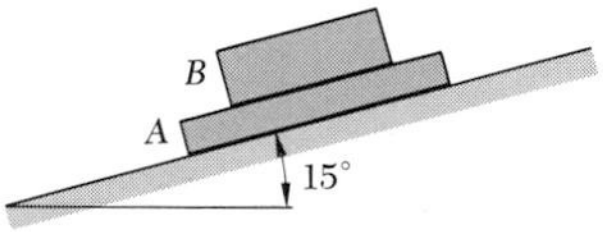

Fig. P12.120

CHAPTER 13

Kinetics of Particles: Energy and Momentum Methods

13.1. Introduction. In the preceding chapter, most problems dealing with the motion of particles were solved through the use of the fundamental equation of motion $\mathbf{F} = m\mathbf{a}$. Given a particle acted upon by a force $\mathbf{F}$, we could solve this equation for the acceleration $\mathbf{a}$; then, by applying the principles of kinematics, we could determine from $\mathbf{a}$ the velocity and position of the particle at any time.

If the equation $\mathbf{F} = m\mathbf{a}$ and the principles of kinematics are combined, two additional methods of analysis may be obtained, the *method of work and energy* and the *method of impulse and momentum*. The advantage of these methods lies in the fact that they make the determination of the acceleration unnecessary. Indeed, the method of work and energy relates directly force, mass, velocity, and displacement, while the method of impulse and momentum relates force, mass, velocity, and time.

The method of work and energy will be considered first. It is based on two important concepts, the concept of the *work of a force* and the concept of the *kinetic energy of a particle*. These concepts are defined in the following sections.

13.2. Work of a Force. We shall first define the terms *displacement* and *work* as they are used in mechanics.† Consider a particle which moves from a point A to a neighboring point A'

† The definition of work was given in Sec. 10.1, and the basic properties of the work of a force were outlined in Secs. 10.1 and 10.5. For convenience, we repeat here the portions of this material which relate to the kinetics of particles.

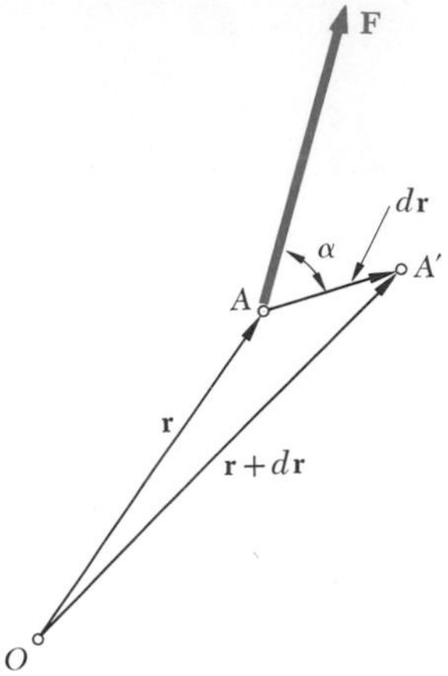

Fig. 13.1

(Fig. 13.1). If $\mathbf{r}$ denotes the position vector corresponding to point A, the small vector joining A and A' may be denoted by the differential $d\mathbf{r}$; the vector $d\mathbf{r}$ is called the *displacement* of the particle. Now, let us assume that a force $\mathbf{F}$ is acting on the particle. The *work of the force* $\mathbf{F}$ *corresponding to the displacement* $d\mathbf{r}$ is defined as the quantity

$$dU = \mathbf{F} \cdot d\mathbf{r} \qquad (13.1)$$

obtained by forming the scalar product of the force $\mathbf{F}$ and of the displacement $d\mathbf{r}$. Denoting respectively by F and ds the magnitudes of the force and of the displacement, and by α the angle formed by $\mathbf{F}$ and $d\mathbf{r}$, and recalling the definition of the scalar product of two vectors (Sec. 3.8), we write

$$dU = F\,ds \cos\alpha \qquad (13.1')$$

Using formula (3.30), we may also express the work dU in terms of the rectangular components of the force and of the displacement:

$$dU = F_x\,dx + F_y\,dy + F_z\,dz \qquad (13.1'')$$

Being a *scalar quantity*, work has a magnitude and a sign, but no direction. We also note that work should be expressed in units obtained by multiplying units of length by units of force. Thus, if U.S. customary units are used, work should be expressed in ft · lb or in · lb. If SI units are used, work should be expressed in N · m. The unit of work N · m is called a *joule* (J).† Recalling the conversion factors indicated in Sec. 12.3, we write

$$1 \text{ ft} \cdot \text{lb} = (1 \text{ ft})(1 \text{ lb}) = (0.3048 \text{ m})(4.448 \text{ N}) = 1.356 \text{ J}$$

It follows from (13.1′) that the work dU is positive if the angle α is acute, and negative if α is obtuse. Three particular cases are of special interest. If the force $\mathbf{F}$ has the same direction as $d\mathbf{r}$, the work dU reduces to $F\,ds$. If $\mathbf{F}$ has a direction opposite to that of $d\mathbf{r}$, the work is $dU = -F\,ds$. Finally, if $\mathbf{F}$ is perpendicular to $d\mathbf{r}$, the work dU is zero.

The work of $\mathbf{F}$ during a *finite* displacement of the particle from A_1 to A_2 (Fig. 13.2*a*) is obtained by integrating Eq. (13.1) along the path described by the particle. This work, denoted by $U_{1\to2}$, is

†The joule (J) is the SI unit of *energy*, whether in mechanical form (work, potential energy, kinetic energy) or in chemical, electrical, or thermal form. We should note that, even though N · m = J, the moment of a force must be expressed in N · m, and not in joules, since the moment of a force is not a form of energy.

$$U_{1\to 2} = \int_{A_1}^{A_2} \mathbf{F} \cdot d\mathbf{r} \tag{13.2}$$

Using the alternate expression (13.1′) for the elementary work dU, and observing that $F \cos \alpha$ represents the tangential component F_t of the force, we may also express the work $U_{1\to 2}$ as

$$U_{1\to 2} = \int_{s_1}^{s_2} (F \cos \alpha)\, ds = \int_{s_1}^{s_2} F_t\, ds \tag{13.2′}$$

where the variable of integration s measures the distance traveled by the particle along the path. The work $U_{1\to 2}$ is represented by the area under the curve obtained by plotting $F_t = F \cos \alpha$ against s (Fig. 13.2*b*).

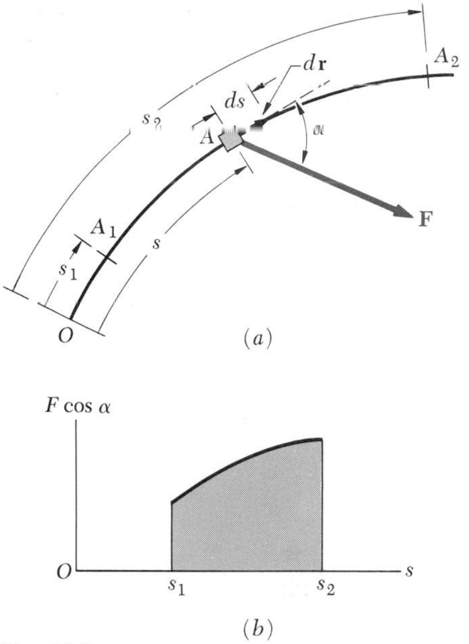

Fig. 13.2

When the force $\mathbf{F}$ is defined by its rectangular components, the expression (13.1″) may be used for the elementary work. We write then

$$U_{1\to 2} = \int_{A_1}^{A_2} (F_x\, dx + F_y\, dy + F_z\, dz) \tag{13.2″}$$

where the integration is to be performed along the path described by the particle.

Work of a Constant Force in Rectilinear Motion. When a particle moving in a straight line is acted upon by a force $\mathbf{F}$ of constant magnitude and of constant direction (Fig. 13.3), formula (13.2′) yields

$$U_{1\to 2} = (F \cos \alpha)\, \Delta x \tag{13.3}$$

where α = angle the force forms with direction of motion

Δx = displacement from A_1 to A_2

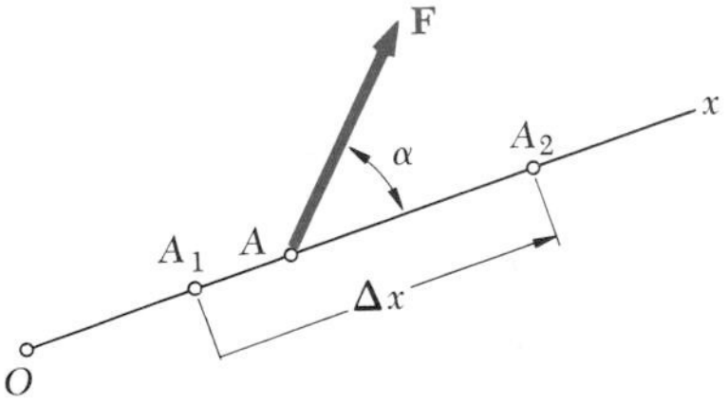

Fig. 13.3

Work of a Weight. The work of the weight $\mathbf{W}$ of a body is obtained by substituting the components of $\mathbf{W}$ into (13.1″) and (13.2″). With the y axis chosen upward (Fig. 13.4), we have $F_x = 0$, $F_y = -W$, $F_z = 0$, and we write

$$dU = -W\, dy$$

$$U_{1\to 2} = -\int_{y_1}^{y_2} W\, dy = Wy_1 - Wy_2 \tag{13.4}$$

or

$$U_{1\to 2} = -W(y_2 - y_1) = -W\, \Delta y \tag{13.4′}$$

where Δy is the vertical displacement from A_1 to A_2. The work of the weight $\mathbf{W}$ is thus equal to *the product of W and of the vertical displacement of the center of gravity of the body*. The work is *positive* when $\Delta y < 0$, that is, *when the body moves down*.

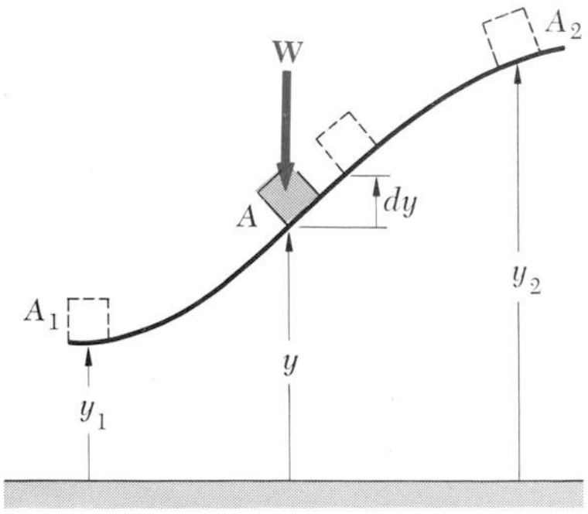

Fig. 13.4

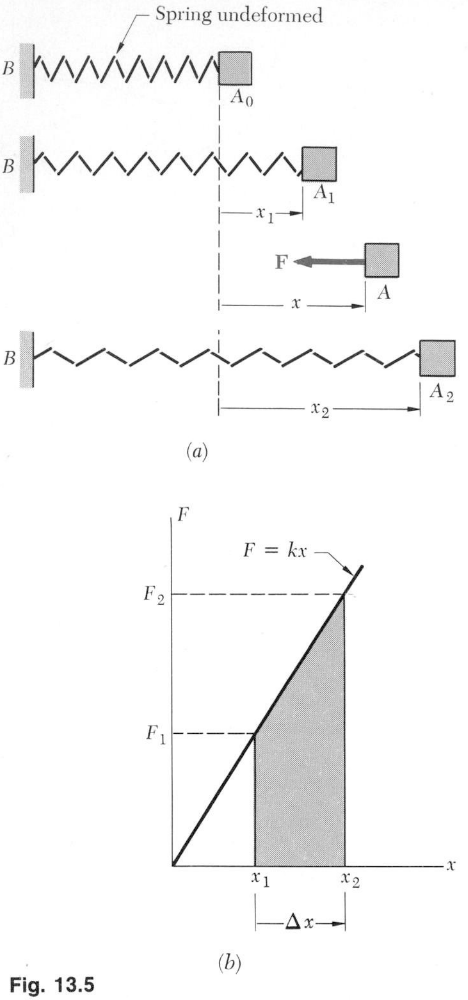

Fig. 13.5

Work of the Force Exerted by a Spring. Consider a body A attached to a fixed point B by a spring; it is assumed that the spring is undeformed when the body is at A_0 (Fig. 13.5*a*). Experimental evidence shows that the magnitude of the force **F** exerted by the spring on body A is proportional to the deflection x of the spring measured from the position A_0. We have

$$F = kx \tag{13.5}$$

where k is the *spring constant,* expressed in N/m or kN/m if SI units are used and in lb/ft or lb/in. if U.S. customary units are used.†

The work of the force **F** exerted by the spring during a finite displacement of the body from $A_1(x = x_1)$ to $A_2(x = x_2)$ is obtained by writing

$$dU = -F\,dx = -kx\,dx$$

$$U_{1\to2} = -\int_{x_1}^{x_2} kx\,dx = \tfrac{1}{2}kx_1^2 - \tfrac{1}{2}kx_2^2 \tag{13.6}$$

Care should be taken to express k and x in consistent units. For example, if U.S. customary units are used, k should be expressed in lb/ft and x in feet, or k in lb/in. and x in inches; in the first case, the work is obtained in ft · lb, in the second case, in in · lb. We note that the work of the force **F** exerted by the spring on the body is *positive* when $x_2 < x_1$, i.e., *when the spring is returning to its undeformed position.*

Since Eq. (13.5) is the equation of a straight line of slope k passing through the origin, the work $U_{1\to2}$ of **F** during the displacement from A_1 to A_2 may be obtained by evaluating the area of the trapezoid shown in Fig. 13.5*b*. This is done by computing F_1 and F_2 and multiplying the base Δx of the trapezoid by its mean height $\frac{1}{2}(F_1 + F_2)$. Since the work of the force **F** exerted by the spring is positive for a negative value of Δx, we write

$$U_{1\to2} = -\tfrac{1}{2}(F_1 + F_2)\,\Delta x \tag{13.6$'$}$$

Formula (13.6′) is usually more convenient to use than (13.6) and affords fewer chances of confusing the units involved.

† The relation $F = kx$ is correct under static conditions only. Under dynamic conditions, formula (13.5) should be modified to take the inertia of the spring into account. However, the error introduced by using the relation $F = kx$ in the solution of kinetics problems is small if the mass of the spring is small compared with the other masses in motion.

Work of a Gravitational Force. We saw in Sec. 12.9 that two particles at distance r from each other and, respectively, of mass M and m, attract each other with equal and opposite forces $\mathbf{F}$ and $-\mathbf{F}$ directed along the line joining the particles, and of magnitude

$$F = G\frac{Mm}{r^2}$$

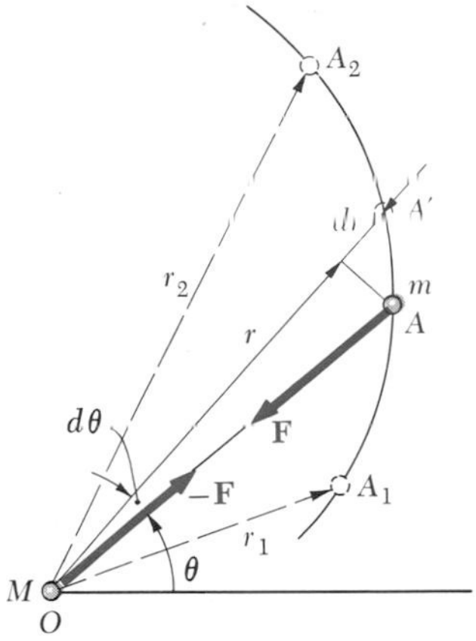

Fig. 13.6

Let us assume that the particle M occupies a fixed position O while the particle m moves along the path shown in Fig. 13.6. The work of the force $\mathbf{F}$ exerted on the particle m during an infinitesimal displacement of the particle from A to A' may be obtained by multiplying the magnitude F of the force by the radial component dr of the displacement. Since $\mathbf{F}$ is directed toward O, the work is negative and we write

$$dU = -F\,dr = -G\frac{Mm}{r^2}\,dr$$

The work of the gravitational force $\mathbf{F}$ during a finite displacement from $A_1(r = r_1)$ to $A_2(r = r_2)$ is therefore

$$U_{1\to 2} = -\int_{r_1}^{r_2} \frac{GMm}{r^2}\,dr = \frac{GMm}{r_2} - \frac{GMm}{r_1} \qquad (13.7)$$

The formula obtained may be used to determine the work of the force exerted by the earth on a body of mass m at a distance r from the center of the earth, when r is larger than the radius R of the earth. The letter M represents then the mass of the earth; recalling the first of the relations (12.29), we may thus replace the product GMm in Eq. (13.7) by WR^2, where R is the radius of the earth ($R = 6.37 \times 10^6$ m or 3960 mi) and W the value of the weight of the body at the surface of the earth.

A number of forces frequently encountered in problems of kinetics *do no work.* They are forces applied to fixed points ($ds = 0$) or acting in a direction perpendicular to the displacement ($\cos\alpha = 0$). Among the forces which do no work are the following: the reaction at a frictionless pin when the body supported rotates about the pin, the reaction at a frictionless surface when the body in contact moves along the surface, the reaction at a roller moving along its track, and the weight of a body when its center of gravity moves horizontally.

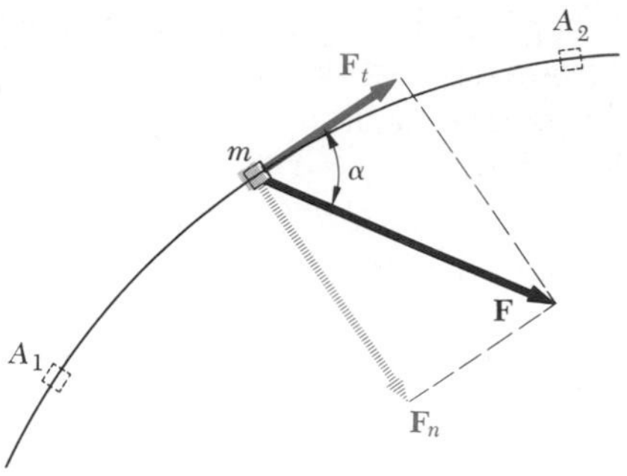

Fig. 13.7

13.3. Kinetic Energy of a Particle. Principle of Work and Energy. Consider a particle of mass m acted upon by a force $\mathbf{F}$ and moving along a path which is either rectilinear or curved (Fig. 13.7). Expressing Newton's second law in terms of the tangential components of the force and of the acceleration (see Sec. 12.4), we write

$$F_t = ma_t \qquad \text{or} \qquad F_t = m\frac{dv}{dt}$$

where v is the speed of the particle. Recalling from Sec. 11.9 that $v = ds/dt$, we obtain

$$F_t = m\frac{dv}{ds}\frac{ds}{dt} = mv\frac{dv}{ds}$$

$$F_t\,ds = mv\,dv$$

Integrating from A_1, where $s = s_1$ and $v = v_1$, to A_2, where $s = s_2$ and $v = v_2$, we write

$$\int_{s_1}^{s_2} F_t\,ds = m\int_{v_1}^{v_2} v\,dv = \tfrac{1}{2}mv_2^2 - \tfrac{1}{2}mv_1^2 \tag{13.8}$$

The left-hand member of Eq. (13.8) represents the work $U_{1\to2}$ of the force $\mathbf{F}$ exerted on the particle during the displacement from A_1 to A_2; as indicated in Sec. 13.2, the work $U_{1\to2}$ is a scalar quantity. The expression $\frac{1}{2}mv^2$ is also a scalar quantity; it is defined as the kinetic energy of the particle and is denoted by T. We write

$$T = \tfrac{1}{2}mv^2 \tag{13.9}$$

Substituting into (13.8), we have

$$U_{1\to2} = T_2 - T_1 \tag{13.10}$$

which expresses that, when a particle moves from A_1 to A_2 under the action of a force **F**, *the work of the force* **F** *is equal to the change in kinetic energy of the particle.* This is known as the *principle of work and energy.* Rearranging the terms in (13.10), we write

$$T_1 + U_{1\to2} = T_2 \tag{13.11}$$

Thus, *the kinetic energy of the particle at* A_2 *may be obtained by adding to its kinetic energy at* A_1 *the work done during the displacement from* A_1 *to* A_2 *by the force* **F** *exerted on the particle.* As Newton's second law from which it is derived, the principle of work and energy applies only with respect to a newtonian frame of reference (Sec. 12.1). The speed v used to determine the kinetic energy T should therefore be measured with respect to a newtonian frame of reference.

Since both work and kinetic energy are scalar quantities, their sum may be computed as an ordinary algebraic sum, the work $U_{1\to2}$ being considered as positive or negative according to the direction of **F**. When several forces act on the particle, the expression $U_{1\to2}$ represents the total work of the forces acting on the particle; it is obtained by adding algebraically the work of the various forces.

As noted above, the kinetic energy of a particle is a scalar quantity. It further appears from the definition $T = \frac{1}{2}mv^2$ that the kinetic energy is always positive, regardless of the direction of motion of the particle. Considering the particular case when $v_1 = 0$, $v_2 = v$, and substituting $T_1 = 0$, $T_2 = T$ into (13.10), we observe that the work done by the forces acting on the particle is equal to T. Thus, the kinetic energy of a particle moving with a speed v represents the work which must be done to bring the particle from rest to the speed v. Substituting $T_1 = T$ and $T_2 = 0$ into (13.10), we also note that, when a particle moving with a speed v is brought to rest, the work done by the forces acting on the particle is $-T$. Assuming that no energy is dissipated into heat, we conclude that the work done by the forces exerted *by the particle* on the bodies which cause it to come to rest is equal to T. Thus, the kinetic energy of a particle also represents *the capacity to do work associated with the speed of the particle.*

The kinetic energy is measured in the same units as work, i.e., in joules if SI units are used, and in ft · lb if U.S. customary units are used. We check that, in SI units,

$$T = \tfrac{1}{2}mv^2 = \text{kg(m/s)}^2 = (\text{kg}\cdot\text{m/s}^2)\text{m} = \text{N}\cdot\text{m} = \text{J}$$

while, in customary units,

$$T = \tfrac{1}{2}mv^2 = (\text{lb}\cdot\text{s}^2/\text{ft})(\text{ft/s})^2 = \text{lb}\cdot\text{ft}$$

13.4. Applications of the Principle of Work and Energy. The application of the principle of work and energy greatly simplifies the solution of many problems involving forces, displacements, and velocities. Consider, for example, the pendulum OA consisting of a bob A of weight W attached to a cord of length l (Fig. 13.8*a*). The pendulum is released with no initial velocity from a horizontal position OA_1 and allowed to swing in a vertical plane. We wish to determine the speed of the bob as it passes through A_2, directly under O.

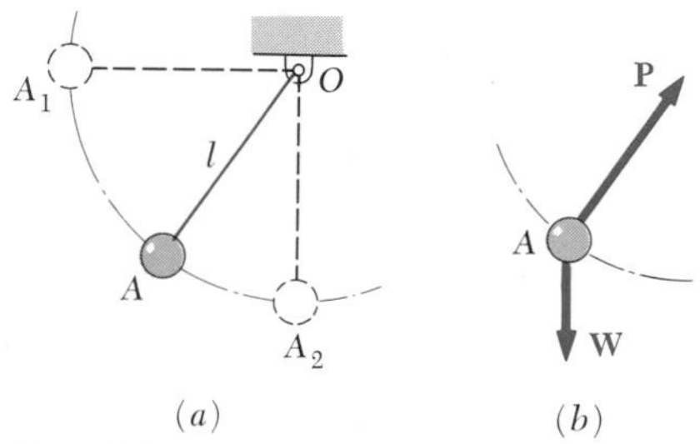

Fig. 13.8

We first determine the work done during the displacement from A_1 to A_2 by the forces acting on the bob. We draw a free-body diagram of the bob, showing all the *actual* forces acting on it, i.e., the weight **W** and the force **P** exerted by the cord (Fig. 13.8*b*). (An inertia vector is not an actual force and *should not* be included in the free-body diagram.) We note that the force **P** does no work, since it is normal to the path; the only force which does work is thus the weight **W.** The work of **W** is obtained by multiplying its magnitude W by the vertical displacement l (Sec. 13.2); since the displacement is downward, the work is positive. We therefore write $U_{1\to2} = Wl$.

Considering, now, the kinetic energy of the bob, we find $T_1 = 0$ at A_1 and $T_2 = \frac{1}{2}(W/g)v_2^2$ at A_2. We may now apply the principle of work and energy; recalling formula (13.11), we write

$$T_1 + U_{1\to2} = T_2 \qquad 0 + Wl = \frac{1}{2}\frac{W}{g}v_2^2$$

Solving for v_2, we find $v_2 = \sqrt{2gl}$. We note that the speed obtained is that of a body falling freely from a height l.

The example we have considered illustrates the following advantages of the method of work and energy:

1. In order to find the speed at A_2, there is no need to determine the acceleration in an intermediate position A and to integrate the expression obtained from A_1 to A_2.
2. All quantities involved are scalars and may be added directly, without using x and y components.
3. Forces which do no work are eliminated from the solution of the problem.

What is an advantage in one problem, however, may become a disadvantage in another. It is evident, for instance, that the method of work and energy cannot be used to directly determine an acceleration. We also note that it should be supplemented by the direct application of Newton's second law in order to determine a force which is normal to the path of the particle, since such a force does no work. Suppose, for example, that we wish to determine the tension in the cord of the pendulum of Fig. 13.8a as the bob passes through A_2. We draw a free-body diagram of the bob in that position (Fig. 13.9) and express

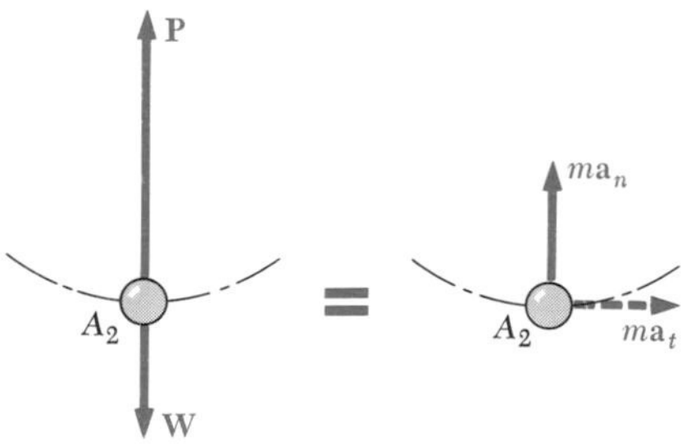

Fig. 13.9

Newton's second law in terms of tangential and normal components. The equations $\Sigma F_t = ma_t$ and $\Sigma F_n = ma_n$ yield, respectively, $a_t = 0$ and

$$P - W = ma_n = \frac{W}{g}\frac{v^2}{l}$$

But the speed at A_2 was determined earlier by the method of work and energy. Substituting $v^2 = 2gl$ and solving for P, we write

$$P = W + \frac{W}{g}\frac{2gl}{l} = 3W$$

When a problem involves two particles or more, each particle may be considered separately and the principle of work and energy may be applied to each particle. Adding the kinetic energies of the various particles, and considering the work of all the forces acting on them, we may also write a single equation of work and energy for all the particles involved. We have

$$T_1 + U_{1\to 2} = T_2 \tag{13.11}$$

where T represents the arithmetic sum of the kinetic energies of the particles involved (all terms are positive) and $U_{1\to 2}$ the work of all the forces acting on the particles, *including the forces of action and reaction exerted by the particles on each other.* In problems involving bodies connected by *inextensible cords or links*, however, the work of the forces exerted by a given cord or link on the two bodies it connects cancels out since the points of application of these forces move through equal distances (see Sample Prob. 13.2).†

13.5. Power and Efficiency. *Power* is defined as the time rate at which work is done. In the selection of a motor or engine, power is a much more important criterion than the actual amount of work to be performed. A small motor or a large power plant may both be used to do a given amount of work; but the small motor may require a month to do the work done by the power plant in a matter of minutes. If ΔU is the work done during the time interval Δt, then the average power during this time interval is

$$\text{Average power} = \frac{\Delta U}{\Delta t}$$

Letting Δt approach zero, we obtain at the limit

$$\text{Power} = \frac{dU}{dt} \tag{13.12}$$

Substituting the scalar product $\mathbf{F} \cdot d\mathbf{r}$ for dU, we may also write

$$\text{Power} = \frac{dU}{dt} = \frac{\mathbf{F} \cdot d\mathbf{r}}{dt}$$

and, recalling that $d\mathbf{r}/dt$ represents the velocity $\mathbf{v}$ of the point of application of $\mathbf{F}$,

$$\text{Power} = \mathbf{F} \cdot \mathbf{v} \tag{13.13}$$

† The application of the method of work and energy to a system of particles is discussed in detail in Chap. 14.

Since power was defined as the time rate at which work is done, it should be expressed in units obtained by dividing units of work by the unit of time. Thus, if SI units are used, power should be expressed in J/s; this unit is called a *watt* (W). We have

$$1\ \mathrm{W} = 1\ \mathrm{J/s} = 1\ \mathrm{N \cdot m/s}$$

If U.S. customary units are used, power should be expressed in ft · lb/s or in *horsepower* (hp), with the latter defined as

$$1\ \mathrm{hp} = 550\ \mathrm{ft \cdot lb/s}$$

Recalling from Sec. 13.2 that 1 ft · lb = 1.356 J, we verify that

$$1\ \mathrm{ft \cdot lb/s} = 1.356\ \mathrm{J/s} = 1.356\ \mathrm{W}$$
$$1\ \mathrm{hp} = 550(1.356\ \mathrm{W}) = 746\ \mathrm{W} = 0.746\ \mathrm{kW}$$

The *mechanical efficiency* of a machine was defined in Sec. 10.4 as the ratio of the output work to the input work:

$$\eta = \frac{\text{output work}}{\text{input work}} \tag{13.14}$$

This definition is based on the assumption that work is done at a constant rate. The ratio of the output to the input work is therefore equal to the ratio of the rates at which output and input work are done, and we have

$$\eta = \frac{\text{power output}}{\text{power input}} \tag{13.15}$$

Because of energy losses due to friction, the output work is always smaller than the input work, and, consequently, the power output is always smaller than the power input. The mechanical efficiency of a machine, therefore, is always less than 1.

When a machine is used to transform mechanical energy into electric energy, or thermal energy into mechanical energy, its *overall efficiency* may be obtained from formula (13.15). The overall efficiency of a machine is always less than 1; it provides a measure of all the various energy losses involved (losses of electric or thermal energy as well as frictional losses). We should note that it is necessary, before using formula (13.15), to express the power output and the power input in the same units.

SAMPLE PROBLEM 13.1

An automobile weighing 4000 lb is driven down a 5° incline at a speed of 60 mi/h when the brakes are applied, causing a constant total braking force (applied by the road on the tires) of 1500 lb. Determine the distance traveled by the automobile as it comes to a stop.

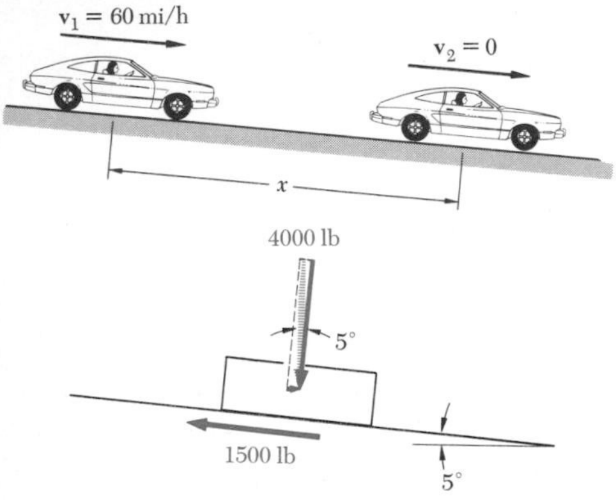

Solution. ***Kinetic Energy***

Position *1*:

$$v_1 = \left(60\frac{\text{mi}}{\text{h}}\right)\left(\frac{5280\text{ ft}}{1\text{ mi}}\right)\left(\frac{1\text{ h}}{3600\text{ s}}\right) = 88\text{ ft/s}$$

$$T_1 = \tfrac{1}{2}mv_1^2 = \tfrac{1}{2}(4000/32.2)(88)^2 = 481{,}000\text{ ft}\cdot\text{lb}$$

Position *2*:

$$v_2 = 0 \qquad T_2 = 0$$

Work $\quad U_{1\to2} = -1500x + (4000\sin 5°)x = -1151x$

Principle of Work and Energy

$$T_1 + U_{1\to2} = T_2$$

$$481{,}000 - 1151x = 0 \qquad x = 418\text{ ft} \blacktriangleleft$$

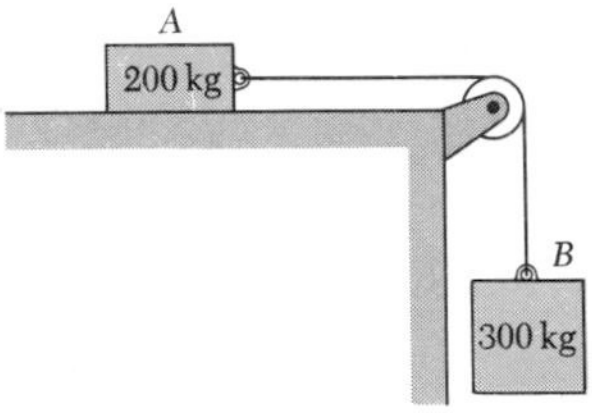

SAMPLE PROBLEM 13.2

Two blocks are joined by an inextensible cable as shown. If the system is released from rest, determine the velocity of block *A* after it has moved 2 m. Assume that μ equals 0.25 between block *A* and the plane and that the pulley is weightless and frictionless.

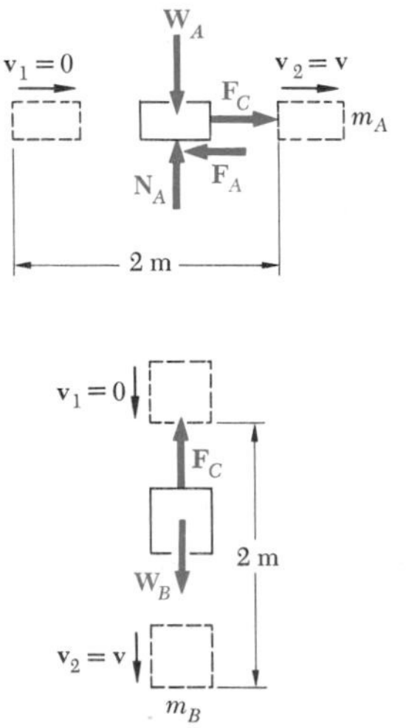

Solution. ***Work and Energy for Block A.*** We denote by $\mathbf{F}_A$ the friction force, by $\mathbf{F}_C$ the force exerted by the cable, and write

$$m_A = 200\text{ kg} \qquad W_A = (200\text{ kg})(9.81\text{ m/s}^2) = 1962\text{ N}$$

$$F_A = \mu N_A = \mu W_A = 0.25(1962\text{ N}) = 490\text{ N}$$

$$T_1 + U_{1\to2} = T_2: \qquad 0 + F_C(2\text{ m}) - F_A(2\text{ m}) = \tfrac{1}{2}m_A v^2$$

$$F_C(2\text{ m}) - (490\text{ N})(2\text{ m}) = \tfrac{1}{2}(200\text{ kg})v^2 \qquad (1)$$

Work and Energy for Block B. We write

$$m_B = 300\text{ kg} \qquad W_B = (300\text{ kg})(9.81\text{ m/s}^2) = 2940\text{ N}$$

$$T_1 + U_{1\to2} = T_2: \qquad 0 + W_B(2\text{ m}) - F_C(2\text{ m}) = \tfrac{1}{2}m_B v^2$$

$$(2940\text{ N})(2\text{ m}) - F_C(2\text{ m}) = \tfrac{1}{2}(300\text{ kg})v^2 \qquad (2)$$

Adding the left-hand and right-hand members of (1) and (2), we observe that the work of the forces exerted by the cable on *A* and *B* cancels out:

$$(2940\text{ N})(2\text{ m}) - (490\text{ N})(2\text{ m}) = \tfrac{1}{2}(200\text{ kg} + 300\text{ kg})v^2$$

$$4900\text{ J} = \tfrac{1}{2}(500\text{ kg})v^2 \qquad v = 4.43\text{ m/s} \blacktriangleleft$$

SAMPLE PROBLEM 13.3

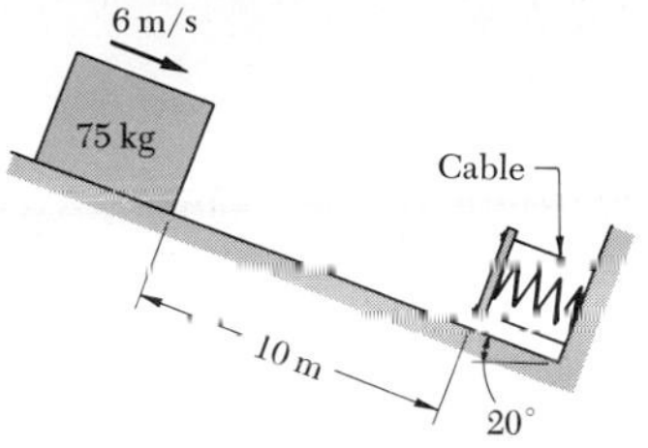

A spring is used to stop a 75-kg package which is moving down a 20° incline. The spring has a constant $k = 25$ kN/m, and is held by cables so that it is initially compressed 100 mm. If the velocity of the package is 6 m/s when it is 10 m from the spring, determine the maximum additional deformation of the spring in bringing the package to rest. Assume $\mu = 0.20$.

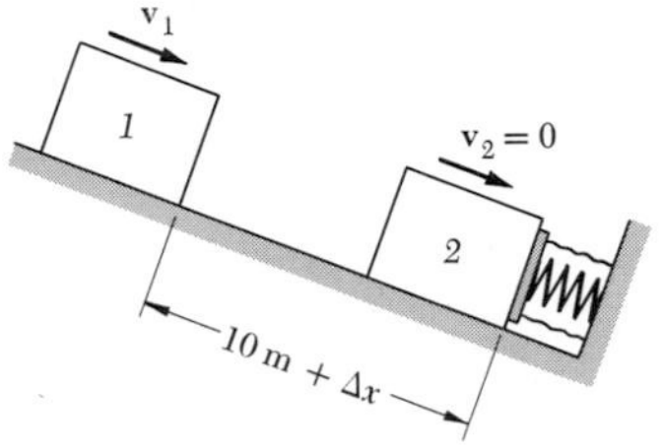

Kinetic Energy

Position *1*: $\qquad v_1 = 6\ \text{m/s}$

$$T_1 = \tfrac{1}{2}mv_1^2 = \tfrac{1}{2}(75\ \text{kg})(6\ \text{m/s})^2 = 1350\ \text{N}\cdot\text{m} = 1350\ \text{J}$$

Position *2* (maximum spring deformation):

$$v_2 = 0 \qquad T_2 = 0$$

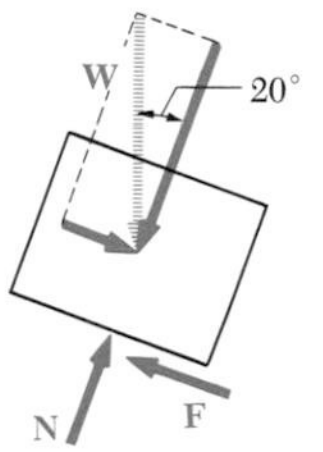

Work. We assume that when the package is brought to rest the additional deflection of the spring is Δx. The component of the weight parallel to the plane and the friction force act through the entire displacement, i.e., through $10\ \text{m} + \Delta x$. The total work done by these forces is

$$\begin{aligned} U_{1\to2} &= W_t(10\ \text{m} + \Delta x) - F(10\ \text{m} + \Delta x) \\ &= (W \sin 20°)(10\ \text{m} + \Delta x) - 0.20(W \cos 20°)(10\ \text{m} + \Delta x) \\ &= 0.1541W(10\ \text{m} + \Delta x) \end{aligned}$$

or, since $W = mg = (75\ \text{kg})(9.81\ \text{m/s}^2) = 736\ \text{N}$,

$$U_{1\to2} = 0.1541(736\ \text{N})(10\ \text{m} + \Delta x) = 1134\ \text{J} + (113.4\ \text{N})\,\Delta x$$

In addition, during the compression of the spring, the variable force **P** exerted by the spring does an amount of negative work equal to the area under the force-deflection curve of the spring force.

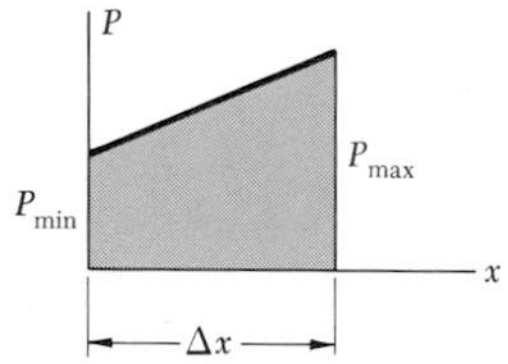

$$P_{min} = kx = (25\ \text{kN/m})(100\ \text{mm}) = (25\,000\ \text{N/m})(0.100\ \text{m}) = 2500\ \text{N}$$

$$\begin{aligned} P_{max} &= P_{min} + k\,\Delta x \\ &= 2500\ \text{N} + (25\,000\ \text{N/m})\,\Delta x \end{aligned}$$

$$U_{1\to2} = -\tfrac{1}{2}(P_{min} + P_{max})\,\Delta x = -(2500\ \text{N})\,\Delta x - (12\,500\ \text{N/m})(\Delta x)^2$$

The total work is thus

$$U_{1\to2} = 1134\ \text{J} + (113.4\ \text{N})\,\Delta x - (2500\ \text{N})\,\Delta x - (12\,500\ \text{N/m})(\Delta x)^2$$

Principle of Work and Energy

$$T_1 + U_{1\to2} = T_2$$

$$1350 + 1134 + 113.4\,\Delta x - 2500\,\Delta x - 12\,500(\Delta x)^2 = 0$$

$$(\Delta x)^2 + 0.1909\,\Delta x - 0.1987 = 0$$

$$\Delta x = 0.360\ \text{m} \qquad \Delta x = 360\ \text{mm} \blacktriangleleft$$

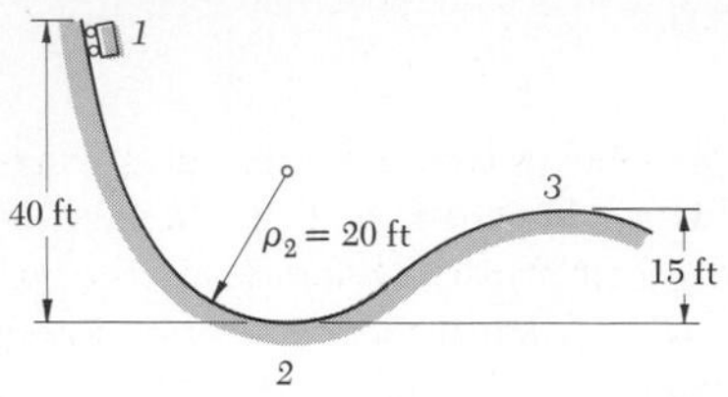

SAMPLE PROBLEM 13.4

A 2000-lb car starts from rest at point *1* and moves without friction down the track shown. (*a*) Determine the force exerted by the track on the car at point *2*, where the radius of curvature of the track is 20 ft. (*b*) Determine the minimum safe value of the radius of curvature at point *3*.

***a.* Force Exerted by the Track at Point *2*.** The principle of work and energy is used to determine the velocity of the car as it passes through point *2*.

Kinetic Energy: $\quad T_1 = 0 \qquad T_2 = \frac{1}{2}mv_2^2 = \frac{1}{2}\frac{W}{g}v_2^2$

Work. The only force which does work is the weight **W**. Since the vertical displacement from point *1* to point *2* is 40 ft downward, the work of the weight is

$$U_{1\to2} = +W(40\text{ ft})$$

Principle of Work and Energy

$$T_1 + U_{1\to2} = T_2 \qquad 0 + W(40\text{ ft}) = \frac{1}{2}\frac{W}{g}v_2^2$$

$$v_2^2 = 80g = 80(32.2) \qquad v_2 = 50.8\text{ ft/s}$$

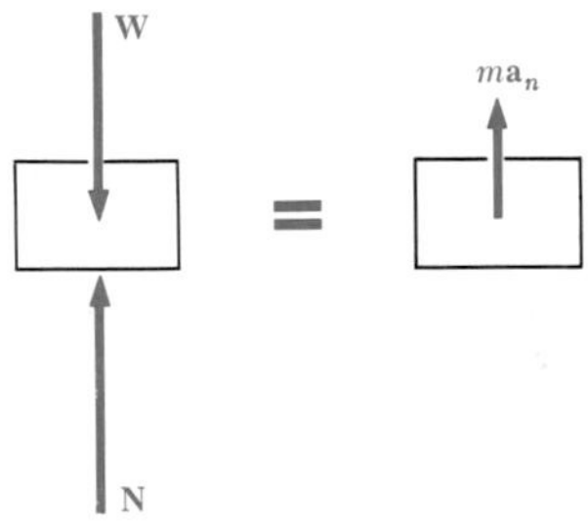

Newton's Second Law at Point 2. The acceleration $\mathbf{a}_n$ of the car at point *2* has a magnitude $a_n = v_2^2/\rho$ and is directed upward. Since the external forces acting on the car are **W** and **N**, we write

$$+\uparrow\Sigma F_n = ma_n: \qquad -W + N = ma_n$$

$$= \frac{W}{g}\frac{v_2^2}{\rho}$$

$$= \frac{W}{g}\frac{80g}{20}$$

$$N = 5W \qquad \mathrm{N} = 10{,}000\text{ lb}\uparrow \quad ◄$$

b.* Minimum Value of ρ at Point *3*.** ***Principle of Work and Energy. Applying the principle of work and energy between point *1* and point *3*, we obtain

$$T_1 + U_{1\to3} = T_3 \qquad 0 + W(25\text{ ft}) = \frac{1}{2}\frac{W}{g}v_3^2$$

$$v_3^2 = 50g = 50(32.2) \qquad v_3 = 40.1\text{ ft/s}$$

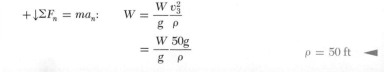

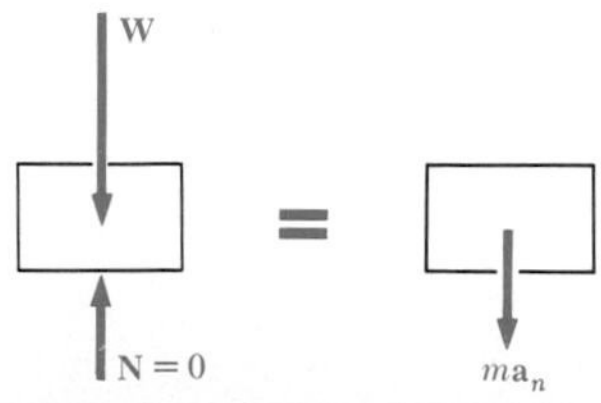

Newton's Second Law at Point 3. The minimum safe value of ρ occurs when $\mathbf{N} = 0$. In this case, the acceleration $\mathbf{a}_n$, of magnitude $a_n = v_3^2/\rho$, is directed downward, and we write

$$+\downarrow\Sigma F_n = ma_n: \qquad W = \frac{W}{g}\frac{v_3^2}{\rho}$$

$$= \frac{W}{g}\frac{50g}{\rho} \qquad \rho = 50\text{ ft} \quad ◄$$

SAMPLE PROBLEM 13.5

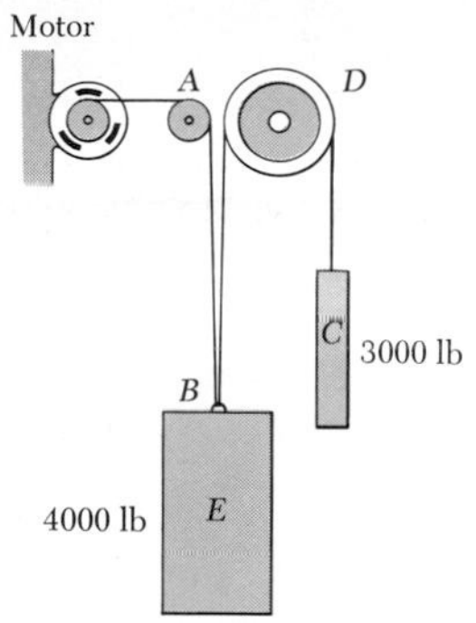

The elevator shown weighs 4000 lb when fully loaded. It is connected to a 3000-lb counterweight C and is powered by an electric motor. Determine the power required when the elevator (a) is moving upward at a constant speed of 20 ft/s, (b) has an instantaneous velocity of 20 ft/s upward and an upward acceleration of 3 ft/s².

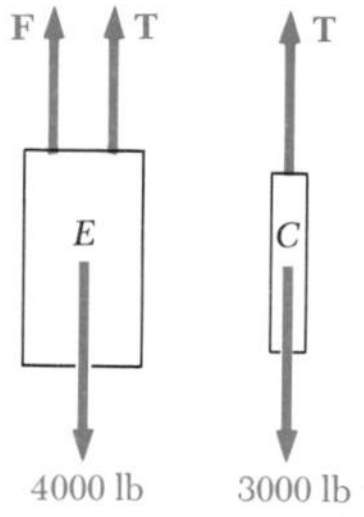

Solution. Since **F** and **v** have the same direction, the power is equal to Fv. We must first determine the force **F** exerted by cable AB on the elevator in each of the two given situations.

Force **F.** The forces acting on the elevator and on the counterweight are shown in the adjoining sketches.

a. Uniform Motion. We have $a = 0$; both bodies are in equilibrium.

Free Body C: $\quad +\uparrow\Sigma F_y = 0: \qquad T - 3000\text{ lb} = 0$

Free Body E: $\quad +\uparrow\Sigma F_y = 0: \qquad F + T - 4000\text{ lb} = 0$

Eliminating T, we find: $\quad F = 1000\text{ lb}$

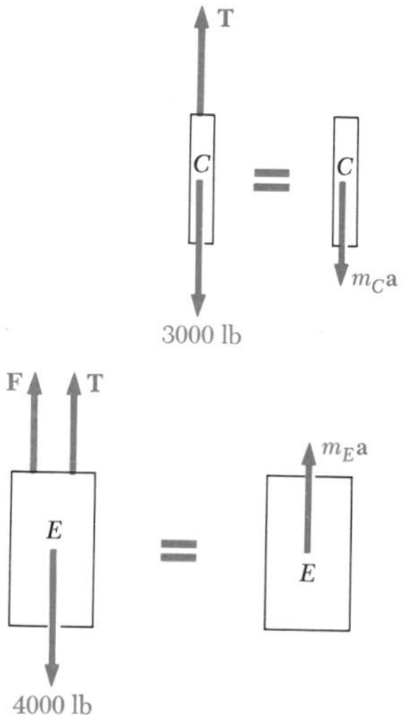

b. Accelerated Motion. We have $a = 3\text{ ft/s}^2$. The equations of motion are

Free Body C: $\quad +\downarrow\Sigma F_y = m_C a: \qquad 3000 - T = \dfrac{3000}{32.2}3$

Free Body E: $\quad +\uparrow\Sigma F_y = m_E a: \qquad F + T - 4000 = \dfrac{4000}{32.2}3$

Eliminating T: $\quad F = 1000 + \dfrac{7000}{32.2}3 = 1652\text{ lb}$

Power. Substituting the given values of v and the values found for F into the expression for the power, we have

a. $Fv = (1000\text{ lb})(20\text{ ft/s}) = 20{,}000\text{ ft}\cdot\text{lb/s}$

$$\text{Power} = (20{,}000\text{ ft}\cdot\text{lb/s})\frac{1\text{ hp}}{550\text{ ft}\cdot\text{lb/s}} = 36.4\text{ hp} \quad ◄$$

b. $Fv = (1652\text{ lb})(20\text{ ft/s}) = 33{,}040\text{ ft}\cdot\text{lb/s}$

$$\text{Power} = \frac{33{,}040}{550} = 60.1\text{ hp} \quad ◄$$

PROBLEMS

13.1 A stone which weighs 8 lb is dropped from a height h and strikes the ground with a velocity of 75 ft/s. (*a*) Find the kinetic energy of the stone as it strikes the ground and the height h from which it was dropped. (*b*) Solve part *a*, assuming that the same stone is dropped on the moon. (Acceleration of gravity on the moon = 5.31 ft/s^2.)

13.2 A 100-kg satellite was placed in a circular orbit 2000 km above the surface of the earth. At this elevation the acceleration of gravity is 5.68 m/s^2. Determine the kinetic energy of the satellite, knowing that its orbital speed is 24.8×10^3 km/h.

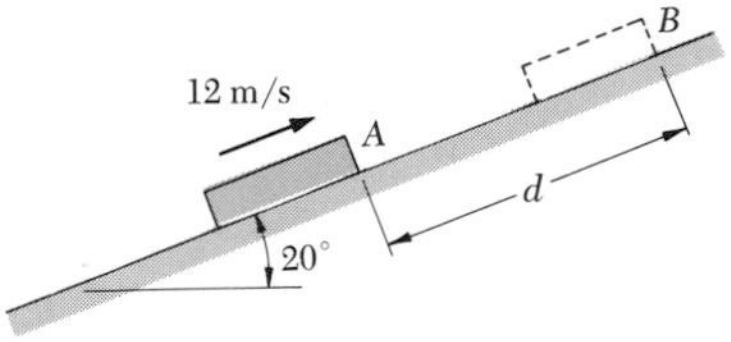

Fig. P13.3

13.3 A 20-kg package is projected up a 20° incline with an initial velocity of 12 m/s. The coefficient of friction between the incline and the package is 0.15. Determine (*a*) the maximum distance that the package will move up the incline, (*b*) the velocity of the package when it returns to its original position.

13.4 Using the method of work and energy, solve Prob. 12.8.

13.5 The conveyor belt shown moves at a constant speed v_0 and discharges packages on to the chute *AB*. The coefficient of friction between the packages and the chute is 0.50. Knowing that the packages must reach point *B* with a speed of 12 ft/s, determine the required speed v_0 of the conveyor belt.

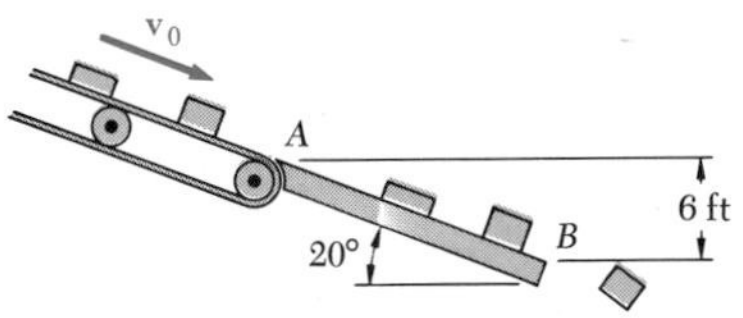

Fig. P13.5

13.6 Solve Prob. 13.5, assuming that the coefficient of friction between the packages and the chute is 0.30.

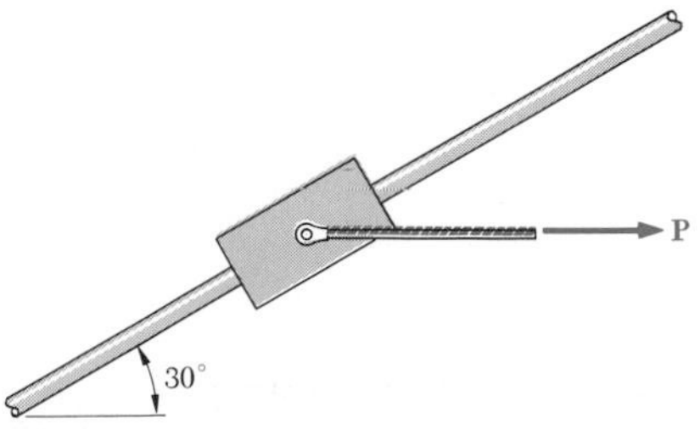

Fig. P13.7

13.7 The 2-kg collar was moving down the rod with a velocity of 3 m/s when a force **P** was applied to the horizontal cable. Assuming negligible friction between the collar and the rod, determine the magnitude of the force **P** if the collar stopped after moving 1.2 m more down the rod.

13.8 Solve Prob. 13.7, assuming a coefficient of friction of 0.20 between the collar and the rod.

13.9 Knowing that the system shown is initially at rest and neglecting the effect of friction, determine the force **P** required if the velocity of collar B is to be 8 ft/s after it has moved 2.5 ft to the right.

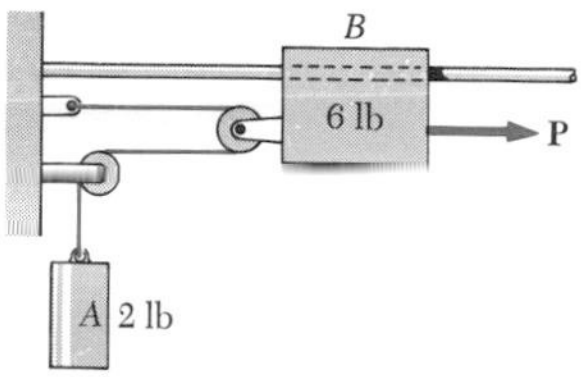

Fig. P13.9

13.10 The system shown is at rest when the 20-lb force is applied to block A. Neglecting the effect of friction, determine the velocity of block A after it has moved 9 ft.

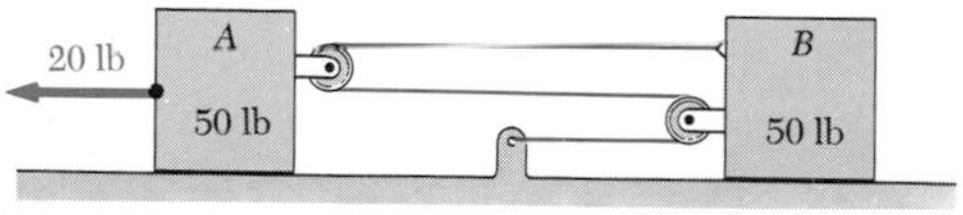

Fig. P13.10

13.11 Solve Prob. 13.10, assuming that the coefficient of friction between the blocks and the horizontal plane is 0.20.

13.12 Three 20-kg packages rest on a belt which passes over a pulley and is attached to a 40-kg block. Knowing that the coefficient of friction between the belt and the horizontal surface and also between the belt and the packages is 0.50, determine the speed of package B as it falls off the belt at E.

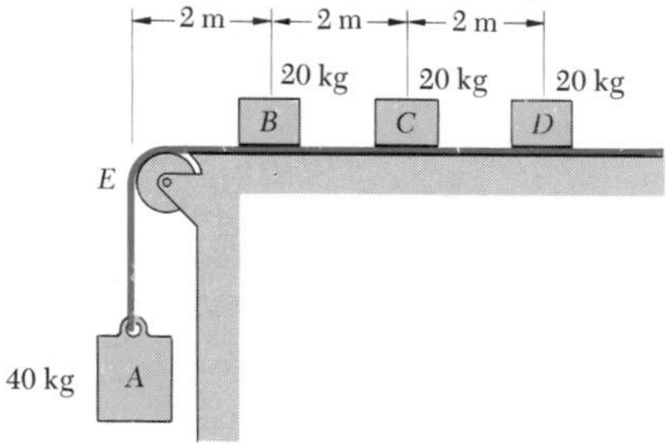

Fig. P13.12

13.13 In Prob. 13.12, determine the speed of package C as it falls off the belt at E.

13.14 Two cylinders are suspended from an inextensible cable as shown. If the system is released from rest, determine (*a*) the maximum velocity attained by the 10-lb cylinder, (*b*) the maximum height above the floor to which the 10-lb cylinder will rise.

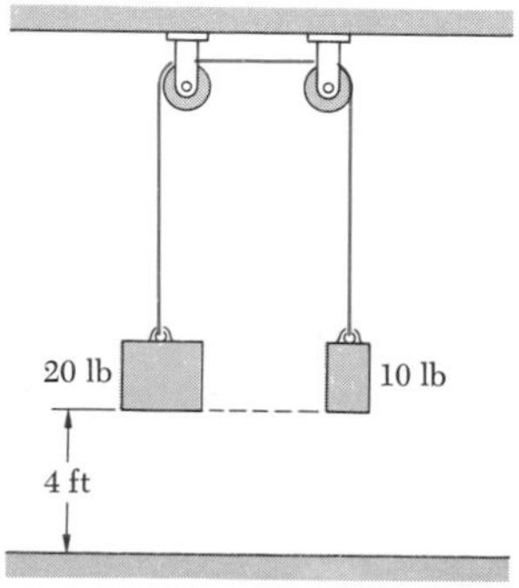

Fig. P13.14

13.15 Solve Prob. 13.14, assuming that the 20-lb cylinder is replaced by a 50-lb cylinder.

13.16 Four packages weighing 50 lb each are placed as shown on a conveyor belt which is disengaged from its drive motor. Package *1* is just to the right of the horizontal portion of the belt. If the system is released from rest, determine the velocity of package *1* as it falls off the belt at point *A*. Assume that the weight of the belt and rollers is small compared to the weight of the packages.

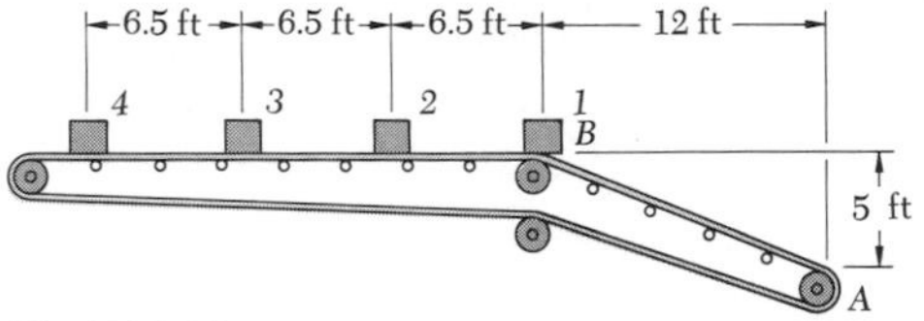

Fig. P13.16

13.17 In Prob. 13.16, determine the velocity of package *2* as it falls from the belt at *A*.

13.18 Using the method of work and energy, solve Prob. 12.16.

13.19 Using the method of work and energy, solve Prob. 12.15.

13.20 Using the method of work and energy, solve Prob. 12.18*c*.

13.21 Using the method of work and energy, solve Prob. 12.17*c*.

13.22 In order to protect it during shipping, a delicate instrument weighing 4 oz is packed in excelsior. From the static test of similar excelsior, the force-deflection curve shown was obtained. Determine the maximum height from which the package may be dropped if the force exerted on the instrument is not to exceed 12 lb.

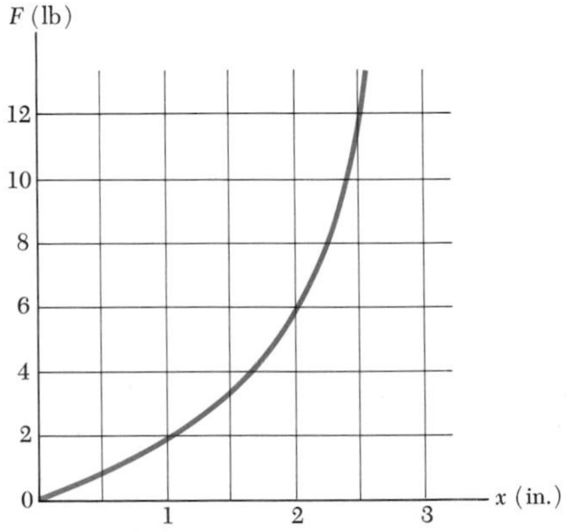

Fig. P13.22

13.23 A 5000-kg airplane lands on an aircraft carrier and is caught by an arresting cable which is characterized by the force-deflection diagram shown. Knowing that the landing speed of the plane is 144 km/h, determine (*a*) the distance required for the plane to come to rest, (*b*) the maximum rate of deceleration of the plane.

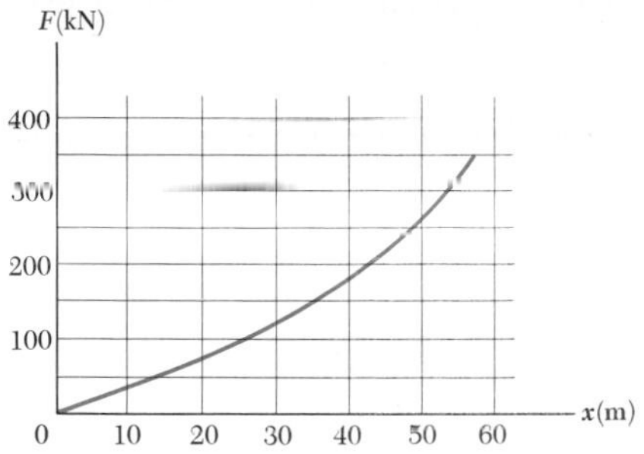

Fig. P13.23

13.24 A 2-kg block is at rest on a spring of constant 400 N/m. A 4-kg block is held above the 2-kg block so that it just touches it, and released. Determine (*a*) the maximum velocity attained by the blocks, (*b*) the maximum force exerted on the blocks.

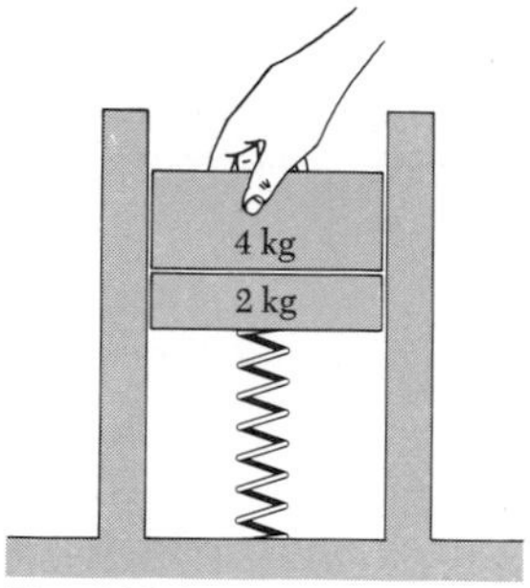

Fig. P13.24

13.25 As the bracket *ABC* is slowly rotated, the 6-kg block starts to slide toward the spring when $\theta = 15°$. The maximum deflection of the spring is observed to be 50 mm. Determine the values of the coefficients of static and kinetic friction.

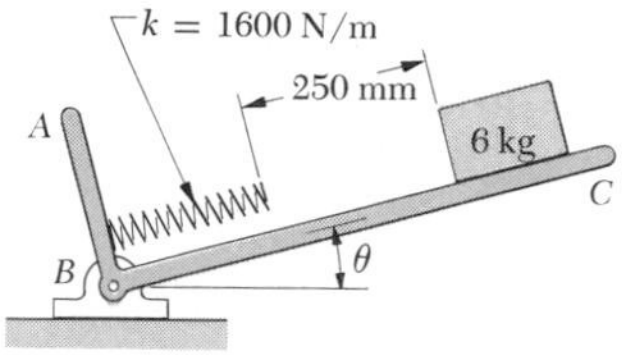

Fig. P13.25

13.26 A 15-lb plunger is released from rest in the position shown and is stopped by two nested springs; the stiffness of the outer spring is 20 lb/in. and the stiffness of the inner spring is 60 lb/in. If the maximum deflection of the outer spring is observed to be 5 in., determine the height *h* from which the plunger was released.

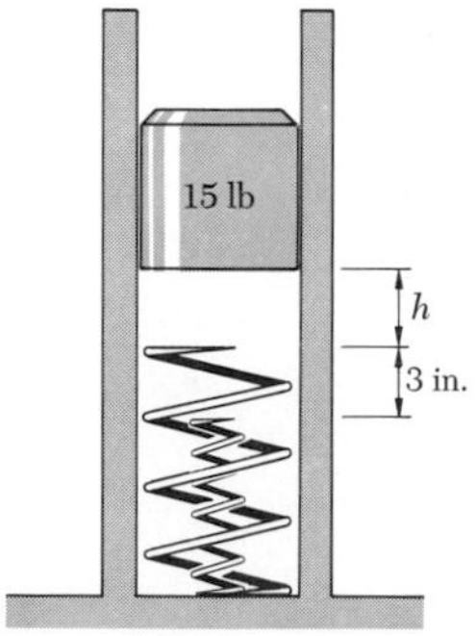

Fig. P13.26

13.27 A railroad car weighing 60,000 lb starts from rest and coasts down a 1-percent incline for a distance of 40 ft. It is stopped by a bumper having a spring constant of 7500 lb/in. (*a*) What is the speed of the car at the bottom of the incline? (*b*) How many inches will the spring be compressed?

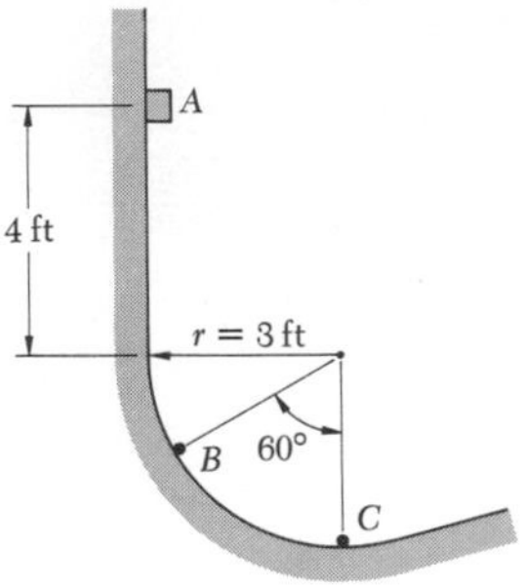

Fig. P13.28

13.28 A 0.5-lb pellet is released from rest at *A* and slides without friction along the surface shown. Determine the force exerted by the surface on the pellet as it passes (*a*) point *B*, (*b*) point *C*.

13.29 A roller coaster is released with no velocity at *A* and rolls down the track shown. The brakes are suddenly applied as the car passes through point *B*, causing the wheels of the car to slide on the track ($\mu = 0.30$). Assuming no energy loss between *A* and *B* and knowing that the radius of curvature of the track at *B* is 80 ft, determine the normal and tangential components of the acceleration of the car just after the brakes have been applied.

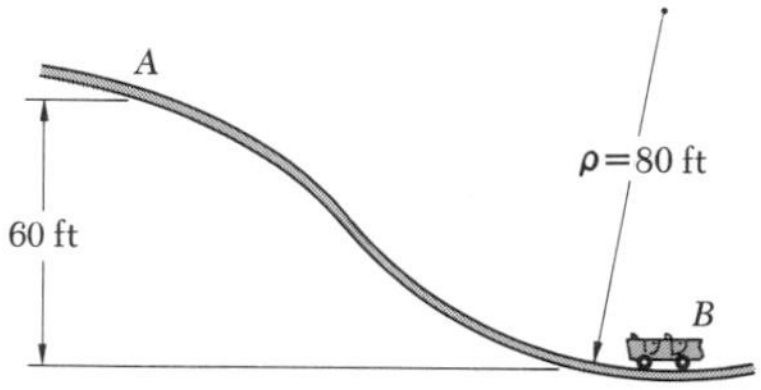

Fig. P13.29

13.30 A small package of mass *m* is projected into a vertical return loop at *A* with a velocity $\mathbf{v}_0$. The package travels without friction along a circle of radius *r* and is deposited on a horizontal surface at *C*. For each of the two loops shown, determine (*a*) the smallest velocity $\mathbf{v}_0$ for which the package will reach the horizontal surface at *C*, (*b*) the corresponding force exerted by the loop on the package as it passes point *B*.

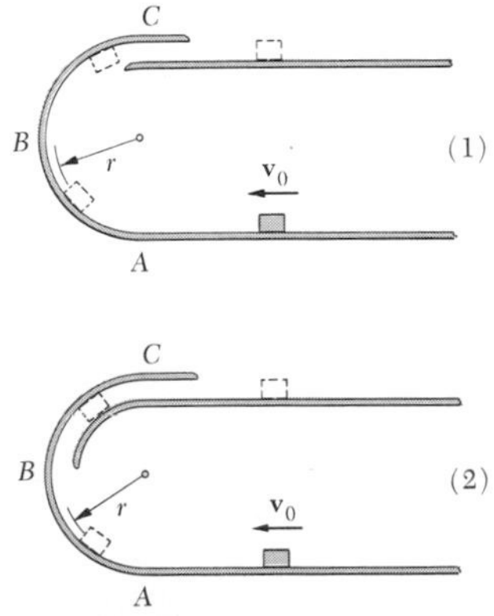

Fig. P13.30

13.31 In Prob. 13.30, it is desired to have the package deposited on the horizontal surface at C with a speed of 2 m/s. Knowing that $r = 0.6$ m, (*a*) show that this requirement cannot be fulfilled by the first loop, (*b*) determine the required initial velocity $\mathbf{v}_0$ when the second loop is used.

13.32 A 6-in.-diameter piston weighing 8 lb slides without friction in a cylinder. When the piston is at a distance $x = 10$ in. from the end of the cylinder, the pressure in the cylinder is atmospheric ($p_a = 14.7$ lb/in^2). If the pressure varies inversely as the volume, find the work done in moving the piston until $x = 4$ in.

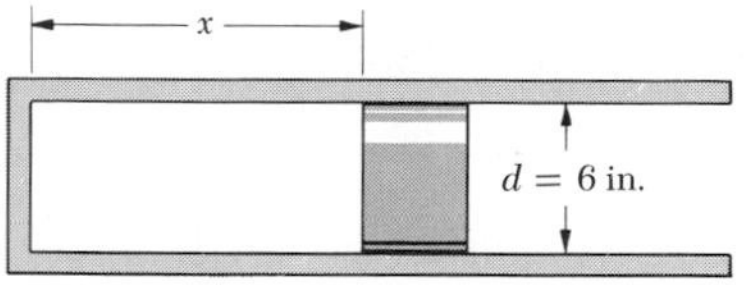

Fig. P13.32

13.33 The piston of Prob. 13.32 is moved to the left and released with no velocity when $x = 4$ in. Neglecting friction, determine (*a*) the maximum velocity attained by the piston, (*b*) the maximum value of the coordinate x.

13.34 An object is released with no velocity at an altitude equal to the radius of the earth. Neglecting air resistance, determine the velocity of the object as it strikes the earth. Give the answer in both SI and U.S. customary units.

13.35 A rocket is fired vertically from the ground. Knowing that at burnout the rocket is 80 km above the ground and has a velocity of 5000 m/s, determine the highest altitude it will reach.

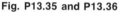

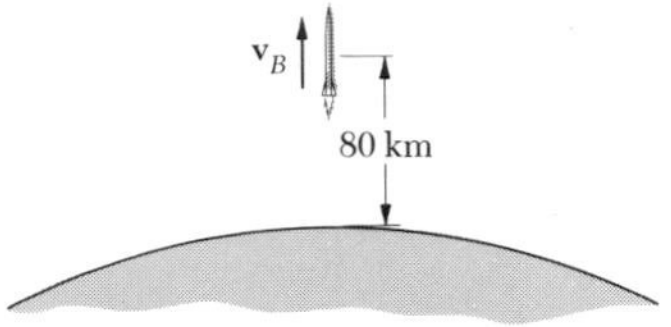

Fig. P13.35 and P13.36

13.36 A rocket is fired vertically from the ground. What should be its velocity $\mathbf{v}_B$ at burnout, 80 km above the ground, if it is to reach an altitude of 1000 km?

13.37 An object is released with no initial velocity at an altitude of 400 mi. (*a*) Neglecting air resistance, determine the velocity of the object as it strikes the ground. (*b*) What percent error is introduced by assuming a uniform gravitational field?

13.38 A 70-kg man and an 80-kg man run up a flight of stairs in 5 s. If the flight of stairs is 4 m high, determine the average power required by each man.

13.39 An industrial hoist can lift its maximum allowable load of 60,000 lb at the rate of 4 ft/min. Knowing that the hoist is run by a 15-hp motor, determine the overall efficiency of the hoist.

13.40 A 1500-kg automobile travels 200 m while being accelerated at a uniform rate from 50 to 75 km/h. During the entire motion, the automobile is traveling on a horizontal road, and the rolling resistance is equal to 2 percent of the weight of the automobile. Determine (*a*) the maximum power required, (*b*) the power required to maintain a constant speed of 75 km/h.

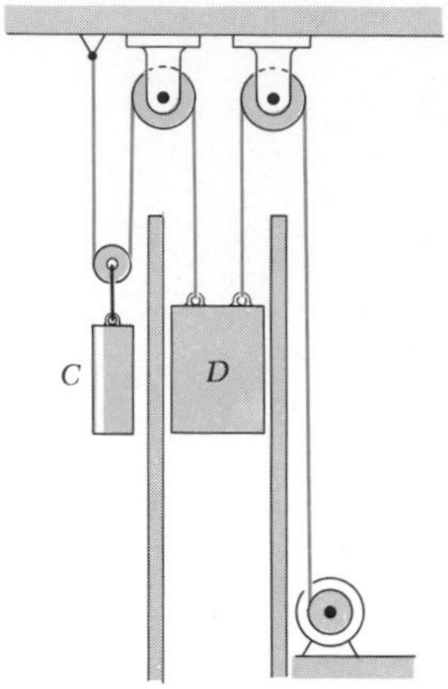

Fig. P13.42 and P13.43

13.41 A train of total weight 600 tons starts from rest and accelerates uniformly to a speed of 30 mi/h in 40 sec. After reaching this speed, the train travels with constant velocity. During the entire motion the train is traveling up a 2 percent grade, and the rolling resistance is 15 lb/ton. Determine the power required as a function of time.

13.42 The dumbwaiter D and its counterweight C weigh 750 lb each. Determine the power required when the dumbwaiter (*a*) is moving upward at a constant speed of 12 ft/s, (*b*) has an instantaneous velocity of 12 ft/s upward and an upward acceleration of 3 ft/s^2.

13.43 The dumbwaiter D and its counterweight C weigh 750 lb each. Knowing that the motor is delivering to the system 9 hp at the instant the speed of the dumbwaiter is 12 ft/s upward, determine the acceleration of the dumbwaiter.

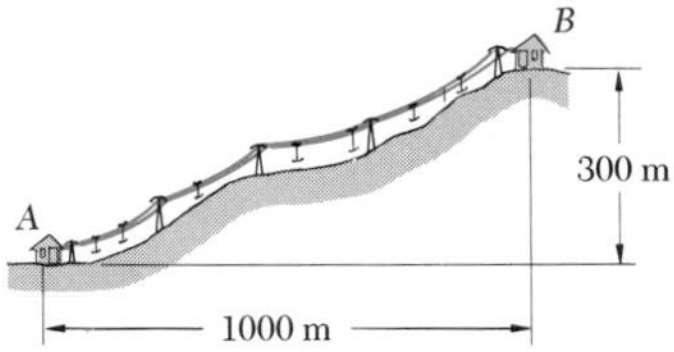

Fig. P13.44

13.44 A chair-lift is designed to transport 900 skiers per hour from the base A to the summit B. The average mass of a skier is 75 kg, and the average speed of the lift is 80 m/min. Determine (*a*) the average power required, (*b*) the required capacity of the motor if the mechanical efficiency is 85 percent and if a 300-percent overload is to be allowed.

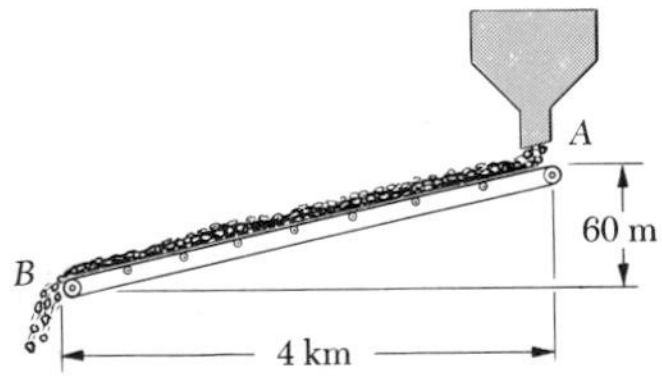

Fig. P13.45

13.45 Crushed stone is moved from a quarry at A to a construction site at B at the rate of 2000 Mg per 8-h period. An electric generator is attached to the system in order to maintain a constant belt speed. Knowing that the efficiency of the belt-generator system is 0.65, determine the average power developed by the generator (*a*) if the belt speed is 0.75 m/s, (*b*) if the belt speed is 2 m/s.

13.46 The fluid transmission of a 15-ton truck permits the engine to deliver an essentially constant power of 60 hp to the driving wheels. Determine the time required and the distance traveled as the speed of the truck is increased (*a*) from 15 to 30 mi/h, (*b*) from 30 to 45 mi/h.

13.47 The fluid transmission of a truck of mass m permits the engine to deliver an essentially constant power P to the driving wheels. Determine the time elapsed and the distance traveled as the speed is increased from v_0 to v_1.

13.48 The frictional resistance of a ship is known to vary directly as the 1.75 power of the speed v of the ship. A single tugboat at full power can tow the ship at a constant speed of 5 km/h by exerting a constant force of 200 kN. Determine (*a*) the power developed by the tugboat, (*b*) the maximum speed at which two tugboats, capable of delivering the same power, can tow the ship.

13.49 Determine the speed at which the single tugboat of Prob. 13.48 will tow the ship if the tugboat is developing half of its maximum power.

13.6. Potential Energy.†

Let us consider again a body of weight **W** which moves along a curved path from a point A_1 of elevation y_1 to a point A_2 of elevation y_2 (Fig. 13.4). We recall from Sec. 13.2 that the work of the weight **W** during this displacement is

$$U_{1\to 2} = Wy_1 - Wy_2 \qquad (13.4)$$

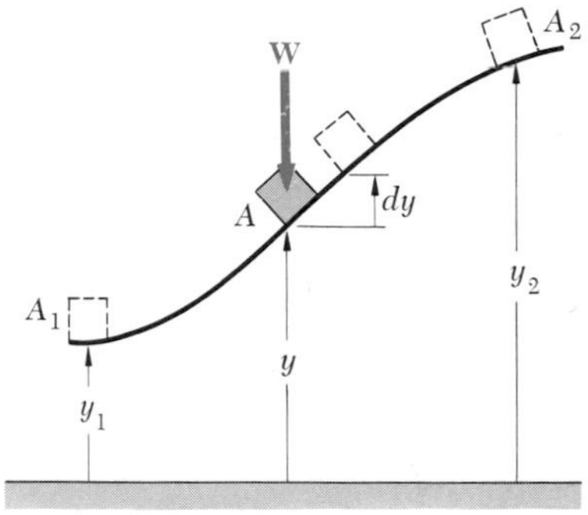

Fig. 13.4 (*repeated*)

The work of **W** may thus be obtained by subtracting the value of the function Wy corresponding to the second position of the body from its value corresponding to the first position. The work of **W** is independent of the actual path followed; it depends only upon the initial and final values of the function Wy. This function is called the *potential energy* of the body with respect to the *force of gravity* **W** and is denoted by V_g. We write

$$U_{1\to 2} = (V_g)_1 - (V_g)_2 \qquad \text{with } V_g = Wy \qquad (13.16)$$

We note that if $(V_g)_2 > (V_g)_1$, i.e., *if the potential energy increases* during the displacement (as in the case considered here), *the work* $U_{1\to 2}$ *is negative.* If, on the other hand, the work of **W** is positive, the potential energy decreases. Therefore, the potential energy V_g of the body provides a measure of the work which may be done by its weight **W**. Since only the *change* in potential energy, and not the actual value of V_g, is involved in formula (13.16), an arbitrary constant may be added to the expression obtained for V_g. In other words, the level, or datum, from which the elevation y is measured may be chosen arbitrarily. Note that potential energy is expressed in the same units as work, i.e., in joules if SI units are used, and in ft · lb or in · lb if U.S. customary units are used.

It should be noted that the expression just obtained for the potential energy of a body with respect to gravity is valid only as long as the weight **W** of the body may be assumed to remain constant, i.e., as long as the displacements of the body are small compared to the radius of the earth. In the case of a space vehicle, however, we should take into consideration the variation of the force of gravity with the distance r from the center of

† Some of the material in this section has already been considered in Sec. 10.6.

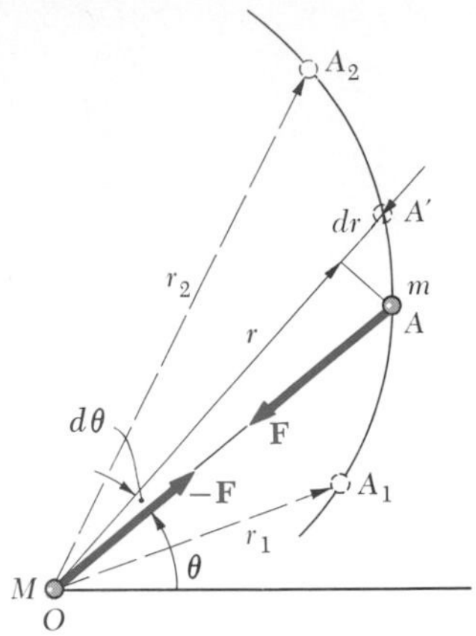

Fig. 13.6 (repeated)

the earth. Using the expression obtained in Sec. 13.2 for the work of a gravitational force, we write (Fig. 13.6)

$$U_{1\to2} = \frac{GMm}{r_2} - \frac{GMm}{r_1} \tag{13.7}$$

The work of the force of gravity may therefore be obtained by subtracting the value of the function $-GMm/r$ corresponding to the second position of the body from its value corresponding to the first position. Thus, the expression which should be used for the potential energy V_g when the variation in the force of gravity cannot be neglected is

$$V_g = -\frac{GMm}{r} \tag{13.17}$$

Taking the first of the relations (12.29) into account, we write V_g in the alternate form

$$V_g = -\frac{WR^2}{r} \tag{13.17$'$}$$

where R is the radius of the earth and W the value of the weight of the body at the surface of the earth. When either of the relations (13.17) and (13.17′) is used to express V_g, the distance r should, of course, be measured from the center of the earth.† Note that V_g is always negative and that it approaches zero for very large values of r.

Consider, now, a body attached to a spring and moving from a position A_1, corresponding to a deflection x_1 of the spring, to a position A_2, corresponding to a deflection x_2 (Fig. 13.5). We recall from Sec. 13.2 that the work of the force **F** exerted by the spring on the body is

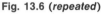

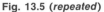

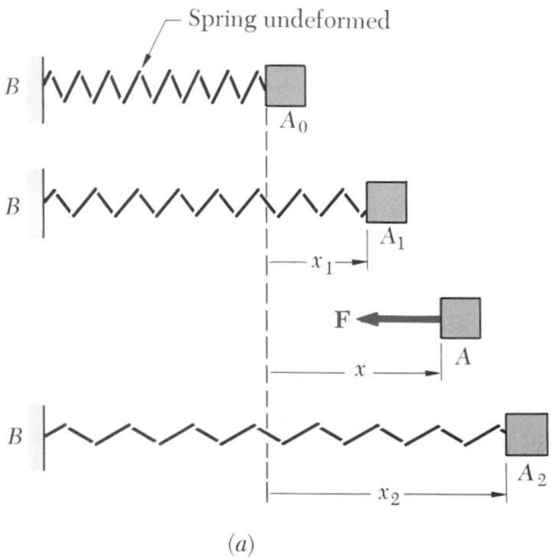

Fig. 13.5 (repeated)

$$U_{1\to2} = \tfrac{1}{2}kx_1^2 - \tfrac{1}{2}kx_2^2 \tag{13.6}$$

The work of the elastic force is thus obtained by subtracting the value of the function $\frac{1}{2}kx^2$ corresponding to the second position of the body from its value corresponding to the first position. This function is denoted by V_e and is called the *potential energy* of the body with respect to the *elastic force* **F**. We write

$$U_{1\to2} = (V_e)_1 - (V_e)_2 \qquad \text{with } V_e = \tfrac{1}{2}kx^2 \tag{13.18}$$

and observe that, during the displacement considered, the work of the force **F** exerted by the spring on the body is negative

† The expressions given for V_g in (13.17) and (13.17′) are valid only when $r \geqq R$, that is, when the body considered is above the surface of the earth.

and the potential energy V_e increases. We should note that the expression obtained for V_e is valid only if the deflection of the spring is measured from its undeformed position. On the other hand, formula (13.18) may be used even when the spring is rotated about its fixed end (Fig. 13.10*a*). The work of the elastic force depends only upon the initial and final deflections of the spring (Fig. 13.10*b*).

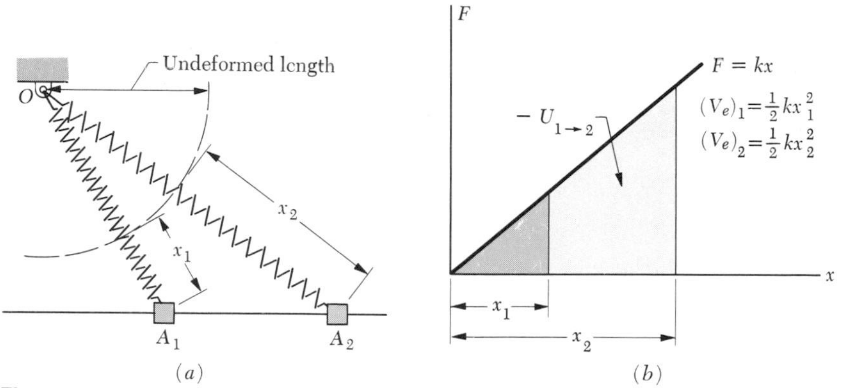

Fig. 13.10

The concept of potential energy may be used when forces other than gravity forces and elastic forces are involved. Indeed, it remains valid as long as the work of the force considered is independent of the path followed by its point of application as this point moves from a given position A_1 to a given position A_2. Such forces are said to be *conservative forces;* the general properties of conservative forces are studied in the following section.

***13.7. Conservative Forces.** As indicated in the preceding section, a force **F** acting on a particle *A* is said to be conservative *if its work $U_{1\to2}$ is independent of the path followed by the particle A as it moves from A_1 to A_2* (Fig. 13.11*a*). We may then write

$$U_{1\to2} = V(x_1,y_1,z_1) - V(x_2,y_2,z_2) \tag{13.19}$$

or, for short,

$$U_{1\to2} = V_1 - V_2 \tag{13.19'}$$

The function $V(x,y,z)$ is called the potential energy, or *potential function,* of **F**.

We note that, if A_2 is chosen to coincide with A_1, i.e., if the particle describes a closed path (Fig. 13.11*b*), we have $V_1 = V_2$

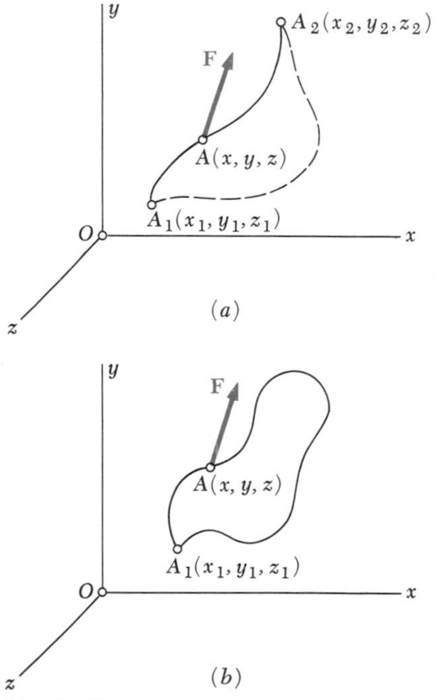

Fig. 13.11

and the work is zero. We may thus write for any conservative force **F**

$$\oint \mathbf{F} \cdot d\mathbf{r} = 0 \tag{13.20}$$

where the circle on the integral sign indicates that the path is closed.

Let us now apply (13.19) between two neighboring points $A(x,y,z)$ and $A'(x + dx, y + dy, z + dz)$. The elementary work dU corresponding to the displacement $d\mathbf{r}$ from A to A' is

$$dU = V(x,y,z) - V(x + dx, y + dy, z + dz)$$

or,

$$dU = -dV(x,y,z) \tag{13.21}$$

Thus, the elementary work of a conservative force is an *exact differential.*

Substituting for dU in (13.21) the expression obtained in (13.1″), and recalling the definition of the differential of a function of several variables, we write

$$F_x\,dx + F_y\,dy + F_z\,dz = -\left(\frac{\partial V}{\partial x}dx + \frac{\partial V}{\partial y}dy + \frac{\partial V}{\partial z}dz\right)$$

from which it follows that

$$F_x = -\frac{\partial V}{\partial x} \qquad F_y = -\frac{\partial V}{\partial y} \qquad F_z = -\frac{\partial V}{\partial z} \tag{13.22}$$

It is clear that the components of **F** must be functions of the coordinates x, y, z. Thus, a *necessary* condition for a conservative force is that it depend only upon the position of its point of application. The relations (13.22) may be expressed more concisely if we write

$$\mathbf{F} = F_x\mathbf{i} + F_y\mathbf{j} + F_z\mathbf{k} = -\left(\frac{\partial V}{\partial x}\mathbf{i} + \frac{\partial V}{\partial y}\mathbf{j} + \frac{\partial V}{\partial z}\mathbf{k}\right)$$

The vector in parentheses is known as the *gradient of the scalar function V* and is denoted by **grad** V. We thus write for any conservative force

$$\mathbf{F} = -\mathbf{grad}\, V \tag{13.23}$$

The relations (13.19) to (13.23) were shown to be satisfied by any conservative force. It may also be shown that if a force **F** satisfies one of these relations, **F** must be a conservative force.

13.8. Conservation of Energy. We saw in the preceding two sections that the work of a conservative force, such as the weight of a particle or the force exerted by a spring, may be expressed as a change in potential energy. When a particle moves under the action of conservative forces, the principle of work and energy stated in Sec. 13.3 may be expressed in a modified form. Substituting for $U_{1\to2}$ from (13.19′) into (13.10), we write

$$V_1 - V_2 = T_2 - T_1$$

$$T_1 + V_1 = T_2 + V_2 \qquad (13.24)$$

Formula (13.24) indicates that, when a particle moves under the action of conservative forces, *the sum of the kinetic energy and of the potential energy of the particle remains constant.* The sum $T + V$ is called the *total mechanical energy* of the particle and is denoted by E.

Consider, for example, the pendulum analyzed in Sec. 13.4, which is released with no velocity from A_1 and allowed to swing in a vertical plane (Fig. 13.12). Measuring the potential energy from the level of A_2, we have, at A_1,

$$T_1 = 0 \qquad V_1 = Wl \qquad T_1 + V_1 = Wl$$

Recalling that, at A_2, the speed of the pendulum is $v_2 = \sqrt{2gl}$, we have

$$T_2 = \tfrac{1}{2} m v_2^2 = \frac{1}{2}\frac{W}{g}(2\,gl) = Wl \qquad V_2 = 0$$
$$T_2 + V_2 = Wl$$

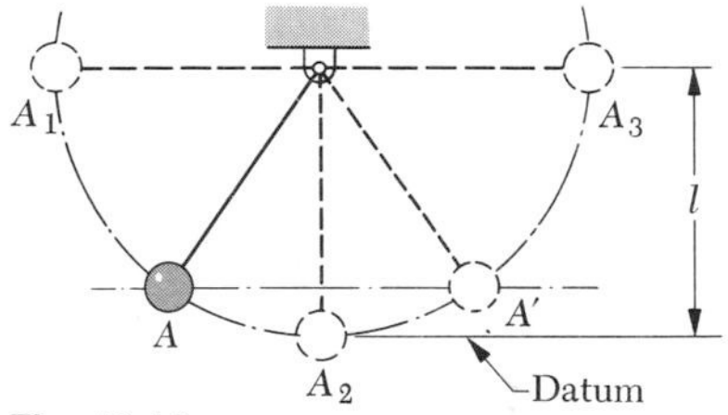

Fig. 13.12

We thus check that the total mechanical energy $E = T + V$ of the pendulum is the same at A_1 and A_2. While the energy is entirely potential at A_1, it becomes entirely kinetic at A_2 and, as the pendulum keeps swinging to the right, the kinetic energy is transformed back into potential energy. At A_3, we shall have $T_3 = 0$ and $V_3 = Wl$.

Since the total mechanical energy of the pendulum remains constant and since its potential energy depends only upon its elevation, the kinetic energy of the pendulum will have the same value at any two points located on the same level. Thus, the speed of the pendulum is the same at A and at A' (Fig. 13.12). This result may be extended to the case of a particle moving along any given path, regardless of the shape of the path, as long as the only forces acting on the particle are its weight and the normal reaction of the path. The particle of Fig. 13.13, for example, which slides in a vertical plane along a frictionless track, will have the same speed at A, A', and A''.

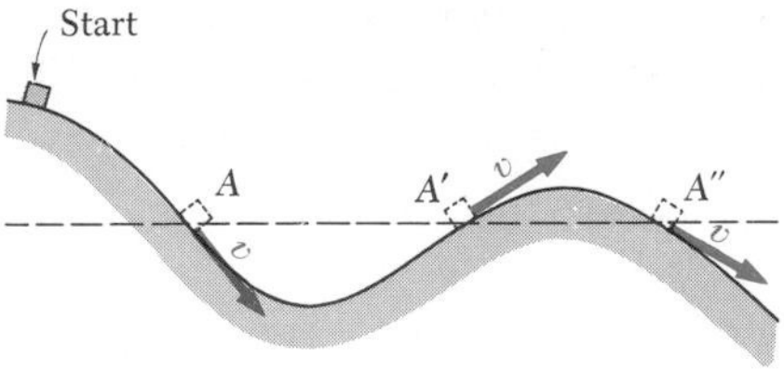

Fig. 13.13

While the weight of a particle and the force exerted by a spring are conservative forces, *friction forces are nonconservative forces*. In other words, *the work of a friction force cannot be expressed as a change in potential energy*. The work of a friction force depends upon the path followed by its point of application; and while the work $U_{1\to2}$ defined by (13.19) is positive or negative according to the sense of motion, *the work of a friction force is always negative*. It follows that, when a mechanical system involves friction, its total mechanical energy does not remain constant but decreases. The mechanical energy of the system, however, is not lost; it is transformed into heat, and the sum of the *mechanical energy* and of the *thermal energy* of the system remains constant.

Other forms of energy may also be involved in a system. For instance, a generator converts mechanical energy into *electric energy;* a gasoline engine converts *chemical energy* into mechanical energy; a nuclear reactor converts *mass* into thermal energy. If all forms of energy are considered, the energy of any system may be considered as constant and the principle of conservation of energy remains valid under all conditions.

13.9. Motion under a Conservative Central Force. Application to Space Mechanics. We saw in Sec. 12.8 that, when a particle P moves under a central force $\mathbf{F}$, the angular momentum $\mathbf{H}_O$ of the particle about the center of force O is constant. If the force $\mathbf{F}$ is also conservative, there exists a potential energy V associated with $\mathbf{F}$, and the total energy $E = T + V$ of the particle is constant (Sec. 13.8). Thus, when a particle moves under a conservative central force, both the principle of conservation of angular momentum and the principle of conservation of energy may be used to study its motion.

Consider, for example, a space vehicle moving under the earth's gravitational force. We shall assume that it begins its free flight at point P_0 at a distance r_0 from the center of the earth, with a velocity $\mathbf{v}_0$ forming an angle ϕ_0 with the radius vector OP_0 (Fig. 13.14). Let P be a point of the trajectory described by the vehicle; we denote by r the distance from O to P, by $\mathbf{v}$ the velocity of the vehicle at P, and by ϕ the angle formed by $\mathbf{v}$ and the radius vector OP. Applying the principle of conservation of angular momentum about O between P_0 and P (Sec. 12.8), we write

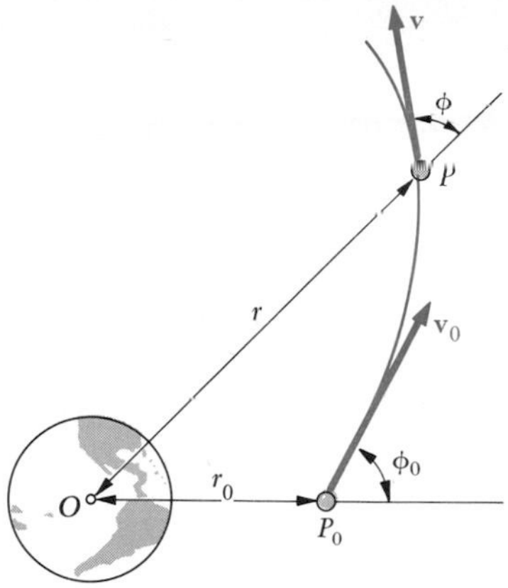

Fig. 13.14

$$r_0 m v_0 \sin \phi_0 = r m v \sin \phi \tag{13.25}$$

Recalling expression (13.17) obtained for the potential energy due to a gravitational force, we apply the principle of conservation of energy between P_0 and P and write

$$T_0 + V_0 = T + V$$

$$\tfrac{1}{2} m v_0^2 - \frac{GMm}{r_0} = \tfrac{1}{2} m v^2 - \frac{GMm}{r} \tag{13.26}$$

where M is the mass of the earth.

Equation (13.26) may be solved for the magnitude v of the velocity of the vehicle at P when the distance r from O to P is known; Eq. (13.25) may then be used to determine the angle ϕ that the velocity forms with the radius vector OP.

Equations (13.25) and (13.26) may also be used to determine the maximum and minimum values of r in the case of a satellite launched from P_0 in a direction forming an angle ϕ_0 with the vertical OP_0 (Fig. 13.15). The desired values of r are obtained by making $\phi = 90°$ in (13.25) and eliminating v between Eqs. (13.25) and (13.26).

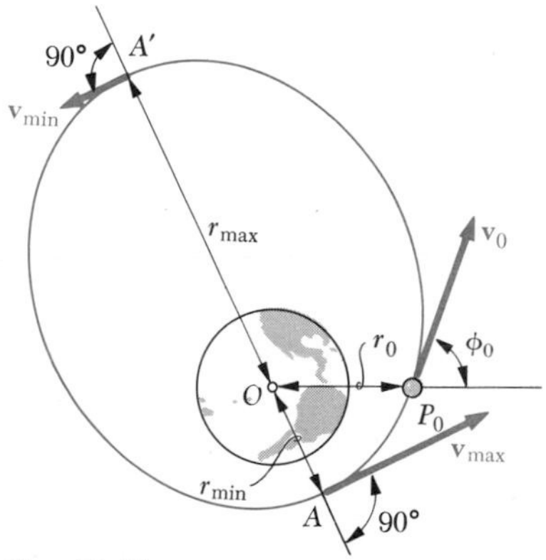

Fig. 13.15

It should be noted that the application of the principles of conservation of energy and of conservation of angular momentum leads to a more fundamental formulation of the problems of space mechanics than the method indicated in Sec. 12.11. In all cases involving oblique launchings, it will also result in much simpler computations. And while the method of Sec. 12.11 must be used when the actual trajectory or the periodic time of a space vehicle is to be determined, the calculations will be simplified if the conservation principles are first used to compute the maximum and minimum values of the radius vector r.

SAMPLE PROBLEM 13.6

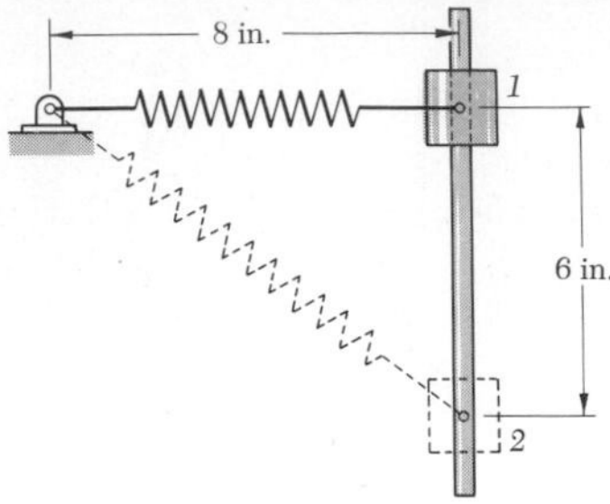

A 20-lb collar slides without friction along a vertical rod as shown. The spring attached to the collar has an undeformed length of 4 in. and a constant of 3 lb/in. If the collar is released from rest in position *1*, determine its velocity after it has moved 6 in. to position *2*.

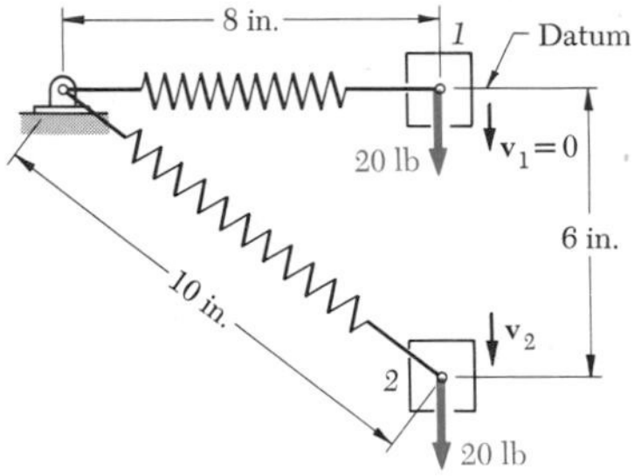

Position *1*. ***Potential Energy.*** The elongation of the spring is $x_1 = 8 \text{ in.} - 4 \text{ in.} = 4 \text{ in.}$, and we have

$$V_e = \tfrac{1}{2}kx_1^2 = \tfrac{1}{2}(3 \text{ lb/in.})(4 \text{ in.})^2 = 24 \text{ in} \cdot \text{lb}$$

Choosing the datum as shown, we have $V_g = 0$. Therefore,

$$V_1 = V_e + V_g = 24 \text{ in} \cdot \text{lb} = 2 \text{ ft} \cdot \text{lb}$$

Kinetic Energy. Since the velocity in position *1* is zero, $T_1 = 0$.

Position *2*. ***Potential Energy.*** The elongation of the spring is $x_2 = 10 \text{ in.} - 4 \text{ in.} = 6 \text{ in.}$, and we have

$$V_e = \tfrac{1}{2}kx_2^2 = \tfrac{1}{2}(3 \text{ lb/in.})(6 \text{ in.})^2 = 54 \text{ in} \cdot \text{lb}$$
$$V_g = Wy = (20 \text{ lb})(-6 \text{ in.}) = -120 \text{ in} \cdot \text{lb}$$

Therefore,

$$V_2 = V_e + V_g = 54 - 120 = -66 \text{ in} \cdot \text{lb} = -5.5 \text{ ft} \cdot \text{lb}$$

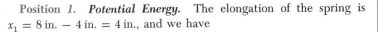

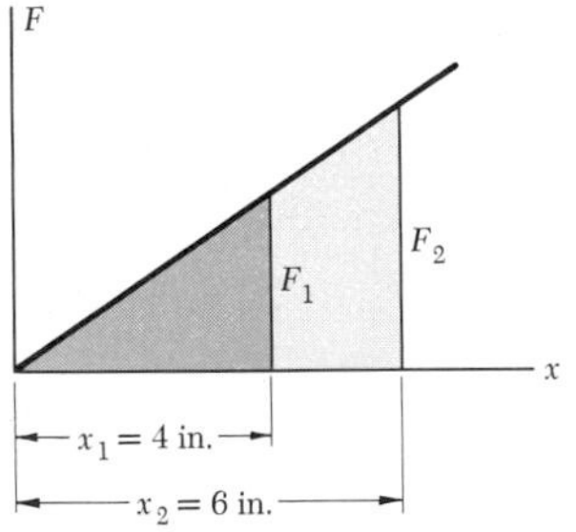

Kinetic Energy

$$T_2 = \tfrac{1}{2} mv_2^2 = \frac{1}{2}\frac{20}{32.2}v_2^2 = 0.311v_2^2$$

Conservation of Energy. Applying the principle of conservation of energy between positions *1* and *2*, we write

$$T_1 + V_1 = T_2 + V_2$$
$$0 + 2 \text{ ft} \cdot \text{lb} = 0.311v_2^2 - 5.5 \text{ ft} \cdot \text{lb}$$
$$v_2 = \pm 4.91 \text{ ft/s}$$

$\mathbf{v}_2 = 4.91 \text{ ft/s} \downarrow$ ◄

SAMPLE PROBLEM 13.7

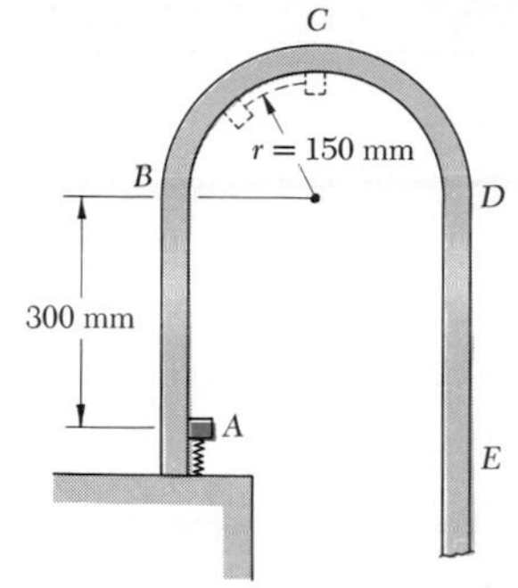

The 200-g pellet is released from rest at A when the spring is compressed 75 mm and travels around the loop $ABCDE$. Determine the smallest value of the spring constant for which the pellet will travel around the loop and will at all times remain in contact with the loop.

Required Speed at Point C. As the pellet passes through the highest point C, its potential energy with respect to gravity is maximum; thus, at the same point its kinetic energy and its speed are minimum. Since the pellet must remain in contact with the loop, the force **N** exerted on the pellet by the loop must be equal to, or greater than, zero. Setting $\mathbf{N} = 0$, we compute the smallest possible speed v_C.

$$+\downarrow \Sigma F_n = ma_n: \qquad W = ma_n \qquad mg = ma_n \qquad a_n = g$$

$$a_n = \frac{v_C^2}{r}: \qquad v_C^2 = ra_n = rg = (0.150\text{ m})(9.81\text{ m/s}^2) = 1.472\text{ m}^2/\text{s}^2$$

Position 1. ***Potential Energy.*** Since the spring is compressed 0.075 m from its undeformed position, we have

$$V_e = \tfrac{1}{2}kx^2 = \tfrac{1}{2}k(0.075\text{ m})^2 = (0.00281\text{ m}^2)k$$

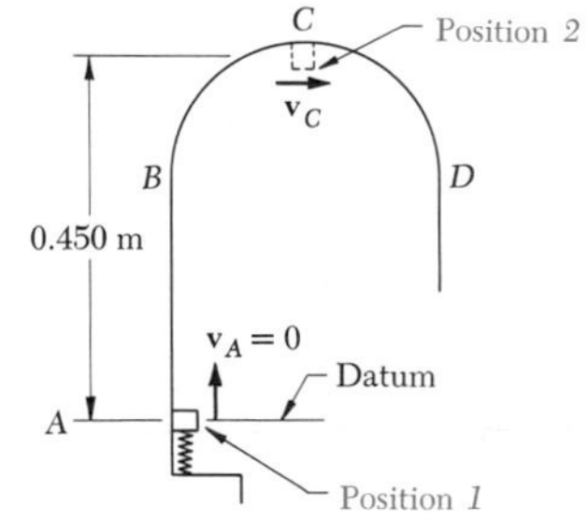

Choosing the datum at A, we have $V_g = 0$; therefore

$$V_1 = V_e + V_g = (0.00281\text{ m}^2)k$$

Kinetic Energy. Since the pellet is released from rest, $v_A = 0$ and we have $T_1 = 0$.

Position 2. ***Potential Energy.*** The spring is now undeformed; thus $V_e = 0$. Since the pellet is 0.450 m above the datum, and since $W = (0.200\text{ kg})(9.81\text{ m/s}^2) = 1.962\text{ N}$, we have

$$V_g = Wy = (1.962\text{ N})(0.450\text{ m}) = 0.883\text{ N}\cdot\text{m} = 0.883\text{ J}$$
$$V_2 = V_e + V_g = 0.883\text{ J}$$

Kinetic Energy. Using the value of v_C^2 obtained above, we write

$$T_2 = \tfrac{1}{2}mv_C^2 = \tfrac{1}{2}(0.200\text{ kg})(1.472\text{ m}^2/\text{s}^2) = 0.1472\text{ N}\cdot\text{m} = 0.1472\text{ J}$$

Conservation of Energy. Applying the principle of conservation of energy between positions *1* and *2*, we write

$$T_1 + V_1 = T_2 + V_2$$
$$0 + (0.00281\text{ m}^2)k = 0.1472\text{ J} + 0.883\text{ J}$$
$$k = 367\text{ J/m}^2 = 367\text{ N/m}$$

The required minimum value of k is therefore

$$k = 367\text{ N/m} \quad ◄$$

SAMPLE PROBLEM 13.8

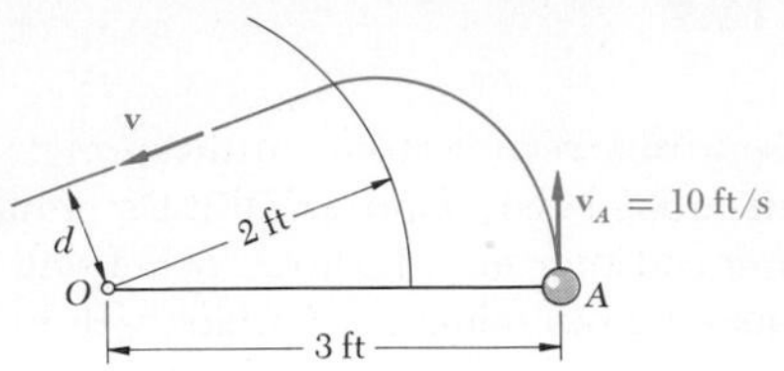

A ball weighing 0.5 lb is attached to a fixed point O by means of an elastic cord of constant $k = 10$ lb/ft and of undeformed length equal to 2 ft. The ball slides on a horizontal frictionless surface. If the ball is placed at point A, 3 ft from O, and is given an initial velocity of 10 ft/s in a direction perpendicular to OA, determine (*a*) the speed of the ball after the cord has become slack, (*b*) the closest distance d that the ball will come to O.

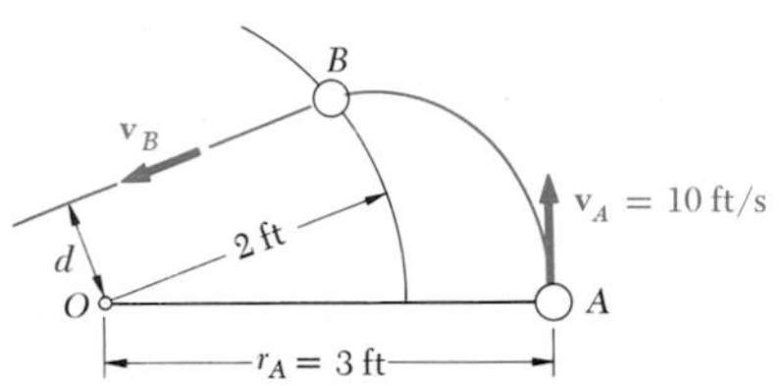

Solution. The force exerted by the cord on the ball passes through the fixed point O, and its work may be expressed as a change in potential energy. It is therefore a conservative central force, and both the total energy of the ball and its angular momentum about O are conserved between points A and B. After the cord has become slack at B, the resultant force acting on the ball is zero. The ball, therefore, will move in a straight line at a constant speed v. The straight line is the line of action of $\mathbf{v}_B$ and the speed v is equal to v_B.

a. Conservation of Energy.

At point A: $\quad T_A = \frac{1}{2}mv_A^2 = \frac{1}{2}\frac{0.5\text{ lb}}{32.2\text{ ft/s}^2}(10\text{ ft/s})^2 = 0.776\text{ ft}\cdot\text{lb}$

$\quad V_A = \frac{1}{2}kx_A^2 = \frac{1}{2}(10\text{ lb/ft})(3\text{ ft} - 2\text{ ft})^2 = 5\text{ ft}\cdot\text{lb}$

At point B: $\quad T_B = \frac{1}{2}mv_B^2 = \frac{1}{2}\frac{0.5}{32.2}v_B^2 = 0.00776v_B^2$

$\quad V_B = 0$

Applying the principle of conservation of energy between points A and B, we write

$$T_A + V_A = T_B + V_B$$
$$0.776 + 5 = 0.00776v_B^2$$
$$v_B^2 = 744 \qquad v = v_B = 27.3\text{ ft/s}$$ ◄

b. Conservation of Angular Momentum About O. Since r_A and d represent the perpendicular distances to $\mathbf{v}_A$ and $\mathbf{v}_B$, respectively, we write

$$r_A(mv_A) = d(mv_B)$$
$$(3\text{ ft})\left(\frac{0.5}{g}\right)(10\text{ ft/s}) = d\left(\frac{0.5}{g}\right)(27.3\text{ ft/s})$$

$$d = 1.099\text{ ft}$$ ◄

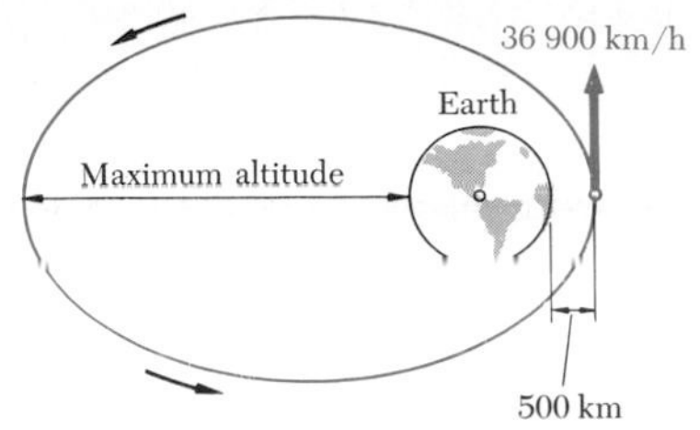

SAMPLE PROBLEM 13.9

A satellite is launched in a direction parallel to the surface of the earth with a velocity of 36 900 km/h from an altitude of 500 km. Determine (*a*) the maximum altitude reached by the satellite, (*b*) the maximum allowable error in the direction of launching if the satellite is to go into orbit and to come not closer than 200 km to the surface of the earth.

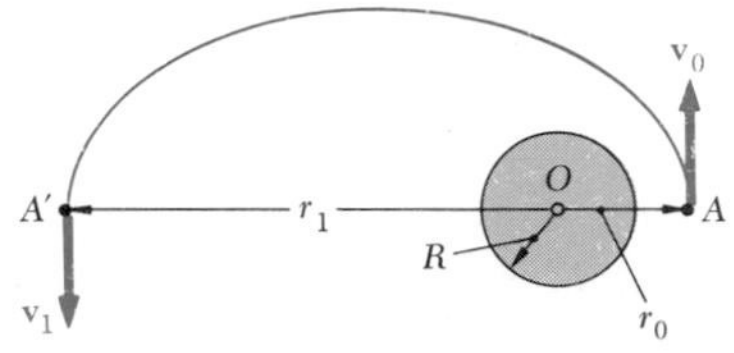

a. **Maximum Altitude.** We denote by A' the point of the orbit farthest from the earth and by r_1 the corresponding distance from the center of the earth. Since the satellite is in free flight between A and A', we apply the principle of conservation of energy.

$$T_A + V_A = T_{A'} + V_{A'}$$

$$\tfrac{1}{2}mv_0^2 - \frac{GMm}{r_0} = \tfrac{1}{2}mv_1^2 - \frac{GMm}{r_1} \tag{1}$$

Since the only force acting on the satellite is the force of gravity, which is a central force, the angular momentum of the satellite about O is conserved. Considering points A and A', we write

$$r_0 m v_0 = r_1 m v_1 \qquad v_1 = v_0 \frac{r_0}{r_1} \tag{2}$$

Substituting this expression for v_1 into Eq. (1) and dividing each term by the mass m, we obtain after rearranging the terms,

$$\tfrac{1}{2}v_0^2\left(1 - \frac{r_0^2}{r_1^2}\right) = \frac{GM}{r_0}\left(1 - \frac{r_0}{r_1}\right) \qquad 1 + \frac{r_0}{r_1} = \frac{2GM}{r_0 v_0^2} \tag{3}$$

Recalling that the radius of the earth is $R = 6370$ km, we compute

$r_0 = 6370\text{ km} + 500\text{ km} = 6870\text{ km} = 6.87 \times 10^6\text{ m}$
$v_0 = 36\,900\text{ km/h} = (3.69 \times 10^7\text{ m})/(3.6 \times 10^3\text{ s}) = 1.025 \times 10^4\text{ m/s}$
$GM = gR^2 = (9.81\text{ m/s}^2)(6.37 \times 10^6\text{ m})^2 = 3.98 \times 10^{14}\text{ m}^3/\text{s}^2$

Substituting these values into (3), we obtain $r_1 = 66.8 \times 10^6$ m

Maximum altitude $= 66.8 \times 10^6\text{ m} - 6.37 \times 10^6\text{ m} = 60.4 \times 10^6\text{ m}$
$= 60\,400$ km ◄

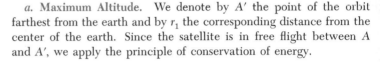

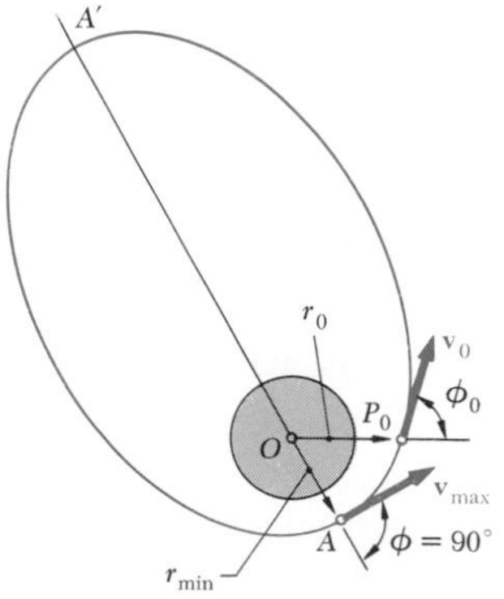

b. **Allowable Error in Direction of Launching.** The satellite is launched from P_0 in a direction forming an angle ϕ_0 with the vertical OP_0. The value of ϕ_0 corresponding to $r_{min} = 6370\text{ km} + 200\text{ km} = 6570$ km is obtained by applying the principles of conservation of energy and of conservation of angular momentum between P_0 and A.

$$\tfrac{1}{2}mv_0^2 - \frac{GMm}{r_0} = \tfrac{1}{2}mv_{max}^2 - \frac{GMm}{r_{min}} \tag{4}$$

$$r_0 m v_0 \sin\phi_0 = r_{min} m v_{max} \tag{5}$$

Solving (5) for v_{max} and then substituting for v_{max} into (4), we may solve (4) for $\sin\phi_0$. Using the values of v_0 and GM computed in part *a* and noting that $r_0/r_{min} = 6870/6570 = 1.0457$, we find

$\sin\phi_0 = 0.9801 \qquad \phi_0 = 90° \pm 11.5° \qquad$ Allowable error $= \pm 11.5°$ ◄

PROBLEMS

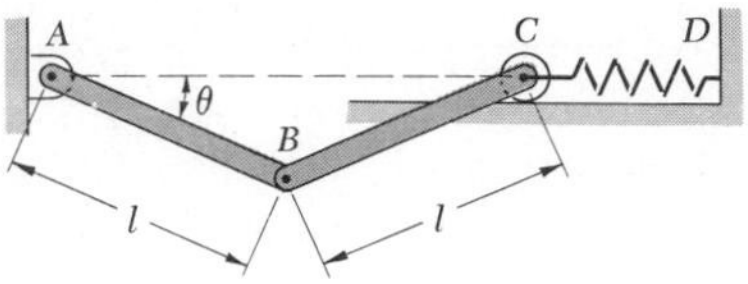

Fig. P13.50

13.50 The uniform rods AB and BC are each of mass m; the spring CD is of constant k and is unstretched when $\theta = 0$. Determine the potential energy of the system with respect to (*a*) the spring, (*b*) gravity. (Place datum at A.)

13.51 A slender rod AB of negligible mass is attached to blocks A and B, each of mass m. The constant of the spring is k and the spring is undeformed when AB is horizontal. Determine the potential energy of the system with respect to (*a*) the spring, (*b*) gravity. (Place datum at B.)

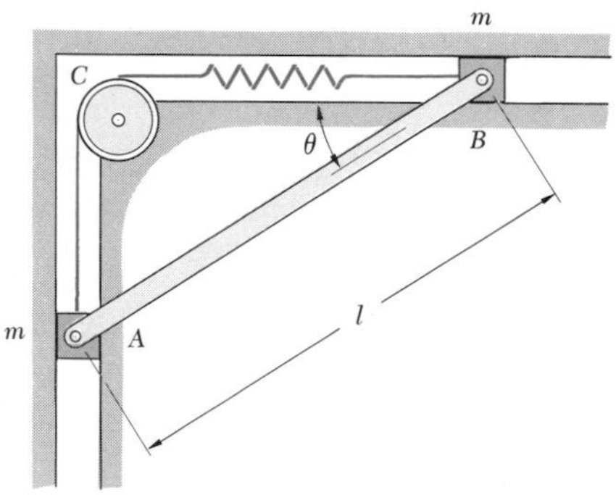

Fig. P13.51

13.52 Prove that a force $\mathbf{F}(x,y,z)$ is conservative if, and only if, the following relations are satisfied:

$$\frac{\partial F_x}{\partial y} = \frac{\partial F_y}{\partial x} \qquad \frac{\partial F_y}{\partial z} = \frac{\partial F_z}{\partial y} \qquad \frac{\partial F_z}{\partial x} = \frac{\partial F_x}{\partial z}$$

13.53 The force $\mathbf{F} = (x\mathbf{i} + y\mathbf{j})/(x^2 + y^2)$ acts on the particle $P(x,y)$ which moves in the xy plane. (*a*) Using the first of the relations derived in Prob. 13.52, prove that $\mathbf{F}$ is a conservative force. (*b*) Determine the potential function $V(x,y)$ associated with $\mathbf{F}$.

13.54 The force $\mathbf{F} = (x\mathbf{i} + y\mathbf{j} + z\mathbf{k})/(x^2 + y^2 + z^2)^{3/2}$ acts on the particle $P(x,y,z)$ which moves in space. (*a*) Using the relations derived in Prob. 13.52, prove that $\mathbf{F}$ is a conservative force. (*b*) Determine the potential function $V(x,y,z)$ associated with $\mathbf{F}$.

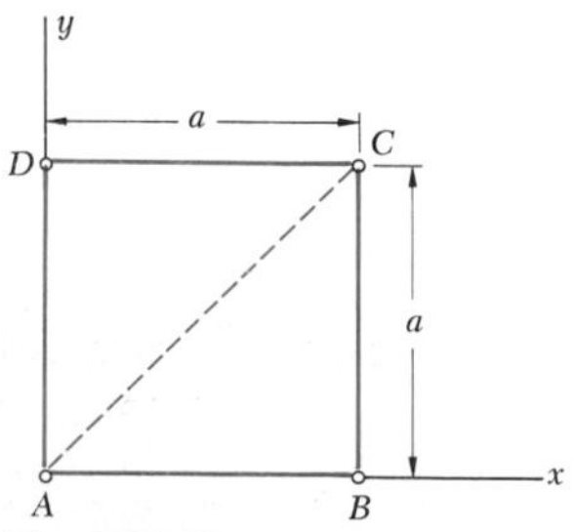

Fig. P13.55

13.55 The force $\mathbf{F} = x^2y\mathbf{i} + xy^2\mathbf{j}$ acts on the particle $P(x,y)$ which moves in the xy plane. Prove that $\mathbf{F}$ is a nonconservative force and determine the work of $\mathbf{F}$ as it moves from A to C along each of the paths ABC, ADC, and AC.

13.56 The spring AB is of constant 6 lb/in. and is attached to the 4-lb collar A which moves freely along the horizontal rod. The unstretched length of the spring is 10 in. If the collar is released from rest in the position shown, determine the maximum velocity attained by the collar.

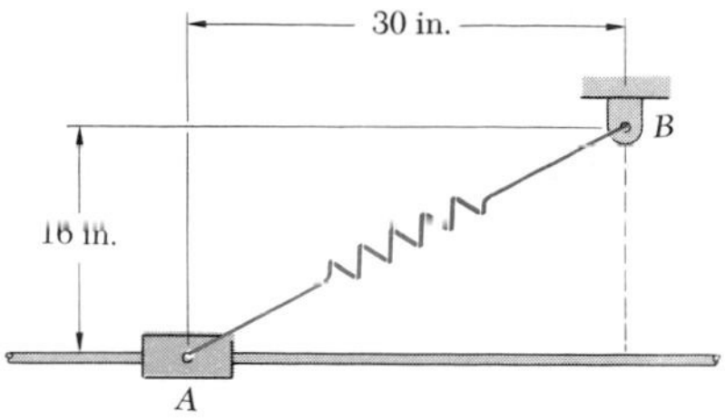

Fig. P13.56

13.57 In Prob. 13.56, determine the weight of the collar A for which the maximum velocity is 30 ft/s.

13.58 A collar of mass 1.5 kg is attached to a spring and slides without friction along a circular rod which lies in a *horizontal* plane. The spring is undeformed when the collar is at C and the constant of the spring is 400 N/m. If the collar is released from rest at B, determine the velocity of the collar as it passes through point C.

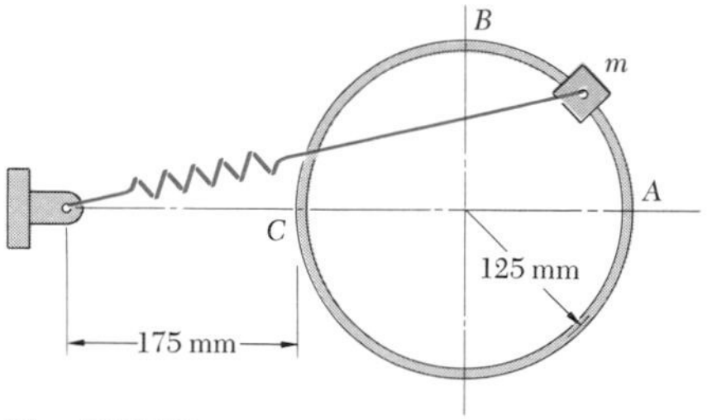

Fig. P13.58

13.59 Plunger A has a mass of 200 g and is to be shot to the right by the mechanism shown. The undeformed length of the spring is 180 mm and it is compressed to a length of 60 mm; it will expand to a length of 110 mm when the plunger is released. Knowing that a force of 36 N is required to hold the plunger in the position shown, determine the velocity attained by the plunger as it leaves the mechanism.

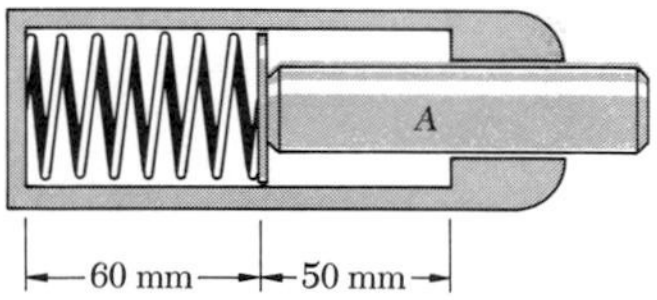

Fig. P13.59

13.60 The collar of Prob. 13.58 has a continuous, although nonuniform, motion along the rod. If the speed of the collar at A is to be half of its speed at C, determine (*a*) the required speed at C, (*b*) the corresponding speed at B.

13.61 The 2-lb collar slides without friction along the horizontal rod. Knowing that the constant of the spring is 3 lb/in. and that $v_0 = 12$ ft/s, determine the required spring tension in the position shown if the speed of the collar is to be 8 ft/s at point C.

13.62 The 2-lb collar slides without friction along the horizontal rod. Knowing that the spring has a constant $k = 3$ lb/in. and is unstretched in the position shown, determine the required speed v_0 if it is to reach point C.

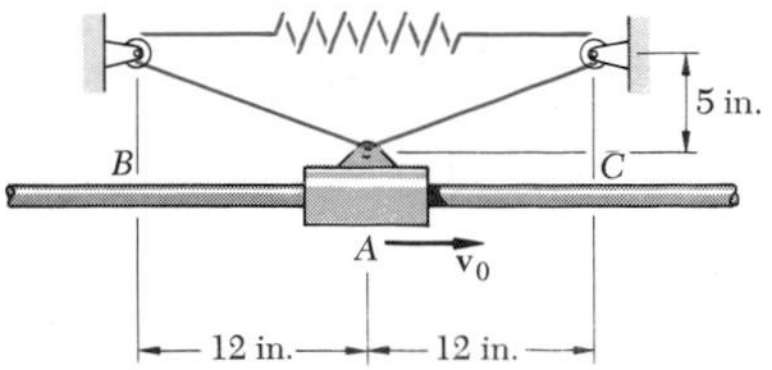

Fig. P13.61 and P13.62

13.63 The 50-kg block is released from rest when $\phi = 0$. If the speed of the block when $\phi = 90°$ is to be 2.5 m/s, determine the required value of the initial tension in the spring.

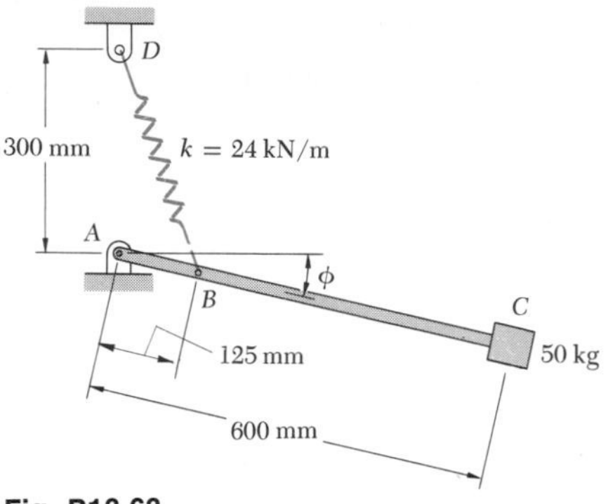

Fig. P13.63

13.64 A sling shot is made by stretching an elastic band between pins A and B located 100 mm apart in the same horizontal plane. The spring constant for the entire length of the elastic band is 600 N/m and the tension in the band is 40 N when it is stretched directly between A and B. Determine the maximum speed attained by a 50-g pellet which is placed at C and released.

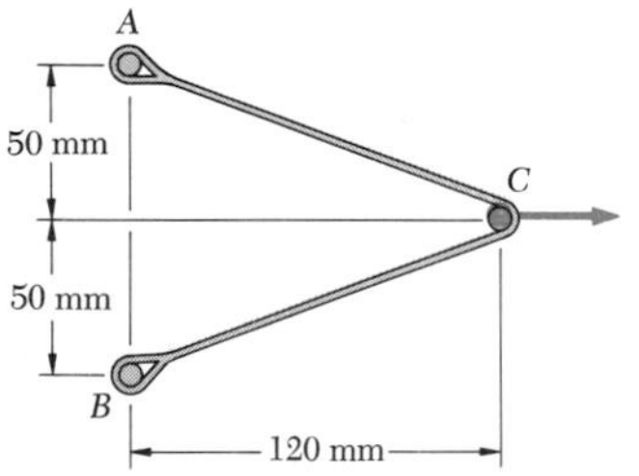

Fig. P13.64

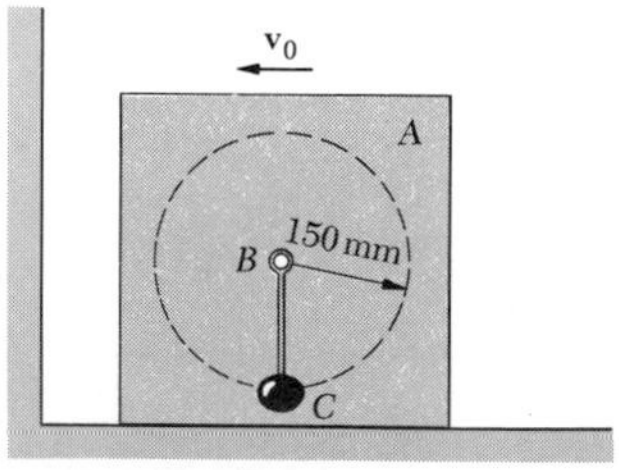

Fig. P13.65

13.65 The sphere C and the block A are both moving to the left with a velocity $\mathbf{v}_0$ when the block is suddenly stopped by the wall. Determine the smallest velocity $\mathbf{v}_0$ for which the sphere C will swing in a full circle about the pivot B (*a*) if BC is a slender rod of negligible weight, (*b*) if BC is a cord.

13.66 The collar of Prob. 13.58 is released from rest at point A. Determine the horizontal component of the force exerted by the rod on the collar as the collar passes through point B. Show that the force component is independent of the mass of the collar.

13.67 A 1.5-lb collar may slide without friction along the semicircular rod *BCD*. The spring is of constant 2 lb/in. and its undeformed length is 12 in. The collar is released from rest at *B*. As the collar passes through point *C*, determine (*a*) the speed of the collar, (*b*) the force exerted by the rod on the collar.

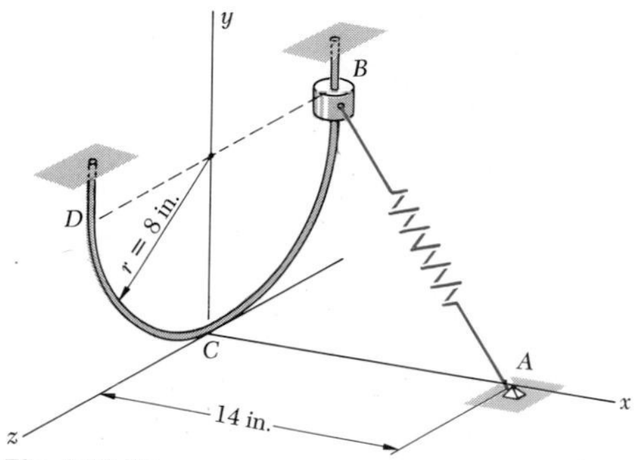

Fig. P13.67

13.68 A small block is released at *A* with zero velocity and moves along the frictionless guide to point *B* where it leaves the guide with a horizontal velocity. Knowing that $h = 8$ ft and $b = 3$ ft, determine (*a*) the speed of the block as it strikes the ground at *C*, (*b*) the corresponding distance *c*.

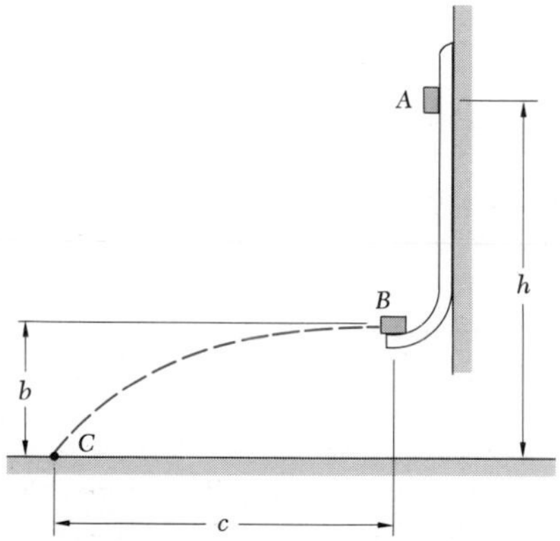

Fig. P13.68

13.69 Assuming a given height *h* in Prob. 13.68, (*a*) show that the speed at *C* is independent of the height *b*, (*b*) determine the height *b* for which the distance *c* is maximum and the corresponding value of *c*.

13.70 A ball of mass m attached to an inextensible cord rotates in a vertical circle of radius r. Show that the difference between the maximum value T_{max} of the tension in the cord and its minimum value T_{min} is independent of the speed v_0 of the ball as measured at the bottom of the circle, and determine $T_{max} - T_{min}$.

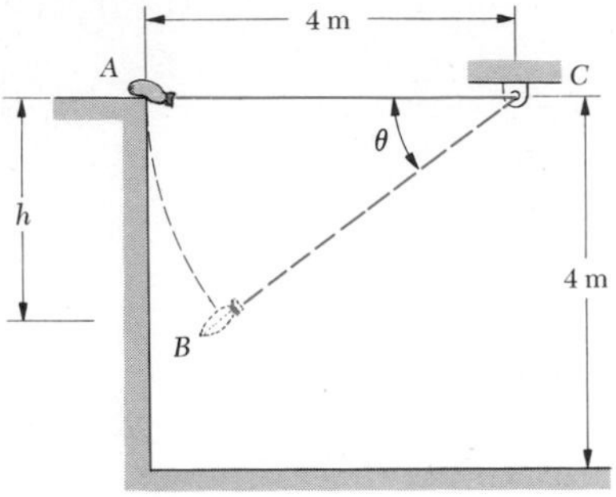

Fig. P13.71

13.71 A bag is gently pushed off the top of a wall at A and swings in a vertical plane at the end of a 4-m rope which can withstand a maximum tension equal to twice the weight of the bag. (*a*) Determine the difference in elevation h between point A and point B where the rope will break. (*b*) How far from the vertical wall will the bag strike the floor?

13.72 A delicate instrument weighing 12 lb is placed on a spring of length l so that its base is just touching the undeformed spring. The instrument is then inadvertently released from that position. Determine the maximum deflection x of the spring and the maximum force exerted by the spring if the constant of the spring is $k = 15$ lb/in.

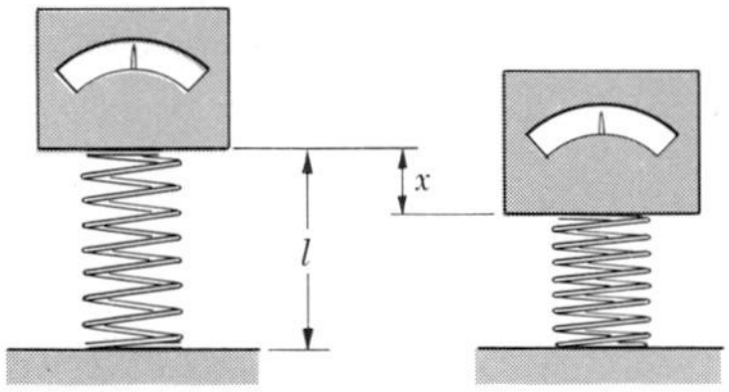

Fig. P13.72

13.73 Nonlinear springs are classified as hard or soft, depending upon the curvature of their force-deflection curves (see figure). Solve Prob. 13.72, assuming (*a*) that a hard spring is used, for which $F = 15x(1 + 0.1x^2)$, (*b*) that a soft spring is used, for which $F = 15x(1 - 0.1x^2)$.

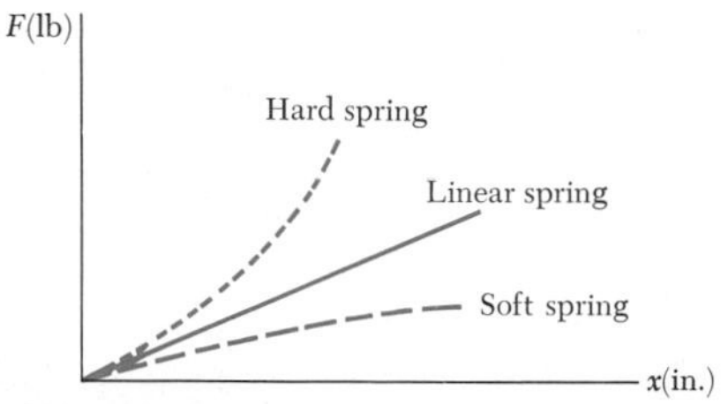

Fig. P13.73

13.74 Determine the escape velocity of a missile, i.e., the velocity with which it should be fired from the surface of the earth if it is to reach an infinite distance from the earth. Give the answer in both SI and U.S. customary units. Show that the result obtained is independent of the firing angle.

13.75 How much energy per kilogram should be imparted to a satellite in order to place it in a circular orbit at an altitude of (*a*) 500 km, (*b*) 5000 km?

13.76 A lunar excursion module (LEM) was used in the Apollo moon-landing missions to save fuel by making it unnecessary to launch the entire Apollo spacecraft from the moon's surface on its return trip to the earth. Check the effectiveness of this approach by computing the energy per pound required for a spacecraft to escape the gravitational field of the moon if the spacecraft starts (*a*) from the moon's surface, (*b*) from a circular orbit 60 mi above the moon's surface. Neglect the effect of the earth's gravitational field. (The radius of the moon is 1080 mi and its mass is 0.01230 times the mass of the earth.)

13.77 Show, by setting $r = R + y$ in formula (13.17′) and expanding in a power series in y/R, that the expression obtained in (13.16) for the potential energy V_g due to gravity is a first-order approximation for the expression given in (13.17′). Using the same expansion, derive a second-order approximation for V_g.

13.78 Show that the ratio of the potential and kinetic energies of an electron, as it enters the plates of the cathode-ray tube of Prob. 12.60, is equal to $d\delta/lL$. (Place the datum at the surface of the positive plate.)

13.79 In Sample Prob. 13.8, determine the required magnitude of $\mathbf{v}_A$ if the ball is to pass at a distance $d = 4$ in. from point O. Assume that the direction of $\mathbf{v}_A$ is not changed.

13.80 A 2-kg sphere is attached to an elastic cord of constant 150 N/m which is undeformed when the sphere is located at the origin O. Knowing that in the position shown $\mathbf{v}_A$ is perpendicular to OP and has a magnitude of 10 m/s, determine (*a*) the maximum distance from the origin attained by the sphere, (*b*) the corresponding speed of the sphere.

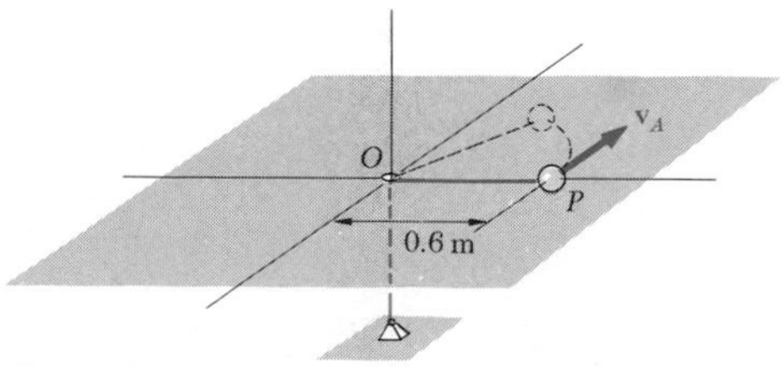

Fig. P13.80

13.81 In Prob. 13.80, determine the required initial speed v_A if the maximum distance from the origin attained by the sphere is to be 1.5 m.

13.82 A 1.5-lb block P rests on a frictionless horizontal table at a distance of 1 ft from a fixed pin O. The block is attached to pin O by an elastic cord of constant $k = 10$ lb/ft and of undeformed length 2 ft. If the block is set in motion to the right as shown, determine (*a*) the speed v_1 for which the distance from O to the block P will reach a maximum value of 3 ft, (*b*) the speed v_2 when $OP = 3$ ft, (*c*) the radius of curvature of the path of the block when $OP = 3$ ft.

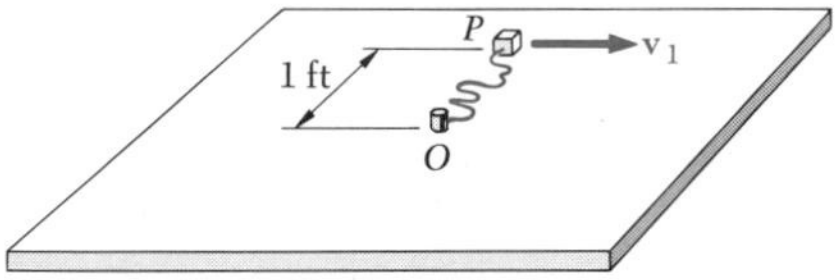

Fig. P13.82

13.83 Collar B weighs 10 lb and is attached to a spring of constant 50 lb/ft and of undeformed length equal to 18 in. The system is set in motion with $r = 12$ in., $v_\theta = 16$ ft/s, and $v_r = 0$. Neglecting the mass of the rod and the effect of friction, determine the radial and transverse components of the velocity of the collar when $r = 21$ in.

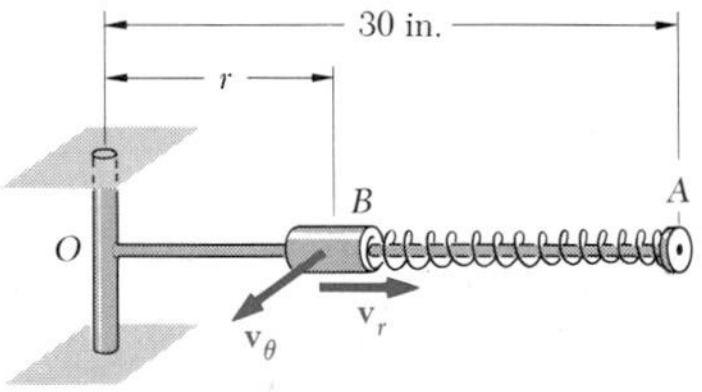

Fig. P13.83

13.84 For the motion described in Prob. 13.83, determine (*a*) the maximum distance between the origin and the collar, (*b*) the corresponding velocity.

13.85 In Sample Prob. 13.8, determine the smallest magnitude of $\mathbf{v}_A$ for which the elastic cord will remain taut at all times.

13.86 through 13.89 Using the principles of conservation of energy and conservation of angular momentum, solve the following problems:

13.86 Prob. 12.88.
13.87 Prob. 12.93.
13.88 Prob. 12.92*a* and *b*.
13.89 Prob. 12.89.

13.90 A space shuttle is to rendezvous with an orbiting laboratory which circles the earth at the constant altitude of 240 mi. The shuttle has reached an altitude of 40 mi when its engine is shut off, and its velocity $\mathbf{v}_0$ forms an angle $\phi_0 = 45°$ with the vertical OB at that time. What magnitude should $\mathbf{v}_0$ have if the shuttle's trajectory is to be tangent at A to the orbit of the laboratory?

13.91 A space shuttle is to rendezvous with an orbiting laboratory which circles the earth at the constant altitude of 240 mi. The shuttle has reached an altitude of 40 mi and a velocity $\mathbf{v}_0$ of magnitude 12,000 ft/s when its engine is shut off. What is the angle ϕ_0 that $\mathbf{v}_0$ should form with the vertical OB if the shuttle's trajectory is to be tangent at A to the orbit of the laboratory?

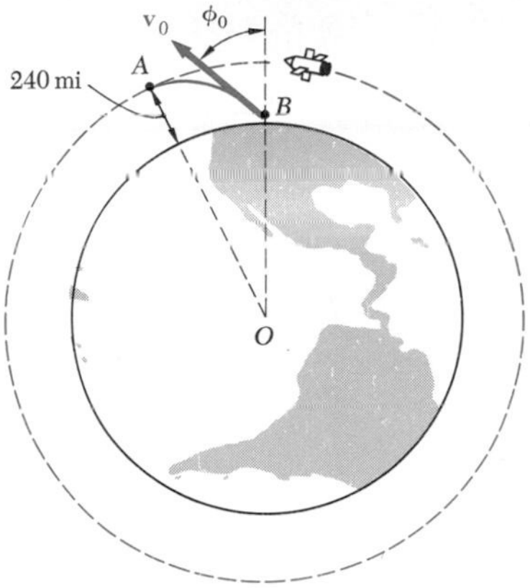

Fig. P13.90 and P13.91

13.92 Determine the magnitude and direction (angle ϕ formed with the vertical OB) of the velocity $\mathbf{v}_B$ of the spacecraft of Prob. 12.100 just before splashdown at B. Neglect the effect of the atmosphere.

13.93 To what value v_0 should the speed of the spacecraft of Prob. 12.101 be reduced preparatory to reentry if its velocity $\mathbf{v}_B$ just before splashdown at B is to form an angle $\phi = 30°$ with the vertical OB? Neglect the effect of the atmosphere.

13.94 Upon the LEM's return to the command module, the Apollo spacecraft of Prob. 12.93 is turned around so that the LEM faces to the rear. The LEM is then cast adrift with a velocity of 600 ft/s relative to the command module. Determine the magnitude and direction (angle ϕ formed with the vertical OC) of the velocity $\mathbf{v}_C$ of the LEM just before it crashes at C on the moon's surface.

13.95 At engine burnout a satellite has reached an altitude of 2400 km and has a velocity $\mathbf{v}_0$ of magnitude 8100 m/s forming an angle $\phi_0 = 76°$ with the vertical. Determine the maximum and minimum heights reached by the satellite.

13.96 At engine burnout a satellite has reached an altitude of 2400 km and has a velocity $\mathbf{v}_0$ of magnitude 8100 m/s. For what range of values of the angle ϕ_0, formed by $\mathbf{v}_0$ and the vertical, will the satellite go into a permanent orbit? (Assume that if the satellite gets closer than 300 km from the earth's surface, it will soon burn up.)

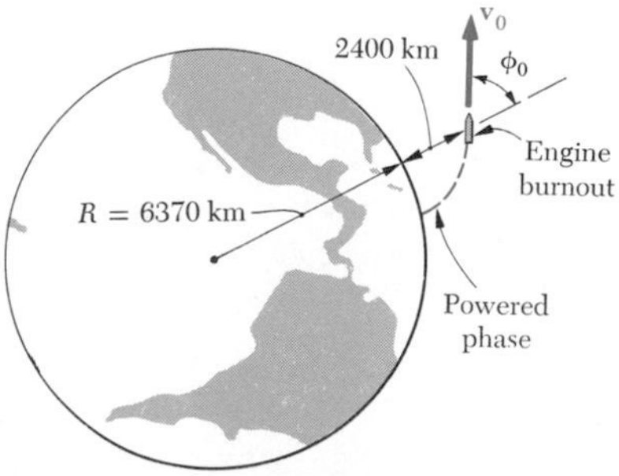

Fig. P13.95 and P13.96

13.97 A satellite is projected into space with a velocity $\mathbf{v}_0$ at a distance r_0 from the center of the earth by the last stage of its launching rocket. The velocity $\mathbf{v}_0$ was designed to send the satellite into a circular orbit of radius r_0. However, owing to a malfunction of control, the satellite is not projected horizontally but at an angle α with the horizontal and, as a result, is propelled into an elliptic orbit. Determine the maximum and minimum values of the distance from the center of the earth to the satellite.

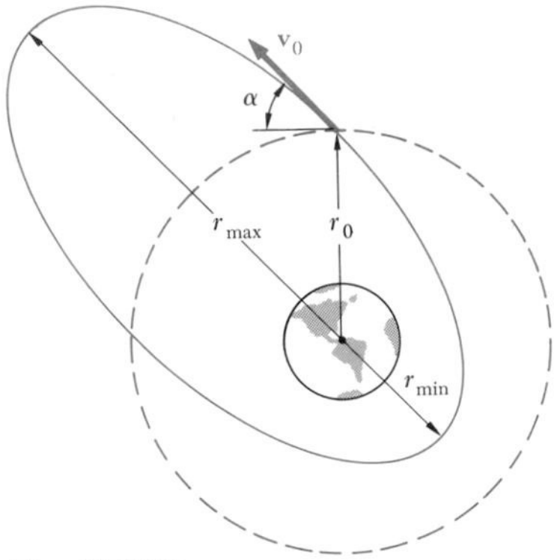

Fig. P13.97

***13.98** Using the answers obtained in Prob. 13.97, show that the intended circular orbit and the resulting elliptic orbit intersect at the ends of the minor axis of the elliptic orbit.

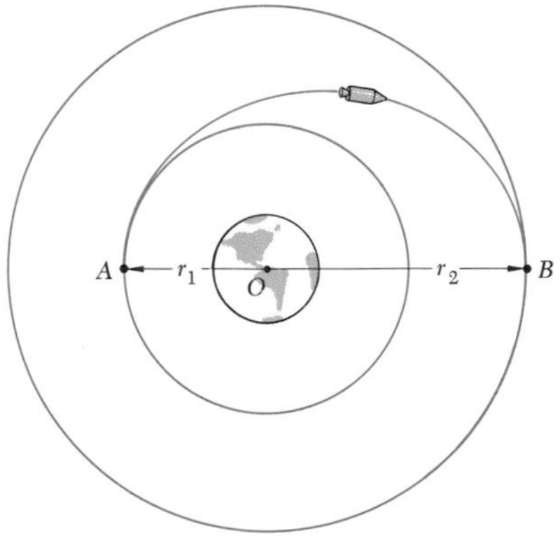

Fig. P13.99

13.99 A spacecraft of mass m describes a circular orbit of radius r_1 around the earth. (*a*) Show that the additional energy ΔE which must be imparted to the spacecraft to transfer it to a circular orbit of larger radius r_2 is

$$\Delta E = \frac{GMm(r_2 - r_1)}{2r_1 r_2}$$

where M is the mass of the earth. (*b*) Further show that, if the transfer from one circular orbit to the other is executed by placing the spacecraft on a transitional semielliptic path AB, the amounts of energy ΔE_A and ΔE_B which must be imparted at A and B are respectively proportional to r_2 and r_1:

$$\Delta E_A = \frac{r_2}{r_1 + r_2}\Delta E \qquad \Delta E_B = \frac{r_1}{r_1 + r_2}\Delta E$$

13.100 Show that the total energy E of a satellite of mass m describing an elliptic orbit is $E = -GMm/(r_1 + r_2)$, where M is the mass of the earth, and r_1 and r_2 represent, respectively, the maximum and minimum distance of the orbit to the center of the earth. (It is recalled that the gravitational potential energy of a satellite was defined as being zero at an infinite distance from the earth.)

***13.101** (*a*) Express the angular momentum per unit mass, h, and the total energy per unit mass, E/m, of a space vehicle moving under the earth's gravitational force in terms of r_{min} and v_{max} (Fig. 13.15). (*b*) Eliminating v_{max} between the equations obtained, derive the formula

$$\frac{1}{r_{min}} = \frac{GM}{h^2} + \sqrt{\left(\frac{GM}{h^2}\right)^2 + \frac{2(E/m)}{h^2}}$$

(*c*) Show that the constant C in Eq. (12.39) of Sec. 12.11 may be expressed as

$$C = \sqrt{\left(\frac{GM}{h^2}\right)^2 + \frac{2(E/m)}{h^2}}$$

(*d*) Further show that the trajectory of the vehicle is a hyperbola, an ellipse, or a parabola, depending on whether E is positive, negative or zero.

***13.102** In Prob. 13.90, determine the distance separating the two points located on the surface of the earth directly below points B and A where engine shut-off and rendezvous with the orbiting laboratory respectively take place. [*Hint.* Use Eq. (12.39) of Sec. 12.11, noting that point A corresponds to $\theta = 180°$.]

***13.103** A missile is fired from the ground with a velocity $\mathbf{v}_0$ of magnitude $v_0 = \sqrt{gR}$, forming an angle ϕ_0 with the vertical. (*a*) Express the maximum height d reached by the missile in terms of ϕ_0. (*b*) Show that the angle 2α subtending the trajectory BAC of the missile is equal to $2\phi_0$ and explain what happens when ϕ_0 approaches 90°. [*Hint.* Use Eq. (12.39) of Sec. 12.11 to solve part *b*, noting that $\theta = 180°$ for point A and $\theta = 180° - \alpha$ for point B.]

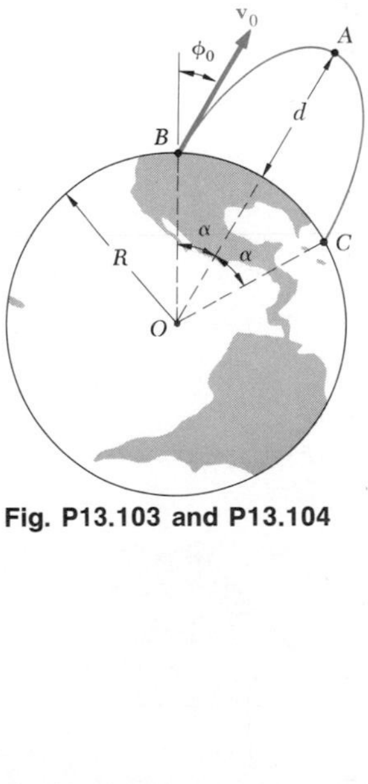

Fig. P13.103 and P13.104

***13.104** A missile is fired from the ground with an initial velocity $\mathbf{v}_0$ forming an angle ϕ_0 with the vertical. If the missile is to reach a maximum altitude equal to the radius of the earth, (*a*) show that the required angle ϕ_0 is defined by the relation

$$\sin\phi_0 = 2\sqrt{1 - \frac{1}{2}\left(\frac{v_{esc}}{v_0}\right)^2}$$

where v_{esc} is the escape velocity, (*b*) determine the maximum and minimum allowable values of v_0.

13.10. Principle of Impulse and Momentum. A third basic method for the solution of problems dealing with the motion of particles will be considered now. This method is based on the principle of impulse and momentum and may be used to solve problems involving force, mass, velocity, and time. It is of particular interest in the solution of problems involving impulsive motion or impact (Secs. 13.11 and 13.12).

Consider a particle of mass m acted upon by a force **F**. As we saw in Sec. 12.2, Newton's second law may be expressed in the form

$$\mathbf{F} = \frac{d}{dt}(m\mathbf{v}) \tag{13.27}$$

where $m\mathbf{v}$ is the linear momentum of the particle. Multiplying both sides of Eq. (13.27) by dt and integrating from a time t_1 to a time t_2, we write

$$\mathbf{F}\,dt = d(m\mathbf{v})$$

$$\int_{t_1}^{t_2} \mathbf{F}\,dt = m\mathbf{v}_2 - m\mathbf{v}_1$$

or, transposing the last term,

$$m\mathbf{v}_1 + \int_{t_1}^{t_2} \mathbf{F}\,dt = m\mathbf{v}_2 \tag{13.28}$$

The integral in Eq. (13.28) is a vector known as the *linear impulse,* or simply the *impulse,* of the force **F** during the interval of time considered. Resolving **F** into rectangular components, we write

$$\begin{aligned}\mathbf{Imp}_{1\to2} &= \int_{t_1}^{t_2} \mathbf{F}\,dt \\ &= \mathbf{i}\int_{t_1}^{t_2} F_x\,dt + \mathbf{j}\int_{t_1}^{t_2} F_y\,dt + \mathbf{k}\int_{t_1}^{t_2} F_z\,dt\end{aligned} \tag{13.29}$$

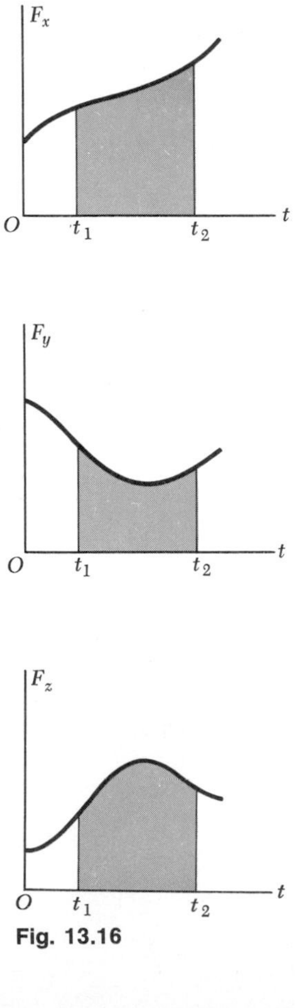

Fig. 13.16

and note that the components of the impulse of the force **F** are, respectively, equal to the areas under the curves obtained by plotting the components F_x, F_y, and F_z against t (Fig. 13.16). In the case of a force **F** of constant magnitude and direction, the impulse is represented by the vector $\mathbf{F}(t_2 - t_1)$, which has the same direction as **F**.

If SI units are used, the magnitude of the impulse of a force is expressed in N · s. But, recalling the definition of the newton, we have

$$\text{N}\cdot\text{s} = (\text{kg}\cdot\text{m/s}^2)\cdot\text{s} = \text{kg}\cdot\text{m/s}$$

which is the unit obtained in Sec. 12.3 for the linear momentum of a particle. We thus check that Eq. (13.28) is dimensionally correct. If U.S. customary units are used, the impulse of a force is expressed in lb · s, which is also the unit obtained in Sec. 12.3 for the linear momentum of a particle.

Equation (13.28) expresses that, when a particle is acted upon by a force $\mathbf{F}$ during a given time interval, *the final momentum* $m\mathbf{v}_2$ *of the particle may be obtained by adding vectorially its initial momentum* $m\mathbf{v}_1$ *and the impulse of the force* $\mathbf{F}$ *during the time interval considered* (Fig. 13.17). We write

$$m\mathbf{v}_1 + \mathbf{Imp}_{1\rightarrow 2} = m\mathbf{v}_2 \tag{13.30}$$

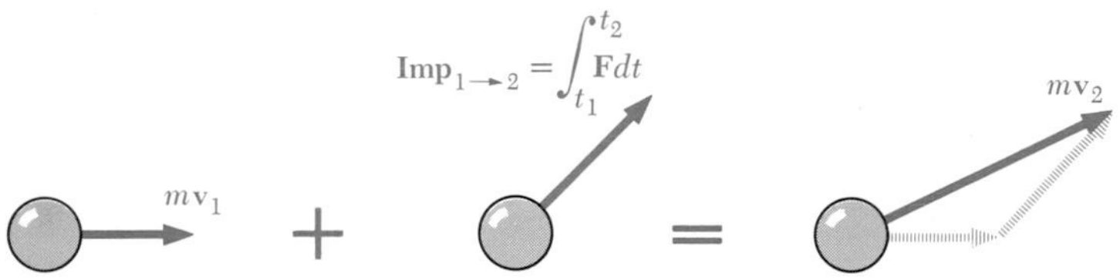

Fig. 13.17

We note that, while kinetic energy and work are scalar quantities, momentum and impulse are vector quantities. To obtain an analytic solution, it is thus necessary to replace Eq. (13.30) by the equivalent component equations

$$\begin{aligned}(mv_x)_1 + \int_{t_1}^{t_2} F_x\,dt &= (mv_x)_2\\ (mv_y)_1 + \int_{t_1}^{t_2} F_y\,dt &= (mv_y)_2\\ (mv_z)_1 + \int_{t_1}^{t_2} F_z\,dt &= (mv_z)_2\end{aligned} \tag{13.31}$$

When several forces act on a particle, the impulse of each of the forces must be considered. We have

$$m\mathbf{v}_1 + \Sigma\,\mathbf{Imp}_{1\rightarrow 2} = m\mathbf{v}_2 \tag{13.32}$$

Again, the equation obtained represents a relation between vector quantities; in the actual solution of a problem, it should be replaced by the corresponding component equations.

When a problem involves two particles or more, each particle may be considered separately and Eq. (13.32) may be written for each particle. We may also add vectorially the momenta of all the particles and the impulses of all the forces involved. We write then

$$\Sigma m\mathbf{v}_1 + \Sigma\,\mathbf{Imp}_{1\to 2} = \Sigma m\mathbf{v}_2 \qquad (13.33)$$

Since the forces of action and reaction exerted by the particles on each other form pairs of equal and opposite forces, and since the time interval from t_1 to t_2 is common to all the forces involved, the impulses of the forces of action and reaction cancel out, and only the impulses of the external forces need be considered.†

If no external force is exerted on the particles or, more generally, if the sum of the external forces is zero, the second term in Eq. (13.33) vanishes, and Eq. (13.33) reduces to

$$\Sigma m\mathbf{v}_1 = \Sigma m\mathbf{v}_2 \qquad (13.34)$$

which expresses that *the total momentum of the particles is conserved.* Consider, for example, two boats, of mass m_A and m_B, initially at rest, which are being pulled together (Fig. 13.18). If the resistance of the water is neglected, the only external forces acting on the boats are their weights and the buoyant forces exerted on them. Since these forces are balanced, we write

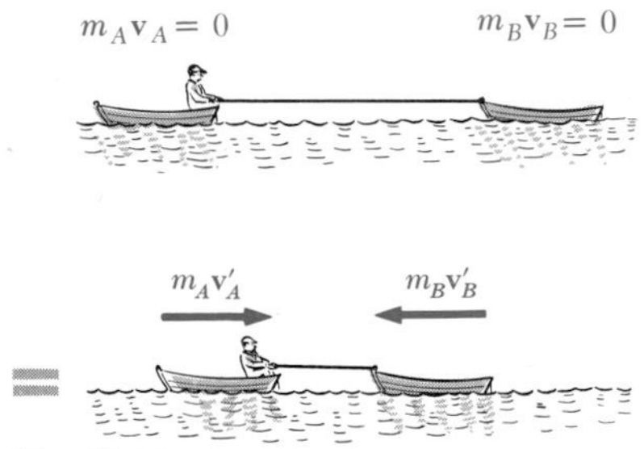

Fig. 13.18

$$\Sigma m\mathbf{v}_1 = \Sigma m\mathbf{v}_2$$
$$0 = m_A\mathbf{v}'_A + m_B\mathbf{v}'_B$$

where $\mathbf{v}'_A$ and $\mathbf{v}'_B$ represent the velocities of the boats after a finite interval of time. The equation obtained indicates that the boats move in opposite directions (toward each other) with velocities inversely proportional to their masses.‡

† We should note the difference between this statement and the corresponding statement made in Sec. 13.4 regarding the work of the forces of action and reaction between several particles. While the sum of the impulses of these forces is always zero, the sum of their work is zero only under special circumstances, e.g., when the various bodies involved are connected by inextensible cords or links and are thus constrained to move through equal distances.

‡ The application of the method of impulse and momentum to a system of particles and the concept of conservation of momentum for a system of particles are discussed in detail in Chap. 14.

13.11. Impulsive Motion. In some problems, a very large force may act during a very short time interval on a particle and produce a definite change in momentum. Such a force is called an *impulsive force* and the resulting motion an *impulsive motion.* For example, when a baseball is struck, the contact between bat and ball takes place during a very short time interval Δt. But the average value of the force $\mathbf{F}$ exerted by the bat on the ball is very large, and the resulting impulse $\mathbf{F}\,\Delta t$ is large enough to change the sense of motion of the ball (Fig. 13.19).

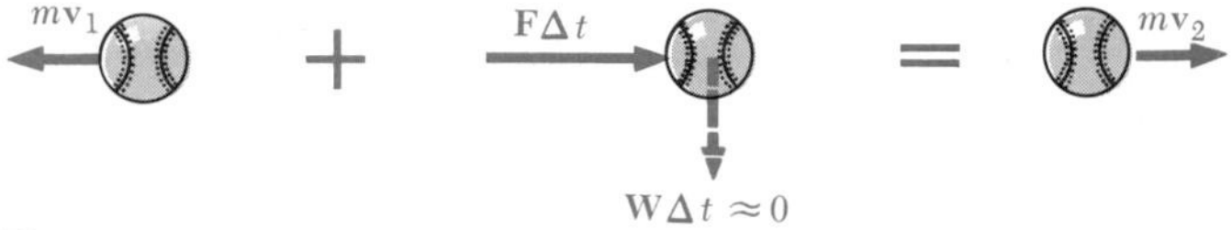

Fig. 13.19

When impulsive forces act on a particle, Eq. (13.32) becomes

$$m\mathbf{v}_1 + \Sigma\mathbf{F}\,\Delta t = m\mathbf{v}_2 \qquad (13.35)$$

Any force which is not an impulsive force may be neglected, since the corresponding impulse $\mathbf{F}\,\Delta t$ is very small. *Nonimpulsive forces* include the weight of the body, the force exerted by a spring, or any other force which is *known* to be small compared with an impulsive force. Unknown reactions may or may not be impulsive; their impulse should therefore be included in Eq. (13.35) as long as it has not been proved negligible. The impulse of the weight of the baseball considered above, for example, may be neglected. If the motion of the bat is analyzed, the impulse of the weight of the bat may also be neglected. The impulses of the reactions of the player's hands on the bat, however, should be included; these impulses will not be negligible if the ball is incorrectly hit.

In the case of the impulsive motion of several particles, Eq. (13.33) may be used. It reduces to

$$\Sigma m\mathbf{v}_1 + \Sigma\mathbf{F}\,\Delta t = \Sigma m\mathbf{v}_2 \qquad (13.36)$$

where the second term involves only impulsive, external forces. If all the external forces are nonimpulsive, the second term vanishes, and Eq. (13.36) reduces to Eq. (13.34); the total momentum of the particles is conserved.

SAMPLE PROBLEM 13.10

An automobile weighing 4000 lb is driven down a 5° incline at a speed of 60 mi/h when the brakes are applied, causing a constant total braking force (applied by the road on the tires) of 1500 lb. Determine the time required for the automobile to come to a stop.

Solution. We apply the principle of impulse and momentum. Since each force is constant in magnitude and direction, each corresponding impulse is equal to the product of the force and of the time interval t.

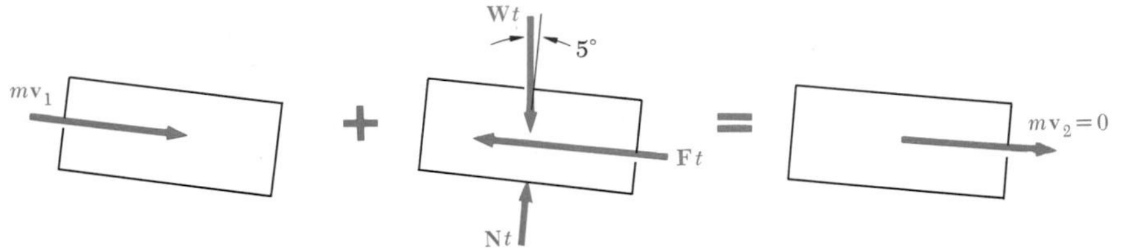

$$m\mathbf{v}_1 + \Sigma\, \mathbf{Imp}_{1\to 2} = m\mathbf{v}_2$$

$+\searrow x$ components: $\quad mv_1 + (W \sin 5^\circ)t - Ft = 0$

$$(4000/32.2)(88 \text{ ft/s}) + (4000 \sin 5^\circ)t - 1500t = 0$$

$$t = 9.49 \text{ s} \quad ◄$$

SAMPLE PROBLEM 13.11

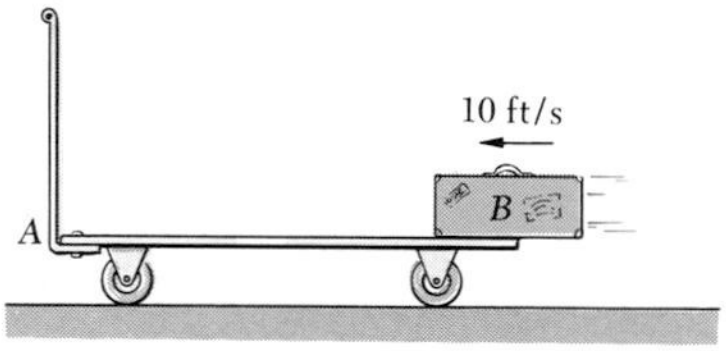

An airline employee tosses a 30-lb suitcase with a horizontal velocity of 10 ft/s onto a 70-lb baggage carrier. Knowing that the carrier can roll freely and is initially at rest, determine the velocity of the carrier after the suitcase has slid to a relative stop on the carrier.

Solution. We apply the principle of impulse and momentum to the carrier-suitcase system. Since the impulses of the internal forces cancel out, and since there are no horizontal external forces, the total momentum of the carrier and suitcase is conserved.

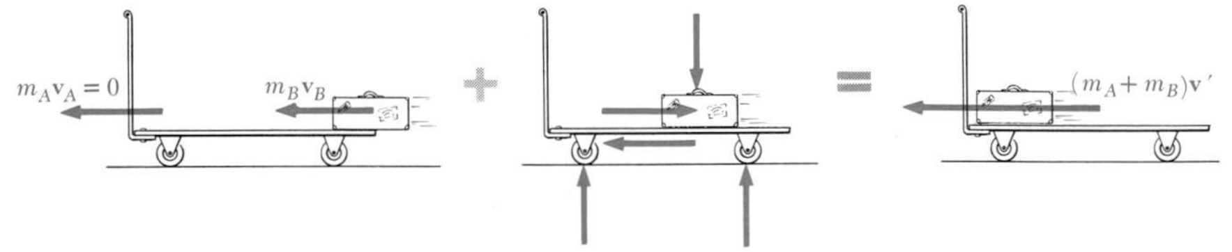

$$m_A\mathbf{v}_A + m_B\mathbf{v}_B = (m_A + m_B)\mathbf{v}'$$

$\overset{+}{\leftarrow} x$ components: $\quad 0 + \dfrac{30}{g}(10 \text{ ft/s}) = \dfrac{70 + 30}{g}v'$

$$v' = 3 \text{ ft/s} \quad ◄$$

SAMPLE PROBLEM 13.12

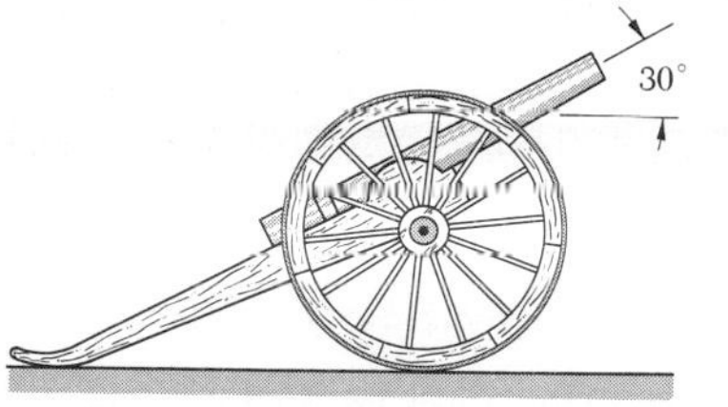

An old 2000-kg gun fires a 10-kg shell with an initial velocity of 600 m/s at an angle of 30°. The gun rests on a horizontal surface and is free to move horizontally. Assuming that the barrel of the gun is rigidly attached to the frame (no recoil mechanism) and that the shell leaves the barrel 6 ms after firing, determine the recoil velocity of the gun and the resultant **R** of the vertical impulsive forces exerted by the ground on the gun.

Solution. We first apply the principle of impulse and momentum to the shell and find the impulse $\mathbf{F}\,\Delta t$ exerted by the gun on the shell. We then apply it to the gun and find the final momentum of the gun and the impulse $\mathbf{R}\,\Delta t$ exerted by the ground on the gun. Since the time interval $\Delta t = 6\text{ ms} = 0.006\text{ s}$ is very short, we neglect all nonimpulsive forces.

Free Body: Shell

$$(m_S\mathbf{v}_S)_1 + \Sigma\,\mathbf{Imp}_{1\to2} = (m_S\mathbf{v}_S)_2$$
$$0 + F\,\Delta t = (10\text{ kg})(600\text{ m/s})$$
$$F\,\Delta t = 6000\text{ kg}\cdot\text{m/s} = 6000\text{ N}\cdot\text{s}$$

Free Body: Gun

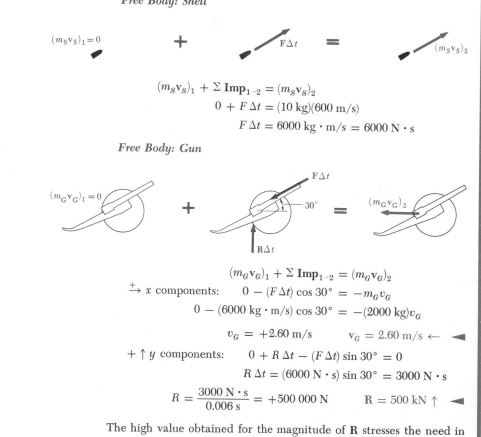

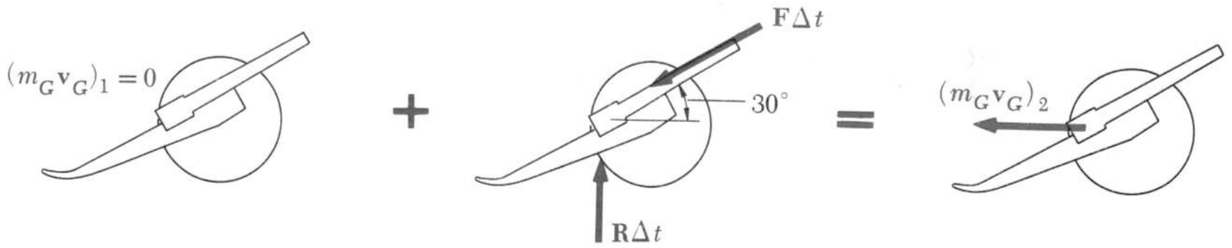

$$(m_G\mathbf{v}_G)_1 + \Sigma\,\mathbf{Imp}_{1\to2} = (m_G\mathbf{v}_G)_2$$

$\overset{+}{\rightarrow}$ x components: $\quad 0 - (F\,\Delta t)\cos 30° = -m_G v_G$

$$0 - (6000\text{ kg}\cdot\text{m/s})\cos 30° = -(2000\text{ kg})v_G$$

$$v_G = +2.60\text{ m/s} \qquad \mathbf{v}_G = 2.60\text{ m/s} \leftarrow \blacktriangleleft$$

$+\uparrow$ y components: $\quad 0 + R\,\Delta t - (F\,\Delta t)\sin 30° = 0$

$$R\,\Delta t = (6000\text{ N}\cdot\text{s})\sin 30° = 3000\text{ N}\cdot\text{s}$$

$$R = \frac{3000\text{ N}\cdot\text{s}}{0.006\text{ s}} = +500\,000\text{ N} \qquad \mathbf{R} = 500\text{ kN}\uparrow \blacktriangleleft$$

The high value obtained for the magnitude of **R** stresses the need in modern guns for a recoil mechanism which allows the barrel to move and brings it to rest over a period of time substantially longer than Δt. Although the total vertical impulse remains the same, the longer time interval results in a smaller value for the magnitude of **R**.

PROBLEMS

13.105 A 2750-lb automobile is moving at a speed of 45 mi/h when the brakes are fully applied, causing all four wheels to skid. Determine the time required to stop the automobile (*a*) on concrete ($\mu = 0.80$), (*b*) on ice ($\mu = 0.10$).

13.106 A tugboat exerts a constant force of 25 tons on a 200,000-ton oil tanker. Neglecting the frictional resistance of the water, determine the time required to increase the speed of the tanker (*a*) from 1 mi/h to 2 mi/h, (*b*) from 2 mi/h to 3 mi/h.

13.107 A 3-lb particle is acted upon by a force **F** of magnitude $F = 14t^2$ (lb) which acts in the direction of the unit vector $\boldsymbol{\lambda} = \frac{2}{7}\mathbf{i} + \frac{3}{7}\mathbf{j} + \frac{6}{7}\mathbf{k}$. Knowing that the velocity of the particle at $t = 0$ is $\mathbf{v} = (400 \text{ ft/s})\mathbf{j} - (250 \text{ ft/s})\mathbf{k}$, determine the velocity when $t = 3$ s.

13.108 A 2-kg particle is acted upon by the force, expressed in newtons, $\mathbf{F} = (8 - 6t)\mathbf{i} + (4 - t^2)\mathbf{j} + (4 + t)\mathbf{k}$. Knowing that the velocity of the particle is $\mathbf{v} = (150 \text{ m/s})\mathbf{i} + (100 \text{ m/s})\mathbf{j} - (250 \text{ m/s})\mathbf{k}$ at $t = 0$, determine (*a*) the time at which the velocity of the particle is parallel to the *yz* plane, (*b*) the corresponding velocity of the particle.

13.109 and 13.110 The initial velocity of the 50-kg car is 5 m/s to the left. Determine the time *t* at which the car has (*a*) no velocity, (*b*) a velocity of 5 m/s to the right.

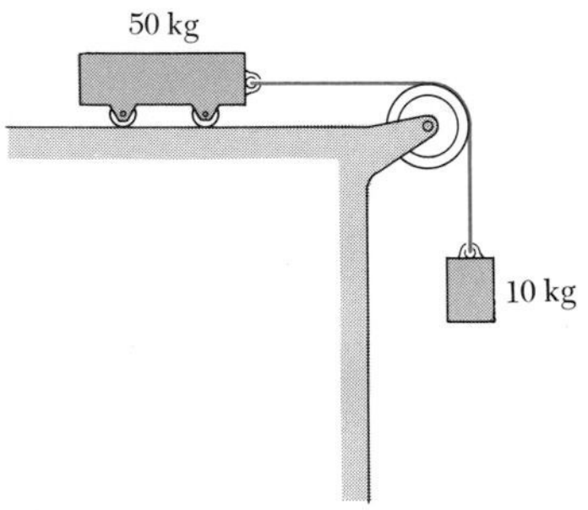

Fig. P13.109

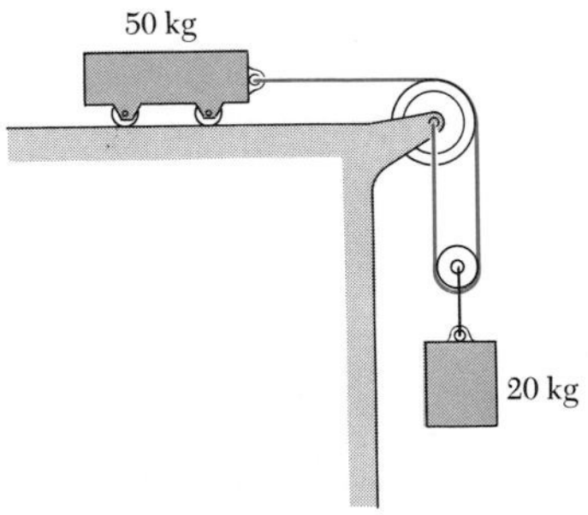

Fig. P13.110

13.111 Using the principle of impulse and momentum, solve Prob. 12.17*b*.

13.112 Using the principle of impulse and momentum, solve Prob. 12.18*b*.

13.113 A light train made of two cars travels at 100 km/h. The mass of car A is 15 Mg, and the mass of car B is 20 Mg. When the brakes are applied, a constant braking force of 25 kN is applied to each car. Determine (a) the time required for the train to stop after the brakes are applied, (b) the force in the coupling between the cars while the train is slowing down.

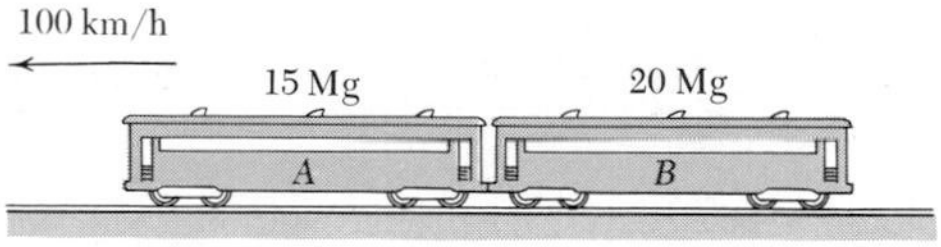

Fig. P13.113

13.114 Solve Prob. 13.113, assuming that a constant braking force of 25 kN is applied to car B but that the brakes on car A are not applied.

13.115 The 3-lb collar is initially at rest and is acted upon by the force **Q** which varies as shown. Knowing that $\mu = 0.25$, determine the velocity of the collar at (a) $t = 1$ s, (b) $t = 2$ s.

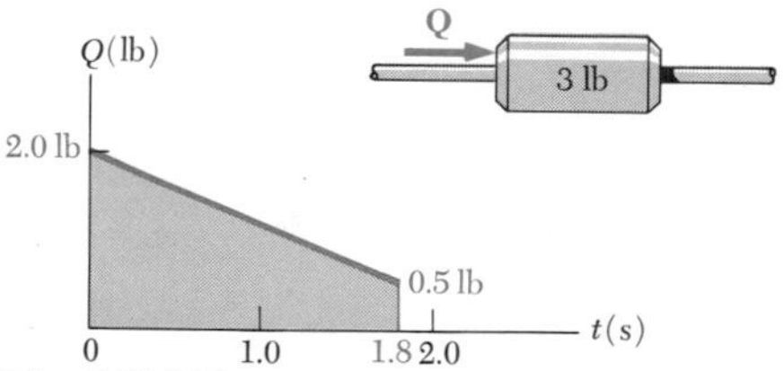

Fig. P13.115

13.116 In Prob. 13.115, determine (a) the maximum velocity reached by the collar and the corresponding time, (b) the time at which the collar comes to rest.

13.117 A 20-kg block is initially at rest and is subjected to a force **P** which varies as shown. Neglecting the effect of friction, determine (a) the maximum speed attained by the block, (b) the speed of the block at $t = 1.5$ s.

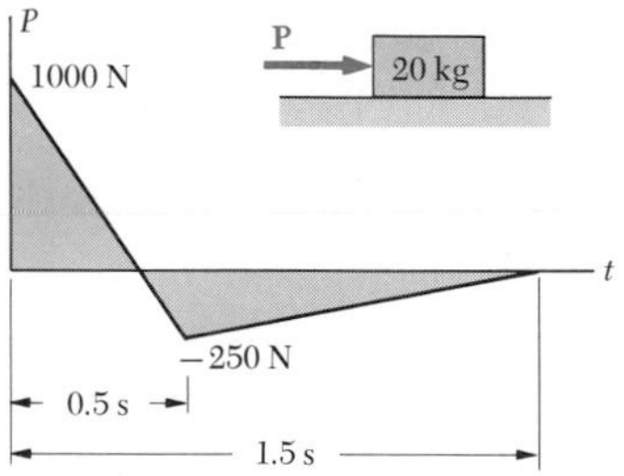

Fig. P13.117

13.118 Solve Prob. 13.117, assuming that $\mu = 0.25$ between the block and the surface.

13.119 A gun of mass 50 Mg is designed to fire a 250-kg shell with an initial velocity of 600 m/s. Determine the average force required to hold the gun motionless if the shell leaves the gun 0.02 s after being fired.

13.120 A 6000-kg plane lands on the deck of an aircraft carrier at a speed of 200 km/h relative to the carrier and is brought to a stop in 3.0 s. Determine the average horizontal force exerted by the carrier on the plane (*a*) if the carrier is at rest, (*b*) if the carrier is moving at a speed of 15 knots in the same direction as the airplane. (1 knot = 0.514 m/s.)

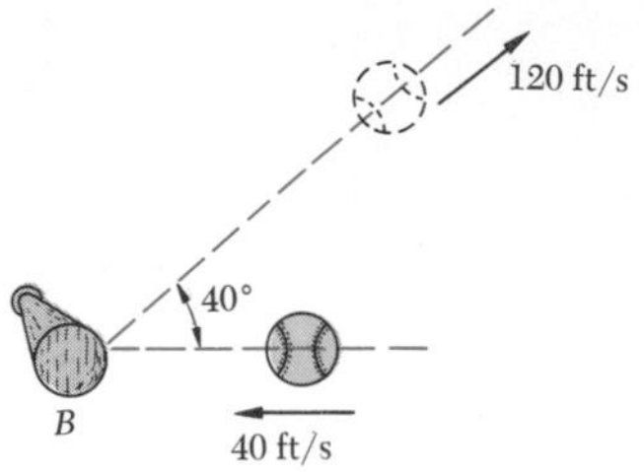

Fig. P13.121

13.121 A 4-oz baseball is pitched with a velocity of 40 ft/s toward a batter. After the ball is hit by the bat *B*, it has a velocity of 120 ft/s in the direction shown. If the bat and ball are in contact 0.025 s, determine the average impulsive force exerted on the ball during the impact.

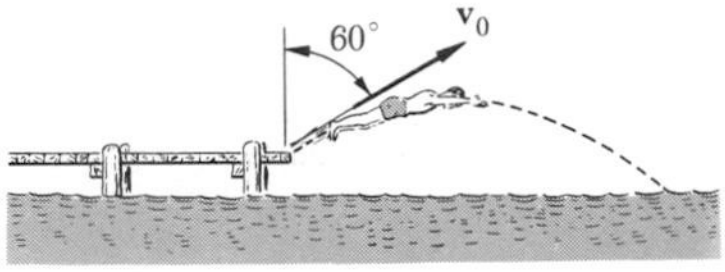

Fig. P13.122

13.122 A 160-lb man dives off the end of a pier with an initial velocity of 9 ft/s in the direction shown. Determine the horizontal and vertical components of the average force exerted on the pier during the 0.8 s that the man takes to leave the pier.

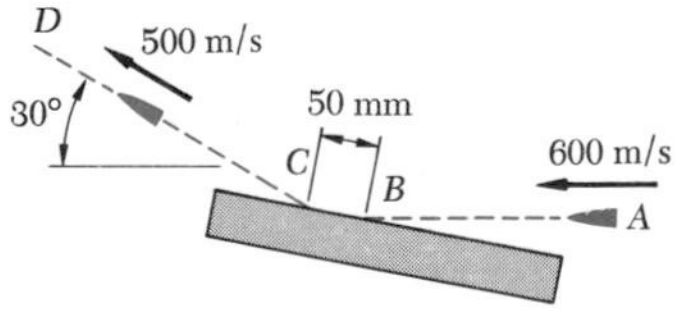

Fig. P13.123

13.123 A steel-jacketed bullet of mass 20 g is fired with a velocity of 600 m/s toward a steel plate; the bullet ricochets along the path *CD* with a velocity of 500 m/s. Knowing that the bullet caused a 50-mm scratch on the surface of the plate, determine the average impulsive force exerted on the bullet during its contact with the plate. (*Hint.* Assume an average speed of 550 m/s during contact.)

13.124 Determine the initial recoil velocity of an 8-lb rifle which fires a $\frac{3}{4}$-oz bullet with a velocity of 1600 ft/s.

13.125 A 2-oz rifle bullet is fired horizontally with a velocity of 1200 ft/s into an 8-lb block of wood which can move freely in the horizontal direction. Determine (*a*) the final velocity of the block, (*b*) the ratio of the final kinetic energy of the block and bullet to the initial kinetic energy of the bullet.

13.126 Collars *A* and *B* are moved toward each other, thus compressing the spring, and are then released from rest. The spring is not attached to the collars. Neglecting the effect of friction and knowing that collar *B* is observed to move to the right with a velocity of 6 m/s, determine (*a*) the corresponding velocity of collar *A*, (*b*) the kinetic energy of each collar.

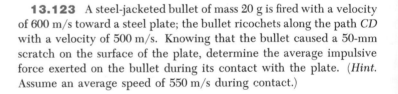

Fig. P13.126

13.127 A barge is initially at rest and carries a 600-kg crate. The barge has a mass of 3000 kg and is equipped with a winch which is used to move the crate along the deck. Neglecting any friction between the crate and the barge, determine (*a*) the velocity of both the barge and the crate when the winch is drawing in rope at the rate of 1.5 m/s, (*b*) the final position of the barge after 12 m of rope has been drawn in by the winch. (*c*) Solve parts *a* and *b* assuming that $\mu = 0.30$ between the crate and the barge.

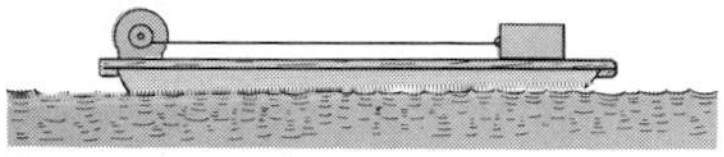

Fig. P13.127

13.128 A 60-ton railroad car is to be coupled to a second car which weighs 40 tons. If initially the speed of the 60-ton car is 1 mi/h and the 40-ton car is at rest, determine (*a*) the final speed of the coupled cars, (*b*) the average impulsive force acting on each car if the coupling is completed in 0.5 s.

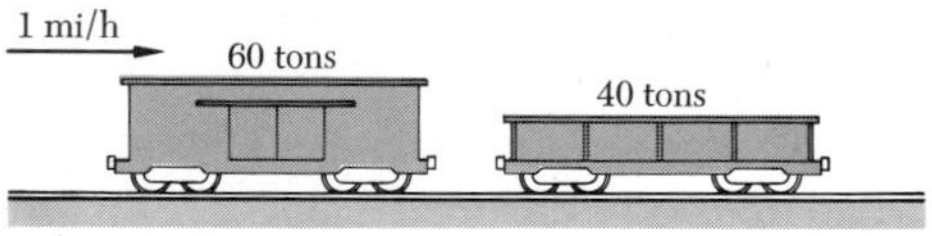

Fig. P13.128

13.129 Solve Prob. 13.128, assuming that, initially, the 60-ton car is at rest and the 40-ton car has a speed of 1 mi/h.

13.130 A 10-kg package is discharged from a conveyor belt with a velocity of 3 m/s and lands in a 25-kg cart. Knowing that the cart is initially at rest and may roll freely, determine the final velocity of the cart.

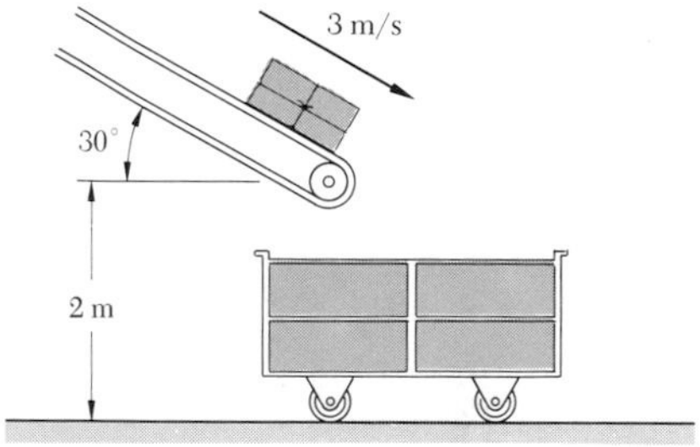

Fig. P13.130

13.131 Solve Prob. 13.130, assuming that the single 10-kg package is replaced by two 5-kg packages. The first 5-kg package comes to relative rest in the cart before the second package strikes the cart.

13.132 In order to test the resistance of a chain to impact, the chain is suspended from a 100-kg block supported by two columns. A rod attached to the last link of the chain is then hit by a 25-kg cylinder dropped from a 1.5-m height. Determine the initial impulse exerted on the chain, assuming that the impact is perfectly plastic and that the columns supporting the dead weight (*a*) are perfectly rigid, (*b*) are equivalent to two perfectly elastic springs. (*c*) Determine the energy absorbed by the chain in parts *a* and *b*.

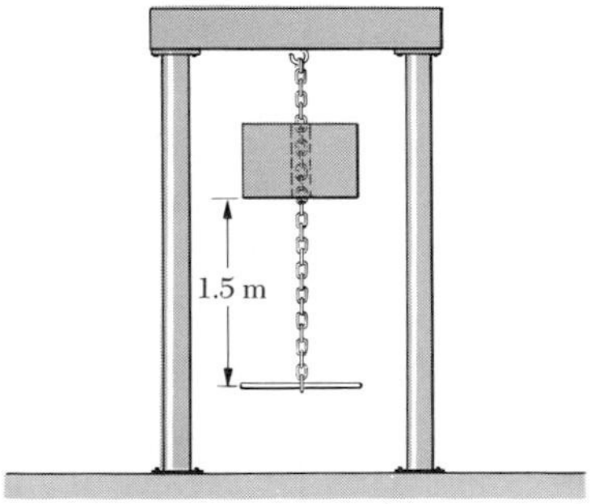

Fig. P13.132

13.133 A machine part is forged in a small drop forge. The hammer weighs 300 lb and is dropped from a height of 4 ft. Determine the initial impulse exerted on the machine part, assuming that the 800-lb anvil (*a*) is resting directly on hard ground, (*b*) is supported by springs.

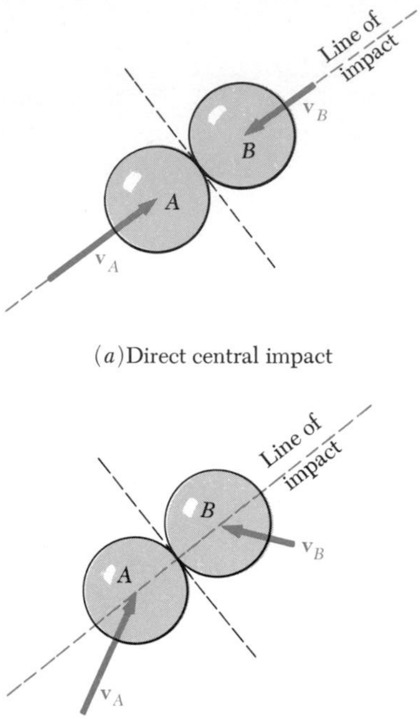

(*a*) Direct central impact

(*b*) Oblique central impact

Fig. 13.20

13.12. Impact.

A collision between two bodies which occurs in a very small interval of time, and during which the two bodies exert on each other relatively large forces, is called an *impact*. The common normal to the surfaces in contact during the impact is called the *line of impact*. If the mass centers of the two colliding bodies are located on this line, the impact is a *central impact*. Otherwise, the impact is said to be *eccentric*. We shall limit our present study to that of the central impact of two particles and postpone until later the analysis of the eccentric impact of two rigid bodies (Sec. 17.11).

If the velocities of the two particles are directed along the line of impact, the impact is said to be a *direct impact* (Fig. 13.20*a*). If, on the other hand, either or both particles move along a line other than the line of impact, the impact is said to be an *oblique impact* (Fig. 13.20*b*).

13.13. Direct Central Impact.

Consider two particles A and B, of mass m_A and m_B, which are moving in the same straight line and to the right with known velocities $\mathbf{v}_A$ and $\mathbf{v}_B$ (Fig. 13.21*a*). If $\mathbf{v}_A$ is larger than $\mathbf{v}_B$, particle A will eventually strike particle B. Under the impact, the two particles will *deform* and, at the end of the period of deformation, they will have the same velocity $\mathbf{u}$ (Fig. 13.21*b*). A period of *restitution* will then take place, at the end of which, depending upon the magnitude of the impact forces and upon the materials involved, the two particles either will have regained their original shape or will stay permanently deformed. Our purpose here is to determine the velocities $\mathbf{v}'_A$ and $\mathbf{v}'_B$ of the particles at the end of the period of restitution (Fig. 13.21*c*).

Considering first the two particles together, we note that there is no impulsive, external force. Thus, the total momentum of the two particles is conserved, and we write

$$m_A\mathbf{v}_A + m_B\mathbf{v}_B = m_A\mathbf{v}'_A + m_B\mathbf{v}'_B$$

Since all the velocities considered are directed along the same axis, we may replace the equation obtained by the following relation involving only scalar components:

$$m_A v_A + m_B v_B = m_A v'_A + m_B v'_B \tag{13.37}$$

A positive value for any of the scalar quantities v_A, v_B, v'_A, or v'_B means that the corresponding vector is directed to the right; a negative value indicates that the corresponding vector is directed to the left.

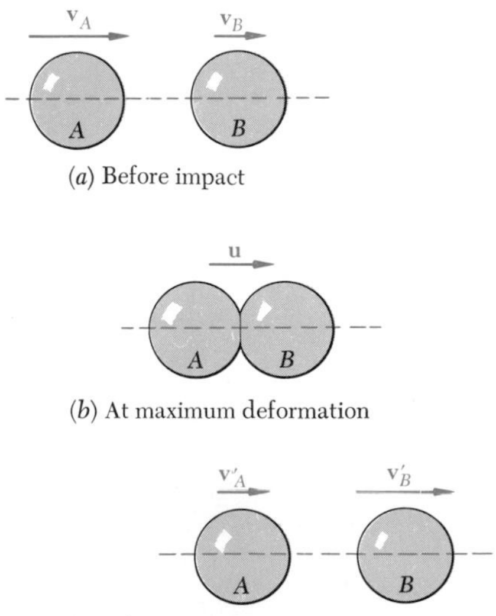

(*a*) Before impact

(*b*) At maximum deformation

(*c*) After impact

Fig. 13.21

To obtain the velocities $\mathbf{v}'_A$ and $\mathbf{v}'_B$, it is necessary to establish a second relation between the scalars v'_A and v'_B. For this purpose, we shall consider now the motion of particle A during the period of deformation and apply the principle of impulse and momentum. Since the only impulsive force acting on A during this period is the force $\mathbf{P}$ exerted by B (Fig. 13.22*a*), we write, using again scalar components,

$$m_A v_A - \int P\,dt = m_A u \tag{13.38}$$

where the integral extends over the period of deformation.

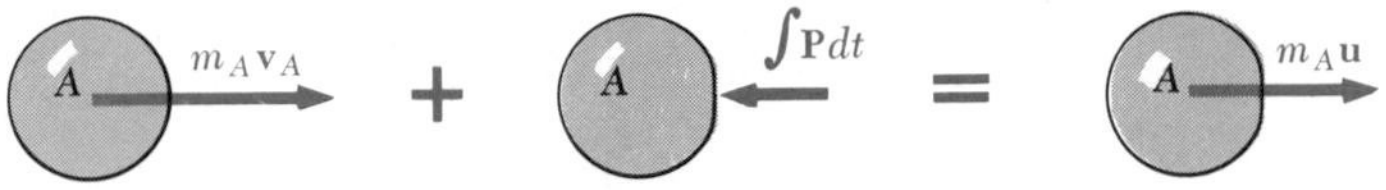

(*a*) Period of deformation

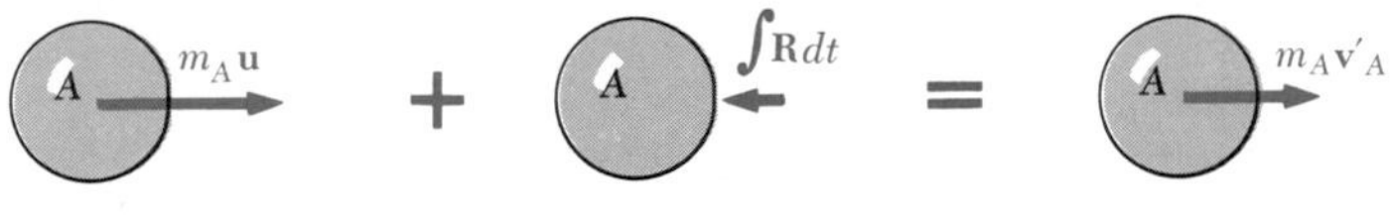

(*b*) Period of restitution

Fig. 13.22

Considering now the motion of A during the period of restitution, and denoting by $\mathbf{R}$ the force exerted by B on A during this period (Fig. 13.22*b*), we write

$$m_A u - \int R\,dt = m_A v'_A \tag{13.39}$$

where the integral extends over the period of restitution.

In general, the force $\mathbf{R}$ exerted on A during the period of restitution differs from the force $\mathbf{P}$ exerted during the period of deformation, and the magnitude $\int R\,dt$ of its impulse is smaller than the magnitude $\int P\,dt$ of the impulse of $\mathbf{P}$. The ratio of the magnitudes of the impulses corresponding respectively to the period of restitution and to the period of deformation is called the *coefficient of restitution* and is denoted by e. We write

$$e = \frac{\int R\,dt}{\int P\,dt} \tag{13.40}$$

The value of the coefficient e is always between 0 and 1 and depends to a large extent on the two materials involved. However, it also varies considerably with the impact velocity and the shape and size of the two colliding bodies.

Solving Eqs. (13.38) and (13.39) for the two impulses and substituting into (13.40), we write

$$e = \frac{u - v'_A}{v_A - u} \tag{13.41}$$

A similar analysis of particle B leads to the relation

$$e = \frac{v'_B - u}{u - v_B} \tag{13.42}$$

Since the quotients in (13.41) and (13.42) are equal, they are also equal to the quotient obtained by adding, respectively, their numerators and their denominators. We have, therefore,

$$e = \frac{(u - v'_A) + (v'_B - u)}{(v_A - u) + (u - v_B)} = \frac{v'_B - v'_A}{v_A - v_B}$$

and

$$v'_B - v'_A = e(v_A - v_B) \tag{13.43}$$

Since $v'_B - v'_A$ represents the relative velocity of the two particles after impact and $v_A - v_B$ their relative velocity before impact, formula (13.43) expresses that *the relative velocity of the two particles after impact may be obtained by multiplying their relative velocity before impact by the coefficient of restitution.* This property is used to determine experimentally the value of the coefficient of restitution of two given materials.

The velocities of the two particles after impact may now be obtained by solving Eqs. (13.37) and (13.43) simultaneously for v'_A and v'_B. It is recalled that the derivation of Eqs. (13.37) and (13.43) was based on the assumption that particle B is located to the right of A, and that both particles are initially moving to the right. If particle B is initially moving to the left, the scalar v_B should be considered negative. The same sign convention holds for the velocities after impact: a positive sign for v'_A will indicate that particle A moves to the right after impact, and a negative sign that it moves to the left.

Two particular cases of impact are of special interest:

1. $e = 0$, *Perfectly Plastic Impact.* When $e = 0$, Eq. (13.43) yields $v'_B = v'_A$. There is no period of restitution, and both particles stay together after impact. Substituting $v'_B = v'_A = v'$ into Eq. (13.37), which expresses that the total momentum of the particles is conserved, we write

$$m_A v_A + m_B v_B = (m_A + m_B)v' \tag{13.44}$$

This equation may be solved for the common velocity v' of the two particles after impact.

2. $e = 1$, *Perfectly Elastic Impact.* When $e = 1$, Eq. (13.43) reduces to

$$v'_B - v'_A = v_A - v_B \tag{13.45}$$

which expresses that the relative velocities before and after impact are equal. The impulses received by each particle during the period of deformation and during the period of restitution are equal. The particles move away from each other after impact with the same velocity with which they approached each other before impact. The velocities v'_A and v'_B may be obtained by solving Eqs. (13.37) and (13.45) simultaneously.

It is worth noting that, *in the case of a perfectly elastic impact, the total energy of the two particles,* as well as their total momentum, *is conserved.* Equations (13.37) and (13.45) may be written as follows:

$$m_A(v_A - v'_A) = m_B(v'_B - v_B) \tag{13.37'}$$

$$v_A + v'_A = v_B + v'_B \tag{13.45'}$$

Multiplying (13.37′) and (13.45′) member by member, we have

$$m_A(v_A - v'_A)(v_A + v'_A) = m_B(v'_B - v_B)(v'_B + v_B)$$

$$m_A v_A^2 - m_A(v'_A)^2 = m_B(v'_B)^2 - m_B v_B^2$$

Rearranging the terms in the equation obtained, and multiplying by $\frac{1}{2}$, we write

$$\tfrac{1}{2}m_A v_A^2 + \tfrac{1}{2}m_B v_B^2 = \tfrac{1}{2}m_A(v'_A)^2 + \tfrac{1}{2}m_B(v'_B)^2$$

which expresses that the kinetic energy of the particles is conserved. It should be noted, however, that *in the general case of impact,* i.e., when e is not equal to 1, *the total energy of the particles is not conserved.* This may be shown in any given case by comparing the kinetic energies before and after impact. The lost kinetic energy is in part transformed into heat and in part spent in generating elastic waves within the two colliding bodies.

13.14. Oblique Central Impact. Let us now consider the case when the velocities of the two colliding particles are *not* directed along the line of impact (Fig. 13.23). As indicated in Sec. 13.12, the impact is said to be *oblique.* Since the velocities $\mathbf{v}'_A$ and $\mathbf{v}'_B$ of the particles after impact are unknown in direction as well as in magnitude, their determination will require the use of four independent equations.

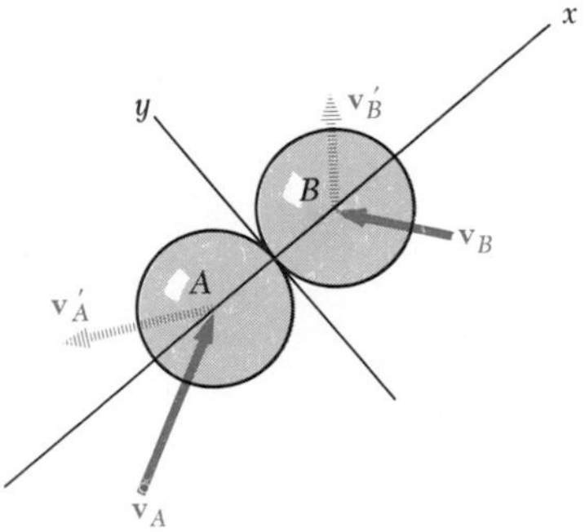

Fig. 13.23

We choose x and y axes, respectively, along the line of impact and along the common tangent to the surfaces in contact. Assuming that the particles are perfectly *smooth and frictionless,*

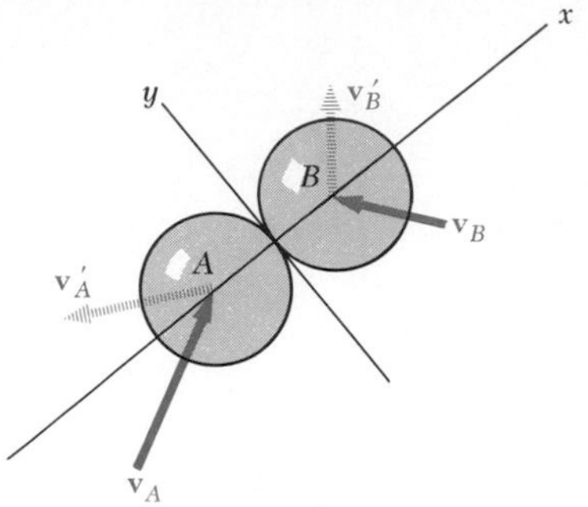

Fig. 13.23 (*repeated*)

we observe that the only impulsive forces acting on the particles during the impact are internal forces directed along the x axis. We may therefore express that:

1. The y component of the momentum of particle A is conserved.
2. The y component of the momentum of particle B is conserved.
3. The x component of the total momentum of the particles is conserved.
4. The x component of the relative velocity of the two particles after impact is obtained by multiplying the x component of their relative velocity before impact by the coefficient of restitution.

We thus obtain four independent equations which may be solved for the components of the velocities of A and B after impact. This method of solution is illustrated in Sample Prob. 13.15.

13.15. Problems Involving Energy and Momentum. We have now at our disposal three different methods for the solution of kinetics problems: the direct application of Newton's second law, $\Sigma \mathbf{F} = m\mathbf{a}$, the method of work and energy, and the method of impulse and momentum. To derive maximum benefit from these three methods, we should be able to choose the method best suited for the solution of a given problem. We should also be prepared to use different methods for solving the various parts of a problem when such a procedure seems advisable.

We have already seen that the method of work and energy is in many cases more expeditious than the direct application of Newton's second law. As indicated in Sec. 13.4, however, the method of work and energy has limitations, and it must sometimes be supplemented by the use of $\Sigma \mathbf{F} = m\mathbf{a}$. This is the case, for example, when we wish to determine an acceleration or a normal force.

There is generally no great advantage in using the method of impulse and momentum for the solution of problems involving no impulsive forces. It will usually be found that the equation $\Sigma \mathbf{F} = m\mathbf{a}$ yields a solution just as fast and that the method of work and energy, if it applies, is more rapid and more convenient. However, the method of impulse and momentum is the only practicable method in problems of impact. A solution based on the direct application of $\Sigma \mathbf{F} = m\mathbf{a}$ would be unwieldy, and the method of work and energy cannot be used since impact (unless perfectly elastic) involves a loss of mechanical energy.

Many problems involve only conservative forces, except for a short impact phase during which impulsive forces act. The solution of such problems may be divided into several parts. While the part corresponding to the impact phase calls for the use of the method of impulse and momentum and of the relation between relative velocities, the other parts may usually be solved by the method of work and energy. The use of the equation $\Sigma \mathbf{F} = m\mathbf{a}$ will be necessary, however, if the problem involves the determination of a normal force.

Consider, for example, a pendulum A, of mass m_A and length l, which is released with no velocity from a position A_1 (Fig. 13.24*a*). The pendulum swings freely in a vertical plane and hits a second pendulum B, of mass m_B and same length l, which is initially at rest. After the impact (with coefficient of restitution e), pendulum B swings through an angle θ that we wish to determine.

The solution of the problem may be divided into three parts:

1. *Pendulum A Swings from A_1 to A_2.* The principle of conservation of energy may be used to determine the velocity $(\mathbf{v}_A)_2$ of the pendulum at A_2 (Fig. 13.24*b*).
2. *Pendulum A Hits Pendulum B.* Using the fact that the total momentum of the two pendulums is conserved and the relation between their relative velocities, we determine the velocities $(\mathbf{v}_A)_3$ and $(\mathbf{v}_B)_3$ of the two pendulums after impact (Fig. 13.24*c*).
3. *Pendulum B Swings from B_3 to B_4.* Applying the principle of conservation of energy, we determine the maximum elevation y_4 reached by pendulum B (Fig. 13.24*d*). The angle θ may then be determined by trigonometry.

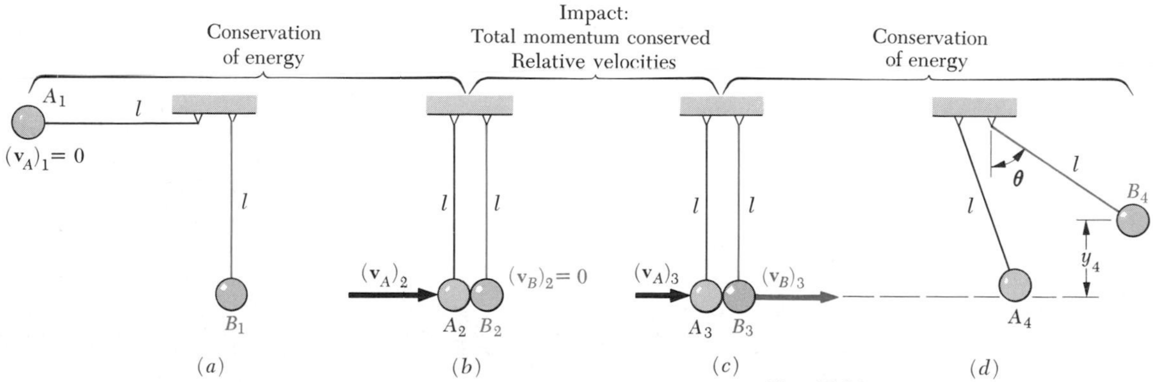

Fig. 13.24

We note that the method of solution just described should be supplemented by the use of $\Sigma \mathbf{F} = m\mathbf{a}$ if the tensions in the cords holding the pendulums are to be determined.

SAMPLE PROBLEM 13.13

A 20-Mg railroad car moving at a speed of 0.5 m/s to the right collides with a 35-Mg car which is at rest. If after the collision the 35-Mg car is observed to move to the right at a speed of 0.3 m/s, determine the coefficient of restitution between the two cars.

Solution. We express that the total momentum of the two cars is conserved.

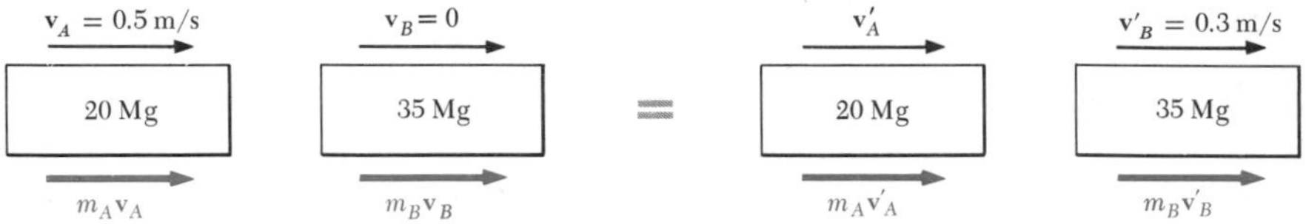

$$m_A \mathbf{v}_A + m_B \mathbf{v}_B = m_A \mathbf{v}'_A + m_B \mathbf{v}'_B$$

$$(20\text{ Mg})(+0.5\text{ m/s}) + (35\text{ Mg})(0) = (20\text{ Mg})v'_A + (35\text{ Mg})(+0.3\text{ m/s})$$

$$v'_A = -0.025\text{ m/s} \qquad \mathbf{v}'_A = 0.025\text{ m/s} \leftarrow$$

The coefficient of restitution is obtained by writing

$$e = \frac{v'_B - v'_A}{v_A - v_B} = \frac{+0.3 - (-0.025)}{+0.5 - 0} = \frac{0.325}{0.5} \qquad e = 0.65 \blacktriangleleft$$

SAMPLE PROBLEM 13.14

A ball is thrown against a frictionless vertical wall. Immediately before the ball strikes the wall, its velocity has a magnitude v and forms an angle of 30° with the horizontal. Knowing that $e = 0.90$, determine the magnitude and direction of the velocity of the ball as it rebounds from the wall.

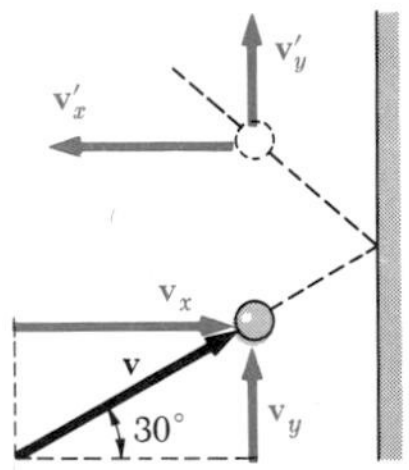

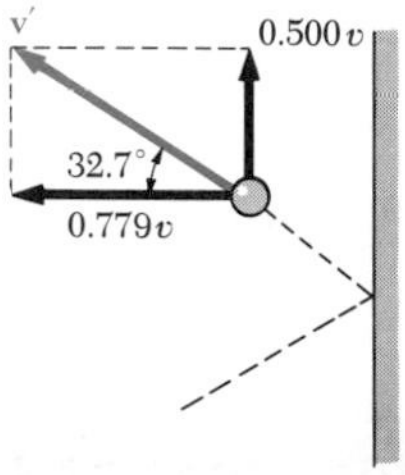

Solution. We resolve the initial velocity of the ball into components

$$v_x = v \cos 30° = 0.866v \qquad v_y = v \sin 30° = 0.500v$$

Vertical Motion. Since the wall is frictionless, no vertical impulsive force acts on the ball during the time it is in contact with the wall. The vertical component of the momentum, and hence the vertical component of the velocity, of the ball is thus unchanged:

$$\mathbf{v}'_y = \mathbf{v}_y = 0.500v \uparrow$$

Horizontal Motion. Since the mass of the wall (and earth) is essentially infinite, there is no point in expressing that the total momentum of the ball and the wall is conserved. Using the relation between relative velocities, we write

$$0 - v'_x = e(v_x - 0)$$
$$v'_x = -0.90(0.866v) = -0.779v \qquad \mathbf{v}'_x = 0.779v \leftarrow$$

Resultant Motion. Adding vectorially the components $\mathbf{v}'_x$ and $\mathbf{v}'_y$,

$$\mathbf{v}' = 0.926v \;\angle\; 32.7° \blacktriangleleft$$

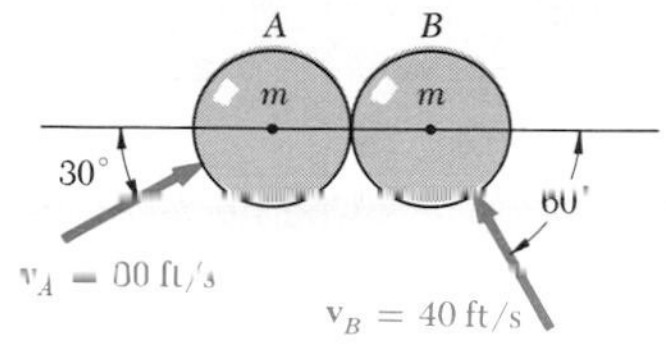

SAMPLE PROBLEM 13.15

The magnitude and direction of the velocities of two identical frictionless balls before they strike each other are as shown. Assuming $e = 0.90$, determine the magnitude and direction of the velocity of each ball after the impact.

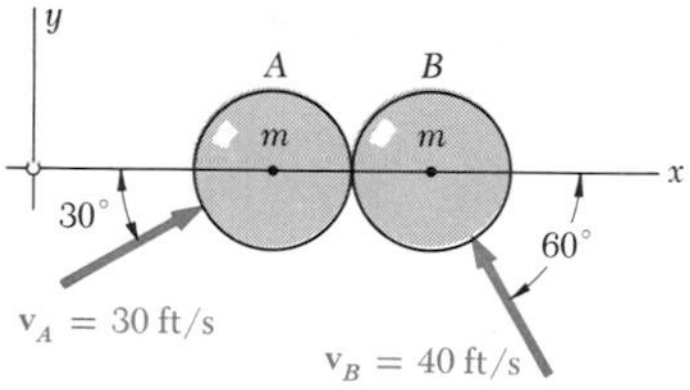

Solution. The impulsive forces acting between the balls during the impact are directed along a line joining the centers of the balls called the *line of impact.* Choosing x and y axes, respectively, parallel and perpendicular to the line of impact and directed as shown, we write

$$(v_A)_x = v_A \cos 30° = +26.0 \text{ ft/s}$$
$$(v_A)_y = v_A \sin 30° = +15.0 \text{ ft/s}$$
$$(v_B)_x = -v_B \cos 60° = -20.0 \text{ ft/s}$$
$$(v_B)_y = v_B \sin 60° = +34.6 \text{ ft/s}$$

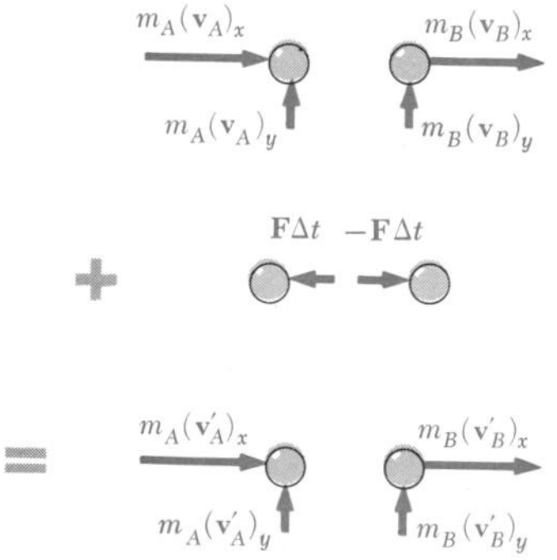

Principle of Impulse and Momentum. In the adjoining sketches we show in turn the initial momenta, the impulsive reactions, and the final momenta.

Motion Perpendicular to the Line of Impact. Considering only the y components, we apply the principle of impulse and momentum to each ball *separately.* Since no vertical impulsive force acts during the impact, the vertical component of the momentum, and hence the vertical component of the velocity, of each ball is unchanged.

$$(\mathbf{v}'_A)_y = 15.0 \text{ ft/s} \uparrow \qquad (\mathbf{v}'_B)_y = 34.6 \text{ ft/s} \uparrow$$

Motion Parallel to the Line of Impact. In the x direction, we consider the two balls together and note that, by Newton's third law, the internal impulses are, respectively, $\mathbf{F}\,\Delta t$ and $-\mathbf{F}\,\Delta t$ and cancel. We thus write that the total momentum of the balls is conserved:

$$m_A(v_A)_x + m_B(v_B)_x = m_A(v'_A)_x + m_B(v'_B)_x$$
$$m(26.0) + m(-20.0) = m(v'_A)_x + m(v'_B)_x$$
$$(v'_A)_x + (v'_B)_x = 6.0 \qquad (1)$$

Using the relation between relative velocities, we write

$$(v'_B)_x - (v'_A)_x = e[(v_A)_x - (v_B)_x]$$
$$(v'_B)_x - (v'_A)_x = (0.90)[26.0 - (-20.0)]$$
$$(v'_B)_x - (v'_A)_x = 41.4 \qquad (2)$$

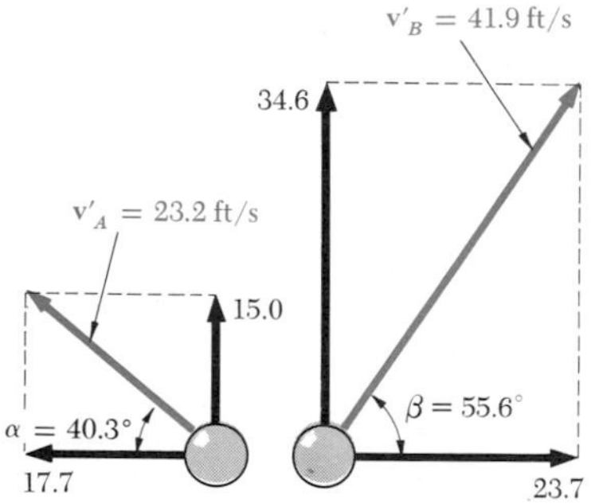

Solving Eqs. (1) and (2) simultaneously, we obtain

$$(v'_A)_x = -17.7 \qquad (v'_B)_x = +23.7$$
$$(\mathbf{v}'_A)_x = 17.7 \text{ ft/s} \leftarrow \qquad (\mathbf{v}'_B)_x = 23.7 \text{ ft/s} \rightarrow$$

Resultant Motion. Adding vectorially the velocity components of each ball, we obtain

$$\mathbf{v}'_A = 23.2 \text{ ft/s} \measuredangle 40.3° \qquad \mathbf{v}'_B = 41.9 \text{ ft/s} \measuredangle 55.6° \blacktriangleleft$$

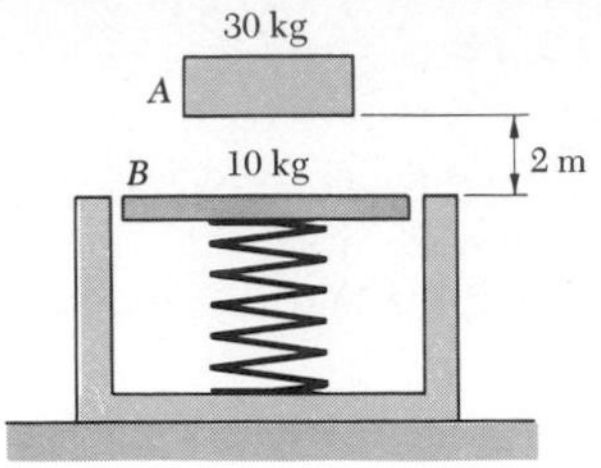

SAMPLE PROBLEM 13.16

A 30-kg block is dropped from a height of 2 m onto the 10-kg pan of a spring scale. Assuming the impact to be perfectly plastic, determine the maximum deflection of the pan. The constant of the spring is $k = 20$ kN/m.

Solution. The impact between the block and the pan *must* be treated separately; therefore we divide the solution into three parts.

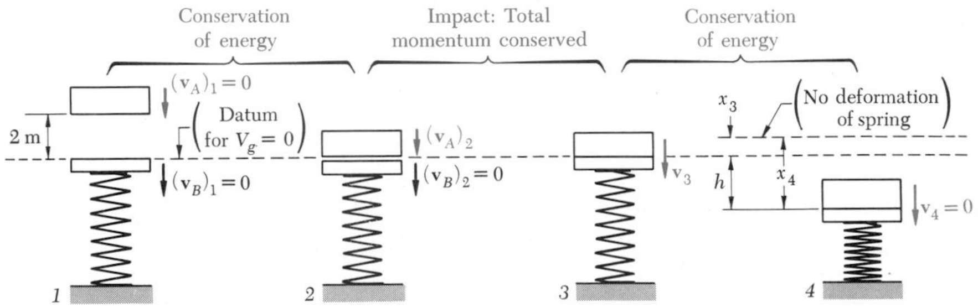

Conservation of Energy. Block: $W_A = (30 \text{ kg})(9.81 \text{ m/s}^2) = 294 \text{ N}$

$$T_1 = \tfrac{1}{2}m_A(v_A)_1^2 = 0 \qquad V_1 = W_A y = (294 \text{ N})(2 \text{ m}) = 588 \text{ J}$$
$$T_2 = \tfrac{1}{2}m_A(v_A)_2^2 = \tfrac{1}{2}(30 \text{ kg})(v_A)_2^2 \qquad V_2 = 0$$
$$T_1 + V_1 = T_2 + V_2: \qquad 0 + 588 \text{ J} = \tfrac{1}{2}(30 \text{ kg})(v_A)_2^2 + 0$$
$$(v_A)_2 = +6.26 \text{ m/s} \qquad (\mathbf{v}_A)_2 = 6.26 \text{ m/s} \downarrow$$

Impact: Conservation of Momentum. Since the impact is perfectly plastic, $e = 0$; the block and pan move together after the impact.

$$m_A(v_A)_2 + m_B(v_B)_2 = (m_A + m_B)v_3$$
$$(30 \text{ kg})(6.26 \text{ m/s}) + 0 = (30 \text{ kg} + 10 \text{ kg})v_3$$
$$v_3 = +4.70 \text{ m/s} \qquad \mathbf{v}_3 = 4.70 \text{ m/s} \downarrow$$

Conservation of Energy. Initially the spring supports the weight W_B of the pan; thus the initial deflection of the spring is

$$x_3 = \frac{W_B}{k} = \frac{(10 \text{ kg})(9.81 \text{ m/s}^2)}{20 \times 10^3 \text{ N/m}} = \frac{98.1 \text{ N}}{20 \times 10^3 \text{ N/m}} = 4.91 \times 10^{-3} \text{ m}$$

Denoting by x_4 the total maximum deflection of the spring, we write

$$T_3 = \tfrac{1}{2}(m_A + m_B)v_3^2 = \tfrac{1}{2}(30 \text{ kg} + 10 \text{ kg})(4.70 \text{ m/s})^2 = 442 \text{ J}$$
$$V_3 = V_g + V_e = 0 + \tfrac{1}{2}kx_3^2 = \tfrac{1}{2}(20 \times 10^3)(4.91 \times 10^{-3})^2 = 0.241 \text{ J}$$
$$T_4 = 0$$
$$V_4 = V_g + V_e = (W_A + W_B)(-h) + \tfrac{1}{2}kx_4^2 = -(392)h + \tfrac{1}{2}(20 \times 10^3)x_4^2$$

Noting that the displacement of the pan is $h = x_4 - x_3$, we write

$$T_3 + V_3 = T_4 + V_4:$$
$$442 + 0.241 = 0 - 392(x_4 - 4.91 \times 10^{-3}) + \tfrac{1}{2}(20 \times 10^3)x_4^2$$
$$x_4 = 0.230 \text{ m} \qquad h = x_4 - x_3 = 0.230 \text{ m} - 4.91 \times 10^{-3} \text{ m}$$
$$h = 0.225 \text{ m} \qquad h = 225 \text{ mm} ◄$$

PROBLEMS

13.134 The coefficient of restitution between the two collars is known to be 0.75; determine (*a*) their velocities after impact, (*b*) the energy loss during the impact.

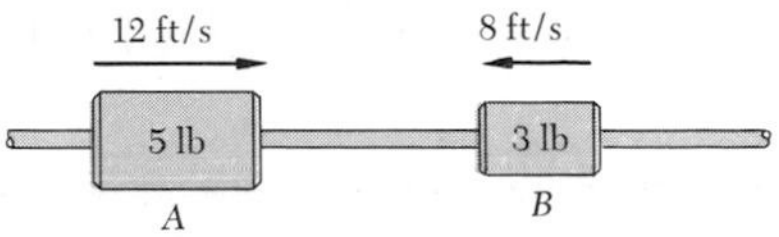

Fig. P13.134

13.135 Solve Prob. 13.134, assuming that the velocity of collar *B* is 4 ft/s to the right.

13.136 Two steel blocks slide without friction on a horizontal surface; immediately before impact their velocities are as shown. Knowing that $e = 0.75$, determine (*a*) their velocities after impact, (*b*) the energy loss during impact.

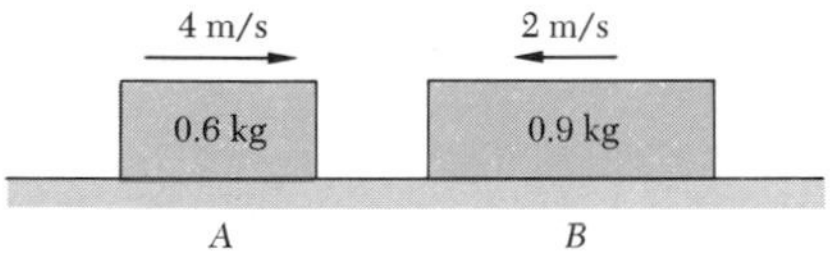

Fig. P13.136 and P13.137

13.137 The velocities of two steel blocks before impact are as shown. If after the impact the velocity of block *B* is observed to be 2.5 m/s to the right, determine the coefficient of restitution between the two blocks.

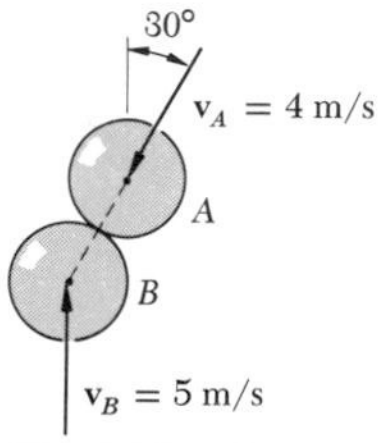

Fig. P13.138

13.138 As ball *A* is falling, a juggler tosses an identical ball *B* which strikes ball *A*. The line of impact forms an angle of 30° with the vertical. Assuming the balls frictionless and $e = 0.8$, determine the velocity of each ball immediately after impact.

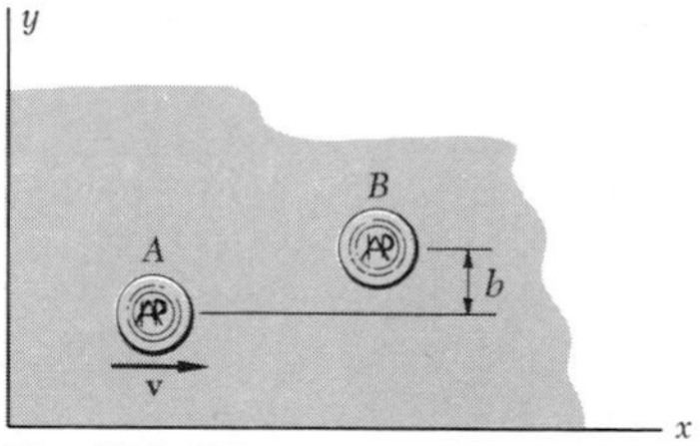

Fig. P13.139

13.139 Two identical pucks *A* and *B*, of 80-mm diameter, may move freely on a hockey rink. Puck *B* is at rest and puck *A* has an initial velocity **v** as shown. (*a*) Knowing that $b = 40$ mm and $e = 0.80$, and assuming no friction, determine the velocity of each puck after the impact. (*b*) Show that if $e = 1$, the final velocities of the pucks form a right angle for all values of *b*.

13.140 Assuming perfectly elastic impact, determine the velocity imparted to a quarter-dollar coin which is at rest and is struck squarely by (*a*) a dime moving with a velocity $\mathbf{v}_0$, (*b*) a half-dollar moving with a velocity $\mathbf{v}_0$. (*Masses:* half-dollar, 192.9 grains; quarter dollar, 96.45 grains; dime, 38.58 grains.)

13.141 A dime which is at rest on a *rough* surface is struck squarely by a half dollar moving to the right. After the impact, each coin slides and comes to rest; the dime slides 19.2 in. to the right, and the half dollar slides 3.8 in. to the right. Assuming the coefficient of friction is the same for each coin, determine the value of the coefficient of restitution between the coins. (See Prob. 13.140 for the mass of United States coins.)

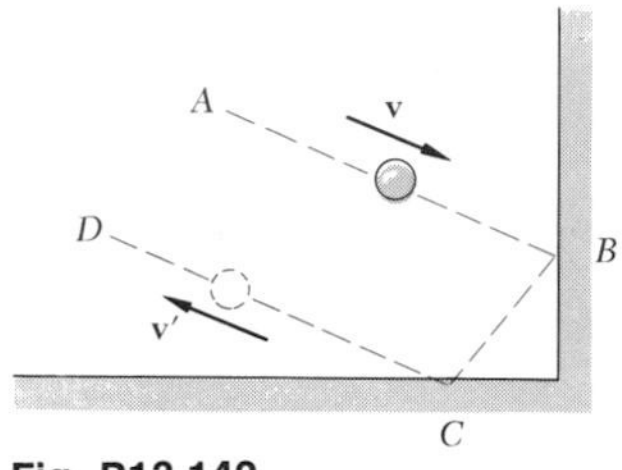

Fig. P13.142

13.142 A ball is thrown into a 90° corner with an initial velocity **v**. Denoting the coefficient of restitution by e and assuming no friction, show that the final velocity **v**′ is of magnitude ev and that the initial and final paths AB and CD are parallel.

13.143 A steel ball falling vertically strikes a rigid plate A and rebounds horizontally as shown. Denoting by e the coefficient of restitution and assuming no friction, determine (*a*) the required angle θ, (*b*) the magnitude of the velocity $\mathbf{v}_1$.

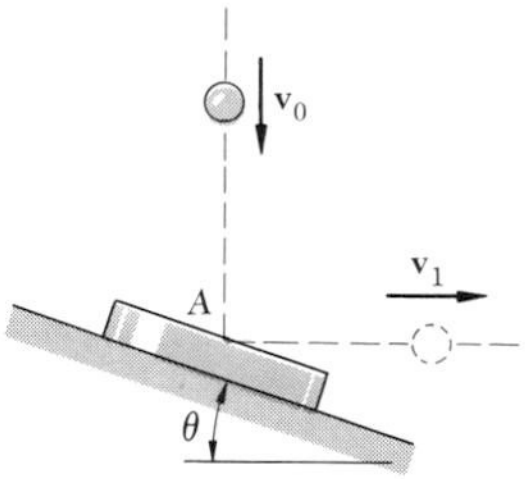

Fig. P13.143

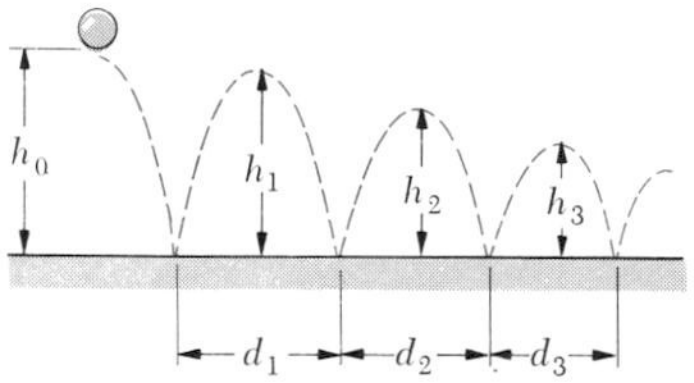

Fig. P13.144, P13.145, and P13.146

13.144 A ball is dropped from a height $h_0 = 36$ in. onto a frictionless floor. Knowing that for the first bounce $h_1 = 32$ in. and $d_1 = 16$ in., determine (*a*) the coefficient of restitution, (*b*) the height and length of the second bounce.

13.145 A ball is dropped onto a frictionless floor and allowed to bounce several times as shown. Derive an expression for the coefficient of restitution in terms of (*a*) the heights of two successive bounces h_n and h_{n+1}, (*b*) the lengths of two successive bounces d_n and d_{n+1}, (*c*) the durations of two successive bounces t_n and t_{n+1}.

13.146 A ball is dropped onto a frictionless floor and bounces as shown. The lengths of the first two bounces are measured and found to be $d_1 = 14.5$ in. and $d_2 = 12.8$ in. Determine (*a*) the coefficient of restitution, (*b*) the expected length d_3 of the third bounce.

13.147 A ball is dropped from a height h above the landing and bounces down a flight of stairs. Denoting by e the coefficient of restitution, determine the value of h for which the ball will bounce to the same height above each step.

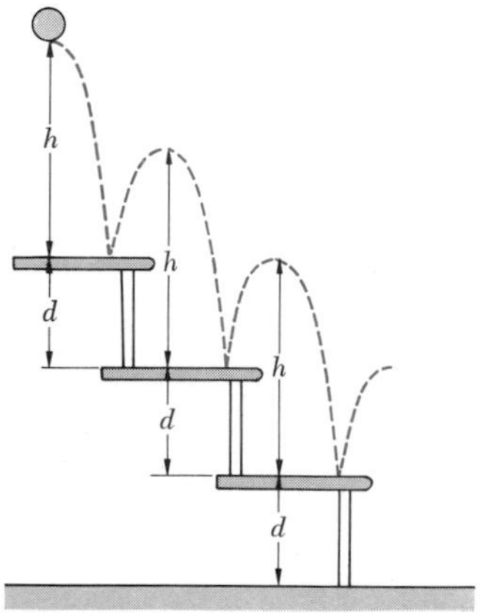

Fig. P13.147

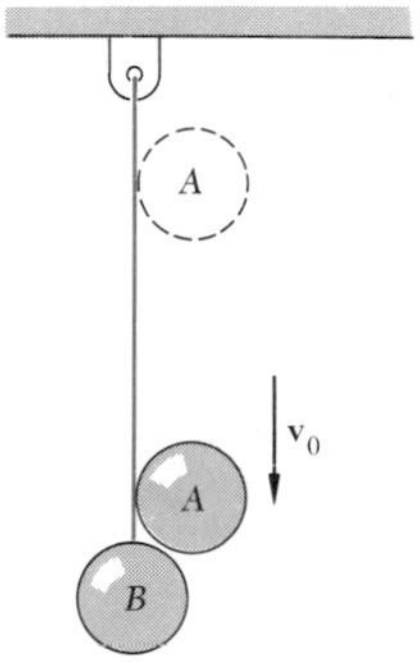

Fig. P13.148

***13.148** Ball B is suspended by an inextensible cord. An identical ball A is released from rest when it is just touching the cord and acquires a velocity $\mathbf{v}_0$ before striking ball B. Assuming $e = 1$ and no friction, determine the velocity of each ball immediately after impact.

***13.149** A 2-kg sphere moving to the left with a velocity of 10 m/s strikes the frictionless, inclined surface of a 5-kg block which is at rest. The block rests on rollers and may move freely in the horizontal direction. Knowing that $e = 0.75$, determine the velocities of the block and of the sphere immediately after impact.

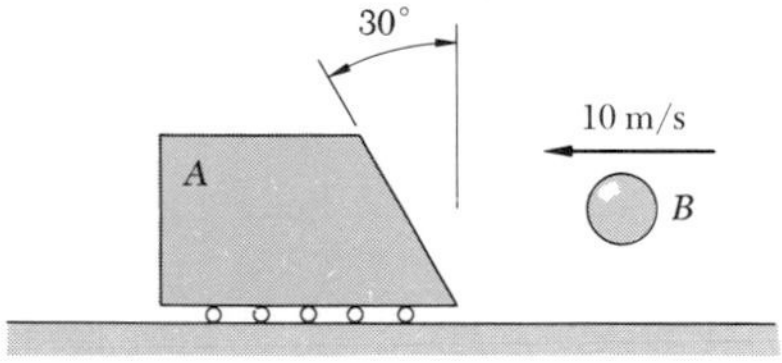

Fig. P13.149

13.150 The 4.5-kg sphere A strikes the 1.5-kg sphere B. Knowing that $e = 0.90$, determine the angle θ_A at which A must be released if the maximum angle θ_B reached by B is to be 90°.

13.151 The 4.5-kg sphere A is released from rest when $\theta_A = 60°$ and strikes the 1.5-kg sphere B. Knowing that $e = 0.90$, determine (*a*) the highest position to which sphere B will rise, (*b*) the maximum tension which will occur in the cord holding B.

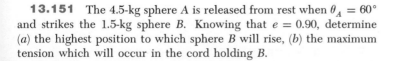

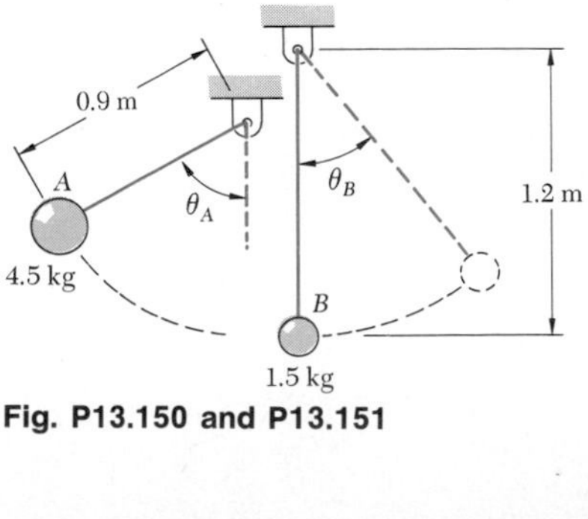

Fig. P13.150 and P13.151

13.152 Block A is released when $\theta_A = 90°$ and slides without friction until it strikes ball B. Knowing that $e = 0.90$, determine (*a*) the velocity of B immediately after impact, (*b*) the maximum tension in the cord holding B, (*c*) the maximum height to which ball B will rise.

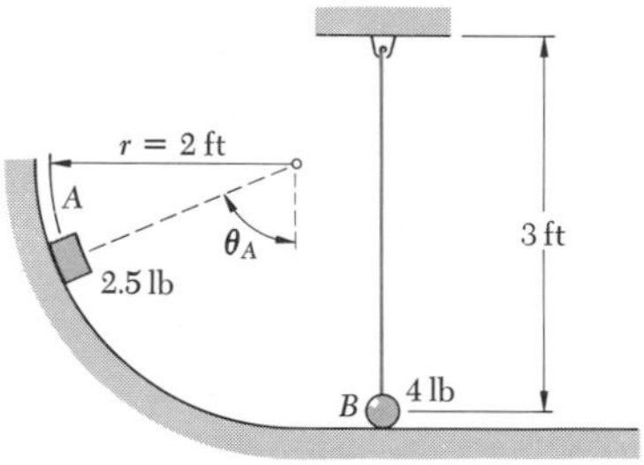

Fig. P13.152

13.153 What should be the value of the angle θ_A in Prob. 13.152 if the maximum angle between the cord holding ball B and the vertical is to be 45°?

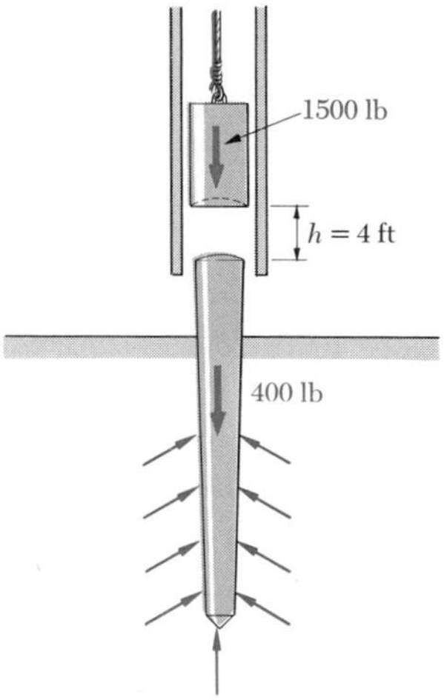

Fig. P13.154 and P13.155

13.154 It is desired to drive the 400-lb pile into the ground until the resistance to its penetration is 24,000 lb. Each blow of the 1500-lb hammer is the result of a 4-ft free fall onto the top of the pile. Determine how far the pile will be driven into the ground by a single blow when the 24,000-lb resistance is achieved. Assume that the impact is perfectly plastic.

13.155 The 1500-lb hammer of a drop-hammer pile driver falls from a height of 4 ft onto the top of a 400-lb pile. The pile is driven 4 in. into the ground. Assuming perfectly plastic impact, determine the average resistance of the ground to penetration.

13.156 Cylinder A is dropped 2 m onto cylinder B, which is resting on a spring of constant $k = 3$ kN/m. Assuming a perfectly plastic impact, determine (*a*) the maximum deflection of cylinder B, (*b*) the energy loss during the impact.

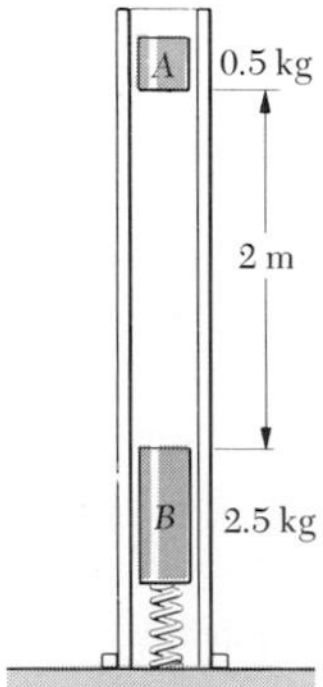

Fig. P13.156

13.157 The efficiency η of a drop-hammer pile driver may be defined as the ratio of the kinetic energy available after impact to the kinetic energy immediately before impact. Denoting by r the ratio of the pile mass m_p to the hammer mass m_h, and assuming perfectly plastic impact, show that $\eta = 1/(1 + r)$.

13.158 A bumper is designed to protect a 1600-kg automobile from damage when it hits a rigid wall at speeds up to 12 km/h. Assuming perfectly plastic impact, determine (*a*) the energy absorbed by the bumper during the impact, (*b*) the speed at which the automobile can hit another 1600-kg automobile without incurring any damage, if the other automobile is at rest and is similarly protected.

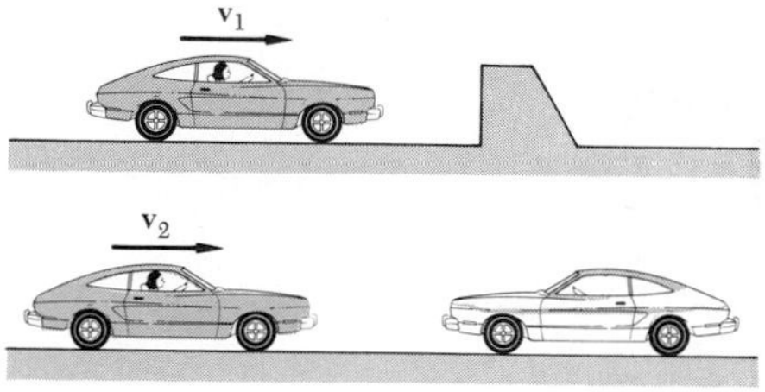

Fig. P13.158

13.159 Solve Prob. 13.158, assuming a coefficient of restitution $e = 0.50$. Show that the answer to part *b* is independent of e.

13.160 A small rivet connecting two pieces of sheet metal is being clinched by hammering. Determine the energy absorbed by the rivet under each blow, knowing that the head of the hammer weighs 1.5 lb and that it strikes the rivet with a velocity of 20 ft/s. Assume that the anvil is supported by springs and (*a*) is infinite in weight (rigid support), (*b*) weighs 10 lb.

***13.161** A ball of mass m_A moving to the right with a velocity $\mathbf{v}_A$ strikes a second ball of mass m_B which is at rest. Derive an expression for the kinetic-energy loss during impact. Assume that the balls strike each other squarely, and denote the coefficient of restitution by e.

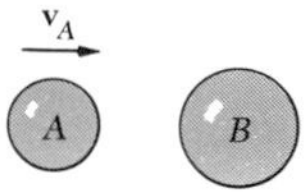

Fig. P13.161

REVIEW PROBLEMS

13.162 Collar B has an initial velocity of 2 m/s. It strikes collar A causing a series of impacts involving the collars and the fixed support at C. Assuming $e = 1$ for all impacts and neglecting friction, determine (*a*) the number of impacts which will occur, (*b*) the final velocity of B, (*c*) the final position of A.

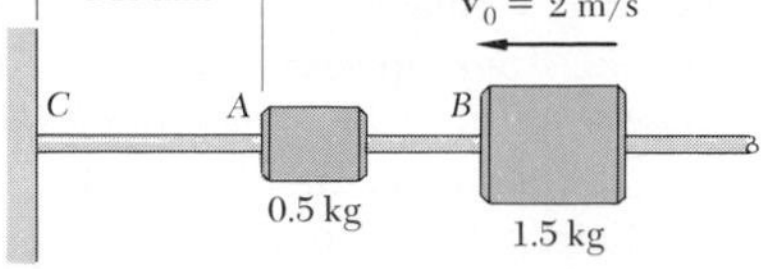

Fig. P13.162

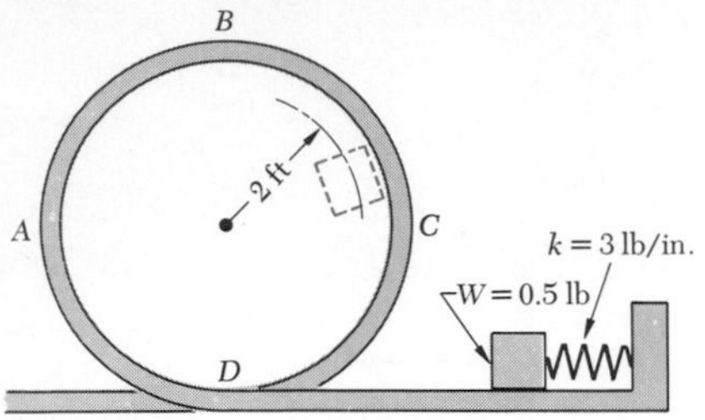

Fig. P13.163

13.163 The 0.5-lb pellet is released when the spring is compressed 6 in. and travels without friction around the vertical loop *ABCD*. Determine the force exerted by the loop on the pellet (*a*) at point *A*, (*b*) at point *B*, (*c*) at point *C*.

13.164 In Prob. 13.163, determine the smallest allowable deflection of the spring if the pellet is to travel around the entire loop without leaving the track.

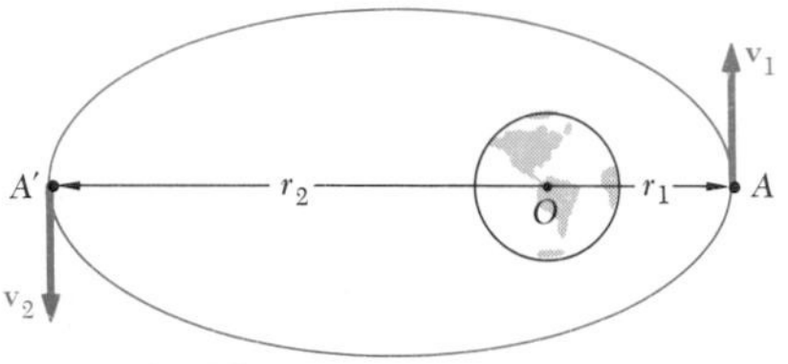

Fig. P13.165

13.165 Show that the values v_1 and v_2 of the speed of a satellite at the perigee *A* and the apogee *A'* of an elliptic orbit are defined by the relations

$$v_1^2 = \frac{2GM}{r_1 + r_2}\frac{r_2}{r_1} \qquad v_2^2 = \frac{2GM}{r_1 + r_2}\frac{r_1}{r_2}$$

13.166 The 20-Mg truck and the 40-Mg railroad flatcar are both at rest with their brakes released. An engine bumps the flatcar and causes the flatcar alone to start moving with a velocity of 1 m/s to the right. Assuming $e = 1$ between the truck and the ends of the flatcar, determine the velocities of the truck and of the flatcar after end *A* strikes the truck. Describe the subsequent motion of the system. Neglect the effect of friction.

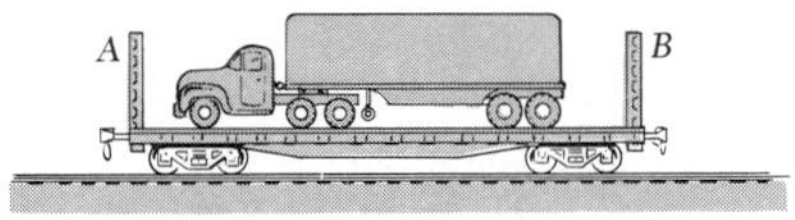

Fig. P13.166

13.167 An elevator travels upward at a constant speed of 2 m/s. A boy riding the elevator throws a 0.8-kg stone upward with a speed of 4 m/s *relative* to the elevator. Determine (*a*) the work done by the boy in throwing the stone, (*b*) the difference in the values of the kinetic energy of the stone before and after it was thrown. (*c*) Why are the values obtained in parts *a* and *b* not the same?

13.168 Two blocks are joined by an inextensible cable as shown. If the system is released from rest, determine the velocity of block A after it has moved 2 m. Assume that μ equals 0.25 between block A and the plane and neglect the mass and friction of the pulleys.

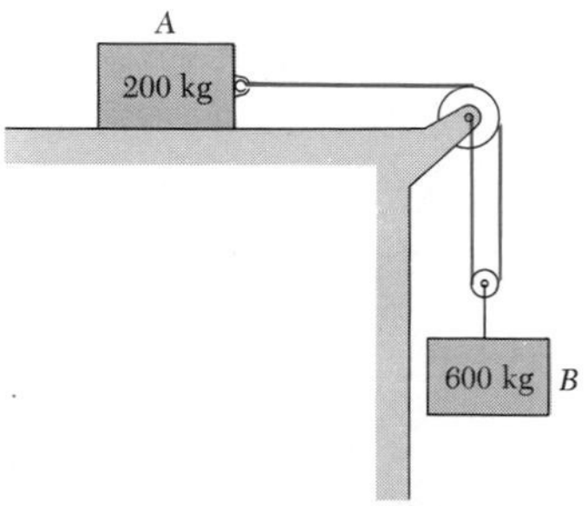

Fig. P13.168 and P13.169

13.169 Knowing that the system is released from rest, determine the additional mass that must be added to block B if its velocity is to be 4 m/s half a second after release. Assume that $\mu = 0.25$ between block A and the plane and neglect the mass and friction of the pulley.

13.170 A steel ball is dropped from A, strikes a rigid, frictionless steel plate at B, and bounces to point C. Knowing that the coefficient of restitution is 0.80, determine the distance d.

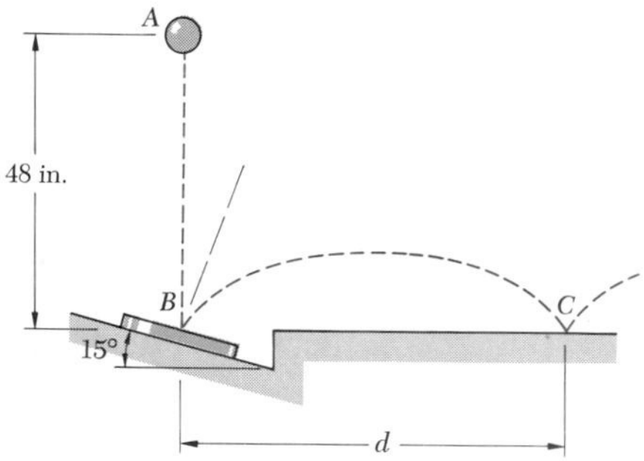

Fig. P13.170

13.171 Two portions AB and BC of the same elastic cord are connected as shown. The portion of cord BC supports a load W while, initially, the portion AB is under no tension. Determine the maximum tension which will develop in the entire cord after the stick DE suddenly breaks. (Assume that the tensions in AB and BC are instantaneously equalized after the stick breaks and that the elongation of the cord is small compared to L.)

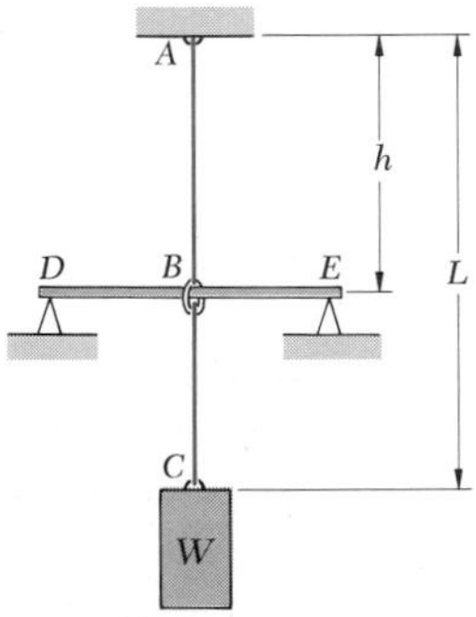

Fig. P13.171

13.172 A 5-kg collar slides without friction along a rod which forms an angle of 30° with the vertical. The spring is unstretched when the collar is at A. If the collar is released from rest at A, determine the value of the spring constant k for which the collar has zero velocity at B.

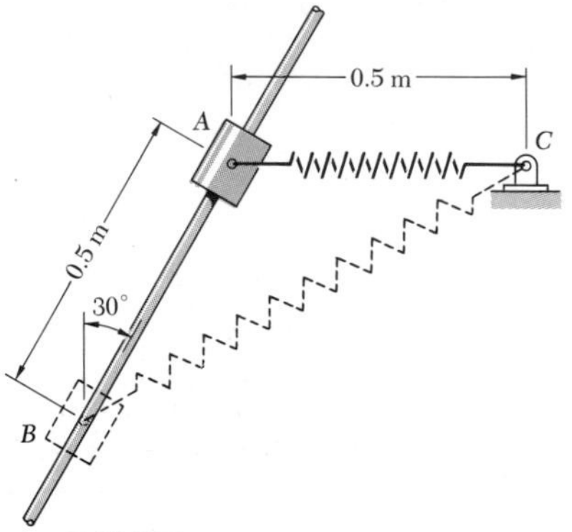

Fig. P13.172

13.173 In Prob. 13.172, determine the value of the spring constant k for which the velocity of the collar at B is 1.5 m/s.

Systems of Particles

CHAPTER

14

14.1. Application of Newton's Laws to the Motion of a System of Particles. Effective Forces. We shall be concerned in this chapter with the application of Newton's laws of motion to a *system of particles*, i.e., to a large number of particles considered together. The results obtained will enable us to analyze the effect of streams of particles on vanes or ducts and will provide us with the basic principles underlying the theory of jet and rocket propulsion (Secs. 14.9 through 14.11). Since a rigid body may be assumed to consist of a very large number of particles, the principles developed in this chapter will also provide us with a basis for the study of the kinetics of rigid bodies.

In order to derive the equations of motion for a system of n particles, we shall begin by writing Newton's second law for each individual particle of the system. Consider the particle P_i, where $1 \leq i \leq n$. Let m_i be the mass of P_i and $\mathbf{a}_i$ its acceleration with respect to the newtonian frame of reference $Oxyz$. We shall denote by $\mathbf{f}_{ij}$ the force exerted on P_i by another particle P_j of the system (Fig. 14.1); this force is called an *internal force*. The resultant of the internal forces exerted on P_i by all the other particles of the system is thus $\sum_{j=1}^{n} \mathbf{f}_{ij}$ (where $\mathbf{f}_{ii}$ has no meaning and is assumed equal to zero). Denoting, on the other hand, by $\mathbf{F}_i$

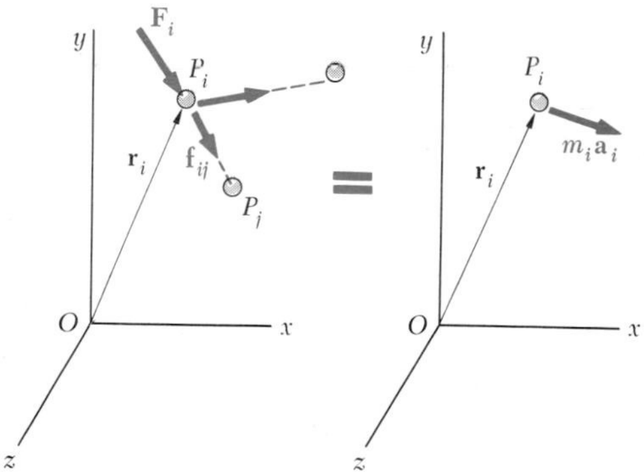

Fig. 14.1

the resultant of all the *external forces* acting on P_i, we write Newton's second law for the particle P_i as follows:

$$\mathbf{F}_i + \sum_{j=1}^{n} \mathbf{f}_{ij} = m_i\mathbf{a}_i \tag{14.1}$$

Denoting by $\mathbf{r}_i$ the position vector of P_i and taking the moments about O of the various terms in Eq. (14.1), we also write

$$\mathbf{r}_i \times \mathbf{F}_i + \sum_{j=1}^{n} (\mathbf{r}_i \times \mathbf{f}_{ij}) = \mathbf{r}_i \times m_i\mathbf{a}_i \tag{14.2}$$

Repeating this procedure for each particle P_i of the system, we obtain n equations of the type (14.1) and n equations of the type (14.2), where i takes successively the values 1, 2, . . . , n. The vectors $m_i\mathbf{a}_i$ are referred to as the *effective forces* of the particles. Thus the equations obtained express the fact that the external forces $\mathbf{F}_i$ and the internal forces $\mathbf{f}_{ij}$ acting on the various particles form a system equivalent to the system of the effective forces $m_i\mathbf{a}_i$ (i.e., one system may be replaced by the other) (Fig. 14.2).

Before proceeding further with our derivation, let us examine the internal forces $\mathbf{f}_{ij}$. We note that these forces occur in pairs $\mathbf{f}_{ij}$, $\mathbf{f}_{ji}$, where $\mathbf{f}_{ij}$ represents the force exerted by the particle P_j on the particle P_i and $\mathbf{f}_{ji}$ the force exerted by P_i on P_j (Fig. 14.2). Now,

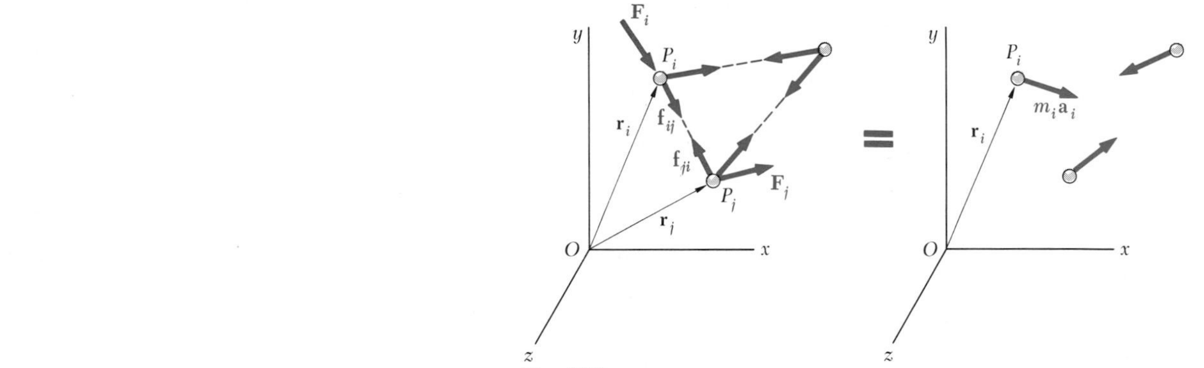

Fig. 14.2

according to Newton's third law (Sec. 6.1), as extended by Newton's law of gravitation to particles acting at a distance (Sec. 12.9), the forces $\mathbf{f}_{ij}$ and $\mathbf{f}_{ji}$ are equal and opposite and have the same line of action. Their sum is therefore $\mathbf{f}_{ij} + \mathbf{f}_{ji} = 0$ and the sum of their moments about O is

$$\mathbf{r}_i \times \mathbf{f}_{ij} + \mathbf{r}_j \times \mathbf{f}_{ji} = \mathbf{r}_i \times (\mathbf{f}_{ij} + \mathbf{f}_{ji}) + (\mathbf{r}_j - \mathbf{r}_i) \times \mathbf{f}_{ji} = 0$$

since the vectors $\mathbf{r}_j - \mathbf{r}_i$ and $\mathbf{f}_{ji}$ in the last term are collinear.

Adding all the internal forces of the system, and summing their moments about O, we obtain the equations

$$\sum_{i=1}^{n}\sum_{j=1}^{n} \mathbf{f}_{ij} = 0 \qquad \sum_{i=1}^{n}\sum_{j=1}^{n} (\mathbf{r}_i \times \mathbf{f}_{ij}) = 0 \tag{14.3}$$

which express the fact that the resultant and the moment resultant of the internal forces of the system are zero.

Returning now to the n equations (14.1), where $i = 1, 2, \ldots, n$, we add them member by member. Taking into account the first of Eqs. (14.3), we obtain

$$\sum_{i=1}^{n} \mathbf{F}_i = \sum_{i=1}^{n} m_i \mathbf{a}_i \tag{14.4}$$

Proceeding similarly with Eqs. (14.2), and taking into account the second of Eqs. (14.3), we have

$$\sum_{i=1}^{n} (\mathbf{r}_i \times \mathbf{F}_i) = \sum_{i=1}^{n} (\mathbf{r}_i \times m_i \mathbf{a}_i) \tag{14.5}$$

Equations (14.4) and (14.5) express the fact that the system of the external forces $\mathbf{F}_i$ and the system of the effective forces $m_i\mathbf{a}_i$ have the same resultant and the same moment resultant. Referring to the definition given in Sec. 3.18 for two equipollent systems of vectors, we may therefore state that *the system of the external forces acting on the particles and the system of the effective forces of the particles are equipollent*† (Fig. 14.3).

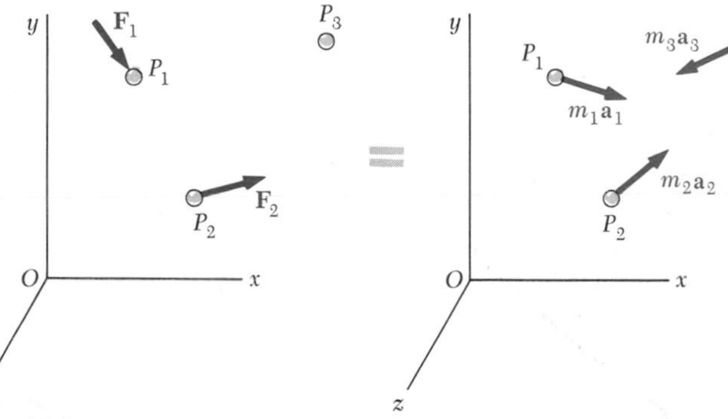

Fig. 14.3

† The result just obtained is often referred to as *D'Alembert's principle,* after the French mathematician Jean le Rond d'Alembert (1717–1783). However, d'Alembert's original statement refers to the motion of a system of connected bodies, with $\mathbf{f}_{ij}$ representing constraint forces which, if applied by themselves, will not cause the system to move. Since, as it will now be shown, this is in general not the case for the internal forces acting on a system of free particles, we shall postpone the consideration of D'Alembert's principle until the study of the motion of rigid bodies (Chap. 16).

We may note that Eqs. (14.3) express the fact that the system of the internal forces $\mathbf{f}_{ij}$ is equipollent to zero. It does *not* follow, however, that the internal forces have no effect on the particles under consideration. Indeed, the gravitational forces that the sun and the planets exert on each other are internal to the solar system and equipollent to zero. Yet these forces are alone responsible for the motion of the planets about the sun.

Similarly, it does not follow from Eqs. (14.4) and (14.5) that two systems of external forces which have the same resultant and the same moment resultant will have the same effect on a given system of particles. Clearly, the systems shown in Figs. 14.4*a* and 14.4*b* have the same resultant and the same moment resultant; yet the first system accelerates particle *A* and leaves particle *B* unaffected, while the second accelerates *B* and does not affect *A*.

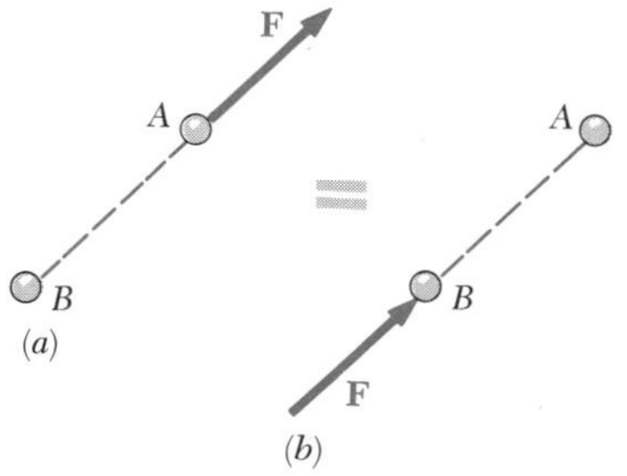

Fig. 14.4

It is important to recall that, when we stated in Sec. 3.18 that two equipollent systems of forces acting on a rigid body are also equivalent, we specifically noted that this property could *not* be extended to a system of forces acting on a set of independent particles such as those considered in this chapter.

In order to avoid any confusion, we shall use gray equals signs to connect equipollent systems of vectors, such as those shown in Figs. 14.3 and 14.4. These signs will indicate that the two systems of vectors have the same resultant and the same moment resultant. Blue equals signs will continue to be used to indicate that two systems of vectors are equivalent, i.e., that one system may actually be replaced by the other (Fig. 14.2).

14.2. Linear and Angular Momentum of a System of Particles. Equations (14.4) and (14.5), obtained in the preceding section for the motion of a system of particles, may be expressed in a more condensed form if we introduce the linear and the angular momentum of the system of particles. Defining the linear momentum **L** of the system of particles as the sum of the linear momenta of the various particles of the system (Sec. 12.2), we write

$$\mathbf{L} = \sum_{i=1}^{n} m_i \mathbf{v}_i \tag{14.6}$$

Defining the angular momentum $\mathbf{H}_O$ about O of the system of particles in a similar way (Sec. 12.6), we have

$$\mathbf{H}_O = \sum_{i=1}^{n} (\mathbf{r}_i \times m_i \mathbf{v}_i) \tag{14.7}$$

Differentiating both members of Eqs. (14.6) and (14.7) with respect to t, we write

$$\dot{\mathbf{L}} = \sum_{i=1}^{n} m_i \dot{\mathbf{v}}_i = \sum_{i=1}^{n} m_i \mathbf{a}_i \tag{14.8}$$

and

$$\dot{\mathbf{H}}_O = \sum_{i=1}^{n} (\dot{\mathbf{r}}_i \times m_i \mathbf{v}_i) + \sum_{i=1}^{n} (\mathbf{r}_i \times m_i \dot{\mathbf{v}}_i)$$

$$= \sum_{i=1}^{n} (\mathbf{v}_i \times m_i \mathbf{v}_i) + \sum_{i=1}^{n} (\mathbf{r}_i \times m_i \mathbf{a}_i)$$

which reduces to

$$\dot{\mathbf{H}}_O = \sum_{i=1}^{n} (\mathbf{r}_i \times m_i \mathbf{a}_i) \tag{14.9}$$

since the vectors $\mathbf{v}_i$ and $m_i \mathbf{v}_i$ are collinear.

We observe that the right-hand members of Eqs. (14.8) and (14.9) are, respectively, identical with the right-hand members of Eqs. (14.4) and (14.5). It follows that the left-hand members of these equations are respectively equal. Recalling that the left-hand member of Eq. (14.5) represents the sum of the moments $\mathbf{M}_O$ about O of the external forces acting on the particles of the system, and omitting the subscript i from the sums, we write

$$\Sigma \mathbf{F} = \dot{\mathbf{L}} \tag{14.10}$$

$$\Sigma \mathbf{M}_O = \dot{\mathbf{H}}_O \tag{14.11}$$

These equations express that *the resultant and the moment resultant about the fixed point O of the external forces are respectively equal to the rates of change of the linear momentum and of the angular momentum about O of the system of particles.*

14.3. Motion of the Mass Center of a System of Particles. Equation (14.10) may be written in an alternate form if the *mass center* of the system of particles is considered. The mass center of the system is the point G defined by the position vector $\bar{\mathbf{r}}$ which satisfies the relation

$$m\bar{\mathbf{r}} = \sum_{i=1}^{n} m_i \mathbf{r}_i \tag{14.12}$$

where m represents the total mass $\sum_{i=1}^{n} m_i$ of the particles. Resolving the position vectors $\bar{\mathbf{r}}$ and $\mathbf{r}_i$ into rectangular components, we obtain the following three scalar equations, which may be used to determine the coordinates $\bar{x}$, $\bar{y}$, $\bar{z}$ of the mass center:

$$m\bar{x} = \sum_{i=1}^{n} m_i x_i \qquad m\bar{y} = \sum_{i=1}^{n} m_i y_i \qquad m\bar{z} = \sum_{i=1}^{n} m_i z_i \tag{14.12$'$}$$

Since $m_i g$ represents the weight of the particle P_i, and mg the total weight of the particles, we note that G is also the center of gravity of the system of particles. However, in order to avoid any confusion, we shall call G the *mass center* of the system of particles when discussing properties of the system associated with the *mass* of the particles, and we shall refer to it as the *center of gravity* of the system when considering properties associated with the *weight* of the particles. Particles located outside the gravitational field of the earth, for example, have a mass but no weight. We may then properly refer to their mass center, but obviously not to their center of gravity.†

Differentiating both members of Eq. (14.12) with respect to t, we write

$$m\dot{\bar{\mathbf{r}}} = \sum_{i=1}^{n} m_i \dot{\mathbf{r}}_i$$

or

$$m\bar{\mathbf{v}} = \sum_{i=1}^{n} m_i \mathbf{v}_i \tag{14.13}$$

† It may also be pointed out that the mass center and the center of gravity of a system of particles do not exactly coincide, since the weights of the particles are directed toward the center of the earth and thus do not truly form a system of parallel forces.

where $\overline{\mathbf{v}}$ represents the velocity of the mass center G of the system of particles. But the right-hand member of Eq. (14.13) is, by definition, the linear momentum $\mathbf{L}$ of the system (Sec. 14.2). We have therefore

$$\mathbf{L} = m\overline{\mathbf{v}} \tag{14.14}$$

and, differentiating both members with respect to t,

$$\dot{\mathbf{L}} = m\overline{\mathbf{a}} \tag{14.15}$$

where $\overline{\mathbf{a}}$ represents the acceleration of the mass center G. Substituting for $\dot{\mathbf{L}}$ from (14.15) into (14.10), we write the equation

$$\Sigma\mathbf{F} = m\overline{\mathbf{a}} \tag{14.16}$$

which defines the motion of the mass center G of the system of particles.

We note that Eq. (14.16) is identical with the equation we would obtain for a particle of mass m equal to the total mass of the particles of the system, acted upon by all the external forces. We state therefore: *The mass center of a system of particles moves as if the entire mass of the system and all the external forces were concentrated at that point.*

This principle is best illustrated by the motion of an exploding shell. We know that, if the resistance of the air is neglected, a shell may be assumed to travel along a parabolic path. After the shell has exploded, the mass center G of the fragments of shell will continue to travel along the same path. Indeed, point G must move as if the mass and the weight of all fragments were concentrated at G; it must move, therefore, as if the shell had not exploded.

It should be noted that the preceding derivation does not involve the moments of the external forces. Therefore, *it would be wrong to assume* that the external forces are equipollent to a vector $m\overline{\mathbf{a}}$ attached at the mass center G. This is, in general, not the case, since, as we shall see in the next section, the sum of the moments about G of the external forces is, in general, not equal to zero.

14.4. Angular Momentum of a System of Particles about Its Mass Center. In some applications (for example, in the analysis of the motion of a rigid body) it is convenient to consider the motion of the particles of the system with respect to a centroidal frame of reference $Gx'y'z'$ which translates with respect to the newtonian frame of reference $Oxyz$ (Fig. 14.5).

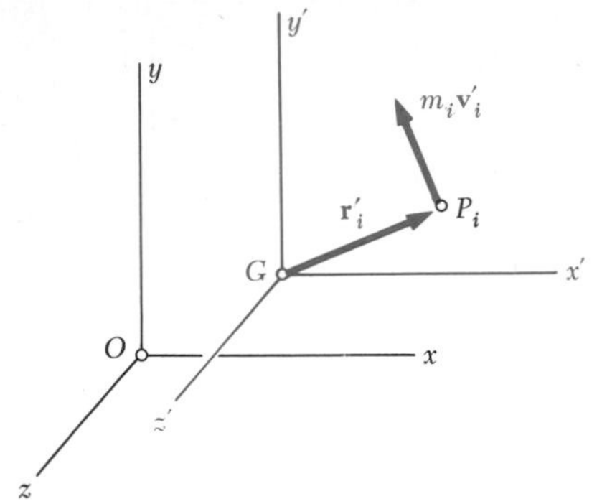

Fig. 14.5

While such a frame is not, in general, a newtonian frame of reference, we shall see that the fundamental relation (14.11) still holds when the frame $Oxyz$ is replaced by $Gx'y'z'$.

Denoting respectively by $\mathbf{r}'_i$ and $\mathbf{v}'_i$ the position vector and the velocity of the particle P_i relative to the moving frame of reference $Gx'y'z'$, we define the *angular momentum* $\mathbf{H}'_G$ of the system of particles *about the mass center* G as follows:

$$\mathbf{H}'_G = \sum_{i=1}^{n} (\mathbf{r}'_i \times m_i\mathbf{v}'_i) \tag{14.17}$$

We now differentiate both members of Eq. (14.17) with respect to t. This operation being similar to that performed in Sec. 14.2 on Eq. (14.7), we write immediately

$$\dot{\mathbf{H}}'_G = \sum_{i=1}^{n} (\mathbf{r}'_i \times m_i\mathbf{a}'_i) \tag{14.18}$$

where $\mathbf{a}'_i$ denotes the acceleration of P_i relative to the moving frame of reference. Referring to Sec. 11.12, we write

$$\mathbf{a}_i = \bar{\mathbf{a}} + \mathbf{a}'_i$$

where $\mathbf{a}_i$ and $\bar{\mathbf{a}}$ denote, respectively, the accelerations of P_i and G relative to the frame $Oxyz$. Solving for $\mathbf{a}'_i$ and substituting into (14.18), we have

$$\dot{\mathbf{H}}'_G = \sum_{i=1}^{n} (\mathbf{r}'_i \times m_i\mathbf{a}_i) - \left(\sum_{i=1}^{n} m_i\mathbf{r}'_i\right) \times \bar{\mathbf{a}} \tag{14.19}$$

But, by (14.12), the second sum in Eq. (14.19) is equal to $m\bar{\mathbf{r}}'$ and, thus, to zero, since the position vector $\bar{\mathbf{r}}'$ of G relative to the frame $Gx'y'z'$ is clearly zero. On the other hand, since $\mathbf{a}_i$ represents the acceleration of P_i relative to a newtonian frame, we may use Eq. (14.1) and replace $m_i\mathbf{a}_i$ by the sum of the internal forces $\mathbf{f}_{ij}$ and of the resultant $\mathbf{F}_i$ of the external forces acting on P_i. But a reasoning similar to that used in Sec. 14.1 shows that the moment resultant about G of the internal forces $\mathbf{f}_{ij}$ of the entire system is zero. The first sum in Eq. (14.19) reduces therefore to the moment resultant about G of the external forces acting on the particles of the system, and we write

$$\Sigma\mathbf{M}_G = \dot{\mathbf{H}}'_G \tag{14.20}$$

which expresses that *the moment resultant about G of the exter-*

nal forces is equal to the rate of change of the angular momentum about G of the system of particles.

It should be noted that, in Eq. (14.17), we defined the angular momentum $\mathbf{H}'_G$ as the sum of the moments about G of the momenta of the particles $m_i\mathbf{v}'_i$ *in their motion relative to the centroidal frame of reference* $Gx'y'z'$. We may sometimes want to compute the sum $\mathbf{H}_G$ of the moments about G of the momenta of the particles $m_i\mathbf{v}_i$ *in their absolute motion,* i.e., in their motion as observed from the newtonian frame of reference $Oxyz$ (Fig. 14.6):

$$\mathbf{H}_G = \sum_{i=1}^{n} (\mathbf{r}'_i \times m_i\mathbf{v}_i) \tag{14.21}$$

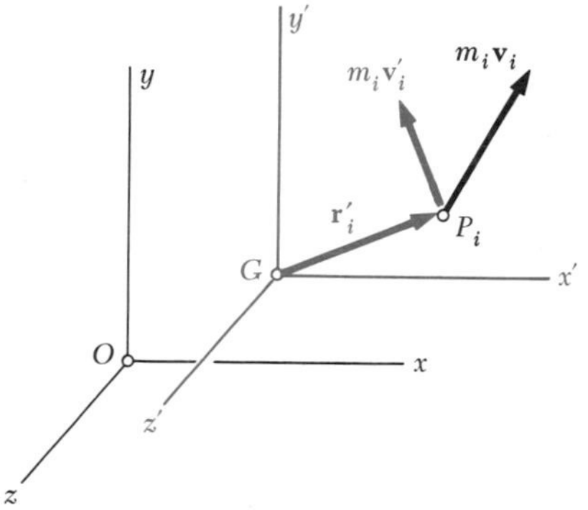

Fig. 14.6

Remarkably, the angular momenta $\mathbf{H}'_G$ and $\mathbf{H}_G$ are identically equal. This may be verified by referring to Sec. 11.12 and writing

$$\mathbf{v}_i = \overline{\mathbf{v}} + \mathbf{v}'_i \tag{14.22}$$

Substituting for $\mathbf{v}_i$ from (14.22) into Eq. (14.21), we have

$$\mathbf{H}_G = \left(\sum_{i=1}^{n} m_i\mathbf{r}'_i\right) \times \overline{\mathbf{v}} + \sum_{i=1}^{n} (\mathbf{r}'_i \times m_i\mathbf{v}'_i)$$

But, as observed earlier, the first sum is equal to zero. Thus $\mathbf{H}_G$ reduces to the second sum which, by definition, is equal to $\mathbf{H}'_G$.†

†Note that this property is peculiar to the centroidal frame $Gx'y'z'$ and does not hold, in general, for other frames of reference (see Prob. 14.19).

Taking advantage of the property we have just established, we shall simplify our notation by dropping the prime (′) from Eq. (14.20). We therefore write

$$\Sigma \mathbf{M}_G = \dot{\mathbf{H}}_G \tag{14.23}$$

where it is understood that the angular momentum $\mathbf{H}_G$ may be computed by forming the moments about G of the momenta of the particles in their motion with respect to either the newtonian frame $Oxyz$ or the centroidal frame $Gx'y'z'$:

$$\mathbf{H}_G = \sum_{i=1}^{n} (\mathbf{r}'_i \times m_i \mathbf{v}_i) = \sum_{i=1}^{n} (\mathbf{r}'_i \times m_i \mathbf{v}'_i) \tag{14.24}$$

14.5. Conservation of Momentum for a System of Particles. If no external force acts on the particles of a system, the left-hand members of Eqs. (14.10) and (14.11) are equal to zero and these equations reduce to $\dot{\mathbf{L}} = 0$ and $\dot{\mathbf{H}}_O = 0$. We conclude that

$$\mathbf{L} = \text{constant} \qquad \mathbf{H}_O = \text{constant} \tag{14.25}$$

The equations obtained express that the linear momentum of the system of particles and its angular momentum about the fixed point O are conserved.

In some applications, such as problems involving central forces, the moment about a fixed point O of each of the external forces may be zero, without any of the forces being zero. In such cases, the second of Eqs. (14.25) still holds; the angular momentum of the system of particles about O is conserved.

The concept of conservation of momentum may also be applied to the analysis of the motion of the mass center G of a system of particles and to the analysis of the motion of the system about G. For example, if the sum of the external forces is zero, the first of Eqs. (14.25) applies. Recalling Eq. (14.14), we write

$$\overline{\mathbf{v}} = \text{constant} \tag{14.26}$$

which expresses that the mass center G of the system moves in a straight line and at a constant speed. On the other hand, if the sum of the moments about G of the external forces is zero, it follows from Eq. (14.23) that the angular momentum of the system about its mass center is conserved:

$$\mathbf{H}_G = \text{constant} \tag{14.27}$$

SAMPLE PROBLEM 14.1

A 200-kg space vehicle is observed at $t = 0$ to pass through the origin of a newtonian reference frame $Oxyz$ with the velocity $\mathbf{v}_0 = (150 \text{ m/s})\mathbf{i}$ relative to the frame. Following the detonation of explosive charges, the vehicle separates into three parts, A, B, and C, of mass 100 kg, 60 kg, and 40 kg, respectively. Knowing that, at $t = 2.5$ s, the positions of parts A and B are observed to be $A(555, -180, 240)$ and $B(255, 0, -120)$, where the coordinates are expressed in meters, determine the position of part C at that time.

Solution. Since there is no external force, the mass center G of the system moves with the constant velocity $\mathbf{v}_0 = (150 \text{ m/s})\mathbf{i}$. At $t = 2.5$ s, its position is

$$\bar{\mathbf{r}} = \mathbf{v}_0 t = (150 \text{ m/s})\mathbf{i}(2.5 \text{ s}) = (375 \text{ m})\mathbf{i}$$

Recalling Eq. (14.12), we write

$$m\bar{\mathbf{r}} = m_A\mathbf{r}_A + m_B\mathbf{r}_B + m_C\mathbf{r}_C$$

$$(200 \text{ kg})(375 \text{ m})\mathbf{i} = (100 \text{ kg})[(555 \text{ m})\mathbf{i} - (180 \text{ m})\mathbf{j} + (240 \text{ m})\mathbf{k}]$$
$$+ (60 \text{ kg})[(255 \text{ m})\mathbf{i} - (120 \text{ m})\mathbf{k}] + (40 \text{ kg})\mathbf{r}_C$$
$$\mathbf{r}_C = (105 \text{ m})\mathbf{i} + (450 \text{ m})\mathbf{j} - (420 \text{ m})\mathbf{k}$$ ◄

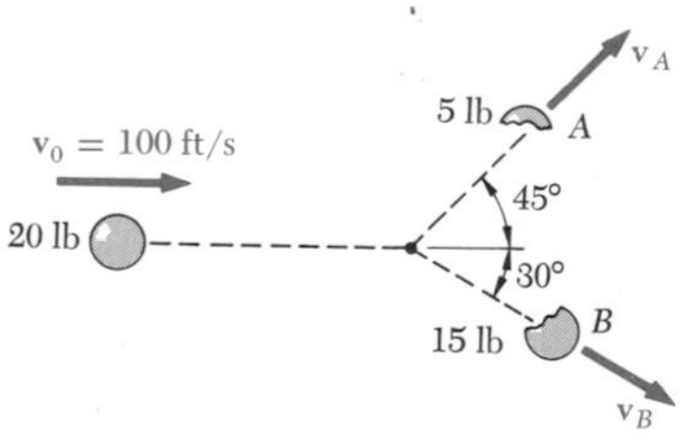

SAMPLE PROBLEM 14.2

A 20-lb projectile is moving with a velocity of 100 ft/sec when it explodes into two fragments A and B, weighing 5 lb and 15 lb, respectively. Knowing that immediately after the explosion the fragments travel in the directions shown, determine the velocity of each fragment.

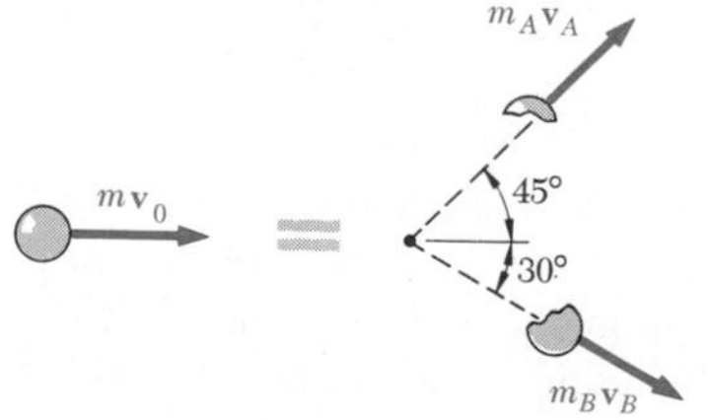

Solution. Since there is no external force, the linear momentum of the system is conserved, and we write

$$m_A\mathbf{v}_A + m_B\mathbf{v}_B = m\mathbf{v}_0$$
$$(5/g)\mathbf{v}_A + (15/g)\mathbf{v}_B = (20/g)\mathbf{v}_0$$

$\xrightarrow{+}$ x components: $\quad 5v_A \cos 45° + 15v_B \cos 30° = 20(100)$

$+\uparrow y$ components: $\quad 5v_A \sin 45° - 15v_B \sin 30° = 0$

Solving simultaneously the two equations for v_A and v_B, we have

$$v_A = 207 \text{ ft/s} \qquad v_B = 97.6 \text{ ft/s}$$
$$\mathbf{v}_A = 207 \text{ ft/s} \angle 45° \qquad \mathbf{v}_B = 97.6 \text{ ft/s} \searrow 30°$$ ◄

PROBLEMS

14.1 Two men dive horizontally and to the right off the end of a 300-lb boat. The boat is initially at rest, and each man weighs 150 lb. If each man dives so that his relative horizontal velocity with respect to the boat is 12 ft/s, determine (*a*) the velocity of the boat after the men dive simultaneously, (*b*) the velocity of the boat after one man dives and the velocity of the boat after the second man dives.

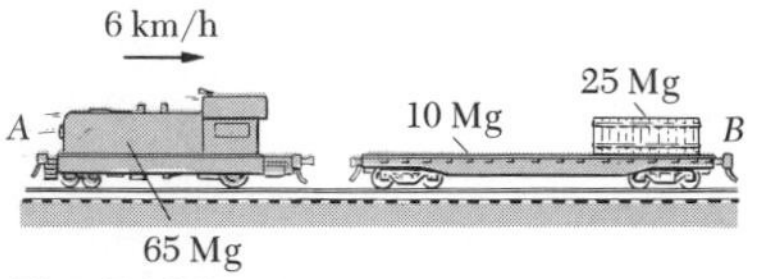

Fig. P14.2

14.2 A 65-Mg engine coasting at 6 km/h strikes, and is automatically coupled with, a 10-Mg flat car which carries a 25-Mg load. The load is *not* securely fastened to the car but may slide along the floor ($\mu = 0.20$). Knowing that the car was at rest with its brakes released and that the coupling takes place instantaneously, determine the velocity of the engine (*a*) immediately after the coupling, (*b*) after the load has slid to a stop relative to the car.

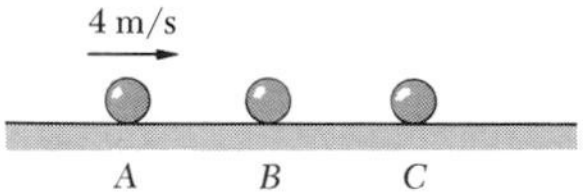

Fig. P14.3

14.3 Two identical balls B and C are at rest when ball B is struck by a ball A of the same mass, moving with a velocity of 4 m/s. This causes a series of collisions between the various balls. Knowing that $e = 0.40$, determine the velocity of each ball after *all* collisions have taken place.

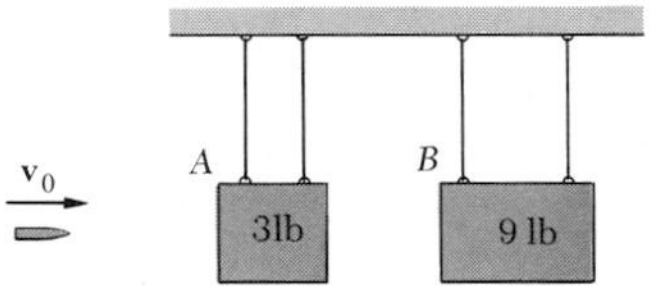

Fig. P14.4

14.4 A $\frac{3}{4}$-oz bullet is fired in a horizontal direction through block A and becomes embedded in block B. The bullet causes A and B to start moving with velocities of 8 and 6 ft/s, respectively. Determine (*a*) the initial velocity $\mathbf{v}_0$ of the bullet, (*b*) the velocity of the bullet as it travels from block A to block B.

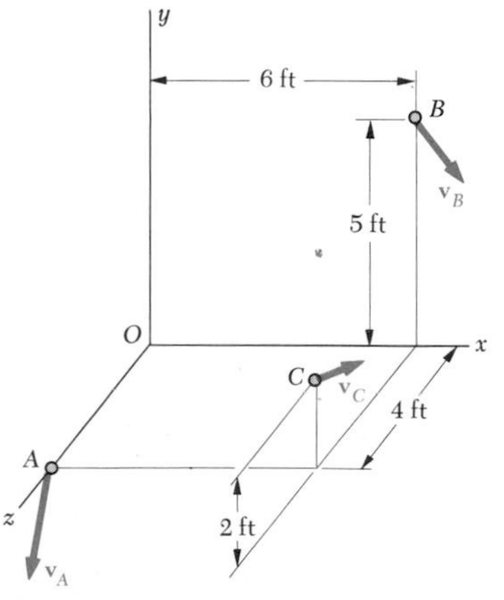

Fig. P14.5

14.5 A system consists of three particles A, B, and C. We know that $W_A = 2$ lb, $W_B = 3$ lb, and $W_C = 4$ lb and that the velocities of the particles expressed in feet per second are, respectively, $\mathbf{v}_A = -10\mathbf{j} + 5\mathbf{k}$, $\mathbf{v}_B = 8\mathbf{i} - 6\mathbf{j} + 4\mathbf{k}$, and $\mathbf{v}_C = v_x\mathbf{i} + v_y\mathbf{j} + 10\mathbf{k}$. Determine (*a*) the components v_x and v_y of the velocity of particle C for which the angular momentum $\mathbf{H}_O$ of the system about O is parallel to the z axis, (*b*) the corresponding value of $\mathbf{H}_O$.

14.6 For the system of particles of Prob. 14.5, determine (*a*) the components v_x and v_y of the velocity of particle C for which the angular momentum $\mathbf{H}_O$ of the system about O is parallel to the x axis, (*b*) the corresponding value of $\mathbf{H}_O$.

14.7 A system consists of three particles A, B, and C. We know that $m_A = 1$ kg, $m_B = 2$ kg, and $m_C = 3$ kg and that the velocities of the particles expressed in meters per second are, respectively, $\mathbf{v}_A = 3\mathbf{i} - 2\mathbf{j} + 4\mathbf{k}$, $\mathbf{v}_B = 4\mathbf{i} + 3\mathbf{j}$, and $\mathbf{v}_C = 2\mathbf{i} + 5\mathbf{j} - 3\mathbf{k}$. (*a*) Determine the angular momentum $\mathbf{H}_O$ of the system about O. (*b*) Using the result of part *a* and the answers to Prob. 14.8, check that the relation given in Prob. 14.17 is satisfied.

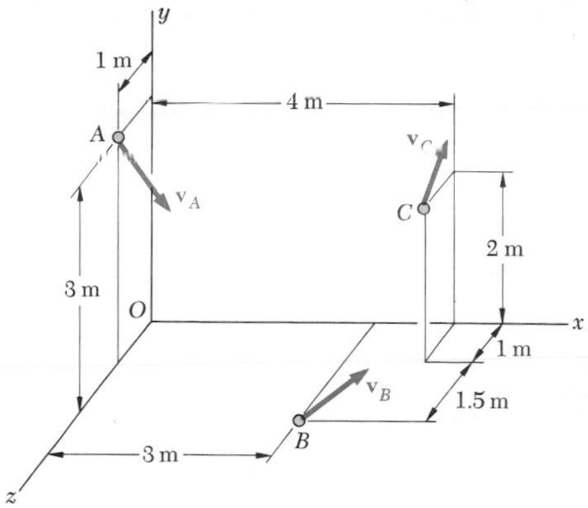

Fig. P14.7

14.8 For the system of particles of Prob. 14.7, determine (*a*) the position vector $\bar{\mathbf{r}}$ of the mass center G of the system, (*b*) the linear momentum $m\bar{\mathbf{v}}$ of the system, (*c*) the angular momentum $\mathbf{H}_G$ of the system about G.

14.9 A 240-kg space vehicle traveling with the velocity $\mathbf{v}_0 = (500 \text{ m/s})\mathbf{k}$ passes through the origin O at $t = 0$. Explosive charges then separate the vehicle into three parts, A, B, and C, of mass 40 kg, 80 kg, and 120 kg, respectively. Knowing that at $t = 3$ s the positions of parts B and C are observed to be $B(375, 825, 2025)$ and $C(-300, -600, 1200)$, where the coordinates are expressed in meters, determine the corresponding position of part A. Neglect the effect of gravity.

14.10 Two 30-lb cannon balls are chained together and fired horizontally with a velocity of 500 ft/s from the top of a 45-ft wall. The chain breaks during the flight of the cannon balls and one of them strikes the ground at $t = 1.5$ s, at a distance of 720 ft from the foot of the wall, and 21 ft to the right of the line of fire. Determine the position of the other cannon ball at that instant. Neglect the resistance of the air.

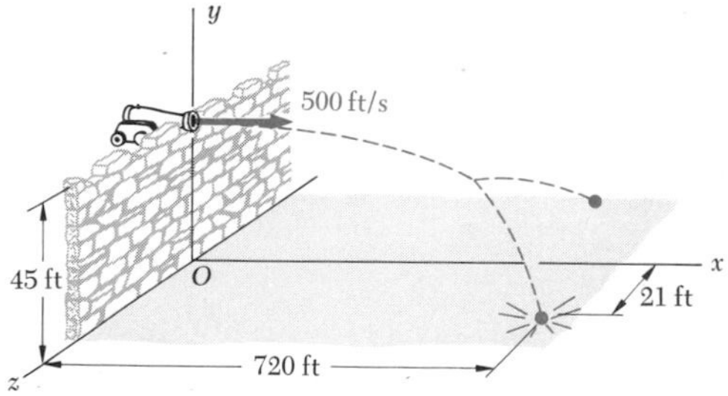

Fig. P14.10

14.11 Solve Prob. 14.10, if the cannon ball which first strikes the ground weighs 24 lb and the other 36 lb. Assume that the time of flight and the point of impact of the first cannon ball remain the same.

14.12 A 10-kg projectile is passing through the origin O with a velocity $\mathbf{v}_0 = (60 \text{ m/s})\mathbf{i}$ when it explodes into two fragments, A and B, of mass 4 kg and 6 kg, respectively. Knowing that, 2 s later, the position of the first fragment is $A(150 \text{ m}, 12 \text{ m}, -24 \text{ m})$, determine the position of fragment B at the same instant. Assume $g = 9.81 \text{ m/s}^2$ and neglect the resistance of the air.

14.13 An archer hits a game bird flying in a horizontal straight line 30 ft above the ground with a 500-grain wooden arrow [1 grain = (1/7000) lb]. Knowing that the arrow strikes the bird from behind with a velocity of 350 ft/s at an angle of 30° with the vertical, and that the bird falls to the ground in 1.5 s and 48 ft beyond the point where it was hit, determine (*a*) the weight of the bird, (*b*) the speed at which it was flying when it was hit.

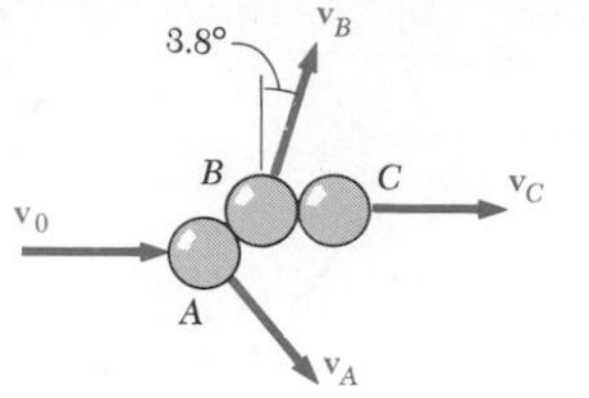

Fig. P14.14

14.14 In a game of billiards, ball A is moving with the velocity $\mathbf{v}_0 = (10 \text{ ft/s})\mathbf{i}$ when it strikes balls B and C which are at rest side by side. After the collision, A is observed to move with the velocity $\mathbf{v}_A = (3.92 \text{ ft/s})\mathbf{i} - (4.56 \text{ ft/s})\mathbf{j}$, while B and C move in the directions shown. Determine the magnitudes of the velocities $\mathbf{v}_B$ and $\mathbf{v}_C$.

14.15 A 5-kg object is falling vertically when, at point D, it explodes into three fragments A, B, and C, weighing, respectively, 1.5 kg, 2.5 kg, and 1 kg. Immediately after the explosion the velocity of each fragment is directed as shown and the speed of fragment A is observed to be 70 m/s. Determine the velocity of the 5-kg object immediately before the explosion.

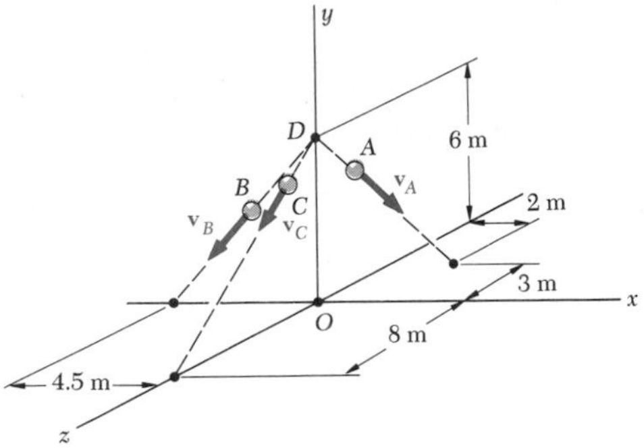

Fig. P14.15

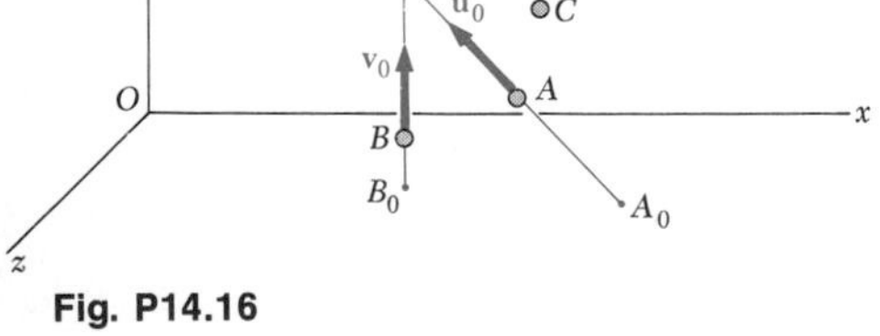

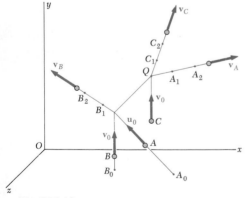

Fig. P14.16

14.16 In a scattering experiment, an alpha particle A is projected with the velocity $\mathbf{u}_0 = -(600 \text{ m/s})\mathbf{i} + (750 \text{ m/s})\mathbf{j} - (800 \text{ m/s})\mathbf{k}$ into a stream of oxygen nuclei moving with the common velocity $\mathbf{v}_0 = (600 \text{ m/s})\mathbf{j}$. After colliding successively with the nuclei B and C, particle A is observed to move along the path defined by the points $A_1(280, 240, 120)$, $A_2(360, 320, 160)$, while nuclei B and C are observed to move along paths defined, respectively, by $B_1(147, 220, 130)$, $B_2(114, 290, 120)$ and by $C_1(240, 232, 90)$, $C_2(240, 280, 75)$. All paths are along straight lines and all coordinates are expressed in millimeters. Knowing that the mass of an oxygen nucleus is four times that of an alpha particle, determine the speed of each of the three particles after the collisions.

14.17 Derive the relation

$$\mathbf{H}_O = \bar{\mathbf{r}} \times m\bar{\mathbf{v}} + \mathbf{H}_G$$

between the angular momenta $\mathbf{H}_O$ and $\mathbf{H}_G$ defined in Eqs. (14.7) and (14.24), respectively. The vectors $\bar{\mathbf{r}}$ and $\bar{\mathbf{v}}$ define, respectively, the position and velocity of the mass center G of the system of particles relative to the newtonian frame of reference $Oxyz$, and m represents the total mass of the system.

14.18 Show that Eq. (14.23) may be derived directly from Eq. (14.11) by substituting for $\mathbf{H}_O$ the expression given in Prob. 14.17.

14.19 Consider the frame of reference $Ax'y'z'$ in translation with respect to the newtonian frame of reference $Oxyz$. We define the angular momentum $\mathbf{H}'_A$ of a system of n particles about A as the sum

$$\mathbf{H}'_A = \sum_{i=1}^{n} \mathbf{r}'_i \times m_i\mathbf{v}'_i \tag{1}$$

of the moments about A of the momenta $m_i\mathbf{v}'_i$ of the particles in their motion relative to the frame $Ax'y'z'$. Denoting by $\mathbf{H}_A$ the sum

$$\mathbf{H}_A = \sum_{i=1}^{n} \mathbf{r}'_i \times m_i\mathbf{v}_i \tag{2}$$

of the moments about A of the momenta $m_i\mathbf{v}_i$ of the particles in their motion relative to the newtonian frame $Oxyz$, show that $\mathbf{H}_A = \mathbf{H}'_A$ at a given instant if, and only if, one of the following conditions is satisfied at that instant: (*a*) A has zero velocity with respect to the frame $Oxyz$, (*b*) A coincides with the mass center G of the system, (*c*) the velocity $\mathbf{v}_A$ relative to $Oxyz$ is directed along the line AG.

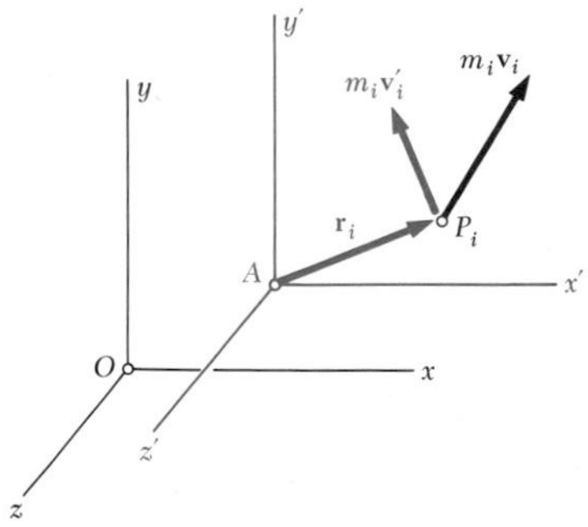

Fig. P14.19

14.20 Show that the relation

$$\Sigma\mathbf{M}_A = \dot{\mathbf{H}}'_A$$

where $\mathbf{H}'_A$ is defined by Eq. (1) of Prob. 14.19 and where $\Sigma\mathbf{M}_A$ represents the sum of the moments about A of the external forces acting on the system of particles, is valid if, and only if, one of the following conditions is satisfied: (*a*) the frame $Ax'y'z'$ is itself a newtonian frame of reference, (*b*) A coincides with the mass center G, (*c*) the acceleration $\mathbf{a}_A$ of A relative to $Oxyz$ is directed along the line AG.

14.6. Kinetic Energy of a System of Particles. The kinetic energy T of a system of particles is defined as the sum of the kinetic energies of the various particles of the system. Referring to Sec. 13.3, we therefore write

$$T = \frac{1}{2}\sum_{i=1}^{n} m_i v_i^2 \tag{14.28}$$

Using a Centroidal Frame of Reference. It is often convenient, when computing the kinetic energy of a system comprising a large number of particles (as in the case of a rigid body), to consider separately the motion of the mass center G of the system and the motion of the system relative to a moving frame of reference attached to G.

Let P_i be a particle of the system, $\mathbf{v}_i$ its velocity relative to the newtonian frame of reference $Oxyz$, and $\mathbf{v}'_i$ its velocity relative to the moving frame $Gx'y'z'$ which is in translation with respect to $Oxyz$ (Fig. 14.7). We recall from the preceding section that

$$\mathbf{v}_i = \overline{\mathbf{v}} + \mathbf{v}'_i \tag{14.22}$$

where $\overline{\mathbf{v}}$ denotes the velocity of the mass center G relative to the newtonian frame $Oxyz$. Observing that v_i^2 is equal to the scalar

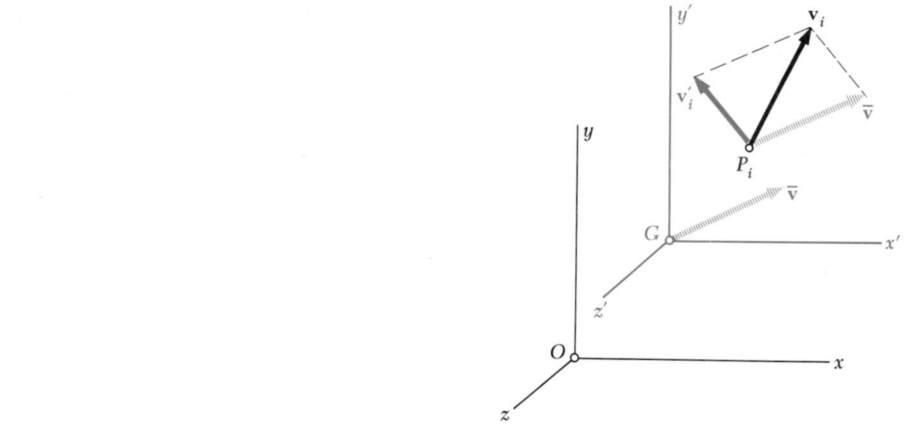

Fig. 14.7

product $\mathbf{v}_i \cdot \mathbf{v}_i$, we express as follows the kinetic energy T of the system relative to the newtonian frame $Oxyz$:

$$T = \frac{1}{2}\sum_{i=1}^{n} m_i v_i^2 = \frac{1}{2}\sum_{i=1}^{n} (m_i \mathbf{v}_i \cdot \mathbf{v}_i)$$

or, substituting for $\mathbf{v}_i$ from (14.22),

$$T = \frac{1}{2}\sum_{i=1}^{n}[m_i(\overline{\mathbf{v}} + \mathbf{v}'_i)\cdot(\overline{\mathbf{v}} + \mathbf{v}'_i)]$$

$$= \frac{1}{2}\left(\sum_{i=1}^{n} m_i\right)\bar{v}^2 + \overline{\mathbf{v}}\cdot\sum_{i=1}^{n} m_i\mathbf{v}'_i + \frac{1}{2}\sum_{i=1}^{n} m_i v_i'^2$$

The first sum represents the total mass m of the system. Recalling Eq. (14.13), we note that the second sum is equal to $m\overline{\mathbf{v}}'$ and thus to zero, since $\overline{\mathbf{v}}'$, which represents the velocity of G relative to the frame $Gx'y'z'$, is clearly zero. We therefore write

$$T = \tfrac{1}{2}m\bar{v}^2 + \frac{1}{2}\sum_{i=1}^{n} m_i v_i'^2 \tag{14.29}$$

This equation shows that the kinetic energy T of a system of particles may be obtained *by adding the kinetic energy of the mass center* G (assuming the entire mass concentrated at G) *and the kinetic energy of the system in its motion relative to the frame* $Gx'y'z'$.

14.7. Work-Energy Principle. Conservation of Energy for a System of Particles. The principle of work and energy may be applied to each particle P_i of a system of particles. We write

$$T_1 + U_{1\to 2} = T_2 \tag{14.30}$$

for each particle P_i, where $U_{1\to 2}$ represents the work done by the internal forces $\mathbf{f}_{ij}$ and the resultant external force $\mathbf{F}_i$ acting on P_i. Adding the kinetic energies of the various particles of the system, and considering the work of all the forces involved, we may apply Eq. (14.30) to the entire system. The quantities T_1 and T_2 now represent the kinetic energy of the entire system and may be computed from either Eq. (14.28) or Eq. (14.29). The quantity $U_{1\to 2}$ represents the work of all the forces acting on the particles of the system. We should note that, while the internal forces $\mathbf{f}_{ij}$ and $\mathbf{f}_{ji}$ are equal and opposite, the work of these forces, in general, will not cancel out, since the particles P_i and P_j on which they act will, in general, undergo different displacements. Therefore, in computing $U_{1\to 2}$, *we should consider the work of the internal forces* $\mathbf{f}_{ij}$ *as well as the work of the external forces* $\mathbf{F}_i$.

If all the forces acting on the particles of the system are conservative, Eq. (14.30) may be replaced by

$$T_1 + V_1 = T_2 + V_2 \tag{14.31}$$

where V represents the potential energy associated with the internal and external forces acting on the particles of the system. Equation (14.31) expresses the principle of *conservation of energy* for the system of particles.

14.8. Principle of Impulse and Momentum for a System of Particles. Integrating Eqs. (14.10) and (14.11) in t from a time t_1 to a time t_2, we write

$$\sum \int_{t_1}^{t_2} \mathbf{F}\, dt = \mathbf{L}_2 - \mathbf{L}_1 \tag{14.32}$$

$$\sum \int_{t_1}^{t_2} \mathbf{M}_O\, dt = (\mathbf{H}_O)_2 - (\mathbf{H}_O)_1 \tag{14.33}$$

Recalling the definition of the linear impulse of a force given in Sec. 13.10, we observe that the integrals in Eq. (14.32) represent the linear impulses of the external forces acting on the particles of the system. We shall refer in a similar way to the integrals in Eq. (14.33) as the *angular impulses* about O of the external forces. Thus, Eq. (14.32) expresses that the sum of the linear impulses of the external forces acting on the system is equal to the change in linear momentum of the system. Similarly, Eq. (14.33) expresses that the sum of the angular impulses about O of the external forces is equal to the change in angular momentum about O of the system.

In order to understand the physical significance of Eqs. (14.32) and (14.33), we shall rearrange the terms in these equations and write

$$\mathbf{L}_1 + \sum \int_{t_1}^{t_2} \mathbf{F}\, dt = \mathbf{L}_2 \tag{14.34}$$

$$(\mathbf{H}_O)_1 + \sum \int_{t_1}^{t_2} \mathbf{M}_O\, dt = (\mathbf{H}_O)_2 \tag{14.35}$$

We have sketched in parts a and c of Fig. 14.8 the momenta of the particles of the system at times t_1 and t_2, respectively, and we have shown in part b of the same figure a vector equal to the sum of the linear impulses of the external forces and a couple of moment equal to the sum of the angular impulses about O of the external forces. For simplicity, the particles have been assumed to move in the plane of the figure, but the present discussion remains valid in the case of particles moving in space. Recalling from Eq. (14.6) that $\mathbf{L}$, by definition, is the resultant of the momenta $m_i\mathbf{v}_i$, we note that Eq. (14.34) expresses that the resultant of the vectors shown in parts a and b of Fig. 14.8 is equal to

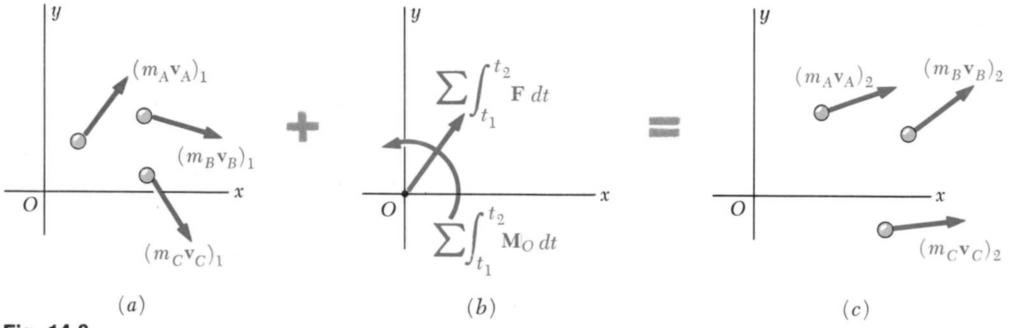

Fig. 14.8

the resultant of the vectors shown in part c of the same figure. Recalling from Eq. (14.7) that $\mathbf{H}_O$ is the moment resultant of the momenta $m_i\mathbf{v}_i$, we note that Eq. (14.35) similarly expresses that the moment resultant of the vectors in parts a and b of Fig. 14.8 is equal to the moment resultant of the vectors in part c. Together, Eqs. (14.34) and (14.35) thus express that *the momenta of the particles at time t_1 and the impulses of the external forces from t_1 to t_2 form a system of vectors equipollent to the system of the momenta of the particles at time t_2*. This has been indicated in Fig. 14.8 by the use of gray plus and equals signs.

If no external force acts on the particles of the system, the integrals in Eqs. (14.34) and (14.35) are zero, and these equations reduce to

$$\mathbf{L}_1 = \mathbf{L}_2 \tag{14.36}$$

$$(\mathbf{H}_O)_1 = (\mathbf{H}_O)_2 \tag{14.37}$$

We thus check the result obtained in Sec. 14.5: If no external force acts on the particles of a system, the linear momentum and the angular momentum about O of the system of particles are conserved. The system of the initial momenta is equipollent to the system of the final momenta, and it follows that the angular momentum of the system of particles about *any* fixed point is conserved.

SAMPLE PROBLEM 14.3

For the 200-kg space vehicle of Sample Prob. 14.1, it is known that, at $t = 2.5$ s, the velocity of part A is $\mathbf{v}_A = (270 \text{ m/s})\mathbf{i} - (120 \text{ m/s})\mathbf{j} + (160 \text{ m/s})\mathbf{k}$ and the velocity of part B is parallel to the xz plane. Determine the velocity of part C.

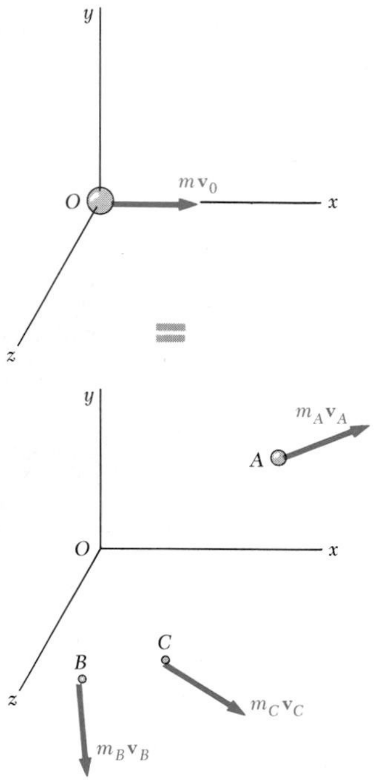

Solution. Since there is no external force, the initial momentum $m\mathbf{v}_0$ is equipollent to the system of the final momenta. Equating first the sums of the vectors in both parts of the adjoining sketch, and then the sums of their moments about O, we write

$$\mathbf{L}_1 = \mathbf{L}_2: \qquad m\mathbf{v}_0 = m_A\mathbf{v}_A + m_B\mathbf{v}_B + m_C\mathbf{v}_C \tag{1}$$

$$(\mathbf{H}_O)_1 = (\mathbf{H}_O)_2: \qquad 0 = \mathbf{r}_A \times m_A\mathbf{v}_A + \mathbf{r}_B \times m_B\mathbf{v}_B + \mathbf{r}_C \times m_C\mathbf{v}_C \tag{2}$$

Recalling from Sample Prob. 14.1 that $\mathbf{v}_0 = (150 \text{ m/s})\mathbf{i}$,

$$m_A = 100 \text{ kg} \qquad m_B = 60 \text{ kg} \qquad m_C = 40 \text{ kg}$$
$$\mathbf{r}_A = (555 \text{ m})\mathbf{i} - (180 \text{ m})\mathbf{j} + (240 \text{ m})\mathbf{k}$$
$$\mathbf{r}_B = (255 \text{ m})\mathbf{i} - (120 \text{ m})\mathbf{k}$$
$$\mathbf{r}_C = (105 \text{ m})\mathbf{i} + (450 \text{ m})\mathbf{j} - (420 \text{ m})\mathbf{k}$$

and using the information given in the statement of this problem, we rewrite Eqs. (1) and (2) as follows:

$$200(150\mathbf{i}) = 100(270\mathbf{i} - 120\mathbf{j} + 160\mathbf{k}) + 60[(v_B)_x\mathbf{i} + (v_B)_z\mathbf{k}] + 40[(v_C)_x\mathbf{i} + (v_C)_y\mathbf{j} + (v_C)_z\mathbf{k}] \tag{1'}$$

$$0 = 100\begin{vmatrix} \mathbf{i} & \mathbf{j} & \mathbf{k} \\ 555 & -180 & 240 \\ 270 & -120 & 160 \end{vmatrix} + 60\begin{vmatrix} \mathbf{i} & \mathbf{j} & \mathbf{k} \\ 255 & 0 & -120 \\ (v_B)_x & 0 & (v_B)_z \end{vmatrix} + 40\begin{vmatrix} \mathbf{i} & \mathbf{j} & \mathbf{k} \\ 105 & 450 & -420 \\ (v_C)_x & (v_C)_y & (v_C)_z \end{vmatrix} \tag{2'}$$

Equating to zero the coefficient of $\mathbf{j}$ in (1′) and the coefficients of $\mathbf{i}$ and $\mathbf{k}$ in (2′), we write, after reductions, the three scalar equations

$$(v_C)_y - 300 = 0$$
$$450(v_C)_z + 420(v_C)_y = 0$$
$$105(v_C)_y - 450(v_C)_x - 45\,000 = 0$$

which yield, respectively,

$$(v_C)_y = 300 \qquad (v_C)_z = -280 \qquad (v_C)_x = -30$$

The velocity of part C is thus

$$\mathbf{v}_C = -(30 \text{ m/s})\mathbf{i} + (300 \text{ m/s})\mathbf{j} - (280 \text{ m/s})\mathbf{k} \quad ◀$$

SAMPLE PROBLEM 14.4

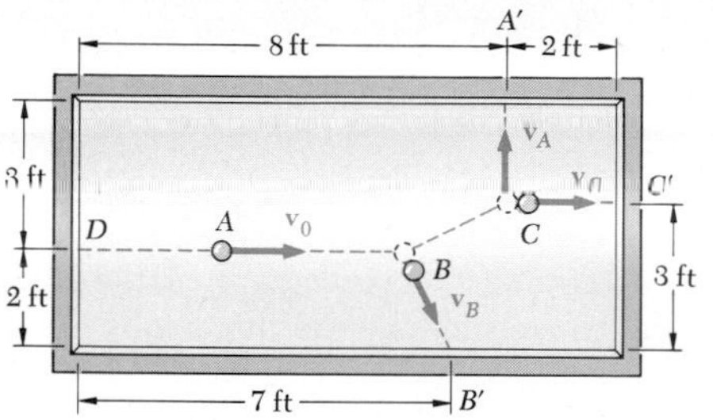

In a game of billiards, ball A is given an initial velocity $\mathbf{v}_0$ of magnitude $v_0 = 10$ ft/s along line DA parallel to the axis of the table. It hits ball B and then ball C, which are both at rest. Knowing that A and C hit the sides of the table squarely at points A' and C', respectively, that B hits the side obliquely at B', and assuming frictionless surfaces and perfectly elastic impacts, determine the velocities $\mathbf{v}_A$, $\mathbf{v}_B$, and $\mathbf{v}_C$ with which the balls hit the sides of the table. (*Remark.* In this sample problem and in several of the problems which follow, the billiard balls are assumed to be particles moving freely in a horizontal plane, rather than the rolling and sliding spheres they actually are.)

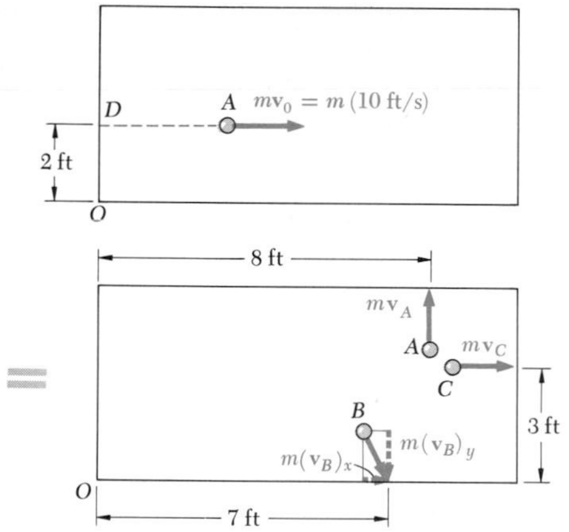

Solution. ***Conservation of Momentum.*** Since there is no external force, the initial momentum $m\mathbf{v}_0$ is equipollent to the system of momenta after the two collisions (and before any of the balls hits the side of the table). Referring to the adjoining sketch, we write

$$\xrightarrow{} x \text{ components:} \qquad m(10\text{ ft/s}) = m(v_B)_x + mv_C \qquad (1)$$

$$+\uparrow y \text{ components:} \qquad 0 = mv_A - m(v_B)_y \qquad (2)$$

$$+\circlearrowleft \text{ moments about } O\text{:} \quad -(2\text{ ft})m(10\text{ ft/s}) = (8\text{ ft})mv_A - (7\text{ ft})m(v_B)_y - (3\text{ ft})mv_C \qquad (3)$$

Solving the three equations for v_A, $(v_B)_x$, and $(v_B)_y$ in terms of v_C:

$$v_A = (v_B)_y = 3v_C - 20 \qquad (v_B)_x = 10 - v_C \qquad (4)$$

Conservation of Energy. Since the surfaces are frictionless and the impacts are perfectly elastic, the initial kinetic energy $\frac{1}{2}mv_0^2$ is equal to the final kinetic energy of the system:

$$\tfrac{1}{2}mv_0^2 = \tfrac{1}{2}m_Av_A^2 + \tfrac{1}{2}m_Bv_B^2 + \tfrac{1}{2}m_Cv_C^2$$

$$v_A^2 + (v_B)_x^2 + (v_B)_y^2 + v_C^2 = (10\text{ ft/s})^2 \qquad (5)$$

Substituting for v_A, $(v_B)_x$, and $(v_B)_y$ from (4) into (5), we have

$$2(3v_C - 20)^2 + (10 - v_C)^2 + v_C^2 = 100$$

$$20v_C^2 - 260v_C + 800 = 0$$

Solving for v_C, we find $v_C = 5$ ft/s and $v_C = 8$ ft/s. Since only the second root yields a positive value for v_A after substitution into Eqs. (4), we conclude that $v_C = 8$ ft/s and

$$v_A = (v_B)_y = 3(8) - 20 = 4\text{ ft/s} \qquad (v_B)_x = 10 - 8 = 2\text{ ft/s}$$

$\mathbf{v}_A = 4\text{ ft/s}\uparrow \qquad \mathbf{v}_B = 4.47\text{ ft/s} ⦩ 63.4° \qquad \mathbf{v}_C = 8\text{ ft/s}\rightarrow$ ◀

PROBLEMS

14.21 In Prob. 14.13, determine the amount of energy lost as the arrow hits the game bird.

14.22 In Prob. 14.14, determine the percentage of the initial kinetic energy lost due to the impacts among the three balls.

14.23 In Prob. 14.15, determine the work done by the internal forces during the explosion.

14.24 In Prob. 14.16, determine the percentage of the initial kinetic energy lost due to the collisions between the alpha particle and the two oxygen nuclei and check that, taking into account the numerical accuracy of the given data and of the calculations, the result obtained suggests conservation of energy.

14.25 A 5-lb weight slides without friction on the xy plane. At $t = 0$ it passes through the origin with a velocity $\mathbf{v}_0 = (20 \text{ ft/s})\mathbf{i}$. Internal springs then separate the weight into the three parts shown. Knowing that, at $t = 3$ s, $\mathbf{r}_A = (42 \text{ ft})\mathbf{i} + (27 \text{ ft})\mathbf{j}$ and $\mathbf{r}_B = (60 \text{ ft})\mathbf{i} - (6 \text{ ft})\mathbf{j}$, that $\mathbf{v}_A = (14 \text{ ft/s})\mathbf{i} + (9 \text{ ft/s})\mathbf{j}$, and that $\mathbf{v}_B$ is parallel to the x axis, determine the corresponding position and velocity of part C.

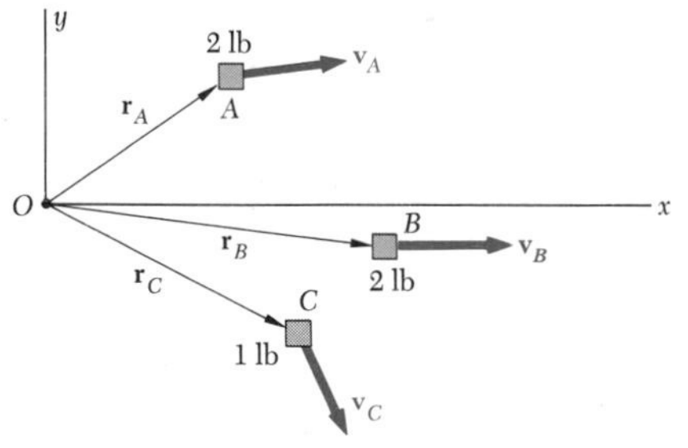

Fig. P14.25

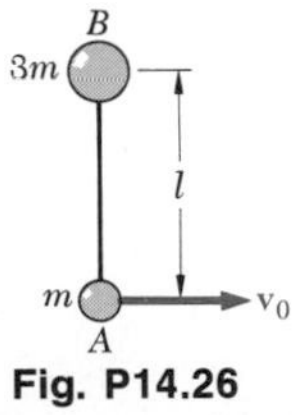

Fig. P14.26

14.26 Two small spheres A and B, respectively of mass m and $3m$, are connected by a rigid rod of length l and negligible mass. The two spheres are resting on a horizontal, frictionless surface when A is suddenly given the velocity $\mathbf{v}_0 = v_0\mathbf{i}$. Determine ($a$) the linear momentum of the system and its angular momentum about its mass center G, (b) the velocities of A and B after the rod AB has rotated through 90°, (c) the velocities of A and B after the rod AB has rotated through 180°.

14.27 A 240-kg space vehicle traveling with the velocity $\mathbf{v}_0 = (500\text{ m/s})\mathbf{k}$ passes through the origin O at $t = 0$. Explosive charges then separate the vehicle into three parts, A, B, and C, of mass 40 kg, 80 kg, and 120 kg, respectively. Knowing that at $t = 3$ s the positions of the three parts are, respectively, $A(150, 150, 1350)$, $B(375, 825, 2025)$, and $C(-300, -600, 1200)$, where the coordinates are expressed in meters, that the velocity of C is $\mathbf{v}_C = -(100\text{ m/s})\mathbf{i} - (200\text{ m/s})\mathbf{j} + (400\text{ m/s})\mathbf{k}$, and that the y component of the velocity of B is $+350$ m/s, determine the velocity of part A.

14.28 In the scattering experiment of Prob. 14.16, it is known that the alpha particle is projected from $A_0(300, 0, 300)$ and that it collides with the oxygen nucleus C at $Q(240, 200, 100)$, where all coordinates are expressed in millimeters. Determine the coordinates of point B_0 where the original path of nucleus B intersects the xz plane. (*Hint.* Express that the angular momentum of the three particles about Q is conserved.)

14.29 In a game of billiards, ball A is moving with the velocity $\mathbf{v}_0 = v_0\mathbf{i}$ when it strikes balls B and C which are at rest side by side. After the collision, the three balls are observed to move in the directions shown. Assuming frictionless surfaces and perfectly elastic impacts (i.e., conservation of energy), determine the magnitudes of the velocities $\mathbf{v}_A$, $\mathbf{v}_B$, and $\mathbf{v}_C$ in terms of v_0 and θ.

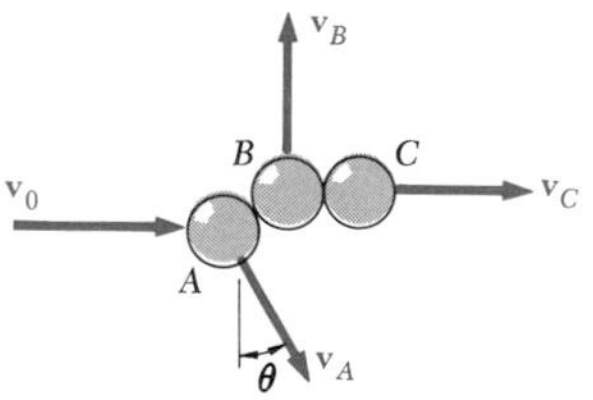

Fig. P14.29 and P14.30

14.30 In a game of billiards, ball A is moving with the velocity $\mathbf{v}_0 = (3\text{ m/s})\mathbf{i}$ when it strikes balls B and C which are at rest side by side. After the collision, the three balls are observed to move in the directions shown, with $\theta = 30°$. Assuming frictionless surfaces and perfectly elastic impacts (i.e., conservation of energy), determine the magnitudes of the velocities $\mathbf{v}_A$, $\mathbf{v}_B$, and $\mathbf{v}_C$.

14.31 In a scattering experiment, an alpha particle A is projected with the velocity $\mathbf{u}_0 = (960\text{ m/s})\mathbf{i} + (1200\text{ m/s})\mathbf{j} + (1280\text{ m/s})\mathbf{k}$ into a stream of oxygen nuclei moving with the common velocity $\mathbf{v}_0 = v_0\mathbf{j}$. After colliding successively with the nuclei B and C, particle A is observed to move in the direction defined by the unit vector $\boldsymbol{\lambda}_A = -0.463\mathbf{i} + 0.853\mathbf{j} - 0.241\mathbf{k}$, while nuclei B and C are observed to move in directions defined, respectively, by $\boldsymbol{\lambda}_B = 0.939\mathbf{j} + 0.344\mathbf{k}$ and $\boldsymbol{\lambda}_C = 0.628\mathbf{i} + 0.778\mathbf{j}$. Knowing that the mass of an oxygen nucleus is four times that of an alpha particle, and assuming conservation of energy, determine (*a*) the speed v_0 of the oxygen nuclei before the collisions, (*b*) the speed of each of the three particles after the collisions.

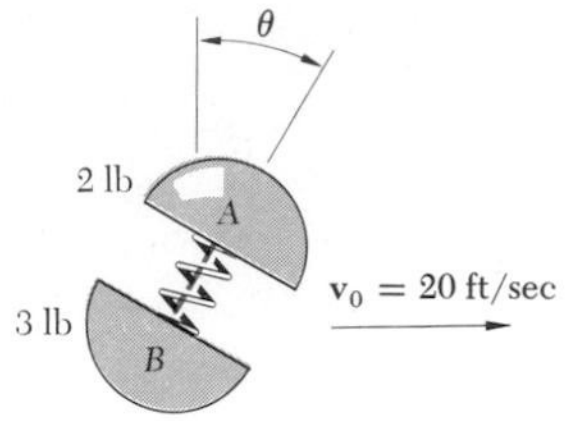

Fig. P14.32

14.32 When the cord connecting particles A and B is severed, the compressed spring causes the particles to fly apart (the spring is not connected to the particles). The potential energy of the compressed spring is known to be 20 ft · lb and the assembly has an initial velocity $\mathbf{v}_0$ as shown. If the cord is severed when $\theta = 30°$, determine the resulting velocity of each particle.

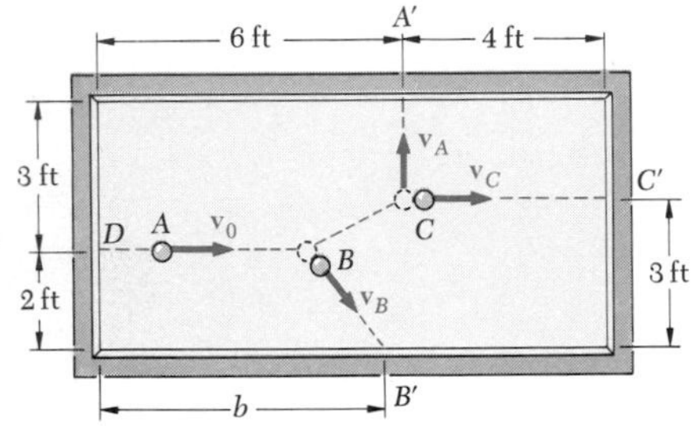

Fig. P14.33

14.33 In a game of billiards, ball A is given an initial velocity $\mathbf{v}_0$ along line DA parallel to the axis of the table. It hits ball B and then ball C, which are at rest. Knowing that A and C hit the sides of the table squarely at points A' and C', respectively, with velocities of magnitude $v_A = 4$ ft/s and $v_C = 6$ ft/s, and assuming frictionless surfaces and perfectly elastic impacts (i.e., conservation of energy), determine (a) the initial velocity $\mathbf{v}_0$ of ball A, (b) the velocity $\mathbf{v}_B$ of ball B, (c) the point B' where B hits the side of the table.

14.34 Solve Prob. 14.33 if $v_A = 6$ ft/s and $v_C = 4$ ft/s.

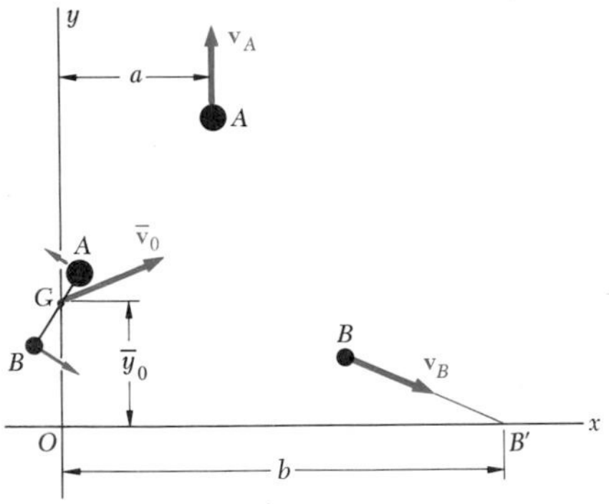

Fig. P14.35 and P14.36

14.35 Two small disks A and B, of mass 2 kg and 1 kg, respectively, may slide on a horizontal and frictionless surface. They are connected by a cord of negligible mass and spin about their mass center G. At $t = 0$, the coordinates of G are $\bar{x}_0 = 0$, $\bar{y}_0 = 1.6$ m, and its velocity is $\bar{\mathbf{v}}_0 = (1.5 \text{ m/s})\mathbf{i} + (1.2 \text{ m/s})\mathbf{j}$. Shortly thereafter, the cord breaks and disk A is observed to move along a path parallel to the y axis at a distance $a = 1.96$ m from that axis. Knowing that, initially, the angular momentum of the two disks about G was 3 kg · m²/s counterclockwise and that their kinetic energy relative to a centroidal frame was 18.75 J, determine (a) the velocities of A and B after the cord breaks, (b) the abscissa b of the point B' where the path of B intersects the x axis.

14.36 Two small disks A and B, of mass 2 kg and 1 kg, respectively, may slide on a horizontal and frictionless surface. They are connected by a cord of negligible mass and spin about their mass center G. At $t = 0$, G is moving with the velocity $\bar{\mathbf{v}}_0$ and its coordinates are $\bar{x}_0 = 0$, $\bar{y}_0 = 1.89$ m. Shortly thereafter, the cord breaks and disk A is observed to move with the velocity $\mathbf{v}_A = (5 \text{ m/s})\mathbf{j}$ in a straight line and at a distance $a = 2.56$ m from the y axis, while B moves with the velocity $\mathbf{v}_B = (7.2 \text{ m/s})\mathbf{i} - (4.6 \text{ m/s})\mathbf{j}$ along a path intersecting the x axis at a distance $b = 7.48$ m from the origin O. Determine (a) the initial velocity $\bar{\mathbf{v}}_0$ of the mass center G of the two disks, (b) the angular momentum $\mathbf{H}_G$ of the system about G and its kinetic energy relative to a centroidal frame before the cord broke, (c) the length of the cord initially connecting the two disks, (d) the rate in rad/s at which the disks were spinning about G.

***14.9. Variable Systems of Particles.** All the systems of particles considered so far consisted of well-defined particles. These systems did not gain or lose any particles during their motion. In a large number of engineering applications, however, it is necessary to consider *variable systems of particles*, i.e., systems which are continuously gaining or losing particles, or doing both at the same time. Consider, for example, a hydraulic turbine. Its analysis involves the determination of the forces exerted by a stream of water on rotating blades, and we note that the particles of water in contact with the blades form an everchanging system which continuously acquires and loses particles. Rockets furnish another example of variable systems, since their propulsion depends upon the continuous ejection of fuel particles.

We recall that all the kinetics principles established so far were derived for constant systems of particles, which neither gain nor lose particles. We must therefore find a way to reduce the analysis of a variable system of particles to that of an auxiliary constant system. The procedure to follow is indicated in Secs. 14.10 and 14.11 for two broad categories of applications.

***14.10. Steady Stream of Particles.** Consider a steady stream of particles, such as a stream of water diverted by a fixed vane or a flow of air through a duct or through a blower. In order to determine the resultant of the forces exerted on the particles in contact with the vane, duct, or blower, we isolate these particles and denote by S the system thus defined (Fig. 14.9). We observe that S is a variable system of particles, since it continuously gains particles flowing in and loses an equal number of particles flowing out. Therefore, the kinetics principles that have been established so far cannot be directly applied to S.

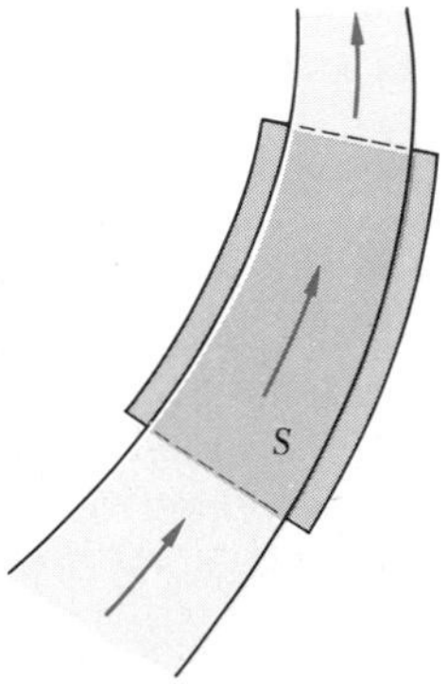

Fig. 14.9

However, we may easily define an auxiliary system of particles which does remain constant for a short interval of time Δt. Consider at time t the system S *plus* the particles which will enter S during the interval of time Δt (Fig. 14.10*a*). Next, consider at time $t + \Delta t$ the system S *plus* the particles which have left S during the interval Δt (Fig. 14.10*c*). Clearly, *the same particles are involved in both cases,* and we may apply to these particles the principle of impulse and momentum. Since the total mass m of the system S remains constant, the particles entering the system and those leaving the system in the time Δt must have the same mass Δm. Denoting by $\mathbf{v}_A$ and $\mathbf{v}_B$, respectively, the velocities of the particles entering S at A and leaving S at B, we represent the momentum of the particles entering S by $(\Delta m)\mathbf{v}_A$ (Fig. 14.10*a*) and the momentum of the particles leaving S by $(\Delta m)\mathbf{v}_B$ (Fig. 14.10*c*). We also represent the momenta $m_i\mathbf{v}_i$ of the particles forming S and the impulses of the forces exerted on S by the appropriate vectors, and indicate by gray plus and equals signs that the system of the momenta and impulses in parts a and b of Fig. 14.10 is equipollent to the system of the momenta in part c of the same figure.

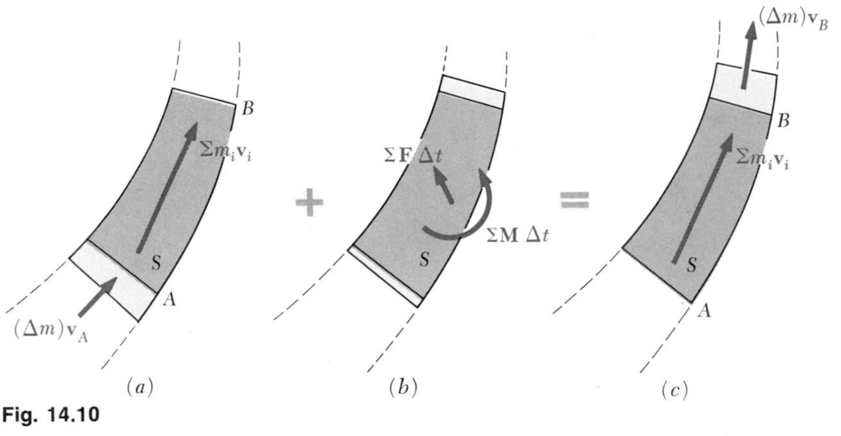

Fig. 14.10

Since the resultant $\Sigma m_i\mathbf{v}_i$ of the momenta of the particles of S is found on both sides of the equals sign, it may be omitted. We conclude that *the system formed by the momentum* $(\Delta m)\mathbf{v}_A$ *of the particles entering S in the time* Δt *and the impulses of the forces exerted on S during that time is equipollent to the momentum* $(\Delta m)\mathbf{v}_B$ *of the particles leaving S in the same time* Δt. We may therefore write

$$(\Delta m)\mathbf{v}_A + \Sigma\mathbf{F}\,\Delta t = (\Delta m)\mathbf{v}_B \qquad (14.38)$$

A similar equation may be obtained by taking the moments of the vectors involved (see Sample Prob. 14.5). Dividing all terms of Eq. (14.38) by Δt and letting Δt approach zero, we obtain at the limit

$$\Sigma \mathbf{F} = \frac{dm}{dt}(\mathbf{v}_B - \mathbf{v}_A) \tag{14.39}$$

where $\mathbf{v}_B - \mathbf{v}_A$ represents the difference between the *vectors* $\mathbf{v}_B$ and $\mathbf{v}_A$.

If SI units are used, dm/dt is expressed in kg/s and the velocities in m/s; we check that both members of Eq. (14.39) are expressed in the same units (newtons). If U.S. customary units are used, dm/dt must be expressed in slugs/s and the velocities in ft/s; we check again that both members of the equation are expressed in the same units (pounds).†

The principle we have established may be used to analyze a large number of engineering applications. Some of the most common are indicated below.

Fluid Stream Diverted by a Vane. If the vane is fixed, the method of analysis given above may be applied directly to find the force **F** exerted by the vane on the stream. We note that **F** is the only force which needs to be considered since the pressure in the stream is constant (atmospheric pressure). The force exerted by the stream on the vane will be equal and opposite to **F**. If the vane moves with a constant velocity, the stream is not steady. However, it will appear steady to an observer moving with the vane. We should therefore choose a system of axes moving with the vane. Since this system of axes is not accelerated, Eq. (14.38) may still be used, but $\mathbf{v}_A$ and $\mathbf{v}_B$ must be replaced by the *relative velocities* of the stream with respect to the vane (see Sample Prob. 14.6).

Fluid Flowing through a Pipe. The force exerted by the fluid on a pipe transition such as a bend or a contraction may be determined by considering the system of particles S in contact with the transition. Since, in general, the pressure in the flow will vary, we should also consider the forces exerted on S by the adjoining portions of the fluid.

† It is often convenient to express the mass rate of flow dm/dt as the product ρQ, where ρ is the density of the stream (mass per unit volume) and Q its volume rate of flow (volume per unit time). If SI units are used, ρ is expressed in kg/m^3 (for instance, $\rho = 1000$ kg/m^3 for water) and Q in m^3/s. However, if U.S. customary units are used, ρ will generally have to be computed from the corresponding specific weight γ (weight per unit volume), $\rho = \gamma/g$. Since γ is expressed in lb/ft^3 (for instance, $\gamma = 62.4$ lb/ft^3 for water), ρ is obtained in slugs/ft^3. The volume rate of flow Q is expressed in ft^3/s.

Jet Engine. In a jet engine, air enters with no velocity through the front of the engine and leaves through the rear with a high velocity. The energy required to accelerate the air particles is obtained by burning fuel. While the exhaust gases contain burned fuel, the mass of the fuel is small compared with the mass of the air flowing through the engine and usually may be neglected. Thus, the analysis of a jet engine reduces to that of an air stream. This stream may be considered as a steady stream if all velocities are measured with respect to the airplane. The air stream shall be assumed, therefore, to enter the engine with a velocity **v** of magnitude equal to the speed of the airplane and to leave with a velocity **u** equal to the relative velocity of the exhaust gases (Fig. 14.11). Since the intake and exhaust

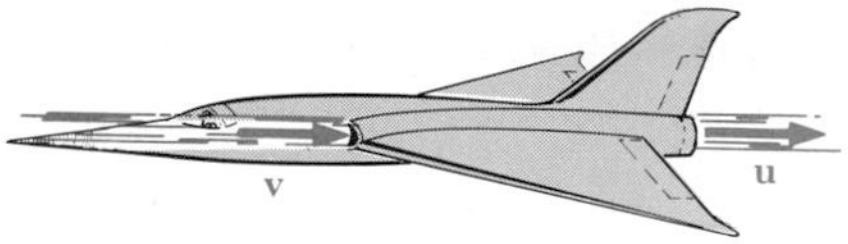

Fig. 14.11

pressures are nearly atmospheric, the only external force which needs to be considered is the force exerted by the engine on the air stream. This force is equal and opposite to the thrust.†

Fan. We consider the system of particles S shown in Fig. 14.12. The velocity $\mathbf{v}_A$ of the particles entering the system is assumed equal to zero, and the velocity $\mathbf{v}_B$ of the particles leaving the system is the velocity of the *slipstream*. The rate of flow may be obtained by multiplying v_B by the cross-sectional

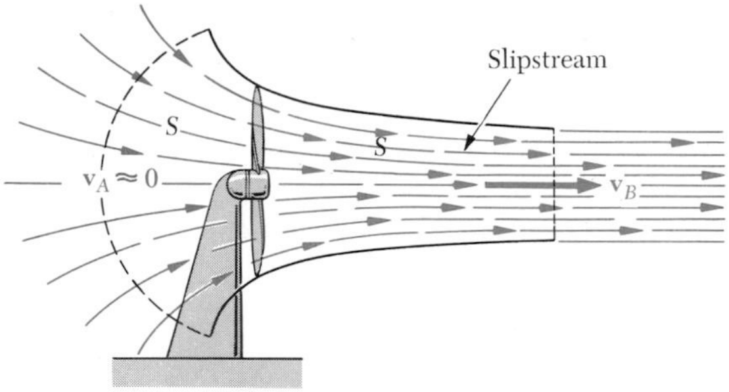

Fig. 14.12

†Note that, if the airplane is accelerated, it cannot be used as a newtonian frame of reference. The same result will be obtained for the thrust, however, by using a reference frame at rest with respect to the atmosphere, since the air particles will then be observed to enter the engine with no velocity and to leave it with a velocity of magnitude $u - v$.

area of the slipstream. Since the pressure all around S is atmospheric, the only external force acting on S is the thrust of the fan.

Airplane Propeller. In order to obtain a steady stream of air, velocities should be measured with respect to the airplane. Thus, the air particles will be assumed to enter the system with a velocity **v** of magnitude equal to the speed of the airplane and to leave with a velocity **u** equal to the relative velocity of the slipstream.

***14.11. Systems Gaining or Losing Mass.** We shall now analyze a different type of variable system of particles, namely, a system which gains mass by continuously absorbing particles or loses mass by continuously expelling particles. Consider the system S shown in Fig. 14.13. Its mass, equal to m at the instant t, increases by Δm in the interval of time Δt. In order to apply the principle of impulse and momentum to the analysis of this system, we must consider at time t the system

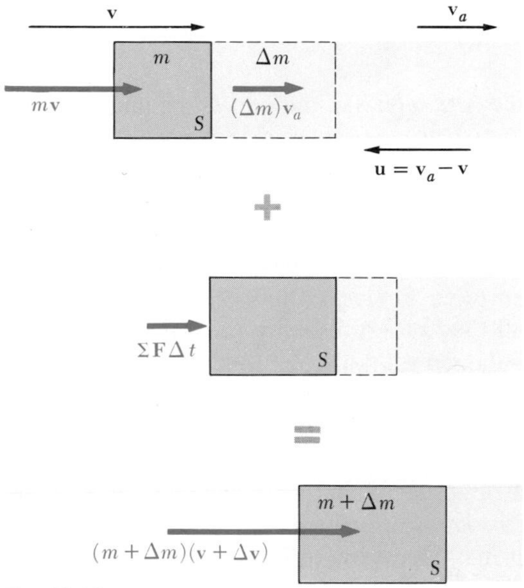

Fig. 14.13

S *plus* the particles of mass Δm which S absorbs during the time interval Δt. The velocity of S at time t is denoted by **v**, and its velocity at time $t + \Delta t$ is denoted by $\mathbf{v} + \Delta\mathbf{v}$, while the absolute velocity of the particles which are absorbed is denoted by $\mathbf{v}_a$. Applying the principle of impulse and momentum, we write

$$m\mathbf{v} + (\Delta m)\mathbf{v}_a + \Sigma\mathbf{F}\,\Delta t = (m + \Delta m)(\mathbf{v} + \Delta\mathbf{v})$$

Solving for the sum $\Sigma\mathbf{F}\,\Delta t$ of the impulses of the external forces acting on S (excluding the forces exerted by the particles being absorbed), we have

$$\Sigma\mathbf{F}\,\Delta t = m\,\Delta\mathbf{v} + \Delta m(\mathbf{v} - \mathbf{v}_a) + (\Delta m)(\Delta\mathbf{v}) \tag{14.40}$$

Introducing the *relative velocity* $\mathbf{u}$ with respect to S of the particles which are absorbed, we write $\mathbf{u} = \mathbf{v}_a - \mathbf{v}$ and note, since $v_a < v$, that the relative velocity $\mathbf{u}$ is directed to the left, as shown in Fig. 14.13. Neglecting the last term in Eq. (14.40), which is of the second order, we write

$$\Sigma\mathbf{F}\,\Delta t = m\,\Delta\mathbf{v} - (\Delta m)\mathbf{u}$$

Dividing through by Δt and letting Δt approach zero, we have at the limit†

$$\Sigma\mathbf{F} = m\frac{d\mathbf{v}}{dt} - \frac{dm}{dt}\mathbf{u} \tag{14.41}$$

Rearranging the terms, we obtain the equation

$$\Sigma\mathbf{F} + \frac{dm}{dt}\mathbf{u} = m\frac{d\mathbf{v}}{dt} \tag{14.42}$$

which shows that the action on S of the particles being absorbed is equivalent to a thrust of magnitude $(dm/dt)u$ which tends to slow down the motion of S, since the relative velocity $\mathbf{u}$ of the particles is directed to the left. If SI units are used, dm/dt is expressed in kg/s, the relative velocity u in m/s, and the corresponding thrust in newtons. If U.S. customary units are used, dm/dt must be expressed in slugs/s and u in ft/s; the corresponding thrust will then be expressed in pounds.‡

The equations obtained may also be used to determine the motion of a system S losing mass. In this case, the rate of change of mass is negative, and the action on S of the particles being expelled is equivalent to a thrust in the direction of $-\mathbf{u}$, that is, in the direction opposite to that in which the particles are being expelled. A *rocket* represents a typical case of a system continuously losing mass (see Sample Prob. 14.7).

†When the absolute velocity $\mathbf{v}_a$ of the particles absorbed is zero, we have $\mathbf{u} = -\mathbf{v}$, and formula (14.41) becomes

$$\Sigma\mathbf{F} = \frac{d}{dt}(m\mathbf{v})$$

Comparing the formula obtained to Eq. (12.3) of Sec. 12.2, we observe that Newton's second law may be applied to a system gaining mass, *provided that the particles absorbed are initially at rest.* It may also be applied to a system losing mass, *provided that the velocity of the particles expelled is zero* with respect to the frame of reference selected.

‡See footnote on page 637.

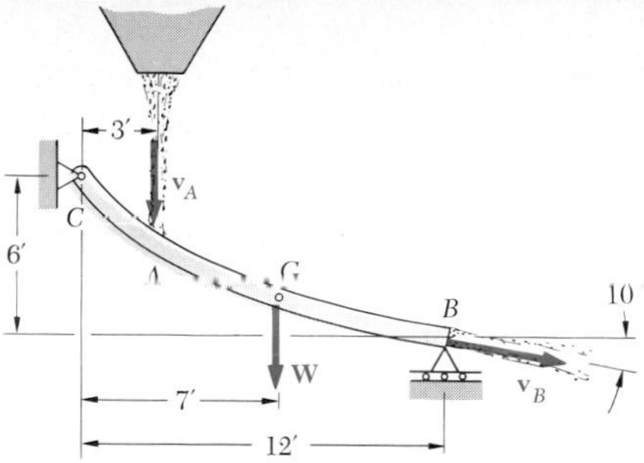

SAMPLE PROBLEM 14.5

Grain falls from a hopper onto a chute CB at the rate of 240 lb/s. It hits the chute at A with a velocity of 20 ft/s and leaves at B with a velocity of 15 ft/s, forming an angle of 10° with the horizontal. Knowing that the combined weight of the chute and of the grain it supports is a force **W** of magnitude 600 lb applied at G, determine the reaction at the roller-support B and the components of the reaction at the hinge C.

Solution. We apply the principle of impulse and momentum for the time interval Δt to the system consisting of the chute, the grain it supports, and the amount of grain which hits the chute in the interval Δt. Since the chute does not move, it has no momentum. We also note that the sum $\Sigma m_i \mathbf{v}_i$ of the momenta of the particles supported by the chute is the same at t and $t + \Delta t$ and thus may be omitted.

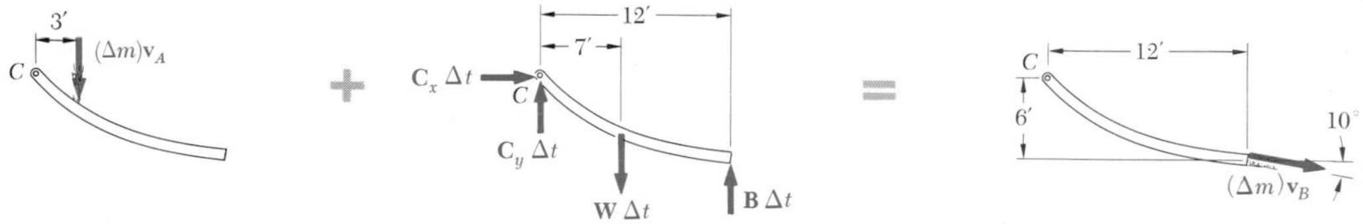

Since the system formed by the momentum $(\Delta m)\mathbf{v}_A$ and the impulses is equipollent to the momentum $(\Delta m)\mathbf{v}_B$, we write

$\xrightarrow{+}$ x components: $\quad C_x\,\Delta t = (\Delta m)v_B \cos 10° \qquad (1)$

$+\uparrow y$ components: $\quad -(\Delta m)v_A + C_y\,\Delta t - W\,\Delta t + B\,\Delta t = -(\Delta m)v_B \sin 10° \qquad (2)$

$+\circlearrowleft$ moments about C: $\quad -3(\Delta m)v_A - 7(W\,\Delta t) + 12(B\,\Delta t) = 6(\Delta m)v_B \cos 10° - 12(\Delta m)v_B \sin 10° \qquad (3)$

Using the given data, $W = 600$ lb, $v_A = 20$ ft/s, $v_B = 15$ ft/s, $\Delta m/\Delta t = 240/32.2 = 7.45$ slugs/s, and solving Eq. (3) for B and Eq. (1) for C_x:

$$12B = 7(600) + 3(7.45)(20) + 6(7.45)(15)(\cos 10° - 2 \sin 10°)$$

$$12B = 5075 \qquad B = 423 \text{ lb} \qquad \mathbf{B} = 423 \text{ lb} \uparrow \blacktriangleleft$$

$$C_x = (7.45)(15) \cos 10° = 110.1 \text{ lb} \qquad \mathbf{C}_x = 110.1 \text{ lb} \rightarrow \blacktriangleleft$$

Substituting for B and solving Eq. (2) for C_y:

$$C_y = 600 - 423 + (7.45)(20 - 15 \sin 10°) = 307 \text{ lb}$$

$$\mathbf{C}_y = 307 \text{ lb} \uparrow \blacktriangleleft$$

SAMPLE PROBLEM 14.6

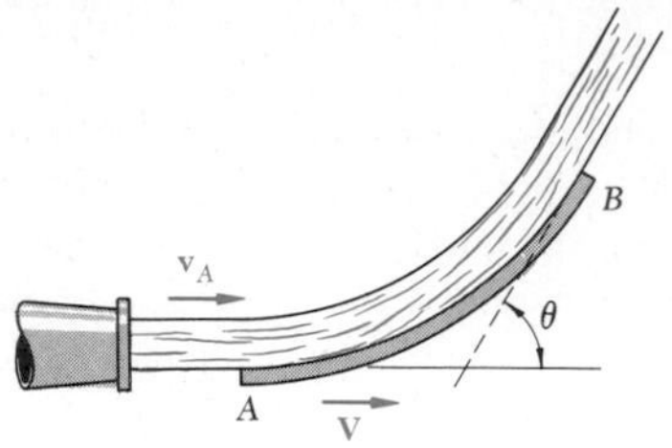

A nozzle discharges a stream of water of cross-sectional area A with a velocity $\mathbf{v}_A$. The stream is deflected by a *single* blade which moves to the right with a constant velocity $\mathbf{V}$. Assuming that the water moves along the blade at constant speed, determine (*a*) the components of the force $\mathbf{F}$ exerted by the blade on the stream, (*b*) the velocity $\mathbf{V}$ for which maximum power is developed.

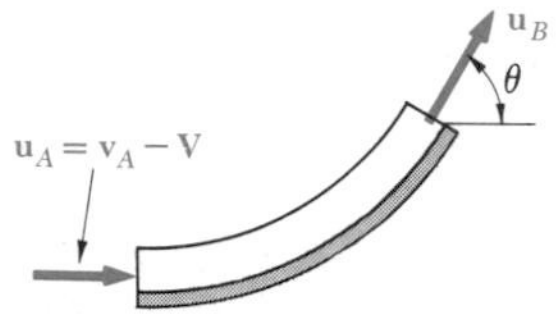

***a.* Components of Force Exerted on Stream.** We choose a coordinate system which moves with the blade at a constant velocity $\mathbf{V}$. The particles of water strike the blade with a relative velocity $\mathbf{u}_A = \mathbf{v}_A - \mathbf{V}$ and leave the blade with a relative velocity $\mathbf{u}_B$. Since the particles move along the blade at a constant speed, the relative velocities $\mathbf{u}_A$ and $\mathbf{u}_B$ have the same magnitude u. Denoting the density of water by ρ, the mass of the particles striking the blade during the time interval Δt is $\Delta m = A\rho(v_A - V)\,\Delta t$; an equal mass of particles leaves the blade during Δt. We apply the principle of impulse and momentum to the system formed by the particles in contact with the blade and by those striking the blade in the time Δt.

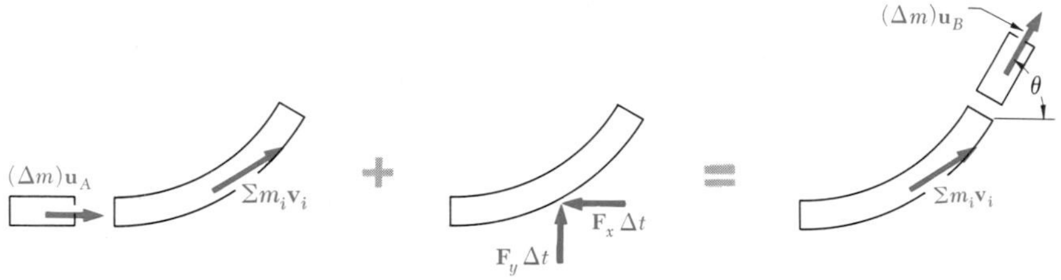

Recalling that $\mathbf{u}_A$ and $\mathbf{u}_B$ have the same magnitude u, and omitting the momentum $\Sigma m_i \mathbf{v}_i$ which appears on both sides, we write

$\xrightarrow{+}$ x components: $\quad (\Delta m)u - F_x\,\Delta t = (\Delta m)u\cos\theta$

$+\uparrow y$ components: $\quad +F_y\,\Delta t = (\Delta m)u\sin\theta$

Substituting $\Delta m = A\rho(v_A - V)\,\Delta t$ and $u = v_A - V$, we obtain

$$\mathbf{F}_x = A\rho(v_A - V)^2(1 - \cos\theta) \leftarrow \qquad \mathbf{F}_y = A\rho(v_A - V)^2 \sin\theta \uparrow \quad ◄$$

***b.* Velocity of Blade for Maximum Power.** The power is obtained by multiplying the velocity V of the blade by the component F_x of the force exerted by the stream on the blade.

$$\text{Power} = F_x V = A\rho(v_A - V)^2(1 - \cos\theta)V$$

Differentiating the power with respect to V and setting the derivative equal to zero, we obtain

$$\frac{d(\text{power})}{dV} = A\rho(v_A^2 - 4v_A V + 3V^2)(1 - \cos\theta) = 0$$

$$V = v_A \qquad V = \tfrac{1}{3}v_A \qquad \text{For maximum power } \mathbf{V} = \tfrac{1}{3}v_A \rightarrow \quad ◄$$

Note. These results are valid only when a *single* blade deflects the stream. Different results are obtained when a series of blades deflects the stream, as in a Pelton-wheel turbine. (See Prob. 14.90.)

SAMPLE PROBLEM 14.7

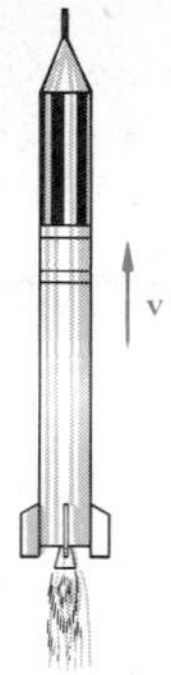

A rocket of initial mass m_0 (including shell and fuel) is fired vertically at time $t = 0$. The fuel is consumed at a constant rate $q = dm/dt$ and is expelled at a constant speed u relative to the rocket. Derive an expression for the velocity of the rocket at time t, neglecting the resistance of the air.

Solution. At time t, the mass of the rocket shell and remaining fuel is $m = m_0 - qt$, and the velocity is $\mathbf{v}$. During the time interval Δt, a mass of fuel $\Delta m = q\,\Delta t$ is expelled with a speed u relative to the rocket. Denoting by $\mathbf{v}_e$ the absolute velocity of the expelled fuel, we apply the principle of impulse and momentum between time t and time $t + \Delta t$.

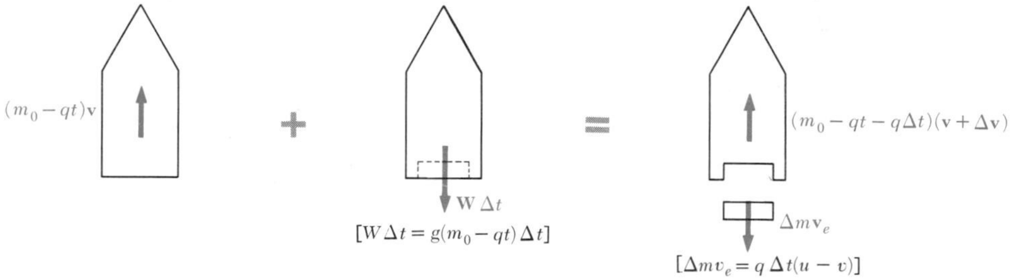

We write

$$(m_0 - qt)v - g(m_0 - qt)\,\Delta t = (m_0 - qt - q\,\Delta t)(v + \Delta v) - q\,\Delta t(u - v)$$

Dividing through by Δt, and letting Δt approach zero, we obtain

$$-g(m_0 - qt) = (m_0 - qt)\frac{dv}{dt} - qu$$

Separating variables and integrating from $t = 0$, $v = 0$ to $t = t$, $v = v$,

$$dv = \left(\frac{qu}{m_0 - qt} - g\right)dt \qquad \int_0^v dv = \int_0^t \left(\frac{qu}{m_0 - qt} - g\right)dt$$

$$v = [-u \ln (m_0 - qt) - gt]_0^t \qquad v = u \ln \frac{m_0}{m_0 - qt} - gt \blacktriangleleft$$

Remark. The mass remaining at time t_f, after all the fuel has been expended, is equal to the mass of the rocket shell $m_s = m_0 - qt_f$, and the maximum velocity attained by the rocket is $v_m = u \ln (m_0/m_s) - gt_f$. Assuming that the fuel is expelled in a relatively short period of time, the term gt_f is small and we have $v_m \approx u \ln (m_0/m_s)$. In order to escape the gravitational field of the earth, a rocket must reach a velocity of 11 180 m/s. Assuming $u = 2200$ m/s and $v_m = 11\,180$ m/s, we obtain $m_0/m_s = 161$. Thus, to project each kilogram of the rocket shell into space, it is necessary to consume more than 161 kg of fuel if a propellant yielding $u = 2200$ m/s is used.

PROBLEMS

Note. In the following problems use $\rho = 1000\ \text{kg/m}^3$ for the density of water in SI units, and $\gamma = 62.4\ \text{lb/ft}^3$ for its specific weight in U.S. customary units. (See footnote on page 637.)

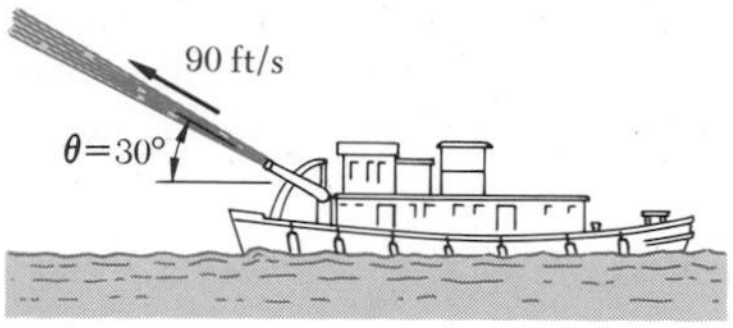

Fig. P14.37

14.37 A hose discharges 2000 gal/min from the stern of a 20-ton fireboat. If the velocity of the water stream is 90 ft/s, determine the reaction on the boat.

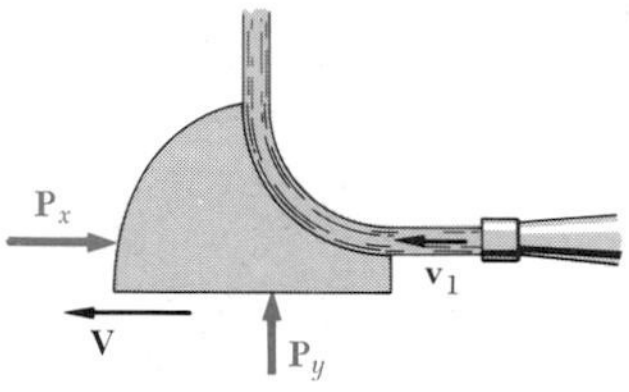

Fig. P14.38 and P14.39

14.38 A stream of water of cross-sectional area A and velocity $\mathbf{v}_1$ strikes the curved surface of a block which is held motionless ($V = 0$) by the forces $\mathbf{P}_x$ and $\mathbf{P}_y$. Determine the magnitudes of $\mathbf{P}_x$ and $\mathbf{P}_y$ when $A = 500\ \text{mm}^2$ and $v_1 = 40\ \text{m/s}$.

14.39 A stream of water of cross-sectional area A and velocity $\mathbf{v}_1$ strikes the curved surface of a block which moves to the left with a velocity $\mathbf{V}$. Determine the magnitudes of the forces $\mathbf{P}_x$ and $\mathbf{P}_y$ required to hold the block when $A = 3\ \text{in}^2$, $v_1 = 90\ \text{ft/s}$, and $V = 25\ \text{ft/s}$.

14.40 Sand is discharged at the rate m (kg/s) from a conveyor belt moving with a velocity $\mathbf{v}_0$. The sand is deflected by a plate at B so that it falls in a vertical stream. After falling a distance h, the sand is again deflected as shown by a curved plate at C. Neglecting the friction between the sand and the plates, determine the force required to hold each plate in the position shown.

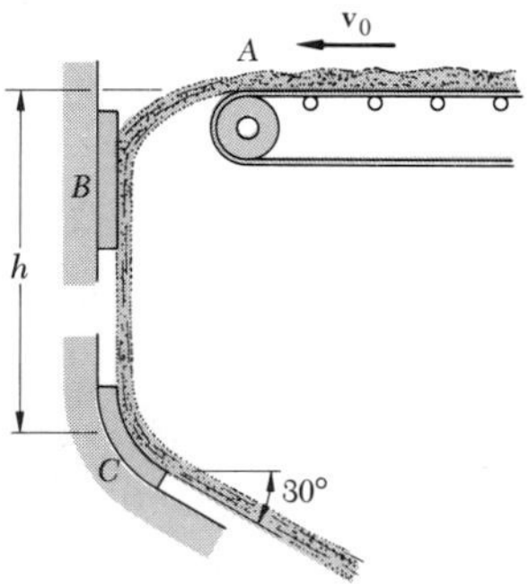

Fig. P14.40

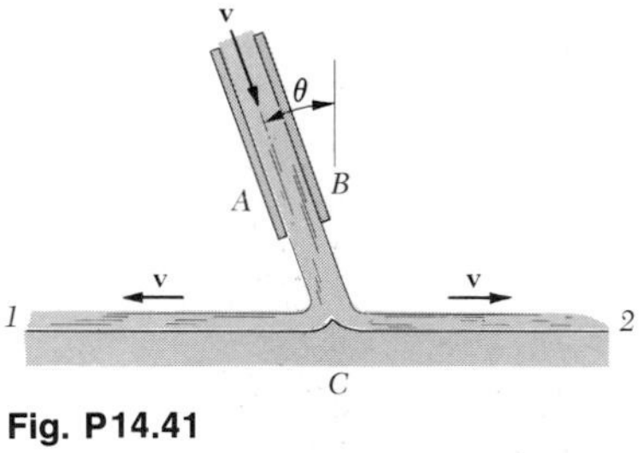

Fig. P14.41

14.41 Water flows in a continuous sheet from between two plates A and B with a velocity $\mathbf{v}$. The stream is split into two equal streams *1* and *2* by a vane attached to plate C. Denoting the total rate of flow by Q, determine the force exerted by the stream on plate C.

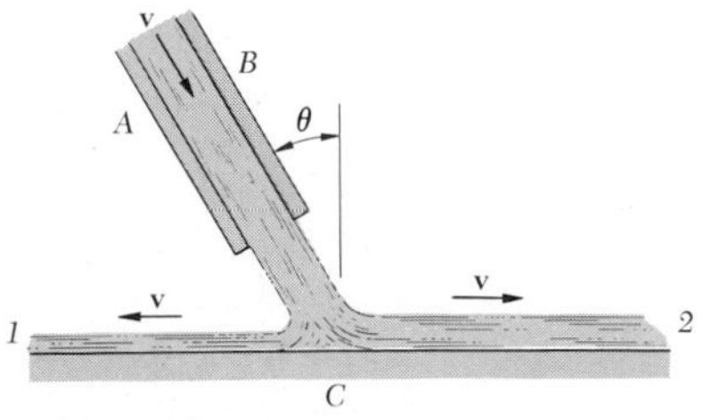

Fig. P14.42

14.42 Water flows in a continuous sheet from between two plates A and B with a velocity $\mathbf{v}$. The stream is split into two parts by a smooth horizontal plate C. Denoting the total rate of flow by Q, determine the rate of flow of each of the resulting streams. (*Hint.* The plate C can exert only a vertical force on the water.)

14.43 A stream of water of cross-sectional area A and velocity $\mathbf{v}_A$ is deflected by a vane AB in the shape of an arc of circle of radius R. Knowing that the vane is welded to a fixed support at A, determine the components of the force-couple system exerted by the support on the vane.

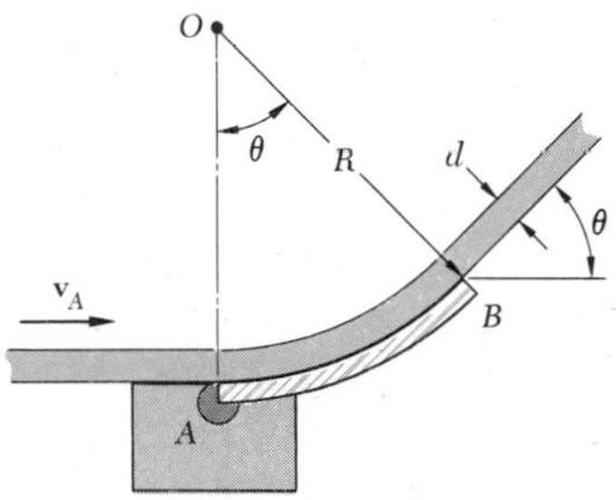

Fig. P14.43

14.44 The nozzle shown discharges 250 gal/min of water with a velocity $\mathbf{v}_A$ of 120 ft/s. The stream is deflected by the fixed vane AB. Determine the force-couple system which must be applied at C in order to hold the vane in place (1 ft^3 = 7.48 gal).

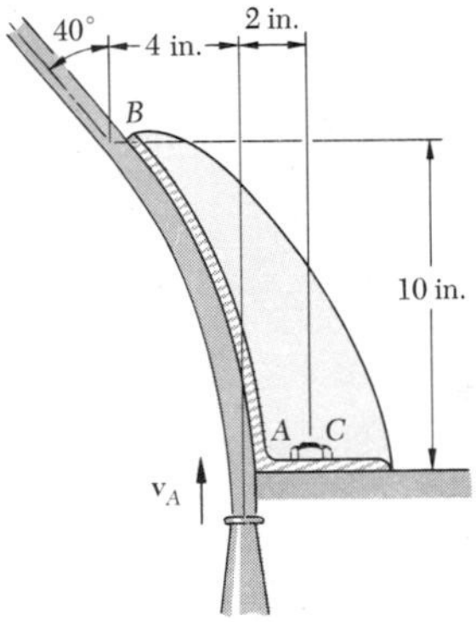

Fig. P14.44

14.45 Knowing that the blade AB of Sample Prob. 14.6 is in the shape of an arc of circle, show that the resultant force $\mathbf{F}$ exerted by the blade on the stream is applied at the midpoint C of the arc AB. (*Hint.* First show that the line of action of $\mathbf{F}$ must pass through the center O of the circle.)

14.46 The stream of water shown flows at the rate of 0.9 m^3/min and moves with a velocity of magnitude 30 m/s at both A and B. The vane is supported by a pin connection at C and by a load cell at D which can exert only a horizontal force. Neglecting the weight of the vane, determine the reactions at C and D.

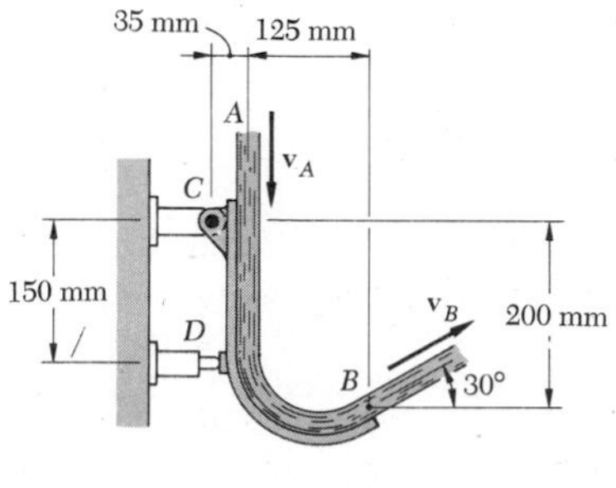

Fig. P14.46

14.47 The nozzle shown discharges water at the rate of $1.2\ m^3/min$. Knowing that at both A and B the stream of water moves with a velocity of magnitude 25 m/s and neglecting the weight of the vane, determine the components of the reactions at C and D.

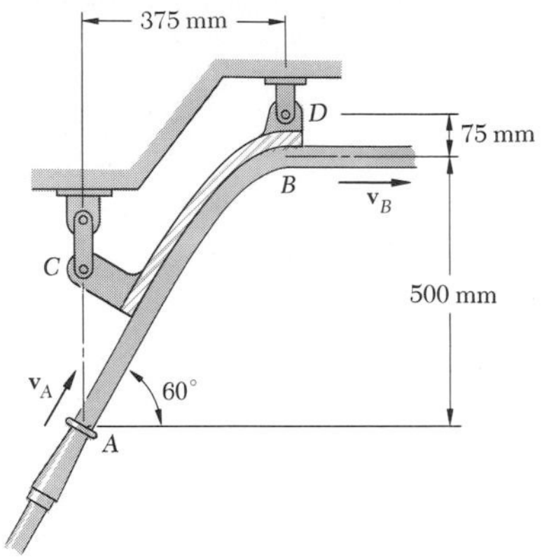

Fig. P14.47

14.48 The final component of a conveyor system receives sand at the rate of 180 lb/s at A and discharges it at B. The sand is moving horizontally at A and B with a velocity of magnitude $v_A = v_B =$ 12 ft/s. Knowing that the combined weight of the component and of the sand it supports is $W = 800$ lb, determine the reactions at C and D.

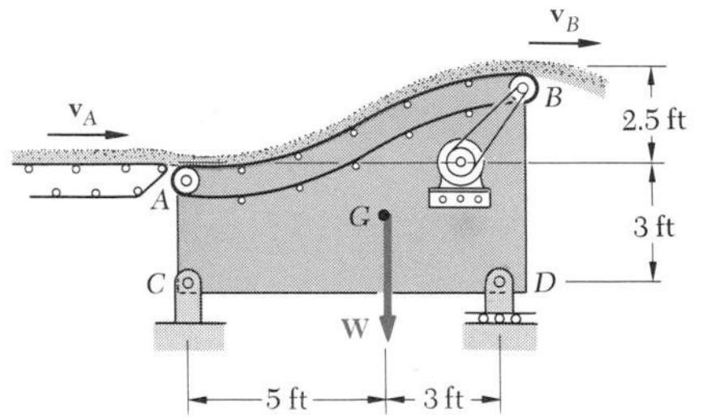

Fig. P14.48

14.49 Solve Prob. 14.48, assuming the velocity of the belt of the final component of the conveyor system is increased in such a way that, while the sand is still received with a velocity $\mathbf{v}_A$ of 12 ft/s, it is discharged with a velocity $\mathbf{v}_B$ of 24 ft/s.

14.50 A jet airplane scoops in air at the rate of 250 lb/s and discharges it with a velocity of 2200 ft/s relative to the airplane. If the speed of the airplane is 600 mi/h, determine (*a*) the propulsive force developed, (*b*) the horsepower actually used to propel the airplane, (*c*) the horsepower developed by the engine.

14.51 The total drag due to air friction on a jet airplane traveling at 1000 km/h is 16 kN. Knowing that the exhaust velocity is 600 m/s relative to the airplane, determine the mass of air which must pass through the engine per second to maintain the speed of 1000 km/h in level flight.

14.52 While cruising in horizontal flight at a speed of 800 km/h, a 9000-kg jet airplane scoops in air at the rate of 70 kg/s and discharges it with a velocity of 600 m/s relative to the airplane. (*a*) Determine the total drag due to air friction. (*b*) Assuming that the drag is proportional to the square of the speed, determine the horizontal cruising speed if the flow of air through the jet is increased by 10 percent, i.e., to 77 kg/s.

14.53 The cruising speed of a jet airliner is 600 mi/h. Each of the four engines discharges air with a velocity of 2000 ft/s relative to the plane. Assuming that the drag due to air resistance is proportional to the square of the speed, determine the speed of the airliner when only two of the engines are in operation.

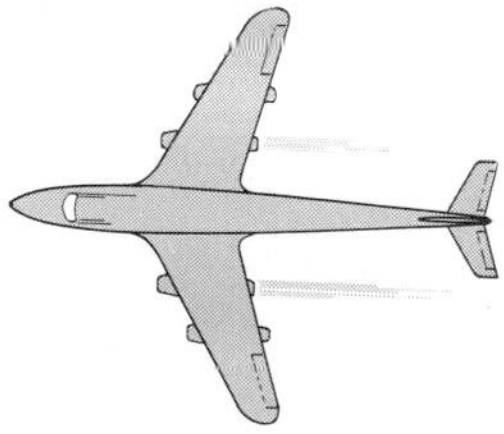

Fig. P14.53

14.54 For use in shallow water the pleasure boat shown is powered by a water jet. Water enters the engine through orifices located in the bow and is discharged through a horizontal pipe at the stern. Knowing that the water is discharged at the rate of 10 m^3/min with a velocity of 15 m/s relative to the boat, determine the propulsive force developed when the speed of the boat is (*a*) 6 m/s, (*b*) zero.

Fig. P14.54

14.55 In order to shorten the distance required for landing, a jet airplane is equipped with movable vanes which partially reverse the direction of the air discharged by each of its engines. Each engine scoops in air at the rate of 200 lb/s and discharges it with a velocity of 2000 ft/s relative to the engine. At an instant when the speed of the airplane is 120 mi/h, determine the reversed thrust provided by each of the engines.

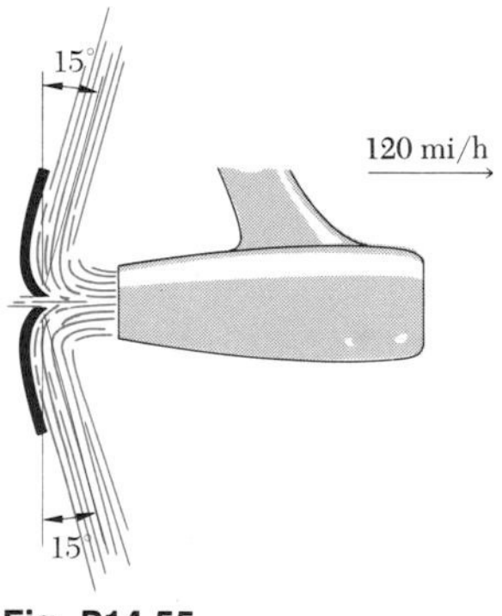

Fig. P14.55

14.56 An unloaded helicopter of weight 5000 lb produces a slipstream of 38-ft diameter. Assuming that air weighs 0.076 lb/ft^3, determine the vertical component of the velocity of the air in the slipstream when the helicopter is hovering in midair.

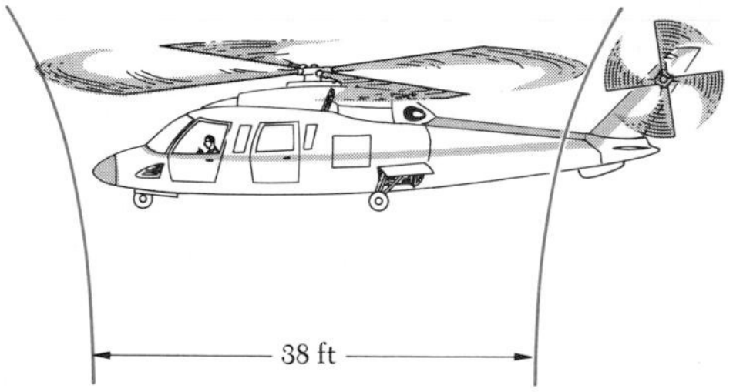

Fig. P14.56 and P14.57

14.57 The helicopter shown weighs 5000 lb and can produce a maximum downward air speed of 60 ft/s in the 38-ft diameter slipstream. Determine the maximum load which the helicopter can carry while hovering in midair.

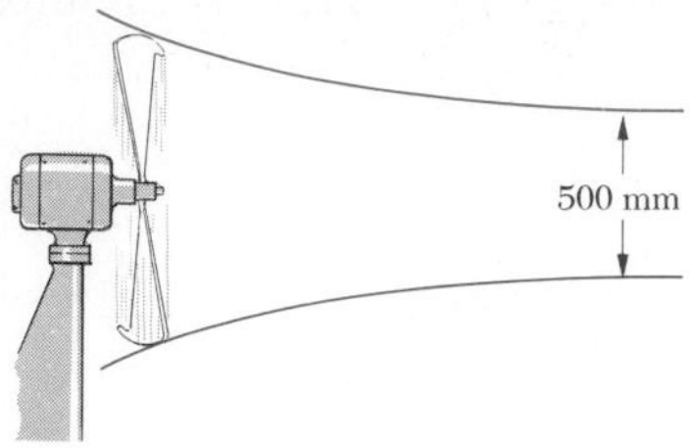

Fig. P14.58

14.58 The slipstream of a fan has a diameter of 500 mm and a velocity of 10 m/s relative to the fan. Assuming $\rho = 1.21\ \mathrm{kg/m^3}$ for air and neglecting the velocity of approach of the air, determine the force required to hold the fan motionless.

14.59 The propeller of an airplane produces a thrust of 4000 N when the airplane is at rest on the ground and has a slipstream of 2-m diameter. Assuming $\rho = 1.21\ \mathrm{kg/m^3}$ for air, determine (*a*) the speed of the air in the slipstream, (*b*) the volume of air passing through the propeller per second, (*c*) the kinetic energy imparted per second to the air of the slipstream.

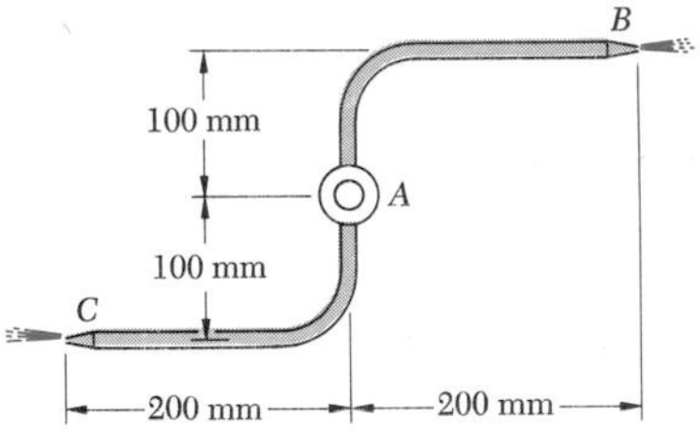

Fig. P14.60

14.60 Each arm of the sprinkler shown discharges water at the rate of 10 liters per minute with a velocity of 12 m/s relative to the arm. Neglecting the effect of friction, determine (*a*) the constant rate at which the sprinkler will rotate, (*b*) the couple **M** which must be applied to the sprinkler to hold it stationary.

14.61 A circular reentrant orifice (also called Borda's mouthpiece) of diameter D is placed at a depth h below the surface of a tank. Knowing that the speed of the issuing stream is $v = \sqrt{2gh}$ and assuming that the speed of approach v_1 is zero, show that the diameter of the stream is $d = D/\sqrt{2}$. (*Hint.* Consider the section of water indicated, and note that P is equal to the pressure at a depth h multiplied by the area of the orifice.)

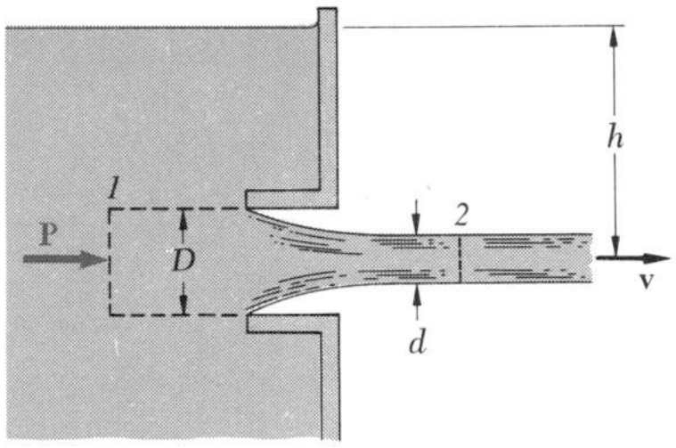

Fig. P14.61

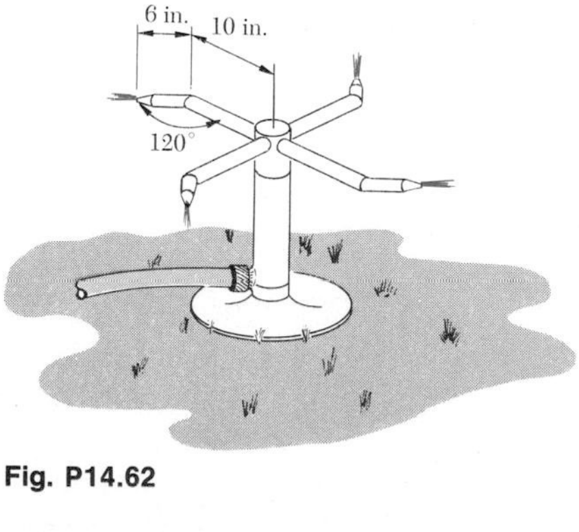

Fig. P14.62

14.62 A garden sprinkler has four rotating arms, each of which consists of two horizontal straight sections of pipe forming an angle of 120°. Each arm discharges water at the rate of 3 gal/min with a velocity of 48 ft/s relative to the arm. Knowing that the friction between the moving and stationary parts of the sprinkler is equivalent to a couple of magnitude $M = 0.200\ \mathrm{lb \cdot ft}$, determine the constant rate at which the sprinkler rotates ($1\ \mathrm{ft^3} = 7.48$ gal).

14.63 Each of the two conveyor belts shown discharges sand at a constant rate of 5 lb/s. The sand falls through a height h and is deflected by a stationary vane. Knowing that the velocity of the sand is horizontal as it leaves the vane, determine the force **P** required to hold the vane when (*a*) $h = 6$ ft, (*b*) $h = 12$ ft.

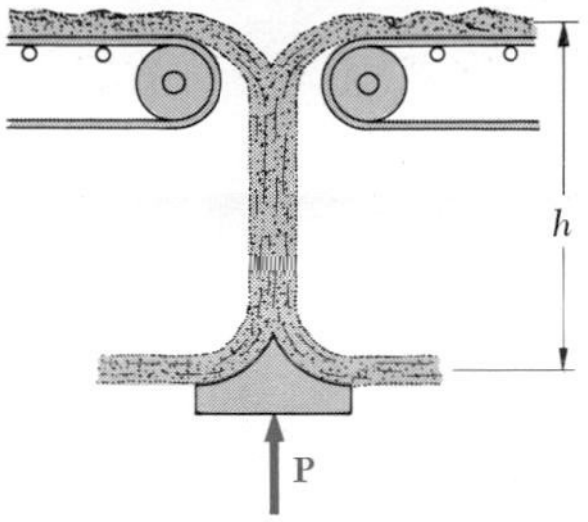

Fig. P14.63

14.64 Gravel falls with practically zero velocity onto a conveyor belt at the constant rate $q = dm/dt$. A force **P** is applied to the belt to maintain a constant speed v. Derive an expression for the angle θ for which the force **P** is zero.

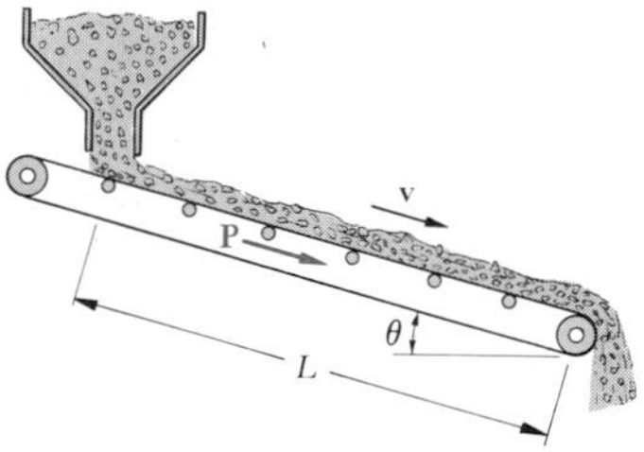

Fig. P14.64 and P14.65

14.65 Gravel falls with practically zero velocity onto a conveyor belt at the constant rate $q = dm/dt$. (*a*) Determine the magnitude of the force **P** required to maintain a constant belt speed v, when $\theta = 0$. (*b*) Show that the kinetic energy acquired by the gravel in a given time interval is equal to half the work done in that interval by the force **P**. Explain what happens to the other half of the work done by **P**.

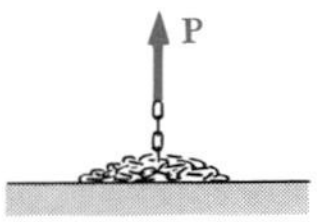

Fig. P14.66

14.66 A chain of mass m per unit length and total length l lies in a pile on the floor. At time $t = 0$ a force **P** is applied and the chain is raised with a constant velocity **v**. Express the required magnitude of the force **P** as a function of the time t.

14.67 A chain of length l and mass m per unit length falls through a small hole in a plate. Initially, when y is very small, the chain is at rest. In each case shown, determine (*a*) the acceleration of the first link A as a function of y, (*b*) the velocity of the chain as the last link passes through the hole. In case *1* assume that the individual links are at rest until they fall through the hole; in case 2 assume that at any instant all links have the same speed. Ignore the effect of friction.

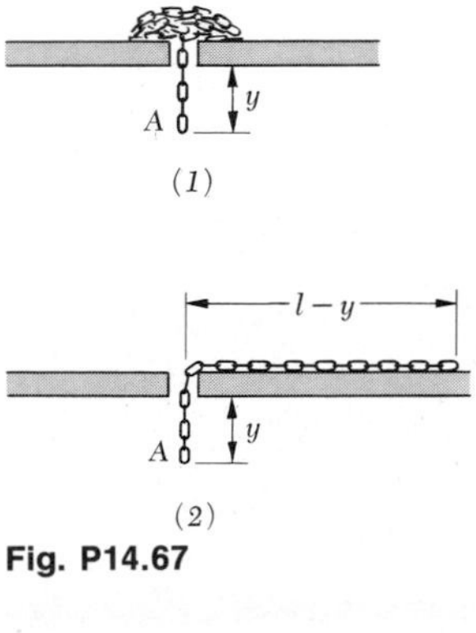

Fig. P14.67

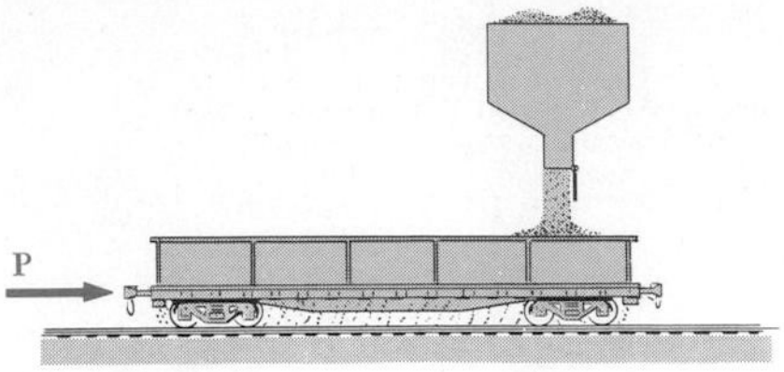

Fig. P14.68

14.68 A moving railroad car, of mass m_0 when empty, is loaded by dropping sand vertically into it from a stationary chute at the rate $q = dm/dt$. At the same time, however, sand is leaking out through the floor of the car at the lesser rate q'. Determine the magnitude of the horizontal force **P** required to keep the car moving at a constant speed v while being loaded.

14.69 For the car and loading conditions of Prob. 14.68, express as a function of t the magnitude of the horizontal force **P** required to keep the car moving with a constant acceleration **a** while being loaded. Denote by v_0 the speed of the car at $t = 0$, when the loading operation begins.

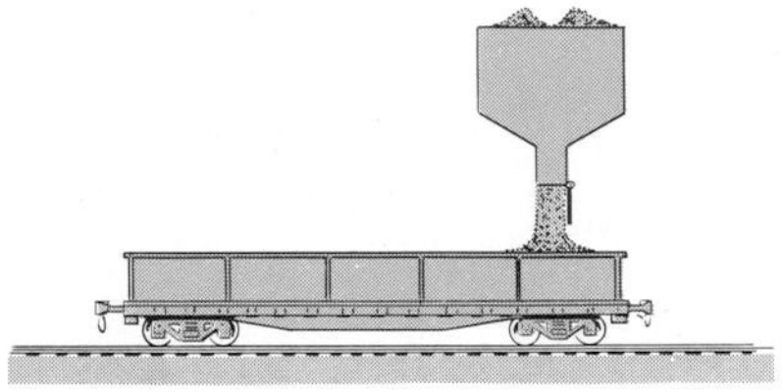

Fig. P14.70

14.70 A railroad car, of mass m_0 when empty and moving freely on a horizontal track, is loaded by dropping sand vertically into it from a stationary chute at the rate $q = dm/dt$. Determine the velocity and acceleration of the car as functions of t. Denote by v_0 the speed of the car at $t = 0$, when the loading operation begins.

14.71 If the car of Prob. 14.68 moves freely ($\mathbf{P} = 0$), determine its velocity and acceleration as functions of t. Denote by v_0 the speed of the car at $t = 0$, when the loading operation begins.

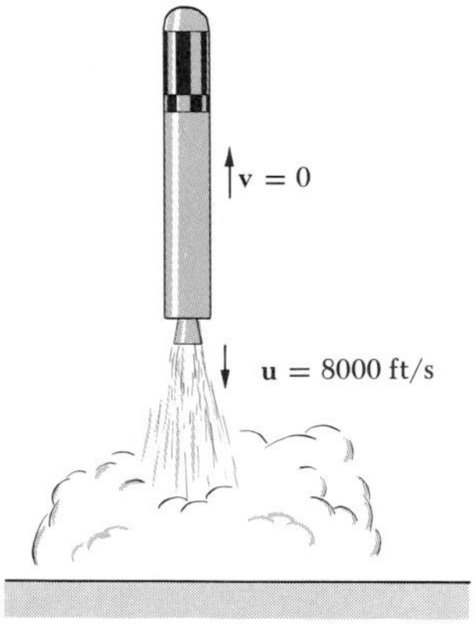

Fig. P14.72

14.72 A test rocket is designed to hover motionless above the ground. The shell of the rocket weighs 2500 lb, and the initial fuel load is 7500 lb. The fuel is burned and ejected with a velocity of 8000 ft/s. Determine the required rate of fuel consumption (*a*) when the rocket is fired, (*b*) as the last particle of fuel is being consumed.

14.73 The main engine installation of a space shuttle consists of three identical rocket engines which are required to provide a total thrust of 6000 kN. Knowing that the hydrogen-oxygen propellent is burned and ejected with a velocity of 3900 m/s, determine the required total rate of fuel consumption.

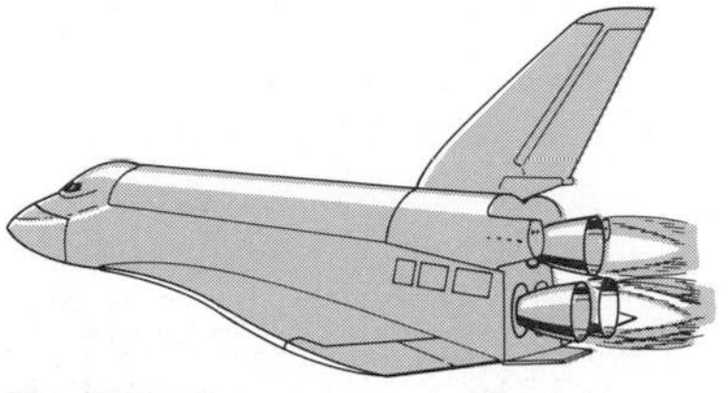

Fig. P14.73

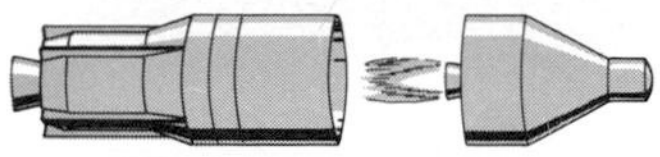

Fig. P14.74

14.74 A space vehicle describing a circular orbit at a speed of 15,000 mi/h releases a capsule which has a gross weight of 1000 lb, including 750 lb of fuel. If the fuel is consumed at the constant rate of 30 lb/s and is ejected with a relative velocity of 8000 ft/s, determine the tangential acceleration of the capsule (*a*) as the engine is fired, (*b*) as the last particle of fuel is being consumed.

14.75 A rocket of gross mass 1000 kg, including 900 kg of fuel, is fired vertically when $t = 0$. Knowing that fuel is consumed at the rate of 10 kg/s and ejected with a relative velocity of 3500 m/s, determine the acceleration and velocity of the rocket when (*a*) $t = 0$, (*b*) $t = 45$ s, (*c*) $t = 90$ s.

14.76 A space tug describing a low-level circular orbit is to be transferred to a high-level orbit. The maneuver is started by firing the rocket engines to increase the speed of the tug from 7370 to 9850 m/s. The initial mass of the tug, fuel, and payload is 14.1 Mg. Knowing that the hydrogen-oxygen propellent is consumed at the rate of 20 kg/s and is ejected with a velocity of 3750 m/s, determine (*a*) the mass of fuel which must be expended to initiate the maneuver, (*b*) the time interval for which the engines must be fired.

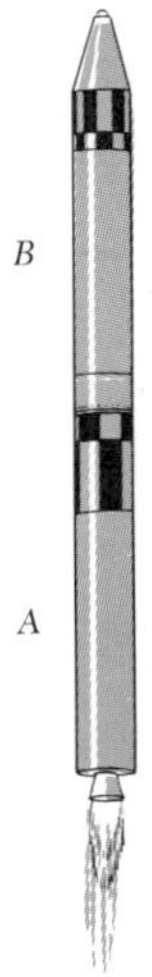

Fig. P14.77

14.77 The rocket of Prob. 14.75 is redesigned as a two-stage rocket consisting of rockets *A* and *B*, each of gross mass 500 kg, including 450 kg of fuel. The fuel is again consumed at the rate of 10 kg/s and is ejected with a relative velocity of 3500 m/s. Knowing that, when rocket *A* expells its last particle of fuel, its shell is released and rocket *B* is fired, determine (*a*) the speed when rocket *A* is released, (*b*) the maximum speed attained by rocket *B*.

14.78 A spacecraft is launched vertically by a two-stage rocket. When the speed is 10,000 mi/h the first-stage-rocket casing is released and the second-stage rocket is fired. Fuel is consumed at the rate of 200 lb/s and ejected with a relative velocity of 8000 ft/s. Knowing that the combined weight of the second-stage rocket and spacecraft is 20,000 lb, including 17,000 lb of fuel, determine the maximum speed which can be attained by the spacecraft.

14.79 For the rocket of Sample Prob. 14.7, derive an expression for the height of the rocket as a function of the time *t*.

14.80 Determine the distance between the spacecraft and the first-stage-rocket casing of Prob. 14.78 as the last particle of fuel is being expelled by the second-stage rocket.

14.81 Determine the distance between the capsule and the space vehicle of Prob. 14.74 as the last particle of fuel is being ejected by the rocket of the capsule. Both the capsule and the space vehicle may be considered to move in a straight line during the time interval considered.

14.82 In a jet airplane, the kinetic energy imparted to the exhaust gases is wasted as far as propelling the airplane is concerned. The useful power is equal to the product of the force available to propel the airplane and the speed of the airplane. If v is the speed of the airplane and u is the relative speed of the expelled gases, show that the efficiency is $\eta = 2v/(u + v)$. Explain why $\eta = 1$ when $u = v$.

14.83 In a rocket, the kinetic energy imparted to the consumed and ejected fuel is wasted as far as propelling the rocket is concerned. The useful power is equal to the product of the force available to propel the rocket and the speed of the rocket. If v is the speed of the rocket and u is the relative speed of the expelled fuel, show that the efficiency is $\eta = 2uv/(u^2 + v^2)$. Explain why $\eta = 1$ when $u = v$.

REVIEW PROBLEMS

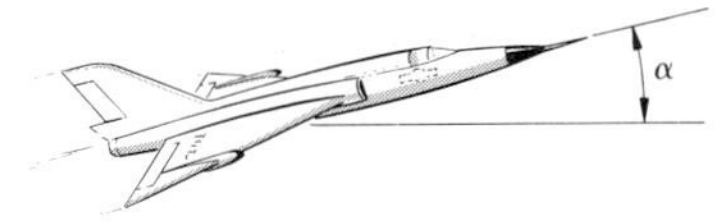

Fig. P14.84

14.84 A 9000-kg jet airplane maintains a constant speed of 900 km/h while climbing at an angle $\alpha = 5°$. The airplane scoops in air at the rate of 80 kg/s and discharges it with a velocity of 700 m/s relative to the airplane. If the pilot changes to a horizontal flight and the same engine conditions are maintained, determine (*a*) the initial acceleration of the plane, (*b*) the maximum horizontal speed attained. Assume that the drag due to air friction is proportional to the square of the speed.

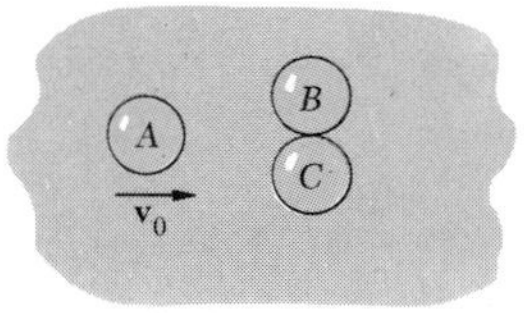

Fig. P14.85

14.85 Three identical balls A, B, and C may roll freely on a horizontal surface. Balls B and C are at rest and in contact when struck by ball A, which was moving to the right with a velocity $\mathbf{v}_0$. Assuming $e = 1$ and no friction, determine the final velocity of ball A if (*a*) the path of A is perfectly centered and A strikes B and C simultaneously, (*b*) the path of A is not perfectly centered and A strikes B slightly before it strikes C.

14.86 A 1-oz bullet is fired with a velocity of 1600 ft/s into block A, which weighs 10 lb. The coefficient of friction between block A and the cart BC is 0.50. Knowing that the cart weighs 8 lb and can roll freely, determine (*a*) the final velocity of the cart and block, (*b*) the final position of the block on the cart.

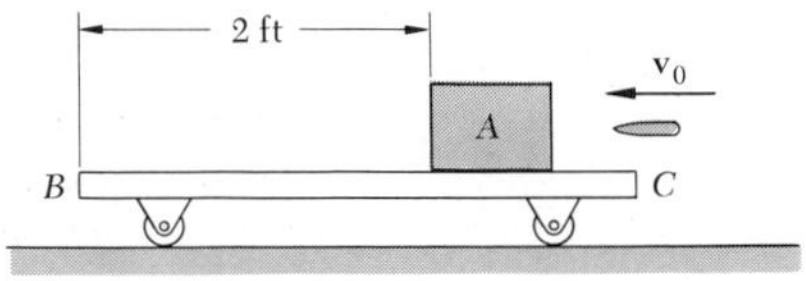

Fig. P14.86

14.87 The ends of a chain of mass m per unit length lie in piles at A and at C; when released, the chain moves over the pulley at B. Determine the required initial speed v for which the chain will move at a constant speed. Neglect axle friction.

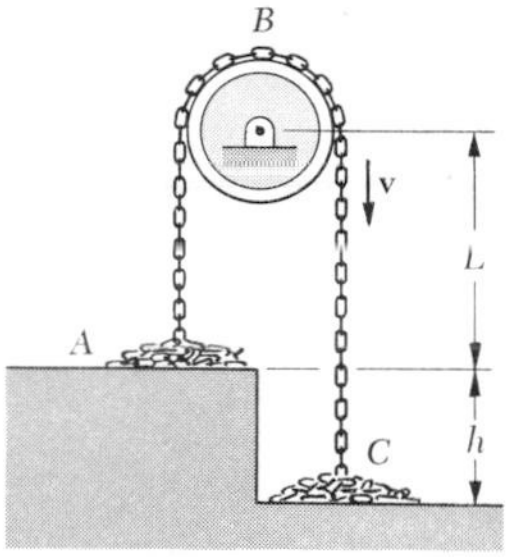

Fig. P14.87

14.88 Two railroad freight cars move with a velocity $\mathbf{v}$ through a switchyard. Car B hits a third car C, which was at rest with its brakes released, and it automatically couples with C. Knowing that all three cars have the same mass, determine their common velocity after they are all coupled together, as well as the percentage of their total initial kinetic energy which is absorbed by each coupling mechanism, assuming (*a*) that cars A and B were originally coupled, (*b*) that cars A and B were moving a few feet apart and that the coupling operation between B and C is completed before A hits B and becomes coupled with it.

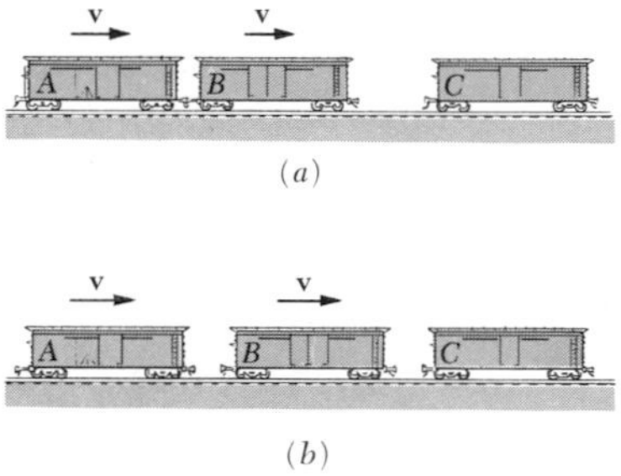

Fig. P14.88

14.89 A 5-kg sphere is moving with a velocity of 60 m/s when it explodes into two fragments. Immediately after the explosion the fragments are observed to travel in the directions shown and the speed of fragment A is observed to be 90 m/s. Determine (*a*) the mass of fragment A, (*b*) the speed of fragment B.

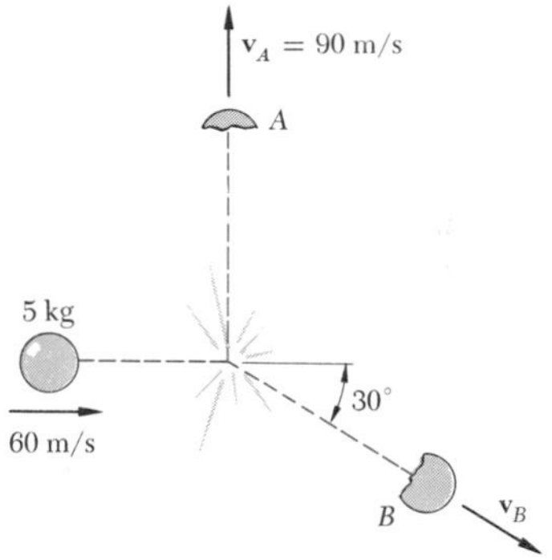

Fig. P14.89

14.90 In a Pelton-wheel turbine, a stream of water is deflected by a series of blades so that the rate at which water is deflected by the blades is equal to the rate at which water issues from the nozzle $(\Delta m/\Delta t = A\rho v_A)$. Using the same notation as in Sample Prob. 14.6, (*a*) determine the velocity $\mathbf{V}$ of the blades for which maximum power is developed, (*b*) derive an expression for the maximum power, (*c*) derive an expression for the mechanical efficiency.

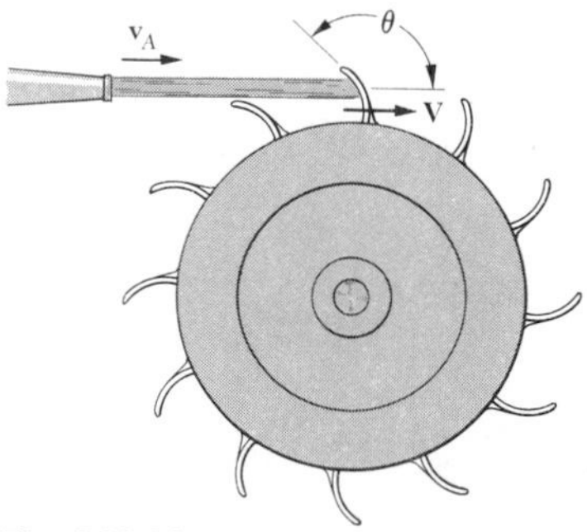

Fig. P14.90

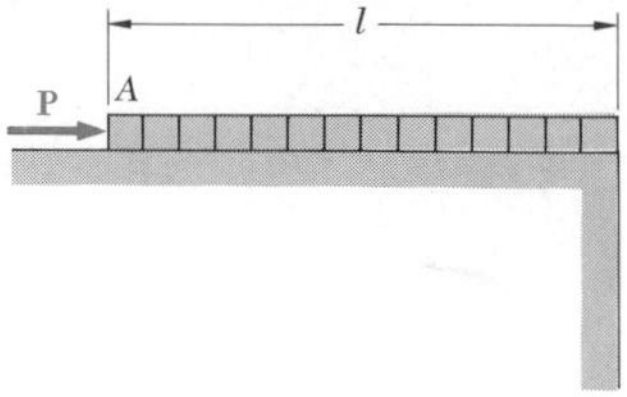

Fig. P14.91

14.91 A large number of small blocks of total mass m are at rest on a table when a constant force **P** is applied to block A. Knowing that the blocks are in contact with each other but not connected, determine the speed of block A after half of the blocks have been pushed off the table, (a) neglecting the effect of friction, (b) assuming a coefficient of friction μ between the table and the blocks.

14.92 A jet of water having a cross-sectional area $A = 600\ \text{mm}^2$ and moving with a velocity of magnitude $v_A = v_B = 20\ \text{m/s}$ is deflected by the two vanes shown, which are welded to a vertical plate. Knowing that the combined mass of the plate and vanes is 5 kg, determine the reactions at C and D.

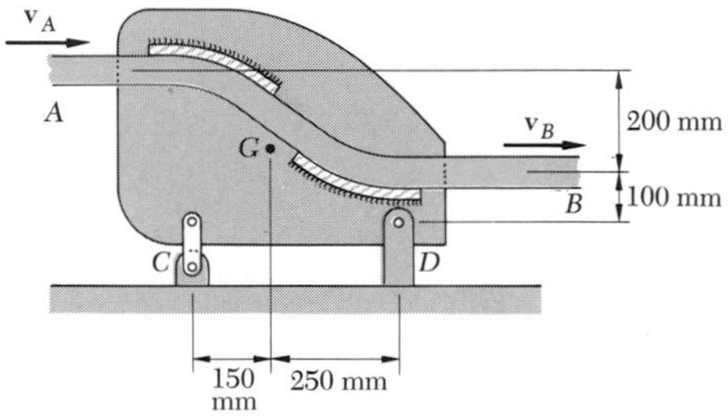

Fig. P14.92

14.93 A space vehicle equipped with a retrorocket, which may expel fuel with a relative velocity **u**, is moving with a velocity $\mathbf{v}_0$. Denoting by m_s the net mass of the vehicle and by m_f the mass of the unexpended fuel, determine the minimum ratio m_f/m_s for which the velocity of the vehicle can be reduced to zero.

14.94 The jet engine shown scoops in air at A at the rate of 165 lb/s and discharges it at B with a velocity of 2500 ft/s relative to the airplane. Determine the magnitude and line of action of the propulsive thrust developed by the engine when the speed of the airplane is (a) 300 mi/h, (b) 600 mi/h.

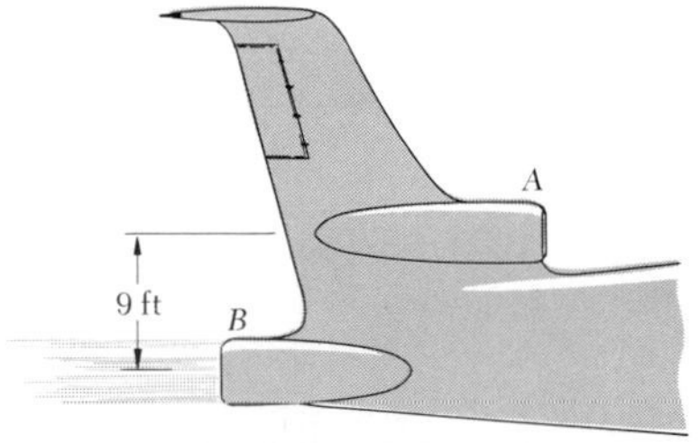

Fig. P14.94

14.95 Solve Prob. 14.94, including the effect of the fuel which is consumed by the engine at the rate of 3 lb/s.

Kinematics of Rigid Bodies

CHAPTER

15

15.1. Introduction. In this chapter, we shall study the kinematics of *rigid bodies*. We shall investigate the relations existing between the time, the positions, the velocities, and the accelerations of the various particles forming a rigid body. As we shall see, the various types of rigid-body motion may be conveniently grouped as follows:

1. *Translation.* A motion is said to be a translation if any straight line inside the body keeps the same direction during the motion. It may also be observed that in a translation all the particles forming the body move along parallel paths. If these paths are straight lines, the motion is said to be a *rectilinear translation* (Fig. 15.1); if the paths are curved lines, the motion is a *curvilinear translation* (Fig. 15.2).

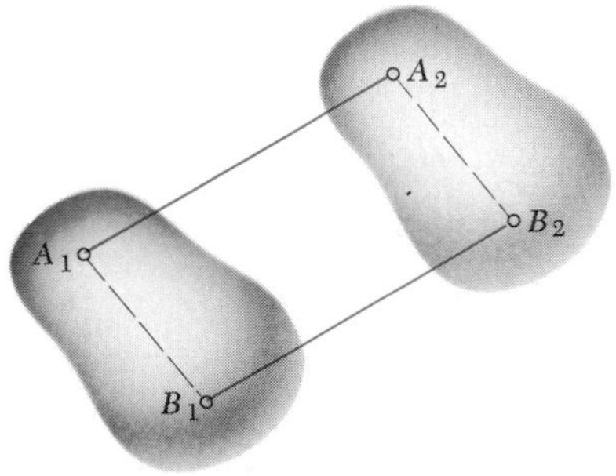

Fig. 15.1

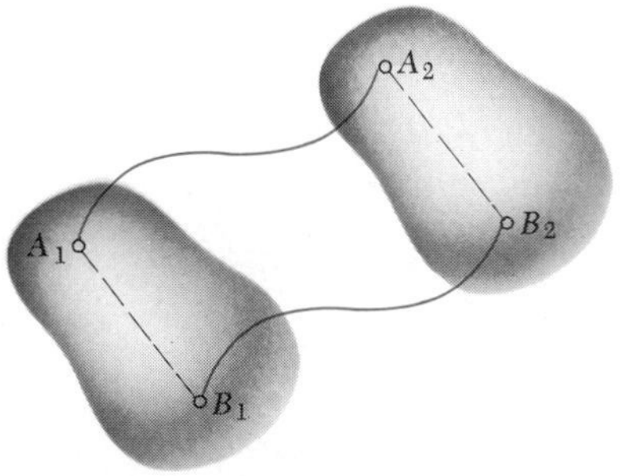

Fig. 15.2

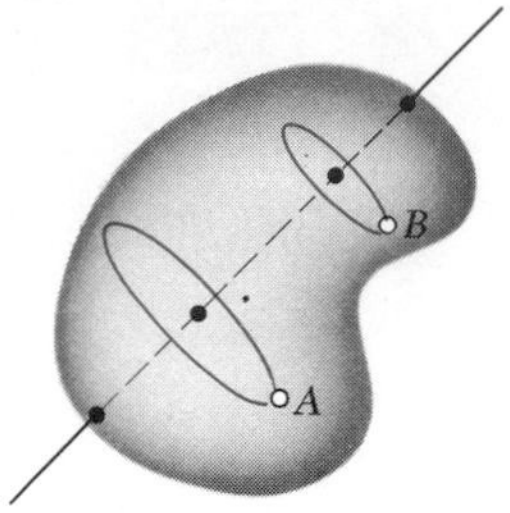

Fig. 15.3

2. *Rotation about a Fixed Axis.* In this motion, the particles forming the rigid body move in parallel planes along circles centered on the same fixed axis (Fig. 15.3). If this axis, called the *axis of rotation*, intersects the rigid body, the particles located on the axis have zero velocity and zero acceleration.

Rotation should not be confused with certain types of curvilinear translation. For example, the plate shown in Fig. 15.4*a* is in curvilinear translation, with all its particles moving along *parallel* circles, while the plate shown in Fig. 15.4*b* is in rotation, with all its particles moving along *concentric* circles. In the first case, any given straight line drawn on the plate will maintain the same direction, while, in the second case, point O remains fixed.

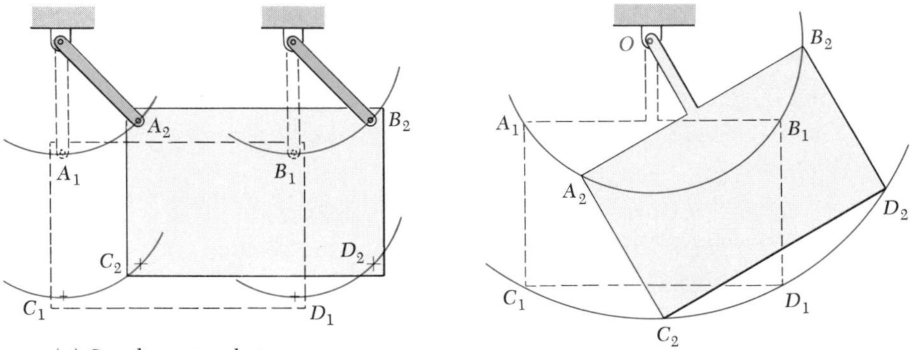

(*a*) Curvilinear translation

(*b*) Rotation

Fig. 15.4

Because each particle moves in a given plane, the rotation of a body about a fixed axis is said to be a *plane motion.*

3. *General Plane Motion.* There are many other types of plane motion, i.e., motions in which all the particles of the body move in parallel planes. Any plane motion which is neither a rotation nor a translation is referred to as a general plane motion. Two examples of general plane motion are given in Fig. 15.5.

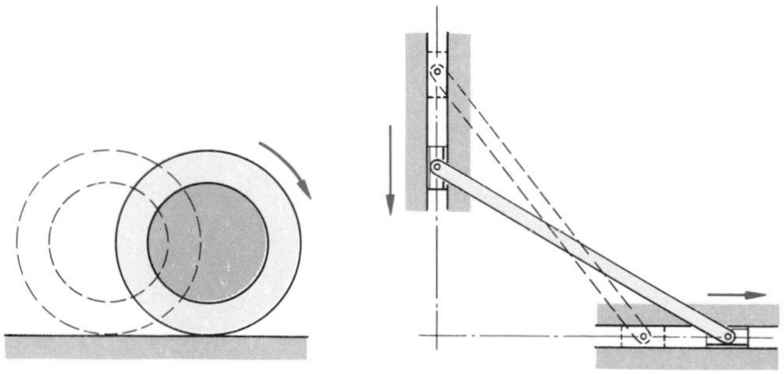
(*a*) Rolling wheel

(*b*) Sliding rod

Fig. 15.5 Examples of general plane motion

4. *Motion about a Fixed Point.* This is the three-dimensional motion of a rigid body attached at a fixed point O. An example of motion about a fixed point is provided by the motion of a top on a rough floor (Fig. 15.6).
5. *General Motion.* Any motion of a rigid body which does not fall in any of the above categories is referred to as a general motion.

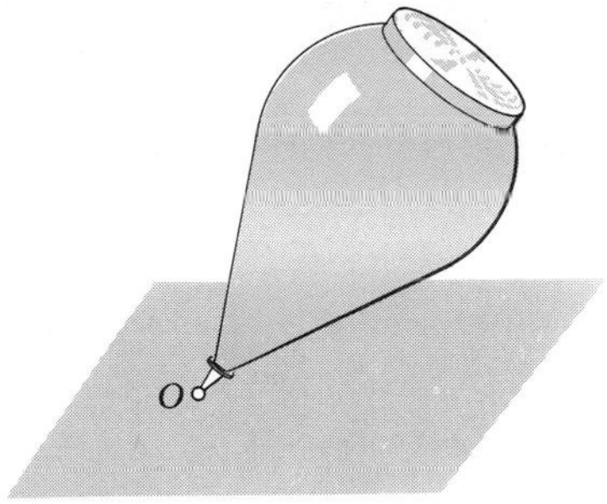

Fig. 15.6

15.2. Translation. Consider a rigid body in translation (either rectilinear or curvilinear translation), and let A and B be any two of its particles (Fig. 15.7*a*). Denoting respectively by $\mathbf{r}_A$ and $\mathbf{r}_B$ the position vectors of A and B with respect to a fixed frame of reference, and by $\mathbf{r}_{B/A}$ the vector joining A and B, we write

$$\mathbf{r}_B = \mathbf{r}_A + \mathbf{r}_{B/A} \tag{15.1}$$

Let us differentiate this relation with respect to t. We note that, from the very definition of a translation, the vector $\mathbf{r}_{B/A}$ must maintain a constant direction; its magnitude must also be constant, since A and B belong to the same rigid body. Thus, the derivative of $\mathbf{r}_{B/A}$ is zero and we have

$$\mathbf{v}_B = \mathbf{v}_A \tag{15.2}$$

Differentiating once more, we write

$$\mathbf{a}_B = \mathbf{a}_A \tag{15.3}$$

Thus, *when a rigid body is in translation, all the points of the body have the same velocity and the same acceleration at any given instant* (Fig. 15.7*b* and *c*). In the case of curvilinear translation, the velocity and acceleration change in direction as well as in magnitude at every instant. In the case of rectilinear

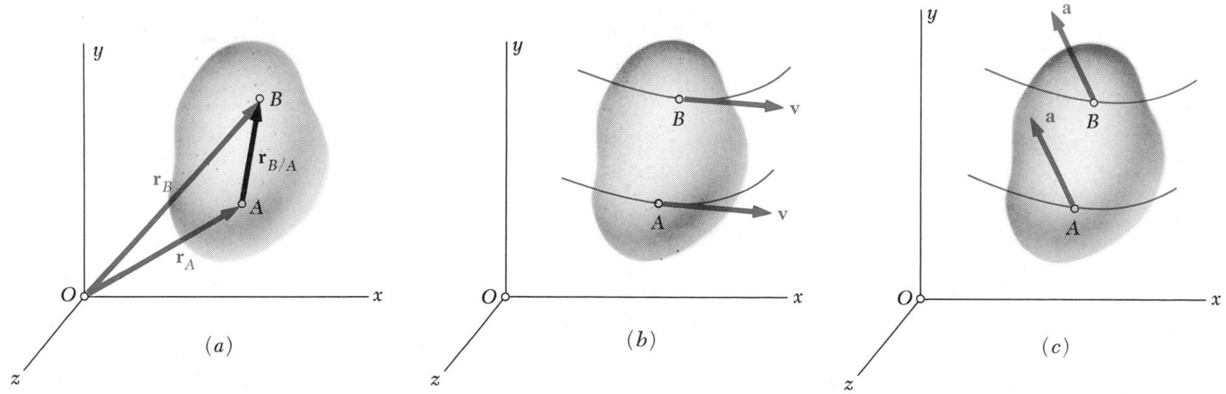

Fig. 15.7

translation, all particles of the body move along parallel straight lines, and their velocity and acceleration keep the same direction during the entire motion.

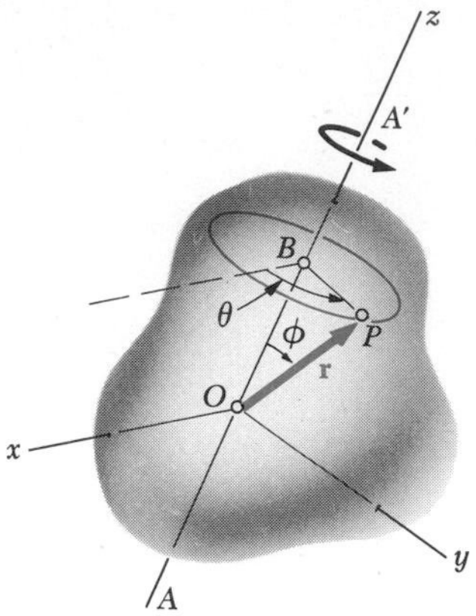

Fig. 15.8

15.3. Rotation about a Fixed Axis. Consider a rigid body which rotates about a fixed axis AA'. Let P be a point of the body and $\mathbf{r}$ its position vector with respect to a fixed frame of reference. For convenience, we shall assume that the frame is centered at point O on AA' and that the z axis coincides with AA' (Fig. 15.8). Let B be the projection of P on AA'; since P must remain at a constant distance from B, it will describe a circle of center B and of radius $r \sin \phi$, where ϕ denotes the angle formed by $\mathbf{r}$ and AA'.

The position of P and of the entire body is completely defined by the angle θ the line BP forms with the zx plane. The angle θ is known as the *angular coordinate* of the body. The angular coordinate is defined as positive when counterclockwise as viewed from A' and will be expressed in radians (rad) or, occasionally, in degrees (°) or revolutions (rev). We recall that

$$1 \text{ rev} = 2\pi \text{ rad} = 360°$$

We recall from Sec. 11.9 that the velocity $\mathbf{v} = d\mathbf{r}/dt$ of a particle P is a vector tangent to the path of P and of magnitude $v = ds/dt$. Observing that the length Δs of the arc described by P when the body rotates through $\Delta\theta$ is

$$\Delta s = (BP)\,\Delta\theta = (r \sin \phi)\,\Delta\theta$$

and dividing both members by Δt, we obtain at the limit, as Δt approaches zero,

$$v = \frac{ds}{dt} = r\dot{\theta} \sin \phi \tag{15.4}$$

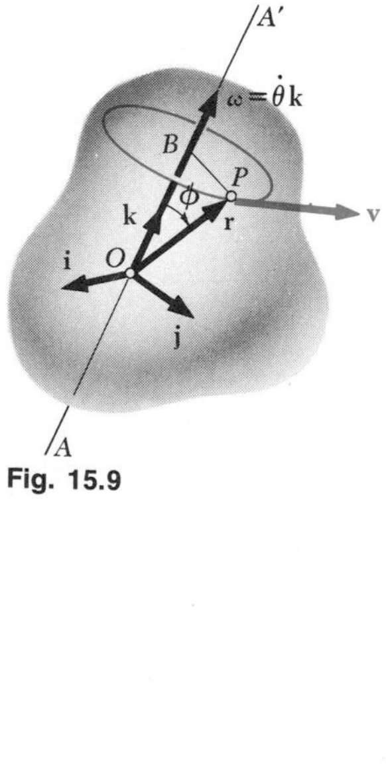

Fig. 15.9

where $\dot{\theta}$ denotes the time derivative of θ. (Note that, while the angle θ depends upon the position of P within the body, the rate of change $\dot{\theta}$ is itself independent of P.) We conclude that the velocity $\mathbf{v}$ of P is a vector perpendicular to the plane containing AA' and $\mathbf{r}$, and of magnitude v defined by (15.4). But this is precisely the result we would obtain if we drew along AA' a vector $\boldsymbol{\omega} = \dot{\theta}\mathbf{k}$ and formed the vector product $\boldsymbol{\omega} \times \mathbf{r}$ (Fig. 15.9). We thus write

$$\mathbf{v} = \frac{d\mathbf{r}}{dt} = \boldsymbol{\omega} \times \mathbf{r} \tag{15.5}$$

The vector

$$\boldsymbol{\omega} = \omega\mathbf{k} = \dot{\theta}\mathbf{k} \tag{15.6}$$

is called the *angular velocity* of the body. It is directed along the axis of rotation, it is equal in magnitude to the rate of change $\dot{\theta}$ of the angular coordinate, and its sense may be obtained by the right-hand rule (Sec. 3.5) from the sense of rotation of the body.†

We shall now determine the acceleration **a** of the particle *P*. Differentiating (15.5) and recalling the rule for the differentiation of a vector product (Sec. 11.10), we write

$$\mathbf{a} = \frac{d\mathbf{v}}{dt} = \frac{d}{dt}(\boldsymbol{\omega} \times \mathbf{r})$$

$$= \frac{d\boldsymbol{\omega}}{dt} \times \mathbf{r} + \boldsymbol{\omega} \times \frac{d\mathbf{r}}{dt}$$

$$= \frac{d\boldsymbol{\omega}}{dt} \times \mathbf{r} + \boldsymbol{\omega} \times \mathbf{v} \qquad (15.7)$$

The vector $d\boldsymbol{\omega}/dt$ is denoted by $\boldsymbol{\alpha}$ and called the *angular acceleration* of the body. Substituting also for **v** from (15.5), we have

$$\mathbf{a} = \boldsymbol{\alpha} \times \mathbf{r} + \boldsymbol{\omega} \times (\boldsymbol{\omega} \times \mathbf{r}) \qquad (15.8)$$

Differentiating (15.6), and recalling that **k** is constant in magnitude and direction, we have

$$\boldsymbol{\alpha} = \alpha\mathbf{k} = \dot{\omega}\mathbf{k} = \ddot{\theta}\mathbf{k} \qquad (15.9)$$

Thus, the angular acceleration of a body rotating about a fixed axis is a vector directed along the axis of rotation, and equal in magnitude to the rate of change $\dot{\omega}$ of the angular velocity. Returning to (15.8), we note that the acceleration of *P* is the sum of two vectors. The first vector is equal to the vector product $\boldsymbol{\alpha} \times \mathbf{r}$; it is tangent to the circle described by *P* and represents, therefore, the tangential component of the acceleration. The second vector is equal to the *vector triple product* $\boldsymbol{\omega} \times (\boldsymbol{\omega} \times \mathbf{r})$ obtained by forming the vector product of $\boldsymbol{\omega}$ and $\boldsymbol{\omega} \times \mathbf{r}$; since $\boldsymbol{\omega} \times \mathbf{r}$ is tangent to the circle described by *P*, the vector triple product is directed toward the center *B* of the circle and represents, therefore, the normal component of the acceleration.

Rotation of a Representative Slab. The rotation of a rigid body about a fixed axis may be defined by the motion of a representative slab in a reference plane perpendicular to the axis of rotation. Let us choose the *xy* plane as the reference plane and assume that it coincides with the plane of the figure, with the

†It will be shown in Sec. 15.12 in the more general case of a rigid body rotating simultaneously about axes having different directions, that angular velocities obey the parallelogram law of addition and, thus, are actually vector quantities.

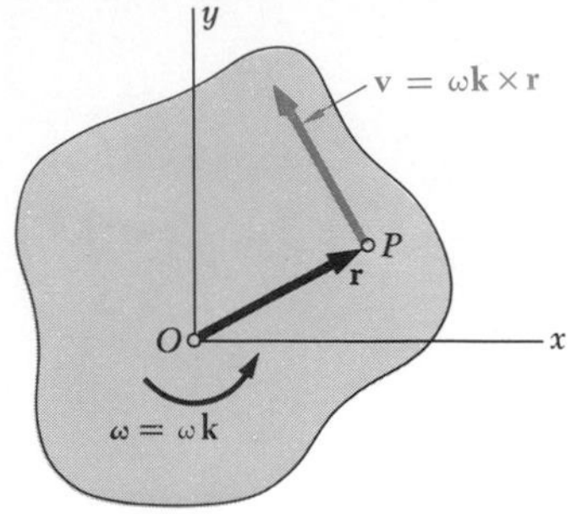

Fig. 15.10

z axis pointing out of the paper (Fig. 15.10). Recalling from (15.6) that $\boldsymbol{\omega} = \omega\mathbf{k}$, we note that a positive value of the scalar ω corresponds to a counterclockwise rotation of the representative slab, and a negative value to a clockwise rotation. Substituting $\omega\mathbf{k}$ for $\boldsymbol{\omega}$ into Eq. (15.5), we express the velocity of any given point P of the slab as

$$\mathbf{v} = \omega\mathbf{k} \times \mathbf{r} \tag{15.10}$$

Since the vectors $\mathbf{k}$ and $\mathbf{r}$ are mutually perpendicular, the magnitude of the velocity $\mathbf{v}$ is

$$v = r\omega \tag{15.10$'$}$$

and its direction may be obtained by rotating $\mathbf{r}$ through 90° in the sense of rotation of the slab.

Substituting $\boldsymbol{\omega} = \omega\mathbf{k}$ and $\boldsymbol{\alpha} = \alpha\mathbf{k}$ into Eq. (15.8), and observing that cross-multiplying $\mathbf{r}$ twice by $\mathbf{k}$ results in a 180° rotation of the vector $\mathbf{r}$, we express the acceleration of point P as

$$\mathbf{a} = \alpha\mathbf{k} \times \mathbf{r} - \omega^2\mathbf{r} \tag{15.11}$$

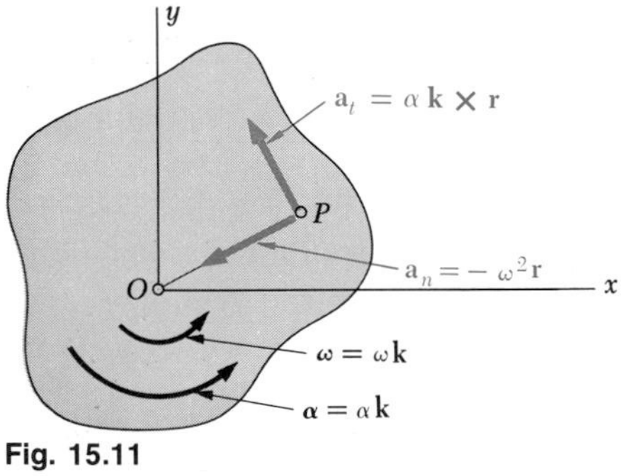

Fig. 15.11

Resolving $\mathbf{a}$ into tangential and normal components (Fig. 15.11), we write

$$\begin{aligned} \mathbf{a}_t &= \alpha\mathbf{k} \times \mathbf{r} \qquad & a_t &= r\alpha \\ \mathbf{a}_n &= -\omega^2\mathbf{r} \qquad & a_n &= r\omega^2 \end{aligned} \tag{15.11$'$}$$

The tangential component $\mathbf{a}_t$ points in the counterclockwise direction if the scalar α is positive, and in the clockwise direction if α is negative. The normal component $\mathbf{a}_n$ always points in the direction opposite to that of $\mathbf{r}$, i.e., toward O.

15.4. Equations Defining the Rotation of a Rigid Body about a Fixed Axis. The motion of a rigid body rotating about a fixed axis AA' is said to be *known* when its angular coordinate θ may be expressed as a known function of t.

In practice, however, the rotation of a rigid body is seldom defined by a relation between θ and t. More often, the conditions of motion will be specified by the type of angular acceleration that the body possesses. For example, α may be given as a function of t, or as a function of θ, or as a function of ω. Recalling the relations (15.6) and (15.9), we write

$$\omega = \frac{d\theta}{dt} \tag{15.12}$$

$$\alpha = \frac{d\omega}{dt} = \frac{d^2\theta}{dt^2} \tag{15.13}$$

or, solving (15.12) for dt and substituting into (15.13),

$$\alpha = \omega \frac{d\omega}{d\theta} \tag{15.14}$$

Since these equations are similar to those obtained in Chap. 11 for the rectilinear motion of a particle, their integration may be performed by following the procedure outlined in Sec. 11.3.

Two particular cases of rotation are frequently encountered:

1. *Uniform Rotation.* This case is characterized by the fact that the angular acceleration is zero. The angular velocity is thus constant, and the angular coordinate is given by the formula

$$\theta = \theta_0 + \omega t \tag{15.15}$$

2. *Uniformly Accelerated Rotation.* In this case, the angular acceleration is constant. The following formulas relating angular velocity, angular coordinate, and time may then be derived in a manner similar to that described in Sec. 11.5. The similitude between the formulas derived here and those obtained for the rectilinear uniformly accelerated motion of a particle is easily noted.

$$\begin{aligned} \omega &= \omega_0 + \alpha t \\ \theta &= \theta_0 + \omega_0 t + \tfrac{1}{2}\alpha t^2 \\ \omega^2 &= \omega_0^2 + 2\alpha(\theta - \theta_0) \end{aligned} \tag{15.16}$$

It should be emphasized that formula (15.15) may be used only when $\alpha = 0$, and formulas (15.16) only when $\alpha =$ constant. In any other case, the general formulas (15.12) to (15.14) should be used.

SAMPLE PROBLEM 15.1

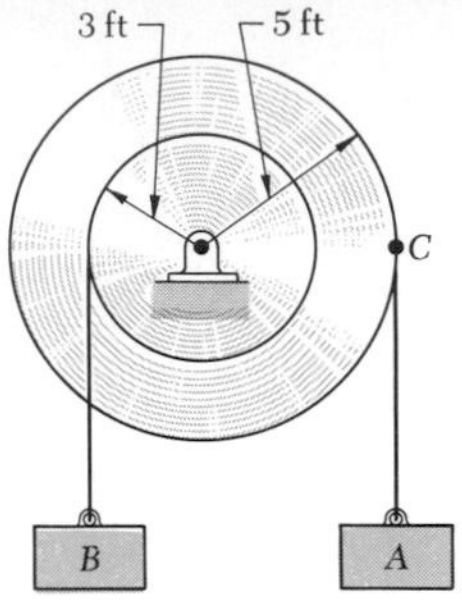

A pulley and two loads are connected by inextensible cords as shown. Load A has a constant acceleration of 10 ft/s^2 and an initial velocity of 15 ft/s, both directed upward. Determine (*a*) the number of revolutions executed by the pulley in 3 s, (*b*) the velocity and position of load B after 3 s, (*c*) the acceleration of point C on the rim of the pulley at $t = 0$.

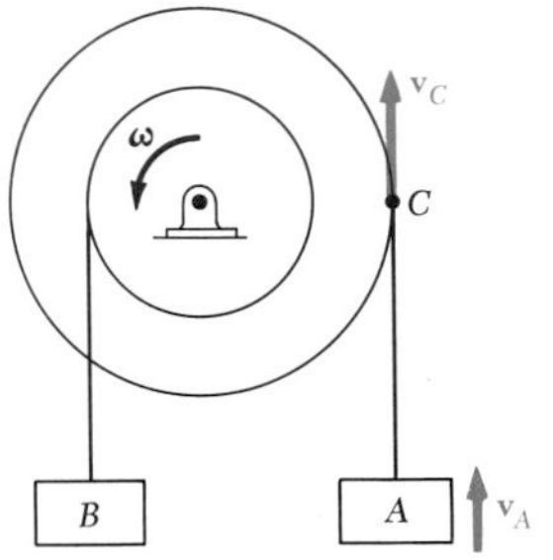

a. **Motion of Pulley.** Since the cord connecting the pulley to load A is inextensible, the velocity of C is equal to the velocity of A and the tangential component of the acceleration of C is equal to the acceleration of A.

$$(\mathbf{v}_C)_0 = (\mathbf{v}_A)_0 = 15 \text{ ft/s} \uparrow \qquad (\mathbf{a}_C)_t = \mathbf{a}_A = 10 \text{ ft/s}^2 \uparrow$$

Noting that the distance from C to the center of the pulley is 5 ft, we write

$$(v_C)_0 = r\omega_0 \qquad 15 \text{ ft/s} = (5 \text{ ft})\omega_0 \qquad \boldsymbol{\omega}_0 = 3 \text{ rad/s} \circlearrowleft$$
$$(a_C)_t = r\alpha \qquad 10 \text{ ft/s}^2 = (5 \text{ ft})\alpha \qquad \boldsymbol{\alpha} = 2 \text{ rad/s}^2 \circlearrowleft$$

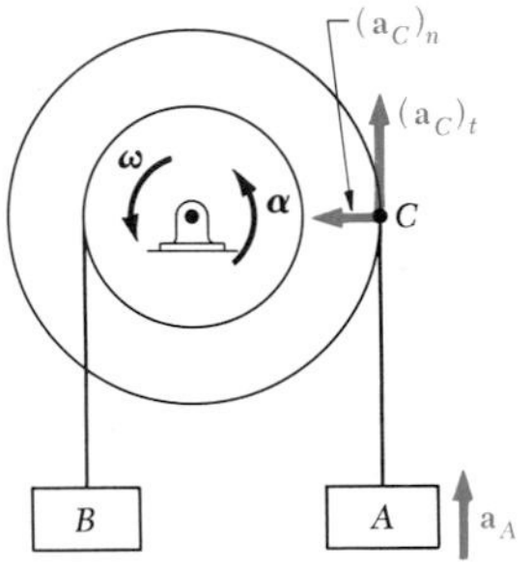

From the equations for uniformly accelerated motion, we obtain, for $t = 3$ s,

$$\omega = \omega_0 + \alpha t = 3 \text{ rad/s} + (2 \text{ rad/s}^2)(3 \text{ s}) = 9 \text{ rad/s}$$
$$\boldsymbol{\omega} = 9 \text{ rad/s} \circlearrowleft$$
$$\theta = \omega_0 t + \tfrac{1}{2}\alpha t^2 = (3 \text{ rad/s})(3 \text{ s}) + \tfrac{1}{2}(2 \text{ rad/s}^2)(3 \text{ s})^2$$
$$\theta = 18 \text{ rad}$$

$$\text{Number of revolutions} = (18 \text{ rad})\left(\frac{1 \text{ rev}}{2\pi \text{ rad}}\right) = 2.86 \text{ rev} \quad ◀$$

b. **Motion of Load *B*.** Using the following relations between the linear and angular motion, with $r = 3$ ft, we write

$$v_B = r\omega = (3 \text{ ft})(9 \text{ rad/s}) \qquad \mathbf{v}_B = 27 \text{ ft/s} \downarrow \quad ◀$$
$$s_B = r\theta = (3 \text{ ft})(18 \text{ rad}) \qquad s_B = 54 \text{ ft} \downarrow \quad ◀$$

c. **Acceleration of Point *C* at $t = 0$.** The tangential component of the acceleration is

$$(\mathbf{a}_C)_t = \mathbf{a}_A = 10 \text{ ft/s}^2 \uparrow$$

Since, at $t = 0$, $\omega_0 = 3$ rad/s, the normal component of the acceleration is

$$(a_C)_n = r_C\omega_0^2 = (5 \text{ ft})(3 \text{ rad/s})^2 \qquad (\mathbf{a}_C)_n = 45 \text{ ft/s}^2 \leftarrow$$

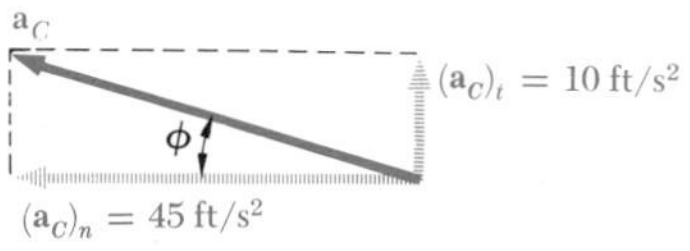

The magnitude and direction of the total acceleration are obtained by writing

$$\tan\phi = (10 \text{ ft/s}^2)/(45 \text{ ft/s}^2) \qquad \phi = 12.5°$$
$$a_C \sin 12.5° = 10 \text{ ft/s}^2 \qquad a_C = 46.1 \text{ ft/s}^2$$
$$\mathbf{a}_C = 46.1 \text{ ft/s}^2 ⦪ 12.5° \quad ◀$$

PROBLEMS

15.1 The motion of a cam is defined by the relation $\theta = t^3 - 2t^2 - 4t + 10$, where θ is expressed in radians and t in seconds. Determine the angular coordinate, the angular velocity, and the angular acceleration of the cam when (*a*) $t = 0$, (*b*) $t = 3$ s.

15.2 The rotor of a steam turbine is rotating at a speed of 7200 rpm when the steam supply is suddenly cut off. It is observed that 5 min are required for the rotor to come to rest. Assuming uniformly accelerated motion, determine (*a*) the angular acceleration, (*b*) the total number of revolutions that the rotor executes before coming to rest.

15.3 A small grinding wheel is attached to the shaft of an electric motor which has a rated speed of 1800 rpm. When the power is turned on, the unit reaches its rated speed in 4 s, and when the power is turned off, the unit coasts to rest in 50 s. Assuming uniformly accelerated motion, determine the number of revolutions that the motor executes (*a*) in reaching its rated speed, (*b*) in coasting to rest.

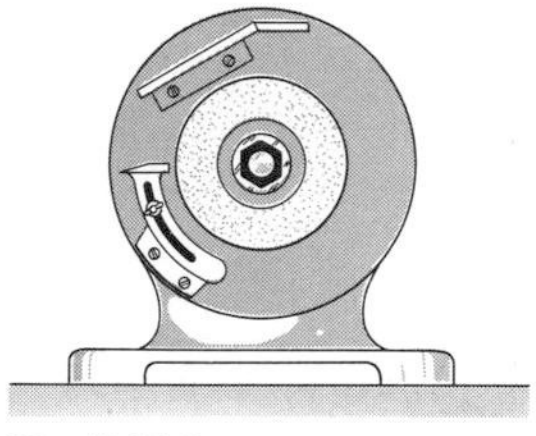

Fig. P15.3

15.4 The rotor of an electric motor has a speed of 1200 rpm when the power is cut off. The rotor is then observed to come to rest after executing 520 revolutions. Assuming uniformly accelerated motion, determine (*a*) the angular acceleration, (*b*) the time required for the rotor to come to rest.

15.5 The assembly shown consists of the straight rod *ABC* which passes through and is welded to the rectangular plate *DEFH*. The assembly rotates about the axis *AC* with a constant angular velocity of 18 rad/s. Knowing that the motion when viewed from *C* is counterclockwise, determine the velocity and acceleration of corner *F*.

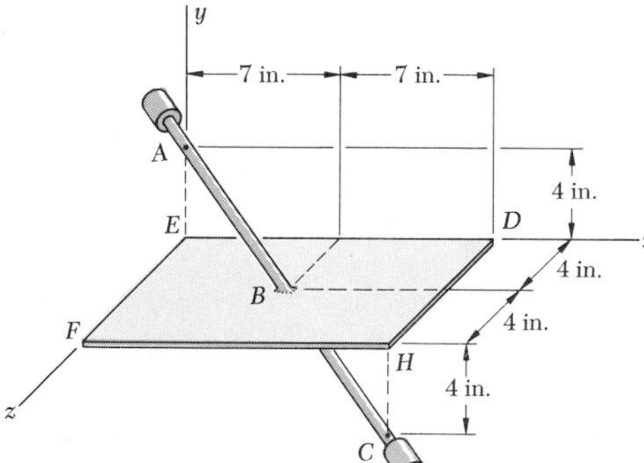

Fig. P15.5

15.6 In Prob. 15.5, assuming that the angular velocity is 18 rad/s and decreases at the rate of 45 rad/s², determine the velocity and acceleration of corner *H*.

15.7 The assembly shown rotates about the rod *AC* with a constant angular velocity of 5 rad/s. Knowing that at the instant considered, the velocity of corner *D* is downward, determine the velocity and acceleration of corner *D*.

15.8 In Prob. 15.7, determine the velocity and acceleration of corner *E*, assuming that the angular velocity is 5 rad/s and increases at the rate of 25 rad/s².

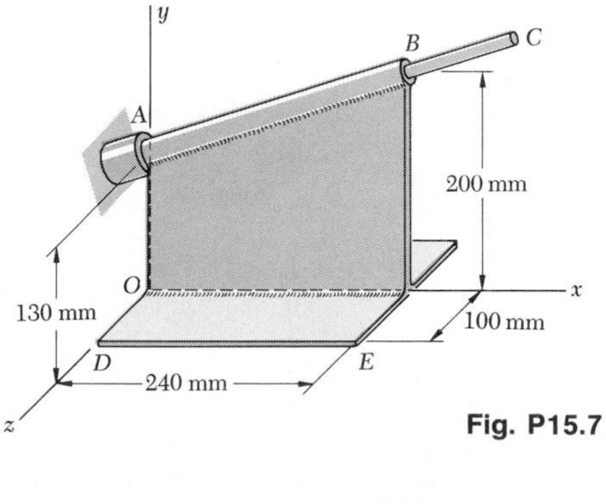

Fig. P15.7

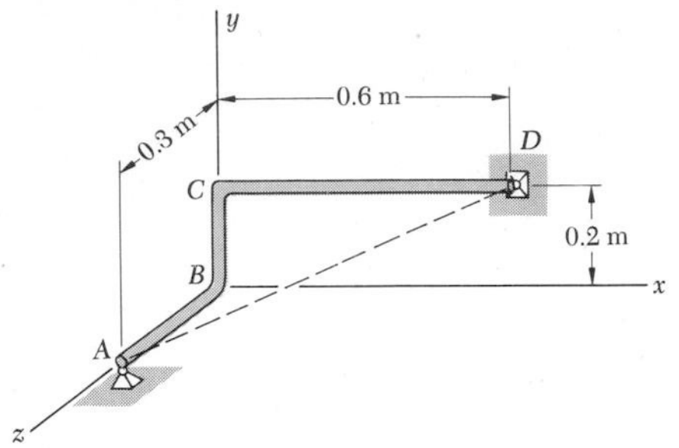

Fig. P15.9 and P15.10

15.9 The rod $ABCD$ has been bent as shown and may rotate about the line joining points A and D. Knowing that the rod starts from rest in the position shown with a constant angular acceleration of 14 rad/s^2 and that the initial acceleration of point B is upward, determine the initial acceleration of point C.

15.10 The bent rod $ABCD$ rotates about the line joining points A and D. At the instant shown, the angular velocity of the rod is 7 rad/s and the angular acceleration is 21 rad/s^2, both counterclockwise when viewed from end A of line AD. Determine the velocity and acceleration of point C.

15.11 The earth makes one complete revolution on its axis in 23.93 h. Knowing that the mean radius of the earth is 3960 mi, determine the linear velocity and acceleration of a point on the surface of the earth (*a*) at the equator, (*b*) at Philadelphia, latitude 40° north, (*c*) at the North Pole.

15.12 The earth makes one complete revolution about the sun in 365.24 days. Assuming that the orbit of the earth is circular and has a radius of 93,000,000 mi, determine the velocity and acceleration of the earth.

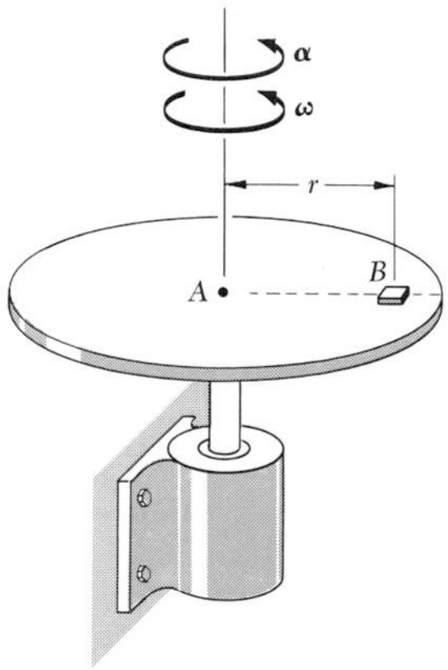

Fig. P15.13 and P15.14

15.13 A small block B rests on a horizontal plate which rotates about a fixed vertical axis. If the plate is initially at rest at $t = 0$ and is accelerated at the constant rate α, derive an expression (*a*) for the total acceleration of the block at time t, (*b*) for the angle between the total acceleration and the radius AB at time t.

15.14 It is known that the static-friction force between block B and the plate will be exceeded and that the block will start sliding on the plate when the total acceleration of the block reaches 5 m/s^2. If the plate starts from rest at $t = 0$ and is accelerated at the constant rate of 6 rad/s^2, determine the time t and the angular velocity of the plate when the block starts sliding, assuming $r = 100$ mm.

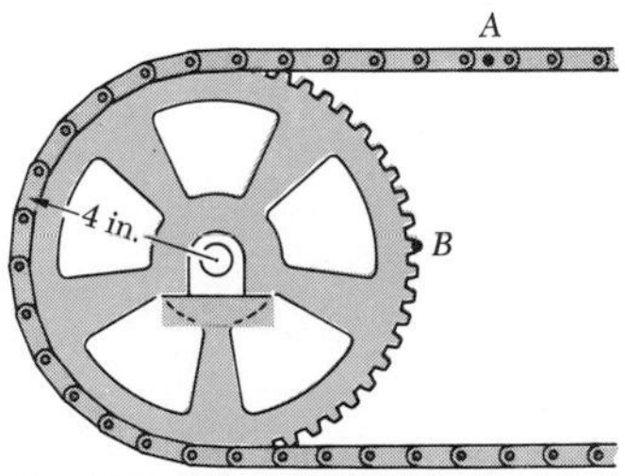

Fig. P15.15 and P15.16

15.15 The sprocket wheel and chain are initially at rest. If the acceleration of point A of the chain has a constant magnitude of 5 in./s^2 and is directed to the left, determine (*a*) the angular velocity of the wheel after it has completed three revolutions, (*b*) the time required for the wheel to reach an angular velocity of 100 rpm.

15.16 At the instant shown the velocity of point A is 8 in./s directed to the right and its acceleration is 12 in./s^2 directed to the left. Determine (*a*) the angular velocity and angular acceleration of the sprocket wheel, (*b*) the total acceleration of sprocket B.

15.17 The friction wheel B executes 100 revolutions about its fixed shaft during the time interval t, while its angular velocity is being increased uniformly from 200 to 600 rpm. Knowing that wheel B rolls without slipping on the inside rim of wheel A, determine (*a*) the angular acceleration of wheel A, (*b*) the time interval t.

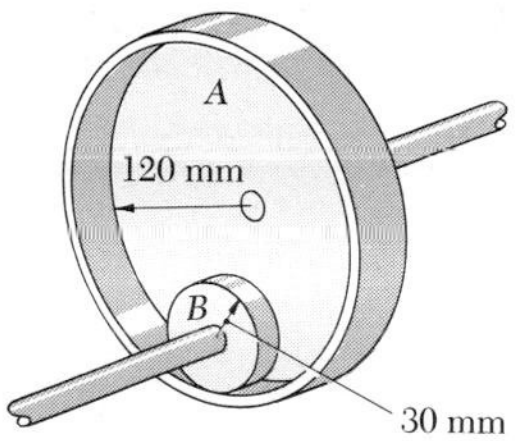

Fig. P15.17

15.18 Ring C has an inside diameter of 120 mm and hangs from the 40-mm-diameter shaft which rotates with a constant angular velocity of 30 rad/s. Knowing that no slipping occurs between the shaft and the ring, determine (*a*) the angular velocity of the ring, (*b*) the acceleration of the points of B and C which are in contact.

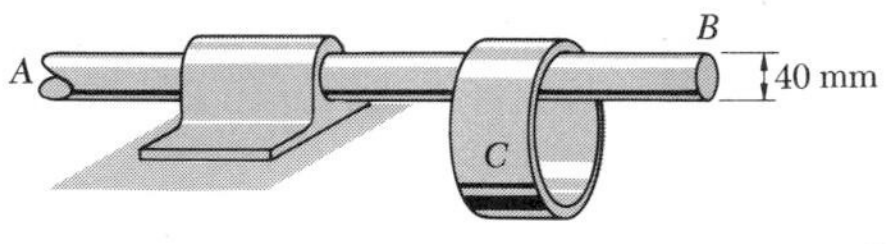

Fig. P15.18

15.19 The system shown starts from rest at $t = 0$ and accelerates uniformly. Knowing that at $t = 4$ s the velocity of the load is 4.8 m/s downward, determine (*a*) the angular acceleration of gear A, (*b*) the number of revolutions executed by gear A during the 4-s interval.

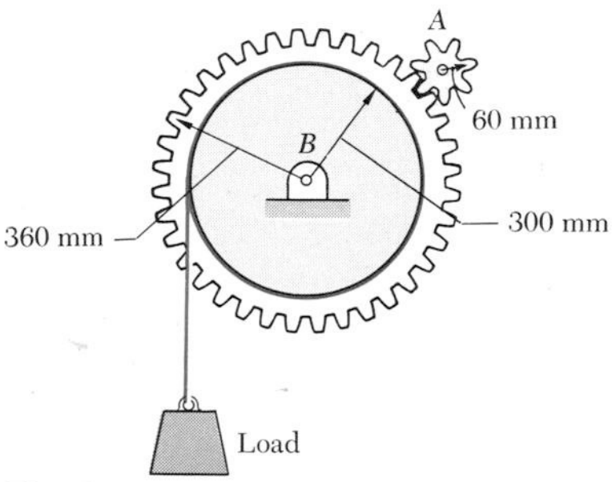

Fig. P15.19

15.20 The two pulleys shown may be operated with the V belt in any of three positions. If the angular acceleration of shaft A is 6 rad/s^2 and if the system is initially at rest, determine the time required for shaft B to reach a speed of 400 rpm with the belt in each of the three positions.

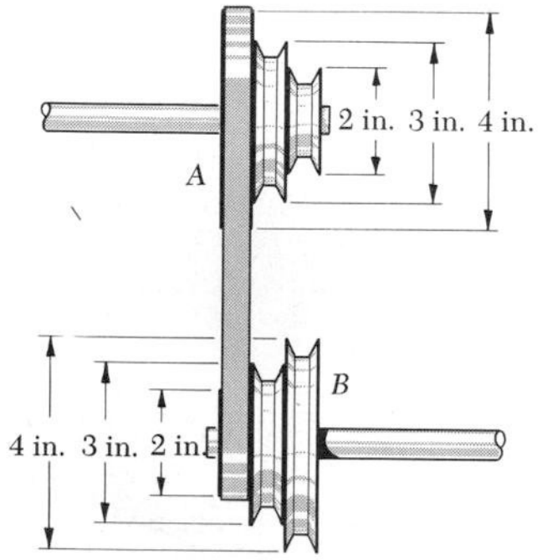

Fig. P15.20

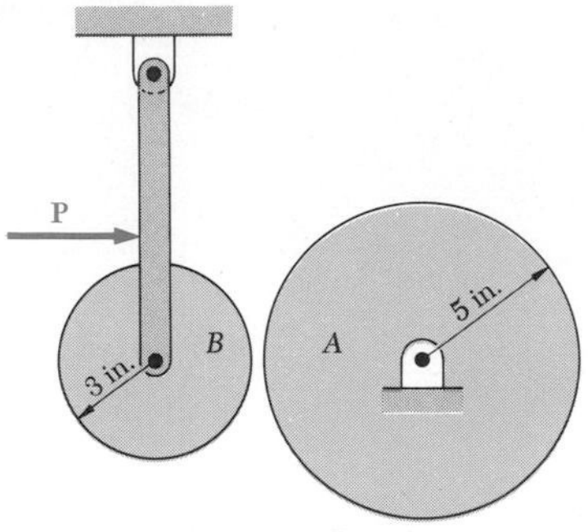

Fig. P15.21 and P15.22

15.21 The two friction wheels A and B are to be brought together. Wheel A has an initial angular velocity of 600 rpm clockwise and will coast to rest in 40 s, while wheel B is initially at rest and is given a constant counterclockwise angular acceleration of 2 rad/s^2. Determine (*a*) at what time the wheels may be brought together if they are not to slip, (*b*) the angular velocity of each wheel as contact is made.

15.22 Two friction wheels A and B are both rotating freely at 300 rpm clockwise when they are brought into contact. After 6 s of slippage, during which each wheel has a constant angular acceleration, wheel A reaches a final angular velocity of 60 rpm clockwise. Determine (*a*) the angular acceleration of each wheel during the period of slippage, (*b*) the time at which the angular velocity of wheel B is equal to zero.

***15.23** The motion of the circular plate of Prob. 15.13 is defined by the relation $\theta = \theta_0 \sin(2\pi t/T)$, where θ is expressed in radians and t in seconds. Derive expressions (*a*) for the magnitude of the total acceleration of B, (*b*) for the values of θ at which the total acceleration of B reaches its maximum and minimum values, and for the corresponding values of the total acceleration of B.

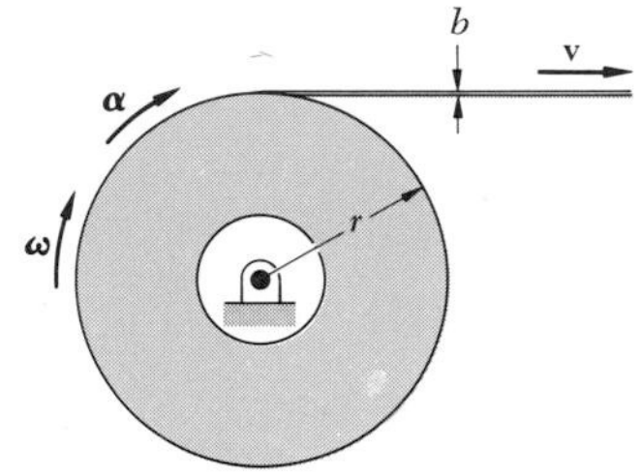

Fig. P15.24

***15.24** In a continuous printing process, paper is drawn into the presses at a constant speed v. Denoting by r the radius of paper on the roll at any given time and by b the thickness of the paper, derive an expression for the angular acceleration of the paper roll.

15.5. General Plane Motion.

As indicated in Sec. 15.1, we understand by general plane motion a plane motion which is neither a translation nor a rotation. As we shall presently see, however, *a general plane motion may always be considered as the sum of a translation and a rotation.*

Consider, for example, a wheel rolling on a straight track (Fig. 15.12). Over a certain interval of time, two given points A and

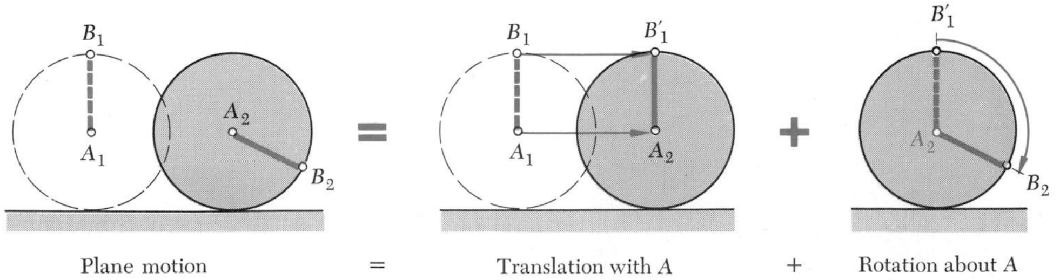

Fig. 15.12

B will have moved, respectively, from A_1 to A_2 and from B_1 to B_2. The same result could be obtained through a translation which would bring A and B into A_2 and B_1' (the line AB remaining vertical), followed by a rotation about A bringing B into B_2. Although the original rolling motion differs from the combination of translation and rotation when these motions are taken in succession, the original motion may be completely duplicated by a combination of simultaneous translation and rotation.

Another example of plane motion is given in Fig. 15.13, which represents a rod whose extremities slide, respectively, along a horizontal and a vertical track. This motion may be replaced by a translation in a horizontal direction and a rotation about A (Fig. 15.13a) or by a translation in a vertical direction and a rotation about B (Fig. 15.13b).

| Plane motion | = | Translation with A | + | Rotation about A |

(a)

| Plane motion | = | Translation with B | + | Rotation about B |

(b)

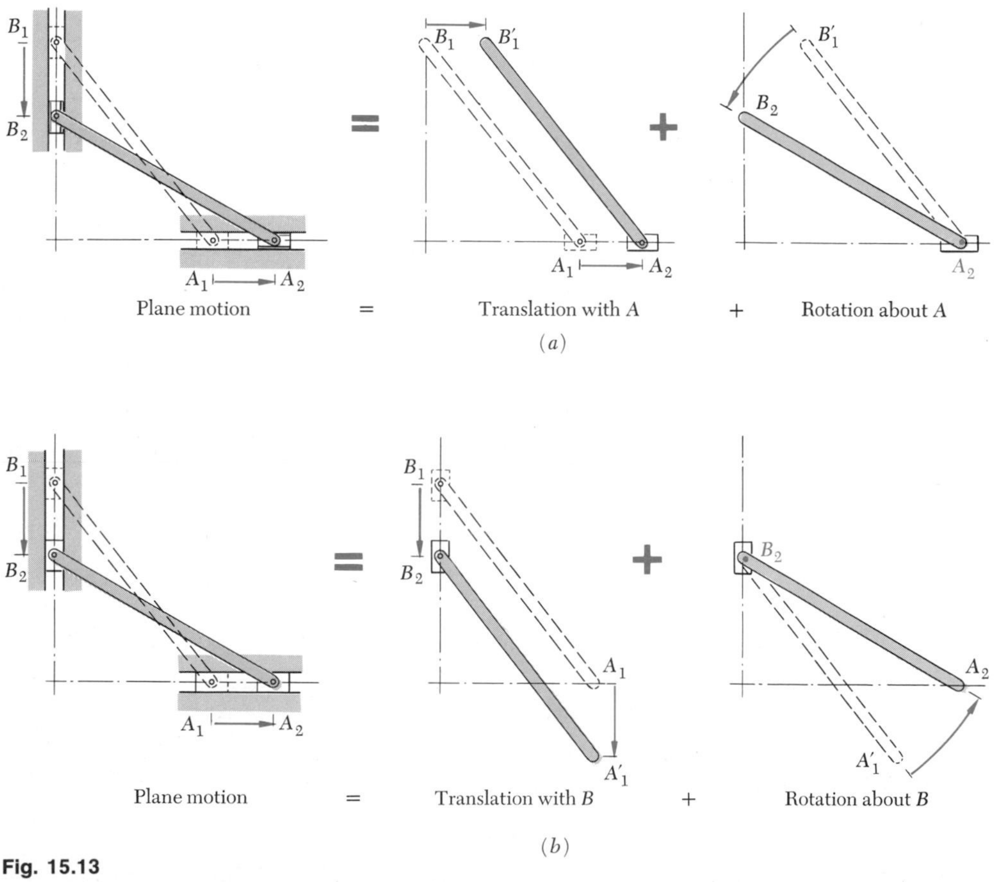

Fig. 15.13

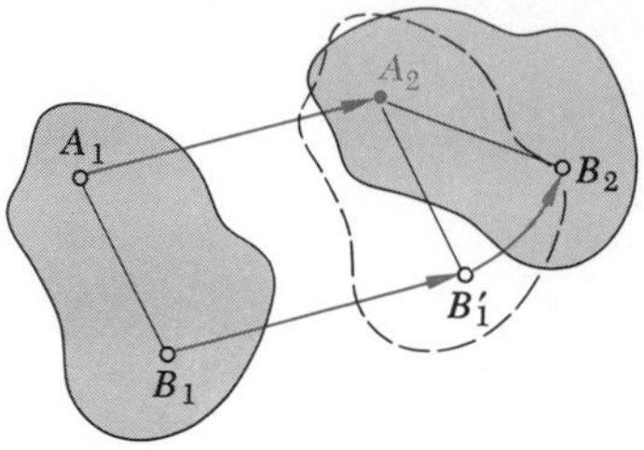

Fig. 15.14

In general, we shall consider a small displacement which brings two particles A and B of a representative slab, respectively, from A_1 and B_1 into A_2 and B_2 (Fig. 15.14). This displacement may be divided into two parts, one in which the particles move into A_2 and B_1' while the line AB maintains the same direction, the other in which B moves into B_2 while A remains fixed. Clearly, the first part of the motion is a translation and the second part a rotation about A.

Recalling from Sec. 11.12 the definition of the "relative motion" of a particle with respect to a moving frame of reference—as opposed to its "absolute motion" with respect to a fixed frame of reference—we may restate as follows the result obtained above: Given two particles A and B of a rigid slab in plane motion, the relative motion of B with respect to a frame attached to A and of fixed orientation is a rotation. To an observer moving with A, but not rotating, particle B will appear to describe an arc of circle centered at A.

15.6. Absolute and Relative Velocity in Plane Motion. We saw in the preceding section that any plane motion of a slab may be replaced by a translation defined by the motion of an arbitrary reference point A, and by a rotation about A. The absolute velocity $\mathbf{v}_B$ of a particle B of the slab is obtained from the relative-velocity formula derived in Sec. 11.12,

$$\mathbf{v}_B = \mathbf{v}_A + \mathbf{v}_{B/A} \tag{15.17}$$

where the right-hand member represents a vector sum. The velocity $\mathbf{v}_A$ corresponds to the translation of the slab with A, while the relative velocity $\mathbf{v}_{B/A}$ is associated with the rotation of the slab about A and is measured with respect to axes centered at A and of fixed orientation (Fig. 15.15). Denoting by $\mathbf{r}_{B/A}$ the position vector of B relative to A, and by $\omega\mathbf{k}$ the angular velocity of the slab with respect to axes of fixed orientation, we have from (15.10) and (15.10′)

$$\mathbf{v}_{B/A} = \omega\mathbf{k} \times \mathbf{r}_{B/A} \qquad v_{B/A} = r\omega \tag{15.18}$$

where r is the distance from A to B. Substituting for $\mathbf{v}_{B/A}$ from (15.18) into (15.17), we may also write

$$\mathbf{v}_B = \mathbf{v}_A + \omega\mathbf{k} \times \mathbf{r}_{B/A} \tag{15.17′}$$

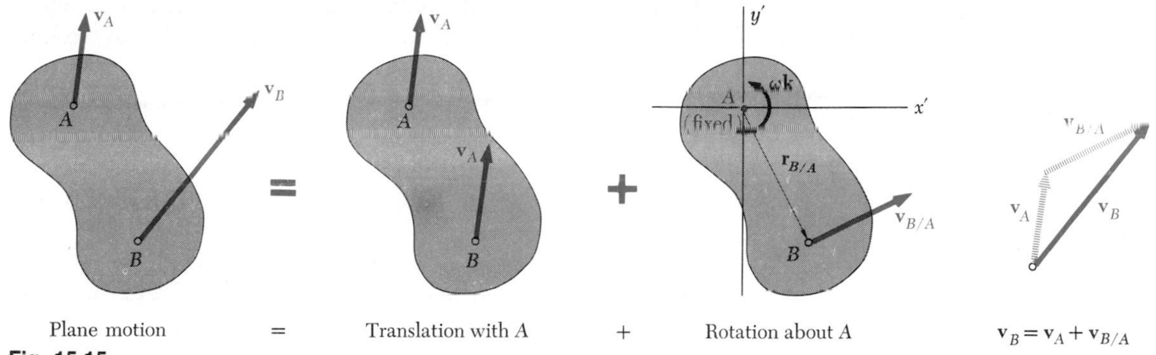

Fig. 15.15

As an example, we shall consider again the rod AB of Fig. 15.13. Assuming that the velocity $\mathbf{v}_A$ of end A is known, we propose to find the velocity $\mathbf{v}_B$ of end B and the angular velocity ω of the rod, in terms of the velocity $\mathbf{v}_A$, the length l, and the angle θ. Choosing A as reference point, we express that the given motion is equivalent to a translation with A and a rotation about A (Fig. 15.16). The absolute velocity of B must therefore be equal to the vector sum

$$\mathbf{v}_B = \mathbf{v}_A + \mathbf{v}_{B/A} \tag{15.17}$$

We note that, while the direction of $\mathbf{v}_{B/A}$ is known, its magnitude $l\omega$ is unknown. However, this is compensated by the fact that the direction of $\mathbf{v}_B$ is known. We may therefore complete the diagram of Fig. 15.16. Solving for the magnitudes v_B and ω, we write

$$v_B = v_A \tan\theta \qquad \omega = \frac{v_{B/A}}{l} = \frac{v_A}{l\cos\theta} \tag{15.19}$$

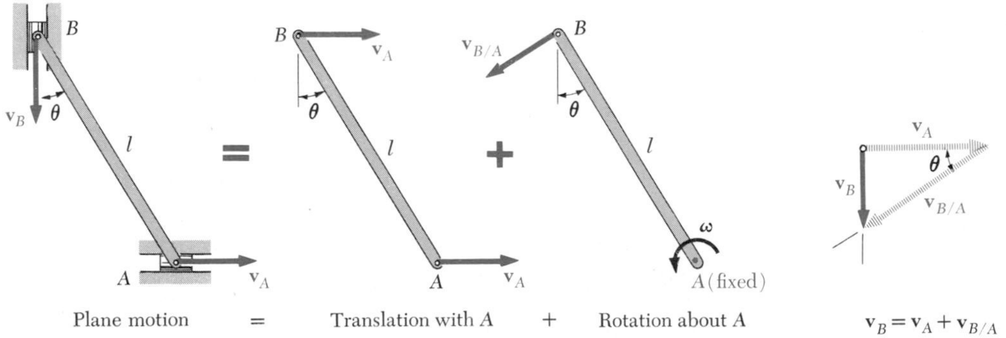

Fig. 15.16

The same result may be obtained by using B as a point of reference. Resolving the given motion into a translation with B and a rotation about B (Fig. 15.17), we write the equation

$$\mathbf{v}_A = \mathbf{v}_B + \mathbf{v}_{A/B} \tag{15.20}$$

which is represented graphically in Fig. 15.17. We note that $\mathbf{v}_{A/B}$ and $\mathbf{v}_{B/A}$ have the same magnitude $l\omega$ but opposite sense. The sense of the relative velocity depends, therefore, upon the point of reference which has been selected and should be carefully ascertained from the appropriate diagram (Fig. 15.16 or 15.17).

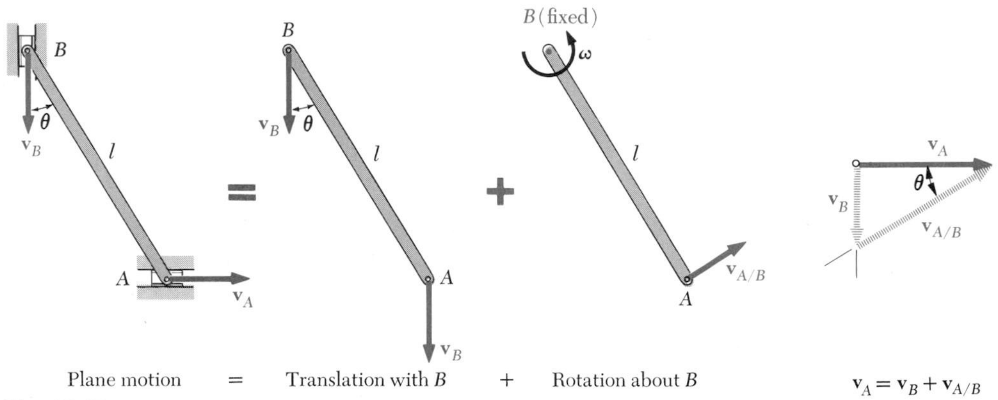

Fig. 15.17

Finally, we observe that the angular velocity ω of the rod in its rotation about B is the same as in its rotation about A. It is measured in both cases by the rate of change of the angle θ. This result is quite general; we should therefore bear in mind that *the angular velocity ω of a rigid body in plane motion is independent of the reference point.*

Most mechanisms consist, not of one, but of *several* moving parts. When the various parts of a mechanism are pin-connected, its analysis may be carried out by considering each part as a rigid body, while keeping in mind that the points where two parts are connected must have the same absolute velocity (see Sample Prob. 15.3). A similar analysis may be used when gears are involved, since the teeth in contact must also have the same absolute velocity. However, when a mechanism contains parts which slide on each other, the relative velocity of the parts in contact must be taken into account (see Secs. 15.10 and 15.11).

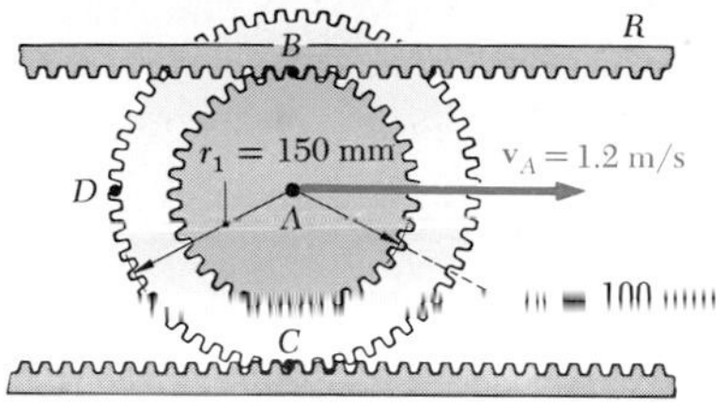

SAMPLE PROBLEM 15.2

The double gear shown rolls on the stationary lower rack; the velocity of its center A is 1.2 m/s directed to the right. Determine (*a*) the angular velocity of the gear, (*b*) the velocities of the upper rack R and of point D of the gear.

a. Angular Velocity of the Gear. Since the gear rolls on the lower rack, its center A moves through a distance equal to the outer circumference $2\pi r_1$ for each full revolution of the gear. Noting that 1 rev = 2π rad, and that when A moves to the right ($x_A > 0$) the gear rotates clockwise ($\theta < 0$), we write

$$\frac{x_A}{2\pi r_1} = -\frac{\theta}{2\pi} \qquad x_A = -r_1\theta$$

Differentiating with respect to the time t and substituting the known values $v_A = 1.2$ m/s and $r_1 = 150$ mm $= 0.150$ m, we obtain

$$v_A = -r_1\omega \qquad 1.2\text{ m/s} = -(0.150\text{ m})\omega \qquad \omega = -8\text{ rad/s}$$

$$\boldsymbol{\omega} = \omega\mathbf{k} = -(8\text{ rad/s})\mathbf{k} \quad ◄$$

where $\mathbf{k}$ is a unit vector pointing out of the paper.

b. Velocities. The rolling motion is resolved into two component motions: a translation with the center A and a rotation about the center A. In the translation, all points of the gear move with the same velocity $\mathbf{v}_A$. In the rotation, each point P of the gear moves about A with a relative velocity $\mathbf{v}_{P/A} = \omega\mathbf{k} \times \mathbf{r}_{P/A}$, where $\mathbf{r}_{P/A}$ is the position vector of P relative to A.

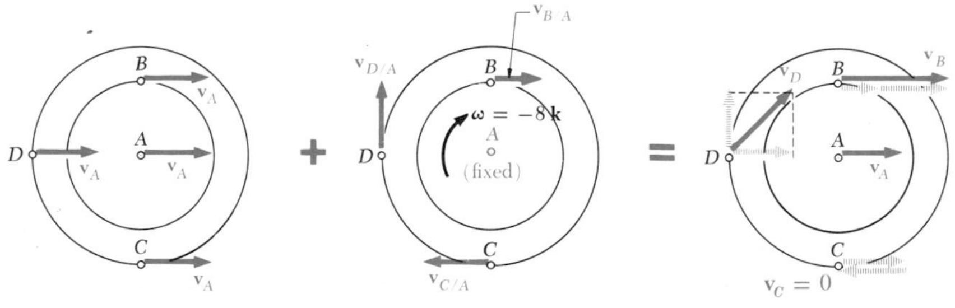

Translation + Rotation = Rolling motion

Velocity of Upper Rack. The velocity of the upper rack is equal to the velocity of point B; we write

$$\begin{aligned}\mathbf{v}_R = \mathbf{v}_B &= \mathbf{v}_A + \mathbf{v}_{B/A} = \mathbf{v}_A + \omega\mathbf{k} \times \mathbf{r}_{B/A} \\ &= (1.2\text{ m/s})\mathbf{i} - (8\text{ rad/s})\mathbf{k} \times (0.100\text{ m})\mathbf{j} \\ &= (1.2\text{ m/s})\mathbf{i} + (0.8\text{ m/s})\mathbf{i} = (2\text{ m/s})\mathbf{i}\end{aligned}$$

$$\mathbf{v}_R = 2\text{ m/s} \rightarrow \quad ◄$$

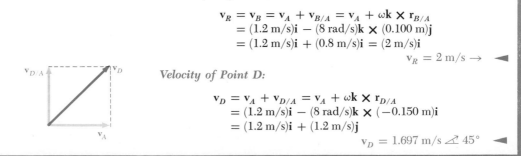

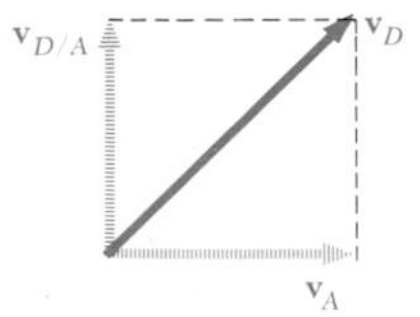

Velocity of Point D:

$$\begin{aligned}\mathbf{v}_D &= \mathbf{v}_A + \mathbf{v}_{D/A} = \mathbf{v}_A + \omega\mathbf{k} \times \mathbf{r}_{D/A} \\ &= (1.2\text{ m/s})\mathbf{i} - (8\text{ rad/s})\mathbf{k} \times (-0.150\text{ m})\mathbf{i} \\ &= (1.2\text{ m/s})\mathbf{i} + (1.2\text{ m/s})\mathbf{j}\end{aligned}$$

$$\mathbf{v}_D = 1.697\text{ m/s} \measuredangle 45° \quad ◄$$

SAMPLE PROBLEM 15.3

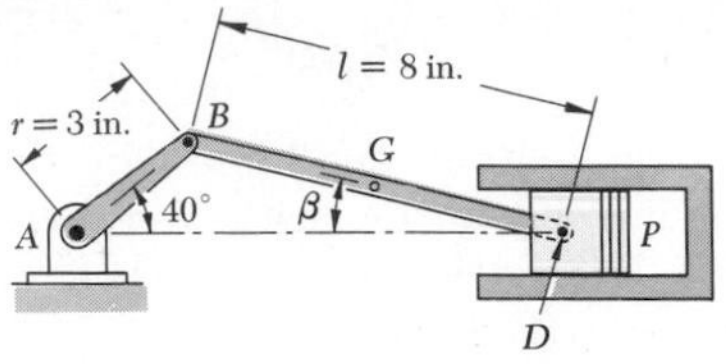

In the engine system shown, the crank AB has a constant clockwise angular velocity of 2000 rpm. For the crank position indicated, determine (*a*) the angular velocity of the connecting rod BD, (*b*) the velocity of the piston P.

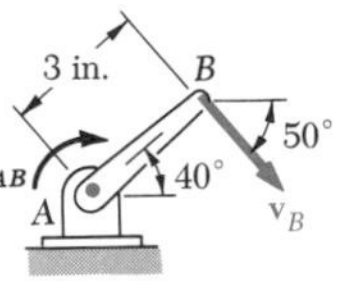

Motion of Crank AB. The crank AB rotates about point A. Expressing ω_{AB} in rad/s and writing $v_B = r\omega_{AB}$, we obtain

$$\omega_{AB} = \left(2000 \frac{\text{rev}}{\text{min}}\right)\left(\frac{1 \text{ min}}{60 \text{ s}}\right)\left(\frac{2\pi \text{ rad}}{1 \text{ rev}}\right) = 209 \text{ rad/s}$$

$$v_B = (AB)\omega_{AB} = (3 \text{ in.})(209 \text{ rad/s}) = 627 \text{ in./s}$$

$$\mathbf{v}_B = 627 \text{ in./s} \searrow 50°$$

Motion of Connecting Rod BD. We consider this motion as a general plane motion. Using the law of sines, we compute the angle β between the connecting rod and the horizontal,

$$\frac{\sin 40°}{8 \text{ in.}} = \frac{\sin \beta}{3 \text{ in.}} \qquad \beta = 13.9°$$

The velocity $\mathbf{v}_D$ of the point D where the rod is attached to the piston must be horizontal, while the velocity of point B is equal to the velocity $\mathbf{v}_B$ obtained above. Resolving the motion of BD into a translation with B and a rotation about B, we obtain

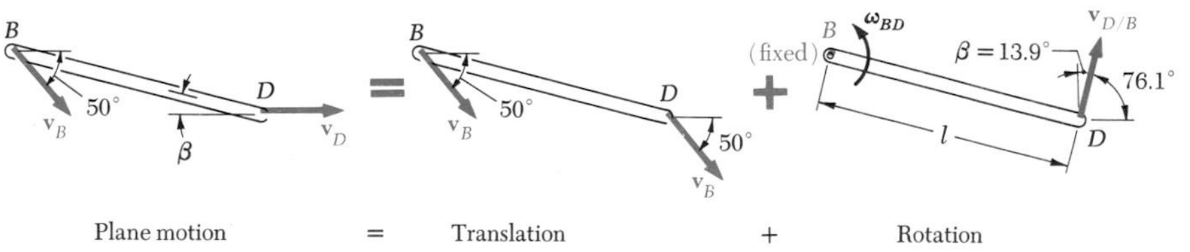

Expressing the relation between the velocities $\mathbf{v}_D$, $\mathbf{v}_B$, and $\mathbf{v}_{D/B}$, we write

$$\mathbf{v}_D = \mathbf{v}_B + \mathbf{v}_{D/B}$$

We draw the vector diagram corresponding to this equation. Recalling that $\beta = 13.9°$, we determine the angles of the triangle and write

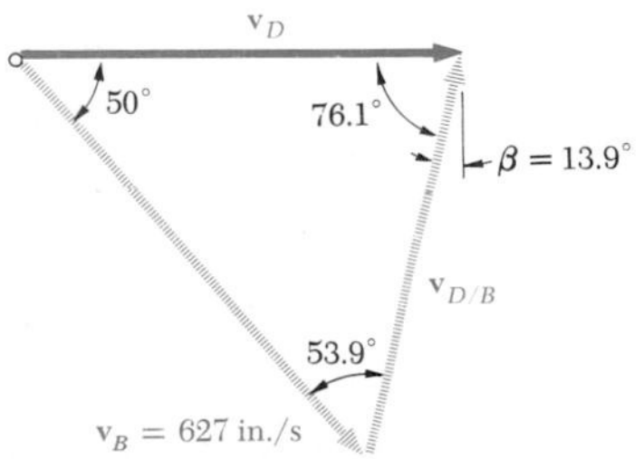

$$\frac{v_D}{\sin 53.9°} = \frac{v_{D/B}}{\sin 50°} = \frac{627 \text{ in./s}}{\sin 76.1°}$$

$$v_{D/B} = 495 \text{ in./s} \qquad \mathbf{v}_{D/B} = 495 \text{ in./s} \measuredangle 76.1°$$

$$v_D = 522 \text{ in./s} = 43.5 \text{ ft/s} \qquad \mathbf{v}_D = 43.5 \text{ ft/s} \rightarrow$$

$$\mathbf{v}_P = \mathbf{v}_D = 43.5 \text{ ft/s} \rightarrow \blacktriangleleft$$

Since $v_{D/B} = l\omega_{BD}$, we have

$$495 \text{ in./s} = (8 \text{ in.})\omega_{BD} \qquad \omega_{BD} = 61.9 \text{ rad/s} \circlearrowleft \blacktriangleleft$$

PROBLEMS

15.25 An automobile travels to the right at a constant speed of 50 km/h. (*a*) If the diameter of a wheel is 610 mm, determine the velocity of points *B*, *C*, *D*, and *E* on the rim of the wheel. (*b*) Solve part *a* assuming that the diameter of the wheel is reduced to 560 mm.

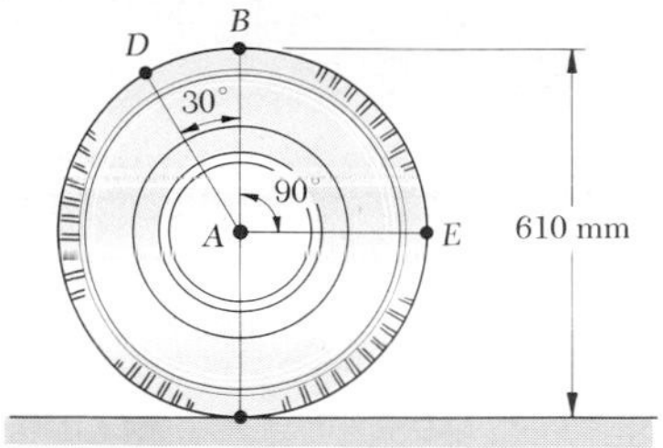

Fig. P15.25

15.26 Collar *B* moves with a constant velocity of 25 in./s to the left. At the instant when $\theta = 30°$, determine (*a*) the angular velocity of rod *AB*, (*b*) the velocity of collar *A*.

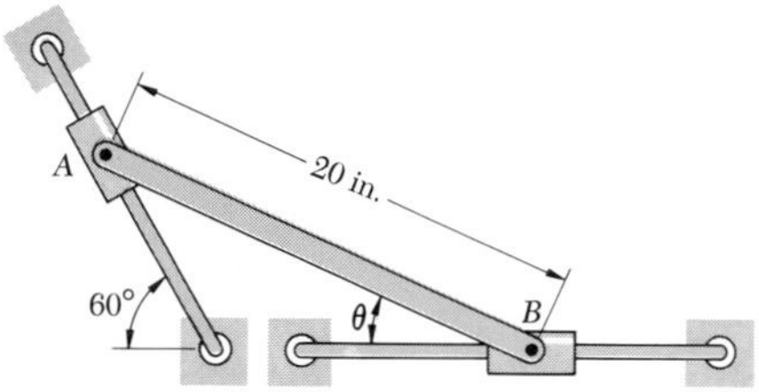

Fig. P15.26

15.27 Solve Prob. 15.26, assuming that $\theta = 45°$.

15.28 The plate shown moves in the *xy* plane. Knowing that $(v_A)_x = 80$ mm/s, $(v_B)_y = 200$ mm/s, and $(v_C)_y = -40$ mm/s, determine (*a*) the angular velocity of the plate, (*b*) the velocity of point *A*.

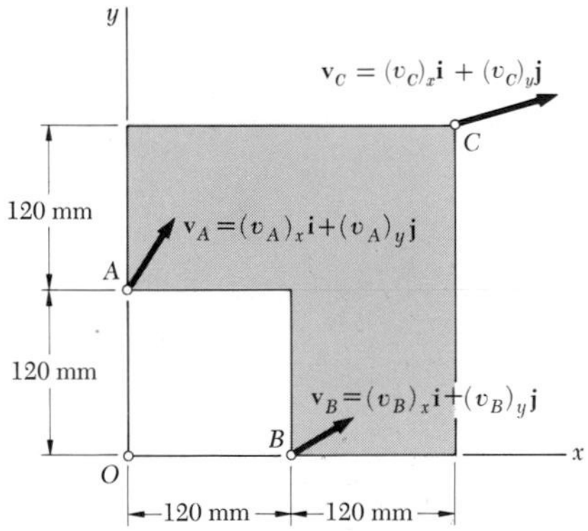

Fig. P15.28

15.29 In Prob. 15.28, determine the equation of the locus of the points of the plate for which the magnitude of the velocity is 100 mm/s.

15.30 In Prob. 15.28, determine (*a*) the velocity of point *B*, (*b*) the point of the plate with zero velocity.

Fig. P15.31

15.31 In the planetary gear system shown, the radius of gears *A*, *B*, *C*, and *D* is *a* and the radius of the outer gear *E* is 3*a*. Knowing that the angular velocity of gear *A* is ω_A clockwise and that the outer gear *E* is stationary, determine (*a*) the angular velocity of the spider connecting the planetary gears, (*b*) the angular velocity of each planetary gear.

15.32 Two rollers *A* and *B* of radius *r* are joined by a link *AB* and roll along a horizontal surface. A drum *C* of radius 2*r* is placed on the rollers as shown. If the link moves to the right with a constant velocity **v**, determine (*a*) the angular velocity of the rollers and of the drum, (*b*) the velocity of points *D*, *E*, and *F* of the drum.

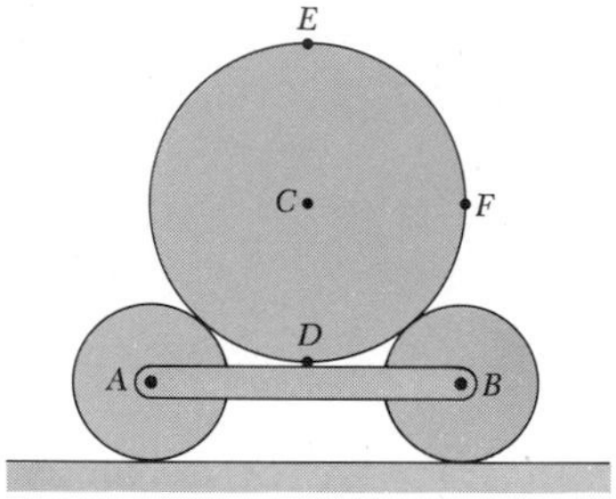

Fig. P15.32

15.33 Gear *A* rotates clockwise with a constant angular velocity of 60 rpm. Knowing that at the same time the arm *AB* rotates counterclockwise with a constant angular velocity of 30 rpm, determine the angular velocity of gear *B*.

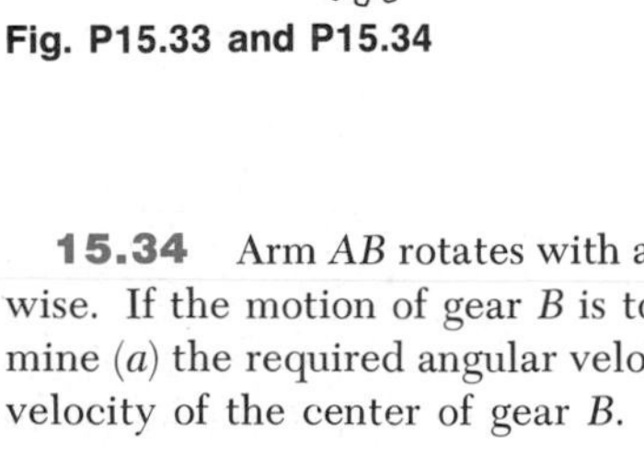

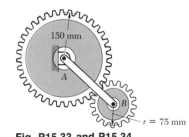

Fig. P15.33 and P15.34

15.34 Arm *AB* rotates with an angular velocity of 120 rpm clockwise. If the motion of gear *B* is to be a curvilinear translation, determine (*a*) the required angular velocity of gear *A*, (*b*) the corresponding velocity of the center of gear *B*.

15.35 Crank AB has a constant angular velocity of 12 rad/s clockwise. Determine the angular velocity of rod BD and the velocity of collar D when (a) $\theta = 0$, (b) $\theta = 90°$, (c) $\theta = 180°$.

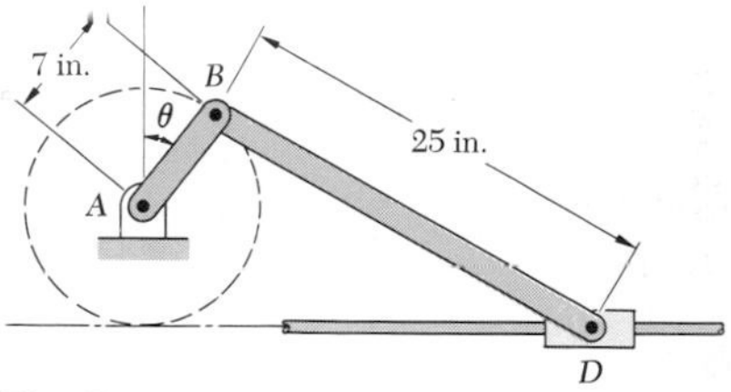

Fig. P15.35

15.36 In the engine system shown, $l = 160$ mm and $b = 60$ mm; the crank AB rotates with a constant angular velocity of 1000 rpm clockwise. Determine the velocity of the piston P and the angular velocity of the connecting rod for the position corresponding to (a) $\theta = 0$, (b) $\theta = 90°$, (c) $\theta = 180°$.

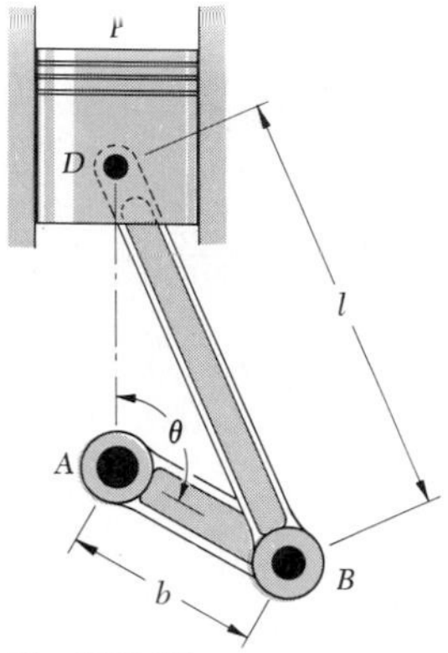

Fig. P15.36

15.37 Solve Prob. 15.36 for the position corresponding to $\theta = 60°$.

15.38 Solve Prob. 15.35 for the position corresponding to $\theta = 120°$.

15.39 through 15.42 In the position shown, bar AB has a constant angular velocity of 3 rad/s counterclockwise. Determine the angular velocity of bars BD and DE.

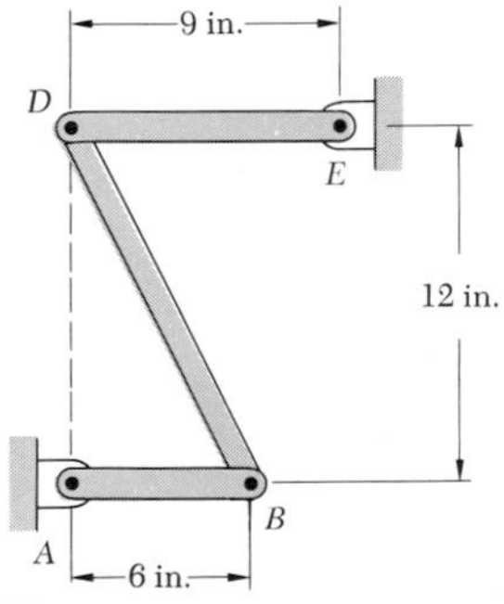

Fig. P15.39

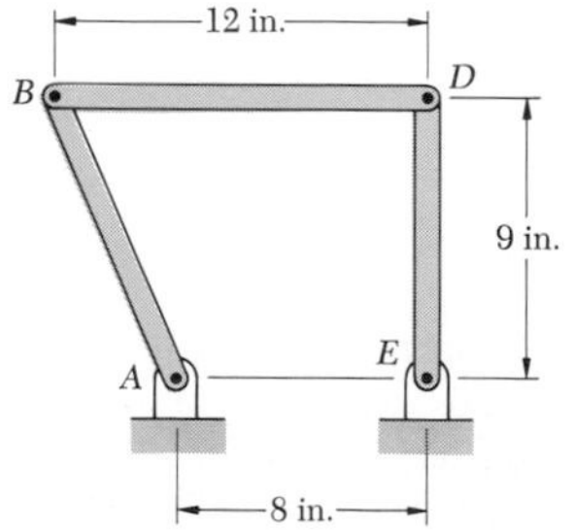

Fig. P15.40

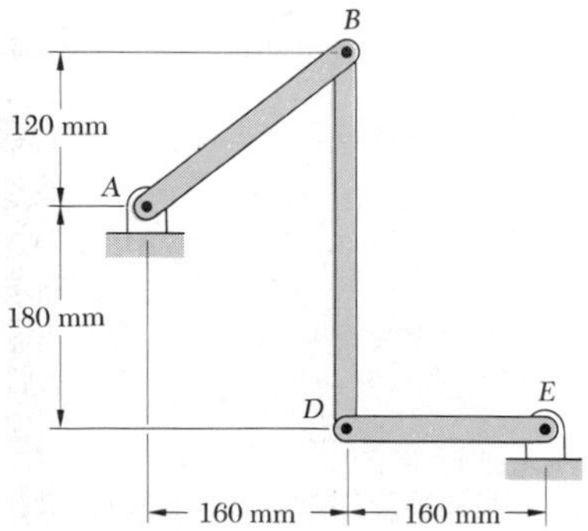

Fig. P15.41

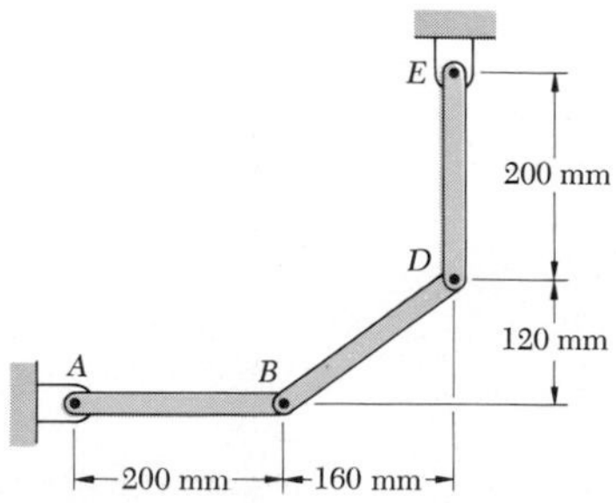

Fig. P15.42

15.43 Two gears, each of 12-in. diameter, are connected by an 18-in. rod AC. Knowing that the center of gear B has a constant velocity of 30 in./s to the right, determine the velocity of the center of gear A and the angular velocity of the connecting rod (*a*) when $\beta = 0$, (*b*) when $\beta = 60°$.

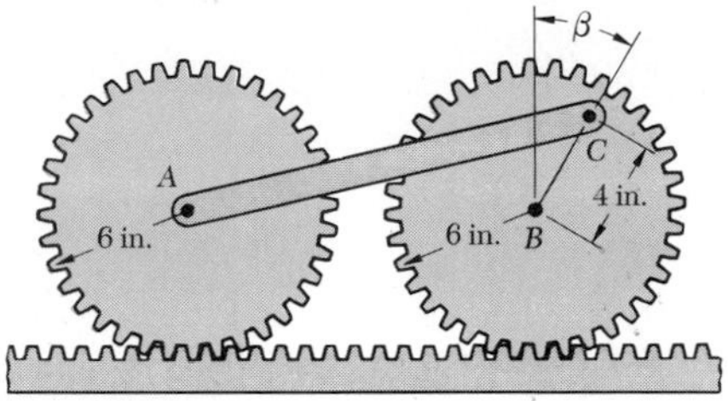

Fig. P15.43

15.44 Solve Prob. 15.43, assuming (*a*) $\beta = 180°$, (*b*) $\beta = 30°$.

15.45 Two collars C and D move along the vertical rod shown. Knowing that the velocity of collar D is 0.210 m/s downward, determine (*a*) the velocity of collar C, (*b*) the angular velocity of member AB.

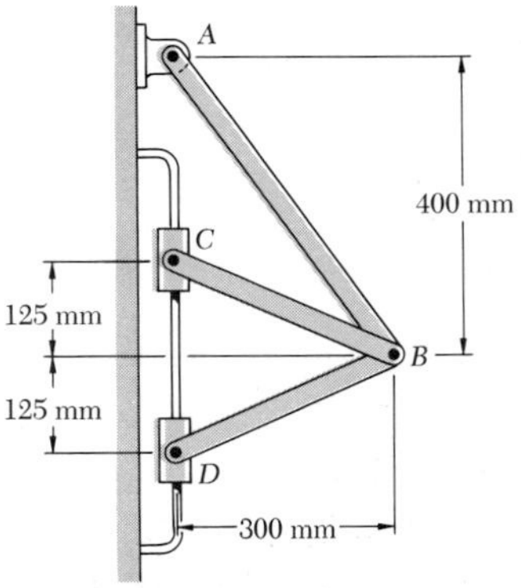

Fig. P15.45

***15.46** Prove for any given position of the mechanism of Prob. 15.45 that the ratio of the magnitudes of the velocities of collars C and D is equal to the ratio of the distances AC and AD.

***15.47** Assuming that the crank AB of Prob. 15.36 rotates with a constant clockwise angular velocity ω and that $\theta = 0$ at $t = 0$, derive an expression for the velocity of the piston P in terms of the time t.

15.7. Instantaneous Center of Rotation in Plane Motion. Consider the general plane motion of a slab. We shall show that at any given instant the velocities of the various particles of the slab are the same as if the slab were rotating about a certain axis perpendicular to the plane of the slab, called the *instantaneous axis of rotation*. This axis intersects the plane of the slab at a point C, called the *instantaneous center of rotation* of the slab.

To prove our statement, we first recall that the plane motion of a slab may always be replaced by a translation defined by the motion of an arbitrary reference point A, and by a rotation about A. As far as the velocities are concerned, the translation is characterized by the velocity $\mathbf{v}_A$ of the reference point A and the rotation is characterized by the angular velocity $\boldsymbol{\omega}$ of the slab (which is independent of the choice of A). Thus, the velocity $\mathbf{v}_A$ of point A and the angular velocity $\boldsymbol{\omega}$ of the slab define completely the velocities of all the other particles of the slab (Fig. 15.18*a*). Now let us assume that $\mathbf{v}_A$ and $\boldsymbol{\omega}$ are known and

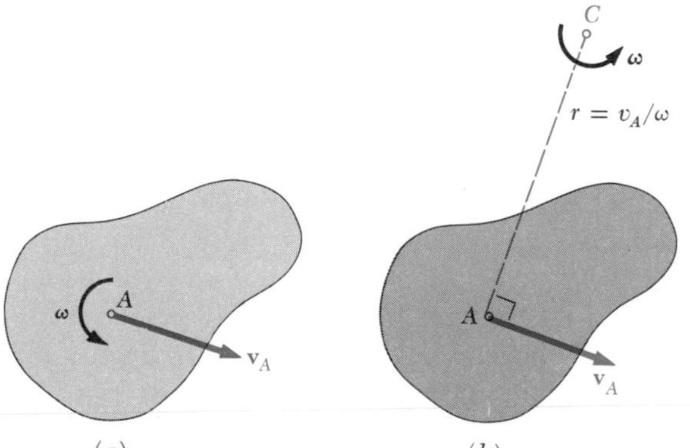

Fig. 15.18

that they are both different from zero. (If $\mathbf{v}_A = 0$, point A is itself the instantaneous center of rotation, and if $\boldsymbol{\omega} = 0$, all the particles have the same velocity $\mathbf{v}_A$.) These velocities could be obtained by letting the slab rotate with the angular velocity $\boldsymbol{\omega}$ about a point C located on the perpendicular to $\mathbf{v}_A$ at a distance $r = v_A/\omega$ from A as shown in Fig. 15.18*b*. We check that the velocity of A would be perpendicular to AC and that its magnitude would be $r\omega = (v_A/\omega)\omega = v_A$. Thus the velocities of all the other particles of the slab would be the same as originally defined. Therefore, *as far as the velocities are concerned, the slab seems to rotate about the instantaneous center* C at the instant considered.

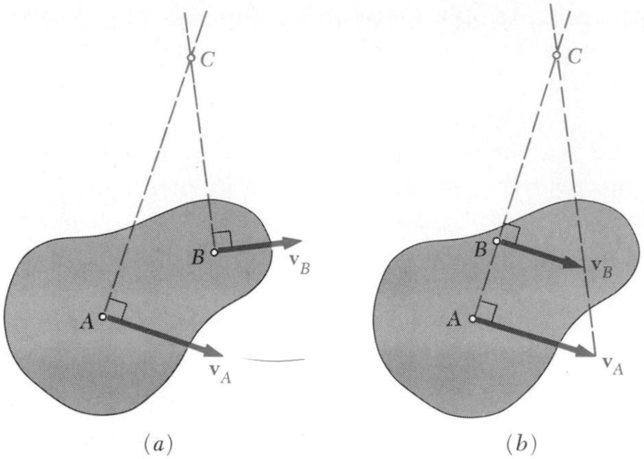

Fig. 15.19

The position of the instantaneous center may be defined in two other ways. If the directions of the velocities of two particles A and B of the slab are known, and if they are different, the instantaneous center C is obtained by drawing the perpendicular to $\mathbf{v}_A$ through A and the perpendicular to $\mathbf{v}_B$ through B and determining the point in which these two lines intersect (Fig. 15.19*a*). If the velocities $\mathbf{v}_A$ and $\mathbf{v}_B$ of two particles A and B are perpendicular to the line AB, and if their magnitudes are known, the instantaneous center may be found by intersecting the line AB with the line joining the extremities of the vectors $\mathbf{v}_A$ and $\mathbf{v}_B$ (Fig. 15.19*b*). Note that, if $\mathbf{v}_A$ and $\mathbf{v}_B$ were parallel in Fig. 15.19*a*, or if $\mathbf{v}_A$ and $\mathbf{v}_B$ had the same magnitude in Fig. 15.19*b*, the instantaneous center C would be at an infinite distance and $\boldsymbol{\omega}$ would be zero; all points of the slab would have the same velocity.

To see how the concept of instantaneous center of rotation may be put to use, let us consider again the rod of Sec. 15.6. Drawing the perpendicular to $\mathbf{v}_A$ through A and the perpendicular to $\mathbf{v}_B$ through B (Fig. 15.20), we obtain the instantaneous center C. At the instant considered, the velocities of all the particles of the rod are thus the same as if the rod rotated about C. Now, if the magnitude v_A of the velocity of A is known, the magnitude ω of the angular velocity of the rod may be obtained by writing

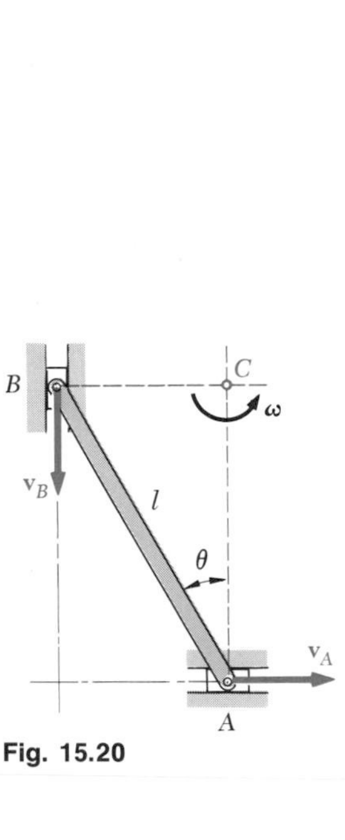

Fig. 15.20

$$\omega = \frac{v_A}{AC} = \frac{v_A}{l\cos\theta}$$

The magnitude of the velocity of B may then be obtained by writing

$$v_B = (BC)\omega = l \sin\theta \frac{v_A}{l \cos\theta} = v_A \tan\theta$$

Note that only *absolute* velocities are involved in the computation.

The instantaneous center of a slab in plane motion may be located either on the slab or outside the slab. If it is located on the slab, the particle C coinciding with the instantaneous center at a given instant t must have zero velocity at that instant. However, it should be noted that the instantaneous center of rotation is valid only at a given instant. Thus, the particle C of the slab which coincides with the instantaneous center at time t will generally not coincide with the instantaneous center at time $t + \Delta t$; while its velocity is zero at time t, it will probably be different from zero at time $t + \Delta t$. This means that, in general, the particle C *does not have zero acceleration,* and therefore that the *accelerations* of the various particles of the slab *cannot* be determined as if the slab were rotating about C.

As the motion of the slab proceeds, the instantaneous center moves in space. But it was just pointed out that the position of the instantaneous center on the slab keeps changing. Thus, the instantaneous center describes one curve in space, called the *space centrode,* and another curve on the slab, called the *body centrode* (Fig. 15.21). It may be shown that, at any instant, these two curves are tangent at C and that, as the slab moves, the body centrode appears to *roll* on the space centrode.

Fig. 15.21

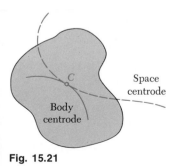

SAMPLE PROBLEM 15.4

Solve Sample Prob. 15.2, using the method of the instantaneous center of rotation.

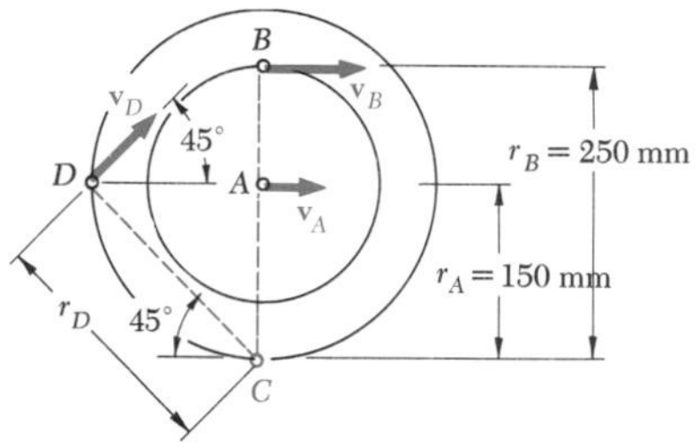

a. **Angular Velocity of the Gear.** Since the gear rolls on the stationary lower rack, the point of contact C of the gear with the rack has no velocity; point C is therefore the instantaneous center of rotation. We write

$$v_A = r_A\omega \qquad 1.2\text{ m/s} = (0.150\text{ m})\omega \qquad \omega = 8\text{ rad/s} \circlearrowright \blacktriangleleft$$

b. **Velocities.** All points of the gear seem to rotate about the instantaneous center as far as velocities are concerned.

Velocity of Upper Rack. Recalling that $v_R = v_B$, we write

$$v_R = v_B = r_B\omega \qquad v_R = (0.250\text{ m})(8\text{ rad/s}) = 2\text{ m/s}$$

$$\mathbf{v}_R = 2\text{ m/s} \rightarrow \blacktriangleleft$$

Velocity of Point D. Since $r_D = (0.150\text{ m})\sqrt{2} = 0.212$ m, we write

$$v_D = r_D\omega \qquad v_D = (0.212\text{ m})(8\text{ rad/s}) = 1.696\text{ m/s}$$

$$\mathbf{v}_D = 1.696\text{ m/s} \measuredangle 45° \blacktriangleleft$$

SAMPLE PROBLEM 15.5

Solve Sample Prob. 15.3, using the method of the instantaneous center of rotation.

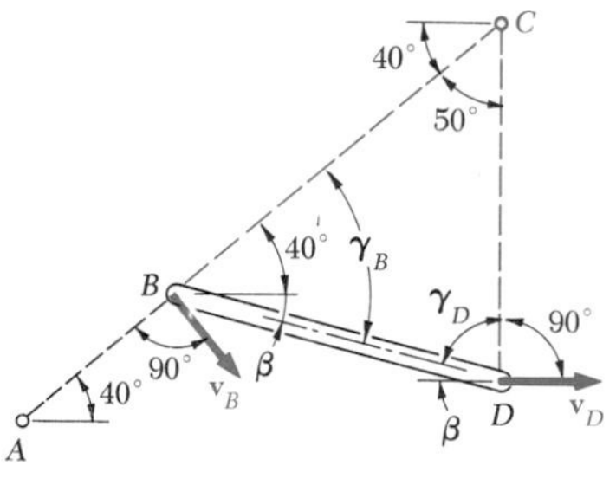

Motion of Crank AB. Referring to Sample Prob. 15.3, we obtain the velocity of point B; $\mathbf{v}_B = 627$ in./s $\searrow$ 50°.

Motion of the Connecting Rod BD. We first locate the instantaneous center C by drawing lines perpendicular to the absolute velocities $\mathbf{v}_B$ and $\mathbf{v}_D$. Recalling from Sample Prob. 15.3 that $\beta = 13.9°$ and that $BD = 8$ in., we solve the triangle BCD.

$$\gamma_B = 40° + \beta = 53.9° \qquad \gamma_D = 90° - \beta = 76.1°$$

$$\frac{BC}{\sin 76.1°} = \frac{CD}{\sin 53.9°} = \frac{8\text{ in.}}{\sin 50°}$$

$$BC = 10.14\text{ in.} \qquad CD = 8.44\text{ in.}$$

Since the connecting rod BD seems to rotate about point C, we write

$$v_B = (BC)\omega_{BD}$$

$$627\text{ in./s} = (10.14\text{ in.})\omega_{BD}$$

$$\omega_{BD} = 61.9\text{ rad/s} \circlearrowleft \blacktriangleleft$$

$$v_D = (CD)\omega_{BD} = (8.44\text{ in.})(61.9\text{ rad/s})$$

$$= 522\text{ in./s} = 43.5\text{ ft/s}$$

$$\mathbf{v}_P = \mathbf{v}_D = 43.5\text{ ft/s} \rightarrow \blacktriangleleft$$

PROBLEMS

15.48 A helicopter moves horizontally in the x direction at a speed of 120 mi/h. Knowing that the main blades rotate clockwise at an angular velocity of 180 rpm, determine the instantaneous axis of rotation of the main blades.

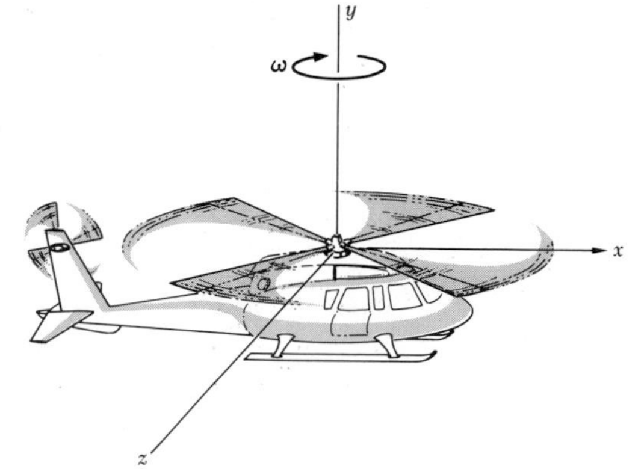

Fig. P15.48

15.49 Denoting by $\mathbf{r}_A$ the position vector of a point A of a rigid slab which moves in plane motion, show that the position vector $\mathbf{r}_C$ of the instantaneous center of rotation is

$$\mathbf{r}_C = \mathbf{r}_A + \frac{\boldsymbol{\omega} \times \mathbf{v}_A}{\omega^2}$$

where $\boldsymbol{\omega}$ is the angular velocity of the slab and $\mathbf{v}_A$ the velocity of point A.

15.50 A drum, of radius 4.5 in., is mounted on a cylinder, of radius 6 in. A cord is wound around the drum, and its extremity D is pulled to the left at a constant velocity of 3 in./s, causing the cylinder to roll without sliding. Determine (*a*) the angular velocity of the cylinder, (*b*) the velocity of the center of the cylinder, (*c*) the length of cord which is wound or unwound per second.

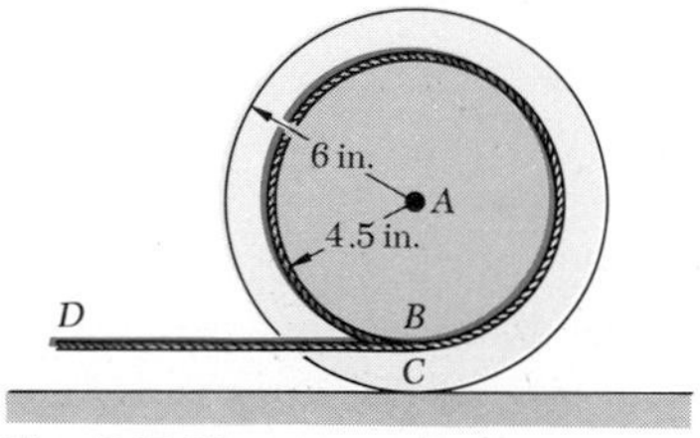

Fig. P15.50

15.51 Solve Sample Prob. 15.2, assuming that the lower rack is not stationary but moves to the left with a velocity of 0.6 m/s.

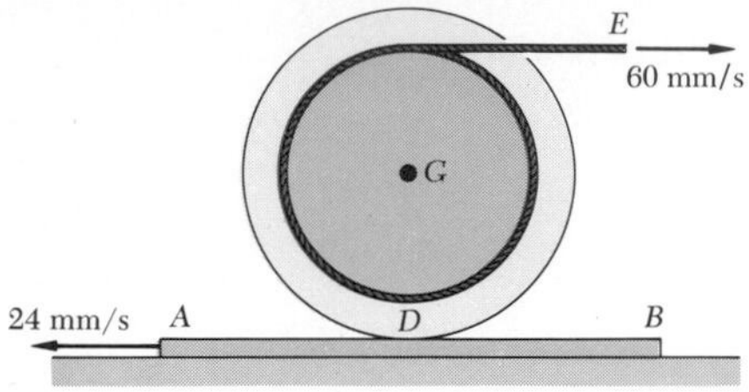

Fig. P15.52

15.52 A double pulley rolls without sliding on the plate *AB*, which moves to the left at a constant speed of 24 mm/s. The 60-mm-radius inner pulley is rigidly attached to the 80-mm-radius outer pulley. Knowing that cord *E* is pulled at a constant speed of 60 mm/s as shown, determine (*a*) the angular velocity of the pulley, (*b*) the velocity of the center *G* of the pulley.

15.53 Knowing that at the instant shown the velocity of collar *D* is 20 in./s upward, determine (*a*) the angular velocity of rod *AD*, (*b*) the velocity of point *B*, (*c*) the velocity of point *A*.

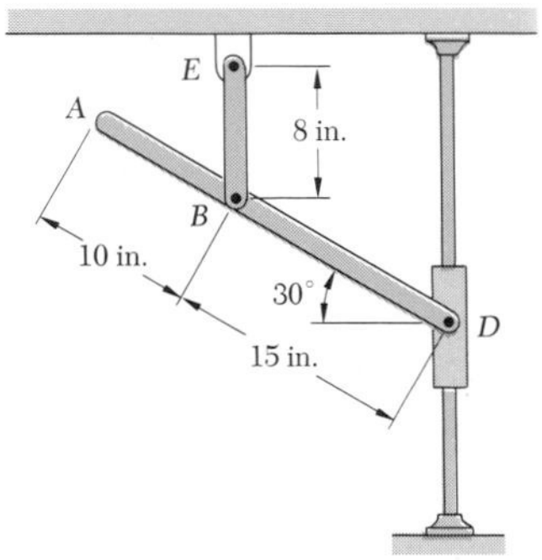

Fig. P15.53

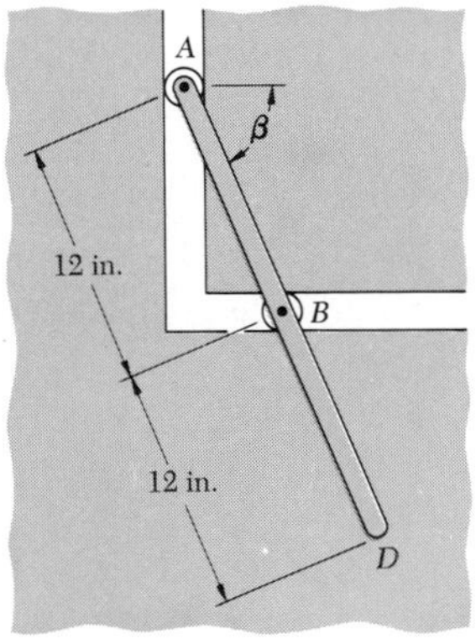

Fig. P15.54

15.54 The rod *ABD* is guided by wheels which roll in the tracks shown. Knowing that $\beta = 60°$ and that the velocity of *A* is 24 in./s downward, determine (*a*) the angular velocity of the rod, (*b*) the velocity of point *D*.

15.55 Solve Prob. 15.54, assuming that $\beta = 30°$.

15.56 Knowing that at the instant shown the angular velocity of crank *AB* is 3 rad/s clockwise, determine (*a*) the angular velocity of link *BD*, (*b*) the velocity of collar *D*, (*c*) the velocity of the midpoint of link *BD*.

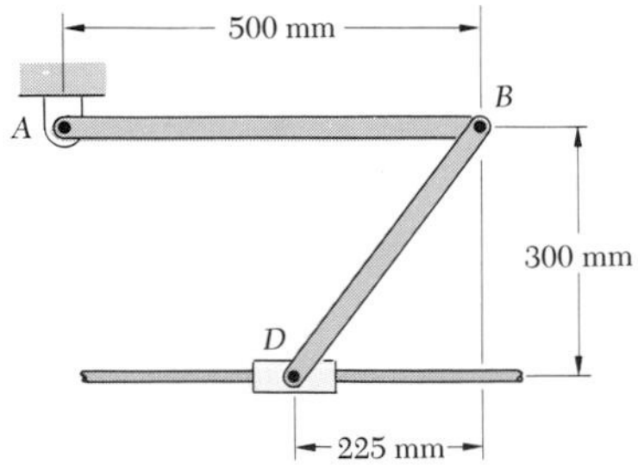

Fig. P15.56 and P15.57

15.57 Knowing that at the instant shown the velocity of collar *D* is 1.5 m/s to the right, determine (*a*) the angular velocities of crank *AB* and link *BD*, (*b*) the velocity of the midpoint of link *BD*.

15.58 Collar A slides downward with a constant velocity $\mathbf{v}_A$. Determine the angle θ corresponding to the position of rod AB for which the velocity of B is horizontal.

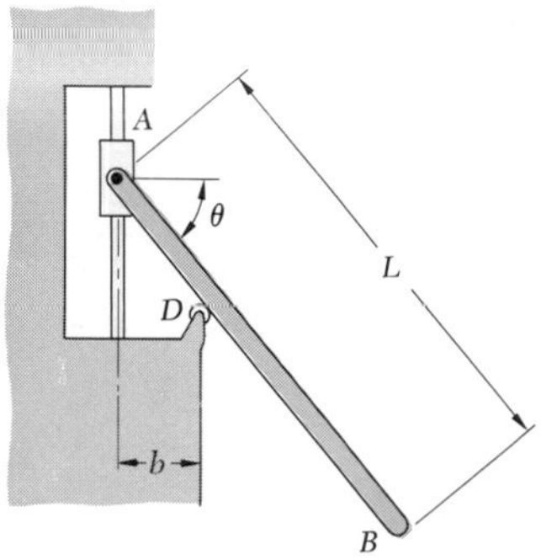

Fig. P15.58 and P15.60

15.59 Two rods AB and BD are connected to three collars as shown. Knowing that collar A moves downward with a constant velocity of 120 mm/s, determine at the instant shown (*a*) the angular velocity of each rod, (*b*) the velocity of collar D.

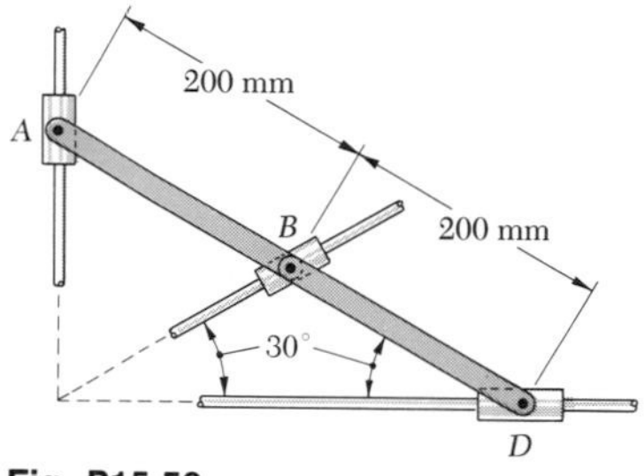

Fig. P15.59

15.60 Collar A slides downward with a constant speed of 16 in./s. Knowing that $b = 2$ in., $L = 10$ in., and $\theta = 60°$, determine (*a*) the angular velocity of rod AB, (*b*) the velocity of B.

15.61 The rectangular plate is supported by two 6-in. links as shown. Knowing that at the instant shown the angular velocity of link AB is 4 rad/s clockwise, determine (*a*) the angular velocity of the plate, (*b*) the velocity of the center of the plate, (*c*) the velocity of corner F.

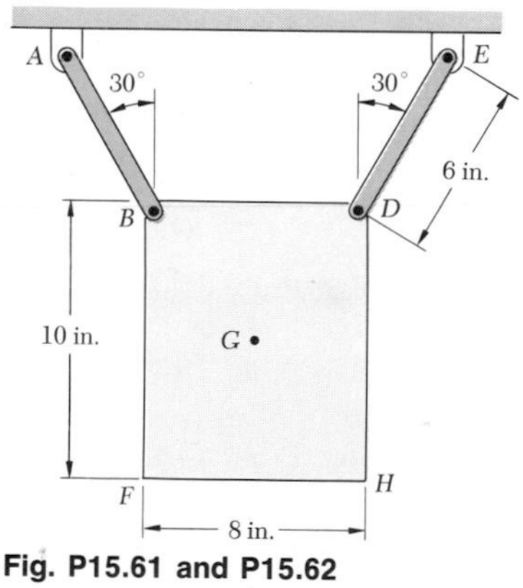

Fig. P15.61 and P15.62

15.62 Knowing that, at the instant shown, the angular velocity of link AB is 4 rad/s clockwise, determine (*a*) the angular velocity of the plate, (*b*) the points of the plate for which the magnitude of the velocity is equal to or less than 6 in./s.

15.63 At the instant shown, the velocity of the center of the gear is 200 mm/s to the right. Determine (*a*) the velocity of point *B*, (*b*) the velocity of collar *D*.

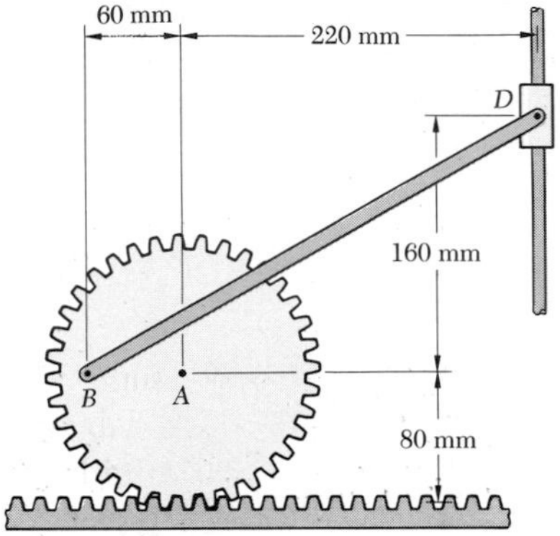

Fig. P15.63 and P15.64

15.64 At the instant shown, the velocity of collar *D* is 360 mm/s downward. Determine (*a*) the angular velocity of rod *BD*, (*b*) the velocity of the center of the gear.

15.65 Describe the space centrode and the body centrode of gear *A* of Prob. 15.63 as the gear rolls on the horizontal rack.

15.66 Describe the space centrode and the body centrode of rod *ABD* of Prob. 15.54 as point *A* moves downward. (*Note.* The body centrode need not lie on a physical portion of the rod.)

15.67 Using the method of Sec. 15.7, solve Prob. 15.35.

15.68 Using the method of Sec. 15.7, solve Prob. 15.36.

15.69 Using the method of Sec. 15.7, solve Prob. 15.39.

15.70 Using the method of Sec. 15.7, solve Prob. 15.40.

15.71 Using the method of Sec. 15.7, solve Prob. 15.41.

15.72 Using the method of Sec. 15.7, solve Prob. 15.42.

15.73 Using the method of Sec. 15.7, solve Prob. 15.43.

15.74 Using the method of Sec. 15.7, solve Prob. 15.32.

15.75 Using the method of Sec. 15.7, solve Prob. 15.33.

15.8. Absolute and Relative Acceleration in Plane Motion.

We saw in Sec. 15.5 that any plane motion may be replaced by a translation defined by the motion of an arbitrary reference point A, and by a rotation about A. This property was used in Sec. 15.6 to determine the velocity of the various points of a moving slab. We shall now use the same property to determine the acceleration of the points of the slab.

We first recall that the absolute acceleration $\mathbf{a}_B$ of a particle of the slab may be obtained from the relative-acceleration formula derived in Sec. 11.12,

$$\mathbf{a}_B = \mathbf{a}_A + \mathbf{a}_{B/A} \qquad (15.21)$$

where the right-hand member represents a vector sum. The acceleration $\mathbf{a}_A$ corresponds to the translation of the slab with A, while the relative acceleration $\mathbf{a}_{B/A}$ is associated with the rotation of the slab about A and is measured with respect to axes centered at A and of fixed orientation. We recall from Sec. 15.3 that the relative acceleration $\mathbf{a}_{B/A}$ may be resolved into two components, a *tangential component* $(\mathbf{a}_{B/A})_t$ perpendicular to the line AB, and a *normal component* $(\mathbf{a}_{B/A})_n$ directed toward A (Fig. 15.22). Denoting by $\mathbf{r}_{B/A}$ the position vector of B relative to A and, respectively, by $\omega\mathbf{k}$ and $\alpha\mathbf{k}$ the angular velocity and angular acceleration of the slab with respect to axes of fixed orientation, we have

$$\begin{aligned} (\mathbf{a}_{B/A})_t &= \alpha\mathbf{k} \times \mathbf{r}_{B/A} & (a_{B/A})_t &= r\alpha \\ (\mathbf{a}_{B/A})_n &= -\omega^2\mathbf{r}_{B/A} & (a_{B/A})_n &= r\omega^2 \end{aligned} \qquad (15.22)$$

where r is the distance from A to B. Substituting into (15.21) the expressions obtained for the tangential and normal components of $\mathbf{a}_{B/A}$, we may also write

$$\mathbf{a}_B = \mathbf{a}_A + \alpha\mathbf{k} \times \mathbf{r}_{B/A} - \omega^2\mathbf{r}_{B/A} \qquad (15.21')$$

| Plane motion | = | Translation with A | + | Rotation about A |

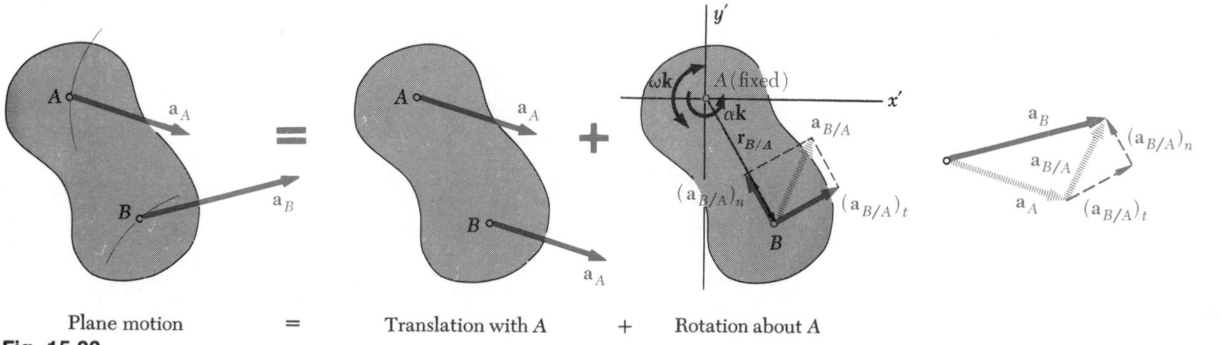

Fig. 15.22

Fig. 15.23

| Plane motion | = | Translation with A | + | Rotation about A |

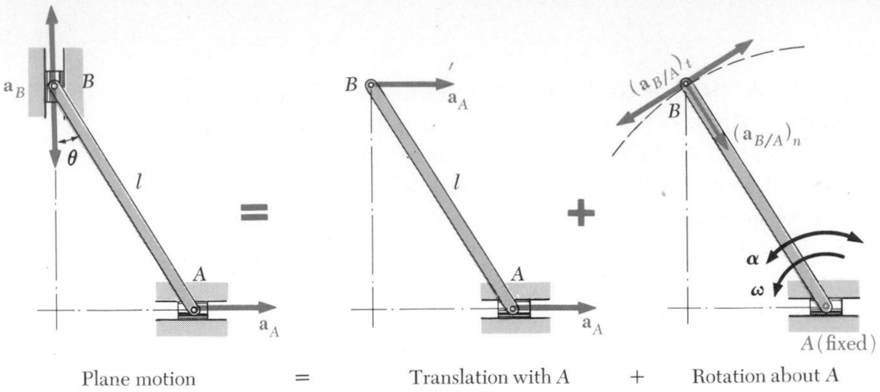

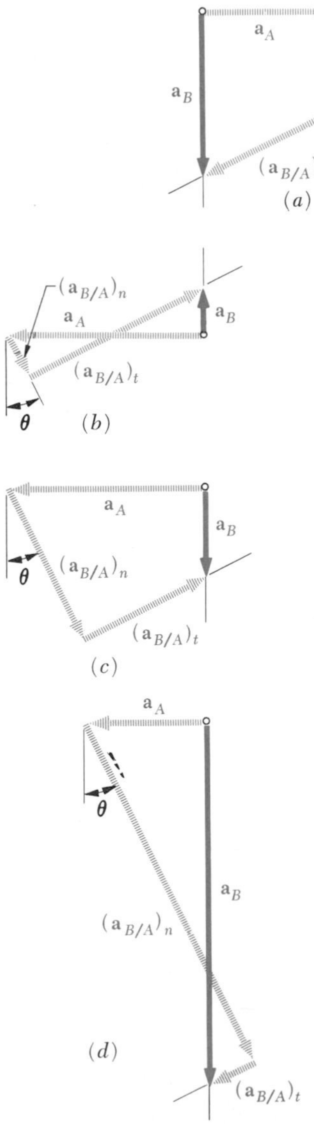

Fig. 15.24

As an example, we shall consider again the rod AB whose extremities slide, respectively, along a horizontal and a vertical track (Fig. 15.23). Assuming that the velocity $\mathbf{v}_A$ and the acceleration $\mathbf{a}_A$ of A are known, we propose to determine the acceleration $\mathbf{a}_B$ of B and the angular acceleration α of the rod. Choosing A as a reference point, we express that the given motion is equivalent to a translation with A and a rotation about A. The absolute acceleration of B must be equal to the sum

$$\mathbf{a}_B = \mathbf{a}_A + \mathbf{a}_{B/A}$$
$$= \mathbf{a}_A + (\mathbf{a}_{B/A})_n + (\mathbf{a}_{B/A})_t \qquad (15.23)$$

where $(\mathbf{a}_{B/A})_n$ has the magnitude $l\omega^2$ and is *directed toward* A, while $(\mathbf{a}_{B/A})_t$ has the magnitude $l\alpha$ and is perpendicular to AB. There is no way of telling at the present time whether the tangential component $(\mathbf{a}_{B/A})_t$ is directed to the left or to the right, and the student should not rely on his "intuition" in this matter. We shall therefore indicate both possible directions for this component in Fig. 15.23. Similarly, we indicate both possible senses for $\mathbf{a}_B$, since we do not know whether point B is accelerated upward or downward.

Equation (15.23) has been expressed geometrically in Fig. 15.24. Four different vector polygons may be obtained, depending upon the sense of $\mathbf{a}_A$ and the relative magnitude of a_A and $(a_{B/A})_n$. If we are to determine a_B and α from one of these diagrams, we must know not only a_A and θ but also ω. The angular velocity of the rod, therefore, should be separately determined by one of the methods indicated in Secs. 15.6 and 15.7. The values of a_B and α may then be obtained by considering successively the x and y components of the vectors shown in Fig. 15.24. In the case of polygon a, for example, we write

$\xrightarrow{+}\,x$ components: $\qquad 0 = a_A + l\omega^2 \sin\theta - l\alpha \cos\theta$

$+\uparrow y$ components: $\qquad -a_B = -l\omega^2 \cos\theta - l\alpha \sin\theta$

and solve for a_B and α. The two unknowns may also be obtained by direct measurement on the vector polygon. In that case, care should be taken to draw first the known vectors $\mathbf{a}_A$ and $(\mathbf{a}_{B/A})_n$.

It is quite evident that the determination of accelerations is considerably more involved than the determination of velocities. Yet, in the example considered here, the extremities A and B of the rod were moving along straight tracks, and the diagrams drawn were relatively simple. If A and B had moved along curved tracks, the accelerations $\mathbf{a}_A$ and $\mathbf{a}_B$ should have been resolved into normal and tangential components and the solution of the problem would have involved six different vectors.

When a mechanism consists of several moving parts which are pin-connected, its analysis may be carried out by considering each part as a rigid body, while keeping in mind that the points where two parts are connected must have the same absolute acceleration (see Sample Prob. 15.7). In the case of meshed gears, the tangential components of the accelerations of the teeth in contact are equal, but their normal components are different.

***15.9. Analysis of Plane Motion in Terms of a Parameter.** In the case of certain mechanisms, it is possible to express the coordinates x and y of all the significant points of the mechanism by means of simple analytic expressions containing a single parameter. It may be advantageous in such a case to determine directly the absolute velocity and the absolute acceleration of the various points of the mechanism, since the components of the velocity and of the acceleration of a given point may be obtained by differentiating the coordinates x and y of that point.

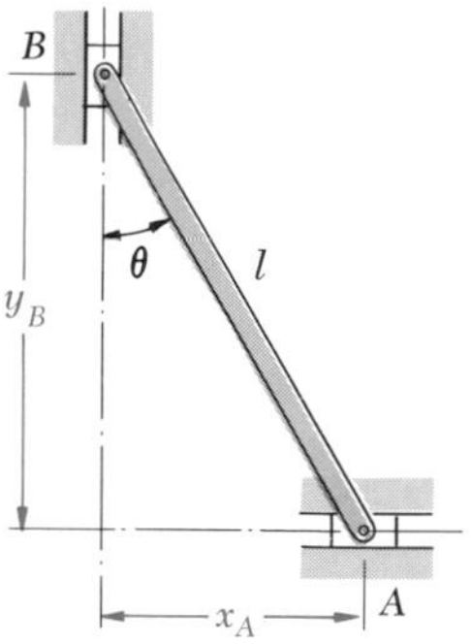

Fig. 15.25

Let us consider again the rod AB whose extremities slide, respectively, in a horizontal and a vertical track (Fig. 15.25). The coordinates x_A and y_B of the extremities of the rod may be expressed in terms of the angle θ the rod forms with the vertical,

$$x_A = l \sin\theta \qquad y_B = l\cos\theta \tag{15.24}$$

Differentiating Eqs. (15.24) twice with respect to t, we write

$$v_A = \dot{x}_A = l\dot{\theta}\cos\theta \qquad v_B = \dot{y}_B = -l\dot{\theta}\sin\theta$$
$$a_A = \ddot{x}_A = -l\dot{\theta}^2\sin\theta + l\ddot{\theta}\cos\theta$$
$$a_B = \ddot{y}_B = -l\dot{\theta}^2\cos\theta - l\ddot{\theta}\sin\theta$$

Recalling that $\dot{\theta} = \omega$ and $\ddot{\theta} = \alpha$, we obtain

$$v_A = l\omega\cos\theta \qquad v_B = -l\omega\sin\theta \tag{15.25}$$
$$a_A = -l\omega^2\sin\theta + l\alpha\cos\theta \qquad a_B = -l\omega^2\cos\theta - l\alpha\sin\theta \tag{15.26}$$

We note that a positive sign for v_A or a_A indicates that the velocity $\mathbf{v}_A$ or the acceleration $\mathbf{a}_A$ is directed to the right; a positive sign for v_B or a_B indicates that $\mathbf{v}_B$ or $\mathbf{a}_B$ is directed upward. Equations (15.25) may be used, for example, to determine v_B and ω when v_A and θ are known. Substituting for ω in (15.26), we may then determine a_B and α if a_A is known.

SAMPLE PROBLEM 15.6

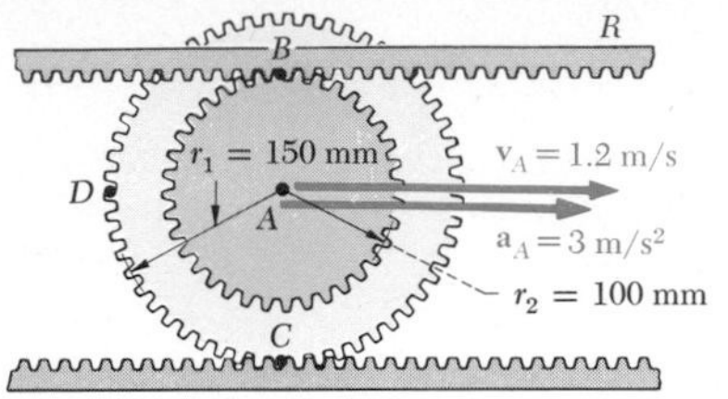

The center of the double gear of Sample Prob. 15.2 has a velocity of 1.2 m/s to the right and an acceleration of 3 m/s² to the right. Determine (*a*) the angular acceleration of the gear, (*b*) the acceleration of points *B*, *C*, and *D* of the gear.

***a.* Angular Acceleration of the Gear.** In Sample Prob. 15.2, we found that $x_A = -r_1\theta$ and $v_A = -r_1\omega$. Differentiating the latter with respect to time, we obtain $a_A = -r_1\alpha$.

$$v_A = -r_1\omega \qquad 1.2\text{ m/s} = -(0.150\text{ m})\omega \qquad \omega = -8\text{ rad/s}$$
$$a_A = -r_1\alpha \qquad 3\text{ m/s}^2 = -(0.150\text{ m})\alpha \qquad \alpha = -20\text{ rad/s}^2$$
$$\boldsymbol{\alpha} = \alpha\mathbf{k} = -(20\text{ rad/s}^2)\mathbf{k} \quad ◀$$

***b.* Accelerations.** The rolling motion of the gear is resolved into a translation with *A* and a rotation about *A*.

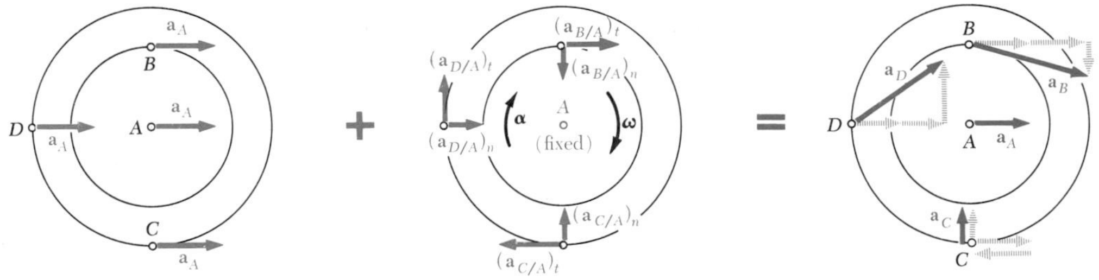

Translation + Rotation = Rolling motion

Acceleration of Point B. Adding vectorially the accelerations corresponding to the translation and to the rotation, we obtain

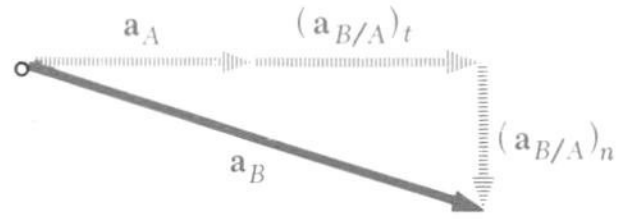

$$\begin{aligned}\mathbf{a}_B &= \mathbf{a}_A + \mathbf{a}_{B/A} = \mathbf{a}_A + (\mathbf{a}_{B/A})_t + (\mathbf{a}_{B/A})_n\\ &= \mathbf{a}_A + \alpha\mathbf{k}\times\mathbf{r}_{B/A} - \omega^2\mathbf{r}_{B/A}\\ &= (3\text{ m/s}^2)\mathbf{i} - (20\text{ rad/s}^2)\mathbf{k}\times(0.100\text{ m})\mathbf{j} - (8\text{ rad/s})^2(0.100\text{ m})\mathbf{j}\\ &= (3\text{ m/s}^2)\mathbf{i} + (2\text{ m/s}^2)\mathbf{i} - (6.40\text{ m/s}^2)\mathbf{j}\end{aligned}$$
$$\mathbf{a}_B = 8.12\text{ m/s}^2 \searrow 52.0° \quad ◀$$

Acceleration of Point C

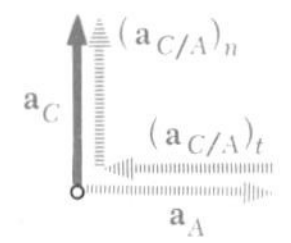

$$\begin{aligned}\mathbf{a}_C &= \mathbf{a}_A + \mathbf{a}_{C/A} = \mathbf{a}_A + \alpha\mathbf{k}\times\mathbf{r}_{C/A} - \omega^2\mathbf{r}_{C/A}\\ &= (3\text{ m/s}^2)\mathbf{i} - (20\text{ rad/s}^2)\mathbf{k}\times(-0.150\text{ m})\mathbf{j} - (8\text{ rad/s})^2(-0.150\text{ m})\mathbf{j}\\ &= (3\text{ m/s}^2)\mathbf{i} - (3\text{ m/s}^2)\mathbf{i} + (9.60\text{ m/s}^2)\mathbf{j}\end{aligned}$$
$$\mathbf{a}_C = 9.60\text{ m/s}^2\uparrow \quad ◀$$

Acceleration of Point D

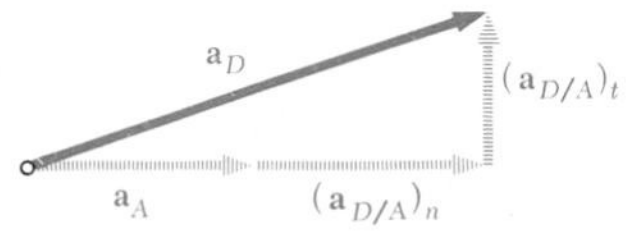

$$\begin{aligned}\mathbf{a}_D &= \mathbf{a}_A + \mathbf{a}_{D/A} = \mathbf{a}_A + \alpha\mathbf{k}\times\mathbf{r}_{D/A} - \omega^2\mathbf{r}_{D/A}\\ &= (3\text{ m/s}^2)\mathbf{i} - (20\text{ rad/s}^2)\mathbf{k}\times(-0.150\text{ m})\mathbf{i} - (8\text{ rad/s})^2(-0.150\text{ m})\mathbf{i}\\ &= (3\text{ m/s}^2)\mathbf{i} + (3\text{ m/s}^2)\mathbf{j} + (9.60\text{ m/s}^2)\mathbf{i}\end{aligned}$$
$$\mathbf{a}_D = 12.95\text{ m/s}^2 \measuredangle 13.4° \quad ◀$$

SAMPLE PROBLEM 15.7

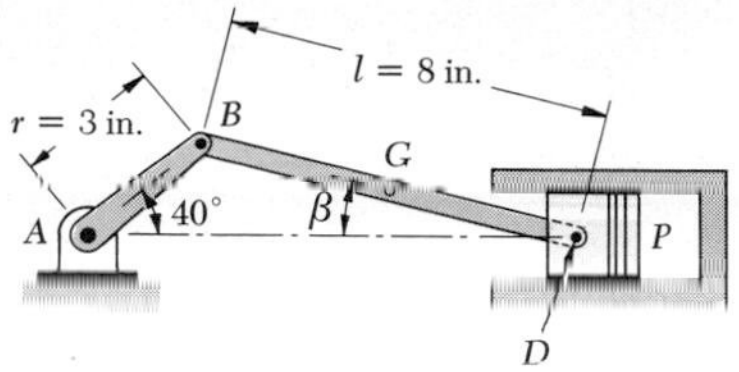

Crank AB of the engine system of Sample Prob. 15.3 has a constant clockwise angular velocity of 2000 rpm. For the crank position shown, determine the angular acceleration of the connecting rod BD and the acceleration of point D.

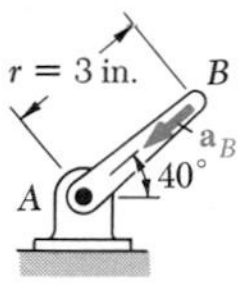

Motion of Crank AB. Since the crank rotates about A with constant $\omega_{AB} = 2000$ rpm $= 209$ rad/s, we have $\alpha_{AB} = 0$. The acceleration of B is therefore directed toward A and has a magnitude

$$a_B = r\omega_{AB}^2 = (\tfrac{3}{12}\text{ ft})(209\text{ rad/s})^2 = 10{,}920\text{ ft/s}^2$$
$$\mathbf{a}_B = 10{,}920\text{ ft/s}^2 \measuredangle\ 40°$$

Motion of the Connecting Rod BD. The angular velocity $\boldsymbol{\omega}_{BD}$ and the value of β were obtained in Sample Prob. 15.3.

$$\omega_{BD} = 61.9\text{ rad/s}\ \circlearrowleft \qquad \beta = 13.9°$$

The motion of BD is resolved into a translation with B and a rotation about B. The relative acceleration $\mathbf{a}_{D/B}$ is resolved into normal and tangential components.

$$(a_{D/B})_n = (BD)\omega_{BD}^2 = (\tfrac{8}{12}\text{ ft})(61.9\text{ rad/s})^2 = 2550\text{ ft/s}^2$$
$$(\mathbf{a}_{D/B})_n = 2550\text{ ft/s}^2 \measuredangle\ 13.9°$$
$$(a_{D/B})_t = (BD)\alpha_{BD} = (\tfrac{8}{12})\alpha_{BD} = 0.667\alpha_{BD}$$
$$(\mathbf{a}_{D/B})_t = 0.667\alpha_{BD} \measuredangle\ 76.1°$$

While $(\mathbf{a}_{B/D})_t$ must be perpendicular to BD, its sense is not known.

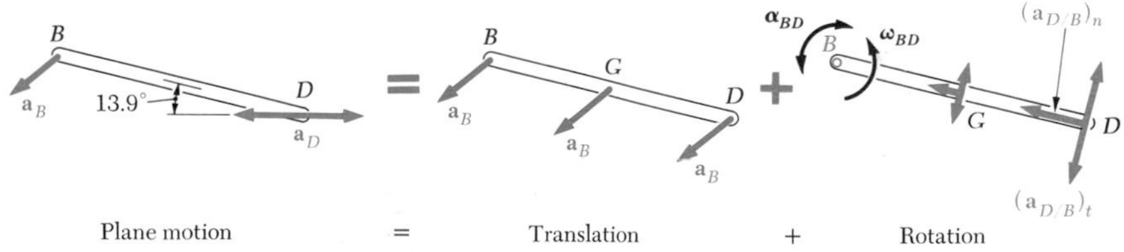

Noting that the acceleration $\mathbf{a}_D$ must be horizontal, we write

$$\mathbf{a}_D = \mathbf{a}_B + \mathbf{a}_{D/B} = \mathbf{a}_B + (\mathbf{a}_{D/B})_n + (\mathbf{a}_{D/B})_t$$
$$[a_D \leftrightarrow] = [10{,}920 \measuredangle\ 40°] + [2560 \measuredangle\ 13.9°] + [0.667\alpha_{BD} \measuredangle\ 76.1°]$$

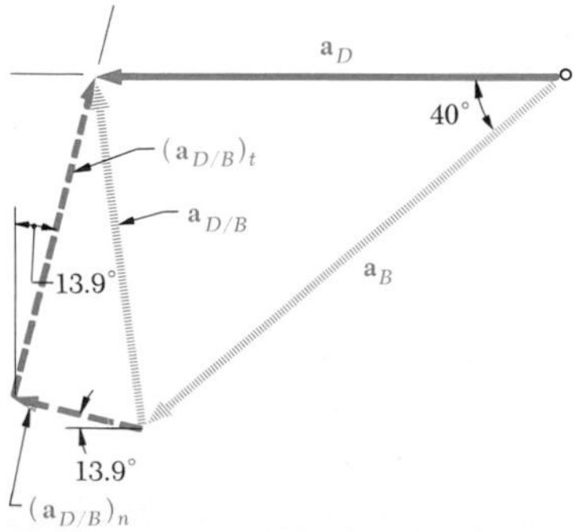

Equating x and y components, we obtain the following scalar equations:

$\xrightarrow{+}$ x components:

$$-a_D = -10{,}920\cos 40° - 2550\cos 13.9° + 0.667\alpha_{BD}\sin 13.9°$$

$+\uparrow y$ components:

$$0 = -10{,}920\sin 40° + 2550\sin 13.9° + 0.667\alpha_{BD}\cos 13.9°$$

Solving the equations simultaneously, we obtain $\alpha_{BD} = +9890$ rad/s² and $a_D = +9260$ ft/s². The positive signs indicate that the senses shown on the vector polygon are correct; we write

$$\boldsymbol{\alpha}_{BD} = 9890\text{ rad/s}^2\ \circlearrowleft \quad ◀$$
$$\mathbf{a}_D = 9260\text{ ft/s}^2 \leftarrow \quad ◀$$

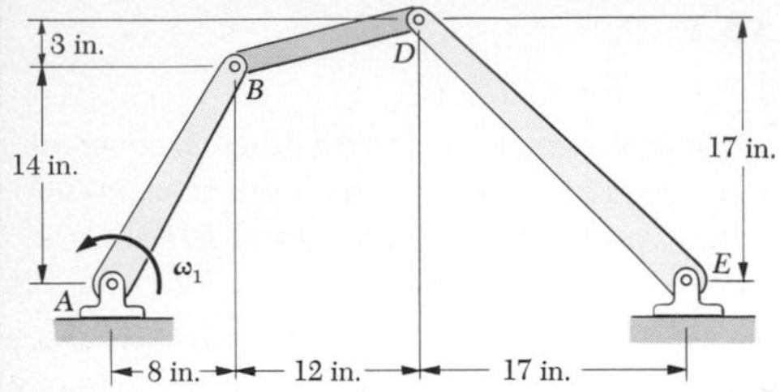

SAMPLE PROBLEM 15.8

The linkage *ABDE* moves in the vertical plane. Knowing that in the position shown crank *AB* has a constant angular velocity ω_1 of 20 rad/s counterclockwise, determine the angular velocities and angular accelerations of the connecting rod *BD* and of the crank *DE*.

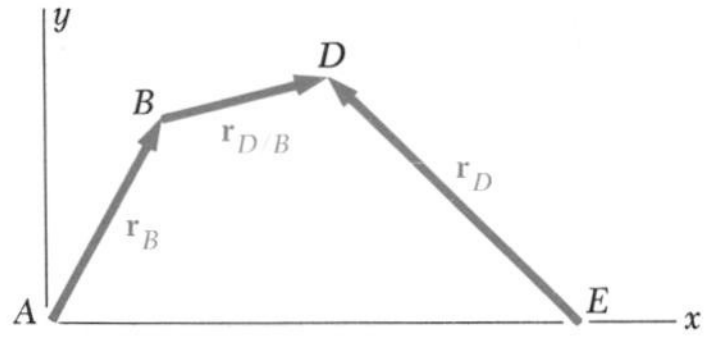

$\mathbf{r}_B = 8\mathbf{i} + 14\mathbf{j}$
$\mathbf{r}_D = -17\mathbf{i} + 17\mathbf{j}$
$\mathbf{r}_{D/B} = 12\mathbf{i} + 3\mathbf{j}$

Solution. While this problem could be solved by the method used in Sample Prob. 15.7, we shall make full use of the vector approach in the present case. The position vectors $\mathbf{r}_B$, $\mathbf{r}_D$, and $\mathbf{r}_{D/B}$ are chosen as shown in the sketch.

Velocities. Since the motion of each element of the linkage is contained in the plane of the figure, we have

$$\boldsymbol{\omega}_{AB} = \omega_{AB}\mathbf{k} = (20 \text{ rad/s})\mathbf{k} \qquad \boldsymbol{\omega}_{BD} = \omega_{BD}\mathbf{k} \qquad \boldsymbol{\omega}_{DE} = \omega_{DE}\mathbf{k}$$

where $\mathbf{k}$ is a unit vector pointing out of the paper. We now write

$$\mathbf{v}_D = \mathbf{v}_B + \mathbf{v}_{D/B}$$
$$\omega_{DE}\mathbf{k} \times \mathbf{r}_D = \omega_{AB}\mathbf{k} \times \mathbf{r}_B + \omega_{BD}\mathbf{k} \times \mathbf{r}_{D/B}$$
$$\omega_{DE}\mathbf{k} \times (-17\mathbf{i} + 17\mathbf{j}) = 20\mathbf{k} \times (8\mathbf{i} + 14\mathbf{j}) + \omega_{BD}\mathbf{k} \times (12\mathbf{i} + 3\mathbf{j})$$
$$-17\omega_{DE}\mathbf{j} - 17\omega_{DE}\mathbf{i} = 160\mathbf{j} - 280\mathbf{i} + 12\omega_{BD}\mathbf{j} - 3\omega_{BD}\mathbf{i}$$

Equating the coefficients of the unit vectors $\mathbf{i}$ and $\mathbf{j}$, we obtain the following two scalar equations:

$$-17\omega_{DE} = -280 - 3\omega_{BD}$$
$$-17\omega_{DE} = +160 + 12\omega_{BD}$$
$$\boldsymbol{\omega}_{BD} = -(29.3 \text{ rad/s})\mathbf{k} \qquad \boldsymbol{\omega}_{DE} = (11.29 \text{ rad/s})\mathbf{k} \quad ◀$$

Accelerations. Noting that at the instant considered crank *AB* has a constant angular velocity, we write

$$\boldsymbol{\alpha}_{AB} = 0 \qquad \boldsymbol{\alpha}_{BD} = \alpha_{BD}\mathbf{k} \qquad \boldsymbol{\alpha}_{DE} = \alpha_{DE}\mathbf{k}$$
$$\mathbf{a}_D = \mathbf{a}_B + \mathbf{a}_{D/B} \tag{1}$$

Each term of Eq. (1) is evaluated separately:

$$\begin{aligned}
\mathbf{a}_D &= \alpha_{DE}\mathbf{k} \times \mathbf{r}_D - \omega_{DE}^2\mathbf{r}_D \\
&= \alpha_{DE}\mathbf{k} \times (-17\mathbf{i} + 17\mathbf{j}) - (11.29)^2(-17\mathbf{i} + 17\mathbf{j}) \\
&= -17\alpha_{DE}\mathbf{j} - 17\alpha_{DE}\mathbf{i} + 2170\mathbf{i} - 2170\mathbf{j} \\
\mathbf{a}_B &= \alpha_{AB}\mathbf{k} \times \mathbf{r}_B - \omega_{AB}^2\mathbf{r}_B = 0 - (20)^2(8\mathbf{i} + 14\mathbf{j}) \\
&= -3200\mathbf{i} - 5600\mathbf{j} \\
\mathbf{a}_{D/B} &= \alpha_{BD}\mathbf{k} \times \mathbf{r}_{D/B} - \omega_{BD}^2\mathbf{r}_{D/B} \\
&= \alpha_{BD}\mathbf{k} \times (12\mathbf{i} + 3\mathbf{j}) - (29.3)^2(12\mathbf{i} + 3\mathbf{j}) \\
&= 12\alpha_{BD}\mathbf{j} - 3\alpha_{BD}\mathbf{i} - 10{,}320\mathbf{i} - 2580\mathbf{j}
\end{aligned}$$

Substituting into Eq. (1) and equating the coefficients of $\mathbf{i}$ and $\mathbf{j}$, we obtain

$$-17\alpha_{DE} + 3\alpha_{BD} = -15{,}690$$
$$-17\alpha_{DE} - 12\alpha_{BD} = -6010$$
$$\boldsymbol{\alpha}_{BD} = -(645 \text{ rad/s}^2)\mathbf{k} \qquad \boldsymbol{\alpha}_{DE} = (809 \text{ rad/s}^2)\mathbf{k} \quad ◀$$

PROBLEMS

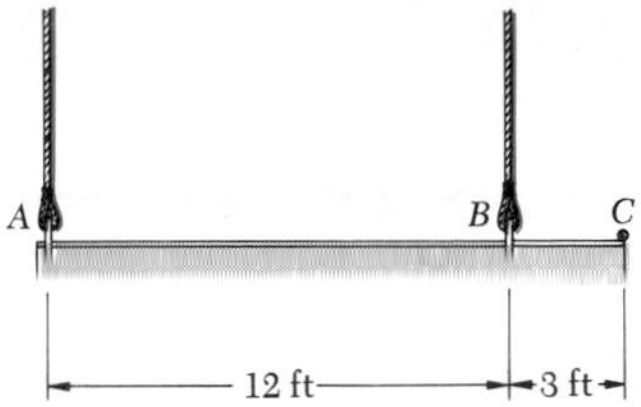

Figs. P15.76 and P15.77

15.76 A 15-ft steel beam is lowered by means of two cables unwinding at the same speed from overhead cranes. As the beam approaches the ground, the crane operators apply brakes to slow down the unwinding motion. At the instant considered the deceleration of the cable attached at A is 13 ft/s^2, while that of the cable attached at B is 7 ft/s^2. Determine (*a*) the angular acceleration of the beam, (*b*) the acceleration of point C.

15.77 The acceleration of point C is 5 ft/s^2 downward and the angular acceleration of the beam is 2 rad/s^2 clockwise. Knowing that the angular velocity of the beam is zero at the instant considered, determine the acceleration of each cable.

15.78 A 600-mm rod rests on a smooth horizontal table. A force **P** applied as shown produces the following accelerations: $\mathbf{a}_A = 0.8$ m/s^2 to the right, $\boldsymbol{\alpha} = 2$ rad/s^2 clockwise as viewed from above. Determine the acceleration (*a*) of point B, (*b*) of point G.

NOTES

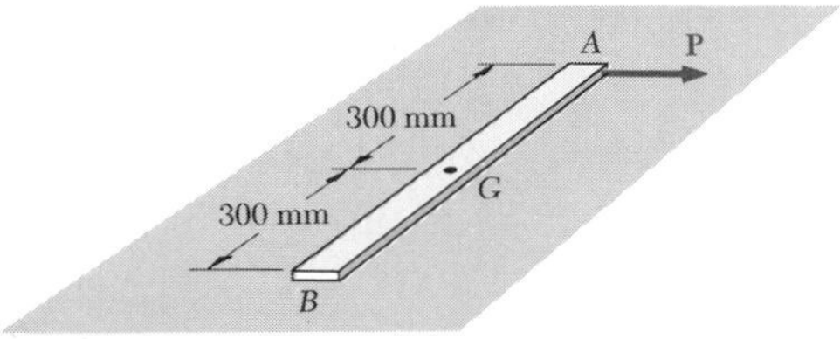

Fig. P15.78

15.79 In Prob. 15.78, determine the point of the rod which (*a*) has no acceleration, (*b*) has an acceleration of 0.350 m/s^2 to the right.

15.80 Determine the accelerations of points C and D of the 610-mm-diameter wheel of Prob. 15.25, knowing that the automobile moves at a constant speed of 50 km/h.

15.81 Determine the accelerations of points B and E of the wheel of Prob. 15.25, knowing that the automobile moves at a constant speed of 50 km/h and assuming the diameter of the wheel is reduced to 560 mm.

15.82 The flanged wheel rolls without slipping on the horizontal rail. If at a given instant the velocity and acceleration of the center of the wheel are as shown, determine the acceleration (*a*) of point B, (*b*) of point C, (*c*) of point D.

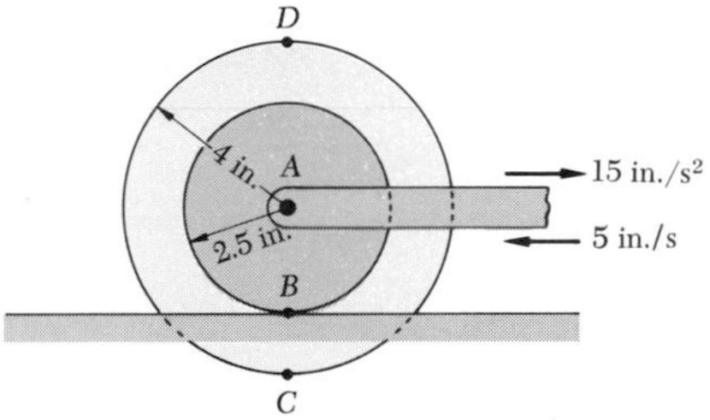

Fig. P15.82

15.83 The moving carriage is supported by two casters A and C, each of $\frac{1}{2}$-in. diameter, and by a $\frac{1}{2}$-in.-diameter ball B. If at a given instant the velocity and acceleration of the carriage are as shown, determine (*a*) the angular accelerations of the ball and of each caster, (*b*) the accelerations of the center of the ball and of each caster.

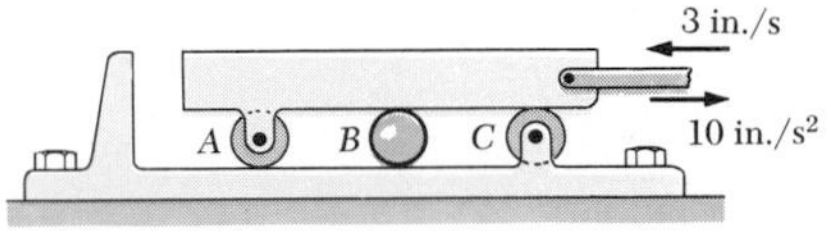

Fig. P15.83

15.84 and 15.85 At the instant shown, the disk rotates with a constant angular velocity ω_0 clockwise. Determine the angular velocities and the angular accelerations of the rods AB and BC.

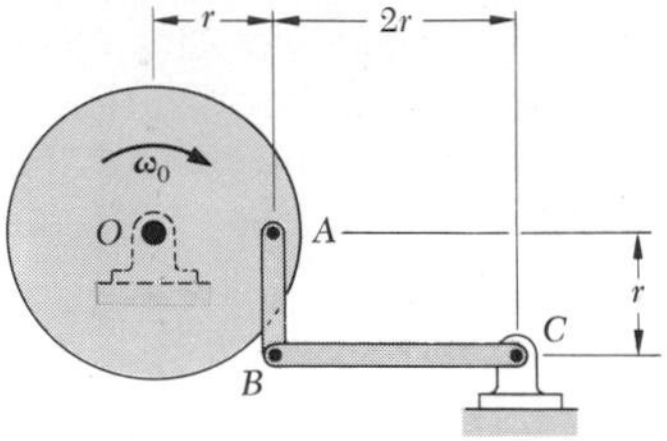

Fig. P15.84

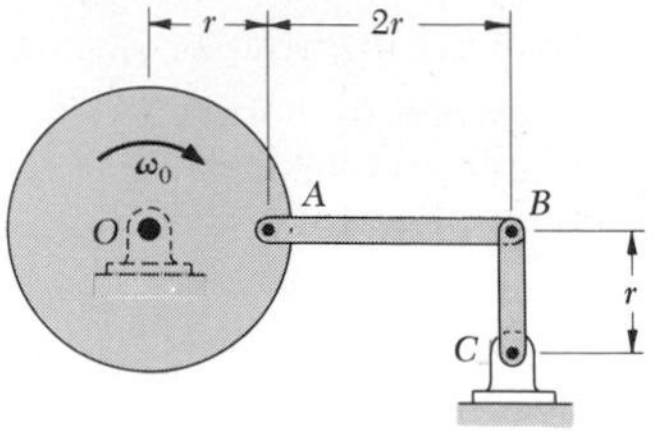

Fig. P15.85

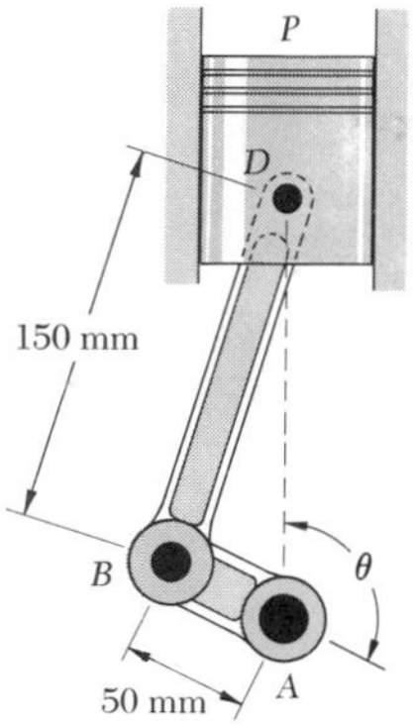

Fig. P15.86

15.86 Crank AB rotates about A with a constant angular velocity of 900 rpm clockwise. Determine the acceleration of the piston P when (*a*) $\theta = 90°$, (*b*) $\theta = 180°$.

15.87 Solve Prob. 15.86 when (*a*) $\theta = 0$, (*b*) $\theta = 270°$.

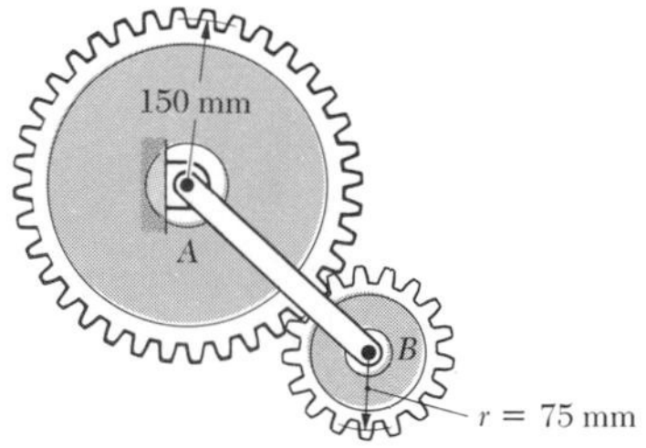

Fig. P15.88

15.88 Arm AB rotates with a constant angular velocity of 120 rpm clockwise. Knowing that gear A does not rotate, determine the acceleration of the tooth of gear B which is in contact with gear A.

15.89 and 15.90 For the linkage indicated, determine the angular acceleration (*a*) of bar BD, (*b*) of bar DE.

15.89 Linkage of Prob. 15.41.
15.90 Linkage of Prob. 15.40.

15.91 and 15.92 The end A of the rod AB moves downward with a constant velocity of 9 in./s. For the position shown, determine (*a*) the angular acceleration of the rod, (*b*) the acceleration of the midpoint G of the rod.

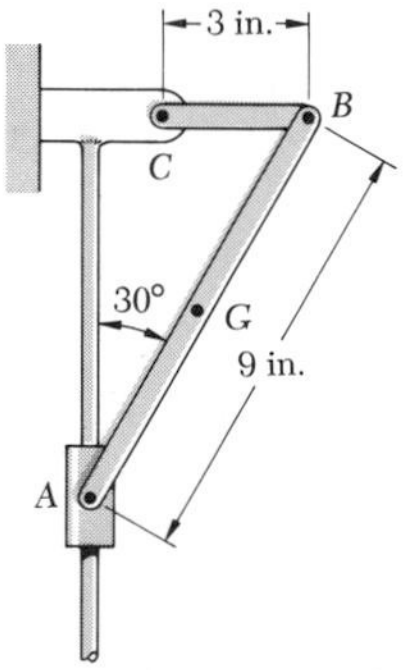

Fig. P15.91 and P15.93

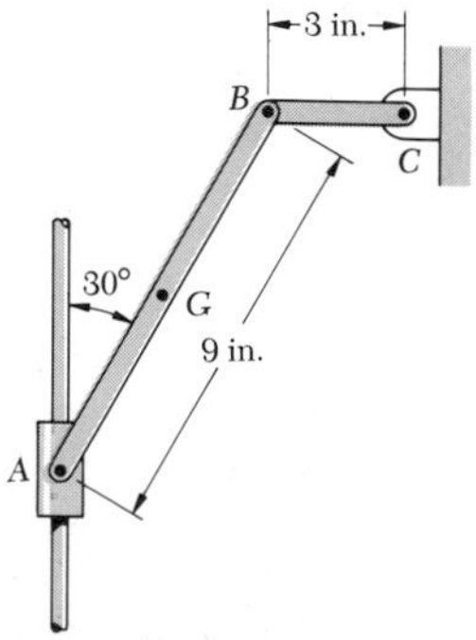

Fig. P15.92 and P15.94

15.93 and 15.94 In the position shown, end A of the rod AB has a velocity of 9 in./s and an acceleration of 6 in./s^2, both directed downward. Determine (*a*) the angular acceleration of the rod, (*b*) the acceleration of the midpoint G of the rod.

15.95 In the position shown, point A of bracket $ABCD$ has a velocity of magnitude $v_A = 250$ mm/s with $dv_A/dt = 0$. Determine (a) the angular acceleration of the bracket, (b) the acceleration of point C.

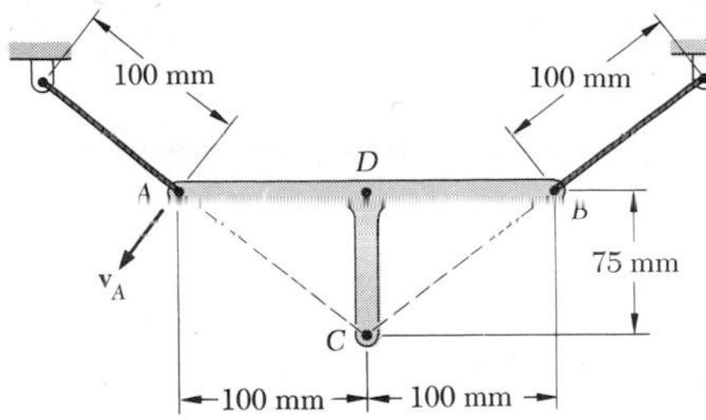

Fig. P15.95

15.96 In Prob. 15.95, determine the acceleration of point D.

15.97 Show that the acceleration of the instantaneous center of rotation of the slab of Prob. 15.49 is zero if, and only if,

$$\mathbf{a}_A = \frac{\alpha}{\omega}\mathbf{v}_A + \boldsymbol{\omega} \times \mathbf{v}_A$$

where $\boldsymbol{\alpha} = \alpha\mathbf{k}$ is the angular acceleration of the slab.

***15.98** Rod AB slides with its ends in contact with the floor and the inclined plane. Using the method of Sec. 15.9, derive an expression for the angular velocity of the rod in terms of v_B, θ, l, and β.

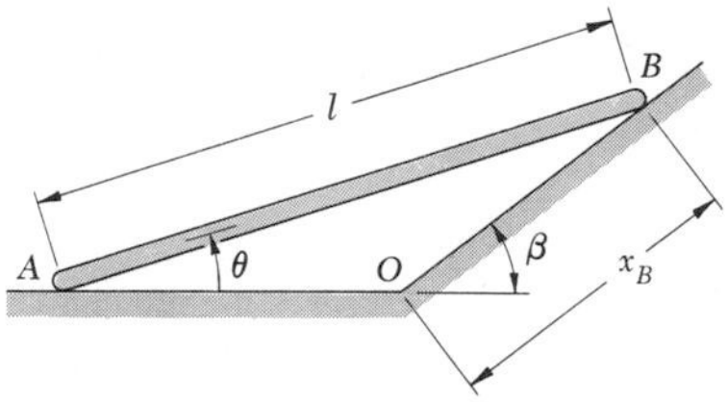

Fig. P15.98 and P15.99

***15.99** Derive an expression for the angular acceleration of the rod AB in terms of v_B, θ, l, and β, knowing that the acceleration of point B is zero.

***15.100** The drive disk of the Scotch crosshead mechanism shown has an angular velocity $\boldsymbol{\omega}$ and an angular acceleration $\boldsymbol{\alpha}$, both directed clockwise. Using the method of Sec. 15.9, derive an expression (a) for the velocity of point B, (b) for the acceleration of point B.

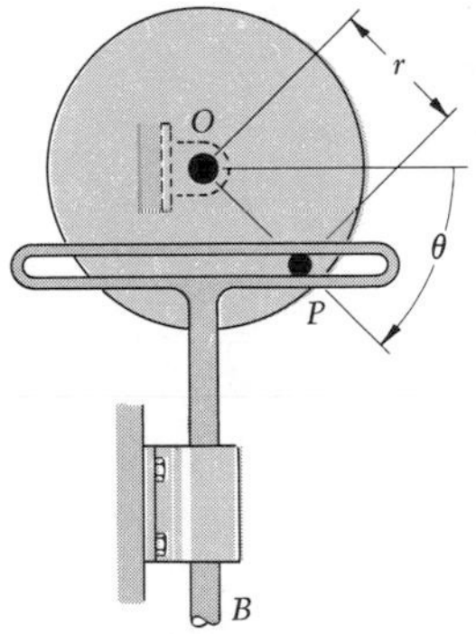

Fig. P15.100

***15.101** A disk of radius r rolls to the right with a constant velocity $\mathbf{v}$. Denoting by P the point of the rim in contact with the ground at $t = 0$, derive expressions for the horizontal and vertical components of the velocity of P at any time t. (The curve described by point P is called a *cycloid*.)

***15.102** Knowing that rod AB rotates with an angular velocity ω and with an angular acceleration $\boldsymbol{\alpha}$, both counterclockwise, derive expressions for the velocity and acceleration of collar D.

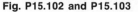

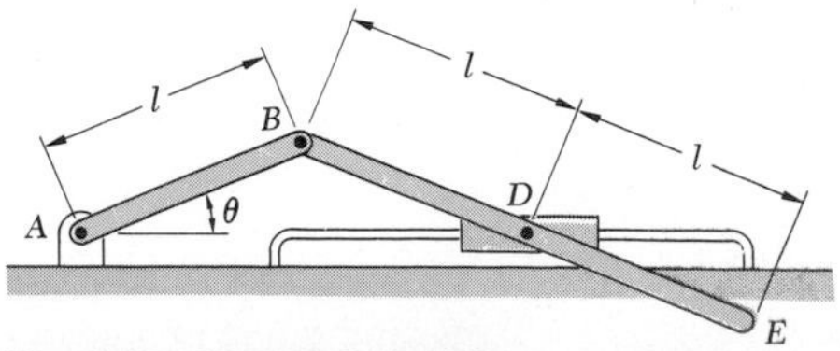

Fig. P15.102 and P15.103

***15.103** Knowing that rod AB rotates with an angular velocity ω and an angular acceleration $\boldsymbol{\alpha}$, both counterclockwise, derive expressions for the components of the velocity and acceleration of point E.

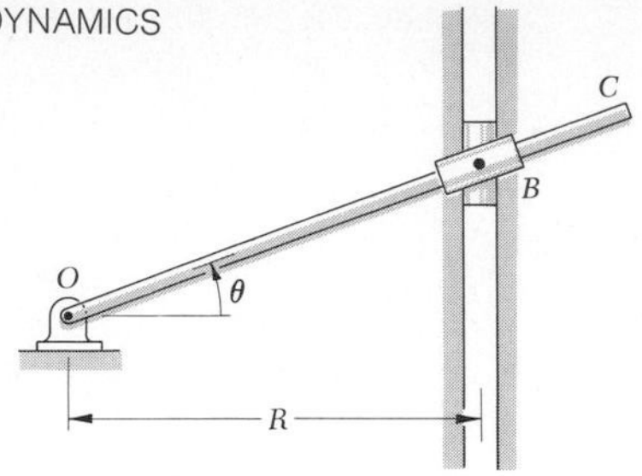

Fig. P15.104 and P15.105

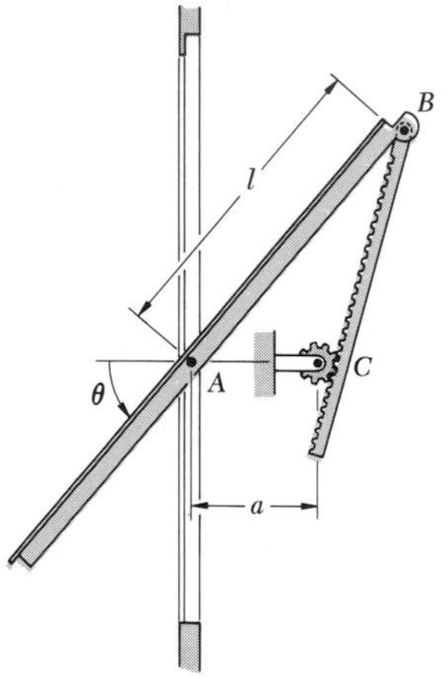

Fig. P15.106

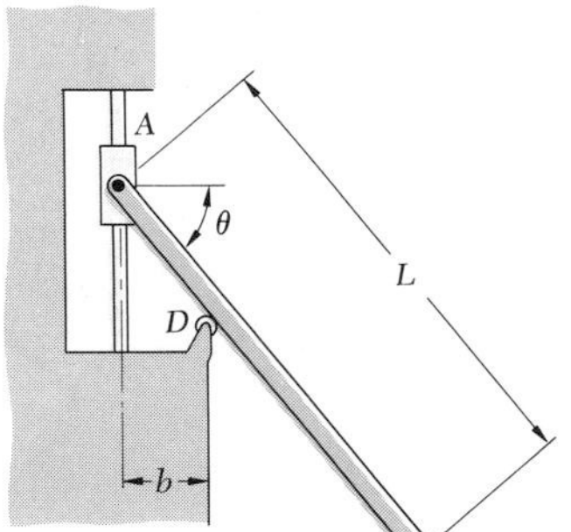

Fig. P15.108

***15.104** Collar B slides along rod OC and is attached to a sliding block which moves in a vertical slot. Knowing that rod OC rotates with an angular velocity $\boldsymbol{\omega}$ and with an angular acceleration $\boldsymbol{\alpha}$, both counterclockwise, derive expressions for the velocity and acceleration of collar B.

***15.105** Collar B slides along rod OC and is attached to a sliding block which moves upward with a constant velocity $\mathbf{v}$ in a vertical slot. Using the method of Sec. 15.9, derive an expression (*a*) for the angular velocity of rod OC, (*b*) for the angular acceleration of rod OC.

***15.106** The position of a factory window is controlled by the rack and pinion shown. Knowing that the pinion C has a radius r and rotates counterclockwise at a constant rate ω, derive an expression for the angular velocity of the window.

***15.107** The crank AB of Prob. 15.36 rotates with a constant clockwise angular velocity $\boldsymbol{\omega}$, and $\theta = 0$ at $t = 0$. Using the method of Sec. 15.9, derive an expression for the velocity of the piston P in terms of the time t.

***15.108** Collar A slides upward with a constant velocity $\mathbf{v}_A$. Using the method of Sec. 15.9, derive an expression for (*a*) the angular velocity of rod AB, (*b*) the components of the velocity of point B.

***15.109** In Prob. 15.108, derive an expression for the angular acceleration of rod AB.

15.10. Rate of Change of a Vector with Respect to a Rotating Frame. We saw in Sec. 11.10 that the rate of change of a vector is the same with respect to a fixed frame and with respect to a frame in translation. In this section, we shall compare the rates of change of a vector $\mathbf{Q}$ with respect to a fixed frame and with respect to a rotating frame of reference.† We shall also learn to determine the rate of change of $\mathbf{Q}$ with respect to one frame of reference when $\mathbf{Q}$ is defined by its components in another frame.

Consider two frames of reference centered at O, a fixed frame $OXYZ$ and a frame $Oxyz$ which rotates about the fixed axis OA; let $\boldsymbol{\Omega}$ denote the angular velocity of the frame $Oxyz$ at a given

† It is recalled that the selection of a fixed frame of reference is arbitrary. Any frame may be designated as "fixed"; all others will then be considered as moving.

instant (Fig. 15.26). Consider now a vector function $\mathbf{Q}(t)$ represented by the vector $\mathbf{Q}$ attached at O; as the time t varies, both the direction and the magnitude of $\mathbf{Q}$ change. Since the variation of $\mathbf{Q}$ is viewed differently by an observer using $OXYZ$ as a frame of reference and by an observer using $Oxyz$, we should expect the rate of change of $\mathbf{Q}$ to depend upon the frame of reference which has been selected. Therefore, we shall denote by $(\dot{\mathbf{Q}})_{OXYZ}$ the rate of change of $\mathbf{Q}$ with respect to the fixed frame $OXYZ$, and by $(\dot{\mathbf{Q}})_{Oxyz}$ its rate of change with respect to the rotating frame $Oxyz$. We propose to determine the relationship existing between these two rates of change.

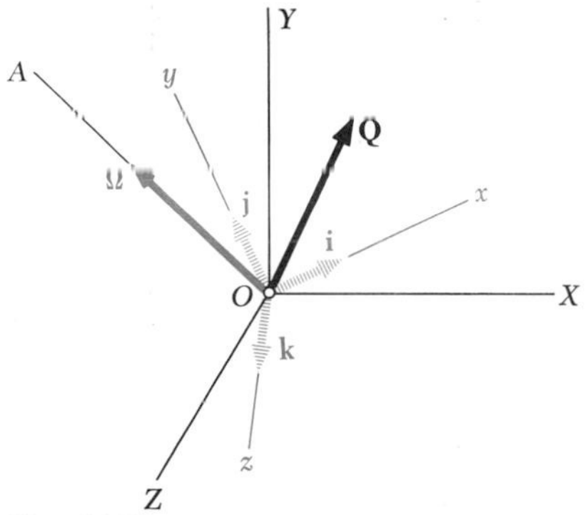

Fig. 15.26

Let us first resolve the vector $\mathbf{Q}$ into components along the x, y, and z axes of the rotating frame. Denoting by $\mathbf{i}$, $\mathbf{j}$, and $\mathbf{k}$ the corresponding unit vectors, we write

$$\mathbf{Q} = Q_x\mathbf{i} + Q_y\mathbf{j} + Q_z\mathbf{k} \tag{15.27}$$

Differentiating (15.27) with respect to t and considering the unit vectors $\mathbf{i}$, $\mathbf{j}$, $\mathbf{k}$ as fixed, we obtain the rate of change of $\mathbf{Q}$ *with respect to the rotating frame Oxyz*:

$$(\dot{\mathbf{Q}})_{Oxyz} = \dot{Q}_x\mathbf{i} + \dot{Q}_y\mathbf{j} + \dot{Q}_z\mathbf{k} \tag{15.28}$$

To obtain the rate of change of $\mathbf{Q}$ *with respect to the fixed frame OXYZ*, we must consider the unit vectors $\mathbf{i}$, $\mathbf{j}$, $\mathbf{k}$ as variable when differentiating (15.27). We therefore write

$$(\dot{\mathbf{Q}})_{OXYZ} = \dot{Q}_x\mathbf{i} + \dot{Q}_y\mathbf{j} + \dot{Q}_z\mathbf{k} + Q_x\frac{d\mathbf{i}}{dt} + Q_y\frac{d\mathbf{j}}{dt} + Q_z\frac{d\mathbf{k}}{dt} \tag{15.29}$$

Recalling (15.28), we observe that the sum of the first three terms in the right-hand member of (15.29) represents the rate of change $(\dot{\mathbf{Q}})_{Oxyz}$. We note, on the other hand, that the rate of change $(\dot{\mathbf{Q}})_{OXYZ}$ would reduce to the last three terms in (15.29) if the vector $\mathbf{Q}$ were fixed within the frame $Oxyz$, since $(\dot{\mathbf{Q}})_{Oxyz}$ would then be zero. But, in that case, $(\dot{\mathbf{Q}})_{OXYZ}$ would represent the velocity of a particle located at the tip of $\mathbf{Q}$ and belonging to a body rigidly attached to the frame $Oxyz$. Thus, the last three terms in (15.29) represent the velocity of that particle; since the frame $Oxyz$ has an angular velocity $\boldsymbol{\Omega}$ with respect to $OXYZ$ at the instant considered, we write, by (15.5),

$$Q_x\frac{d\mathbf{i}}{dt} + Q_y\frac{d\mathbf{j}}{dt} + Q_z\frac{d\mathbf{k}}{dt} = \boldsymbol{\Omega} \times \mathbf{Q} \tag{15.30}$$

Substituting from (15.28) and (15.30) into (15.29), we obtain the fundamental relation

$$(\dot{\mathbf{Q}})_{OXYZ} = (\dot{\mathbf{Q}})_{Oxyz} + \boldsymbol{\Omega} \times \mathbf{Q} \tag{15.31}$$

We conclude that the rate of change of the vector **Q** with respect to the fixed frame *OXYZ* is made of two parts: The first part represents the rate of change of **Q** with respect to the rotating frame *Oxyz*; the second part, $\boldsymbol{\Omega} \times \mathbf{Q}$, is induced by the rotation of the frame *Oxyz*.

The use of relation (15.31) simplifies the determination of the rate of change of a vector **Q** with respect to a fixed frame of reference *OXYZ* when the vector **Q** is defined by its components along the axes of a rotating frame *Oxyz*, since this relation does not require the separate computation of the derivatives of the unit vectors defining the orientation of the rotating frame.

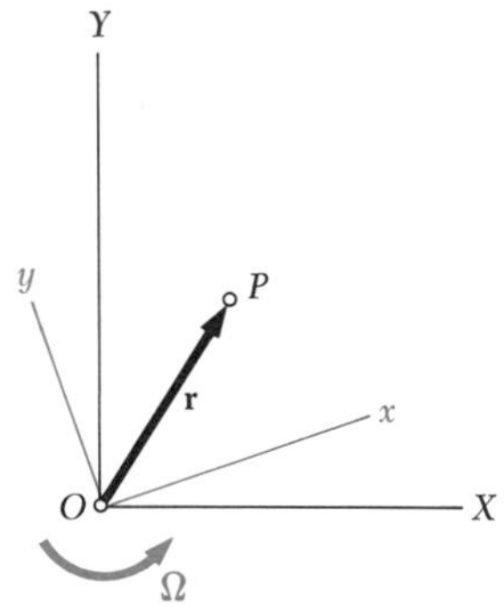

Fig. 15.27

15.11. Plane Motion of a Particle Relative to a Rotating Frame. Coriolis Acceleration. Consider two frames of reference, both centered at *O* and both in the plane of the figure, a fixed frame *OXY*, and a rotating frame *Oxy* (Fig. 15.27). Let *P* be a particle moving in the plane of the figure. While the position vector **r** of *P* is the same in both frames, its rate of change depends upon the frame of reference which has been selected.

The absolute velocity $\mathbf{v}_P$ of the particle is defined as the velocity observed from the fixed frame *OXY* and is equal to the rate of change $(\dot{\mathbf{r}})_{OXY}$ of **r** with respect to that frame. We may, however, express $\mathbf{v}_P$ in terms of the rate of change $(\dot{\mathbf{r}})_{Oxy}$ observed from the rotating frame if we make use of Eq. (15.31). Denoting by $\boldsymbol{\Omega}$ the angular velocity of the frame *Oxy* with respect to *OXY* at the instant considered, we write

$$\mathbf{v}_P = (\dot{\mathbf{r}})_{OXY} = \boldsymbol{\Omega} \times \mathbf{r} + (\dot{\mathbf{r}})_{Oxy} \tag{15.32}$$

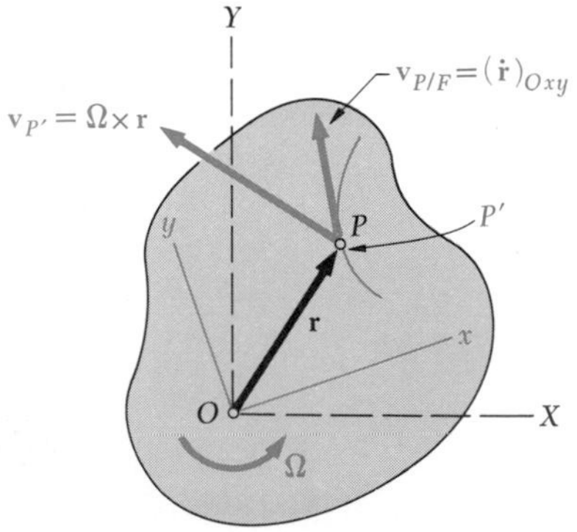

Fig. 15.28

But $(\dot{\mathbf{r}})_{Oxy}$ defines the velocity $\mathbf{v}_{P/F}$ of the particle *P* relative to the frame *Oxy*. If we imagine that a rigid slab has been attached to the rotating frame, $\mathbf{v}_{P/F}$ will represent the velocity of *P* along the path that it describes on that slab (Fig. 15.28). On the other hand, the term $\boldsymbol{\Omega} \times \mathbf{r}$ in (15.32) will represent the velocity $\mathbf{v}_{P'}$ of the point *P'* of the slab—or rotating frame—which coincides with *P* at the instant considered. Thus, we have

$$\mathbf{v}_P = \mathbf{v}_{P'} + \mathbf{v}_{P/F} \tag{15.33}$$

where $\mathbf{v}_P$ = absolute velocity of particle *P*
$\mathbf{v}_{P'}$ = velocity of point *P'* of moving frame coinciding with *P*
$\mathbf{v}_{P/F}$ = velocity of *P* relative to moving frame

The absolute acceleration $\mathbf{a}_P$ of the particle is defined as the rate of change of $\mathbf{v}_P$ with respect to the fixed frame OXY. Computing the rates of change with respect to OXY of the terms in (15.32), we write

$$\mathbf{a}_P = \dot{\mathbf{v}}_P = \dot{\boldsymbol{\Omega}} \times \mathbf{r} + \boldsymbol{\Omega} \times \dot{\mathbf{r}} + \frac{d}{dt}[(\dot{\mathbf{r}})_{Oxy}] \qquad (15.34)$$

where all derivatives are defined with respect to OXY, except where indicated otherwise. Referring to Eq. (15.31), we note that the last term in (15.34) may be expressed as

$$\frac{d}{dt}[(\dot{\mathbf{r}})_{Oxy}] = (\ddot{\mathbf{r}})_{Oxy} + \boldsymbol{\Omega} \times (\dot{\mathbf{r}})_{Oxy}$$

On the other hand, $\dot{\mathbf{r}}$ represents the velocity $\mathbf{v}_P$ and may be replaced by the right-hand member of Eq. (15.32). After completing these two substitutions into (15.34), we write

$$\mathbf{a}_P = \dot{\boldsymbol{\Omega}} \times \mathbf{r} + \boldsymbol{\Omega} \times (\boldsymbol{\Omega} \times \mathbf{r}) + 2\boldsymbol{\Omega} \times (\dot{\mathbf{r}})_{Oxy} + (\ddot{\mathbf{r}})_{Oxy} \qquad (15.35)$$

Referring to the expression (15.8) obtained in Sec. 15.3 for the acceleration of a particle in a rigid body rotating about a fixed axis, we note that the sum of the first two terms represents the acceleration $\mathbf{a}_{P'}$ of the point P' of the rotating frame which coincides with P at the instant considered. On the other hand, the last term defines the acceleration $\mathbf{a}_{P/F}$ of P relative to the rotating frame. If it were not for the third term, which has not been accounted for, a relation similar to (15.33) could be written for the accelerations, and $\mathbf{a}_P$ could be expressed as the sum of $\mathbf{a}_{P'}$ and $\mathbf{a}_{P/F}$. However, it is clear that *such a relation would be incorrect* and that we must include the additional term. This term, which we shall denote by $\mathbf{a}_c$, is called the *complementary acceleration*, or *Coriolis acceleration*, after the French mathematician De Coriolis (1792–1843). We write

$$\mathbf{a}_P = \mathbf{a}_{P'} + \mathbf{a}_{P/F} + \mathbf{a}_c \qquad (15.36)$$

where $\mathbf{a}_P$ = absolute acceleration of particle P

$\mathbf{a}_{P'}$ = acceleration of point P' of moving frame coinciding with P

$\mathbf{a}_{P/F}$ = acceleration of P relative to moving frame

$\mathbf{a}_c = 2\boldsymbol{\Omega} \times (\dot{\mathbf{r}})_{Oxy} = 2\boldsymbol{\Omega} \times \mathbf{v}_{P/F}$
= complementary, or Coriolis, acceleration

We note that, since point P' moves in a circle about the origin O, its acceleration $\mathbf{a}_{P'}$ has, in general, two components: a component $(\mathbf{a}_{P'})_t$ tangent to the circle, and a component $(\mathbf{a}_{P'})_n$ directed toward O. Similarly, the acceleration $\mathbf{a}_{P/F}$ generally has two components: a component $(\mathbf{a}_{P/F})_t$ tangent to the path that P describes on the rotating slab, and a component $(\mathbf{a}_{P/F})_n$ directed toward the center of curvature of that path. We further note that, since the vector $\mathbf{\Omega}$ is perpendicular to the plane of motion, and thus to $\mathbf{v}_{P/F}$, the magnitude of the Coriolis acceleration $\mathbf{a}_c = 2\mathbf{\Omega} \times \mathbf{v}_{P/F}$ is equal to $2\Omega v_{P/F}$, and its direction may be obtained by rotating the vector $\mathbf{v}_{P/F}$ through 90° in the sense of rotation of the moving frame (Fig. 15.29). The Coriolis acceleration reduces to zero when either $\mathbf{\Omega}$ or $\mathbf{v}_{P/F}$ is zero.

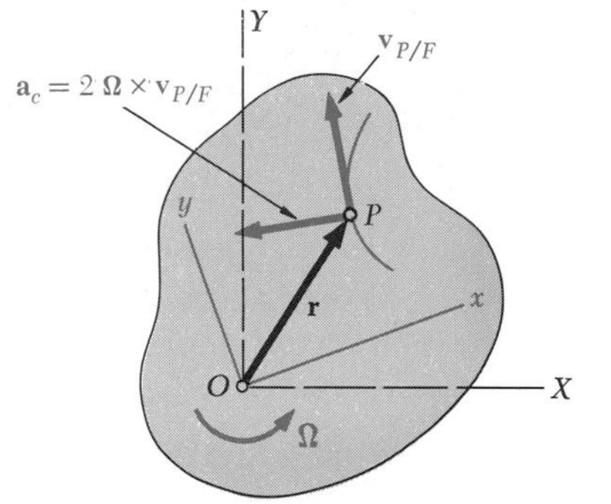

Fig. 15.29

The following example will help in understanding the physical meaning of the Coriolis acceleration. Consider a collar P which is made to slide at a constant relative speed u along a rod OB rotating at a constant angular velocity $\boldsymbol{\omega}$ about O (Fig. 15.30*a*). According to formula (15.36), the absolute acceleration of P may be obtained by adding vectorially the acceleration $\mathbf{a}_A$ of the point A of the rod coinciding with P, the relative acceleration $\mathbf{a}_{P/OB}$ of P with respect to the rod, and the Coriolis acceleration $\mathbf{a}_c$. Since the angular velocity ω of the rod is constant, $\mathbf{a}_A$ reduces to its normal component $(\mathbf{a}_A)_n$ of magnitude $r\omega^2$; and since u is constant, the relative acceleration $\mathbf{a}_{P/OB}$ is zero. According to the definition given above, the Coriolis acceleration is a vector perpendicular to OB, of magnitude $2\omega u$, and directed as shown in the figure. The acceleration of the collar P consists, therefore, of the two vectors shown in Fig. 15.30a. Note that the result obtained may be checked by applying the relation (11.44).

To understand better the significance of the Coriolis acceleration, we shall consider the absolute velocity of P at time t and at time $t + \Delta t$ (Fig. 15.30b). At time t, the velocity may be resolved into its components $\mathbf{u}$ and $\mathbf{v}_A$, and at time $t + \Delta t$ into its components $\mathbf{u}'$ and $\mathbf{v}_{A'}$. Drawing these components from the same origin (Fig. 15.30c), we note that the change in velocity during the time Δt may be represented by the sum of three vectors $\overrightarrow{RR'}$, $\overrightarrow{TT''}$, and $\overrightarrow{T''T'}$. The vector $\overrightarrow{TT''}$ measures the change in direction of the velocity $\mathbf{v}_A$, and the quotient $\overrightarrow{TT''}/\Delta t$ represents the acceleration $\mathbf{a}_A$ when Δt approaches zero. We check that the direction of $\overrightarrow{TT''}$ is that of $\mathbf{a}_A$ when Δt approaches zero and that

$$\lim_{\Delta t \to 0} \frac{TT''}{\Delta t} = \lim_{\Delta t \to 0} v_A \frac{\Delta\theta}{\Delta t} = r\omega\omega = r\omega^2 = a_A$$

The vector $\overrightarrow{RR'}$ measures the change in direction of $\mathbf{u}$ due to the rotation of the rod; the vector $\overrightarrow{T''T'}$ measures the change in magnitude of $\mathbf{v}_A$ due to the motion of P on the rod. The vectors $\overrightarrow{RR'}$ and $\overrightarrow{T''T'}$ result from the *combined effect* of the relative motion of P and of the rotation of the rod; they would vanish if *either* of these two motions stopped. We may easily verify that the sum of these two vectors defines the Coriolis acceleration. Their direction is that of $\mathbf{a}_c$ when Δt approaches zero and, since $RR' = u\,\Delta\theta$ and $T''T' = v_{A'} - v_A = (r + \Delta r)\omega - r\omega = \omega\,\Delta r$, we check that

$$\lim_{\Delta t \to 0}\left(\frac{RR'}{\Delta t} + \frac{T''T'}{\Delta t}\right) = \lim_{\Delta t \to 0}\left(u\frac{\Delta\theta}{\Delta t} + \omega\frac{\Delta r}{\Delta t}\right)$$
$$= u\omega + \omega u = 2\omega u = a_c$$

Formulas (15.33) and (15.36) may be used to analyze the motion of mechanisms which contain parts sliding on each other. They make it possible, for example, to relate the absolute and relative motions of sliding pins and collars (see Sample Probs. 15.9 and 15.10). The concept of Coriolis acceleration is also very useful in the study of long-range projectiles and of other bodies whose motions are appreciably affected by the rotation of the earth. As was pointed out in Sec. 12.1, a system of axes attached to the earth does not truly constitute a newtonian frame of reference; such a system of axes should actually be considered as rotating. The formulas derived in this section will therefore facilitate the study of the motion of bodies with respect to axes attached to the earth.

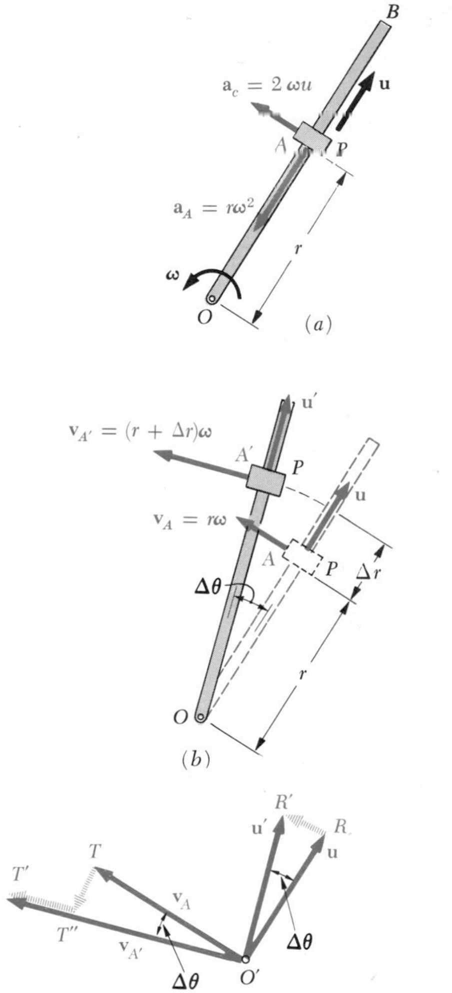

Fig. 15.30

SAMPLE PROBLEM 15.9

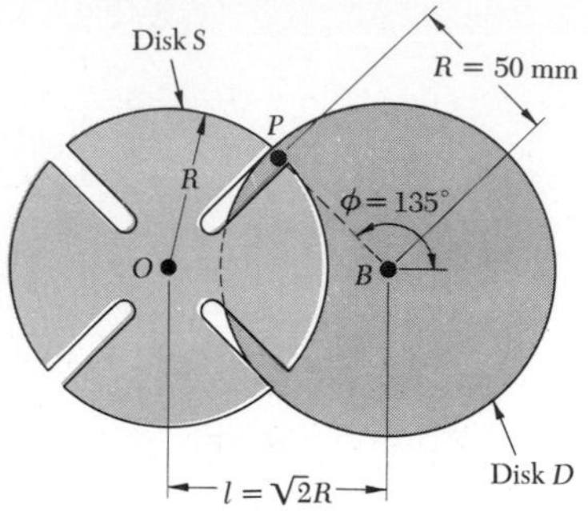

The Geneva mechanism shown is used in many counting instruments and in other applications where an intermittent rotary motion is required. Disk D rotates with a constant counterclockwise angular velocity ω_D of 10 rad/s. A pin P is attached to disk D and slides along one of several slots cut in disk S. It is desirable that the angular velocity of disk S be zero as the pin enters and leaves each slot; in the case of four slots, this will occur if the distance between the centers of the disks is $l = \sqrt{2}\,R$.

At the instant when $\phi = 150°$, determine (*a*) the angular velocity of disk S, (*b*) the velocity of pin P relative to disk S.

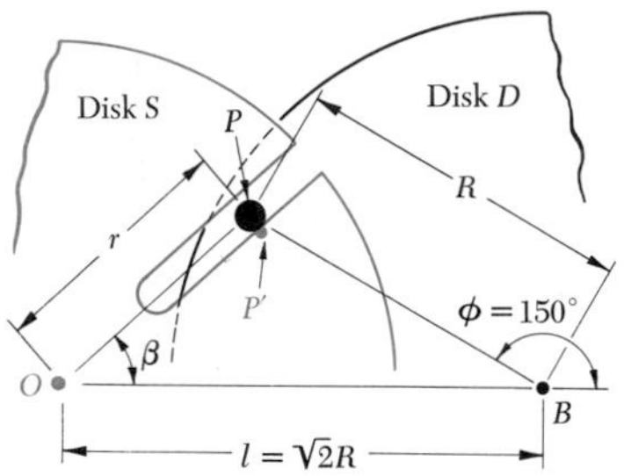

Solution. We solve triangle OPB, which corresponds to the position $\phi = 150°$. Using the law of cosines, we write

$$r^2 = R^2 + l^2 - 2Rl\cos 30° = 0.551R^2 \qquad r = 0.742R = 37.1\text{ mm}$$

From the law of sines

$$\frac{\sin\beta}{R} = \frac{\sin 30°}{r} \qquad \sin\beta = \frac{\sin 30°}{0.742} \qquad \beta = 42.4°$$

Since pin P is attached to disk D, and since disk D rotates about point B, the magnitude of the absolute velocity of P is

$$v_P = R\omega_D = (50\text{ mm})(10\text{ rad/s}) = 500\text{ mm/s}$$
$$\mathbf{v}_P = 500\text{ mm/s} \measuredangle 60°$$

We consider now the motion of pin P along the slot in disk S. Denoting by P' the point of disk S which coincides with P at the instant considered, we write

$$\mathbf{v}_P = \mathbf{v}_{P'} + \mathbf{v}_{P/S}$$

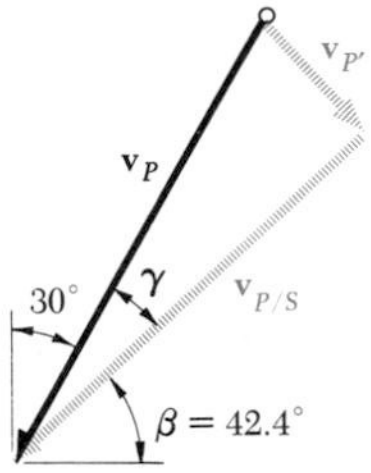

Noting that $\mathbf{v}_{P'}$ is perpendicular to the radius OP and that $\mathbf{v}_{P/S}$ is directed along the slot, we draw the velocity triangle corresponding to the above equation. From the triangle, we compute

$$\gamma = 90° - 42.4° - 30° = 17.6°$$
$$v_{P'} = v_P\sin\gamma = (500\text{ mm/s})\sin 17.6°$$
$$\mathbf{v}_{P'} = 151.2\text{ mm/s} \measuredangle 42.4°$$
$$v_{P/S} = v_P\cos\gamma = (500\text{ mm/s})\cos 17.6°$$
$$\mathbf{v}_{P/S} = 477\text{ mm/s} \measuredangle 42.4° \quad ◀$$

Since $\mathbf{v}_{P'}$ is perpendicular to the radius OP, we write

$$v_{P'} = r\omega_S \qquad 151.2\text{ mm/s} = (37.1\text{ mm})\omega_S$$
$$\omega_S = 4.08\text{ rad/s} \circlearrowright \quad ◀$$

SAMPLE PROBLEM 15.10

In the Geneva mechanism of Sample Prob. 15.9, disk D rotates with a constant counterclockwise angular velocity $\boldsymbol{\omega}_D$ of 10 rad/s. At the instant when $\phi = 150°$, determine the angular acceleration of disk S.

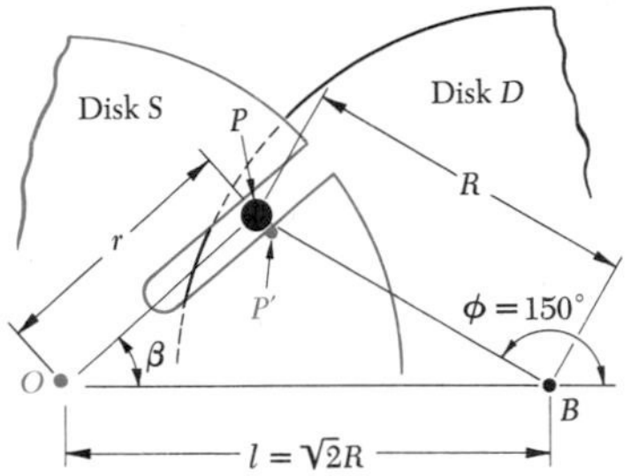

Solution. Referring to Sample Prob. 15.9, we obtain the angular velocity of disk S and the velocity of the pin relative to disk S.

$$\omega_S = 4.08 \text{ rad/s} \curvearrowright$$

$$\beta = 42.4° \qquad \mathbf{v}_{P/S} = 477 \text{ mm/s} \nearrow 42.4°$$

Since pin P moves with respect to the rotating disk S, we write

$$\mathbf{a}_P = \mathbf{a}_{P'} + \mathbf{a}_{P/S} + \mathbf{a}_c \tag{1}$$

Each term of this vector equation is investigated separately.

Absolute Acceleration $\mathbf{a}_P$. Since disk D rotates with constant ω, the absolute acceleration $\mathbf{a}_P$ is directed toward B.

$$a_P = R\omega_D^2 = (50 \text{ mm})(10 \text{ rad/s})^2 = 5000 \text{ mm/s}^2$$

$$\mathbf{a}_P = 5000 \text{ mm/s}^2 \searrow 30°$$

Acceleration $\mathbf{a}_{P'}$ *of the Coinciding Point* P'. The acceleration $\mathbf{a}_{P'}$ of the point P' of disk S which coincides with P at the instant considered is resolved into normal and tangential components. (We recall from Sample Prob. 15.9 that $r = 37.1$ mm.)

$$(a_{P'})_n = r\omega_S^2 = (37.1 \text{ mm})(4.08 \text{ rad/s})^2 = 618 \text{ mm/s}^2$$

$$(\mathbf{a}_{P'})_n = 618 \text{ mm/s}^2 \nearrow 42.4°$$

$$(a_{P'})_t = r\alpha_S = 37.1\alpha_S \qquad (\mathbf{a}_{P'})_t = 37.1\alpha_S \searrow 42.4°$$

Relative Acceleration $\mathbf{a}_{P/S}$. Since the pin P moves in a straight slot cut in disk S, the relative acceleration $\mathbf{a}_{P/S}$ must be parallel to the slot; i.e., its direction must be $\swarrow$ 42.4°.

Coriolis Acceleration $\mathbf{a}_c$. Rotating the relative velocity $\mathbf{v}_{P/S}$ through 90° in the sense of $\boldsymbol{\omega}_S$, we obtain the direction of the Coriolis component of the acceleration.

$$a_c = 2\omega_S v_{P/S} = 2(4.08 \text{ rad/s})(477 \text{ mm/s}) = 3890 \text{ mm/s}^2$$

$$\mathbf{a}_c = 3890 \text{ mm/s}^2 \nwarrow 42.4°$$

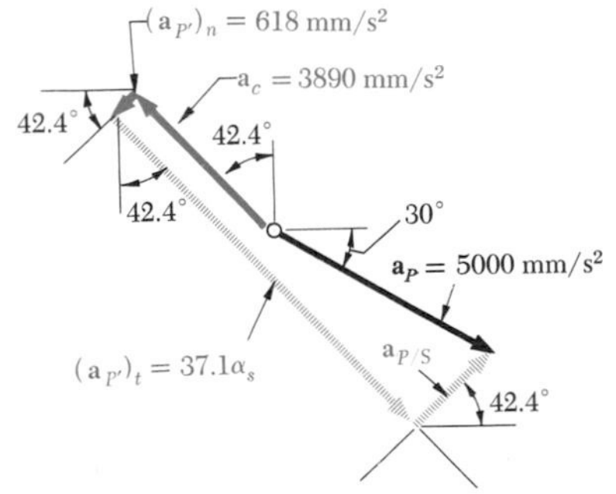

We rewrite Eq. (1) and substitute the accelerations found above.

$$\mathbf{a}_P = (\mathbf{a}_{P'})_n + (\mathbf{a}_{P'})_t + \mathbf{a}_{P/S} + \mathbf{a}_c$$

$$[5000 \searrow 30°] = [618 \nearrow 42.4°] + [37.1\alpha_S \searrow 42.4°]$$
$$+ [a_{P/S} \swarrow 42.4°] + [3890 \nwarrow 42.4°]$$

Equating components in a direction perpendicular to the slot:

$$5000 \cos 17.6° = 37.1\alpha_S - 3890$$

$$\boldsymbol{\alpha}_S = 233 \text{ rad/s}^2 \curvearrowright \blacktriangleleft$$

PROBLEMS

15.110 and 15.111 Two rotating rods are connected by a slider block *P*. The rod attached at *B* rotates with a constant clockwise angular velocity $\boldsymbol{\omega}_B$. For the given data, determine for the position shown (*a*) the angular velocity of the rod attached at *A*, (*b*) the relative velocity of the slider block *P* with respect to the rod on which it slides.

15.110 $b = 10$ in., $\omega_B = 5$ rad/s.
15.111 $b = 200$ mm, $\omega_B = 9$ rad/s.

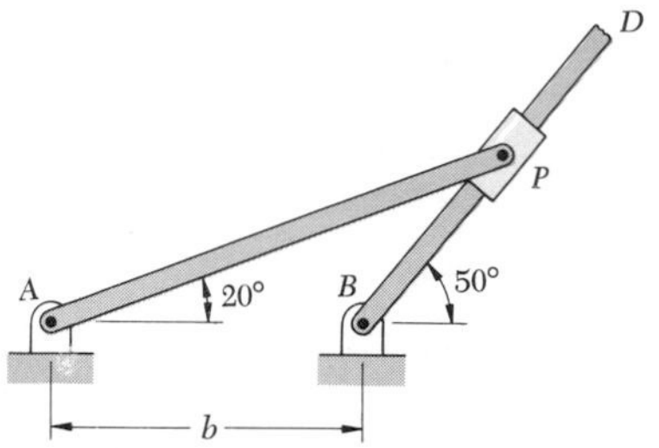

Fig. P15.110 and P15.112

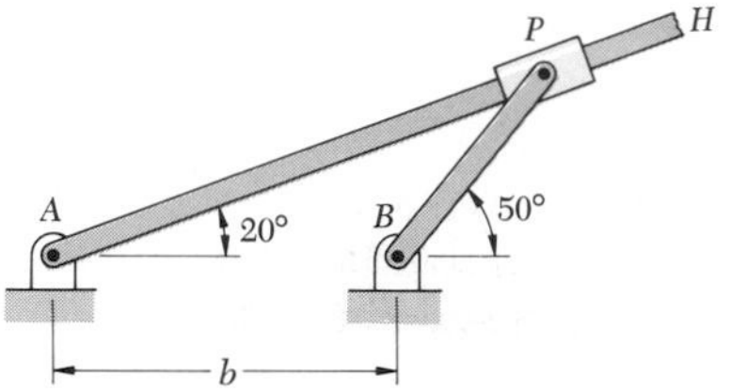

Fig. P15.111 and P15.113

15.112 and 15.113 Two rotating rods are connected by a slider block *P*. The velocity $\mathbf{v}_0$ of the slider block relative to the rod on which it slides is constant and is directed outward. For the given data, determine the angular velocity of each rod for the position shown.

15.112 $b = 200$ mm, $v_0 = 300$ mm/s.
15.113 $b = 10$ in., $v_0 = 15$ in./s.

15.114 Two rods *AH* and *BD* pass through smooth holes drilled in a hexagonal block. (The holes are drilled in different planes so that the rods will not hit each other.) Knowing that rod *AH* rotates counterclockwise at the rate ω, determine the angular velocity of rod *BD* and the relative velocity of the block with respect to each rod when (*a*) $\theta = 30°$, (*b*) $\theta = 15°$.

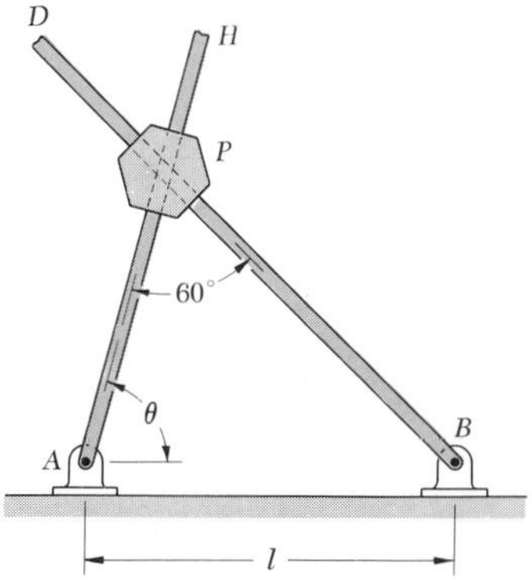

Fig. P15.114

15.115 Solve Prob. 15.114 when (*a*) $\theta = 90°$, (*b*) $\theta = 60°$.

15.116 Four pins slide in four separate slots cut in a circular plate as shown. When the plate is at rest, each pin has a velocity directed as shown and of the same constant magnitude u. If each pin maintains the same velocity in relation to the plate when the plate rotates about O with a constant *clockwise* angular velocity ω, determine the acceleration of each pin.

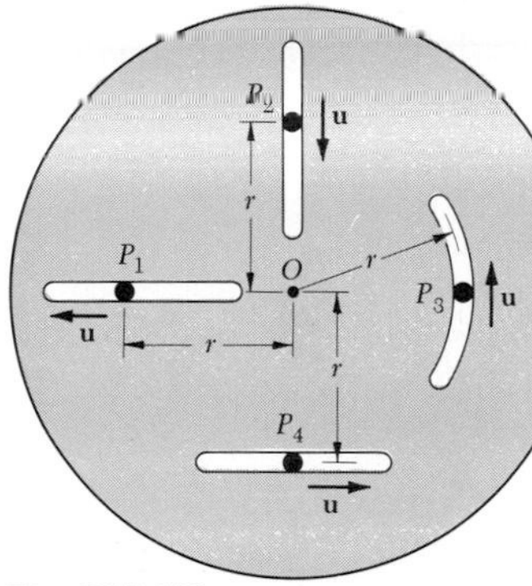

Fig. P15.116

15.117 Solve Prob. 15.116, assuming that the plate rotates about O with a constant *counterclockwise* angular velocity ω.

15.118 At the instant shown the length of the boom is being decreased at the constant rate of 150 mm/s and the boom is being lowered at the constant rate of 0.08 rad/s. Knowing that $\theta = 30°$, determine (*a*) the velocity of point B, (*b*) the acceleration of point B.

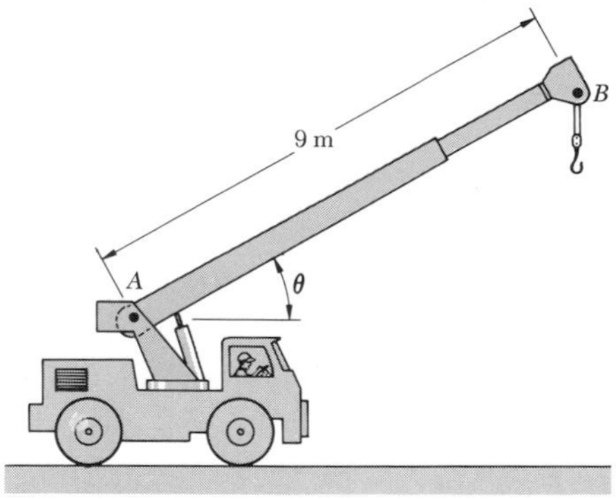

Fig. P15.118

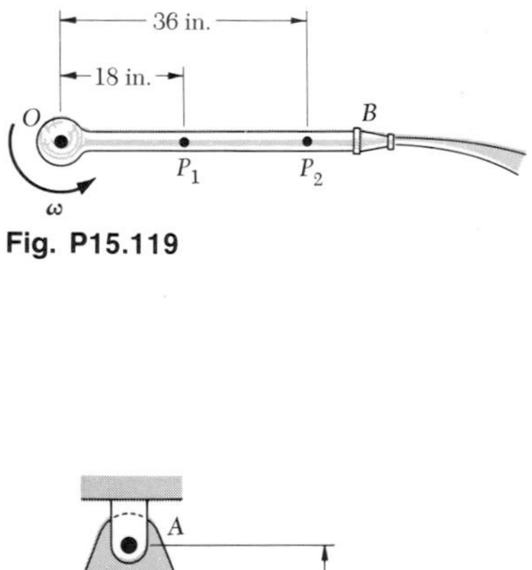

Fig. P15.119

15.119 Water flows through a straight pipe OB which rotates counterclockwise with an angular velocity of 120 rpm. If the velocity of the water relative to the pipe is 20 ft/s, determine the total acceleration (*a*) of the particle of water P_1, (*b*) of the particle of water P_2.

15.120 Pin P slides in the circular slot cut in the plate $ABDE$ at a constant relative speed $u = 0.5$ m/s as the plate rotates about A at the constant rate $\omega = 6$ rad/s. Determine the acceleration of the pin as it passes through (*a*) point B, (*b*) point D, (*c*) point E.

15.121 Solve Prob. 15.120, assuming that at the instant considered the angular velocity ω is being decreased at the rate of 10 rad/s^2 and that the relative velocity $\mathbf{u}$ is being decreased at the rate of 3 m/s^2.

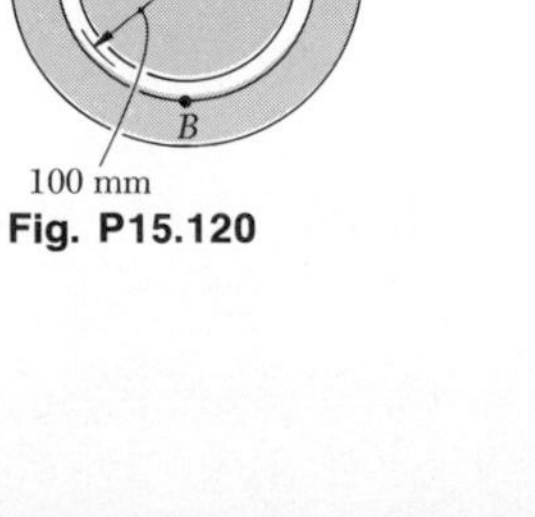

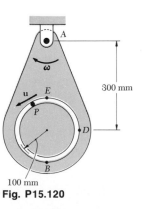

Fig. P15.120

15.122 The cage of a mine elevator moves downward with a constant speed of 40 ft/s. Determine the magnitude and direction of the Coriolis acceleration of the cage if the elevator is located (*a*) at the equator, (*b*) at latitude 40° north, (*c*) at latitude 40° south. (*Hint.* In parts *b* and *c* consider separately the components of the motion parallel and perpendicular to the plane of the equator.)

15.123 A train crosses the parallel 50° north, traveling due north at a constant speed v. Determine the speed of the train if the Coriolis component of its acceleration is 0.01 ft/s^2. (See hint of Prob. 15.122.)

15.124 In Prob. 15.110, determine the angular acceleration of the rod attached at A.

15.125 In Prob. 15.111, determine the angular acceleration of the rod attached at A.

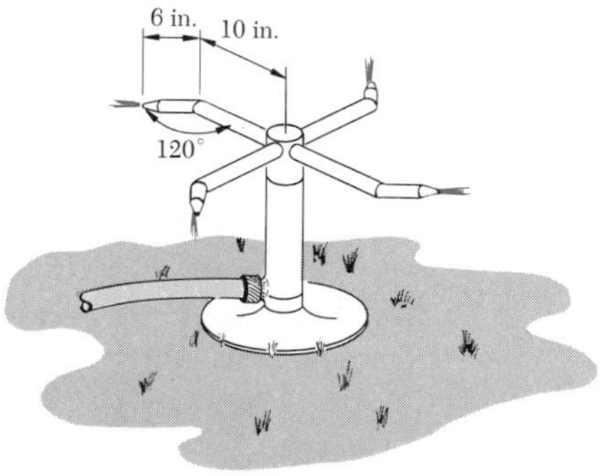

Fig. P15.126

15.126 A garden sprinkler has four rotating arms, each of which consists of two horizontal straight sections of pipe forming an angle of 120°. The sprinkler when operating rotates with a constant angular velocity of 180 rpm. If the velocity of the water relative to the pipe sections is 12 ft/s, determine the magnitude of the total acceleration of a particle of water as it passes the midpoint of (*a*) the 10-in. section of pipe, (*b*) the 6-in. section of pipe.

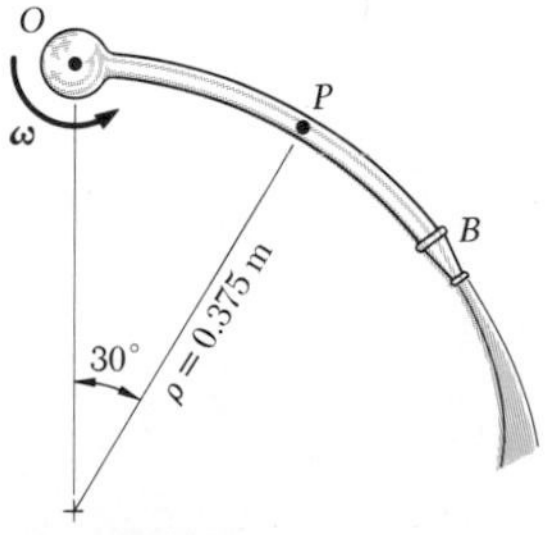

Fig. P15.127

15.127 Water flows through the curved pipe OB, which has a constant radius of 0.375 m and which rotates with a constant counterclockwise angular velocity of 120 rpm. If the velocity of the water relative to the pipe is 12 m/s, determine the total acceleration of the particle of water P.

15.128 The disk shown rotates with a constant clockwise angular velocity of 12 rad/s. At the instant shown, determine (*a*) the angular velocity and angular acceleration of rod *BD*, (*b*) the velocity and acceleration of the point of the rod in contact with collar *E*.

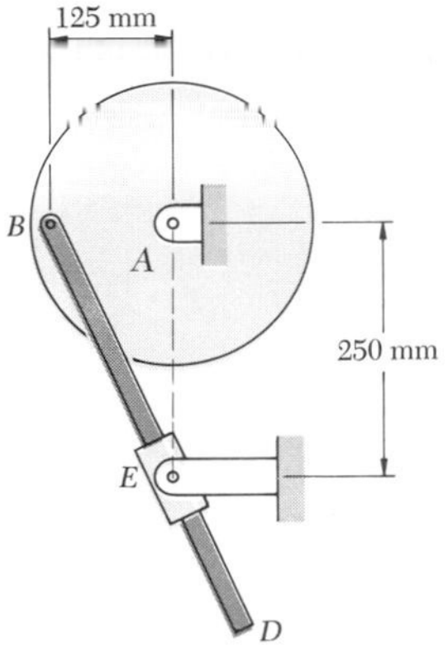

Fig. P15.128

15.129 Solve Prob. 15.128, assuming that the disk rotates with a constant counterclockwise angular velocity of 12 rad/s.

***15.12. Motion about a Fixed Point.** We have studied in Sec. 15.3 the motion of a rigid body constrained to rotate about a fixed axis. We shall now consider the more general case of the motion of a rigid body which has a fixed point *O*.

First, we shall prove that *the most general displacement of a rigid body with a fixed point O is equivalent to a rotation of the body about an axis through O.*† Instead of considering the rigid body itself, we may detach a sphere of center *O* from the body and analyze the motion of that sphere. Clearly, the motion of the sphere completely characterizes the motion of the given body. Since three points define the position of a solid in space, the center *O* and two points *A* and *B* on the surface of the sphere will define the position of the sphere and, thus, the position of the body. Let A_1 and B_1 characterize the position of the sphere at one instant, and A_2 and B_2 its position at a later instant (Fig. 15.31*a*). Since the sphere is rigid, the lengths of the arcs of great circle A_1B_1 and A_2B_2 must be equal, but, except for this requirement, the positions of A_1, A_2, B_1, and B_2 are arbitrary. We propose to prove that the points *A* and *B* may be brought, respectively, from A_1 and B_1 into A_2 and B_2 by a single rotation of the sphere about an axis.

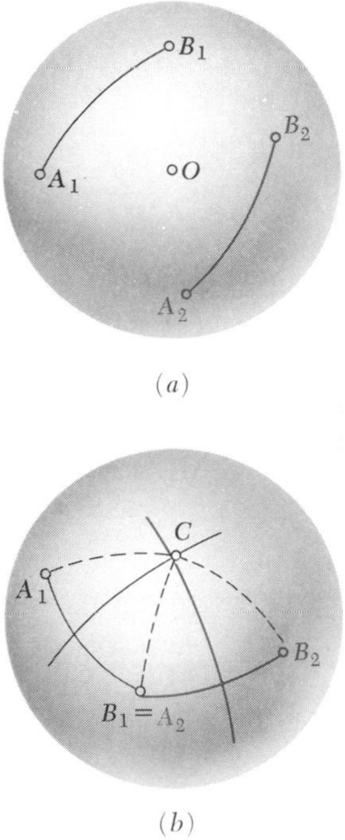

Fig. 15.31

For convenience, and without loss of generality, we may select point *B* so that its initial position coincides with the final position of *A*; thus, $B_1 = A_2$ (Fig. 15.31*b*). We draw the arcs of great circle A_1A_2, A_2B_2 and the arcs bisecting, respectively, A_1A_2 and A_2B_2. Let *C* be the point of intersection of these last two arcs; we complete the construction by drawing A_1C, A_2C, and B_2C. As pointed out above, $A_1B_1 = A_2B_2$ on account of the rigidity of the sphere; on the other hand, since *C* is by construction equidistant from A_1, A_2, and B_2, we have $A_1C = A_2C = B_2C$. As a result, the spherical triangles A_1CA_2 and B_1CB_2 are congruent and the angles A_1CA_2 and B_1CB_2 are equal. Denoting by θ the common value of these angles, we conclude that the sphere may be brought from its initial position into its final position by a single rotation through θ about the axis *OC*.

It follows that the motion during a time interval Δt of a rigid body with a fixed point *O* may be considered as a rotation

† This is known as Euler's theorem.

through $\Delta\theta$ about a certain axis. Drawing along that axis a vector of magnitude $\Delta\theta/\Delta t$ and letting Δt approach zero, we obtain at the limit the *instantaneous axis of rotation* and the angular velocity $\boldsymbol{\omega}$ of the body at the instant considered (Fig. 15.32). The velocity of a particle P of the body may then be obtained, as in Sec. 15.3, by forming the vector product of $\boldsymbol{\omega}$ and of the position vector $\mathbf{r}$ of the particle:

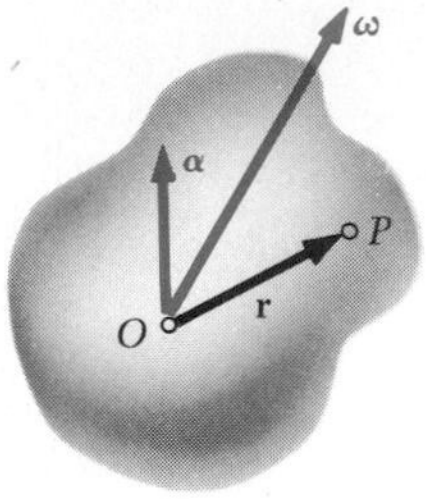

Fig. 15.32

$$\mathbf{v} = \frac{d\mathbf{r}}{dt} = \boldsymbol{\omega} \times \mathbf{r} \tag{15.37}$$

The acceleration of the particle is obtained by differentiating (15.37) with respect to t. As in Sec. 15.3 we have

$$\mathbf{a} = \boldsymbol{\alpha} \times \mathbf{r} + \boldsymbol{\omega} \times (\boldsymbol{\omega} \times \mathbf{r}) \tag{15.38}$$

where the angular acceleration $\boldsymbol{\alpha}$ is defined as the derivative

$$\boldsymbol{\alpha} = \frac{d\boldsymbol{\omega}}{dt} \tag{15.39}$$

of the angular velocity $\boldsymbol{\omega}$.

In the case of the motion of a rigid body with a fixed point, the direction of $\boldsymbol{\omega}$ and of the instantaneous axis of rotation changes from one instant to the next. The angular acceleration $\boldsymbol{\alpha}$, therefore, reflects the change in direction of $\boldsymbol{\omega}$ as well as its change in magnitude and, in general, *is not directed along the instantaneous axis of rotation*. While the particles of the body located on the instantaneous axis of rotation have zero velocity at the instant considered, they do not have zero acceleration. Also, the accelerations of the various particles of the body *cannot* be determined as if the body were rotating permanently about the instantaneous axis.

Recalling the definition of the velocity of a particle with position vector $\mathbf{r}$, we note that the angular acceleration $\boldsymbol{\alpha}$, as expressed in (15.39), represents the velocity of the tip of the vector $\boldsymbol{\omega}$. This property may be useful in the determination of the angular acceleration of a rigid body. For example, it follows that the vector $\boldsymbol{\alpha}$ is tangent to the curve described in space by the tip of the vector $\boldsymbol{\omega}$.

We should note that the vector $\boldsymbol{\omega}$ moves within the body, as well as in space. It thus generates two cones, respectively called the *body cone* and the *space cone* (Fig. 15.33).† It may be shown that, at any given instant, the two cones are tangent along the instantaneous axis of rotation and that, as the body moves, the body cone appears to *roll* on the space cone.

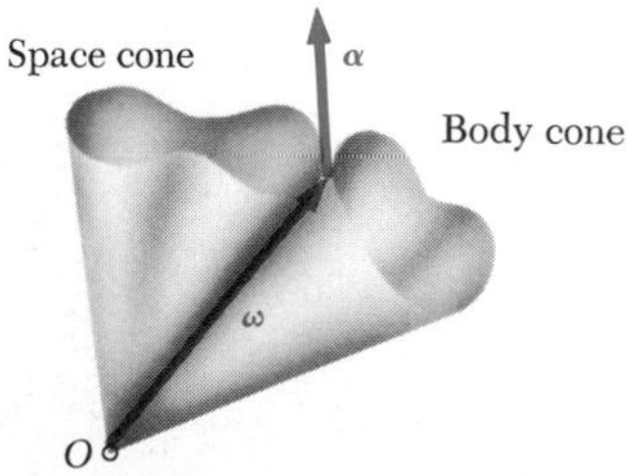

Fig. 15.33

† It is recalled that a *cone* is, by definition, a surface generated by a straight line passing through a fixed point. In general, the cones considered here *will not be circular cones.*

Before concluding our analysis of the motion of a rigid body with a fixed point, we should prove that angular velocities are actually vectors. As it was indicated in Sec. 2.3, some quantities, such as the *finite rotations* of a rigid body, have magnitude and direction, but do not obey the parallelogram law of addition; these quantities cannot be considered as vectors. We shall see presently that angular velocities (and also *infinitesimal rotations*) *do obey* the parallelogram law and, thus, are truly vector quantities.

Consider a rigid body with a fixed point O which, at a given instant, rotates simultaneously about the axes OA and OB with angular velocities $\boldsymbol{\omega}_1$ and $\boldsymbol{\omega}_2$ (Fig. 15.34*a*). We know that this motion must be equivalent at the instant considered to a single rotation of angular velocity $\boldsymbol{\omega}$. We propose to show that

$$\boldsymbol{\omega} = \boldsymbol{\omega}_1 + \boldsymbol{\omega}_2 \tag{15.40}$$

i.e., that the resulting angular velocity may be obtained by adding $\boldsymbol{\omega}_1$ and $\boldsymbol{\omega}_2$ by the parallelogram law (Fig. 15.34*b*).

Consider a particle P of the body, defined by the position vector $\mathbf{r}$. Denoting respectively by $\mathbf{v}_1$, $\mathbf{v}_2$, and $\mathbf{v}$ the velocity of P when the body rotates about OA only, about OB only, and about both axes simultaneously, we write

$$\mathbf{v} = \boldsymbol{\omega} \times \mathbf{r} \qquad \mathbf{v}_1 = \boldsymbol{\omega}_1 \times \mathbf{r} \qquad \mathbf{v}_2 = \boldsymbol{\omega}_2 \times \mathbf{r} \tag{15.41}$$

But the vectorial character of *linear* velocities is well established (since they represent the derivatives of position vectors). We have therefore

$$\mathbf{v} = \mathbf{v}_1 + \mathbf{v}_2$$

where the plus sign indicates vector addition. Substituting from (15.41), we write

$$\boldsymbol{\omega} \times \mathbf{r} = \boldsymbol{\omega}_1 \times \mathbf{r} + \boldsymbol{\omega}_2 \times \mathbf{r}$$
$$\boldsymbol{\omega} \times \mathbf{r} = (\boldsymbol{\omega}_1 + \boldsymbol{\omega}_2) \times \mathbf{r}$$

where the plus sign still indicates vector addition. Since the relation obtained holds for an arbitrary $\mathbf{r}$, we conclude that (15.40) must be true.

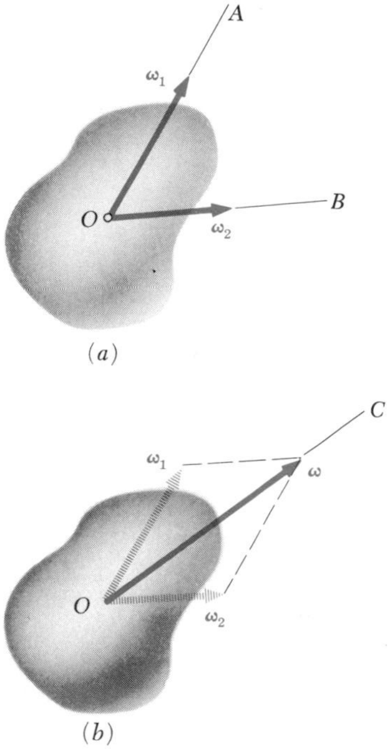

Fig. 15.34

*15.13. General Motion.

We shall now consider the most general motion of a rigid body in space. Let A and B be two particles of the body. We recall from Sec. 11.12 that the velocity of B with respect to the fixed frame of reference $OXYZ$ may be expressed as

$$\mathbf{v}_B = \mathbf{v}_A + \mathbf{v}_{B/A} \tag{15.42}$$

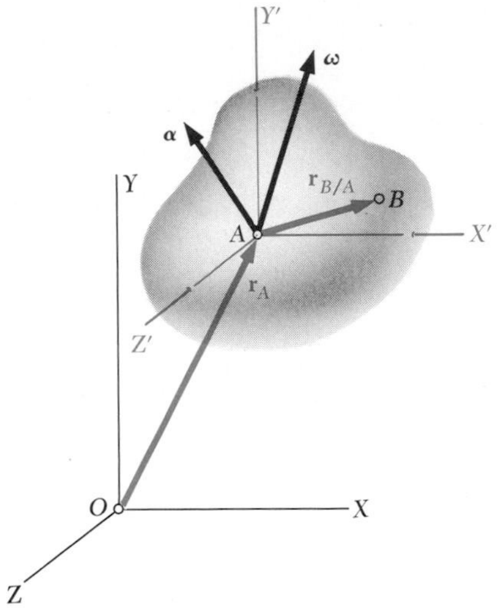

Fig. 15.35

where $\mathbf{v}_{B/A}$ is the velocity of B relative to a frame $AX'Y'Z'$ attached to A and of fixed orientation (Fig. 15.35). Since A is fixed in this frame, the motion of the body relative to $AX'Y'Z'$ is the motion of a body with a fixed point. Therefore, the relative velocity $\mathbf{v}_{B/A}$ may be obtained from (15.37), after $\mathbf{r}$ has been replaced by the position vector $\mathbf{r}_{B/A}$ of B relative to A. Substituting for $\mathbf{v}_{B/A}$ into (15.42), we write

$$\mathbf{v}_B = \mathbf{v}_A + \boldsymbol{\omega} \times \mathbf{r}_{B/A} \tag{15.43}$$

where $\boldsymbol{\omega}$ is the angular velocity of the body at the instant considered.

The acceleration of B is obtained by a similar reasoning. We first write

$$\mathbf{a}_B = \mathbf{a}_A + \mathbf{a}_{B/A}$$

and, recalling Eq. (15.38),

$$\mathbf{a}_B = \mathbf{a}_A + \boldsymbol{\alpha} \times \mathbf{r}_{B/A} + \boldsymbol{\omega} \times (\boldsymbol{\omega} \times \mathbf{r}_{B/A}) \tag{15.44}$$

where $\boldsymbol{\alpha}$ is the angular acceleration of the body at the instant considered.

Equations (15.43) and (15.44) show that *the most general motion of a rigid body is equivalent, at any given instant, to the sum of a translation,* in which all the particles of the body have the same velocity and acceleration as a reference particle A, *and of a motion in which particle A is assumed to be fixed.*†

It may easily be shown, by solving (15.43) and (15.44) for $\mathbf{v}_A$ and $\mathbf{a}_A$, that the motion of the body with respect to a frame attached to B would be characterized by the same vectors $\boldsymbol{\omega}$ and $\boldsymbol{\alpha}$ as its motion relative to $AX'Y'Z'$. Thus, the angular velocity and angular acceleration of a rigid body at a given instant are independent of the choice of the reference point. On the other hand, one should keep in mind that, whether it is attached to A or to B, the moving frame should maintain a fixed orientation; i.e., it should remain parallel to the fixed reference frame $OXYZ$ throughout the motion of the rigid body. In many problems it is found more convenient to use a moving frame which is allowed to rotate as well as to translate. The use of such moving frames will be discussed in Secs. 15.14 and 15.15.

† It is recalled from Sec. 15.12 that, in general, the vectors $\boldsymbol{\omega}$ and $\boldsymbol{\alpha}$ are not collinear, and that the accelerations of the particles of the body in their motion relative to the frame $AX'Y'Z'$ cannot be determined as if the body were rotating permanently about the instantaneous axis through A.

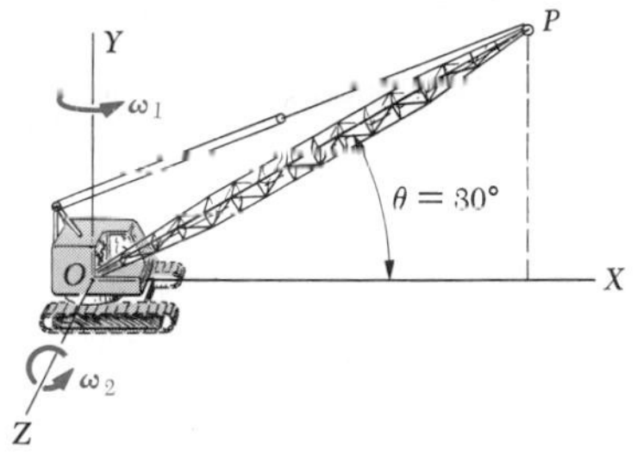

SAMPLE PROBLEM 15.11

The crane shown rotates with a constant angular velocity $\boldsymbol{\omega}_1$ of 0.30 rad/s. Simultaneously, the boom is being raised with a constant angular velocity $\boldsymbol{\omega}_2$ of 0.50 rad/s relative to the cab. Knowing that the length of the boom OP is $l = 12$ m, determine (*a*) the angular velocity $\boldsymbol{\omega}$ of the boom, (*b*) the angular acceleration $\boldsymbol{\alpha}$ of the boom, (*c*) the velocity $\mathbf{v}$ of the tip of the boom, (*d*) the acceleration $\mathbf{a}$ of the tip of the boom.

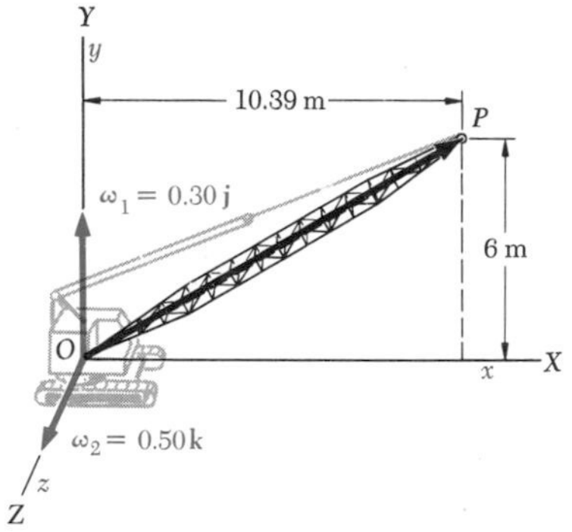

a. **Angular Velocity of Boom.** Adding the angular velocity $\boldsymbol{\omega}_1$ of the cab and the angular velocity $\boldsymbol{\omega}_2$ of the boom relative to the cab, we obtain the angular velocity $\boldsymbol{\omega}$ of the boom at the instant considered:

$$\boldsymbol{\omega} = \boldsymbol{\omega}_1 + \boldsymbol{\omega}_2 \qquad \boldsymbol{\omega} = (0.30 \text{ rad/s})\mathbf{j} + (0.50 \text{ rad/s})\mathbf{k} \quad ◀$$

b. **Angular Acceleration of Boom.** The angular acceleration $\boldsymbol{\alpha}$ of the boom is obtained by differentiating $\boldsymbol{\omega}$. Since the vector $\boldsymbol{\omega}_1$ is constant in magnitude and direction, we have

$$\boldsymbol{\alpha} = \dot{\boldsymbol{\omega}} = \dot{\boldsymbol{\omega}}_1 + \dot{\boldsymbol{\omega}}_2 = 0 + \dot{\boldsymbol{\omega}}_2$$

where the rate of change $\dot{\boldsymbol{\omega}}_2$ is to be computed with respect to the fixed frame $OXYZ$. However, it is more convenient to use a frame $Oxyz$ attached to the cab and rotating with it, since the vector $\boldsymbol{\omega}_2$ also rotates with the cab and, therefore, has zero rate of change with respect to that frame. Using Eq. (15.31) with $\mathbf{Q} = \boldsymbol{\omega}_2$ and $\boldsymbol{\Omega} = \boldsymbol{\omega}_1$, we write

$$(\dot{\mathbf{Q}})_{OXYZ} = (\dot{\mathbf{Q}})_{Oxyz} + \boldsymbol{\Omega} \times \mathbf{Q}$$
$$(\dot{\boldsymbol{\omega}}_2)_{OXYZ} = (\dot{\boldsymbol{\omega}}_2)_{Oxyz} + \boldsymbol{\omega}_1 \times \boldsymbol{\omega}_2$$
$$\boldsymbol{\alpha} = (\dot{\boldsymbol{\omega}}_2)_{OXYZ} = 0 + (0.30 \text{ rad/s})\mathbf{j} \times (0.50 \text{ rad/s})\mathbf{k}$$
$$\boldsymbol{\alpha} = (0.15 \text{ rad/s}^2)\mathbf{i} \quad ◀$$

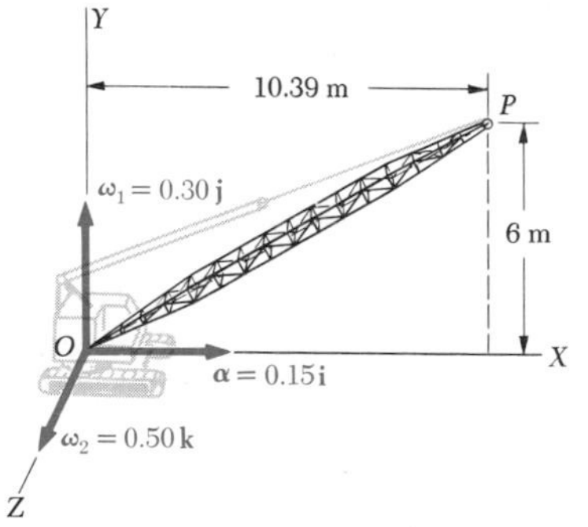

c. **Velocity of Tip of Boom.** Noting that the position vector of point P is $\mathbf{r} = (10.39 \text{ m})\mathbf{i} + (6 \text{ m})\mathbf{j}$ and using the expression found for $\boldsymbol{\omega}$ in part *a*, we write

$$\mathbf{v} = \boldsymbol{\omega} \times \mathbf{r} = \begin{vmatrix} \mathbf{i} & \mathbf{j} & \mathbf{k} \\ 0 & 0.30 \text{ rad/s} & 0.50 \text{ rad/s} \\ 10.39 \text{ m} & 6 \text{ m} & 0 \end{vmatrix}$$

$$\mathbf{v} = -(3 \text{ m/s})\mathbf{i} + (5.20 \text{ m/s})\mathbf{j} - (3.12 \text{ m/s})\mathbf{k} \quad ◀$$

d. **Acceleration of Tip of Boom.** Recalling that $\mathbf{v} = \boldsymbol{\omega} \times \mathbf{r}$, we write

$$\mathbf{a} = \boldsymbol{\alpha} \times \mathbf{r} + \boldsymbol{\omega} \times (\boldsymbol{\omega} \times \mathbf{r}) = \boldsymbol{\alpha} \times \mathbf{r} + \boldsymbol{\omega} \times \mathbf{v}$$

$$\mathbf{a} = \begin{vmatrix} \mathbf{i} & \mathbf{j} & \mathbf{k} \\ 0.15 & 0 & 0 \\ 10.39 & 6 & 0 \end{vmatrix} + \begin{vmatrix} \mathbf{i} & \mathbf{j} & \mathbf{k} \\ 0 & 0.30 & 0.50 \\ -3 & 5.20 & -3.12 \end{vmatrix}$$

$$= 0.90\mathbf{k} - 0.94\mathbf{i} - 2.60\mathbf{i} - 1.50\mathbf{j} + 0.90\mathbf{k}$$

$$\mathbf{a} = -(3.54 \text{ m/s}^2)\mathbf{i} - (1.50 \text{ m/s}^2)\mathbf{j} + (1.80 \text{ m/s}^2)\mathbf{k} \quad ◀$$

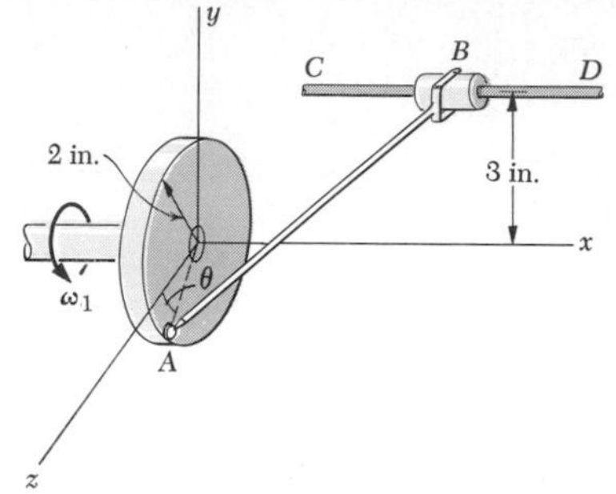

SAMPLE PROBLEM 15.12

The rod AB, of length 7 in., is attached to the disk by a ball-and-socket connection and to the collar B by a clevis. The disk rotates in the yz plane at a constant rate $\omega_1 = 12$ rad/s, while the collar is free to slide along the horizontal rod CD. For the position $\theta = 0$, determine (*a*) the velocity of the collar, (*b*) the angular velocity of the rod.

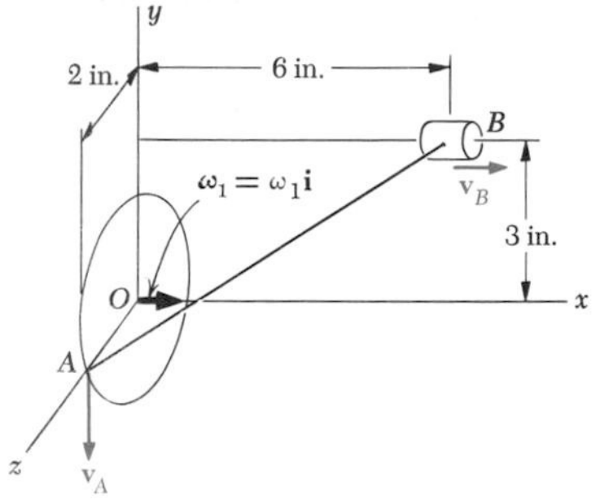

$\boldsymbol{\omega}_1 = 12\mathbf{i}$
$\mathbf{r}_A = 2\mathbf{k}$
$\mathbf{r}_B = 6\mathbf{i} + 3\mathbf{j}$
$\mathbf{r}_{B/A} = 6\mathbf{i} + 3\mathbf{j} - 2\mathbf{k}$

a. Velocity of Collar. Since point A is attached to the disk and since collar B moves parallel to the x axis, we have

$$\mathbf{v}_A = \boldsymbol{\omega}_1 \times \mathbf{r}_A = 12\mathbf{i} \times 2\mathbf{k} = -24\mathbf{j} \qquad \mathbf{v}_B = v_B\mathbf{i}$$

Denoting by $\boldsymbol{\omega}$ the angular velocity of the rod, we write

$$\mathbf{v}_B = \mathbf{v}_A + \mathbf{v}_{B/A} = \mathbf{v}_A + \boldsymbol{\omega} \times \mathbf{r}_{B/A}$$

$$v_B\mathbf{i} = -24\mathbf{j} + \begin{vmatrix} \mathbf{i} & \mathbf{j} & \mathbf{k} \\ \omega_x & \omega_y & \omega_z \\ 6 & 3 & -2 \end{vmatrix}$$

$$v_B\mathbf{i} = -24\mathbf{j} + (-2\omega_y - 3\omega_z)\mathbf{i} + (6\omega_z + 2\omega_x)\mathbf{j} + (3\omega_x - 6\omega_y)\mathbf{k}$$

Equating the coefficients of the unit vectors, we obtain

$$v_B = \quad -2\omega_y \quad -3\omega_z \qquad (1)$$
$$24 = 2\omega_x \quad +6\omega_z \qquad (2)$$
$$0 = 3\omega_x \quad -6\omega_y \qquad (3)$$

Multiplying Eqs. (1), (2), (3), respectively, by 6, 3, −2 and adding, we write

$$6v_B + 72 = 0 \qquad v_B = -12 \qquad \mathbf{v}_B = -(12 \text{ in./s})\mathbf{i} \blacktriangleleft$$

b. Angular Velocity of Rod AB. We note that the angular velocity cannot be determined from Eqs. (1), (2), and (3), since the determinant formed by the coefficients of ω_x, ω_y, and ω_z is zero. We must therefore obtain an additional equation by considering the constraint imposed by the clevis at B.

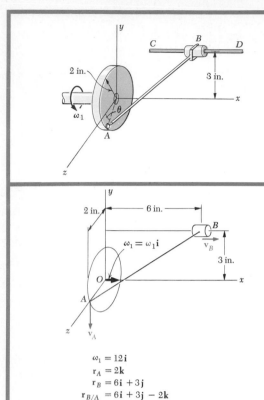

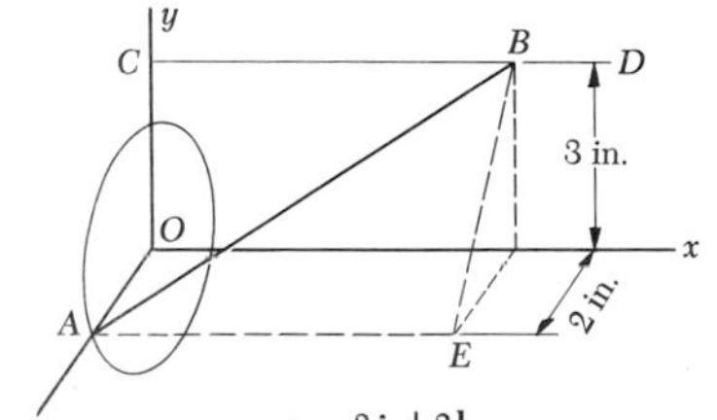

$\mathbf{r}_{E/B} = -3\mathbf{j} + 2\mathbf{k}$

The collar-clevis connection at B permits rotation of AB about the rod CD and also about an axis perpendicular to the plane containing AB and CD. It prevents rotation of AB about the axis EB, which is perpendicular to CD and lies in the plane containing AB and CD. Thus the projection of $\boldsymbol{\omega}$ on $\mathbf{r}_{E/B}$ must be zero and we write†

$$\boldsymbol{\omega} \cdot \mathbf{r}_{E/B} = 0 \qquad (\omega_x\mathbf{i} + \omega_y\mathbf{j} + \omega_z\mathbf{k}) \cdot (-3\mathbf{j} + 2\mathbf{k}) = 0$$
$$-3\omega_y + 2\omega_z = 0 \qquad (4)$$

Solving Eqs. (1) through (4) simultaneously, we obtain

$$v_B = -12 \qquad \omega_x = 3.69 \qquad \omega_y = 1.846 \qquad \omega_z = 2.77$$
$$\boldsymbol{\omega} = (3.69 \text{ rad/s})\mathbf{i} + (1.846 \text{ rad/s})\mathbf{j} + (2.77 \text{ rad/s})\mathbf{k} \blacktriangleleft$$

†We could also note that the direction of EB is that of the vector triple product $\mathbf{r}_{B/C} \times (\mathbf{r}_{B/C} \times \mathbf{r}_{B/A})$ and write $\boldsymbol{\omega} \cdot [\mathbf{r}_{B/C} \times (\mathbf{r}_{B/C} \times \mathbf{r}_{B/A})] = 0$. This formulation would be particularly useful if the rod CD were skew.

PROBLEMS

15.130 The rigid body shown rotates about the origin of coordinates with an angular velocity $\boldsymbol{\omega}$. Denoting the velocity of point A by $\mathbf{v}_A = (v_A)_x\mathbf{i} + (v_A)_y\mathbf{j} + (v_A)_z\mathbf{k}$, and knowing that $(v_A)_x = 40$ mm/s and $(v_A)_y = -200$ mm/s, determine the velocity component $(v_A)_z$.

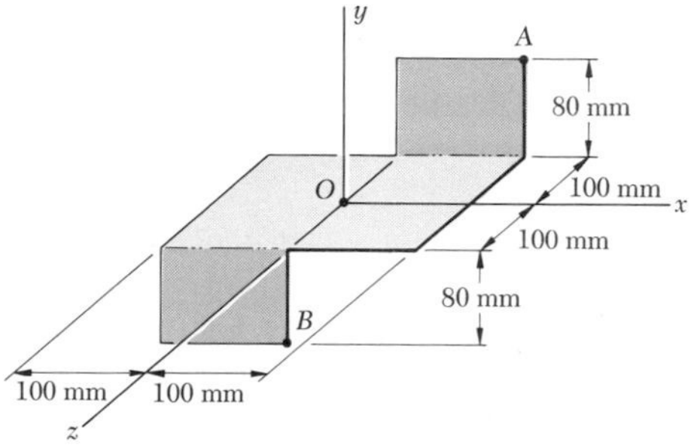

Fig. P15.130 and P15.131

15.131 The rigid body shown rotates about the origin of coordinates with an angular velocity $\boldsymbol{\omega} = \omega_x\mathbf{i} + \omega_y\mathbf{j} + \omega_z\mathbf{k}$. Knowing that $(v_A)_y = 400$ mm/s, $(v_B)_y = -300$ mm/s, and $\omega_y = 2$ rad/s, determine (*a*) the angular velocity of the body, (*b*) the velocities of points A and B.

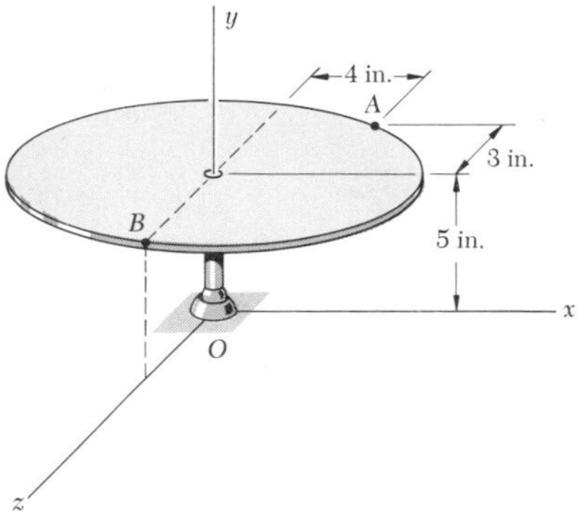

Fig. P15.132

15.132 The circular plate and rod are rigidly connected and rotate about the ball-and-socket joint O with an angular velocity $\boldsymbol{\omega} = \omega_x\mathbf{i} + \omega_y\mathbf{j} + \omega_z\mathbf{k}$. Knowing that $\mathbf{v}_A = -(27 \text{ in./s})\mathbf{i} + (18 \text{ in./s})\mathbf{j} + (v_A)_z\mathbf{k}$ and $\omega_y = 4$ rad/s, determine (*a*) the angular velocity of the assembly, (*b*) the velocity of point B.

15.133 Solve Prob. 15.132, assuming that $\omega_y = 0$.

15.134 The rotor of an electric motor rotates at the constant rate $\omega_1 = 3600$ rpm. Determine the angular acceleration of the rotor as the motor is rotated about the y axis with a constant angular velocity of 6 rpm clockwise when viewed from the positive y axis.

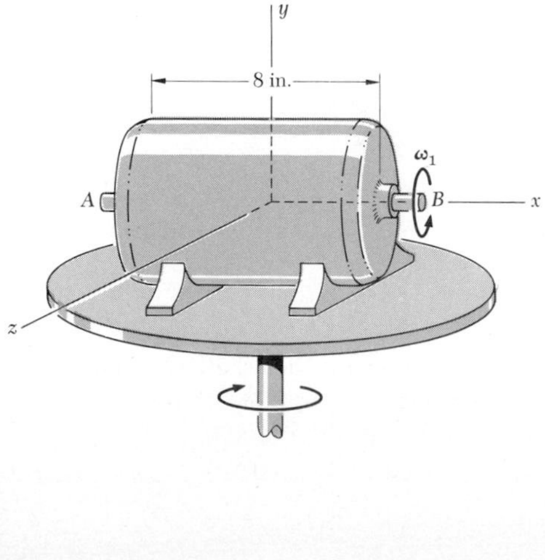

Fig. P15.134

15.135 The propeller of a small airplane rotates at a constant rate of 2200 rpm in a clockwise sense when viewed by the pilot. Knowing that the airplane is turning left along a horizontal circular path of radius 1000 ft, and that the speed of the airplane is 150 mi/h, determine the angular acceleration of the propeller at the instant the airplane is moving due south.

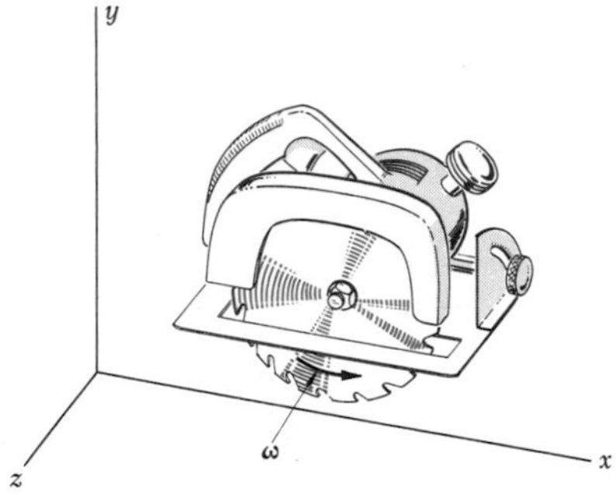

Fig. P15.136

15.136 The blade of a portable saw rotates at a constant rate $\omega = 1800$ rpm as shown. Determine the angular acceleration of the blade as a man rotates the saw about the y axis with an angular velocity of 3 rad/s and an angular acceleration of 5 rad/s^2, both clockwise when viewed from above.

15.137 Knowing that the turbine rotor shown rotates at a constant rate $\omega_1 = 10{,}000$ rpm, determine the angular acceleration of the rotor if the turbine housing has a constant angular velocity of 3 rad/s clockwise as viewed from (*a*) the positive y axis, (*b*) the positive z axis.

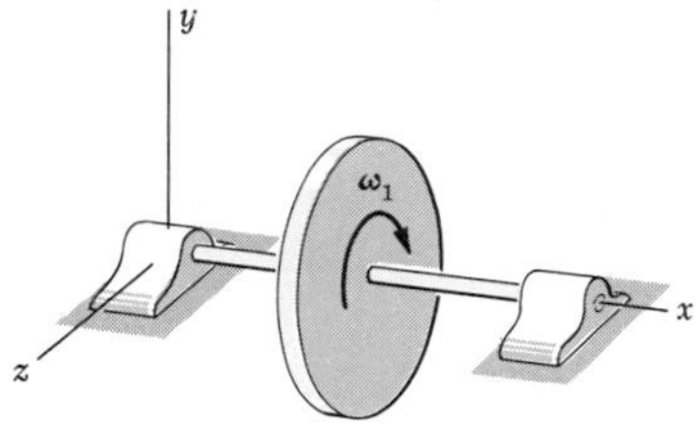

Fig. P15.137

15.138 In the gear system shown, gear A is free to rotate about the horizontal rod OA. Assuming that gear B is fixed and that shaft OC rotates with a constant angular velocity ω_1, determine (*a*) the angular velocity of gear A, (*b*) the angular acceleration of gear A.

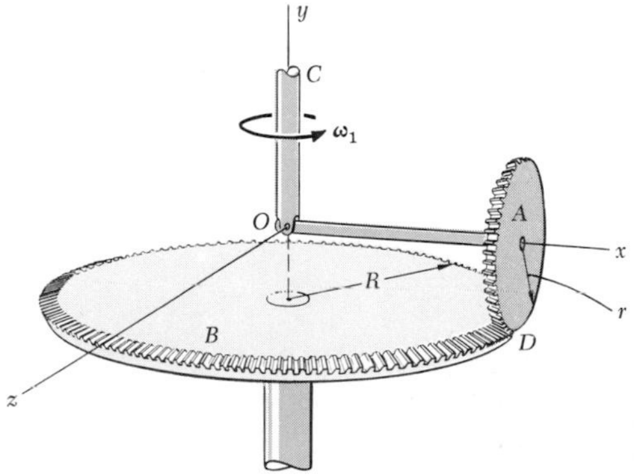

Fig. P15.138

15.139 Solve Prob. 15.138, assuming that shaft OC and gear B rotate with constant angular velocities ω_1 and ω_2, respectively, both counterclockwise as viewed from the positive y axis.

15.140 Two shafts AC and CF, which lie in the vertical xy plane, are connected by a universal joint at C. Shaft CF rotates with a constant angular velocity ω_1 as shown. At a time when the arm of the crosspiece attached to shaft CF is horizontal, determine the angular velocity of shaft AC.

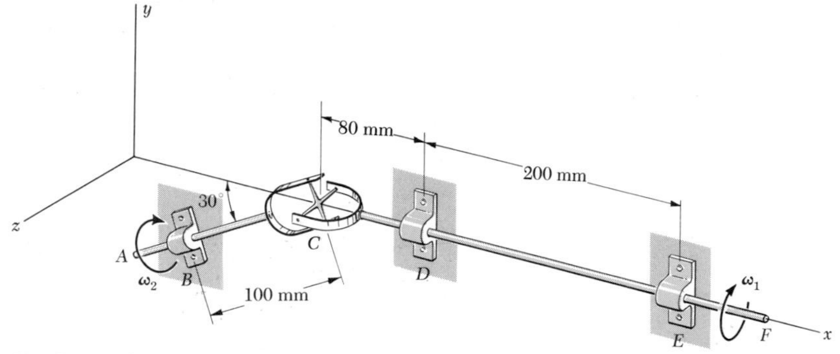

Fig. P15.140

15.141 Solve Prob. 15.140, assuming that the arm of the crosspiece attached to shaft CF is vertical.

15.142 The radar antenna shown rotates with a constant angular velocity ω_1 of 1.5 rad/s about the y axis. At the instant shown the antenna is also rotating about the z axis with an angular velocity ω_2 of 2 rad/s and an angular acceleration α_2 of 2.5 rad/s^2. Determine (*a*) the angular acceleration of the antenna, (*b*) the accelerations of points A and B.

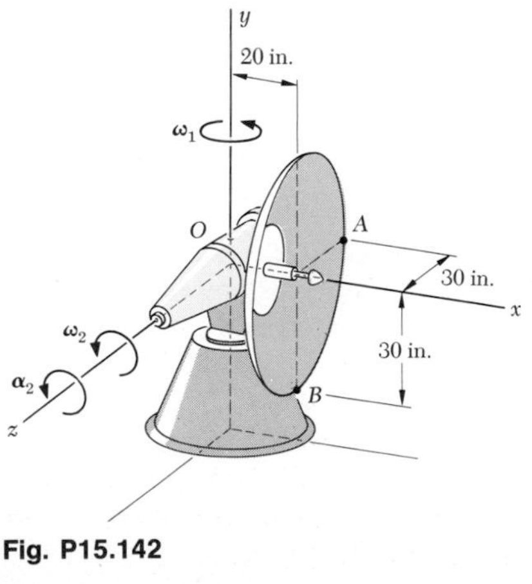

Fig. P15.142

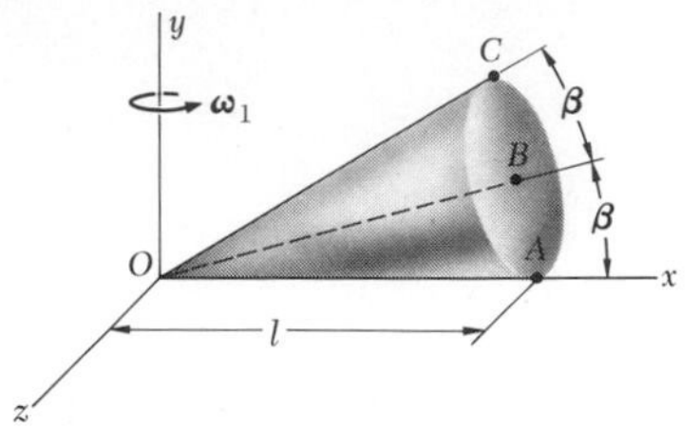

Fig. P15.143

15.143 The cone shown rolls on the zx plane with its apex at the origin of coordinates. Denoting by ω_1 the constant angular velocity of the axis OB of the cone about the y axis, determine (*a*) the rate of spin of the cone about the axis OB, (*b*) the total angular velocity of the cone, (*c*) the angular acceleration of the cone.

15.144 A rod of length $OP = 500$ mm is mounted on a bracket as shown. At the instant considered the angle β is being increased at the constant rate $d\beta/dt = 4$ rad/s and the elevation angle γ is being increased at the constant rate $d\gamma/dt = 1.6$ rad/s. For the position $\beta = 0$ and $\gamma = 30°$, determine (*a*) the angular velocity of the rod, (*b*) the angular acceleration of the rod, (*c*) the velocity and acceleration of point *P*.

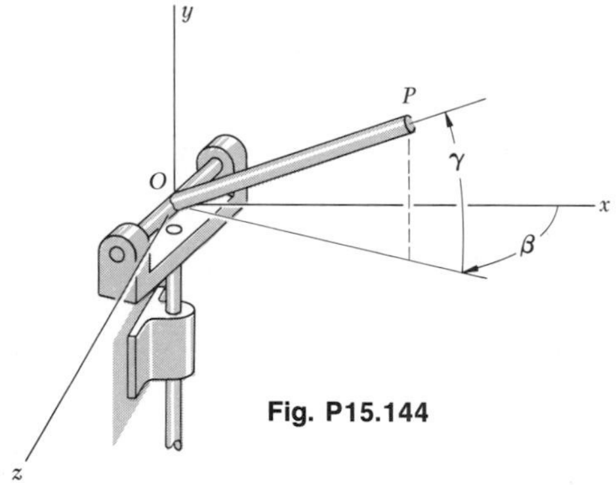

Fig. P15.144

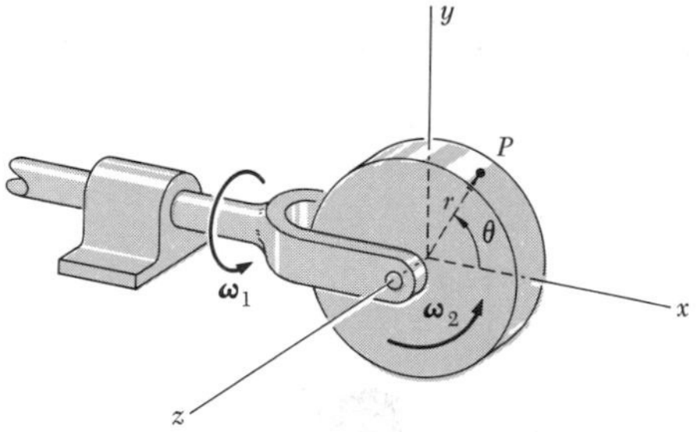

Fig. P15.145 and P15.146

15.145 A disk of radius r spins at the constant rate ω_2 about an axle held by a fork-ended horizontal rod which rotates at the constant rate ω_1. Determine the acceleration of point *P* for an arbitrary value of the angle θ.

15.146 A disk of radius r spins at the constant rate ω_2 about an axle held by a fork-ended horizontal rod which rotates at the constant rate ω_1. Determine (*a*) the angular acceleration of the disk, (*b*) the acceleration of point *P* on the rim of the disk when $\theta = 0$, (*c*) the acceleration of *P* when $\theta = 90°$.

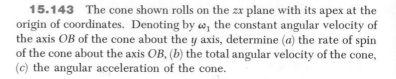

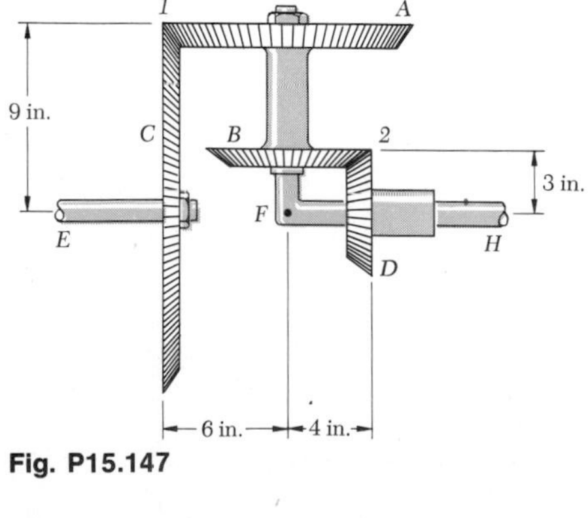

Fig. P15.147

15.147 In the planetary gear system shown, gears *A* and *B* are rigidly connected to each other and rotate as a unit about shaft *FG*. Gears *C* and *D* rotate with constant angular velocities of 15 rad/s and 30 rad/s, respectively (both counterclockwise when viewed from the right). Choosing the x axis to the right, the y axis upward, and the z axis pointing out of the plane of the figure, determine (*a*) the common angular velocity of gears *A* and *B*, (*b*) the angular velocity of shaft *FH*, which is rigidly attached to *FG*.

15.148 In Prob. 15.147, determine (*a*) the common angular acceleration of gears *A* and *B*, (*b*) the acceleration of the tooth of gear *B* which is in contact with gear *D* at point 2.

15.149 Three rods are welded together to form the corner assembly shown which is attached to a fixed ball-and-socket joint at O. The end of rod OA moves on the inclined plane D which is perpendicular to the xy plane. The end of rod OB moves on the horizontal plane E which coincides with the zx plane. Knowing that at the instant shown $\mathbf{v}_B = (1.6 \text{ m/s})\mathbf{k}$, determine (a) the angular velocity of the assembly, (b) the velocity of point C.

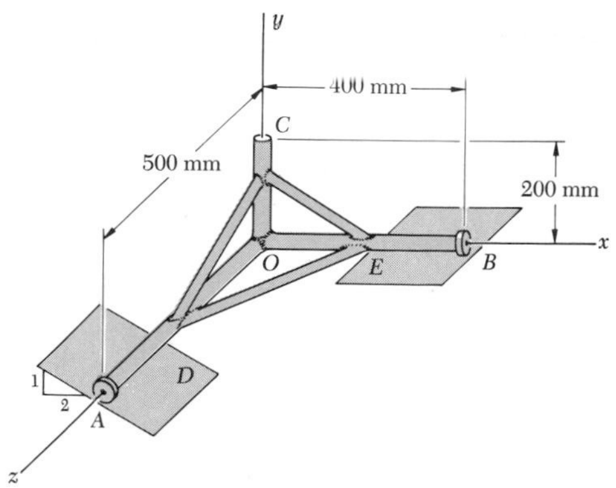

Fig. P15.149

15.150 In Prob. 15.149 the speed of point B is known to be constant. For the position shown, determine (a) the angular acceleration of the assembly, (b) the acceleration of point C.

15.151 In Prob. 15.149 the speed of point B is being decreased at the rate of 0.8 m/s^2. For the position shown, determine (a) the angular acceleration of the assembly, (b) the acceleration of point C.

15.152 Rod AB, of length 220 mm, is connected by ball-and-socket joints to collars A and B, which slide along the two rods shown. Knowing that collar A moves downward with a constant speed of 63 mm/s, determine the velocity of collar B when $c = 120$ mm.

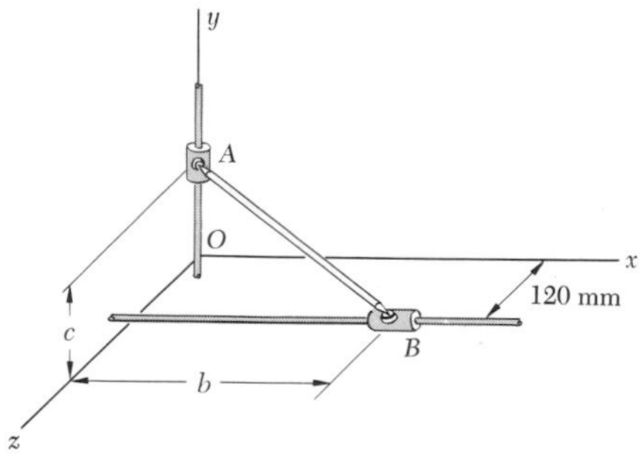

Fig. P15.152

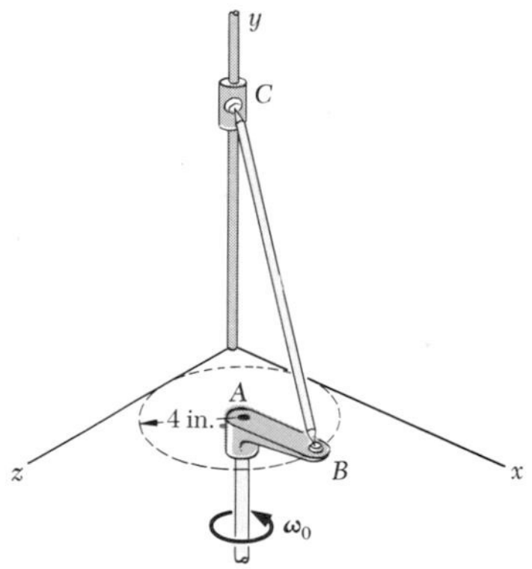

Fig. P15.154

15.153 Solve Prob. 15.152 when $c = 40$ mm.

15.154 Rod BC, of length 21 in., is connected by ball-and-socket joints to the collar C and to the rotating arm AB. Knowing that arm AB rotates in the zx plane at the constant rate $\omega_0 = 38$ rad/s, determine the velocity of collar C.

15.155 In Prob. 15.152, the ball-and-socket joint between the rod and collar A is replaced by the clevis connection shown. Determine (a) the angular velocity of the rod, (b) the velocity of collar B.

15.156 In Prob. 15.154, the ball-and-socket joint between the rod and collar C is replaced by the clevis connection shown. Determine (a) the angular velocity of the rod, (b) the velocity of collar C.

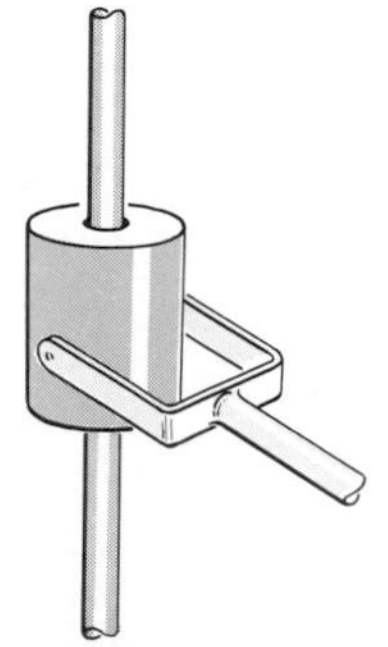
Fig. P15.155 and P15.156

15.157 In the linkage shown, crank BC rotates in the yz plane while crank ED rotates in a plane parallel to the xy plane. Knowing that in the position shown crank BC has an angular velocity $\boldsymbol{\omega}_1$ of 10 rad/s and no angular acceleration, determine the corresponding angular velocity $\boldsymbol{\omega}_2$ of crank ED.

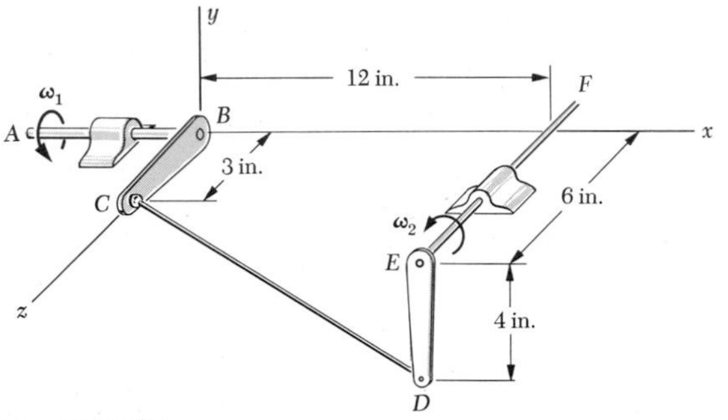

Fig. P15.157

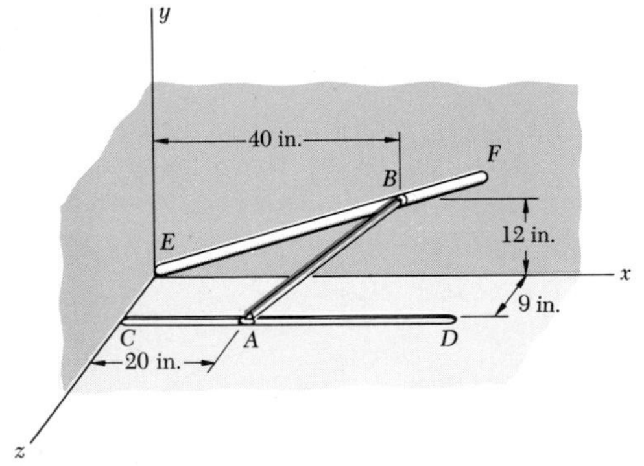

Fig. P15.158

15.158 Rod AB has a length of 25 in. and is guided by pins sliding in the slots CD and EF, which lie in the zx and xy planes, respectively. Knowing that in the position shown end A moves to the left along slot CD with a speed of 17 in./s, determine the velocity of end B of the rod.

***15.159** In Prob. 15.152, determine the acceleration of collar B when $c = 40$ mm.

***15.160** In Prob. 15.152, determine the acceleration of collar B when $c = 120$ mm.

***15.161** In Prob. 15.157, determine the angular acceleration of crank ED.

***15.162** In Prob. 15.154, determine the acceleration of collar C.

***15.14. Three-dimensional Motion of a Particle Relative to a Rotating Frame. Coriolis Acceleration.** We saw in Sec. 15.10 that, given a vector function $\mathbf{Q}(t)$ and two frames of reference centered at O—a fixed frame $OXYZ$ and a rotating frame $Oxyz$—the rates of change of $\mathbf{Q}$ with respect to the two frames satisfy the relation

$$(\dot{\mathbf{Q}})_{OXYZ} = (\dot{\mathbf{Q}})_{Oxyz} + \boldsymbol{\Omega} \times \mathbf{Q} \qquad (15.31)$$

We had assumed at the time that the frame $Oxyz$ was constrained to rotate about a fixed axis OA. However, the derivation given in

Sec. 15.10 remains valid when the frame $Oxyz$ is constrained only to have a fixed point O. Under this more general assumption, the axis OA represents the *instantaneous* axis of rotation of the frame $Oxyz$ (Sec. 15.12), and the vector $\mathbf{\Omega}$ its angular velocity at the instant considered (Fig. 15.36).

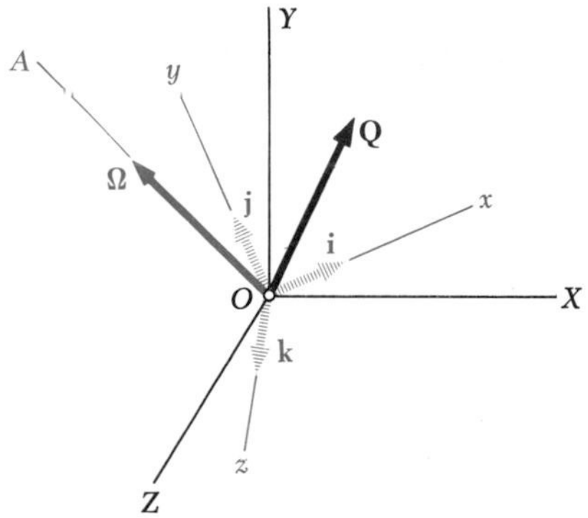

Fig. 15.36

We shall now consider the three-dimensional motion of a particle P relative to a rotating frame $Oxyz$ constrained to have a fixed origin O. Let $\mathbf{r}$ be the position vector of P at a given instant, and $\mathbf{\Omega}$ the angular velocity of the frame $Oxyz$ with respect to the fixed frame $OXYZ$ at the same instant (Fig. 15.37). The derivations given in Sec. 15.11 for the two-dimensional motion of a particle may readily be extended to the three-dimensional case, and we may express the absolute velocity $\mathbf{v}_P$ of P (i.e., its velocity with respect to the fixed frame $OXYZ$) as

$$\mathbf{v}_P = \mathbf{\Omega} \times \mathbf{r} + (\dot{\mathbf{r}})_{Oxyz} \tag{15.45}$$

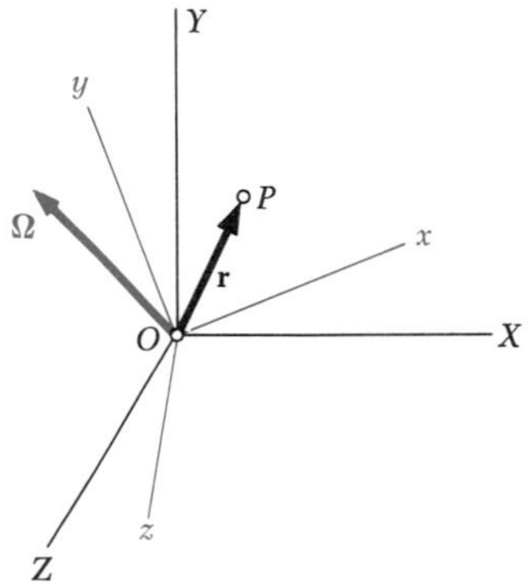

Fig. 15.37

This relation may be written in the alternate form

$$\mathbf{v}_P = \mathbf{v}_{P'} + \mathbf{v}_{P/F} \tag{15.46}$$

where $\mathbf{v}_P$ = absolute velocity of particle P
$\mathbf{v}_{P'}$ = velocity of point P' of moving frame coinciding with P
$\mathbf{v}_{P/F}$ = velocity of P relative to moving frame

The absolute acceleration $\mathbf{a}_P$ of P may be expressed as

$$\mathbf{a}_P = \dot{\mathbf{\Omega}} \times \mathbf{r} + \mathbf{\Omega} \times (\mathbf{\Omega} \times \mathbf{r}) + 2\mathbf{\Omega} \times (\dot{\mathbf{r}})_{Oxyz} + (\ddot{\mathbf{r}})_{Oxyz} \tag{15.47}$$

We may also use the alternate form

$$\mathbf{a}_P = \mathbf{a}_{P'} + \mathbf{a}_{P/F} + \mathbf{a}_c \tag{15.48}$$

where $\mathbf{a}_P$ = absolute acceleration of particle P
$\mathbf{a}_{P'}$ = acceleration of point P' of moving frame coinciding with P
$\mathbf{a}_{P/F}$ = acceleration of P relative to moving frame
$\mathbf{a}_c = 2\mathbf{\Omega} \times (\dot{\mathbf{r}})_{Oxyz} = 2\mathbf{\Omega} \times \mathbf{v}_{P/F}$
= complementary, or Coriolis, acceleration

We note that the Coriolis acceleration is perpendicular to the vectors $\mathbf{\Omega}$ and $\mathbf{v}_{P/F}$. However, since these vectors are usually not perpendicular to each other, the magnitude of $\mathbf{a}_c$, in general, is *not* equal to $2\Omega v_{P/F}$, as was the case for the plane motion of a particle. We further note that the Coriolis acceleration reduces to zero when the vectors $\boldsymbol{\Omega}$ and $\mathbf{v}_{P/F}$ are parallel, or when either of them is zero.

Rotating frames of reference are particularly useful in the study of the three-dimensional motion of rigid bodies. If a rigid body has a fixed point O, as was the case for the crane of Sample Prob. 15.11, we may use a frame $Oxyz$ which is neither fixed nor rigidly attached to the rigid body. Denoting by $\boldsymbol{\Omega}$ the angular velocity of the frame $Oxyz$, we then resolve the angular velocity $\boldsymbol{\omega}$ of the body into the components $\boldsymbol{\Omega}$ and $\boldsymbol{\omega}_{B/F}$, where the second component represents the angular velocity of the body relative to the frame $Oxyz$ (see Sample Prob. 15.14). An appropriate choice of the rotating frame will often lead to a simpler analysis of the motion of the rigid body than would be possible with axes of fixed orientation. This is especially true in the case of the general three-dimensional motion of a rigid body, i.e., when the rigid body under consideration has no fixed point (see Sample Prob. 15.15).

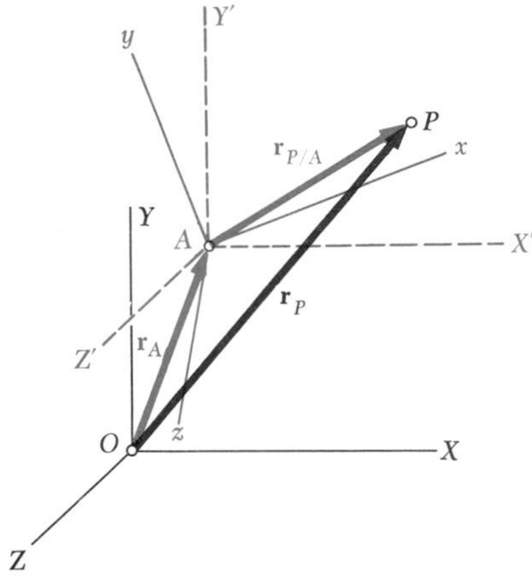

Fig. 15.38

*15.15. Frame of Reference in General Motion.

Consider a fixed frame of reference $OXYZ$ and a frame $Axyz$ which moves in a known, but arbitrary, fashion with respect to $OXYZ$ (Fig. 15.38). Let P be a particle moving in space. The position of P is defined at any instant by the vector $\mathbf{r}_P$ in the fixed frame, and by the vector $\mathbf{r}_{P/A}$ in the moving frame. Denoting by $\mathbf{r}_A$ the position vector of A in the fixed frame, we have

$$\mathbf{r}_P = \mathbf{r}_A + \mathbf{r}_{P/A} \tag{15.49}$$

The absolute velocity $\mathbf{v}_P$ of the particle is obtained by writing

$$\mathbf{v}_P = \dot{\mathbf{r}}_P = \dot{\mathbf{r}}_A + \dot{\mathbf{r}}_{P/A} \tag{15.50}$$

where the derivatives are defined with respect to the fixed frame $OXYZ$. Thus, the first term in the right-hand member of (15.50) represents the velocity $\mathbf{v}_A$ of the origin A of the moving axes. On the other hand, since the rate of change of a vector is the same with respect to a fixed frame and with respect to a frame in translation (Sec. 11.10), the second term may be regarded as the velocity $\mathbf{v}_{P/A}$ of P relative to the frame $AX'Y'Z'$ of the same orientation as $OXYZ$ and the same origin as $Axyz$. We therefore have

$$\mathbf{v}_P = \mathbf{v}_A + \mathbf{v}_{P/A} \tag{15.51}$$

But the velocity $\mathbf{v}_{P/A}$ of P relative to $AX'Y'Z'$ may be obtained from (15.45) by substituting $\mathbf{r}_{P/A}$ for $\mathbf{r}$ in that equation. We write

$$\mathbf{v}_P = \mathbf{v}_A + \boldsymbol{\Omega} \times \mathbf{r}_{P/A} + (\dot{\mathbf{r}}_{P/A})_{Axyz} \tag{15.52}$$

where $\boldsymbol{\Omega}$ is the angular velocity of the frame $Axyz$ at the instant considered.

The absolute acceleration $\mathbf{a}_P$ of the particle is obtained by differentiating (15.51) and writing

$$\mathbf{a}_P = \dot{\mathbf{v}}_P = \dot{\mathbf{v}}_A + \dot{\mathbf{v}}_{P/A} \tag{15.53}$$

where the derivatives are defined with respect to either of the frames $OXYZ$ or $AX'Y'Z'$. Thus, the first term in the right-hand member of (15.53) represents the acceleration $\mathbf{a}_A$ of the origin A of the moving axes, and the second term the acceleration $\mathbf{a}_{P/A}$ of P relative to the frame $AX'Y'Z'$. This acceleration may be obtained from (15.47) by substituting $\mathbf{r}_{P/A}$ for $\mathbf{r}$. We therefore write

$$\begin{aligned}\mathbf{a}_P = \mathbf{a}_A &+ \dot{\boldsymbol{\Omega}} \times \mathbf{r}_{P/A} + \boldsymbol{\Omega} \times (\boldsymbol{\Omega} \times \mathbf{r}_{P/A}) \\ &+ 2\boldsymbol{\Omega} \times (\dot{\mathbf{r}}_{P/A})_{Axyz} + (\ddot{\mathbf{r}}_{P/A})_{Axyz}\end{aligned} \tag{15.54}$$

Formulas (15.52) and (15.54) make it possible to determine the velocity and acceleration of a given particle with respect to a fixed frame of reference, when the motion of the particle is known with respect to a moving frame. These formulas become more significant, and considerably easier to remember, if we note that the sum of the first two terms in (15.52) represents the velocity of the point P' of the moving frame which coincides with P at the instant considered, and that the sum of the first three terms in (15.54) represents the acceleration of the same point. Thus, the relations (15.46) and (15.48) of the preceding section are still valid in the case of a reference frame in general motion, and we write

$$\mathbf{v}_P = \mathbf{v}_{P'} + \mathbf{v}_{P/\mathcal{F}} \tag{15.46}$$
$$\mathbf{a}_P = \mathbf{a}_{P'} + \mathbf{a}_{P/\mathcal{F}} + \mathbf{a}_c \tag{15.48}$$

where the various vectors involved have been defined in Sec. 15.14.

We may note that, if the reference frame $Axyz$ is in translation, the velocity and acceleration of the point P' of the frame which coincides with P become respectively equal to the velocity and acceleration of the origin A of the frame. On the other hand, since the frame maintains a fixed orientation, $\mathbf{a}_c$ is zero, and the relations (15.46) and (15.48) reduce, respectively, to the relations (11.33) and (11.34) derived in Sec. 11.12.

SAMPLE PROBLEM 15.13

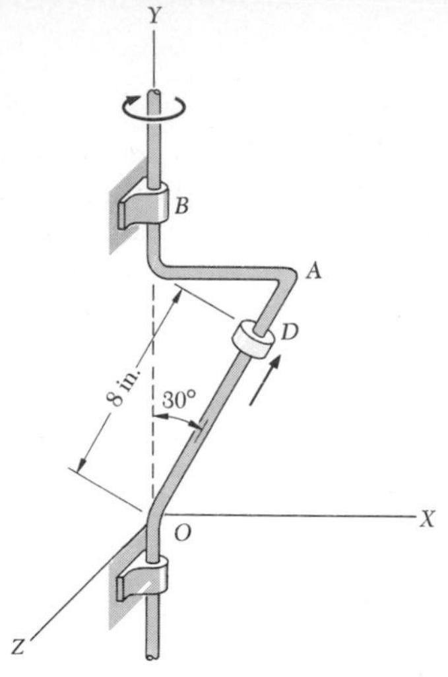

The bent rod OAB rotates about the vertical OB. At the instant considered, its angular velocity and angular acceleration are, respectively, 20 rad/s and 200 rad/s^2, both clockwise when viewed from the positive Y axis. The collar D moves along the rod and, at the instant considered, $OD = 8$ in., and the velocity and acceleration of the collar relative to the rod are, respectively, 50 in./s and 600 in./s^2, both upward. Determine (*a*) the velocity of the collar, (*b*) the acceleration of the collar.

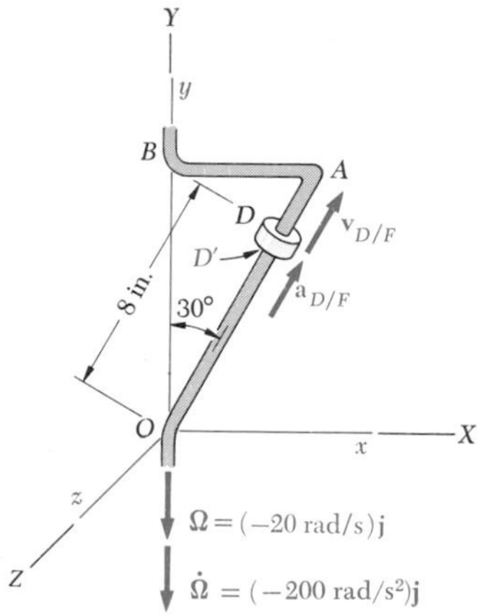

Frames of Reference. The frame $OXYZ$ is fixed. We attach the rotating frame $Oxyz$ to the bent rod. Its angular velocity and angular acceleration relative to $OXYZ$, therefore, are $\boldsymbol{\Omega} = (-20 \text{ rad/s})\mathbf{j}$ and $\dot{\boldsymbol{\Omega}} = (-200 \text{ rad/s}^2)\mathbf{j}$, respectively. The position vector of D is

$$\mathbf{r} = (8 \text{ in.})(\sin 30°\mathbf{i} + \cos 30°\mathbf{j}) = (4 \text{ in.})\mathbf{i} + (6.93 \text{ in.})\mathbf{j}$$

***a.* Velocity $\mathbf{v}_D$.** Denoting by D' the point of the rod which coincides with D, we write from Eq. (15.46)

$$\mathbf{v}_D = \mathbf{v}_{D'} + \mathbf{v}_{D/F} \tag{1}$$

where

$$\mathbf{v}_{D'} = \boldsymbol{\Omega} \times \mathbf{r} = (-20 \text{ rad/s})\mathbf{j} \times [(4 \text{ in.})\mathbf{i} + (6.93 \text{ in.})\mathbf{j}] = (80 \text{ in./s})\mathbf{k}$$
$$\mathbf{v}_{D/F} = (50 \text{ in./s})(\sin 30°\mathbf{i} + \cos 30°\mathbf{j}) = (25 \text{ in./s})\mathbf{i} + (43.3 \text{ in./s})\mathbf{j}$$

Substituting the values obtained for $\mathbf{v}_{D'}$ and $\mathbf{v}_{D/F}$ into (1), we find

$$\mathbf{v}_D = (25 \text{ in./s})\mathbf{i} + (43.3 \text{ in./s})\mathbf{j} + (80 \text{ in./s})\mathbf{k} \quad ◄$$

***b.* Acceleration $\mathbf{a}_D$.** From Eq. (15.48) we write

$$\mathbf{a}_D = \mathbf{a}_{D'} + \mathbf{a}_{D/F} + \mathbf{a}_c \tag{2}$$

where

$$\begin{aligned}
\mathbf{a}_{D'} &= \dot{\boldsymbol{\Omega}} \times \mathbf{r} + \boldsymbol{\Omega} \times (\boldsymbol{\Omega} \times \mathbf{r}) \\
&= (-200 \text{ rad/s}^2)\mathbf{j} \times [(4 \text{ in.})\mathbf{i} + (6.93 \text{ in.})\mathbf{j}] \\
&\qquad - (20 \text{ rad/s})\mathbf{j} \times (80 \text{ in./s})\mathbf{k} \\
&= +(800 \text{ in./s}^2)\mathbf{k} - (1600 \text{ in./s}^2)\mathbf{i} \\
\mathbf{a}_{D/F} &= (600 \text{ in./s}^2)(\sin 30°\mathbf{i} + \cos 30°\mathbf{j}) = (300 \text{ in./s}^2)\mathbf{i} + (520 \text{ in./s}^2)\mathbf{j} \\
\mathbf{a}_c &= 2\boldsymbol{\Omega} \times \mathbf{v}_{D/F} \\
&= 2(-20 \text{ rad/s})\mathbf{j} \times [(25 \text{ in./s})\mathbf{i} + (43.3 \text{ in./s})\mathbf{j}] = (1000 \text{ in./s}^2)\mathbf{k}
\end{aligned}$$

Substituting the values obtained for $\mathbf{a}_{D'}$, $\mathbf{a}_{D/F}$, and $\mathbf{a}_c$ into (2):

$$\mathbf{a}_D = -(1300 \text{ in./s}^2)\mathbf{i} + (520 \text{ in./s}^2)\mathbf{j} + (1800 \text{ in./s}^2)\mathbf{k} \quad ◄$$

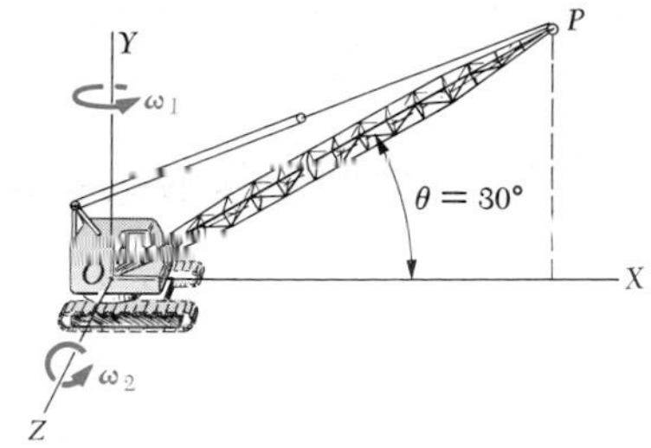

SAMPLE PROBLEM 15.14

The crane shown rotates with a constant angular velocity ω_1 of 0.30 rad/s. Simultaneously, the boom is being raised with a constant angular velocity ω_2 of 0.50 rad/s relative to the cab. Knowing that the length of the boom OP is $l = 12$ m, determine (*a*) the velocity of the tip of the boom, (*b*) the acceleration of the tip of the boom.

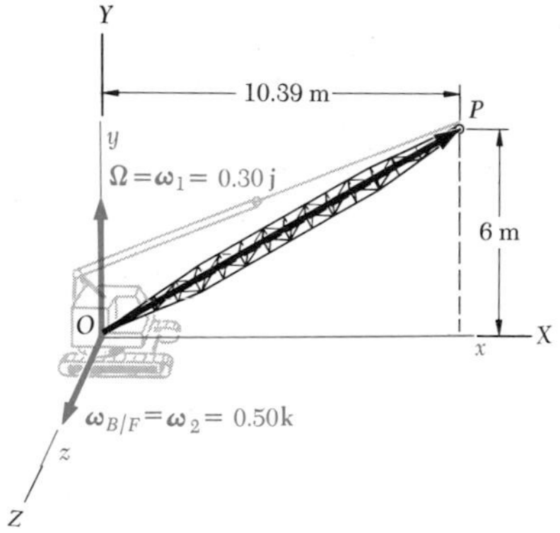

Frames of Reference. The frame $OXYZ$ is fixed. We attach the rotating frame $Oxyz$ to the cab. Its angular velocity with respect to the frame $OXYZ$, therefore, is $\boldsymbol{\Omega} = \boldsymbol{\omega}_1 = (0.30 \text{ rad/s})\mathbf{j}$. The angular velocity of the boom relative to the cab and the rotating frame $Oxyz$ is $\boldsymbol{\omega}_{B/F} = \boldsymbol{\omega}_2 = (0.50 \text{ rad/s})\mathbf{k}$.

a. **Velocity $\mathbf{v}_P$.** From Eq. (15.46) we write

$$\mathbf{v}_P = \mathbf{v}_{P'} + \mathbf{v}_{P/F} \qquad (1)$$

where $\mathbf{v}_{P'}$ is the velocity of the point P' of the frame $Oxyz$ which coincides with P,

$$\mathbf{v}_{P'} = \boldsymbol{\Omega} \times \mathbf{r} = (0.30 \text{ rad/s})\mathbf{j} \times [(10.39 \text{ m})\mathbf{i} + (6 \text{ m})\mathbf{j}] = -(3.12 \text{ m/s})\mathbf{k}$$

and where $\mathbf{v}_{P/F}$ is the velocity of P relative to the rotating frame $Oxyz$. But the angular velocity of the boom relative to $Oxyz$ was found to be $\boldsymbol{\omega}_{B/F} = (0.50 \text{ rad/s})\mathbf{k}$. The velocity of its tip P relative to $Oxyz$ is therefore

$$\begin{aligned}\mathbf{v}_{P/F} = \boldsymbol{\omega}_{B/F} \times \mathbf{r} &= (0.50 \text{ rad/s})\mathbf{k} \times [(10.39 \text{ m})\mathbf{i} + (6 \text{ m})\mathbf{j}] \\ &= -(3 \text{ m/s})\mathbf{i} + (5.20 \text{ m/s})\mathbf{j}\end{aligned}$$

Substituting the values obtained for $\mathbf{v}_{P'}$ and $\mathbf{v}_{P/F}$ into (1), we find

$$\mathbf{v}_P = -(3 \text{ m/s})\mathbf{i} + (5.20 \text{ m/s})\mathbf{j} - (3.12 \text{ m/s})\mathbf{k} \quad ◀$$

b. **Acceleration $\mathbf{a}_P$.** From Eq. (15.48) we write

$$\mathbf{a}_P = \mathbf{a}_{P'} + \mathbf{a}_{P/F} + \mathbf{a}_c \qquad (2)$$

Since $\boldsymbol{\Omega}$ and $\boldsymbol{\omega}_{B/F}$ are both constant, we have

$$\begin{aligned}\mathbf{a}_{P'} &= \boldsymbol{\Omega} \times (\boldsymbol{\Omega} \times \mathbf{r}) = (0.30 \text{ rad/s})\mathbf{j} \times (-3.12 \text{ m/s})\mathbf{k} = -(0.94 \text{ m/s}^2)\mathbf{i} \\ \mathbf{a}_{P/F} &= \boldsymbol{\omega}_{B/F} \times (\boldsymbol{\omega}_{B/F} \times \mathbf{r}) \\ &= (0.50 \text{ rad/s})\mathbf{k} \times [-(3 \text{ m/s})\mathbf{i} + (5.20 \text{ m/s})\mathbf{j}] \\ &= -(1.50 \text{ m/s}^2)\mathbf{j} - (2.60 \text{ m/s}^2)\mathbf{i} \\ \mathbf{a}_c &= 2\boldsymbol{\Omega} \times \mathbf{v}_{P/F} \\ &= 2(0.30 \text{ rad/s})\mathbf{j} \times [-(3 \text{ m/s})\mathbf{i} + (5.20 \text{ m/s})\mathbf{j}] = (1.80 \text{ m/s}^2)\mathbf{k}\end{aligned}$$

Substituting for $\mathbf{a}_{P'}$, $\mathbf{a}_{P/F}$, and $\mathbf{a}_c$ into (2), we find

$$\mathbf{a}_P = -(3.54 \text{ m/s}^2)\mathbf{i} - (1.50 \text{ m/s}^2)\mathbf{j} + (1.80 \text{ m/s}^2)\mathbf{k} \quad ◀$$

SAMPLE PROBLEM 15.15

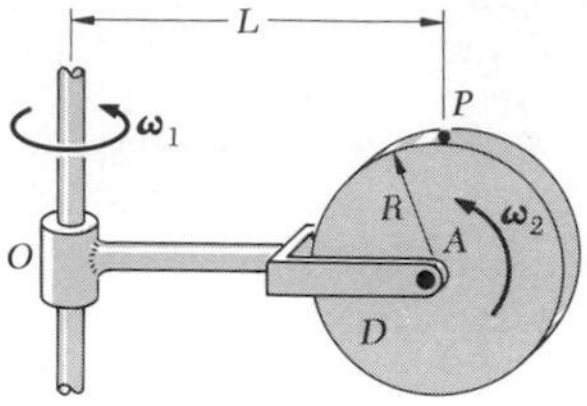

Disk D, of radius R, is pinned to end A of the arm OA of length L located in the plane of the disk. The arm rotates about a vertical axis through O at the constant rate ω_1, and the disk rotates about A at the constant rate ω_2. Determine (*a*) the velocity of point P located directly above A, (*b*) the acceleration of P, (*c*) the angular velocity and angular acceleration of the disk.

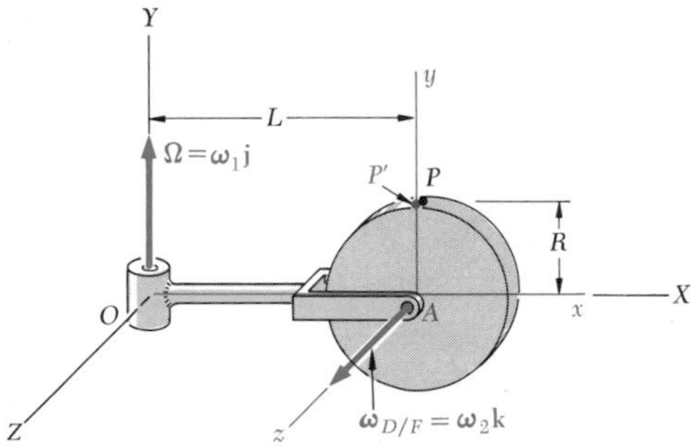

Frames of Reference. The frame $OXYZ$ is fixed. We attach the moving frame $Axyz$ to the arm OA. Its angular velocity with respect to the frame $OXYZ$, therefore, is $\mathbf{\Omega} = \omega_1\mathbf{j}$. The angular velocity of disk D relative to the frame $Axyz$ is $\boldsymbol{\omega}_{D/F} = \omega_2\mathbf{k}$. The position vector of P relative to O is $\mathbf{r} = L\mathbf{i} + R\mathbf{j}$ and its position vector relative to A is $\mathbf{r}_{P/A} = R\mathbf{j}$.

a. **Velocity $\mathbf{v}_P$.** Denoting by P' the point of the frame $Axyz$ which coincides with P, we write from Eq. (15.46)

$$\mathbf{v}_P = \mathbf{v}_{P'} + \mathbf{v}_{P/F} \qquad (1)$$

where
$$\mathbf{v}_{P'} = \mathbf{\Omega} \times \mathbf{r} = \omega_1\mathbf{j} \times (L\mathbf{i} + R\mathbf{j}) = -\omega_1 L\mathbf{k}$$
$$\mathbf{v}_{P/F} = \boldsymbol{\omega}_{D/F} \times \mathbf{r}_{P/A} = \omega_2\mathbf{k} \times R\mathbf{j} = -\omega_2 R\mathbf{i}$$

Substituting the values obtained for $\mathbf{v}_{P'}$ and $\mathbf{v}_{P/F}$ into (1), we find

$$\mathbf{v}_P = -\omega_2 R\mathbf{i} - \omega_1 L\mathbf{k} \quad ◄$$

b. **Acceleration $\mathbf{a}_P$.** From Eq. (15.48) we write

$$\mathbf{a}_P = \mathbf{a}_{P'} + \mathbf{a}_{P/F} + \mathbf{a}_c \qquad (2)$$

Since $\mathbf{\Omega}$ and $\boldsymbol{\omega}_{D/F}$ are both constant, we have

$$\mathbf{a}_{P'} = \mathbf{\Omega} \times (\mathbf{\Omega} \times \mathbf{r}) = \omega_1\mathbf{j} \times (-\omega_1 L\mathbf{k}) = -\omega_1^2 L\mathbf{i}$$
$$\mathbf{a}_{P/F} = \boldsymbol{\omega}_{D/F} \times (\boldsymbol{\omega}_{D/F} \times \mathbf{r}_{P/A}) = \omega_2\mathbf{k} \times (-\omega_2 R\mathbf{i}) = -\omega_2^2 R\mathbf{j}$$
$$\mathbf{a}_c = 2\mathbf{\Omega} \times \mathbf{v}_{P/F} = 2\omega_1\mathbf{j} \times (-\omega_2 R\mathbf{i}) = 2\omega_1\omega_2 R\mathbf{k}$$

Substituting the values obtained into (2), we find

$$\mathbf{a}_P = -\omega_1^2 L\mathbf{i} - \omega_2^2 R\mathbf{j} + 2\omega_1\omega_2 R\mathbf{k} \quad ◄$$

c. **Angular Velocity and Angular Acceleration of Disk.**

$$\boldsymbol{\omega} = \mathbf{\Omega} + \boldsymbol{\omega}_{D/F} \qquad \boldsymbol{\omega} = \omega_1\mathbf{j} + \omega_2\mathbf{k} \quad ◄$$

Using Eq. (15.31) with $\mathbf{Q} = \boldsymbol{\omega}$, we write

$$\boldsymbol{\alpha} = (\dot{\boldsymbol{\omega}})_{OXYZ} = (\dot{\boldsymbol{\omega}})_{Axyz} + \mathbf{\Omega} \times \boldsymbol{\omega}$$
$$= 0 + \omega_1\mathbf{j} \times (\omega_1\mathbf{j} + \omega_2\mathbf{k})$$
$$\boldsymbol{\alpha} = \omega_1\omega_2\mathbf{i} \quad ◄$$

PROBLEMS

15.163 The bent rod ABC rotates at a constant rate $\omega_1 = 8$ rad/s. Knowing that the collar D moves downward along the rod at a constant relative speed $u = 780$ mm/s, determine for the position shown (*a*) the velocity of D, (*b*) the acceleration of D.

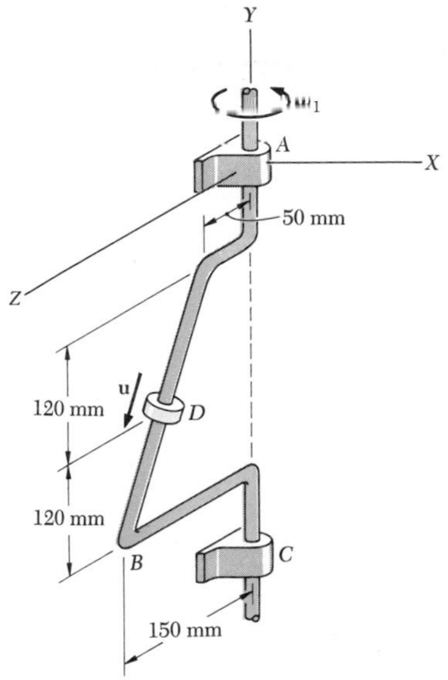

Fig. P15.163

15.164 Solve Prob. 15.163, assuming $\omega_1 = 6$ rad/s and $u = 650$ mm/s.

15.165 The bent rod ABC rotates at a constant rate ω_1. Knowing that the collar D moves downward along the rod at a constant relative speed u, determine for the position shown (*a*) the velocity of D, (*b*) the acceleration of D.

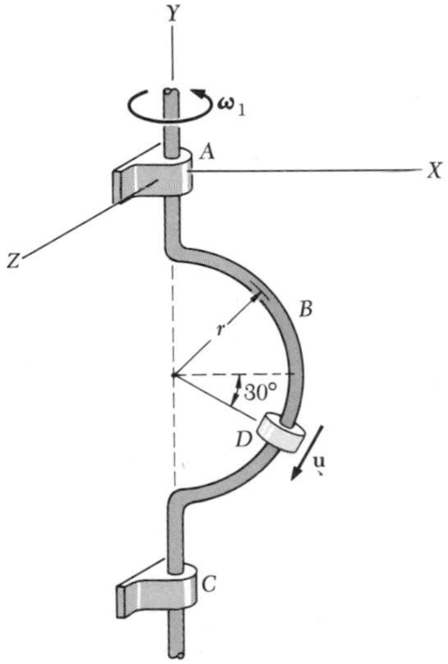

Fig. P15.165 and P15.167

15.166 Solve Prob. 15.165, assuming that $\omega_1 = 9$ rad/s, $u = 40$ in./s, and $r = 6$ in.

15.167 At the instant shown the magnitude of the angular velocity $\boldsymbol{\omega}_1$ of the bent rod ABC is 9 rad/s and is increasing at the rate of 20 rad/s^2, while the relative speed u of collar D is 40 in./s and is increasing at the rate of 100 in./s^2. Knowing that $r = 6$ in., determine the acceleration of D.

15.168 Solve Prob. 15.163, assuming that at the instant shown the angular velocity $\boldsymbol{\omega}_1$ of the rod is 8 rad/s and is decreasing at the rate of 18 rad/s^2, while the relative speed u of the collar is 780 mm/s and is decreasing at the rate of 2.6 m/s^2.

15.169 The cab of the backhoe shown rotates with the constant angular velocity $\boldsymbol{\omega}_1 = (0.4 \text{ rad/s})\mathbf{j}$ about the Y axis. The arm OA is fixed with respect to the cab, while the arm AB rotates about the horizontal axle A at the constant rate $\omega_2 = d\beta/dt = 0.6$ rad/s. Knowing that $\beta = 30°$, determine (*a*) the angular velocity and angular acceleration of AB, (*b*) the velocity and acceleration of point B.

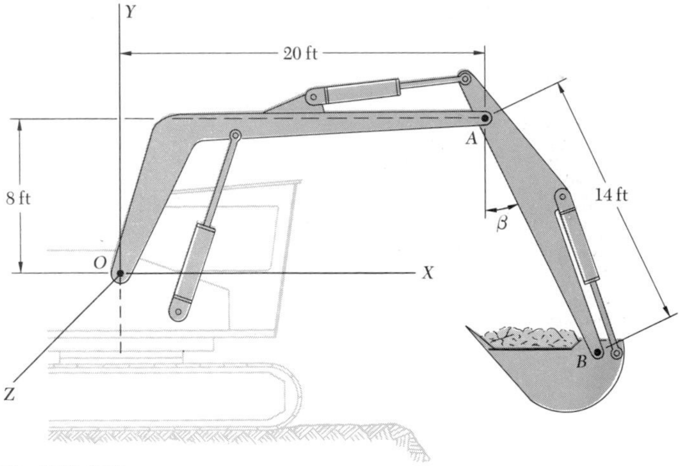

Fig. P15.169

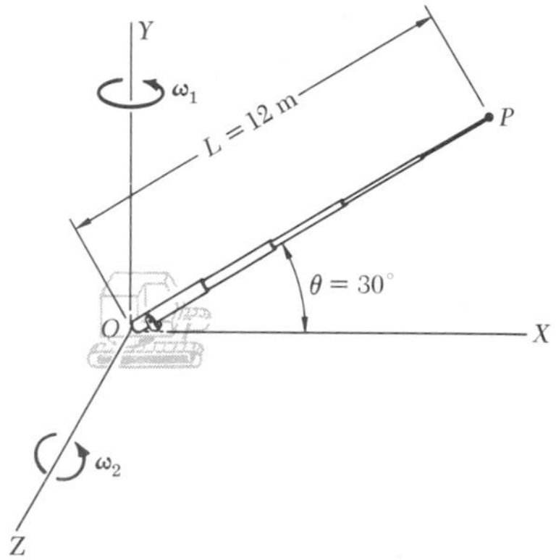

Fig. P15.170

15.170 Solve Sample Prob. 15.14, assuming that the crane has a telescoping boom as shown and that the length of the boom is being increased at the rate $dL/dt = 1.5$ m/s.

15.171 Solve Prob. 15.169, assuming that $\beta = 30°$ and that arms OA and AB rotate as a rigid body with respect to the cab with a constant angular velocity $(0.6 \text{ rad/s})\mathbf{k}$.

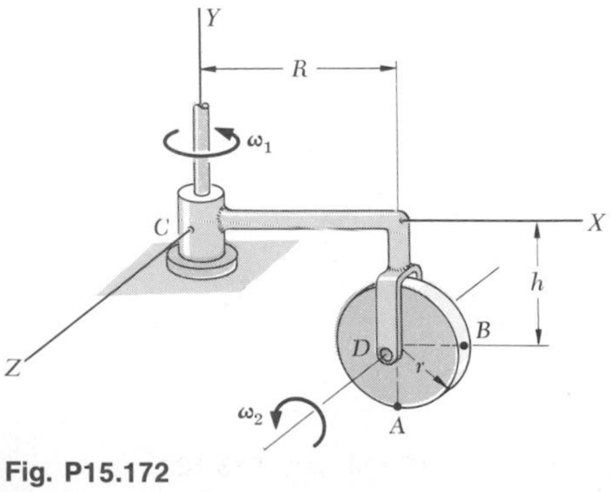

Fig. P15.172

15.172 A disk of radius r rotates at a constant rate ω_2 with respect to the arm CD, which itself rotates at a constant rate ω_1 about the Y axis. Determine (*a*) the angular velocity and angular acceleration of the disk, (*b*) the velocity and acceleration of point B on the rim of the disk.

15.173 In Prob. 15.172, determine the velocity and acceleration of point A on the rim of the disk.

15.174 The 40-ft blades of the experimental wind-turbine generator rotate at a constant rate $\omega = 30$ rpm. Knowing that at the instant shown the entire unit is being rotated about the Y axis at a constant rate $\Omega = 0.1$ rad/s, determine (*a*) the angular acceleration of the blades, (*b*) the velocity and acceleration of blade tip B.

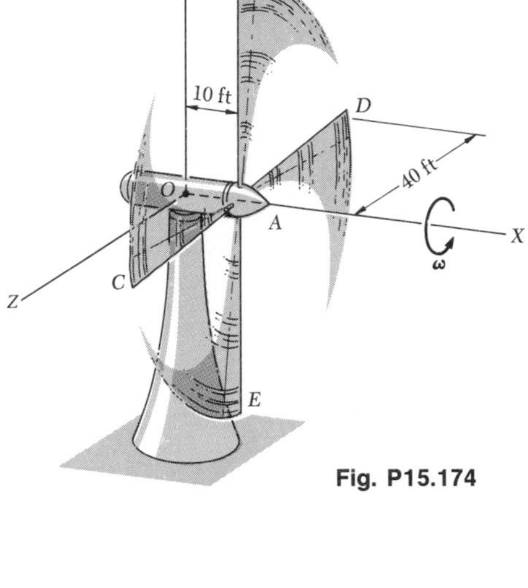

Fig. P15.174

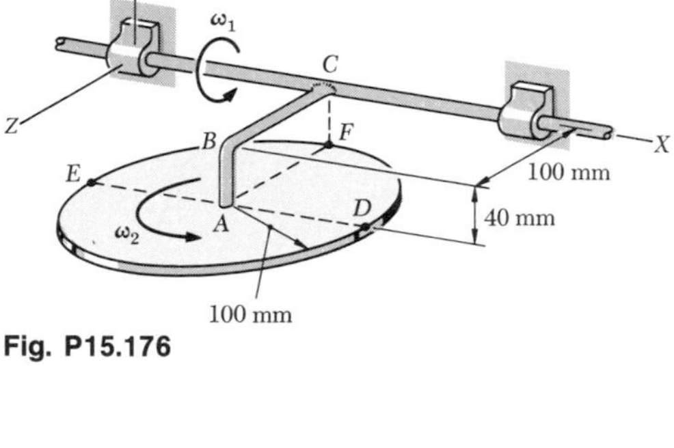

Fig. P15.176

15.175 In Prob. 15.174, determine the velocity and acceleration of (*a*) blade tip C, (*b*) blade tip E.

15.176 A disk of radius 100 mm rotates at a constant rate $\omega_2 = 20$ rad/s with respect to the arm ABC, which itself rotates at a constant rate $\omega_1 = 10$ rad/s about the X axis. Determine (*a*) the angular acceleration of the disk, (*b*) the velocity and acceleration of point D on the rim of the disk.

15.177 In Prob. 15.176, determine the acceleration (*a*) of point E, (*b*) of point F.

15.178 through 15.180 Two collars A and B are connected by a 15-in. rod AB as shown. Knowing that collar A moves downward at a constant speed of 18 in./s, determine the velocities and accelerations of collars A and B for the constant rate of rotation indicated.

15.178 $\omega_1 = 10$ rad/s, $\omega_2 = \omega_3 = 0$.

15.179 $\omega_2 = 10$ rad/s, $\omega_1 = \omega_3 = 0$.

15.180 $\omega_3 = 10$ rad/s, $\omega_1 = \omega_2 = 0$.

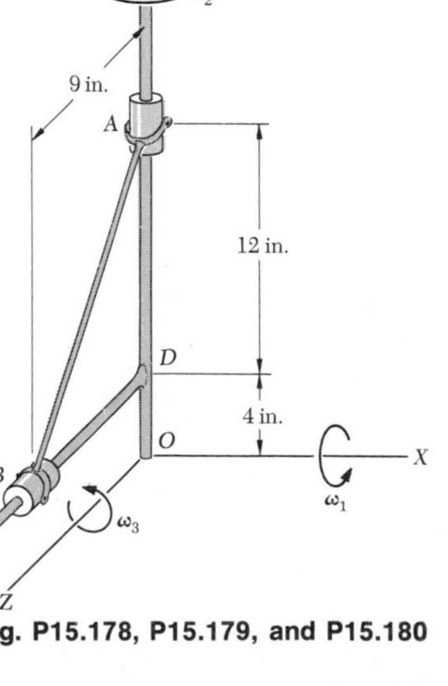

Fig. P15.178, P15.179, and P15.180

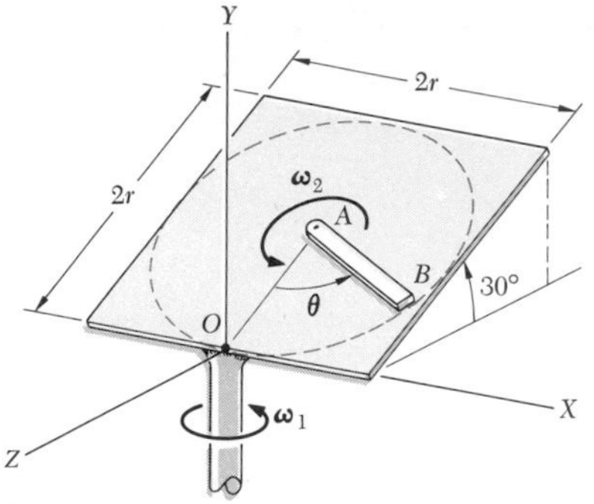

Fig. P15.181

15.181 A square plate of side $2r$ is welded to a vertical shaft which rotates with a constant angular velocity $\boldsymbol{\omega}_1$. At the same time, rod AB of length r rotates about the center of the plate with a constant angular velocity $\boldsymbol{\omega}_2$ with respect to the plate. For the position of the plate shown, determine the acceleration of end B of the rod if (*a*) $\theta = 0$, (*b*) $\theta = 90°$, (*c*) $\theta = 180°$.

15.182 Solve Prob. 15.181, assuming $\omega_1 = 2$ rad/s, $\omega_2 = 3$ rad/s, and $r = 100$ mm.

15.183 In Prob. 15.181, the plate rotates at a constant rate $\omega_1 = 2$ rad/s. At the same time, the magnitude ω_2 is 3 rad/s and is increasing at the rate $\alpha_2 = 5$ rad/s^2. Knowing that $r = 100$ mm, determine the acceleration of end B of the rod if $\theta = 90°$.

15.184 In Prob. 15.181, the magnitude of the angular velocity of the plate is $\omega_1 = 2$ rad/s and is increasing at the rate $\alpha_1 = 8$ rad/s^2. At the same time, the rod AB rotates with respect to the plate at the constant rate $\omega_2 = 3$ rad/s. Knowing that $r = 100$ mm, determine the acceleration of end B of the rod if $\theta = 90°$.

REVIEW PROBLEMS

15.185 It takes 0.8 s for the turntable of a 33-rpm record player to reach full speed after being started. Assuming uniformly accelerated motion, determine (*a*) the angular acceleration of the turntable, (*b*) the normal and tangential components of the acceleration of a point on the rim of the 12-in.-diameter turntable just before the speed of 33 rpm is reached, (*c*) the total acceleration of the same point at that time.

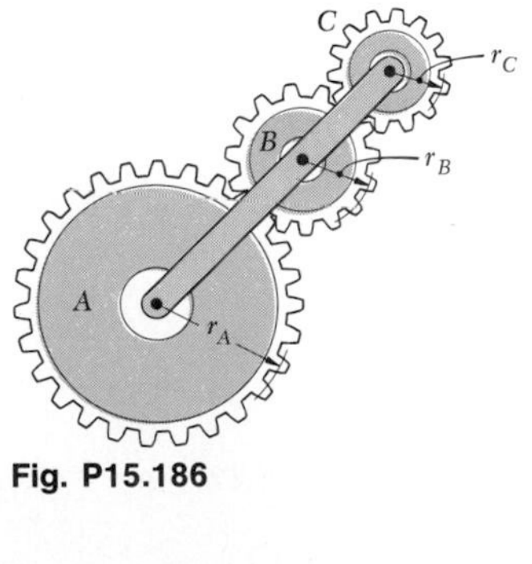

Fig. P15.186

15.186 Three gears A, B, and C are pinned at their centers to rod ABC. Knowing that $r_A = 3r_B = 3r_C$ and that gear A does not rotate, determine the angular velocity of gears B and C when the rod ABC rotates clockwise with a constant angular velocity of 10 rpm.

15.187 In Prob. 15.186 it is known that $r_A = 12$ in., $r_B = r_C = 4$ in. Determine the acceleration of the tooth of gear C which is in contact with gear B.

15.188 Three links AB, BC, and BD are connected by a pin B as shown. Knowing that at the instant shown point D has a velocity of 200 mm/s to the right and no acceleration, determine (*a*) the angular acceleration of each link, (*b*) the accelerations of points A and B.

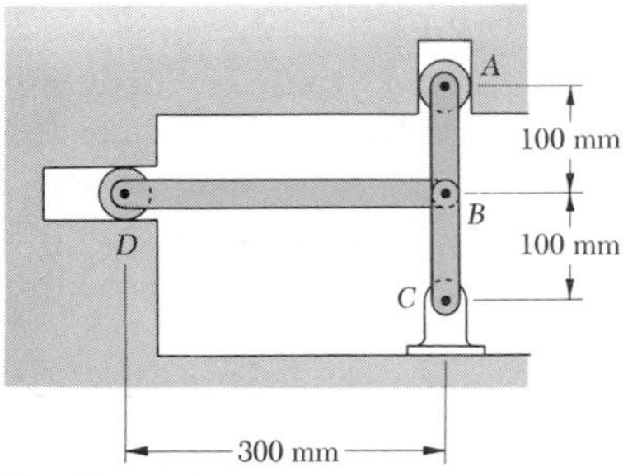

Fig. P15.188

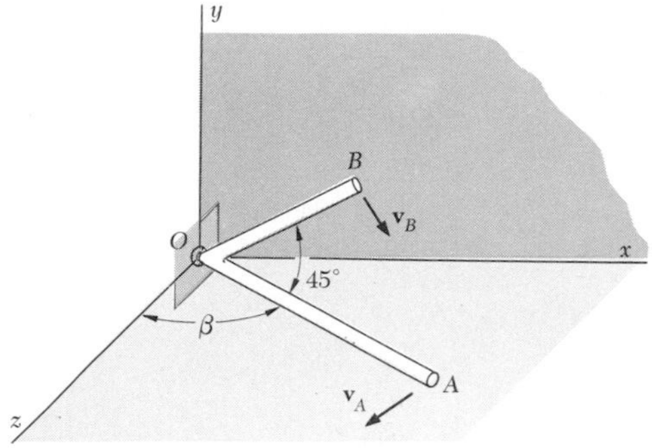

Fig. P15.189

15.189 The bent rod AOB is attached to a fixed ball-and-socket joint at O. The lengths of portions OA and OB are 200 mm and 120 mm, respectively, and the angle formed by the two portions is 45°. As portion OA moves on the horizontal surface, portion OB moves on the vertical wall. Knowing that end A moves at a constant speed of 600 mm/s, determine, at the instant when $\beta = 60°$, (*a*) the angular velocity of the rod, (*b*) the velocity of point B.

15.190 Water flows through the sprinkler arm ABC with a velocity of 16 ft/s relative to the arm. Knowing that the angular velocity of the arm is 90 rpm counterclockwise, determine at the instant shown the total acceleration (*a*) of the particle of water P_1, (*b*) of the particle of water P_2.

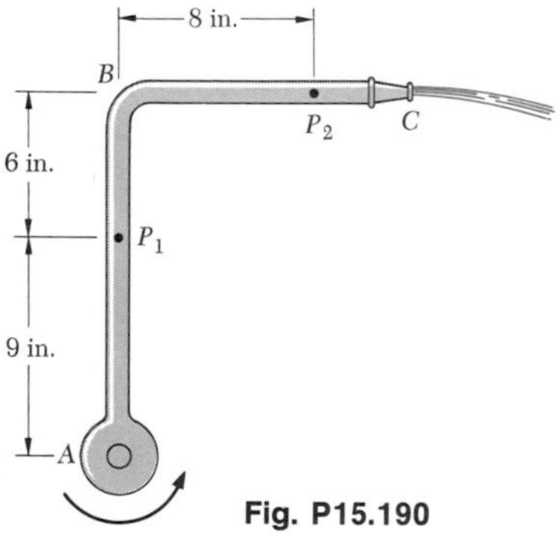

Fig. P15.190

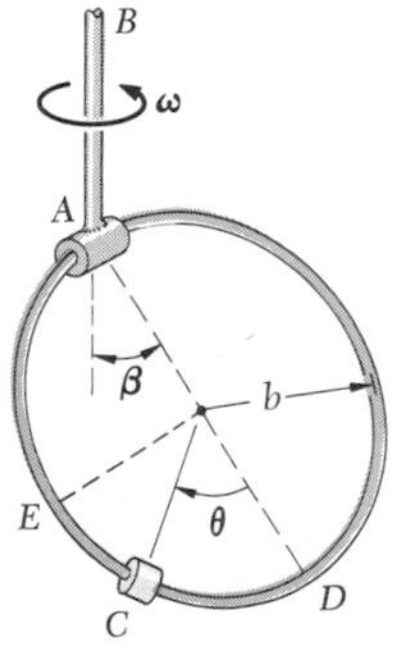

Fig. P15.191

15.191 A thin ring of radius b is attached to a vertical shaft AB which rotates with a constant angular velocity ω. Collar C moves at a constant speed u relative to the ring. For the position $\beta = 30°$, determine the velocity and acceleration of the collar when (*a*) $\theta = 0$, (*b*) $\theta = 90°$.

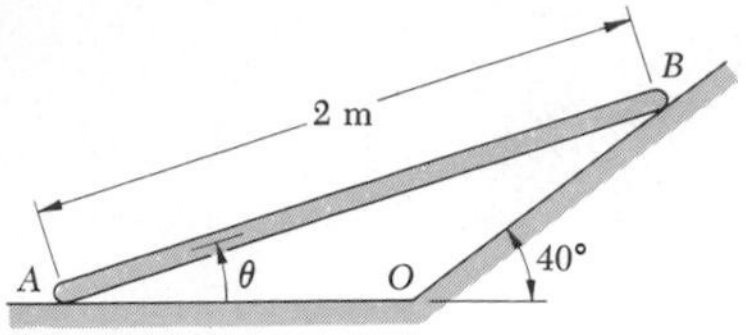

Fig. P15.192

15.192 Rod AB is 2 m long and slides with its ends in contact with the floor and the inclined plane. End A moves with a constant velocity of 6 m/s to the right. At the instant when $\theta = 25°$, determine (*a*) the angular velocity and angular acceleration of the rod, (*b*) the velocity and acceleration of end B.

15.193 Gear A rolls on the fixed gear B and rotates about the axle AD which is rigidly attached at D to the vertical shaft DE. Knowing that shaft DE rotates with a constant angular velocity $\boldsymbol{\omega}_1$, determine (*a*) the rate of spin of gear A about the axle AD, (*b*) the angular acceleration of gear A, (*c*) the acceleration of tooth C of gear A.

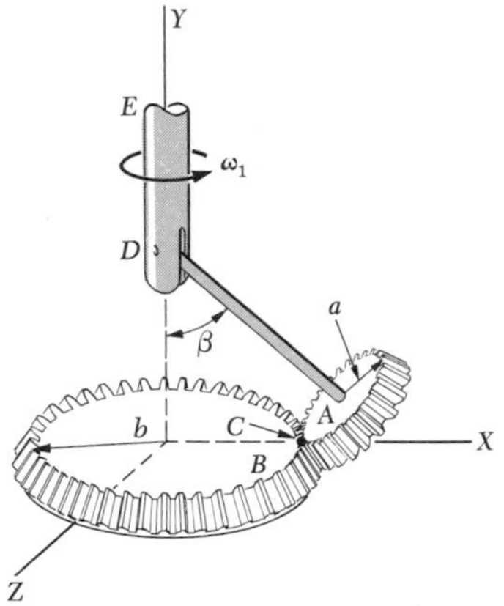

Fig. P15.193

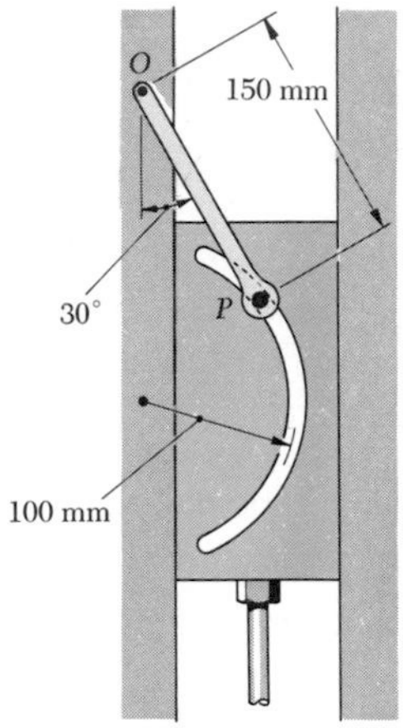

Fig. P15.194

15.194 At the instant shown, the slotted plate slides with a velocity of 0.5 m/s upward and has an acceleration of 2 m/s^2 downward. Determine the angular velocity and the angular acceleration of rod OP.

15.195 Solve Prob. 15.193, assuming $\omega_1 = 90$ rpm, $a = 60$ mm, $b = 160$ mm, and $\beta = 30°$.

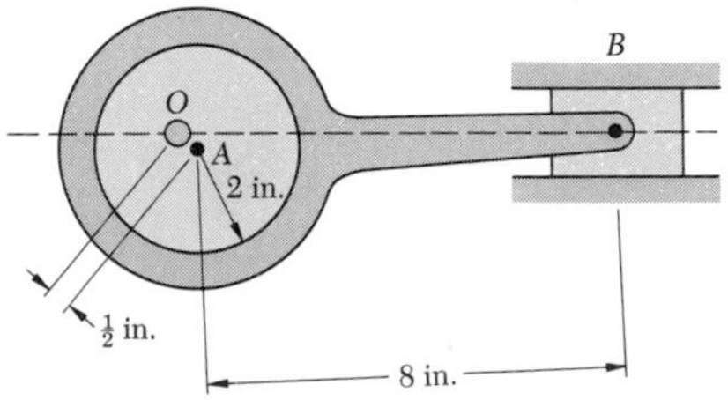

Fig. P15.196

15.196 The eccentric shown consists of a disk of 2-in. radius which revolves about a shaft O located $\frac{1}{2}$ in. from the center of the disk A. Assuming that the disk rotates about O with a constant angular velocity of 1800 rpm clockwise, determine the velocity and acceleration of block B when point A is directly below the shaft O.

CHAPTER 16

Plane Motion of Rigid Bodies: Forces and Accelerations

16.1 Introduction. In this chapter and in Chaps. 17 and 18, we shall study the *kinetics of rigid bodies,* i.e., the relations existing between the forces acting on a rigid body, the shape and mass of the body, and the motion produced. In Chaps. 12 and 13, we studied similar relations, assuming then that the body could be considered as a particle, i.e., that its mass could be concentrated in one point and that all forces acted at that point. We shall now take the shape of the body into account, as well as the exact location of the points of application of the forces. Besides, we shall be concerned not only with the motion of the body as a whole but also with the motion of the body about its mass center.

Our approach will be to consider rigid bodies as made of large numbers of particles and to use the results obtained in Chap. 14 for the motion of systems of particles. In this chapter, we shall use specifically Eq. (14.16), $\Sigma\mathbf{F} = m\bar{\mathbf{a}}$, which relates the resultant of the external forces and the acceleration of the mass center G of the system of particles, and Eq. (14.23), $\Sigma\mathbf{M}_G = \dot{\mathbf{H}}_G$, which relates the moment resultant of the external forces and the angular momentum of the system of particles about G.

Except for Sec. 16.2, which applies to the most general case of the motion of a rigid body, the results derived in this chapter will be limited in two ways: (1) They will be restricted to the *plane motion* of rigid bodies, i.e., to a motion in which each particle of the body remains at a constant distance from a fixed reference plane. (2) The rigid bodies considered will consist only of plane slabs and of bodies which are symmetrical with respect to the reference plane.† The study of the plane motion of nonsymmetrical three-dimensional bodies and, more generally, the motion of rigid bodies in three-dimensional space will be postponed until Chap. 18.

16.2. Equations of Motion for a Rigid Body. Consider a rigid body acted upon by several external forces $\mathbf{F}_1$, $\mathbf{F}_2$, $\mathbf{F}_3$, etc. (Fig. 16.1). We may assume the body to be made of a large number n of particles of mass Δm_i $(i = 1,2, \ldots ,n)$ and apply the results obtained in Chap. 14 for a system of particles (Fig. 16.2).

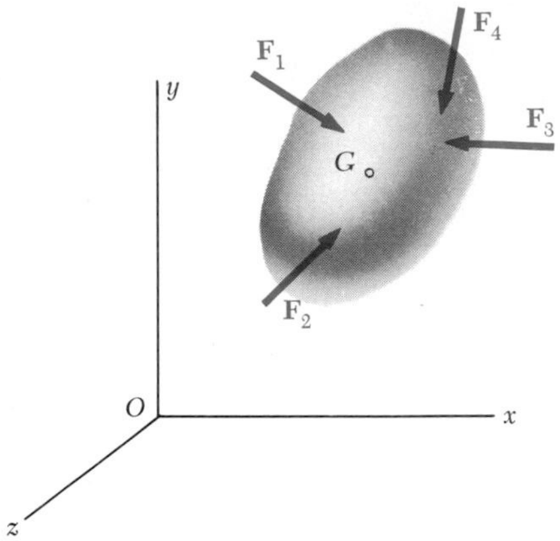

Fig. 16.1

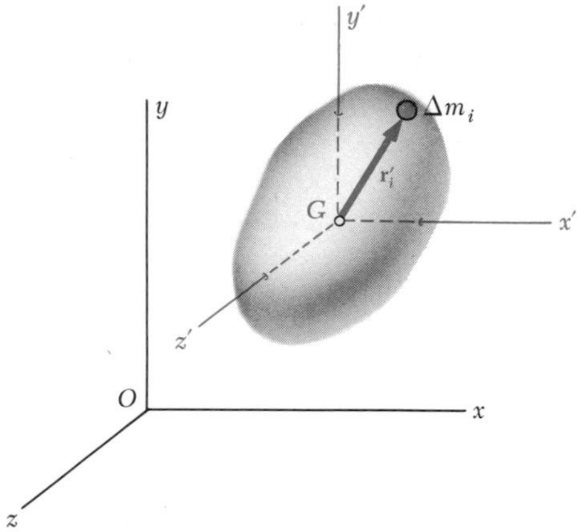

Fig. 16.2

Considering first the motion of the mass center G of the body with respect to the newtonian frame of reference $Oxyz$, we recall Eq. (14.16) and write

$$\Sigma\mathbf{F} = m\bar{\mathbf{a}} \tag{16.1}$$

where m is the mass of the body and $\bar{\mathbf{a}}$ the acceleration of the mass center G. Turning now to the motion of the body relative to the centroidal frame of reference $Gx'y'z'$, we recall Eq. (14.23) and write

$$\Sigma\mathbf{M}_G = \dot{\mathbf{H}}_G \tag{16.2}$$

where $\dot{\mathbf{H}}_G$ represents the rate of change of $\mathbf{H}_G$, the angular momentum about G of the system of particles forming the rigid body. In the following we shall simply refer to $\mathbf{H}_G$ as the *angular momentum of the rigid body about its mass center G*. Together Eqs. (16.1) and (16.2) express that *the system of the external forces*

† Or, more generally, bodies which have a principal centroidal axis of inertia perpendicular to the reference plane.

is equipollent to the system consisting of the vector $m\bar{\mathbf{a}}$ attached at G and the couple of moment $\dot{\mathbf{H}}_G$ (Fig. 16.3).†

Equations (16.1) and (16.2) apply in the most general case of the motion of a rigid body. In the rest of this chapter, however, we shall limit our analysis to the *plane motion* of rigid bodies, i.e., to a motion in which each particle remains at a constant distance from a fixed reference plane, and we shall assume that the rigid bodies considered consist only of plane slabs and of bodies which are symmetrical with respect to the reference plane. Further study of the plane motion of nonsymmetrical three-dimensional bodies and of the motion of rigid bodies in three-dimensional space will be postponed until Chap. 18.

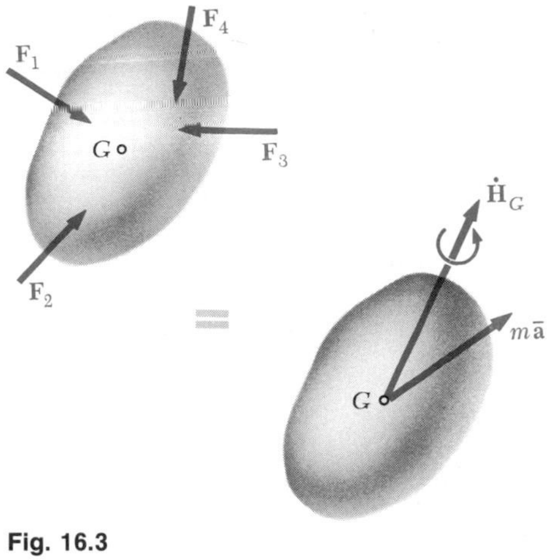

Fig. 16.3

16.3. Angular Momentum of a Rigid Body in Plane Motion. Consider a rigid slab in plane motion. Assuming the slab to be made of a large number n of particles P_i of mass Δm_i and recalling Eq. (14.24) of Sec. 14.4, we note that the angular momentum $\mathbf{H}_G$ of the slab about its mass center G may be computed by taking the moments about G of the momenta of the particles of the slab in their motion with respect to either of the frames Oxy or $Gx'y'$. Choosing the latter course, we write

$$\mathbf{H}_G = \sum_{i=1}^{n} (\mathbf{r}'_i \times \mathbf{v}'_i \, \Delta m_i) \tag{16.3}$$

where $\mathbf{r}'_i$ and $\mathbf{v}'_i \, \Delta m_i$ denote, respectively, the position vector and the linear momentum of the particle P_i relative to the centroidal frame of reference $Gx'y'$ (Fig. 16.4). But, since the particle belongs to the slab, we have $\mathbf{v}'_i = \boldsymbol{\omega} \times \mathbf{r}'_i$, where $\boldsymbol{\omega}$ is the angular velocity of the slab at the instant considered. We write

$$\mathbf{H}_G = \sum_{i=1}^{n} [\mathbf{r}'_i \times (\boldsymbol{\omega} \times \mathbf{r}'_i) \, \Delta m_i]$$

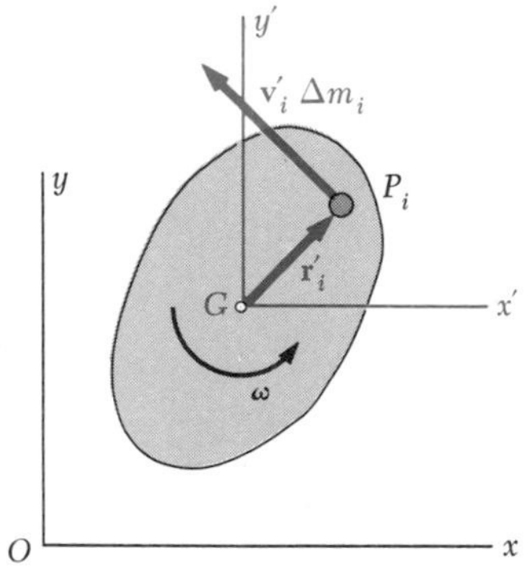

Fig. 16.4

Referring to Fig. 16.4, we easily verify that the expression obtained represents a vector of the same direction as $\boldsymbol{\omega}$ (i.e., perpendicular to the slab) and of magnitude equal to $\omega \Sigma r_i'^2 \, \Delta m_i$. Recalling that the sum $\Sigma r_i'^2 \, \Delta m_i$ represents the moment of inertia $\bar{I}$ of the slab about a centroidal axis perpendicular to the slab, we conclude that the angular momentum $\mathbf{H}_G$ of the slab about its mass center is

$$\mathbf{H}_G = \bar{I}\boldsymbol{\omega} \tag{16.4}$$

† Since the systems involved act on a rigid body, we could conclude at this point, by referring to Sec. 3.18, that the two systems are *equivalent* as well as equipollent and use blue rather than gray equals signs in Fig. 16.3. However, by postponing this conclusion, we shall be able to arrive at it independently (Secs. 16.4 and 18.5), thus eliminating the necessity of including the principle of transmissibility among the axioms of mechanics (Sec. 16.5).

Differentiating both members of Eq. (16.4) we obtain

$$\dot{\mathbf{H}}_G = \bar{I}\dot{\boldsymbol{\omega}} = \bar{I}\boldsymbol{\alpha} \tag{16.5}$$

Thus the rate of change of the angular momentum of the slab is represented by a vector of the same direction as $\boldsymbol{\alpha}$, (i.e., perpendicular to the slab) and of magnitude $\bar{I}\alpha$.

It should be kept in mind that the results obtained in this section have been derived for a rigid slab in plane motion. As we shall see in Chap. 18, they remain valid in the case of the plane motion of rigid bodies which are symmetrical with respect to the reference plane.† However, they do not apply in the case of nonsymmetrical bodies or in the case of three-dimensional motion.

16.4. Plane Motion of a Rigid Body. D'Alembert's Principle. Consider a rigid slab of mass m moving under the action of several external forces $\mathbf{F}_1$, $\mathbf{F}_2$, $\mathbf{F}_3$, etc., contained in the plane of the slab (Fig. 16.5). Substituting for $\dot{\mathbf{H}}_G$ from Eq. (16.5) into Eq. (16.2), and writing the fundamental equations of motion (16.1) and (16.2) in scalar form, we have

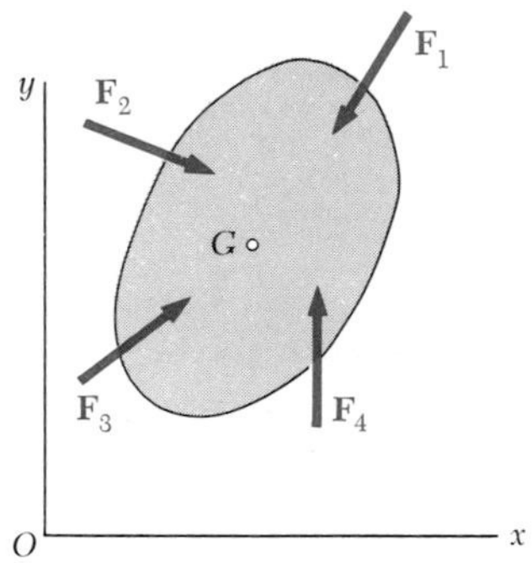

Fig. 16.5

$$\Sigma F_x = m\bar{a}_x \qquad \Sigma F_y = m\bar{a}_y \qquad \Sigma M_G = \bar{I}\alpha \tag{16.6}$$

Equations (16.6) show that the acceleration of the mass center G of the slab and its angular acceleration $\boldsymbol{\alpha}$ may easily be obtained, once the resultant of the external forces acting on the slab and their moment resultant about G have been determined. Given appropriate initial conditions, the coordinates $\bar{x}$ and $\bar{y}$ of the mass center and the angular coordinate θ of the slab may then be obtained at any instant t by integration. Thus *the motion of the slab is completely defined by the resultant and moment resultant about G of the external forces acting on it.*

This property, which will be extended in Chap. 18 to the case of the three-dimensional motion of a rigid body, is characteristic of the motion of a rigid body. Indeed, as we saw in Chap. 14, the motion of a system of particles which are not rigidly connected will in general depend upon the specific external forces, as well as upon the internal forces, acting on the various particles.

Since the motion of a rigid body depends only upon the resultant and moment resultant of the external forces acting on it, it follows that *two systems of forces which are equipollent,* i.e., which have the same resultant and the same moment resultant,

† Or, more generally, bodies which have a principal centroidal axis of inertia perpendicular to the reference plane.

are also equivalent; i.e., they have exactly the same effect on a given rigid body.†

Consider in particular the system of the external forces acting on a rigid body (Fig. 16.6*a*) and the system of the effective forces associated with the particles forming the rigid body (Fig. 16.6*b*). It was shown in Sec. 14.1 that the two systems thus defined are equipollent. But since the particles considered now form a rigid body, it follows from the above discussion that the two systems are also equivalent. We may thus state that *the external forces acting on a rigid body are equivalent to the effective forces of the various particles forming the body.* This statement is referred to as *D'Alembert's principle,* after the French mathematician Jean le Rond d'Alembert (1717–1783), even though D'Alembert's original statement was written in a somewhat different form.

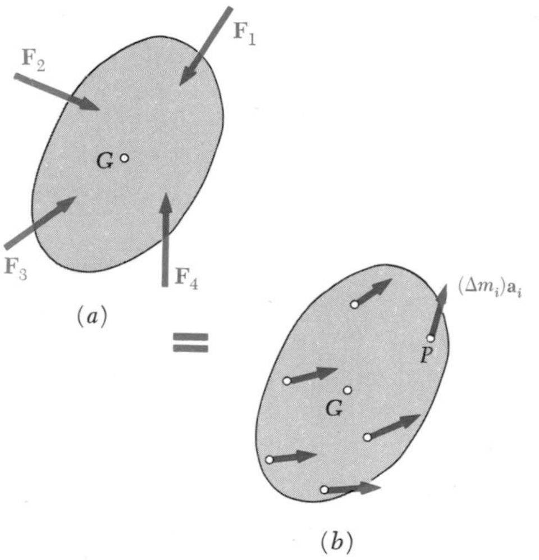

Fig. 16.6

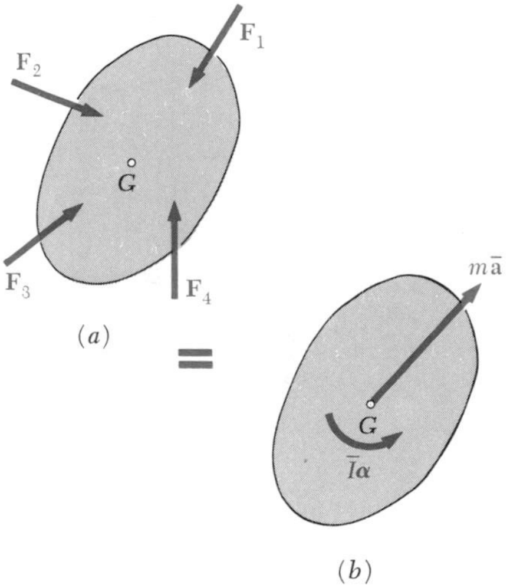

Fig. 16.7

The significance of D'Alembert's principle has been emphasized by the use of a blue equals sign in Fig. 16.6 and also in Fig. 16.7, where, using results obtained earlier in this section, the effective forces have been replaced by a vector $m\bar{\mathbf{a}}$ attached at the mass center G of the slab and a couple of moment $\bar{I}\boldsymbol{\alpha}$.

† This result has already been derived in Sec. 3.18 from the principle of transmissibility (Sec. 3.2). The present derivation, however, is independent of that principle and will make possible its elimination from the axioms of mechanics (Sec. 16.5).

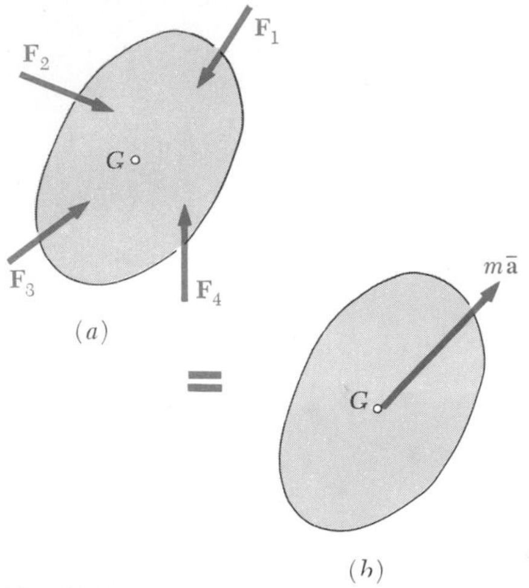

Fig. 16.8

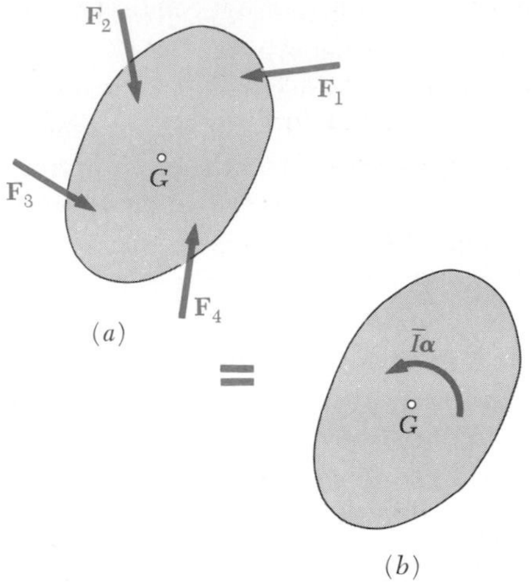

Fig. 16.9

Translation. In the particular case of a body in translation, the angular acceleration of the body is identically equal to zero and its effective forces reduce to the vector $m\bar{\mathbf{a}}$ attached at G (Fig. 16.8). Thus, the resultant of the external forces acting on a rigid body in translation passes through the mass center of the body and is equal to $m\bar{\mathbf{a}}$.

Centroidal Rotation. When a slab, or, more generally, a body symmetrical with respect to the reference plane, rotates about a fixed axis perpendicular to the reference plane and passing through its mass center G, we say that the body is in *centroidal rotation*. Since the acceleration $\bar{\mathbf{a}}$ is identically equal to zero, the effective forces of the body reduce to the couple $\bar{I}\boldsymbol{\alpha}$ (Fig. 16.9). Thus, the external forces acting on a body in centroidal rotation are equivalent to a couple of moment $\bar{I}\boldsymbol{\alpha}$.

General Plane Motion. Comparing Fig. 16.7 with Figs. 16.8 and 16.9, we observe that, from the point of view of *kinetics*, the most general plane motion of a rigid body symmetrical with respect to the reference plane may be replaced by the sum of a translation and a centroidal rotation. We should note that this statement is more restrictive than the similar statement made earlier from the point of view of *kinematics* (Sec. 15.5), since we now require that the mass center of the body be selected as the reference point.

Referring to Eqs. (16.6), we observe that the first two equations are identical with the equations of motion of a particle of mass m acted upon by the given forces $\mathbf{F}_1$, $\mathbf{F}_2$, $\mathbf{F}_3$, etc. We thus check that *the mass center G of a rigid body in plane motion moves as if the entire mass of the body were concentrated at that point, and as if all the external forces acted on it.* We recall that this result

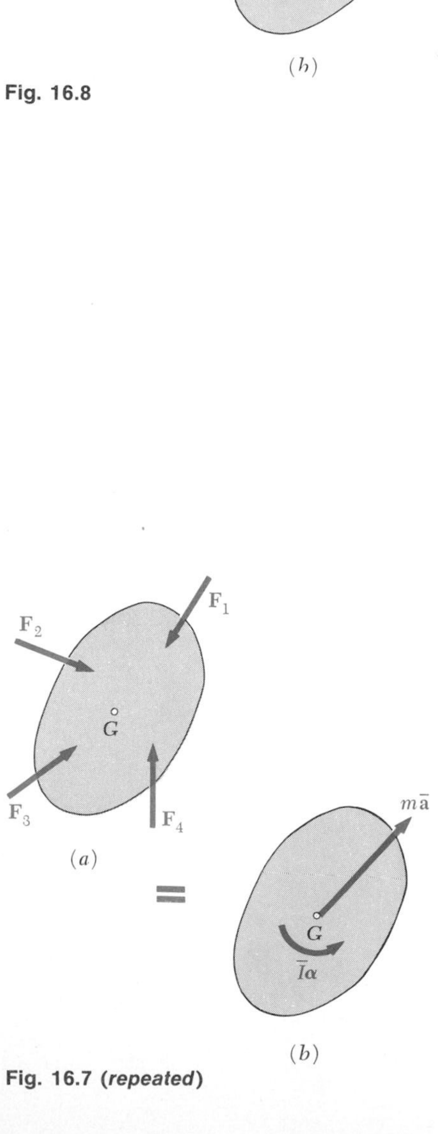

Fig. 16.7 (*repeated*)

has already been obtained in Sec. 14.3 in the general case of a system of particles, the particles being not necessarily rigidly connected. We also note, as we did in Sec. 14.3, that the system of the external forces does not, in general, reduce to a single vector $m\bar{\mathbf{a}}$ attached at G. Therefore, in the general case of the plane motion of a rigid body, *the resultant of the external forces acting on the body does not pass through the mass center of the body.*

Finally, we may observe that the last of Eqs. (16.6) would still be valid if the rigid body, while subjected to the same applied forces, were constrained to rotate about a fixed axis through G. Thus, *a rigid body in plane motion rotates about its mass center as if this point were fixed.*

∗16.5. A Remark on the Axioms of the Mechanics of Rigid Bodies. The fact that two equipollent systems of external forces acting on a rigid body are also equivalent, i.e., have the same effect on that rigid body, has already been established in Sec. 3.18. But there it was derived from the *principle of transmissibility,* one of the axioms used in our study of the statics of rigid bodies. It should be noted that this axiom has not been used in the present chapter, because Newton's second and third laws of motion make its use unnecessary in the study of the dynamics of rigid bodies.

In fact, the principle of transmissibility may now be *derived* from the other axioms used in the study of mechanics. This principle stated, without proof, (Sec. 3.2) that the conditions of equilibrium or motion of a rigid body remain unchanged if a force $\mathbf{F}$ acting at a given point of the rigid body is replaced by a force $\mathbf{F}'$ of the same magnitude and same direction, but acting at a different point, provided that the two forces have the same line of action. But, since $\mathbf{F}$ and $\mathbf{F}'$ have the same moment about any given point, it is clear that they form two equipollent systems of external forces. Thus, we may now *prove,* as a result of what we established in the preceding section, that $\mathbf{F}$ and $\mathbf{F}'$ have the same effect on the rigid body (Fig. 3.3).

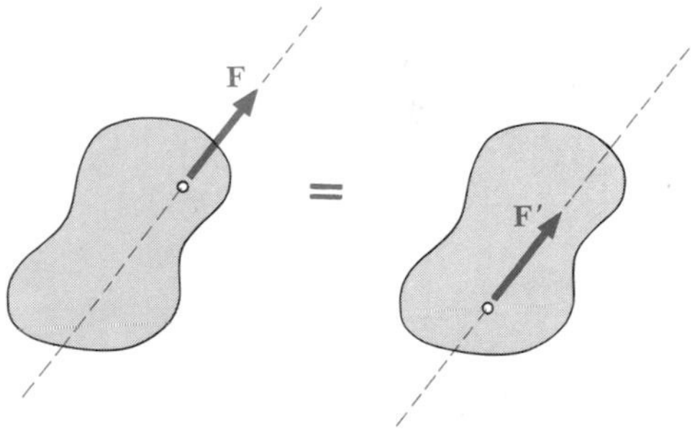

Fig. 3.3 (*repeated*)

The principle of transmissibility may therefore be removed from the list of axioms required for the study of the mechanics of rigid bodies. These axioms are reduced to the parallelogram law of addition of vectors and to Newton's laws of motion.

16.6. Solution of Problems Involving the Motion of a Rigid Body. We saw in Sec. 16.4 that, when a rigid body is in plane motion, there exists a fundamental relation between the forces $\mathbf{F}_1$, $\mathbf{F}_2$, $\mathbf{F}_3$, etc., acting on the body, the acceleration $\bar{\mathbf{a}}$ of its mass center, and the angular acceleration $\boldsymbol{\alpha}$ of the body. This relation, which is represented in Fig. 16.7, may be used to determine the acceleration $\bar{\mathbf{a}}$ and the angular acceleration $\boldsymbol{\alpha}$ produced by a given system of forces acting on a rigid body or, conversely,

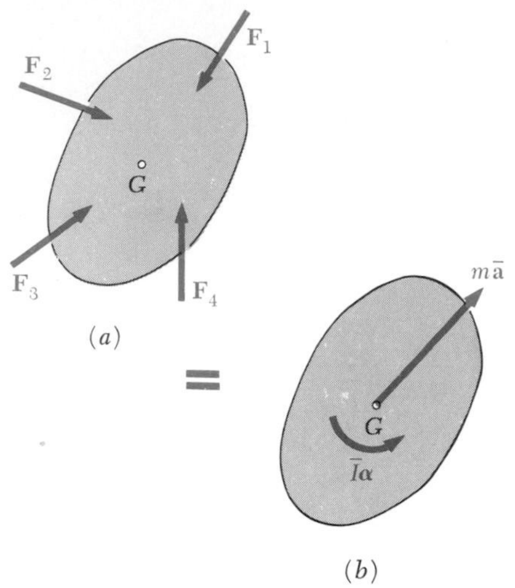

Fig. 16.7 (*repeated*)

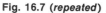

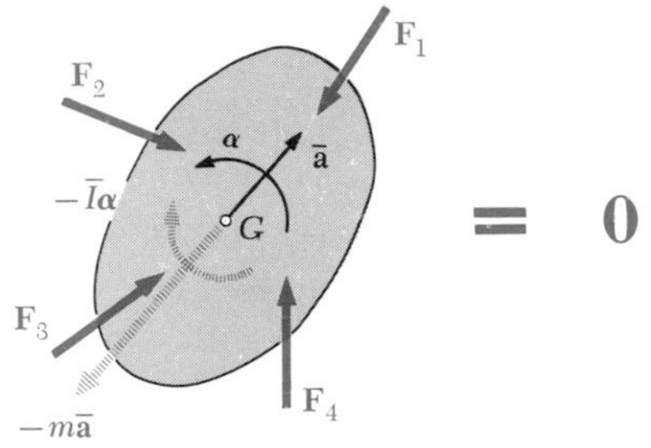

Fig. 16.10

to determine the forces which produce a given motion of the rigid body.

While the three algebraic equations (16.6) may be used to solve problems of plane motion,† our experience in statics suggests that the solution of many problems involving rigid bodies could be simplified by an appropriate choice of the point about which the moments of the forces are computed. It is therefore preferable to remember the relation existing between the forces and the accelerations in the vectorial form shown in Fig. 16.7, and to derive from this fundamental relation the component or moment equations which fit best the solution of the problem under consideration.

The fundamental relation shown in Fig. 16.7 may be presented in an alternate form if we add to the external forces an inertia vector $-m\bar{\mathbf{a}}$ of sense opposite to that of $\bar{\mathbf{a}}$, attached at G, and an inertia couple $-\bar{I}\boldsymbol{\alpha}$ of moment equal in magnitude to $\bar{I}\alpha$ and of sense opposite to that of $\boldsymbol{\alpha}$ (Fig. 16.10). The system obtained is equivalent to zero, and the rigid body is said to be in dynamic equilibrium.

Whether the principle of equivalence of external and effective forces is directly applied, as in Fig. 16.7, or whether the concept of dynamic equilibrium is introduced, as in Fig. 16.10, the use of free-body diagrams showing vectorially the relationship existing between the forces applied on the rigid body and the resulting linear and angular accelerations presents considerable advantages over the blind application of the formulas (16.6). These advantages may be summarized as follows:

1. First of all, a much clearer understanding of the effect of the forces on the motion of the body will result from the use of a pictorial representation.
2. This approach makes it possible to divide the solution of a dynamics problem into two parts: In the first part, the analysis of the kinematic and kinetic characteristics of the problem leads to the free-body diagrams of Fig. 16.7 or 16.10; in the second part, the diagram obtained is used to analyze by the methods of Chap. 3 the various forces and vectors involved.
3. A unified approach is provided for the analysis of the plane motion of a rigid body, regardless of the particular type of motion involved. While the kinematics of the various motions

† We recall that the last of Eqs. (16.6) is valid only in the case of the plane motion of a rigid body symmetrical with respect to the reference plane. In all other cases, the methods of Chap. 18 should be used.

considered may vary from one case to the other, the approach to the kinetics of the motion is consistently the same. In every case we shall draw a diagram showing the external forces, the vector $m\bar{\mathbf{a}}$ associated with the motion of G, and the couple $\bar{I}\boldsymbol{\alpha}$ associated with the rotation of the body about G.

4. The resolution of the plane motion of a rigid body into a translation and a centroidal rotation, which is used here, is a basic concept which may be applied effectively throughout the study of mechanics. We shall use it again in Chap. 17 with the method of work and energy and the method of impulse and momentum.
5. As we shall see in Chap. 18, this approach may be extended to the study of the general three-dimensional motion of a rigid body. The motion of the body will again be resolved into a translation and a rotation about the mass center, and free-body diagrams will be used to indicate the relationship existing between the external forces and the rates of change of the linear and angular momentum of the body.

16.7. Systems of Rigid Bodies. The method described in the preceding section may also be used in problems involving the plane motion of several connected rigid bodies. A diagram similar to Fig. 16.7 or Fig. 16.10 may be drawn for each part of the system. The equations of motion obtained from these diagrams are solved simultaneously.

In some cases, as in Sample Prob. 16.3, a single diagram may be drawn for the entire system. This diagram should include all the external forces, as well as the vectors $m\bar{\mathbf{a}}$ and the couples $\bar{I}\boldsymbol{\alpha}$ associated with the various parts of the system. However, internal forces, such as the forces exerted by connecting cables, may be omitted since they occur in pairs of equal and opposite forces and are thus equipollent to zero. The equations obtained by expressing that the system of the external forces is equipollent to the system of the effective forces may be solved for the remaining unknowns.†

This second approach may not be used in problems involving more than three unknowns, since only three equations of motion are available when a single diagram is used. We shall not elaborate upon this point, since the discussion involved would be completely similar to that given in Sec. 6.11 in the case of the equilibrium of a system of rigid bodies.

† Note that we cannot speak of *equivalent* systems since we are not dealing with a single rigid body.

SAMPLE PROBLEM 16.1

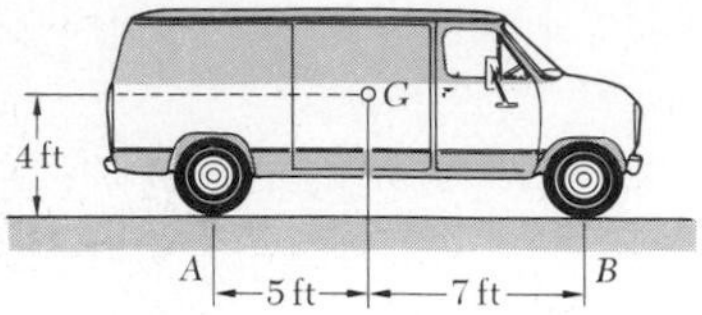

When the forward speed of the truck shown was 30 ft/s, the brakes were suddenly applied, causing all four wheels to stop rotating. It was observed that the truck skidded to rest in 20 ft. Determine the magnitude of the normal reaction and of the friction force at each wheel as the truck skidded to rest.

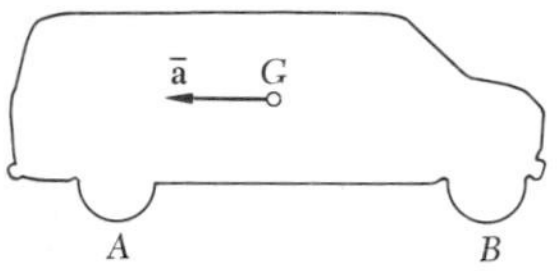

Kinematics of Motion. Choosing the positive sense to the right and using the equations of uniformly accelerated motion, we write

$$\bar{v}_0 = +30\text{ ft/s} \qquad \bar{v}^2 = \bar{v}_0^2 + 2\bar{a}\bar{x} \qquad 0 = (30)^2 + 2\bar{a}(20)$$
$$\bar{a} = -22.5\text{ ft/s}^2 \qquad \bar{\mathbf{a}} = 22.5\text{ ft/s}^2 \leftarrow$$

Equations of Motion. The external forces consist of the weight $\mathbf{W}$ of the truck and of the normal reactions and friction forces at the wheels. (The vectors $\mathbf{N}_A$ and $\mathbf{F}_A$ represent the sum of the reactions at the rear wheels, while $\mathbf{N}_B$ and $\mathbf{F}_B$ represent the sum of the reactions at the front wheels.) Since the truck is in translation, the effective forces reduce to the vector $m\bar{\mathbf{a}}$ attached at G. Three equations of motion are obtained by expressing that the system of the external forces is equivalent to the system of the effective forces.

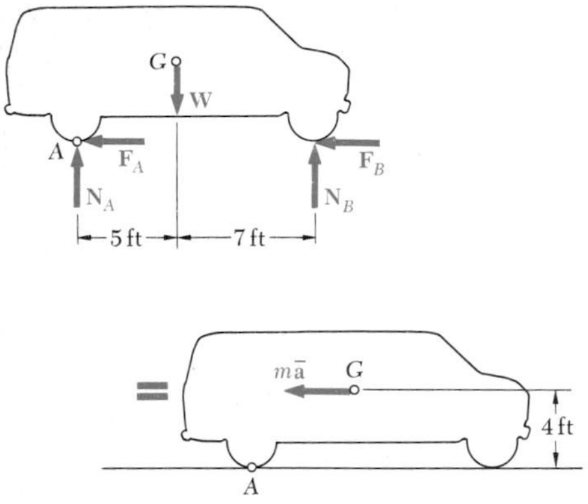

$$+\uparrow \Sigma F_y = \Sigma(F_y)_{\text{eff}}: \qquad N_A + N_B - W = 0$$

Since $F_A = \mu N_A$ and $F_B = \mu N_B$, we find

$$F_A + F_B = \mu(N_A + N_B) = \mu W$$

$$\xrightarrow{+} \Sigma F_x = \Sigma(F_x)_{\text{eff}}: \qquad -(F_A + F_B) = -m\bar{a}$$

$$-\mu W = -\frac{W}{32.2\text{ ft/s}^2}(22.5\text{ ft/s}^2)$$

$$\mu = 0.699$$

$$+\circlearrowleft \Sigma M_A = \Sigma(M_A)_{\text{eff}}: \qquad -W(5\text{ ft}) + N_B(12\text{ ft}) = m\bar{a}(4\text{ ft})$$

$$-W(5\text{ ft}) + N_B(12\text{ ft}) = \frac{W}{32.2\text{ ft/s}^2}(22.5\text{ ft/s}^2)(4\text{ ft})$$

$$N_B = 0.650W$$

$$F_B = \mu N_B = (0.699)(0.650W) \qquad F_B = 0.454W$$

$$+\uparrow \Sigma F_y = \Sigma(F_y)_{\text{eff}}: \qquad N_A + N_B - W = 0$$
$$N_A + 0.650W - W = 0$$
$$N_A = 0.350W$$
$$F_A = \mu N_A = (0.699)(0.350W) \qquad F_A = 0.245W$$

Reactions at Each Wheel. Recalling that the values computed above represent the sum of the reactions at the two front wheels or the two rear wheels, we obtain the magnitude of the reactions at each wheel by writing

$$N_{\text{front}} = \tfrac{1}{2}N_B = 0.325W \qquad N_{\text{rear}} = \tfrac{1}{2}N_A = 0.175W \quad ◀$$
$$F_{\text{front}} = \tfrac{1}{2}F_B = 0.227W \qquad F_{\text{rear}} = \tfrac{1}{2}F_A = 0.122W \quad ◀$$

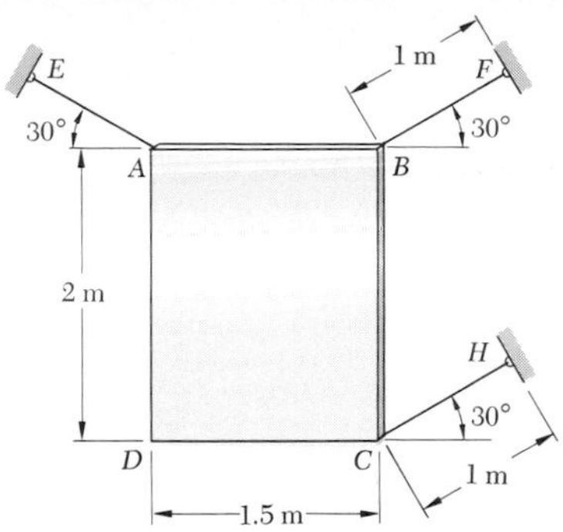

SAMPLE PROBLEM 16.2

The thin plate $ABCD$ has a mass of 50 kg and is held in position by the three inextensible wires AE, BF, and CH. Wire AE is then cut. Determine both (*a*) the acceleration of the plate, (*b*) the tension in wires BF and CH immediately after wire AE has been cut.

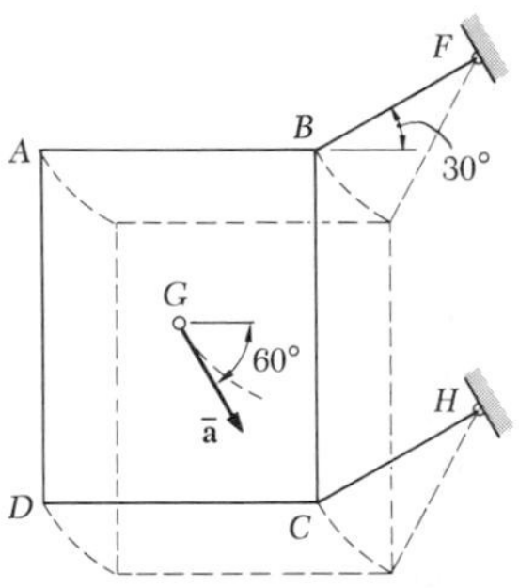

Motion of Plate. After wire AE has been cut, we observe that corners B and C move along parallel circles of radius 1 m centered, respectively, at F and H. The motion of the plate is thus a curvilinear translation; the particles forming the plate move along parallel circles of radius 1 m.

At the instant wire AE is cut, the velocity of the plate is zero; the acceleration $\bar{\mathbf{a}}$ of the mass center G is thus tangent to the circular path which will be described by G.

Equations of Motion. The external forces consist of the weight $\mathbf{W}$ and of the forces $\mathbf{T}_B$ and $\mathbf{T}_C$ exerted by the wires. Since the plate is in translation, the effective forces reduce to the vector $m\bar{\mathbf{a}}$ attached at G and directed along the t axis. Expressing that the system of the external forces is equivalent to the system of the effective forces, we write

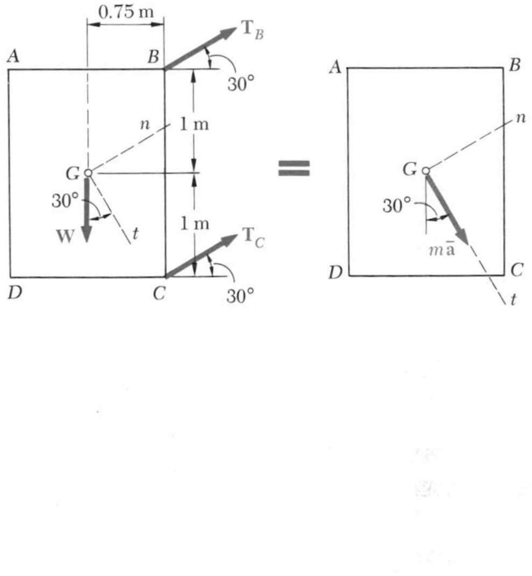

$+\searrow \Sigma F_t = \Sigma (F_t)_{\text{eff}}$: $\quad W \cos 30° = m\bar{a}$

$$mg \cos 30° = m\bar{a} \qquad (1)$$

$$\bar{a} = g \cos 30° = (9.81 \text{ m/s}^2) \cos 30°$$

$$\bar{\mathbf{a}} = 8.50 \text{ m/s}^2 \measuredangle 60° \quad \blacktriangleleft$$

$+\nearrow \Sigma F_n = \Sigma (F_n)_{\text{eff}}$: $\quad T_B + T_C - W \sin 30° = 0 \qquad (2)$

$+\circlearrowleft \Sigma M_G = \Sigma (M_G)_{\text{eff}}$: $\quad (T_B \sin 30°)(0.75 \text{ m}) - (T_B \cos 30°)(1 \text{ m})$

$$+ (T_C \sin 30°)(0.75 \text{ m}) + (T_C \cos 30°)(1 \text{ m}) = 0$$

$$-0.491 T_B + 1.241 T_C = 0$$

$$T_C = 0.396 T_B \qquad (3)$$

Substituting for T_C from (3) into (2), we write

$$T_B + 0.396 T_B - W \sin 30° = 0$$

$$T_B = 0.358 W$$

$$T_C = 0.396(0.358W) = 0.1418W$$

Noting that $W = mg = (50 \text{ kg})(9.81 \text{ m/s}^2) = 491 \text{ N}$, we have

$$T_B = 175.8 \text{ N} \qquad T_C = 69.6 \text{ N} \quad \blacktriangleleft$$

SAMPLE PROBLEM 16.3

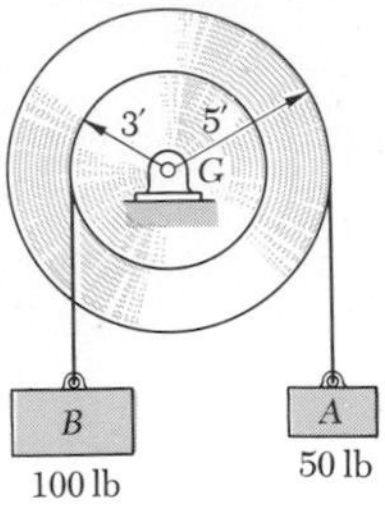

A pulley weighing 120 lb and having a radius of gyration of 4 ft is connected to two blocks as shown. Assuming no axle friction, determine the angular acceleration of the pulley.

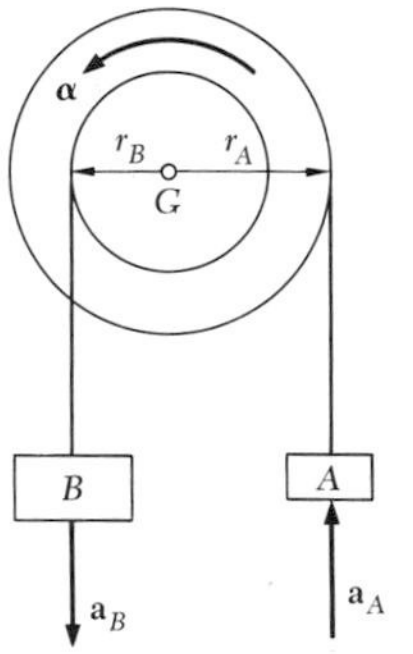

Sense of Motion. Although an arbitrary sense of motion may be assumed (since no friction forces are involved) and later checked by the sign of the answer, we may prefer first to determine the actual sense of rotation of the pulley. We first find the weight of block B required to maintain the equilibrium of the pulley when it is acted upon by the 50-lb block A. We write

$$+\circlearrowleft \Sigma M_G = 0: \qquad W_B(3\text{ ft}) - (50\text{ lb})(5\text{ ft}) = 0 \qquad W_B = 83.3\text{ lb}$$

Since block B actually weighs 100 lb, the pulley will rotate counterclockwise.

Kinematics of Motion. Assuming $\boldsymbol{\alpha}$ counterclockwise and noting that $a_A = r_A\alpha$ and $a_B = r_B\alpha$, we obtain

$$\mathbf{a}_A = 5\alpha \uparrow \qquad \mathbf{a}_B = 3\alpha \downarrow$$

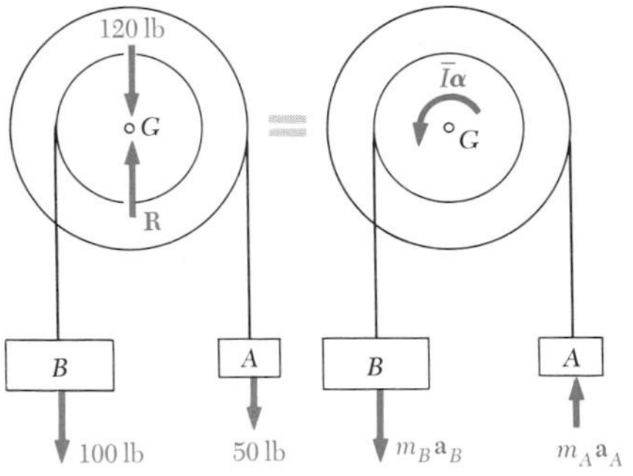

Equations of Motion. A single system consisting of the pulley and the two blocks is considered. Forces external to this system consist of the weights of the pulley and the two blocks and of the reaction at G. (The forces exerted by the cables on the pulley and on the blocks are internal to the system considered and cancel out.) Since the motion of the pulley is a centroidal rotation and the motion of each block is a translation, the effective forces reduce to the couple $\bar{I}\boldsymbol{\alpha}$ and the two vectors $m\mathbf{a}_A$ and $m\mathbf{a}_B$. The centroidal moment of inertia of the pulley is

$$\bar{I} = mk^2 = \frac{W}{g}k^2 = \frac{120\text{ lb}}{32.2\text{ ft/s}^2}(4\text{ ft})^2 = 59.6\text{ lb}\cdot\text{ft}\cdot\text{s}^2$$

Since the system of the external forces is equipollent to the system of the effective forces, we write

$$+\circlearrowleft \Sigma M_G = \Sigma(M_G)_{\text{eff}}:$$

$$(100\text{ lb})(3\text{ ft}) - (50\text{ lb})(5\text{ ft}) = +\bar{I}\alpha + m_B a_B(3\text{ ft}) + m_A a_A(5\text{ ft})$$

$$(100)(3) - (50)(5) = +59.6\alpha + \frac{100}{32.2}(3\alpha)(3) + \frac{50}{32.2}(5\alpha)(5)$$

$$\alpha = +0.396\text{ rad/s}^2 \qquad \boldsymbol{\alpha} = 0.396\text{ rad/s}^2 \circlearrowleft \blacktriangleleft$$

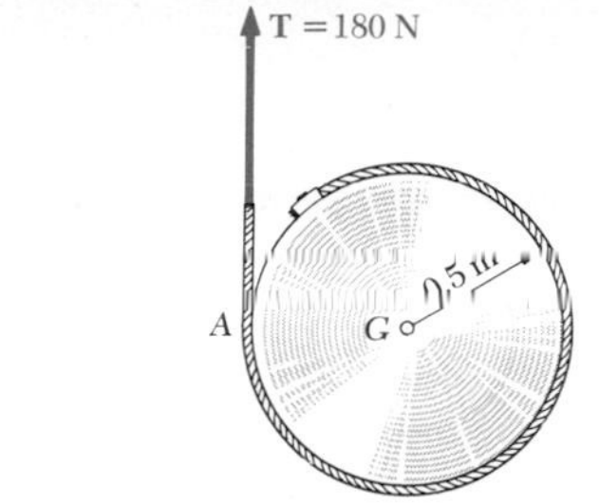

SAMPLE PROBLEM 16.4

A cord is wrapped around a homogeneous disk of radius $r = 0.5$ m and mass $m = 15$ kg. If the cord is pulled upward with a force **T** of magnitude 180 N, determine (*a*) the acceleration of the center of the disk, (*b*) the angular acceleration of the disk, (*c*) the acceleration of the cord.

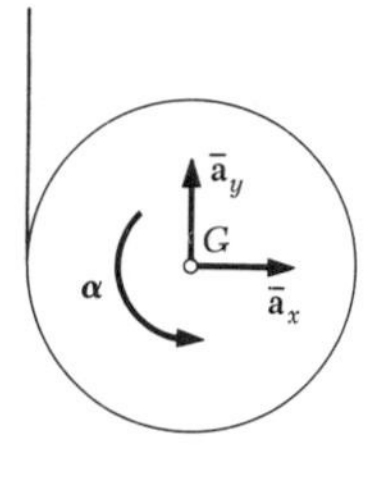

Equations of Motion. We assume that the components $\bar{\mathbf{a}}_x$ and $\bar{\mathbf{a}}_y$ of the acceleration of the center are directed, respectively, to the right and upward and that the angular acceleration of the disk is counterclockwise. The external forces acting on the disk consist of the weight **W** and the force **T** exerted by the cord. This system is equivalent to the system of the effective forces, which consists of a vector of components $m\bar{\mathbf{a}}_x$ and $m\bar{\mathbf{a}}_y$ attached at G and a couple $\bar{I}\boldsymbol{\alpha}$. We write

$$\overset{+}{\rightarrow} \Sigma F_x = \Sigma(F_x)_{\text{eff}}: \qquad 0 = m\bar{a}_x \qquad \bar{\mathbf{a}}_x = 0 \blacktriangleleft$$

$$+\uparrow \Sigma F_y = \Sigma(F_y)_{\text{eff}}: \qquad T - W = m\bar{a}_y$$

$$\bar{a}_y = \frac{T - W}{m}$$

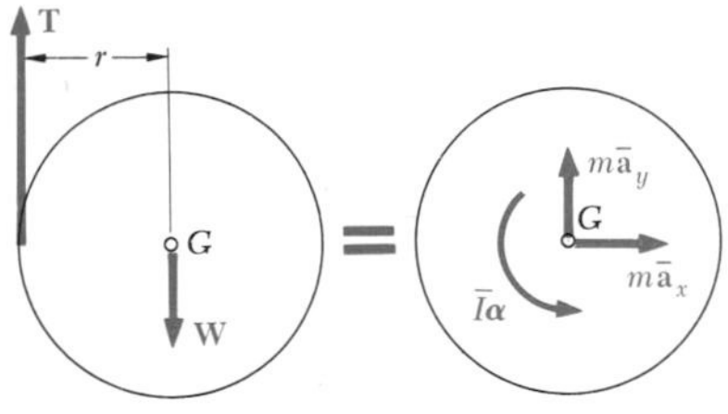

Since $T = 180$ N, $m = 15$ kg, and $W = (15 \text{ kg})(9.81 \text{ m/s}^2) = 147.1$ N, we have

$$\bar{a}_y = \frac{180 \text{ N} - 147.1 \text{ N}}{15 \text{ kg}} = +2.19 \text{ m/s}^2 \qquad \bar{\mathbf{a}}_y = 2.19 \text{ m/s}^2 \uparrow \blacktriangleleft$$

$$+\circlearrowleft \Sigma M_G = \Sigma(M_G)_{\text{eff}}: \qquad -Tr = \bar{I}\alpha$$

$$-Tr = (\tfrac{1}{2}mr^2)\alpha$$

$$\alpha = -\frac{2T}{mr} = -\frac{2(180 \text{ N})}{(15 \text{ kg})(0.5 \text{ m})} = -48.0 \text{ rad/s}^2$$

$$\boldsymbol{\alpha} = 48.0 \text{ rad/s}^2 \circlearrowright \blacktriangleleft$$

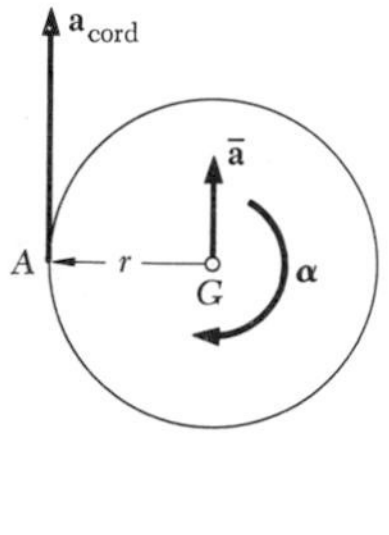

Acceleration of Cord. Since the acceleration of the cord is equal to the tangential component of the acceleration of point A on the disk, we write

$$\mathbf{a}_{\text{cord}} = (\mathbf{a}_A)_t = \bar{\mathbf{a}} + (\mathbf{a}_{A/G})_t$$
$$= [2.19 \text{ m/s}^2 \uparrow] + [(0.5 \text{ m})(48 \text{ rad/s}^2) \uparrow]$$

$$\mathbf{a}_{\text{cord}} = 26.2 \text{ m/s}^2 \uparrow \blacktriangleleft$$

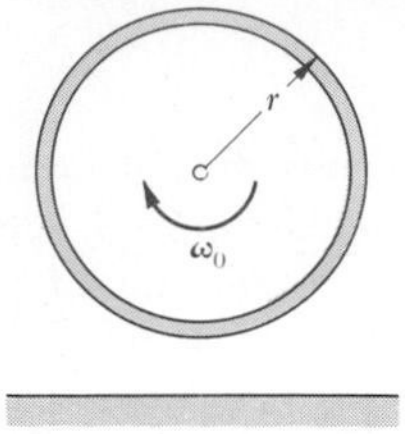

SAMPLE PROBLEM 16.5

A hoop of radius r and mass m is placed on a horizontal surface with no linear velocity but with a clockwise angular velocity $\boldsymbol{\omega}_0$. Denoting by μ the coefficient of friction between the hoop and the floor, determine (*a*) the time t_1 at which the hoop will start rolling without sliding, (*b*) the linear and angular velocities of the hoop at time t_1.

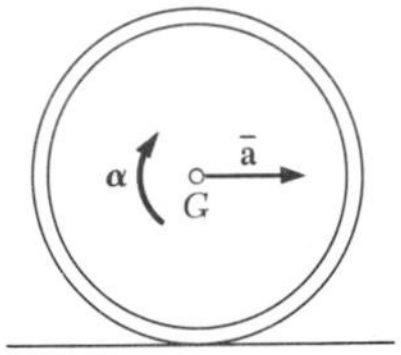

Solution. Since the entire mass is located at a distance r from the center of the hoop, we write $\bar{I} = mr^2$.

Equations of Motion. The positive sense is chosen to the right for $\bar{\mathbf{a}}$ and clockwise for $\boldsymbol{\alpha}$. The external forces acting on the hoop consist of the weight **W**, the normal reaction **N**, and the friction force **F**. While the hoop is sliding, the magnitude of the friction force is $F = \mu N$. The effective forces consist of the vector $m\bar{\mathbf{a}}$ attached at G and the couple $\bar{I}\boldsymbol{\alpha}$. Expressing that the system of the external forces is equivalent to the system of the effective forces, we write

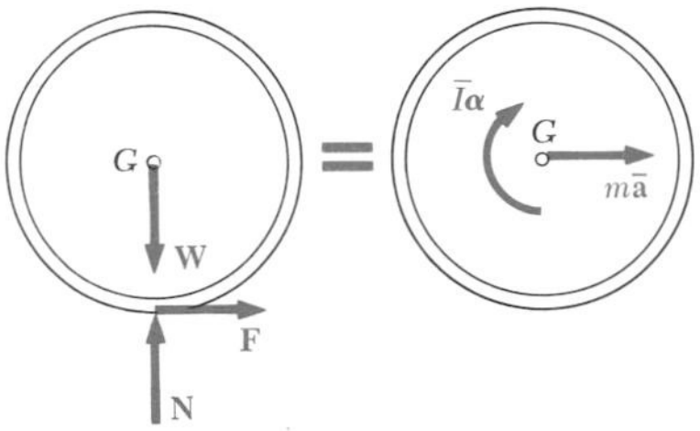

$$+\uparrow \Sigma F_y = \Sigma(F_y)_{\text{eff}}: \quad N - W = 0$$
$$N = W = mg \qquad F = \mu N = \mu mg$$
$$\overset{+}{\rightarrow} \Sigma F_x = \Sigma(F_x)_{\text{eff}}: \quad F = m\bar{a} \qquad \mu mg = m\bar{a} \qquad \bar{a} = +\mu g$$
$$+\circlearrowright \Sigma M_G = \Sigma(M_G)_{\text{eff}}: \quad -Fr = \bar{I}\alpha$$
$$-(\mu mg)r = (mr^2)\alpha \qquad \alpha = -\frac{\mu g}{r}$$

Kinematics of Motion. As long as the hoop both rolls and slides, its linear and angular motions are uniformly accelerated.

$$t = 0,\ \bar{v}_0 = 0 \qquad \bar{v} = \bar{v}_0 + \bar{a}t = 0 + \mu g t \tag{1}$$
$$t = 0,\ \omega = \omega_0 \qquad \omega = \omega_0 + \alpha t = \omega_0 + \left(-\frac{\mu g}{r}\right)t \tag{2}$$

The hoop will start rolling without sliding when the velocity $\mathbf{v}_C$ of the point of contact is zero. At that time, $t = t_1$, point C becomes the instantaneous center of rotation, and we have

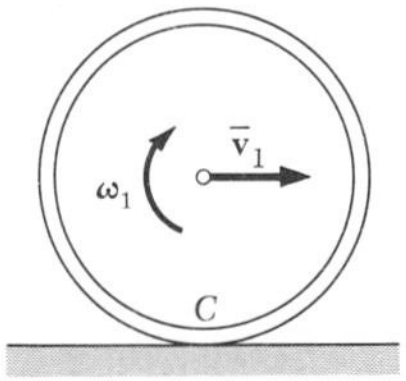

$$\bar{v}_1 = r\omega_1$$
$$\mu g t_1 = r\left(\omega_0 - \frac{\mu g}{r}t_1\right) \qquad t_1 = \frac{r\omega_0}{2\mu g} \quad \blacktriangleleft$$

Substituting for t_1 into (1), we have

$$\bar{v}_1 = \mu g t_1 = \mu g \frac{r\omega_0}{2\mu g} \qquad \bar{v}_1 = \tfrac{1}{2}r\omega_0 \qquad \bar{\mathbf{v}}_1 = \tfrac{1}{2}r\omega_0 \rightarrow \quad \blacktriangleleft$$
$$\bar{v}_1 = r\omega_1 \qquad \tfrac{1}{2}r\omega_0 = r\omega_1 \qquad \omega_1 = \tfrac{1}{2}\omega_0 \qquad \omega_1 = \tfrac{1}{2}\omega_0 \circlearrowright \quad \blacktriangleleft$$

PROBLEMS

16.1 A 6-ft board is placed in a truck so that one end rests against a block on the floor while the other end rests against a vertical wall. Determine the maximum possible uniform acceleration of the truck if the board is to remain in the position shown.

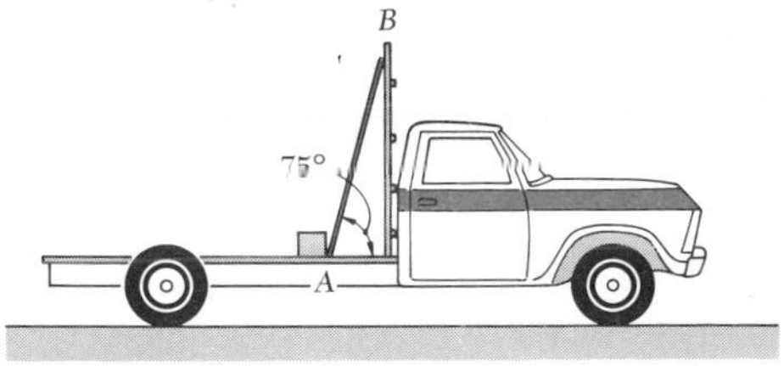

Fig. P16.1

16.2 A uniform rod ABC of mass 8 kg is connected to two collars of negligible mass which slide on smooth horizontal rods located in the same vertical plane. If a force $\mathbf{P}$ of magnitude 40 N is applied at C, determine (a) the acceleration of the rod, (b) the reactions at B and C.

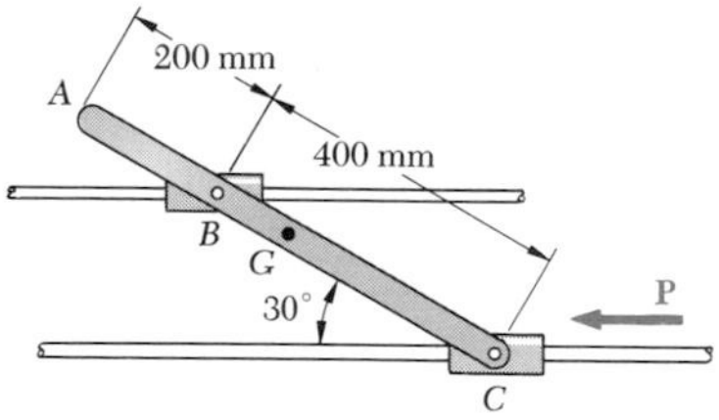

Fig. P16.2

16.3 In Prob. 16.2, determine (a) the required magnitude of $\mathbf{P}$ if the reaction at B is to be 45 N upward, (b) the corresponding acceleration of the rod.

16.4 The motion of a 3-lb semicircular rod is guided by two small wheels which roll freely in a vertical slot. Knowing that the acceleration of the rod is $a = \frac{1}{4}g$ upward, determine (a) the magnitude of the force $\mathbf{P}$, (b) the reactions at A and B.

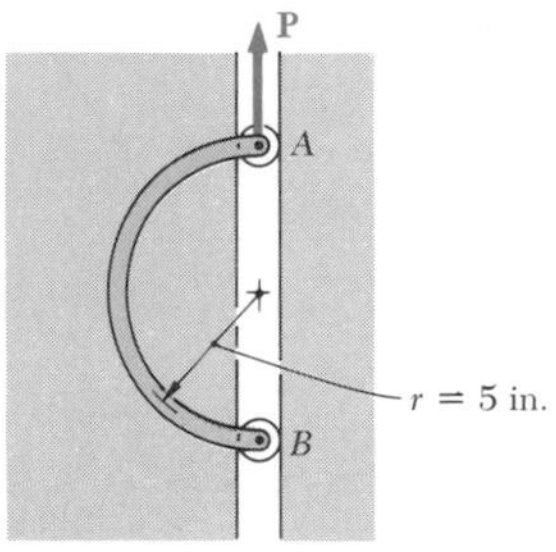

Fig. P16.4

16.5 Cylindrical cans are transported from one elevation to another by the moving horizontal arms shown. Assuming that $\mu = 0.20$ between the cans and the arms, determine (a) the magnitude of the upward acceleration $\mathbf{a}$ for which the cans slide on the horizontal arms, (b) the smallest ratio h/d for which the cans tip before they slide.

16.6 Solve Prob. 16.5, assuming that the acceleration $\mathbf{a}$ of the horizontal arms is directed downward.

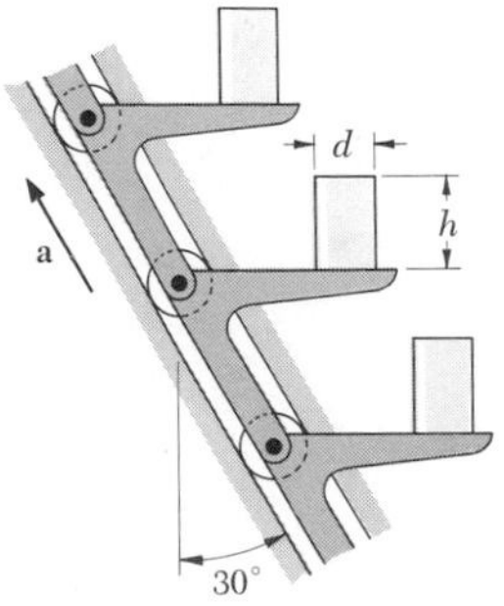

Fig. P16.5

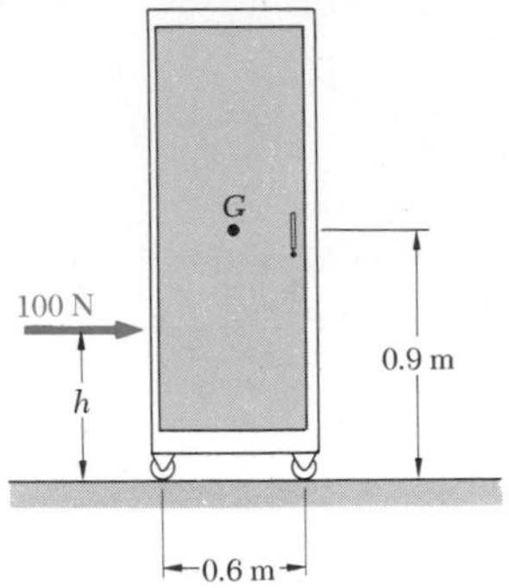

Fig. P16.7

16.7 A 20-kg cabinet is mounted on casters which allow it to move freely ($\mu = 0$) on the floor. If a 100-N force is applied as shown, determine (*a*) the acceleration of the cabinet, (*b*) the range of values of h for which the cabinet will not tip.

16.8 Solve Prob. 16.7, assuming that the casters are locked and slide along the rough floor ($\mu = 0.25$).

16.9 Determine the distance through which the truck of Sample Prob. 16.1 will skid if (*a*) the rear-wheel brakes fail to operate, (*b*) the front-wheel brakes fail to operate.

16.10 A 600-kg fork-lift truck carries the 300-kg crate at the height shown. The truck is moving to the left when the brakes are applied causing a deceleration of 3 m/s^2. Knowing that the coefficient of friction between the crate and the fork lift is 0.5, determine the vertical component of the reaction (*a*) at each of the two wheels *A* (one wheel on each side of the truck), (*b*) at the single steerable wheel *B*.

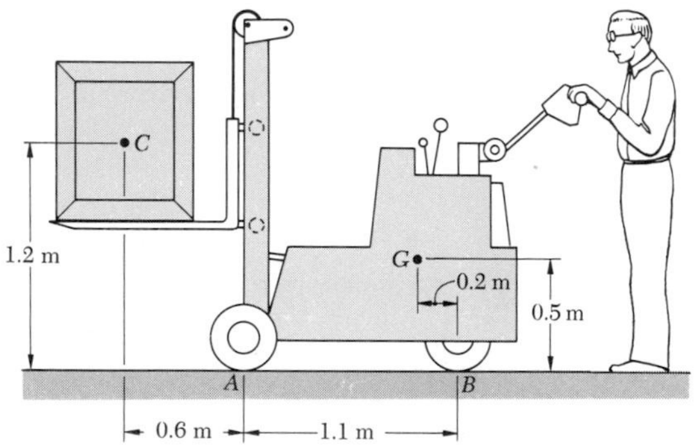

Fig. P16.10

16.11 In Prob. 16.10, determine the maximum deceleration of the truck if the crate is not to slide forward and if the truck is not to tip forward.

16.12 Knowing that the coefficient of friction between the tires and road is 0.80 for the car shown, determine the maximum possible acceleration on a level road, assuming (*a*) four-wheel drive, (*b*) conventional rear-wheel drive, (*c*) front-wheel drive.

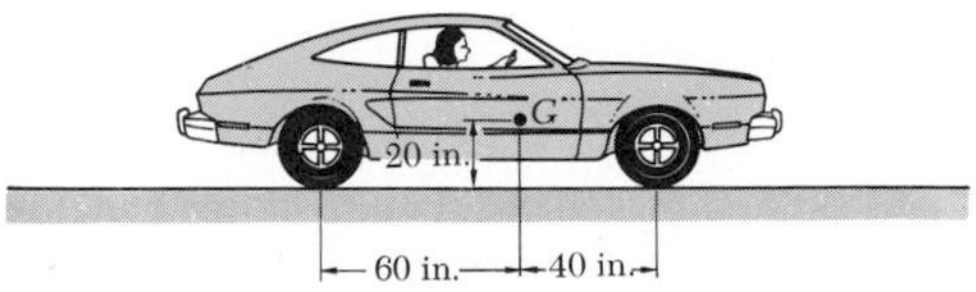

Fig. P16.12

16.13 A man rides a bicycle at a speed of 30 km/h. The distance between axles is 1050 mm, and the mass center of the man and bicycle is located 650 mm behind the front axle and 1000 mm above the ground. If the man applies the brakes on the front wheel only, determine the shortest distance in which he can stop without being thrown over the front wheel.

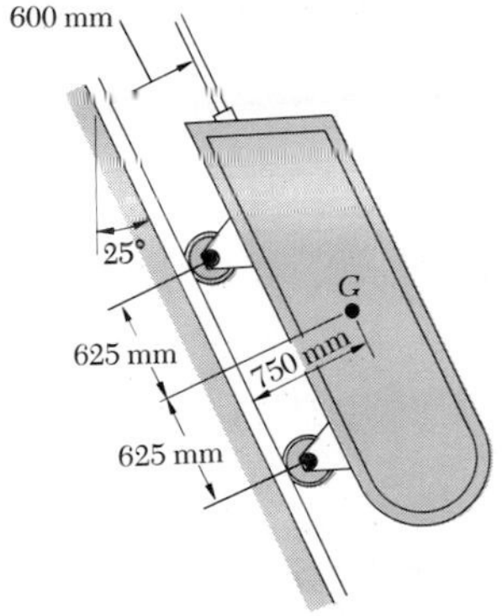

Fig. P16.14

16.14 The total mass of the loading car and its load is 2500 kg. Neglecting the mass and friction of the wheels, determine (*a*) the minimum tension *T* in the cable for which the upper wheels are lifted from the track, (*b*) the corresponding acceleration of the car.

16.15 A 200-lb rectangular panel is suspended from two skids which may slide with no friction on the inclined track shown. If the panel is released from rest, determine (*a*) the acceleration of the panel, (*b*) the reaction at *B*.

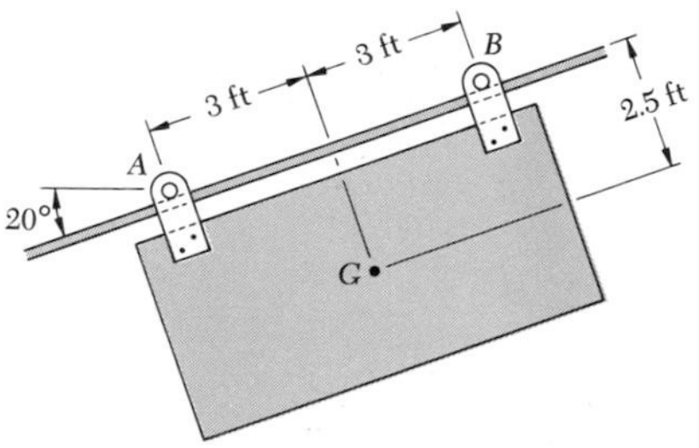

Fig. P16.15

16.16 Solve Prob. 16.15, assuming that the coefficient of friction between each skid and the track is 0.10.

16.17 The 200-kg fire door is supported by wheels *B* and *C* which may roll freely on the horizontal track. The 40-kg counterweight *A* is connected to the door by the cable shown. If the system is released from rest, determine (*a*) the acceleration of the door, (*b*) the reactions at *B* and *C*.

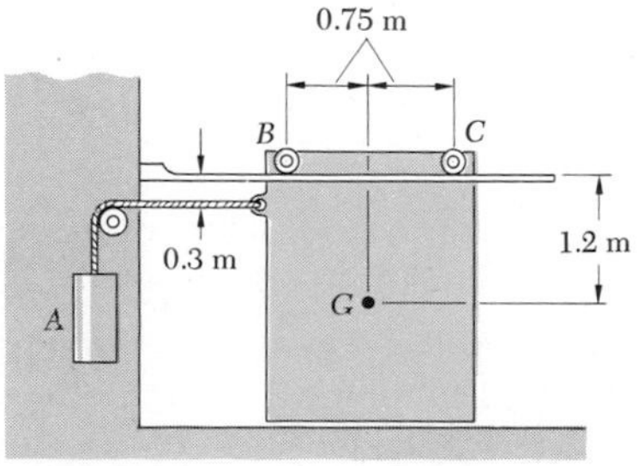

Fig. P16.17

16.18 Two uniform rods *AB* and *CD*, each of mass 2.5 kg, are welded together and are attached to two links *CE* and *DF*. Neglecting the mass of the links, determine the force in each link immediately after the system is released from rest in the position shown.

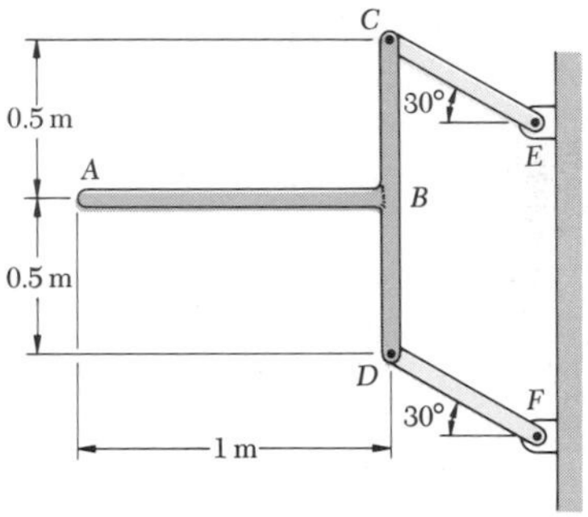

Fig. P16.18

16.19 The retractable shelf shown is supported by two identical linkage-and-spring systems; only one of the systems is shown. A 40-lb machine is placed on the shelf so that half of its weight is supported by the system shown. If the springs are removed and the system is released from rest, determine (*a*) the acceleration of the machine, (*b*) the tension in link *AB*. Neglect the weight of the shelf and links.

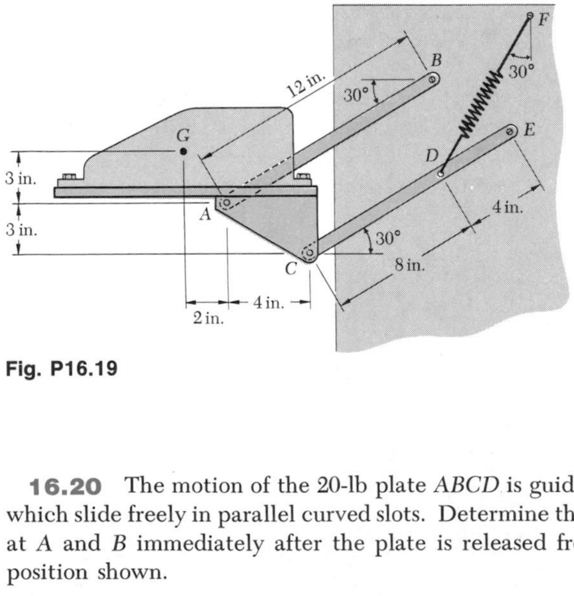

Fig. P16.19

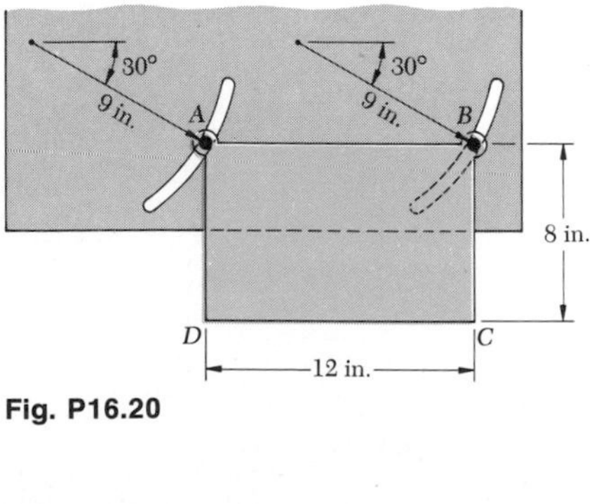

Fig. P16.20

16.20 The motion of the 20-lb plate *ABCD* is guided by two pins which slide freely in parallel curved slots. Determine the pin reactions at *A* and *B* immediately after the plate is released from rest in the position shown.

16.21 The cranks *AB* and *CD* rotate at a constant speed of 240 rpm. For the position $\phi = 30°$, determine the horizontal components of the forces exerted on the 5 kg uniform connecting rod *BC* by the pins *B* and *C*.

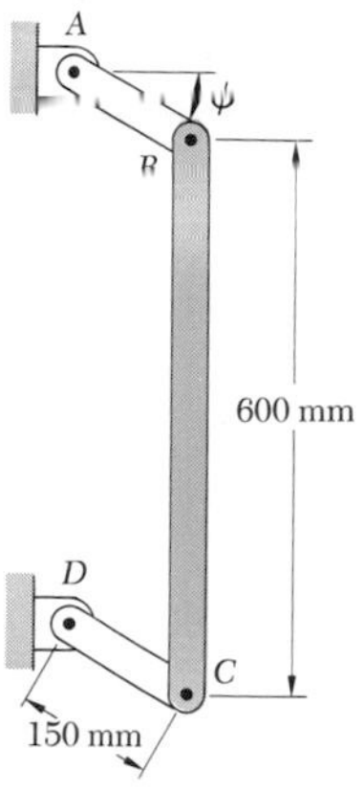

Fig. P16.21

16.22 The control rod *AC* is guided by two pins which slide freely in parallel curved slots of radius 200 mm. The rod has a mass of 10 kg, and its mass center is located at point *G*. Knowing that for the position shown the *vertical* component of the velocity of *C* is 1.25 m/s upward and the *vertical* component of the acceleration of *C* is 5 m/s² upward, determine the magnitude of the force **P**.

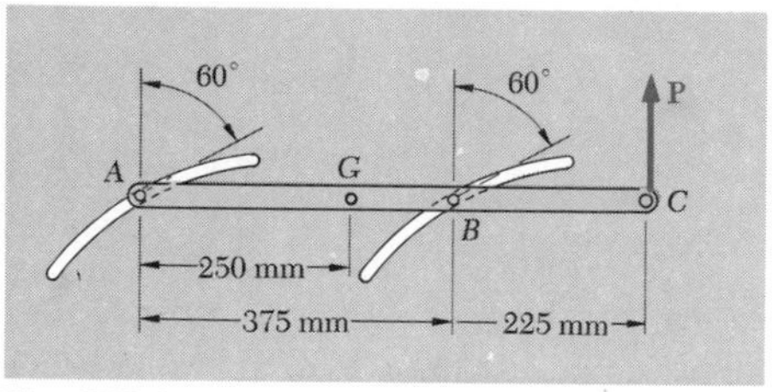

Fig. P16.22

16.23 Assuming that the plate of Prob. 16.20 has acquired a velocity of 4 ft/s in the position shown, determine (*a*) the acceleration of the plate, (*b*) the pin reactions at *A* and *B*.

***16.24** A 12-kg block is placed on a 3-kg platform *BD* which is held in the position shown by three wires. Determine the accelerations of the block and of the platform immediately after wire *AB* has been cut. Assume (*a*) that the block is rigidly attached to *BD*, (*b*) that $\mu = 0$ between the block and *BD*.

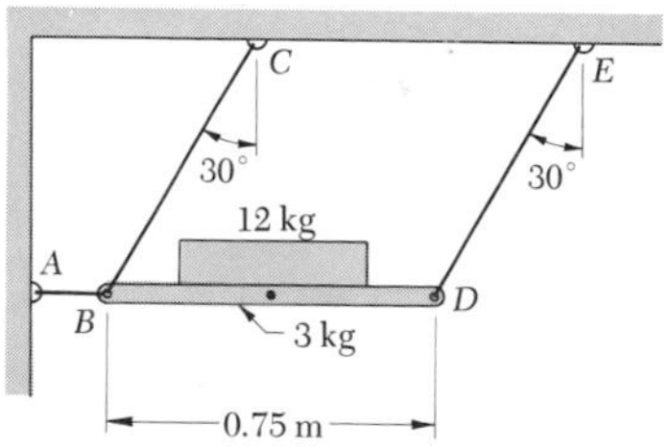

Fig. P16.24 and P16.25

***16.25** The coefficient of friction between the 12-kg block and the 3-kg platform *BD* is 0.50. Determine the accelerations of the block and of the platform immediately after wire *AB* has been cut.

***16.26** Draw the shear and bending moment diagrams for the horizontal rod *AB* of Prob. 16.18.

***16.27** Draw the shear and bending-moment diagrams for the rod *BC* of Prob. 16.21.

16.28 For a rigid slab in translation, show that the system of the effective forces consists of vectors $(\Delta m_i)\bar{\mathbf{a}}$ attached to the various particles of the slab, where $\bar{\mathbf{a}}$ is the acceleration of the mass center G of the slab. Further show, by computing their sum and the sum of their moments about G, that the effective forces reduce to a single vector $m\bar{\mathbf{a}}$ attached at G.

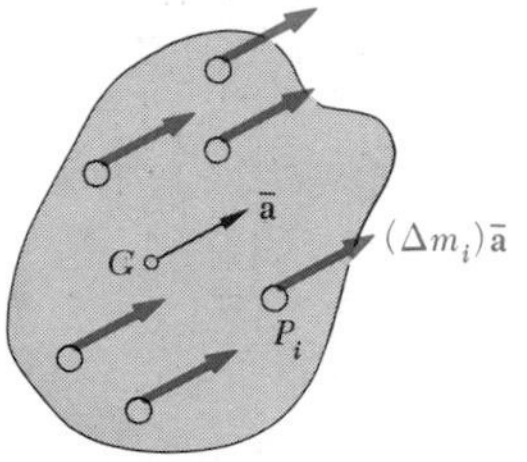

Fig. P16.28

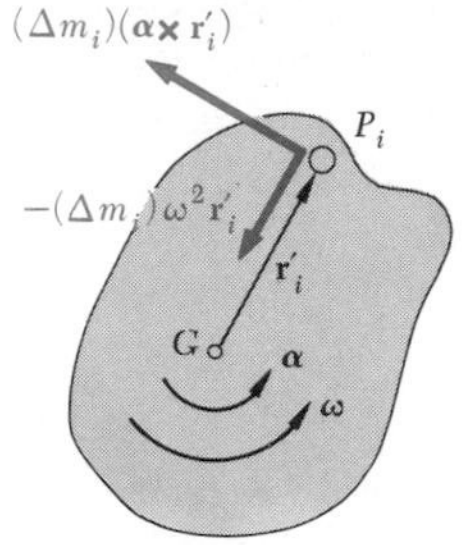

Fig. P16.29

16.29 For a rigid slab in centroidal rotation, show that the system of the effective forces consists of vectors $-(\Delta m_i)\omega^2\mathbf{r}'_i$ and $(\Delta m_i)(\boldsymbol{\alpha} \times \mathbf{r}'_i)$ attached to the various particles P_i of the slab, where $\boldsymbol{\omega}$ and $\boldsymbol{\alpha}$ are the angular velocity and angular acceleration of the slab, and where $\mathbf{r}'_i$ denotes the position vector of the particle P_i relative to the mass center G of the slab. Further show, by computing their sum and the sum of their moments about G, that the effective forces reduce to a couple $\bar{I}\boldsymbol{\alpha}$.

16.30 A turbine-generator unit is shut off when its rotor is rotating at 3600 rpm; it is observed that the rotor coasts to rest in 7.10 min. Knowing that the 1850-kg rotor has a radius of gyration of 234 mm, determine the average magnitude of the couple due to bearing friction.

16.31 An electric motor is rotating at 1200 rpm when the load and power are cut off. The rotor weighs 180 lb and has a radius of gyration of 8 in. If the kinetic friction of the rotor produces a couple of moment 15 lb·in., how many revolutions will the rotor execute before stopping?

16.32 Disk A weighs 12 lb and is at rest when it is placed in contact with a conveyor belt moving at a constant speed. The link AB connecting the center of the disk to the support at B is of negligible weight. Knowing that $r = 6$ in. and $\mu = 0.35$, determine the angular acceleration of the disk while slipping occurs.

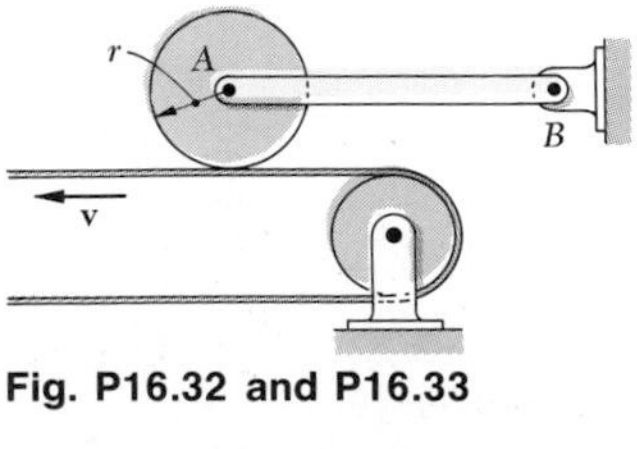

Fig. P16.32 and P16.33

16.33 The uniform disk A is at rest when it is placed in contact with a conveyor belt moving at a constant speed. Neglecting the weight of the link AB, derive an expression for the angular acceleration of the disk while slipping occurs.

16.34 Each of the double pulleys shown has a mass moment of inertia of 10 kg · m² and is initially at rest. The outside radius is 400 mm, and the inner radius is 200 mm. Determine (*a*) the angular acceleration of each pulley, (*b*) the angular velocity of each pulley at $t = 2$ s, (*c*) the angular velocity of each pulley after point A on the cord has moved 2 m.

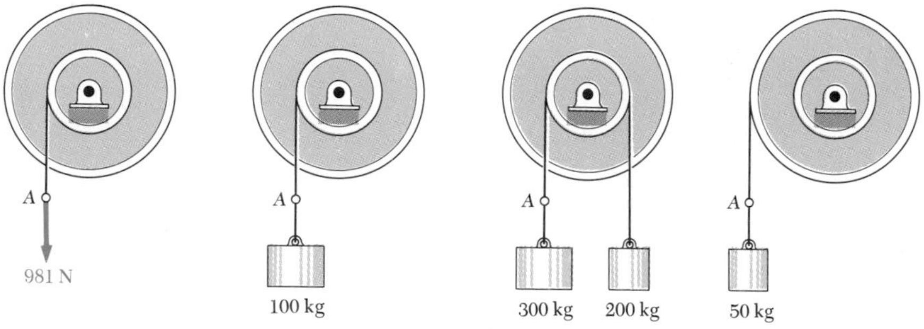

Fig. P16.34

16.35 Solve Prob. 12.17*a* assuming that each pulley is of 8-in. radius and has a centroidal mass moment of inertia 0.25 lb · ft · s².

16.36 The flywheel shown weighs 250 lb and has a radius of gyration of 15 in. A block A of weight 30 lb is attached to a wire wrapped around the rim of radius $r = 20$ in. The system is released from rest. Neglecting the effect of friction, determine (*a*) the acceleration of block A, (*b*) the speed of block A after it has moved 6 ft.

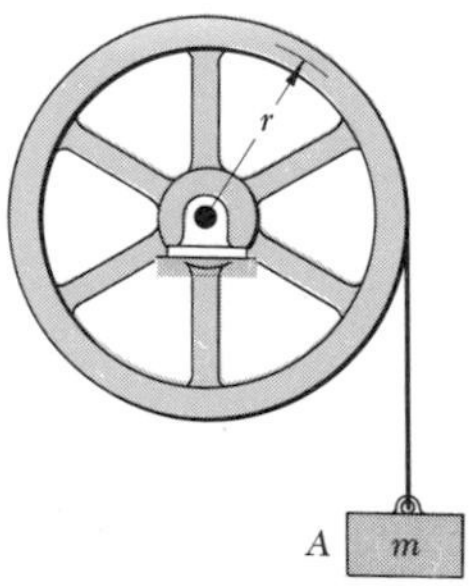

Fig. P16.36 and P16.37

16.37 In order to determine the mass moment of inertia of a flywheel of radius $r = 600$ mm a block of mass 12 kg is attached to a cord which is wrapped around the rim of the flywheel. The block is released from rest and is observed to fall 3 m in 4.6 s. To eliminate bearing friction from the computation, a second block of mass 24 kg is used and is observed to fall 3 m in 3.1 s. Assuming that the moment of the couple due to friction is constant, determine the mass moment of inertia of the flywheel.

16.38 A rope of total mass 10 kg and total length 20 m is wrapped around the drum of a hoist as shown. The mass of the drum and shaft is 18 kg, and they have a combined radius of gyration of 200 mm. Knowing that the system is released from rest when a length $h = 5$ m hangs from the drum, determine the initial angular acceleration of the drum.

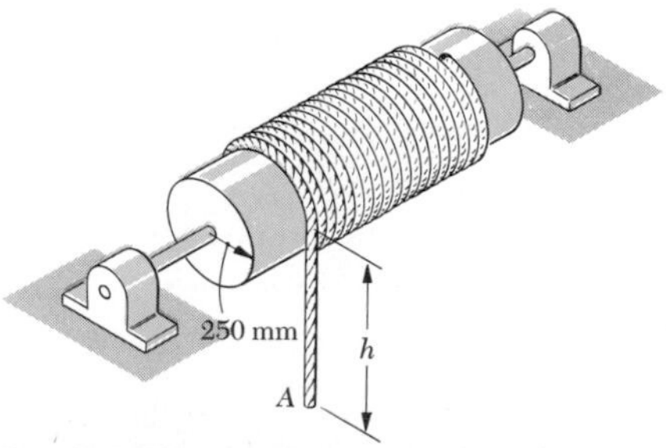

Fig. P16.38

16.39 The flywheel shown consists of a 3-ft-diameter disk which weighs 300 lb. The coefficient of friction between the band and the flywheel is 0.30. If the initial angular velocity of the flywheel is 300 rpm clockwise, determine the magnitude of the force **P** required sto stop the flywheel in 20 revolutions.

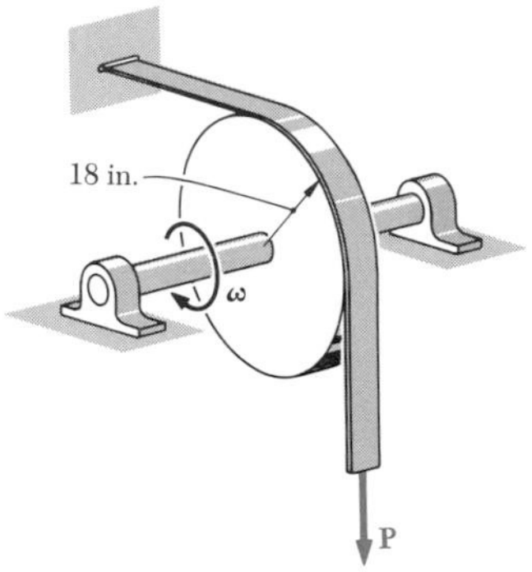

Fig. P16.39

16.40 Solve Prob. 16.39 assuming that the initial angular velocity of the flywheel is 300 rpm counterclockwise.

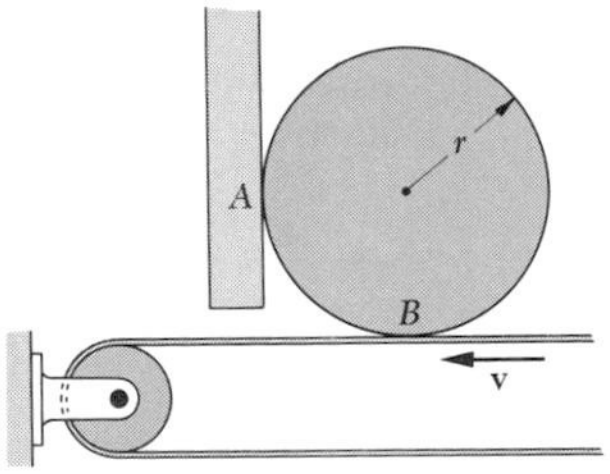

Fig. P16.41

16.41 A cylinder of radius r and mass m is placed with no initial velocity on a belt as shown. Denoting by μ the coefficient of friction at A and at B and assuming that $\mu < 1$, determine the angular acceleration $\boldsymbol{\alpha}$ of the cylinder.

16.42 Shaft A and friction disk B have a combined mass of 15 kg and a combined radius of gyration of 150 mm. Shaft D and friction wheel C rotate with a constant angular velocity of 1000 rpm. Disk B is at rest when it is brought into contact with the rotating wheel. Knowing that disk B accelerates uniformly for 12 s before acquiring its final angular velocity, determine the magnitude of the friction force between the disk and the wheel.

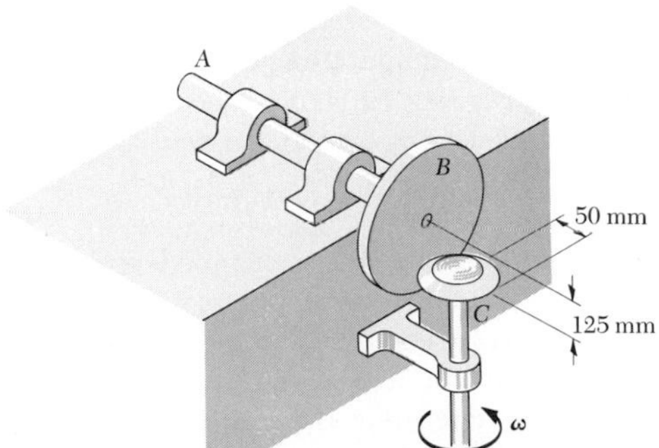

Fig. P16.42

16.43 Each of the gears A and B weighs 4 lb and has a radius of gyration of 3 in., while gear C weighs 20 lb and has a radius of gyration of 9 in. If a couple **M** of constant magnitude 60 lb · in. is applied to gear C, determine (a) the angular acceleration of gear A, (b) the time required for the angular velocity of gear A to increase from 150 to 500 rpm.

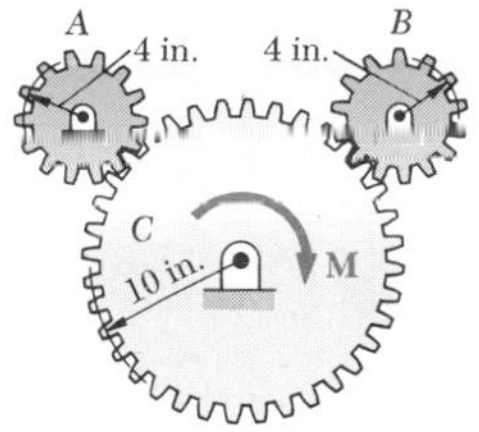

Fig. P16.43

16.44 Disk A is of mass 5 kg and has an initial angular velocity of 300 rpm clockwise. Disk B is of mass 1.8 kg and is at rest when it is placed in contact with disk A. Knowing that $\mu = 0.30$ between the disks and neglecting bearing friction, determine (a) the angular acceleration of each disk, (b) the reaction at the support C.

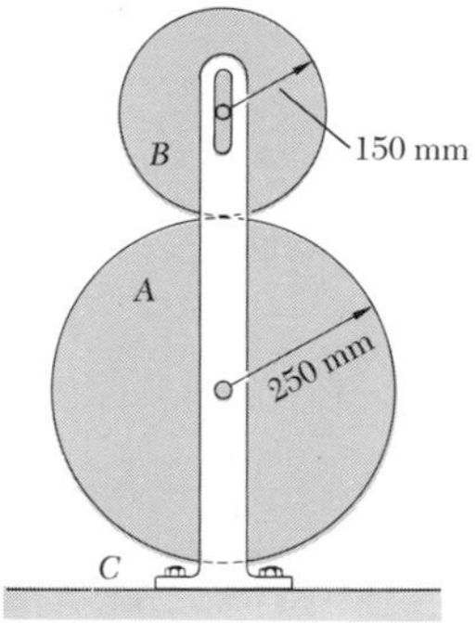

Fig. P16.44

16.45 In Prob. 16.44, (a) determine the final angular velocity of each disk, (b) show that the final angular velocities are independent of μ.

16.46 The two friction disks A and B are brought together by applying the 8-lb force shown. Disk A weighs 6 lb and had an initial angular velocity of 1200 rpm clockwise; disk B weighs 15 lb and was initially at rest. Knowing that $\mu = 0.30$ between the disks and neglecting bearing friction, determine (a) the angular acceleration of each disk, (b) the final angular velocity of each disk.

16.47 Solve Prob. 16.46, assuming that, initially, disk A was at rest and disk B had an angular velocity of 1200 rpm clockwise.

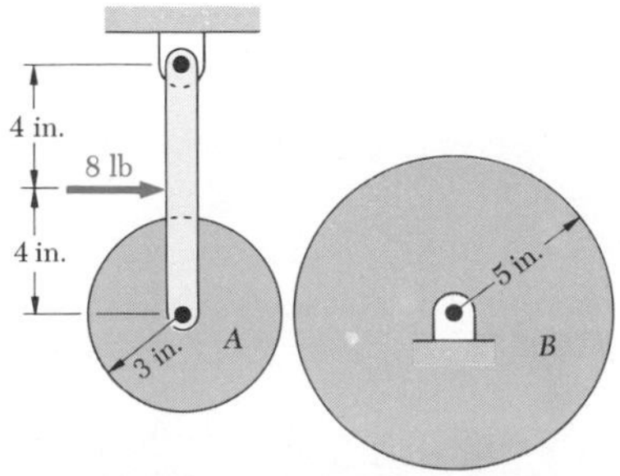

Fig. P16.46

16.48 A coder C, used to record in digital form the rotation of a shaft S, is connected to the shaft by means of the gear train shown, which consists of four gears of the same thickness and of the same material. Two of the gears have a radius r and the other two a radius nr. Let I_R denote the ratio M/α of the moment M of the couple applied to the shaft S and of the resulting angular acceleration α of S. (I_R is sometimes called the "reflected moment of inertia" of the coder and gear train.) Determine I_R in terms of the gear ratio n, the moment of inertia I_0 of the first gear, and the moment of inertia I_C of the coder. Neglect the moments of inertia of the shafts.

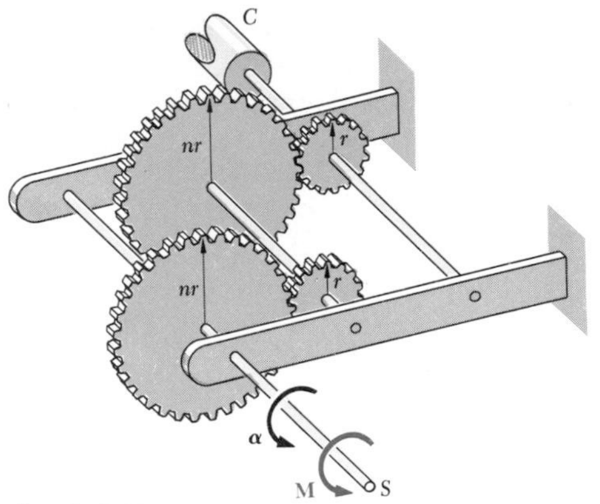

Fig. P16.48

16.49 A 6-kg bar is held between four disks as shown. Each disk has a mass of 3 kg and a diameter of 200 mm. The disks may rotate freely, and the normal reaction exerted by each disk on the bar is sufficient to prevent slipping. If the bar is released from rest, determine (a) its acceleration immediately after release, (b) its velocity after it has dropped 0.75 m

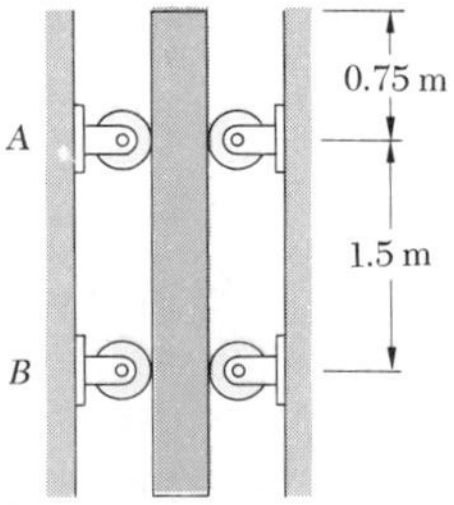

Fig. P16.49

16.50 Show that the system of the effective forces for a rigid slab in plane motion reduces to a single vector, and express the distance from the mass center G of the slab to the line of action of this vector in terms of the centroidal radius of gyration $\bar{k}$ of the slab, the magnitude $\bar{a}$ of the acceleration of G, and the angular acceleration α.

16.51 For a rigid slab in plane motion, show that the system of the effective forces consists of vectors $(\Delta m_i)\bar{\mathbf{a}}$, $-(\Delta m_i)\omega^2\mathbf{r}'_i$, and $(\Delta m_i)(\boldsymbol{\alpha} \times \mathbf{r}'_i)$ attached to the various particles P_i of the slab, where $\bar{\mathbf{a}}$ is the acceleration of the mass center G of the slab, ω the angular velocity of the slab, $\boldsymbol{\alpha}$ its angular acceleration, and where $\mathbf{r}'_i$ denotes the position vector of the particle P_i relative to G. Further show, by computing their sum and the sum of their moments about G, that the effective forces reduce to a vector $m\bar{\mathbf{a}}$ attached at G and a couple $\bar{I}\boldsymbol{\alpha}$.

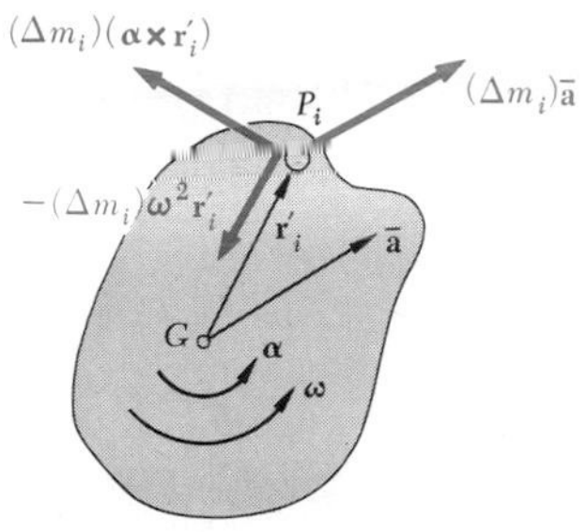

Fig. P16.51

16.52 The uniform slender rod AB weighs 8 lb and is at rest on a frictionless horizontal surface. A force $\mathbf{P}$ of magnitude 2 lb is applied at A in a horizontal direction perpendicular to the rod. Determine (*a*) the angular acceleration of the rod, (*b*) the acceleration of the center of the rod, (*c*) the point of the rod which has no acceleration.

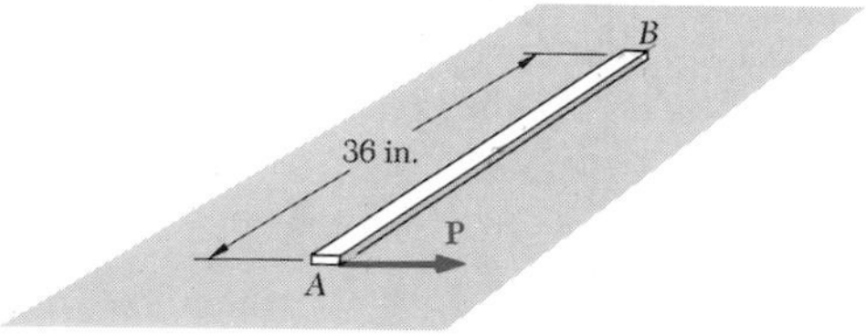

Fig. P16.52

16.53 In Prob. 16.52, determine the point of the rod AB at which the force $\mathbf{P}$ should be applied if the acceleration of point B is to be zero. Knowing that the magnitude of $\mathbf{P}$ is 2 lb, determine the corresponding angular acceleration of the rod and the acceleration of the center of the rod.

16.54 A 50-kg space satellite has a radius of gyration of 450 mm with respect to the y axis, and is symmetrical with respect to the zx plane. The orientation of the satellite is changed by firing four small rockets A, B, C, and D which are equally spaced around the perimeter of the satellite. While being fired, each rocket produces a thrust $\mathbf{T}$ of magnitude 10 N directed as shown. Determine the angular acceleration of the satellite and the acceleration of its mass center G (*a*) when all four rockets are fired, (*b*) when all rockets except rocket D are fired.

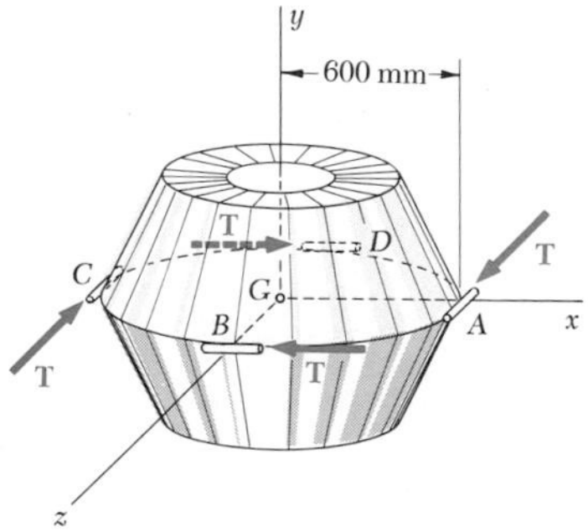

Fig. P16.54

16.55 Solve Prob. 16.54 assuming that only rocket A is fired.

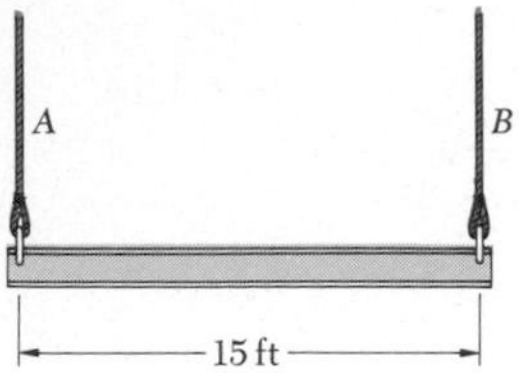

Fig. P16.56 and P16.57

16.56 A 15-ft beam weighing 500 lb is lowered from a considerable height by means of two cables unwinding from overhead cranes. As the beam approaches the ground, the crane operators apply brakes to slow the unwinding motion. The deceleration of cable A is 20 ft/s^2, while that of cable B is 2 ft/s^2. Determine the tension in each cable.

16.57 A 15-ft beam weighing 500 lb is lowered from a considerable height by means of two cables unwinding from overhead cranes. As the beam approaches the ground, the crane operators apply brakes to slow the unwinding motion. Determine the acceleration of each cable at that instant, knowing that $T_A = 360$ lb and $T_B = 320$ lb.

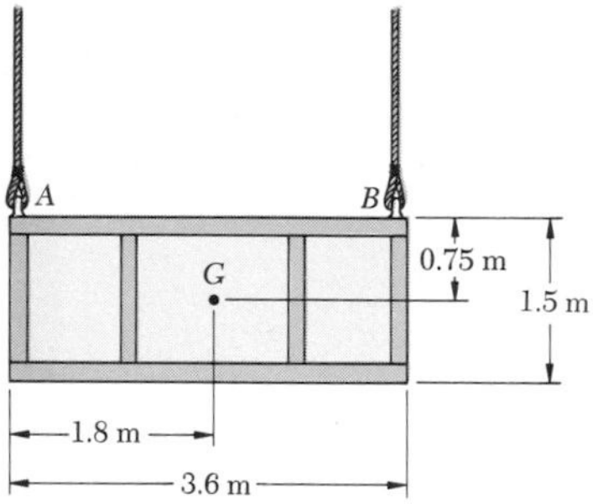

Fig. P16.58 and P16.59

16.58 The 180-kg crate shown is being lowered by means of two overhead cranes. Knowing that at the instant shown the deceleration of cable A is 7 m/s^2, while that of cable B is 1 m/s^2, determine the tension in each cable.

16.59 The 180-kg crate is being lowered by means of two overhead cranes. As the crate approaches the ground, the crane operators apply brakes to slow the motion. Determine the acceleration of each cable at that instant, knowing that $T_A = 1450$ N and $T_B = 1200$ N.

16.60 Solve Sample Prob. 16.4, assuming that the disk rests flat on a frictionless horizontal surface and that the cord is pulled horizontally with a force of magnitude 180 N.

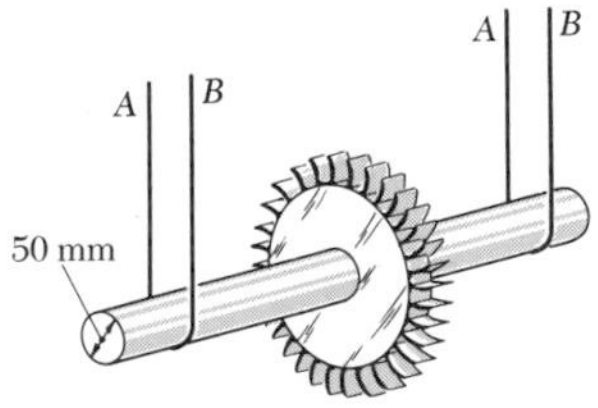

Fig. P16.61

16.61 A turbine disk and shaft have a combined mass of 100 kg and a centroidal radius of gyration of 50 mm. The unit is lifted by two ropes looped around the shaft as shown. Knowing that for each rope $T_A = 270$ N and $T_B = 320$ N, determine (*a*) the angular acceleration of the unit, (*b*) the acceleration of its mass center.

16.62 By pulling on the cord of a yo-yo just fast enough, a man manages to make the yo-yo spin counterclockwise, while remaining at a constant height above the floor. Denoting the weight of the yo-yo by W, the radius of the inner drum on which the cord is wound by r, and the radius of gyration of the yo-yo by $\bar{k}$, determine (*a*) the tension in the cord, (*b*) the angular acceleration of the yo-yo.

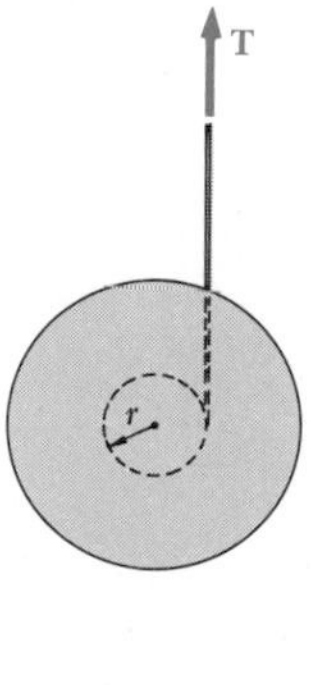

Fig. P16.62

16.63 The 80-lb crate shown rests on four casters which allow it to move without friction in any horizontal direction. A 20-lb horizontal force is applied at the midpoint *A* of edge *CE*. Knowing that the force is perpendicular to side *BCDE*, determine the angular acceleration of the crate and the acceleration of point *A*.

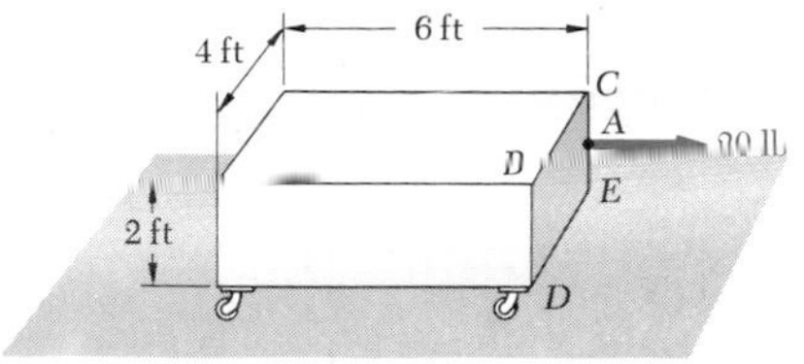

Fig. P16.63

16.64 and 16.65 A uniform slender bar *AB* of mass *m* is suspended from two *springs* as shown. If spring *BC* breaks, determine at that instant (*a*) the angular acceleration of the bar, (*b*) the acceleration of point *A*, (*c*) the acceleration of point *B*.

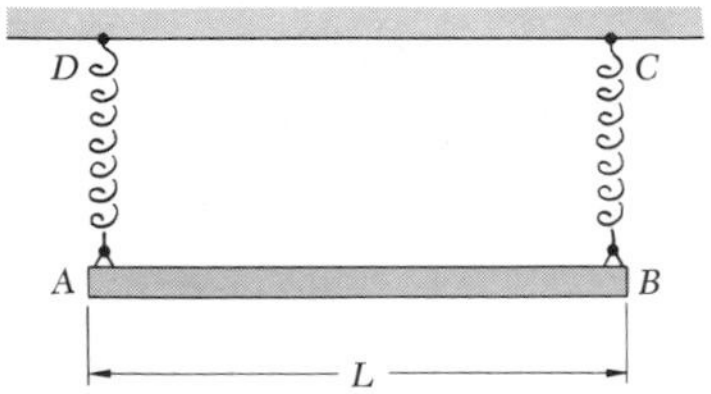

Fig. P16.64

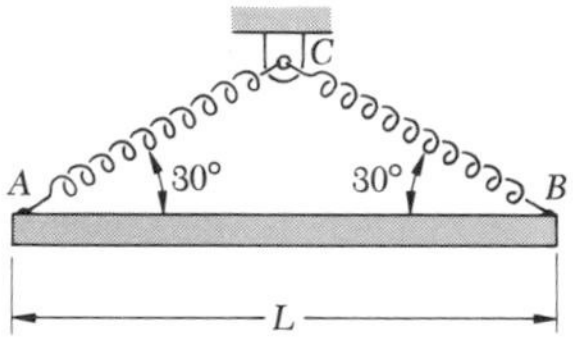

Fig. P16.65

16.66 A sphere of mass *m* and radius *r* is projected along a rough horizontal surface with a linear velocity $\mathbf{v}_0$ and with $\omega_0 = 0$. The sphere will decelerate and then reach a uniform motion. Denoting by μ the coefficient of friction, determine (*a*) the linear and angular acceleration of the sphere before it reaches a uniform motion, (*b*) the time required for the motion to become uniform, (*c*) the distance traveled before the motion becomes uniform, (*d*) the final linear and angular velocities of the sphere.

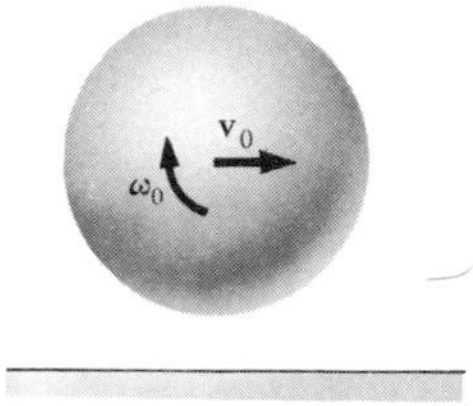

Fig. P16.66

16.67 Solve Prob. 16.66, assuming that the sphere is replaced by a uniform disk of radius *r* and mass *m*.

16.68 A heavy square plate of weight *W*, suspended from four vertical wires, supports a small block *E* of much smaller weight *w*. The coefficient of friction between *E* and the plate is denoted by μ. If the coordinates of *E* are $x = \frac{1}{4}L$ and $z = \frac{1}{4}L$, derive an expression for the magnitude of the force **P** required to cause *E* to slip with respect to the plate. (*Hint.* Neglect *w* in all equations containing *W*.)

***16.69** A square plate of weight $W = 20$ lb and side $L = 3$ ft is suspended from four wires and supports a block *E* of much smaller weight *w*. The coefficient of friction between *E* and the plate is 0.50. If a force **P** of magnitude 10 lb is applied as shown, determine the area of the plate where *E* should be placed if it is not to slip with respect to the plate. (*Hint.* Neglect *w* in all equations containing *W.*)

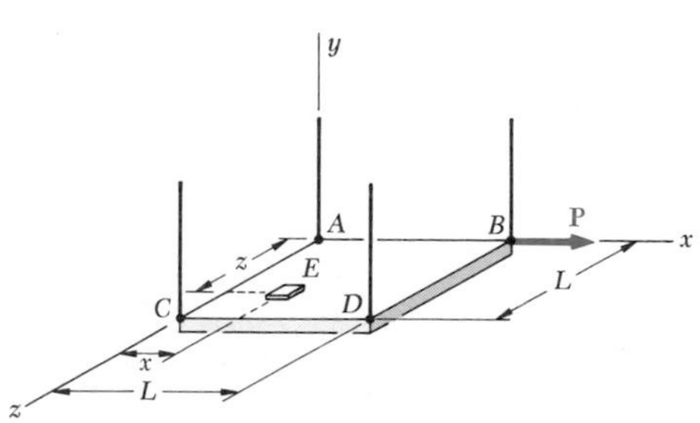

Fig. P16.68 and P16.69

16.8. Constrained Plane Motion. Most engineering applications deal with rigid bodies which are moving under given constraints. Cranks, for example, are constrained to rotate about a fixed axis, wheels roll without sliding, connecting rods must describe certain prescribed motions. In all such cases, definite relations exist between the components of the acceleration $\bar{\mathbf{a}}$ of the mass center G of the body considered and its angular acceleration $\boldsymbol{\alpha}$; the corresponding motion is said to be a *constrained motion.*

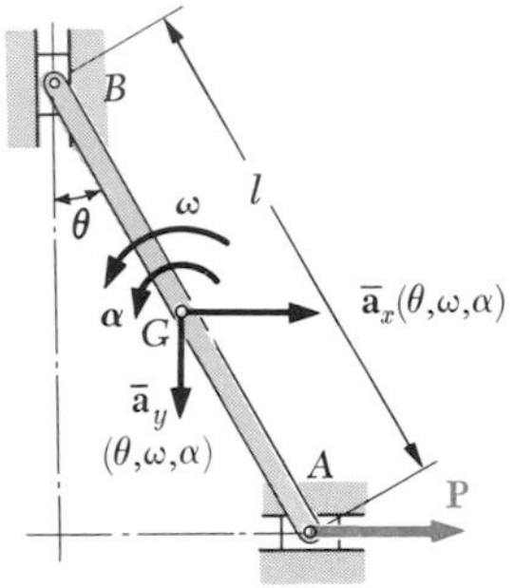

Fig. 16.11

The solution of a problem involving a constrained plane motion calls first for a *kinematic analysis* of the problem. Consider, for example, a slender rod AB of length l and mass m whose extremities are connected to blocks of negligible mass which slide along horizontal and vertical frictionless tracks. The rod is pulled by a force $\mathbf{P}$ applied at A (Fig. 16.11). We know from Sec. 15.8 that the acceleration $\bar{\mathbf{a}}$ of the mass center G of the rod may be determined at any given instant from the position of the rod, its angular velocity, and its angular acceleration at that instant. Suppose, for instance, that the values of θ, ω, and α are known at a given instant and that we wish to determine the corresponding value of the force $\mathbf{P}$, as well as the reactions at A and B. We should first *determine the components* $\bar{a}_x$ *and* $\bar{a}_y$ *of the acceleration of the mass center* G by the method of Sec. 15.8. We next apply D'Alembert's principle (Fig. 16.12), using the expressions obtained for $\bar{a}_x$ and $\bar{a}_y$. The unknown forces $\mathbf{P}$, $\mathbf{N}_A$, and $\mathbf{N}_B$ may then be determined by writing and solving the appropriate equations.

Suppose now that the applied force $\mathbf{P}$, the angle θ, and the angular velocity ω of the rod are known at a given instant and that we wish to find the angular acceleration α of the rod and the components $\bar{a}_x$ and $\bar{a}_y$ of the acceleration of its mass center at that instant, as well as the reactions at A and B. The prelimi-

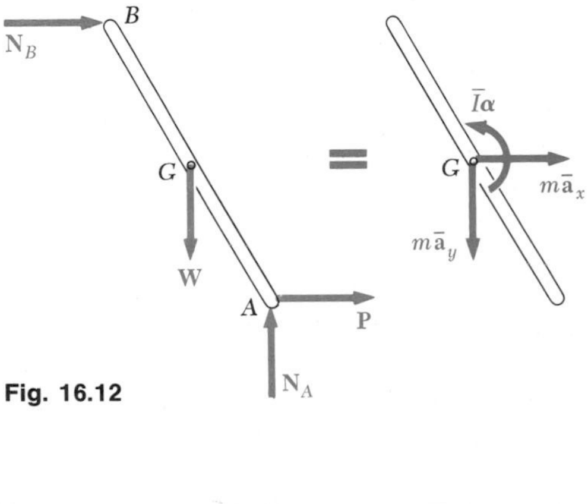

Fig. 16.12

nary kinematic study of the problem will have for its object *to express the components $\bar{a}_x$ and $\bar{a}_y$ of the acceleration of G in terms of the angular acceleration α of the rod.* This will be done by first expressing the acceleration of a suitable reference point such as A in terms of the angular acceleration α. The components $\bar{a}_x$ and $\bar{a}_y$ of the acceleration of G may then be determined in terms of α, and the expressions obtained carried into Fig. 16.12. Three equations may then be derived in terms of α, N_A, and N_B, and solved for the three unknowns (see Sample Prob. 16.10). Note that the method of dynamic equilibrium may also be used to carry out the solution of the two types of problems we have considered (Fig. 16.13).

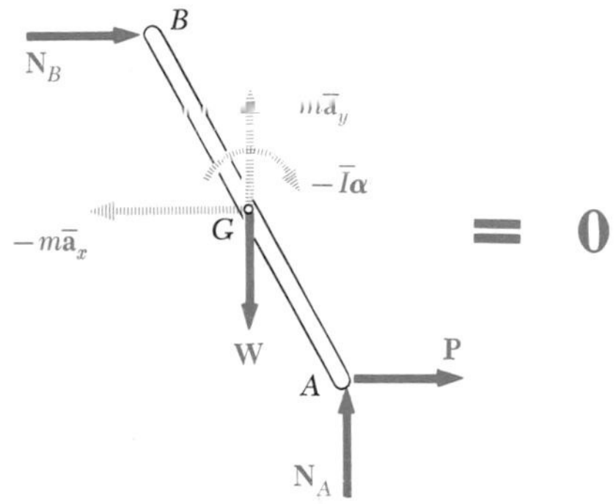

Fig. 16.13

When a mechanism consists of *several moving parts*, the approach just described may be used with each part of the mechanism. The procedure required to determine the various unknowns is then similar to the procedure followed in the case of the equilibrium of a system of connected rigid bodies (Sec. 6.11).

We have analyzed earlier two particular cases of constrained plane motion, the translation of a rigid body, in which the angular acceleration of the body is constrained to be zero, and the centroidal rotation, in which the acceleration $\bar{\mathbf{a}}$ of the mass center of the body is constrained to be zero. Two other particular cases of constrained plane motion are of special interest, the *noncentroidal rotation* of a rigid body and the *rolling motion* of a disk or wheel. These two cases should be analyzed by one of the general methods described above. However, in view of the range of their applications, they deserve a few special comments.

Noncentroidal Rotation. This is the motion of a rigid body constrained to rotate about a fixed axis which does not pass through its mass center. Such a motion is called a *noncentroidal rotation.* The mass center G of the body moves along a circle of radius $\bar{r}$ centered at the point O, where the axis of rotation intersects the plane of reference (Fig. 16.14). Denoting, respectively, by $\boldsymbol{\omega}$ and $\boldsymbol{\alpha}$ the angular velocity and the angular acceleration of the line OG, we obtain the following expressions for the tangential and normal components of the acceleration of G:

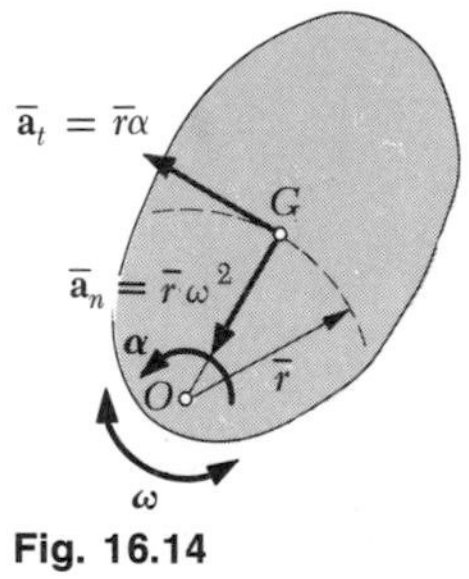

Fig. 16.14

$$\bar{a}_t = \bar{r}\alpha \qquad \bar{a}_n = \bar{r}\omega^2 \tag{16.7}$$

Since line OG belongs to the body, its angular velocity $\boldsymbol{\omega}$ and its angular acceleration $\boldsymbol{\alpha}$ also represent the angular velocity and the angular acceleration of the body in its motion relative to G. Equations (16.7) define, therefore, the kinematic relation existing between the motion of the mass center G and the motion of the body about G. They should be used to eliminate $\bar{a}_t$ and $\bar{a}_n$ from the equations obtained by applying D'Alembert's principle (Fig. 16.15) or the method of dynamic equilibrium (Fig. 16.16).

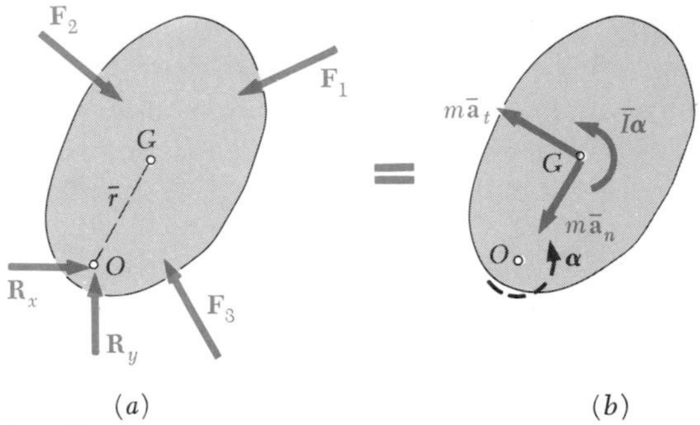

Fig. 16.15

An interesting relation may be obtained by equating the moments about the fixed point O of the forces and vectors shown respectively in parts a and b of Fig. 16.15. We write

$$+\circlearrowleft \Sigma M_O = \bar{I}\alpha + (m\bar{r}\alpha)\bar{r} = (\bar{I} + m\bar{r}^2)\alpha$$

But, according to the parallel-axis theorem, we have $\bar{I} + m\bar{r}^2 = I_O$, where I_O denotes the moment of inertia of the rigid body about the fixed axis. We write, therefore,

$$\Sigma M_O = I_O\alpha \tag{16.8}$$

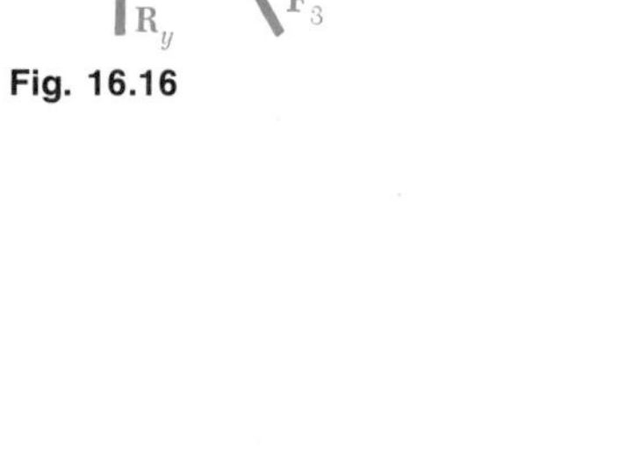

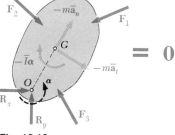

Fig. 16.16

While formula (16.8) expresses an important relation between the sum of the moments of the external forces about the fixed point O and the product $I_O\alpha$, it should be clearly understood that this formula *does not mean* that the system of the external forces is equivalent to a couple of moment $I_O\alpha$. The system of the effective forces, and thus the system of the external forces, reduces to a couple only when O coincides with G, that is, *only when the rotation is centroidal* (Sec. 16.4). In the more general case of noncentroidal rotation, the system of the external forces does not reduce to a couple.

A particular case of noncentroidal rotation is of special interest: the case of *uniform rotation,* in which the angular velocity $\boldsymbol{\omega}$ is constant. Since $\boldsymbol{\alpha}$ is zero, the inertia couple in Fig. 16.16 vanishes and the inertia vector reduces to its normal component. This component (also called *centrifugal force*) represents the tendency of the rigid body to break away from the axis of rotation.

Rolling Motion. Another important case of plane motion is the motion of a disk or wheel rolling on a plane surface. If the disk is constrained to roll without sliding, the acceleration $\bar{\mathbf{a}}$ of its mass center G and its angular acceleration $\boldsymbol{\alpha}$ are not independent. Assuming the disk to be balanced, so that its mass center and its geometric center coincide, we first write that the distance $\bar{x}$ traveled by G during a rotation θ of the disk is $\bar{x} = r\theta$, where r is the radius of the disk. Differentiating this relation twice, we write

$$\bar{a} = r\alpha \tag{16.9}$$

Recalling that the system of the effective forces in plane motion reduces to a vector $m\bar{\mathbf{a}}$ and a couple $\bar{I}\boldsymbol{\alpha}$, we find that, in the particular case of the rolling motion of a balanced disk, the effective forces reduce to a vector of magnitude $mr\alpha$ attached at G and to a couple of magnitude $\bar{I}\alpha$. We may thus express that the external forces are equivalent to the vector and couple shown in Fig. 16.17.

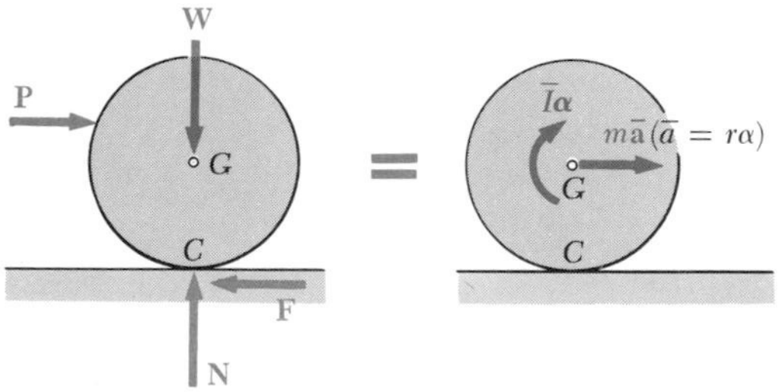

Fig. 16.17

When a disk *rolls without sliding,* there is no relative motion between the point of the disk which is in contact with the ground and the ground itself. As far as the computation of the friction force $\mathbf{F}$ is concerned, a rolling disk may thus be compared with a block at rest on a surface. The magnitude F of the friction force may have any value, as long as it does not exceed the maximum value $F_m = \mu_s N$, where μ_s is the coefficient of static friction and N the magnitude of the normal force. In the case of a rolling disk, the magnitude F of the friction force should therefore be determined independently of N by solving the equation obtained from Fig. 16.17.

When *sliding is impending,* the friction force reaches its maximum value $F_m = \mu_s N$ and may be obtained from N.

When the disk *rotates and slides* at the same time, a relative motion exists between the point of the disk which is in contact with the ground and the ground itself, and the force of friction has the magnitude $F_k = \mu_k N$, where μ_k is the coefficient of kinetic friction. In this case, however, the motion of the mass center G of the disk and the rotation of the disk about G are independent, and $\bar{a}$ is not equal to $r\alpha$.

These three different cases may be summarized as follows:

Rolling, no sliding:	$F \leq \mu_s N$	$\bar{a} = r\alpha$
Rolling, sliding impending:	$F = \mu_s N$	$\bar{a} = r\alpha$
Rotating and sliding:	$F = \mu_k N$	$\bar{a}$ *and* α *independent*

When it is not known whether a disk slides or not, it should first be assumed that the disk rolls without sliding. If F is found smaller than, or equal to, $\mu_s N$, the assumption is proved correct. If F is found larger than $\mu_s N$, the assumption is incorrect and the problem should be started again, assuming rotating and sliding.

When a disk is *unbalanced*, i.e., when its mass center G does not coincide with its geometric center O, the relation (16.9) does not hold between $\bar{a}$ and α. A similar relation will hold, however, between the magnitude a_O of the acceleration of the geometric center and the angular acceleration α,

$$a_O = r\alpha \tag{16.10}$$

To determine $\bar{a}$ in terms of the angular acceleration α and the angular velocity ω of the disk, we may use the relative-acceleration formula,

$$\begin{aligned} \bar{\mathbf{a}} = \mathbf{a}_G &= \mathbf{a}_O + \mathbf{a}_{G/O} \\ &= \mathbf{a}_O + (\mathbf{a}_{G/O})_t + (\mathbf{a}_{G/O})_n \end{aligned} \tag{16.11}$$

where the three component accelerations obtained have the directions indicated in Fig. 16.18 and the magnitudes $a_0 = r\alpha$, $(a_{G/O})_t = (OG)\alpha$, and $(a_{G/O})_n = (OG)\omega^2$.

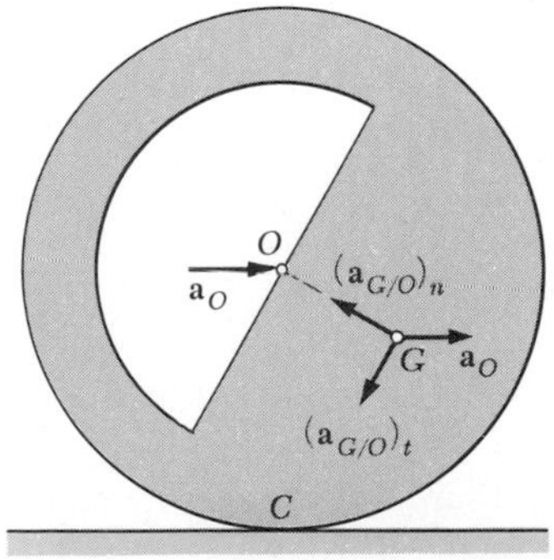

Fig. 16.18

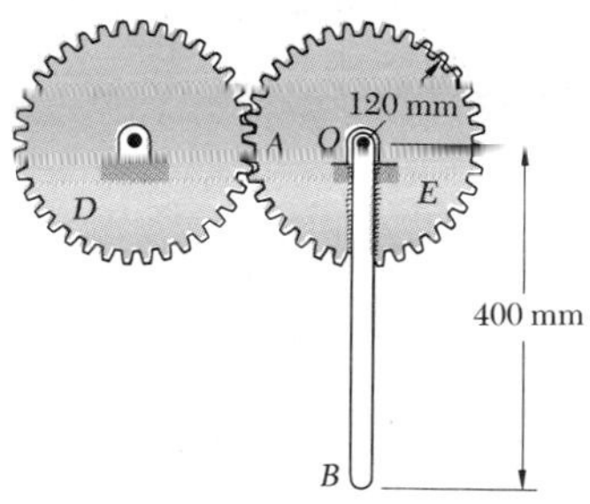

SAMPLE PROBLEM 16.6

The portion AOB of a mechanism consists of a 400-mm steel rod OB welded to a gear E of radius 120 mm which may rotate about a horizontal shaft O. It is actuated by a gear D and, at the instant shown, has a clockwise angular velocity of 8 rad/s and a counterclockwise angular acceleration of 40 rad/s². Knowing that rod OB has a mass of 3 kg and gear E a mass of 4 kg and a radius of gyration of 85 mm, determine (*a*) the tangential force exerted by gear D on gear E, (*b*) the components of the reaction at shaft O.

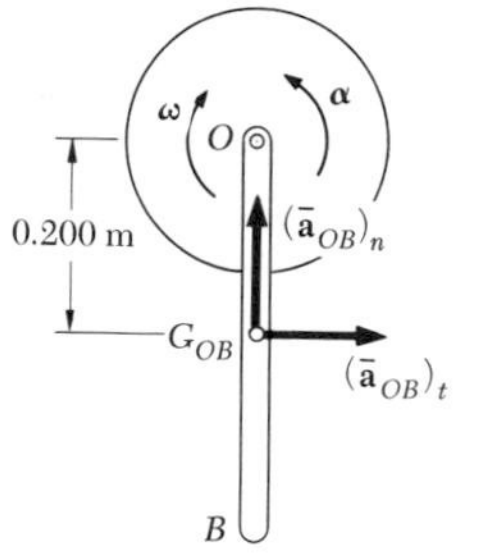

Solution. In determining the effective forces of the rigid body AOB we shall consider separately gear E and rod OB. Therefore, we shall first determine the components of the acceleration of the mass center G_{OB} of the rod:

$$(\bar{a}_{OB})_t = \bar{r}\alpha = (0.200\text{ m})(40\text{ rad/s}^2) = 8\text{ m/s}^2$$
$$(\bar{a}_{OB})_n = \bar{r}\omega^2 = (0.200\text{ m})(8\text{ rad/s})^2 = 12.8\text{ m/s}^2$$

Equations of Motion. Two sketches of the rigid body AOB have been drawn. The first shows the external forces consisting of the weight $\mathbf{W}_E$ of gear E, the weight $\mathbf{W}_{OB}$ of rod OB, the force $\mathbf{F}$ exerted by gear D, and the components $\mathbf{R}_x$ and $\mathbf{R}_y$ of the reaction at O. The magnitudes of the weights are, respectively,

$$W_E = m_E g = (4\text{ kg})(9.81\text{ m/s}^2) = 39.2\text{ N}$$
$$W_{OB} = m_{OB} g = (3\text{ kg})(9.81\text{ m/s}^2) = 29.4\text{ N}$$

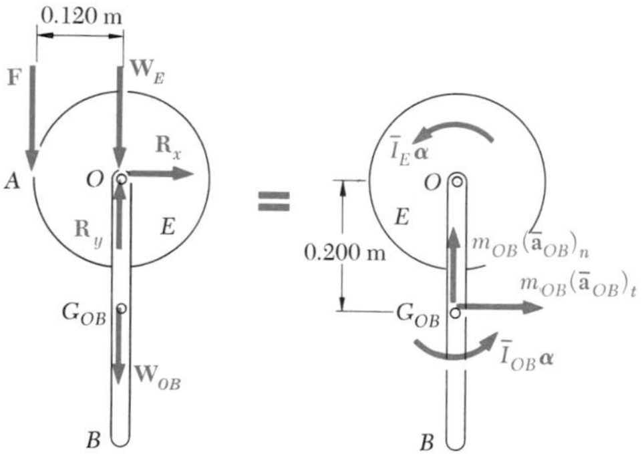

The second sketch shows the effective forces, which consist of a couple $\bar{I}_E\boldsymbol{\alpha}$ (since gear E is in centroidal rotation) and of a couple and two vector components at the mass center of OB. Since the accelerations are known, we compute the magnitudes of these components and couples:

$$\bar{I}_E\alpha = m_E\bar{k}_E^2\alpha = (4\text{ kg})(0.085\text{ m})^2(40\text{ rad/s}^2) = 1.156\text{ N}\cdot\text{m}$$
$$m_{OB}(\bar{a}_{OB})_t = (3\text{ kg})(8\text{ m/s}^2) = 24.0\text{ N}$$
$$m_{OB}(\bar{a}_{OB})_n = (3\text{ kg})(12.8\text{ m/s}^2) = 38.4\text{ N}$$
$$\bar{I}_{OB}\alpha = (\tfrac{1}{12}m_{OB}L^2)\alpha = \tfrac{1}{12}(3\text{ kg})(0.400\text{ m})^2(40\text{ rad/s}^2) = 1.600\text{ N}\cdot\text{m}$$

Expressing that the system of the external forces is equivalent to the system of the effective forces, we write the following equations:

$+\circlearrowleft\ \Sigma M_O = \Sigma(M_O)_{\text{eff}}$:

$$F(0.120\text{ m}) = \bar{I}_E\alpha + m_{OB}(\bar{a}_{OB})_t(0.200\text{ m}) + \bar{I}_{OB}\alpha$$
$$F(0.120\text{ m}) = 1.156\text{ N}\cdot\text{m} + (24.0\text{ N})(0.200\text{ m}) + 1.600\text{ N}\cdot\text{m}$$

$F = 63.0\text{ N}$ $\qquad$ $\mathbf{F} = 63.0\text{ N}\downarrow$ ◄

$\xrightarrow{+}\ \Sigma F_x = \Sigma(F_x)_{\text{eff}}$: $\qquad R_x = m_{OB}(\bar{a}_{OB})_t$

$R_x = 24.0\text{ N}$ $\qquad$ $\mathbf{R}_x = 24.0\text{ N}\rightarrow$ ◄

$+\uparrow\ \Sigma F_y = \Sigma(F_y)_{\text{eff}}$: $\qquad R_y - F - W_E - W_{OB} = m_{OB}(\bar{a}_{OB})_n$

$$R_y - 63.0\text{ N} - 39.2\text{ N} - 29.4\text{ N} = 38.4\text{ N}$$

$R_y = 170.0\text{ N}$ $\qquad$ $\mathbf{R}_y = 170.0\text{ N}\uparrow$ ◄

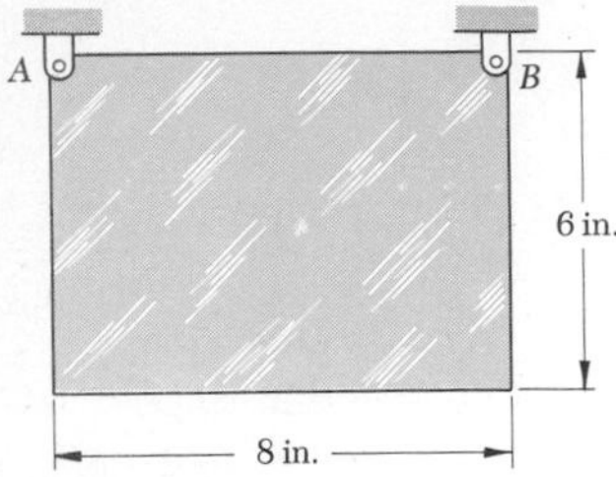

SAMPLE PROBLEM 16.7

A rectangular plate, 6 by 8 in., weighs 60 lb and is suspended from two pins A and B. If pin B is suddenly removed, determine (*a*) the angular acceleration of the plate, (*b*) the components of the reactions at pin A, immediately after pin B has been removed.

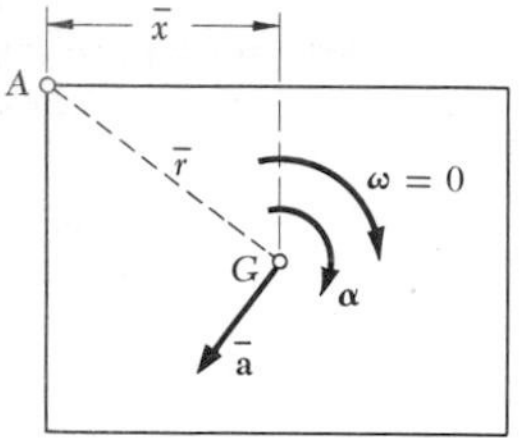

a. **Angular Acceleration.** We observe that as the plate rotates about point A, its mass center G describes a circle of radius $\bar{r}$ with center at A.

Since the plate is released from rest ($\omega = 0$), the normal component of the acceleration of G is zero. The magnitude of the acceleration $\bar{\mathbf{a}}$ of the mass center G is thus $\bar{a} = \bar{r}\alpha$. We draw the diagram shown to express that the external forces are equivalent to the effective forces:

$$+\circlearrowright \Sigma M_A = \Sigma (M_A)_{\text{eff}}: \qquad W\bar{x} = (m\bar{a})\bar{r} + \bar{I}\alpha$$

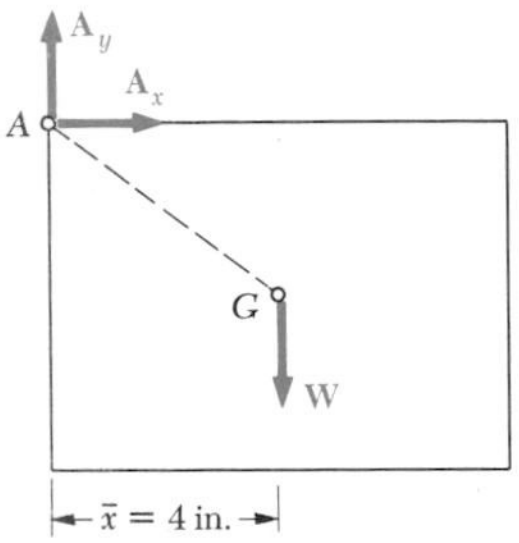

Since $\bar{a} = \bar{r}\alpha$, we have

$$W\bar{x} = m(\bar{r}\alpha)\bar{r} + \bar{I}\alpha \qquad \alpha = \frac{W\bar{x}}{\frac{W}{g}\bar{r}^2 + \bar{I}} \tag{1}$$

The centroidal moment of inertia of the plate is

$$\bar{I} = \frac{m}{12}(a^2 + b^2) = \frac{60 \text{ lb}}{12(32.2 \text{ ft/s}^2)}[(\tfrac{8}{12} \text{ ft})^2 + (\tfrac{6}{12} \text{ ft})^2]$$
$$= 0.1078 \text{ lb} \cdot \text{ft} \cdot \text{s}^2$$

Substituting this value of $\bar{I}$ together with $W = 60$ lb, $\bar{r} = \frac{5}{12}$ ft, and $\bar{x} = \frac{4}{12}$ ft into Eq. (1), we obtain

$$\alpha = +46.4 \text{ rad/s}^2 \qquad \boldsymbol{\alpha} = 46.4 \text{ rad/s}^2 \circlearrowright \quad ◀$$

=

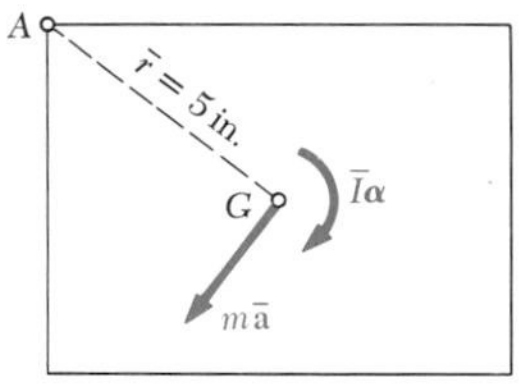

b. **Reaction at A.** Using the computed value of α, we determine the magnitude of the vector $m\bar{\mathbf{a}}$ attached at G,

$$m\bar{a} = m\bar{r}\alpha = \frac{60 \text{ lb}}{32.2 \text{ ft/s}^2}(\tfrac{5}{12} \text{ ft})(46.4 \text{ rad/s}^2) = 36.0 \text{ lb}$$

Showing this result on the diagram, we write the equations of motion

$$\xrightarrow{+} \Sigma F_x = \Sigma (F_x)_{\text{eff}}: \qquad A_x = -\tfrac{3}{5}(36 \text{ lb})$$
$$A_x = -21.6 \text{ lb} \qquad \mathbf{A}_x = 21.6 \text{ lb} \leftarrow \quad ◀$$

$$+\uparrow \Sigma F_y = \Sigma (F_y)_{\text{eff}}: \qquad A_y - 60 \text{ lb} = -\tfrac{4}{5}(36 \text{ lb})$$
$$A_y = +31.2 \text{ lb} \qquad \mathbf{A}_y = 31.2 \text{ lb} \uparrow \quad ◀$$

=

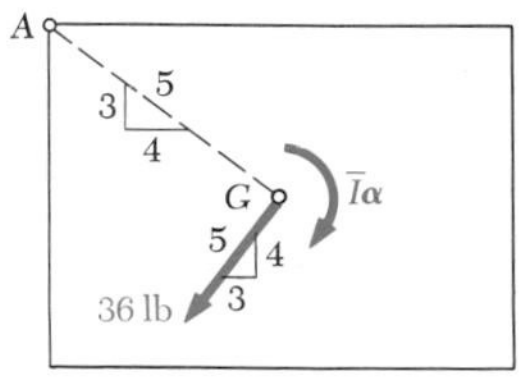

The couple $\bar{I}\boldsymbol{\alpha}$ is not involved in the last two equations; nevertheless, it should be indicated on the diagram.

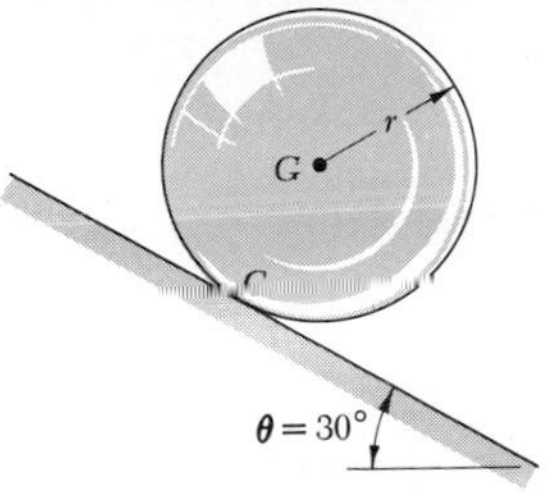

SAMPLE PROBLEM 16.8

A sphere of radius r and weight W is released with no initial velocity on the incline and rolls without slipping. Determine (*a*) the minimum value of the coefficient of friction compatible with the rolling motion, (*b*) the velocity of the center G of the sphere after the sphere has rolled 10 ft, (*c*) the velocity of G if the sphere were to move 10 ft down a frictionless 30° incline.

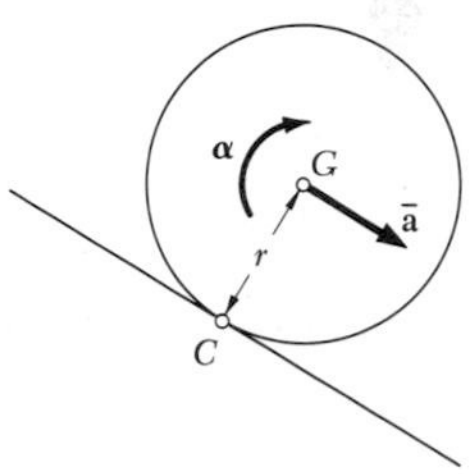

***a.* Minimum μ for Rolling Motion.** The external forces **W, N,** and **F** form a system equivalent to the system of effective forces represented by the vector $m\bar{\mathbf{a}}$ and the couple $\bar{I}\boldsymbol{\alpha}$. Since the sphere rolls without sliding, we have $\bar{a} = r\alpha$.

$$+\downarrow \Sigma M_C = \Sigma(M_C)_{\text{eff}}: \qquad (W \sin\theta)r = (m\bar{a})r + \bar{I}\alpha$$
$$(W \sin\theta)r = (mr\alpha)r + \bar{I}\alpha$$

Noting that $m = W/g$ and $\bar{I} = \frac{2}{5}mr^2$, we write

$$(W\sin\theta)r = \left(\frac{W}{g}r\alpha\right)r + \frac{2}{5}\frac{W}{g}r^2\alpha \qquad \alpha = +\frac{5g\sin\theta}{7r}$$

$$\bar{a} = r\alpha = \frac{5g\sin\theta}{7} = \frac{5(32.2\text{ ft/s}^2)\sin 30°}{7} = 11.50\text{ ft/s}^2$$

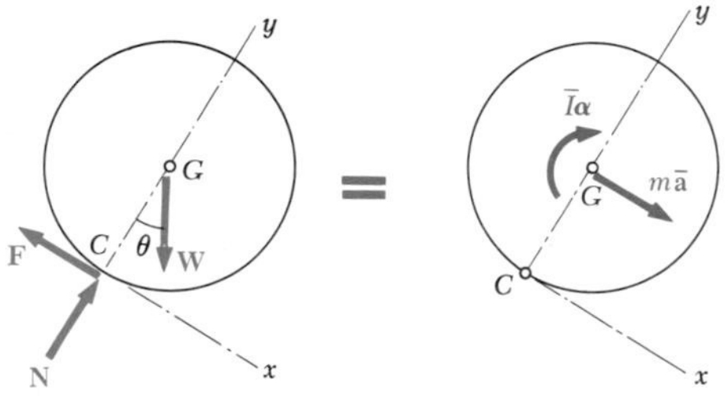

$$+\searrow \Sigma F_x = \Sigma(F_x)_{\text{eff}}: \qquad W\sin\theta - F = m\bar{a}$$
$$W\sin\theta - F = \frac{W}{g}\frac{5g\sin\theta}{7}$$
$$F = +\tfrac{2}{7}W\sin\theta = \tfrac{2}{7}W\sin 30° \qquad \mathbf{F} = 0.143W \;\nwarrow 30°$$

$$+\nearrow \Sigma F_y = \Sigma(F_y)_{\text{eff}}: \qquad N - W\cos\theta = 0$$
$$N = W\cos\theta = 0.866W \qquad \mathbf{N} = 0.866W \;\angle 60°$$

$$\mu_{\min} = \frac{F}{N} = \frac{0.143W}{0.866W} \qquad \mu_{\min} = 0.165 \blacktriangleleft$$

***b.* Velocity of Rolling Sphere.** We have uniformly accelerated motion,

$$\bar{v}_0 = 0 \qquad \bar{a} = 11.50\text{ ft/s}^2 \qquad \bar{x} = 10\text{ ft} \qquad \bar{x}_0 = 0$$
$$\bar{v}^2 = \bar{v}_0^2 + 2\bar{a}(\bar{x} - \bar{x}_0) \qquad \bar{v}^2 = 0 + 2(11.50\text{ ft/s}^2)(10\text{ ft})$$
$$\bar{v} = 15.17\text{ ft/s} \qquad \bar{\mathbf{v}} = 15.17\text{ ft/s} \searrow 30° \blacktriangleleft$$

***c.* Velocity of Sliding Sphere.** Assuming now no friction, we have $F = 0$ and obtain

$$+\downarrow \Sigma M_G = \Sigma(M_G)_{\text{eff}}: \qquad 0 = \bar{I}\alpha \qquad \alpha = 0$$

$$+\searrow \Sigma F_x = \Sigma(F_x)_{\text{eff}}: \qquad W\sin 30° = m\bar{a} \qquad 0.50W = \frac{W}{g}\bar{a}$$

$$\bar{a} = +16.1\text{ ft/s}^2 \qquad \bar{\mathbf{a}} = 16.1\text{ ft/s}^2 \searrow 30°$$

Substituting $\bar{a} = 16.1\text{ ft/s}^2$ into the equations for uniformly accelerated motion, we obtain

$$\bar{v}^2 = \bar{v}_0^2 + 2\bar{a}(\bar{x} - \bar{x}_0) \qquad \bar{v}^2 = 0 + 2(16.1\text{ ft/s}^2)(10\text{ ft})$$
$$\bar{v} = 17.94\text{ ft/s} \qquad \bar{\mathbf{v}} = 17.94\text{ ft/s} \searrow 30° \blacktriangleleft$$

SAMPLE PROBLEM 16.9

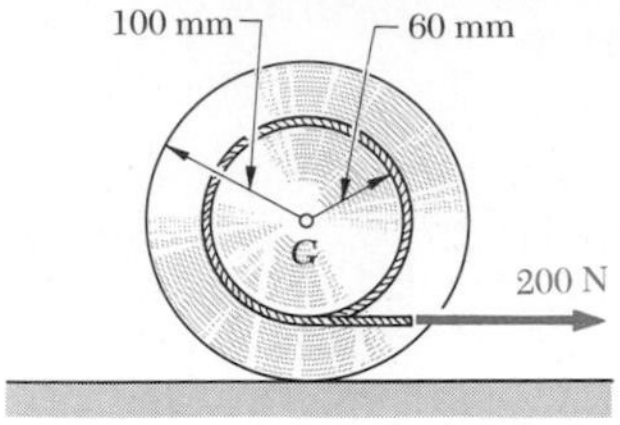

A cord is wrapped around the inner drum of a wheel and pulled horizontally with a force of 200 N. The wheel has a mass of 50 kg and a radius of gyration of 70 mm. Knowing that $\mu_s = 0.20$ and $\mu_k = 0.15$, determine the acceleration of G and the angular acceleration of the wheel.

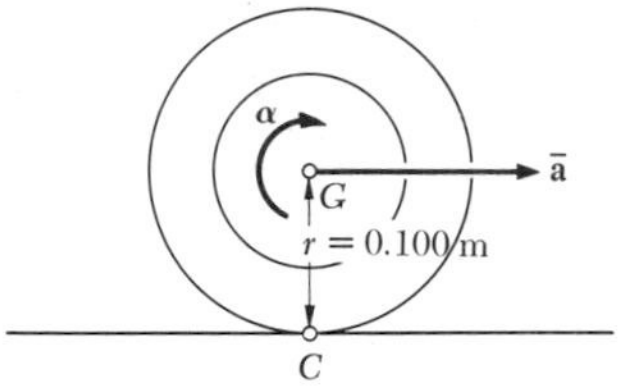

a. **Assume Rolling without Sliding.** In this case, we have

$$\bar{a} = r\alpha = (0.100 \text{ m})\alpha$$

By comparing the friction force obtained with the maximum available friction force, we shall determine whether this assumption is justified. The moment of inertia of the wheel is

$$\bar{I} = m\bar{k}^2 = (50 \text{ kg})(0.070 \text{ m})^2 = 0.245 \text{ kg} \cdot \text{m}^2$$

Equations of Motion

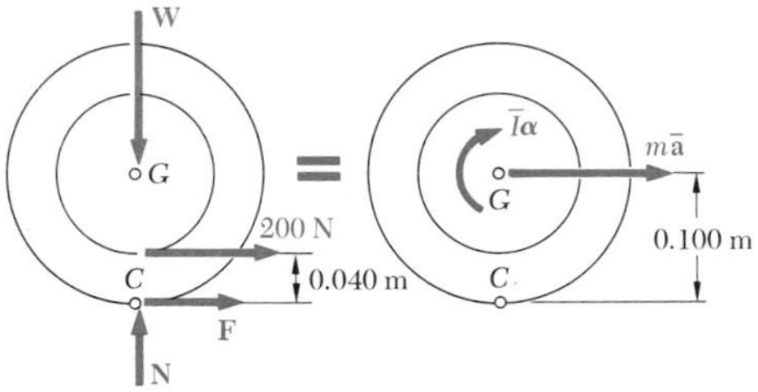

$$+\circlearrowright \Sigma M_C = \Sigma(M_C)_{\text{eff}}: \qquad (200 \text{ N})(0.040 \text{ m}) = m\bar{a}(0.100 \text{ m}) + \bar{I}\alpha$$
$$8.00 \text{ N} = (50 \text{ kg})(0.100 \text{ m})\alpha(0.100 \text{ m}) + (0.245 \text{ kg} \cdot \text{m}^2)\alpha$$
$$\alpha = +10.74 \text{ rad/s}^2$$
$$\bar{a} = r\alpha = (0.100 \text{ m})(10.74 \text{ rad/s}^2) = 1.074 \text{ m/s}^2$$

$$\xrightarrow{+} \Sigma F_x = \Sigma(F_x)_{\text{eff}}: \qquad F + 200 \text{ N} = m\bar{a}$$
$$F + 200 \text{ N} = (50 \text{ kg})(1.074 \text{ m/s}^2)$$
$$F = -146.3 \text{ N} \qquad\qquad \mathbf{F} = 146.3 \text{ N} \leftarrow$$

$$+\uparrow \Sigma F_y = \Sigma(F_y)_{\text{eff}}:$$
$$N - W = 0 \qquad N = W = mg = (50 \text{ kg})(9.81 \text{ m/s}^2) = 490.5 \text{ N}$$
$$\mathbf{N} = 490.5 \text{ N} \uparrow$$

Maximum Available Friction Force

$$F_{\text{max}} = \mu_s N = 0.20\,(490.5 \text{ N}) = 98.1 \text{ N}$$

Since $F > F_{\text{max}}$, the assumed motion is impossible.

b. **Rotating and Sliding.** Since the wheel must rotate and slide at the same time, we draw a new diagram, where $\bar{\mathbf{a}}$ and $\boldsymbol{\alpha}$ are independent and where

$$F = F_k = \mu_k N = 0.15\,(490.5 \text{ N}) = 73.6 \text{ N}$$

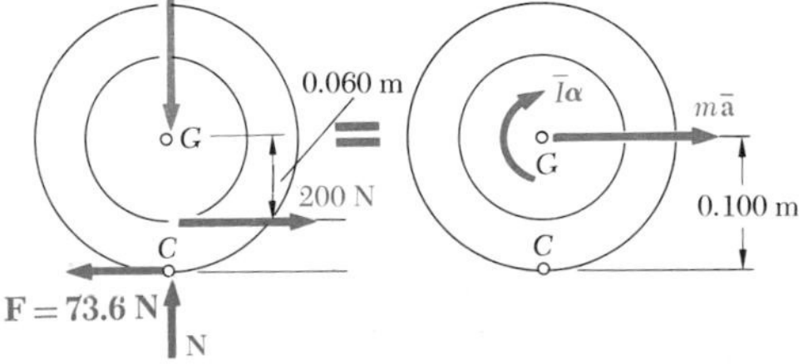

From the computation of part *a*, it appears that **F** should be directed to the left. We write the following equations of motion:

$$\xrightarrow{+} \Sigma F_x = \Sigma(F_x)_{\text{eff}}: \qquad 200 \text{ N} - 73.6 \text{ N} = (50 \text{ kg})\bar{a}$$
$$\bar{a} = +2.53 \text{ m/s}^2 \qquad \bar{\mathbf{a}} = 2.53 \text{ m/s}^2 \rightarrow \quad ◀$$

$$+\circlearrowright \Sigma M_G = \Sigma(M_G)_{\text{eff}}:$$
$$(73.6 \text{ N})(0.100 \text{ m}) - (200 \text{ N})(0.060 \text{ m}) = (0.245 \text{ kg} \cdot \text{m}^2)\alpha$$
$$\alpha = -18.94 \text{ rad/s}^2 \qquad \boldsymbol{\alpha} = 18.94 \text{ rad/s}^2 \circlearrowleft \quad ◀$$

SAMPLE PROBLEM 16.10

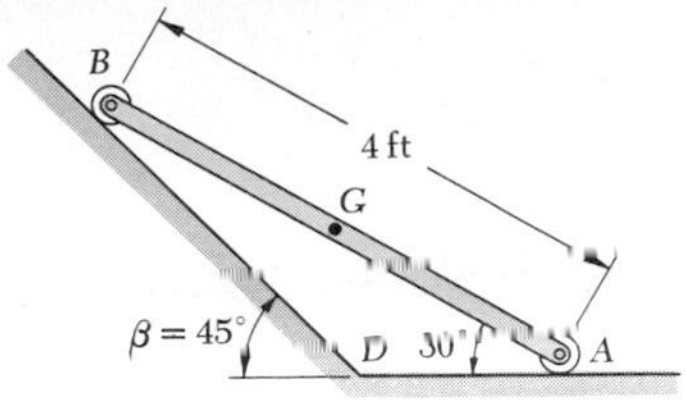

The extremities of a 4-ft rod, weighing 50 lb, may move freely and with no friction along two straight tracks as shown. If the rod is released with no velocity from the position shown, determine (*a*) the angular acceleration of the rod, (*b*) the reactions at *A* and *B*.

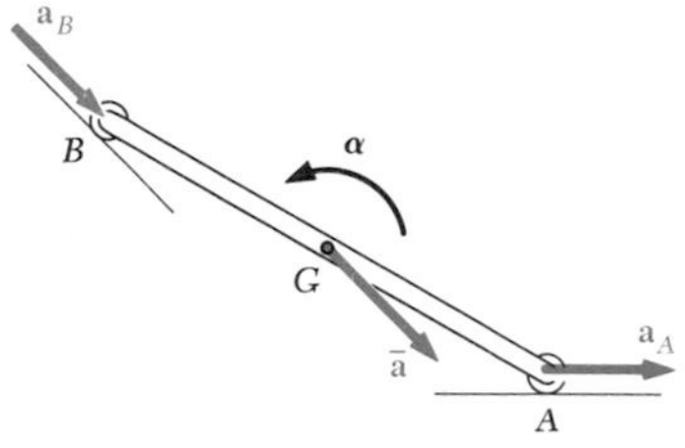

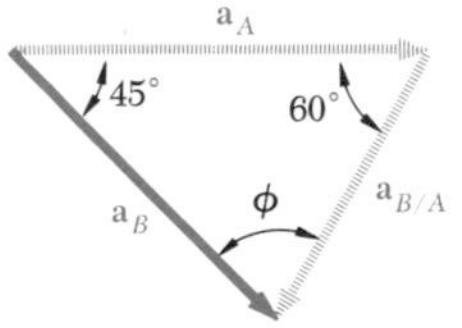

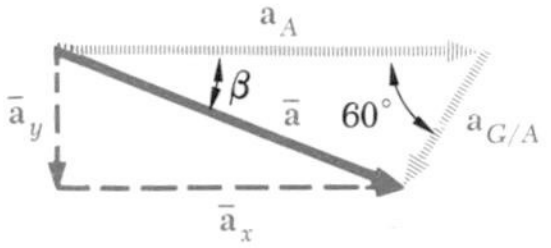

Kinematics of Motion. Since the motion is constrained, the acceleration of G must be related to the angular acceleration $\boldsymbol{\alpha}$. To obtain this relation, we shall first determine the magnitude of the acceleration $\mathbf{a}_A$ of point A in terms of α; assuming $\boldsymbol{\alpha}$ directed counterclockwise and noting that $a_{B/A} = 4\alpha$, we write

$$\mathbf{a}_B = \mathbf{a}_A + \mathbf{a}_{B/A}$$
$$[a_B \searrow 45°] = [a_A \rightarrow] + [4\alpha \nearrow 60°]$$

Noting that $\phi = 75°$ and using the law of sines, we obtain

$$a_A = 5.46\alpha \qquad a_B = 4.90\alpha$$

The acceleration of G is now obtained by writing

$$\bar{\mathbf{a}} = \mathbf{a}_G = \mathbf{a}_A + \mathbf{a}_{G/A}$$
$$\bar{\mathbf{a}} = [5.46\alpha \rightarrow] + [2\alpha \nearrow 60°]$$

Resolving $\bar{\mathbf{a}}$ into x and y components, we obtain

$$\bar{a}_x = 5.46\alpha - 2\alpha \cos 60° = 4.46\alpha \qquad \bar{\mathbf{a}}_x = 4.46\alpha \rightarrow$$
$$\bar{a}_y = -2\alpha \sin 60° = -1.732\alpha \qquad \bar{\mathbf{a}}_y = 1.732\alpha \downarrow$$

Kinetics of Motion. We draw the two sketches shown to express that the system of external forces is equivalent to the system of effective forces represented by the vector of components $m\bar{\mathbf{a}}_x$ and $m\bar{\mathbf{a}}_y$ attached at G and the couple $\bar{I}\boldsymbol{\alpha}$. We compute the following magnitudes:

$$\bar{I} = \frac{1}{12}ml^2 = \frac{1}{12}\frac{50 \text{ lb}}{32.2 \text{ ft/s}^2}(4 \text{ ft})^2 = 2.07 \text{ lb} \cdot \text{ft} \cdot \text{s}^2 \qquad \bar{I}\alpha = 2.07\alpha$$

$$m\bar{a}_x = \frac{50}{32.2}(4.46\alpha) = 6.93\alpha \qquad m\bar{a}_y = -\frac{50}{32.2}(1.732\alpha) = -2.69\alpha$$

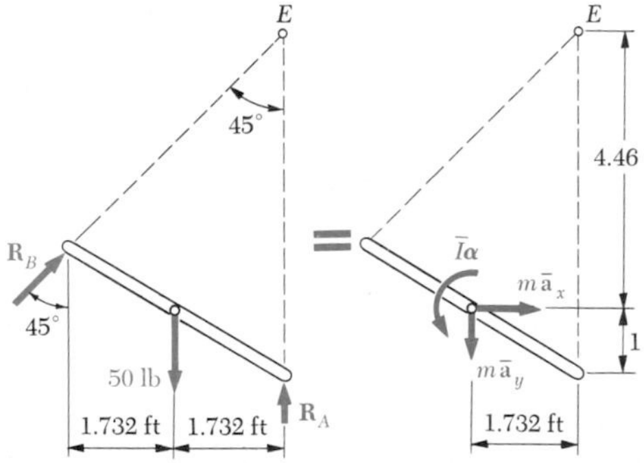

Equations of Motion

$+\circlearrowleft \Sigma M_E = \Sigma(M_E)_{\text{eff}}$:

$$(50)(1.732) = (6.93\alpha)(4.46) + (2.69\alpha)(1.732) + 2.07\alpha$$
$$\alpha = +2.30 \text{ rad/s}^2 \qquad \alpha = 2.30 \text{ rad/s}^2 \circlearrowleft \blacktriangleleft$$

$\xrightarrow{+} \Sigma F_x = \Sigma(F_x)_{\text{eff}}$: $\quad R_B \sin 45° = (6.93)(2.30) = 15.94$

$$R_B = 22.5 \text{ lb} \qquad \mathbf{R}_B = 22.5 \text{ lb} \measuredangle 45° \blacktriangleleft$$

$+\uparrow \Sigma F_y = \Sigma(F_y)_{\text{eff}}$: $\quad R_A + R_B \cos 45° - 50 = -(2.69)(2.30)$

$$R_A = -6.19 - 15.94 + 50 = 27.9 \text{ lb} \qquad \mathbf{R}_A = 27.9 \text{ lb} \uparrow \blacktriangleleft$$

PROBLEMS

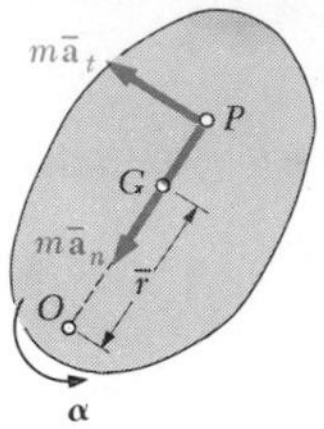

Fig. P16.70

16.70 Show that the couple $\bar{I}\boldsymbol{\alpha}$ of Fig. 16.15 may be eliminated by attaching the vectors $m\bar{\mathbf{a}}_t$ and $m\bar{\mathbf{a}}_n$ at a point P called the *center of percussion*, located on line OG at a distance $GP = \bar{k}^2/\bar{r}$ from the mass center of the body.

16.71 A uniform slender rod, of length $L = 900$ mm and mass $m = 4$ kg, is supported as shown. A horizontal force $\mathbf{P}$ of magnitude 75 N is applied at end B. For $\bar{r} = \frac{1}{4}L = 225$ mm, determine (*a*) the angular acceleration of the rod, (*b*) the components of the reaction at C.

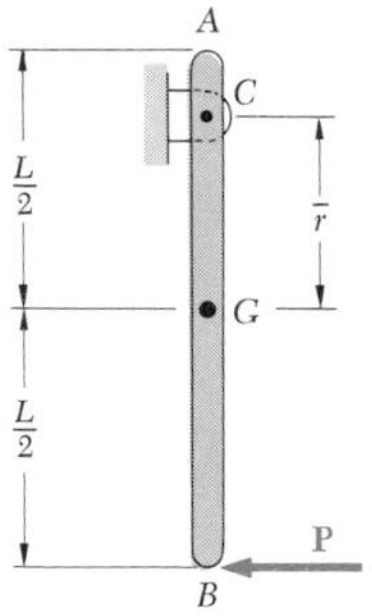

Fig. 16.71

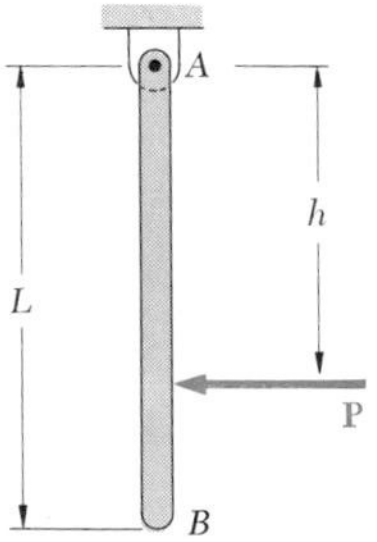

Fig. P16.73 and P16.74

16.72 In Prob. 16.71, determine (*a*) the distance $\bar{r}$ for which the horizontal component of the reaction at C is zero, (*b*) the corresponding angular acceleration of the rod.

16.73 A uniform slender rod, of length L and weight W, hangs freely from a hinge at A. If a horizontal force $\mathbf{P}$ is applied as shown, determine (*a*) the distance h for which the horizontal component of the reaction at A is zero, (*b*) the corresponding angular acceleration of the rod.

16.74 A uniform slender rod, of length L and weight W, hangs freely from a hinge at A. If a force $\mathbf{P}$ is applied at B horizontally to the left ($h = L$), determine (*a*) the angular acceleration of the rod, (*b*) the components of the reaction at A.

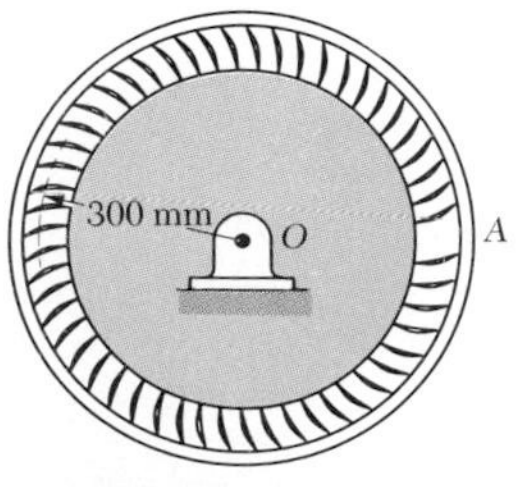

Fig. P16.75

16.75 A turbine disk of mass 75 kg rotates at a constant speed of 9600 rpm; the mass center of the disk coincides with the center of rotation O. Determine the reaction at O after a single vane at A, of mass 45 g, becomes loose and is thrown off.

16.76 A uniform slender rod of length l and mass m rotates about a vertical axis AA' at a constant angular velocity $\boldsymbol{\omega}$. Determine the tension in the rod at a distance x from the axis of rotation.

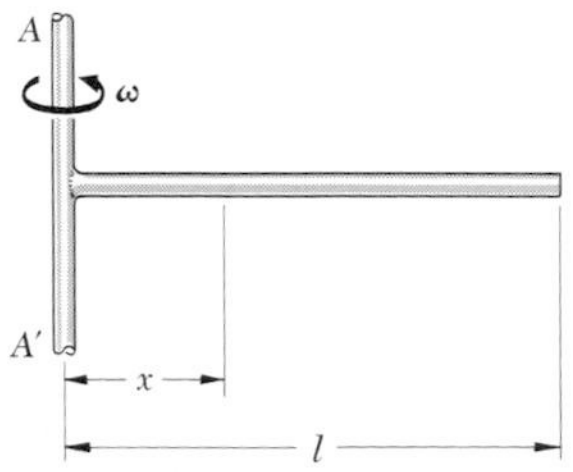

Fig. P16.76

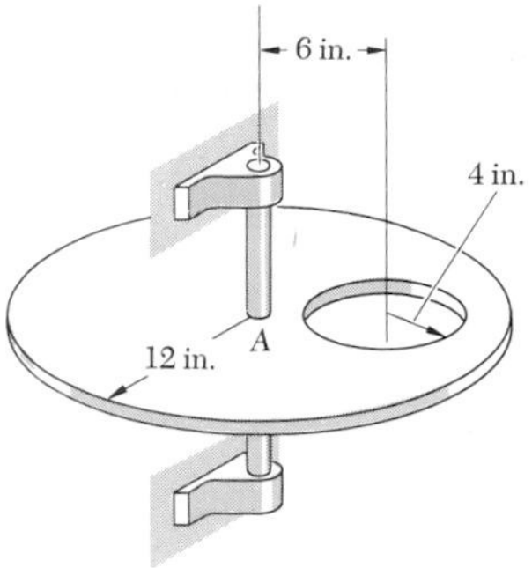

Fig. P16.77

16.77 An 8-in.-diameter hole is cut as shown in a thin disk of diameter 24 in. The disk rotates in a horizontal plane about its geometric center A at a constant angular velocity of 480 rpm. Knowing that the disk weighs 100 lb after the hole has been cut, determine the horizontal component of the force exerted by the shaft on the disk at A.

16.78 A large flywheel is mounted on a horizontal shaft and rotates at a constant rate of 1200 rpm. Experimental data show that the total force exerted by the flywheel on the shaft varies from 55 kN upward to 85 kN downward. Determine (*a*) the mass of the flywheel, (*b*) the distance from the center of the shaft to the mass center of the flywheel.

16.79 and 16.80 A uniform beam of length L and weight W is supported as shown. If the cable suddenly breaks, determine (*a*) the reaction at the pin support, (*b*) the acceleration of point B.

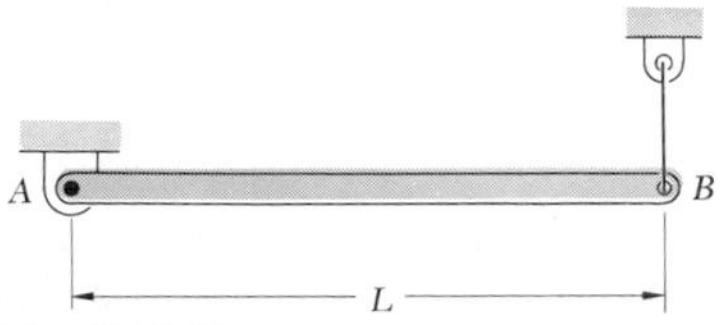

Fig. P16.79

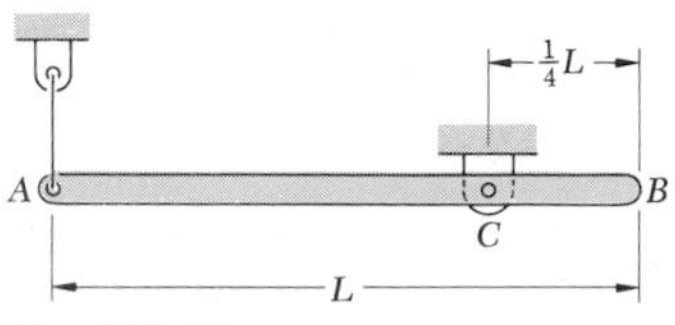

Fig. P16.80

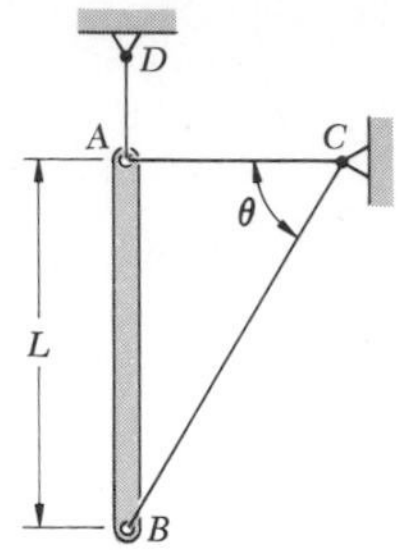

Fig. P16.81

16.81 A uniform slender rod *AB*, of length $L = 4$ ft and weight 10 lb, is held in the position shown by three wires. If $\theta = 60°$, determine the tension in wires *AC* and *BC* immediately after wire *AD* has been cut.

16.82 Two uniform rods, each of mass m, are attached as shown to small gears of negligible mass. If the rods are released from rest in the position shown, determine the angular acceleration of rod *AB* immediately after release, assuming (*a*) $\theta = 0$, (*b*) $\theta = 30°$.

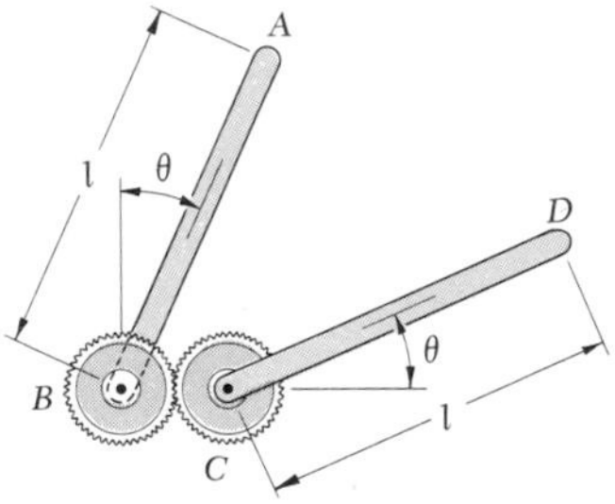

Fig. P16.82

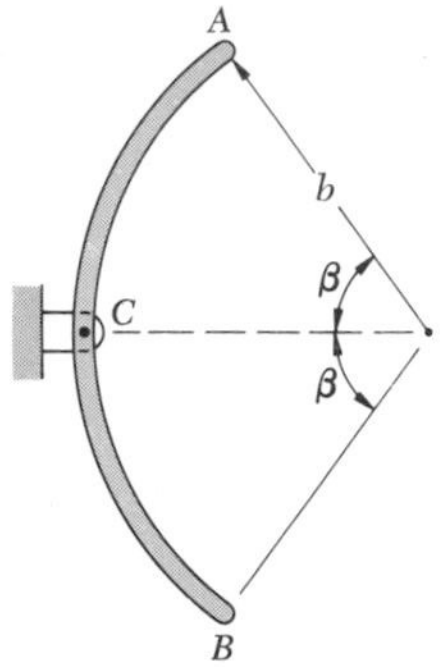

Fig. P16.83

16.83 A uniform rod *AB* is bent in the shape of an arc of circle. Determine the angular acceleration of the rod immediately after it is released from rest and show that it is independent of β.

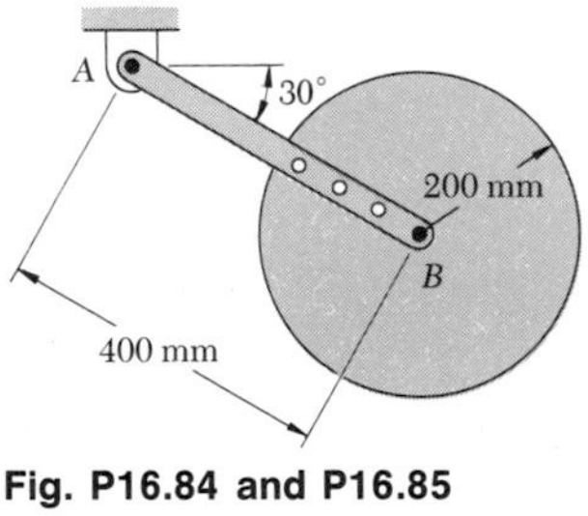

Fig. P16.84 and P16.85

16.84 A 2-kg slender rod is riveted to a 4-kg uniform disk as shown. The assembly swings freely in a vertical plane and, in the position shown, has an angular velocity of 4 rad/s clockwise. Determine (*a*) the angular acceleration of the assembly, (*b*) the components of the reaction at *A*.

16.85 A 2-kg slender rod is riveted to a 4-kg uniform disk as shown. The assembly rotates in a vertical plane under the combined effect of gravity and a couple **M** which is applied to rod *AB*. Knowing that at the instant shown the assembly has an angular velocity of 6 rad/s and an angular acceleration of 10 rad/s^2 both counterclockwise, determine (*a*) the magnitude of the couple **M**, (*b*) the components of the reaction at *A*.

16.86 After being released, the plate of Sample Prob. 16.7 is allowed to swing through 90°. Knowing that at that instant the angular velocity of the plate is 4.82 rad/s, determine (*a*) the angular acceleration of the plate, (*b*) the reaction at *A*.

16.87 Two uniform rods, AB of weight 12 lb and CD of weight 8 lb, are welded together to form the T-shaped assembly shown. The assembly rotates in a vertical plane about a horizontal shaft at E. Knowing that at the instant shown the assembly has an angular velocity of 12 rad/s and an angular acceleration of 36 rad/s^2, both clockwise, determine (*a*) the magnitude of the horizontal force **P**, (*b*) the components of the reaction at E.

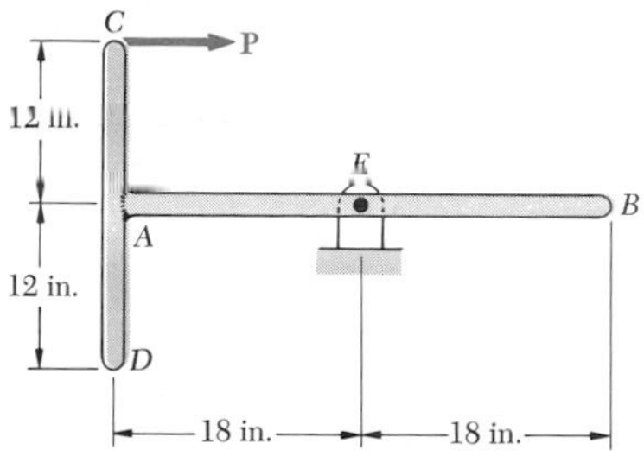

Fig. P16.87

16.88 The uniform rod AB of mass m is released from rest when $\beta = 60°$. Assuming that the friction between end A and the surface is large enough to prevent sliding, determine (*a*) the angular acceleration of the rod just after release, (*b*) the normal reaction and the friction force at A, (*c*) the minimum value of μ compatible with the described motion.

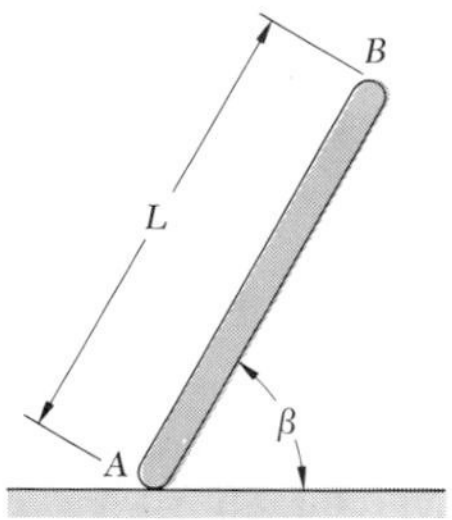

Fig. P16.88 and P16.89

***16.89** Knowing that the coefficient of friction between the rod and the floor is 0.30, determine the range of values of β for which the rod *will* slip immediately after being released from rest.

16.90 Derive the equation $\Sigma M_C = I_C\alpha$ for the rolling disk of Fig. 16.17, where ΣM_C represents the sum of the moments of the external forces about the instantaneous center C and I_C the moment of inertia of the disk about C.

16.91 Show that, in the case of an unbalanced disk, the equation derived in Prob. 16.90 is valid only when the mass center G, the geometric center O, and the instantaneous center C happen to lie in a straight line.

16.92 A homogeneous cylinder C and a section of pipe P are in contact when they are released from rest. Knowing that both the cylinder and the pipe roll without slipping, determine the clear distance between them after 2.5 s.

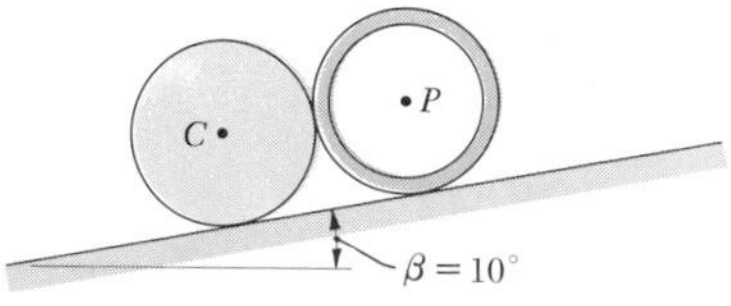

Fig. P16.92

16.93 A flywheel is rigidly attached to a shaft of 40-mm radius which may roll along parallel rails as shown. When released from rest, the system rolls a distance of 3 m in 30 s. Determine the centroidal radius of gyration of the system.

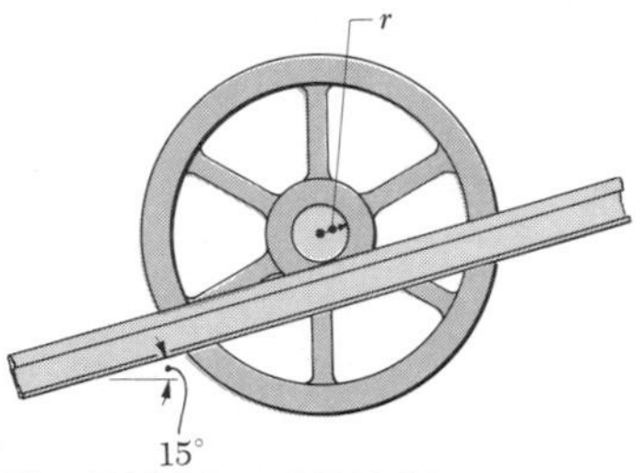

Fig. P16.93 and P16.94

16.94 A flywheel of centroidal radius of gyration $\bar{k} = 600$ mm is rigidly attached to a shaft of radius $r = 30$ mm which may roll along parallel rails. Knowing that the system is released from rest, determine the distance it will roll in 20 s.

16.95 through 16.98 A drum of 80-mm radius is attached to a disk of 160-mm radius. The disk and drum have a total mass of 5 kg and a radius of gyration of 120 mm. A cord is attached as shown and pulled with a force **P** of magnitude 20 N. Knowing that the disk rolls without sliding, determine (*a*) the angular acceleration of the disk and the acceleration of G, (*b*) the minimum value of the coefficient of friction compatible with this motion.

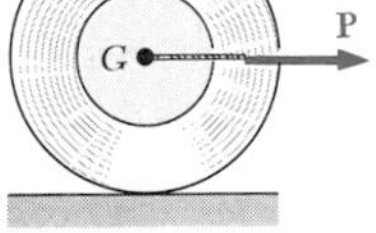

Fig. P16.95 and P16.99

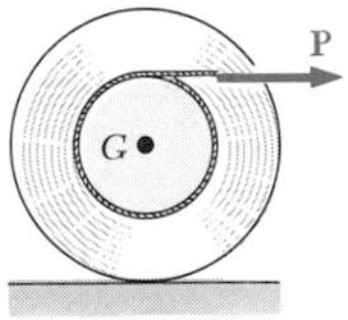

Fig. P16.96 and P16.100

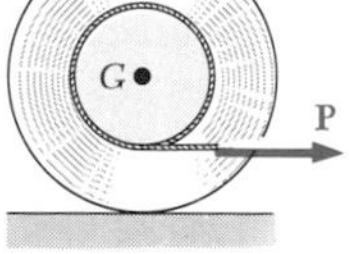

Fig. P16.97 and P16.101

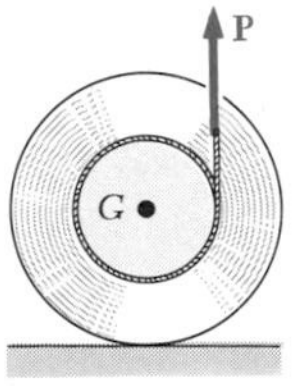

Fig. P16.98 and P16.102

16.99 through 16.102 A drum of 4-in. radius is attached to a disk of 8-in. radius. The disk and drum have a total weight of 10 lb and a radius of gyration of 6 in. A cord is attached as shown and pulled with a force **P** of magnitude 5 lb. Knowing that $\mu = 0.20$, determine (*a*) whether or not the disk slides, (*b*) the angular acceleration of the disk and the acceleration of G.

16.103 and 16.104 The 12-lb carriage is supported as shown by two uniform disks each of weight 8 lb and radius 3 in. Knowing that the disks roll without sliding, determine the acceleration of the carriage when a force of 4 lb is applied to it.

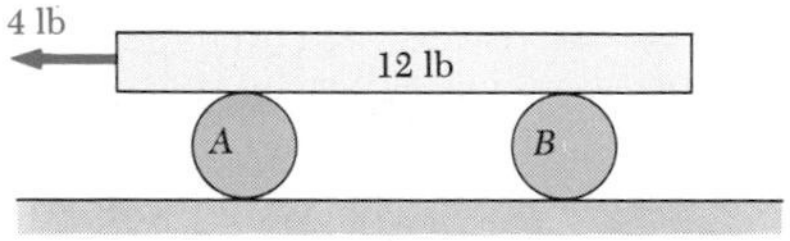

Fig. P16.103

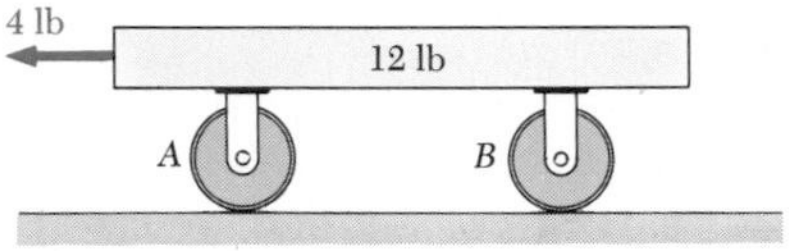

Fig. P16.104

16.105 A half section of pipe of mass m and radius r rests on a rough horizontal surface. A vertical force **P** is applied as shown. Assuming that the section rolls without sliding, derive an expression (*a*) for its angular acceleration, (*b*) for the minimum value of μ compatible with this motion. [*Hint.* Note that $OG = 2r/\pi$ and that, by the parallel-axis theorem, $\bar{I} = mr^2 - m(OG)^2$.]

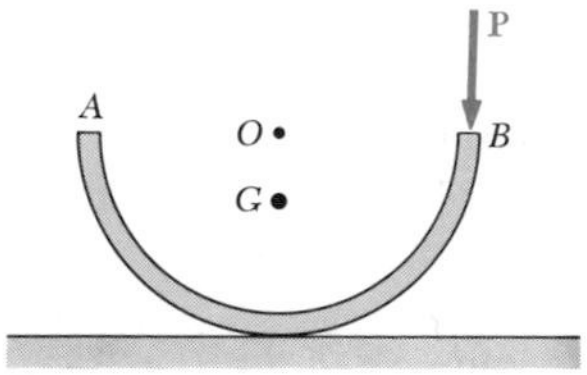

Fig. P16.105

16.106 A small block of mass m is attached at B to a hoop of mass m and radius r. Knowing that when the system is released from rest it starts to roll without sliding, determine (*a*) the angular acceleration of the hoop, (*b*) the acceleration of B.

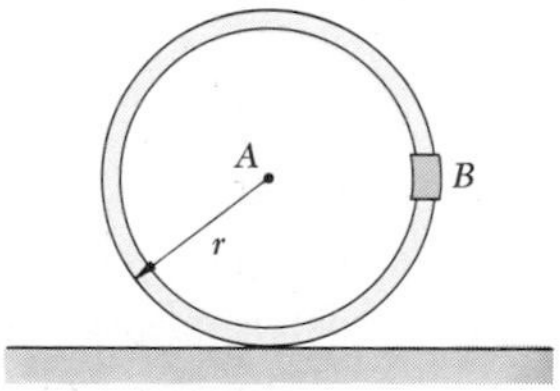

Fig. P16.106

16.107 Solve Prob. 16.105, assuming that the force **P** is applied at B and is directed horizontally to the right.

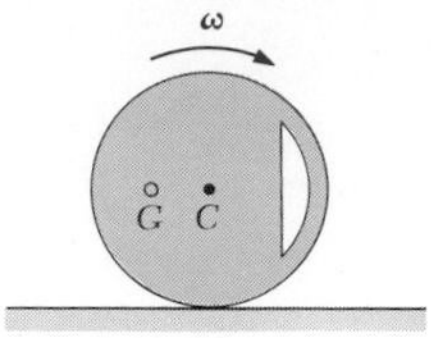

Fig. P16.108

16.108 The mass center G of a 10-lb wheel of radius $R = 12$ in. is located at a distance $r = 4$ in. from its geometric center C. The centroidal radius of gyration is $\bar{k} = 6$ in. As the wheel rolls without sliding, its angular velocity varies and it is observed that $\omega = 8$ rad/s in the position shown. Determine the corresponding angular acceleration of the wheel.

16.109 End A of the 100-lb beam AB moves along the frictionless floor, while end B is supported by a 4-ft cable. Knowing that at the instant shown end A is moving to the left with a constant velocity of 8 ft/s, determine (a) the magnitude of the force **P**, (b) the corresponding tension in the cable.

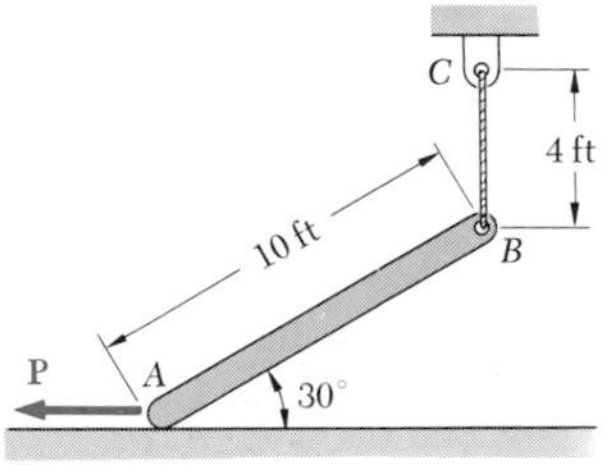

Fig. P16.109

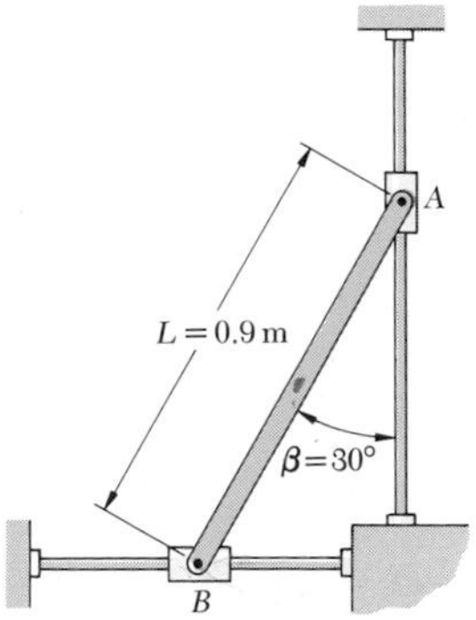

Fig. P16.110 and P16.112

16.110 Ends A and B of a 4-kg slender rod are attached to collars of negligible mass which slide without friction along the rods shown. A horizontal force **P** is applied to collar B, causing the rod to start from rest with a counterclockwise angular acceleration of 12 rad/s^2. Determine (a) the required magnitude of **P**, (b) the reactions at A and B.

16.111 Solve Prob. 16.110, assuming that at the instant considered the angular velocity of the rod is 4 rad/s counterclockwise.

16.112 Ends A and B of a 4-kg slender rod are attached to collars of negligible mass which slide without friction along the rods shown. If the rod is released from rest in the position shown, determine (a) the angular acceleration of the rod, (b) the reactions at A and B.

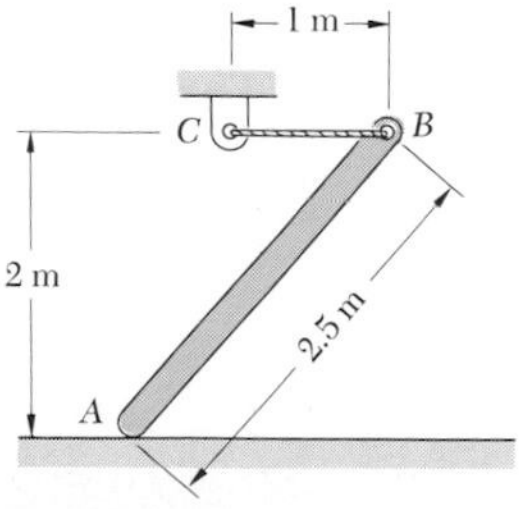

Fig. P16.113

16.113 The 50-kg uniform rod AB is released from rest in the position shown. Knowing that end A may slide freely on the frictionless floor, determine (a) the angular acceleration of the rod, (b) the tension in wire BC, (c) the reaction at A.

16.114 Rod AB weighs 3 lb and is released from rest in the position shown. Assuming that the ends of the rod slide without friction, determine (*a*) the angular acceleration of the rod, (*b*) the reactions at A and B.

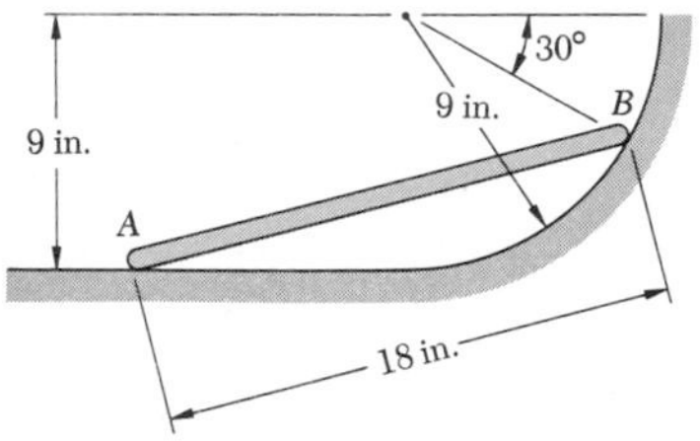

Fig. P16.114

16.115 The 12-lb uniform rod AB is held by the three wires shown. Determine the tension in wires AD and BE immediately after wire AC has been cut.

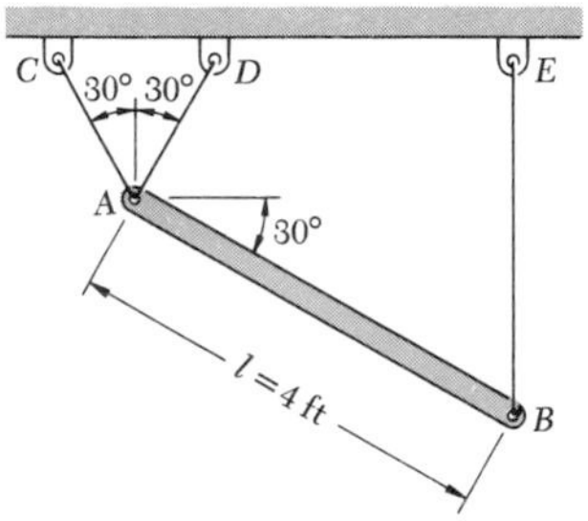

Fig. P16.115

16.116 Show that, for a rigid slab in plane motion, the equation $\Sigma M_A = I_A\alpha$, where ΣM_A represents the sum of the moments of the external forces about point A and I_A the moment of inertia of the slab about the same point A, is verified *if and only if* one of the following conditions is satisfied: (*a*) A is the mass center of the slab, (*b*) A has zero acceleration, (*c*) the acceleration of A is directed along a line joining point A and the mass center G.

16.117 The 6-lb sliding block is connected to the rotating disk by the uniform rod AB which weighs 4 lb. Knowing that the disk has a constant angular velocity of 360 rpm, determine the forces exerted on the connecting rod at A and B when $\beta = 0$. Assume the motion to take place in a horizontal plane.

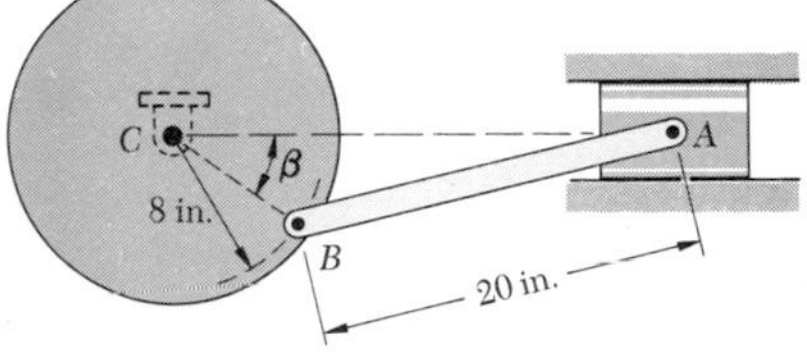

Fig. P16.117

16.118 Solve Prob. 16.117 when $\beta = 180°$.

16.119 Each of the bars shown is 600 mm long and has a mass of 4 kg. A horizontal and variable force **P** is applied at C, causing point C to move to the left with a constant speed of 10 m/s. Determine the force **P** for the position shown.

16.120 The two bars AB and BC are released from rest in the position shown. Each bar is 600 mm long and has a mass of 4 kg. Determine (*a*) the angular acceleration of each bar, (*b*) the reactions at A and C.

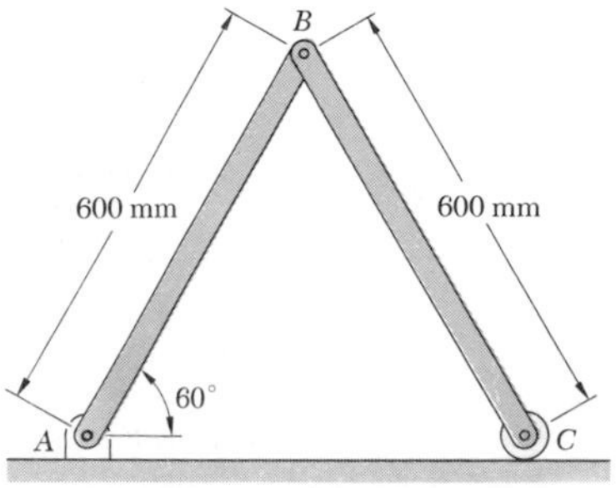

Fig. P16.119 and P16.120

16.121 and 16.122 Two rods AB and BC, of mass m per unit length, are connected as shown to a disk which is made to rotate in a vertical plane at a constant angular velocity ω_0. For the position shown, determine the components of the forces exerted at A and B on rod AB.

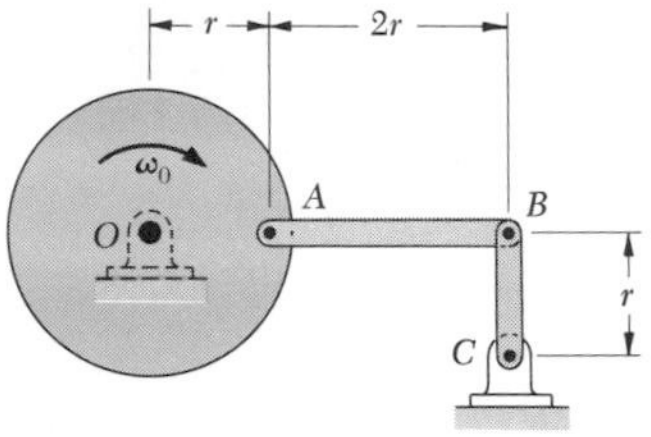

Fig. P16.121

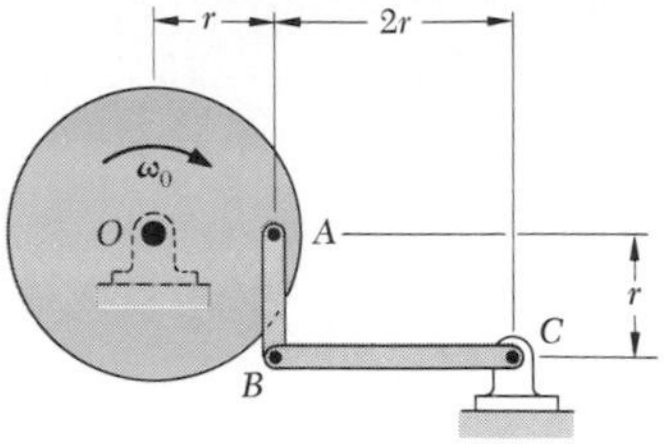

Fig. P16.122

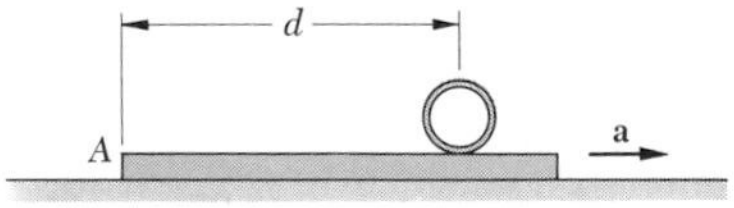

Fig. P16.123

16.123 A section of pipe rests on a plate. The plate is then given a constant acceleration **a** directed to the right. Assuming that the pipe rolls on the plate, determine (*a*) the acceleration of the pipe, (*b*) the distance through which the plate will move before the pipe reaches end A.

16.124 Solve Prob. 16.123, assuming that the pipe is replaced (1) by a solid cylinder, (2) by a sphere.

16.125 and 16.126 Gear C weighs 6 lb and has a centroidal radius of gyration of 3 in. The uniform bar AB weighs 5 lb and gear D is stationary. If the system is released from rest in the position shown, determine (*a*) the angular acceleration of gear C, (*b*) the acceleration of point B.

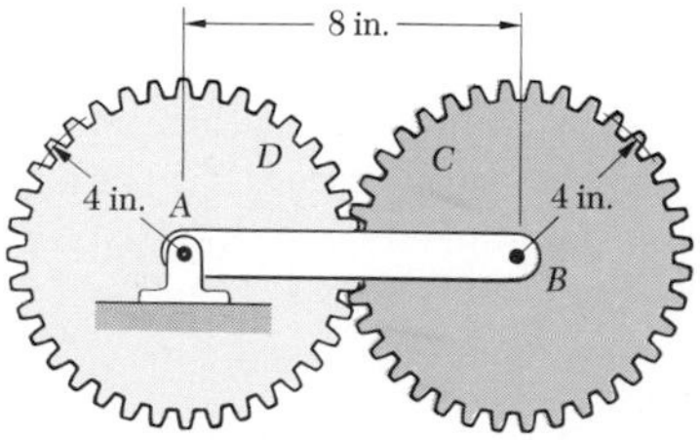

Fig. P16.125

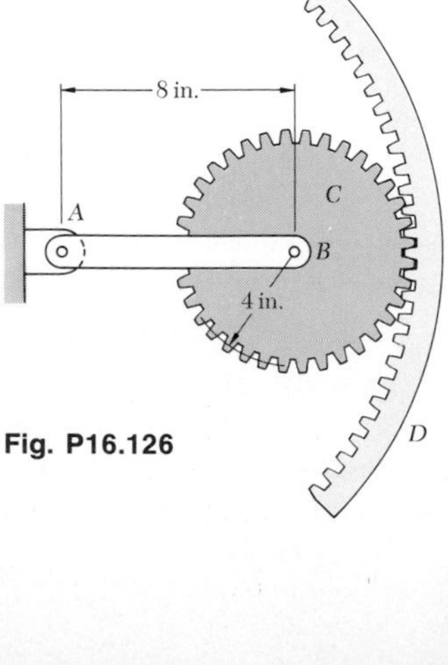

Fig. P16.126

∗16.127 The disk shown rotates with a constant counterclockwise angular velocity of 12 rad/s. The uniform rod *BD* is 450 mm long and has a mass of 3 kg. Knowing that the system moves in a *horizontal* plane, determine the reaction at *E*.

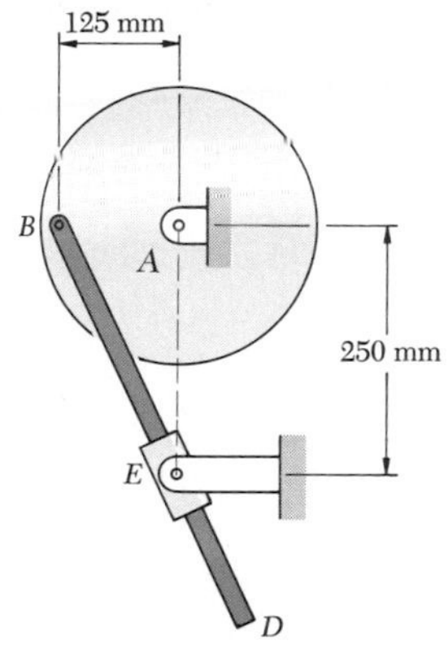

Fig. P16.127

∗16.128 Solve Prob. 16.127, assuming that the disk rotates with a constant clockwise angular velocity of 12 rad/s.

∗16.129 A uniform slender rod of length L and mass m is released from rest in the position shown. Derive an expression for (*a*) the angular acceleration of the rod, (*b*) the acceleration of end *A*, (*c*) the reaction at *A*, immediately after release. Neglect the mass and friction of the roller at *A*.

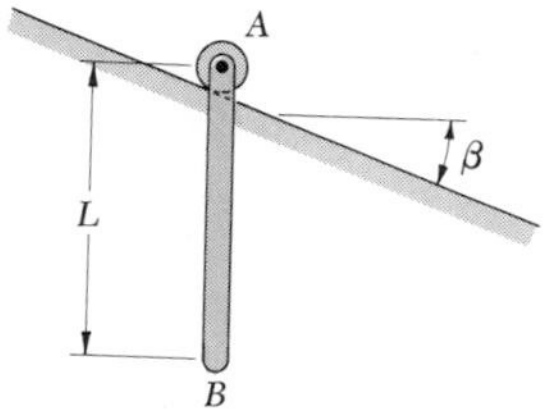

Fig. P16.129 and P16.130

∗16.130 A uniform rod *AB*, of mass 3 kg and length $L = 1.2$ m, is released from rest in the position shown. Knowing that $\beta = 30°$, determine the values immediately after release of (*a*) the angular acceleration of the rod, (*b*) the acceleration of end *A*, (*c*) the reaction at *A*. Neglect the mass and friction of the roller at *A*.

∗16.131 Each of the bars *AB* and *BC* is of length $L = 18$ in. and weight 3 lb. A couple $\mathbf{M}_0$ of magnitude 6 lb · ft is applied to bar *BC*. Determine the angular acceleration of each bar.

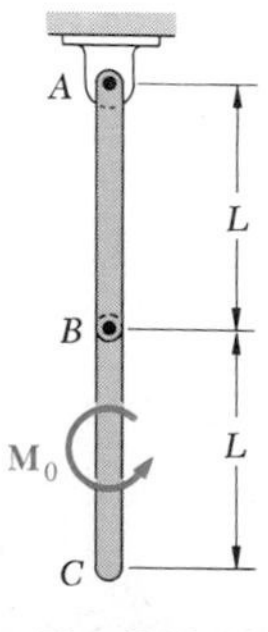

Fig. P16.131

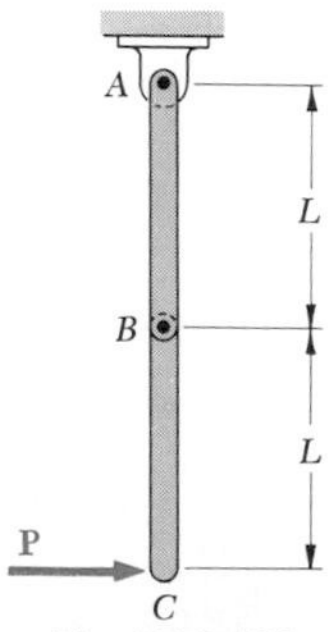

Fig. P16.132

∗16.132 Each of the bars *AB* and *BC* is of length $L = 18$ in. and weight 3 lb. A horizontal force **P** of magnitude 4 lb is applied at *C*. Determine the angular acceleration of each bar.

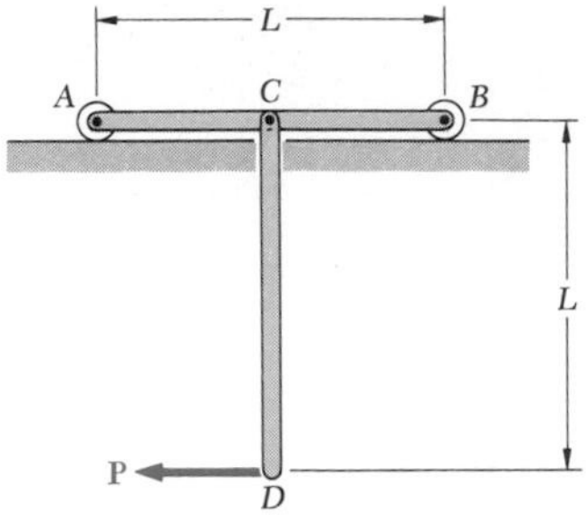

Fig. P16.133

∗16.133 Two uniform slender rods, each of mass m, are connected by a pin at C. Determine the acceleration of points C and D immediately after the horizontal force **P** has been applied at D.

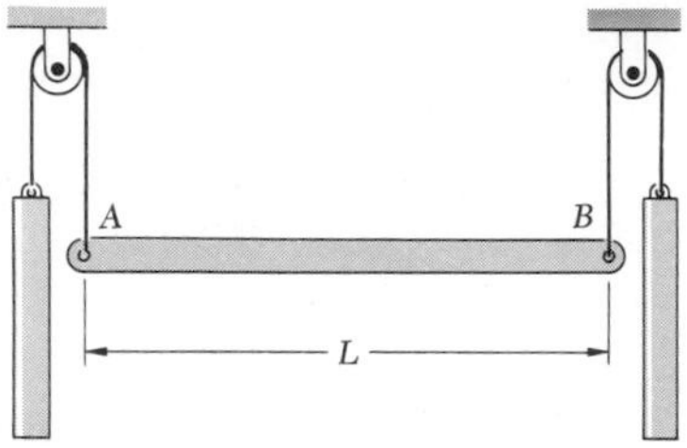

Fig. P16.134

∗16.134 The slender bar AB is of length L and mass m. It is held in equilibrium by two counterweights, each of mass $\frac{1}{2}m$. If the wire at B is cut, determine at that instant the acceleration of (*a*) point A, (*b*) point B.

∗16.135 (*a*) Determine the magnitude and the location of the maximum bending moment in the rod of Prob. 16.74. (*b*) Show that the answer to part *a* is independent of the weight W of the rod.

∗16.136 In Prob. 16.132 the pin at B is severely rusted and the bars rotate as a single rigid body. Determine the bending moment which occurs at B.

∗16.137 Draw the shear and bending-moment diagrams for the beam of Prob. 16.79 immediately after the cable at B breaks.

∗16.138 Draw the shear and bending-moment diagrams for the bar of Prob. 16.134 immediately after the wire at B has been cut.

REVIEW PROBLEMS

16.139 and 16.140 A uniform plate of mass m is suspended in each of the ways shown. For each case determine the acceleration of the center of the plate immediately after the connection at B has been released.

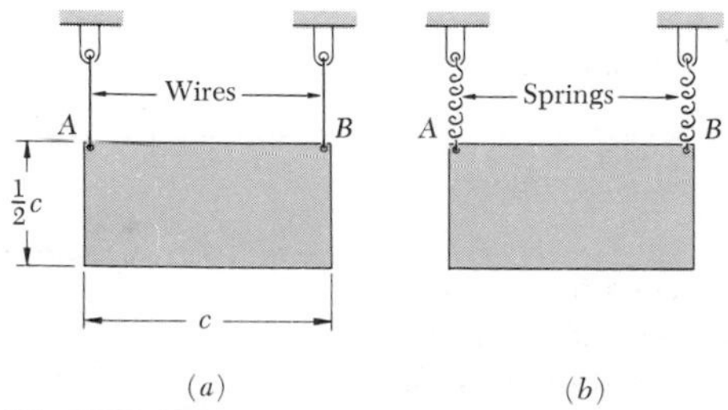

Fig. P16.139

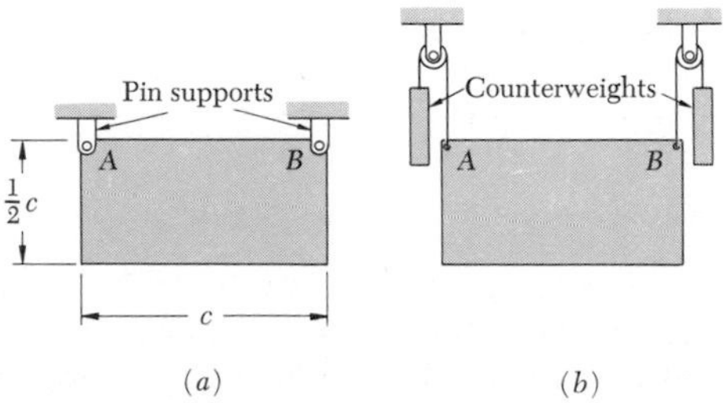

Fig. P16.140

16.141 The flanged wheel shown rolls to the right with a constant velocity of 1.5 m/s. The rod AB is 1.2 m long and has a mass of 5 kg. Knowing that point A slides without friction on the horizontal surface, determine the reaction at A (*a*) when $\beta = 0$, (*b*) when $\beta = 180°$.

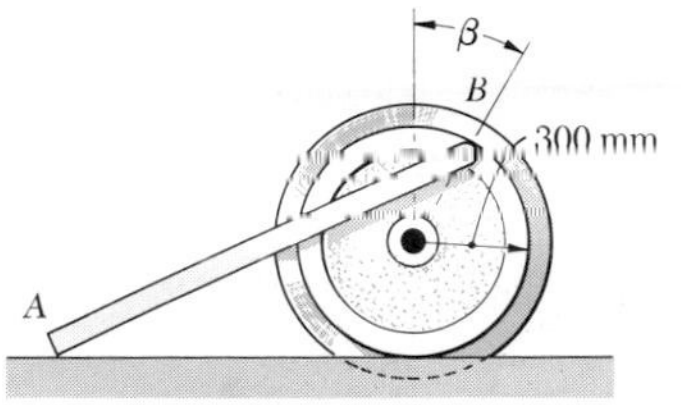

Fig. P16.141

16.142 A 15-lb uniform disk is suspended from a link AB of negligible weight. If a 10-lb force is applied at B, determine the acceleration of B (*a*) if the connection at B is a frictionless pin, (*b*) if the connection at B is "frozen" and the system rotates about A as a rigid body.

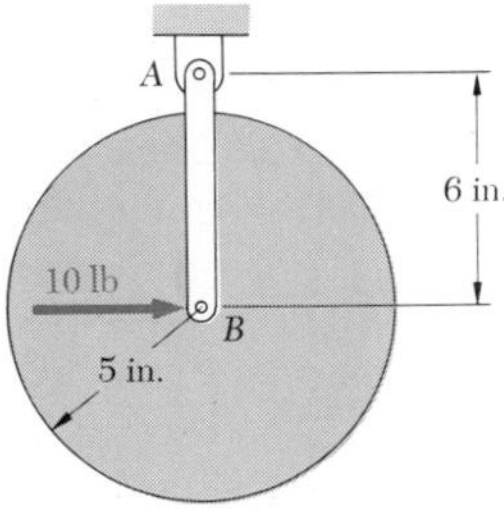

Fig. P16.142

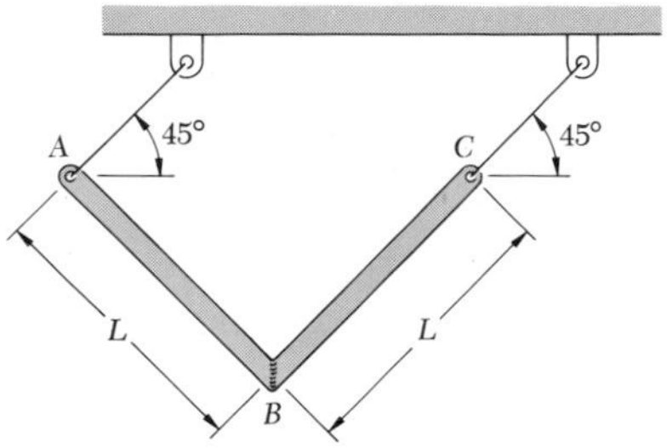

Fig. P16.143

16.143 Two uniform bars AB and BC, each of length $L = 10$ in., are welded together to form an L-shaped rigid body. Knowing that each bar weighs 3 lb, determine the tension in each wire immediately after the body has been released from rest.

16.144 A slender rod of mass m per unit length is placed inside a shallow drum of radius r which rotates at a constant angular velocity ω about a vertical shaft through O. (*a*) Determine the ratio L/r for which the maximum bending moment in the rod is as large as possible. (*b*) Derive an expression for the corresponding value of the maximum bending moment.

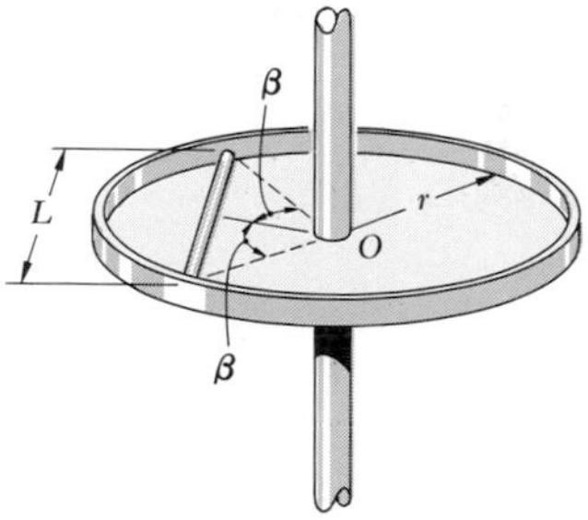

Fig. P16.144

16.145 A collar C of weight W_C is rigidly attached to a uniform slender rod AB of length L and weight W. If the rod is released from rest in the position shown, determine the ratio d/L for which the reaction at B is independent of W_C.

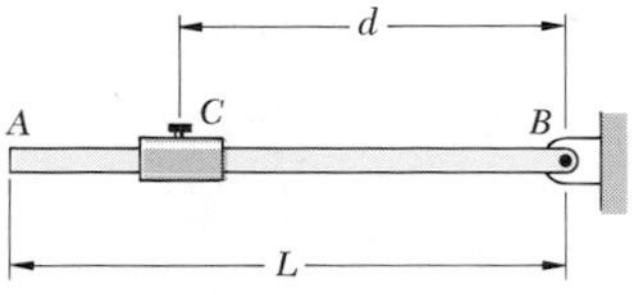

Fig. P16.145 and P16.146

16.146 A collar C of weight 2 lb is rigidly attached to a uniform slender rod AB of weight 12 lb and length $L = 20$ in. If the rod is released from rest in the position shown, determine the distance d for which the angular acceleration of the rod is maximum.

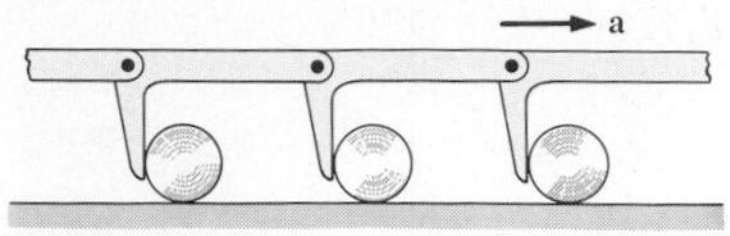

Fig. P16.147

16.147 Identical cylinders of mass m and radius r are pushed by a series of moving arms. Assuming the coefficient of friction between all surfaces to be $\mu < 1$, and denoting by a the magnitude of the acceleration of the arms, derive an expression for (*a*) the maximum allowable value of a if each cylinder is to roll without sliding, (*b*) the minimum allowable value of a if each cylinder is to move to the right without rotating.

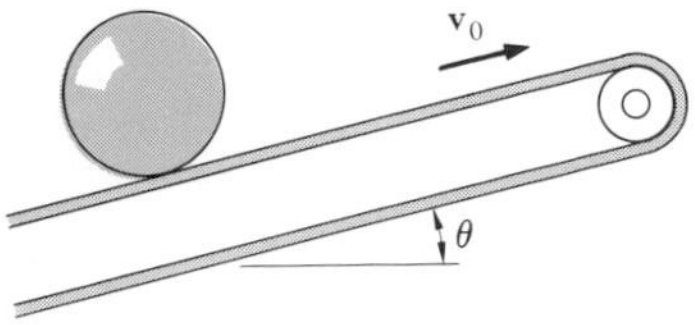

Fig. P16.148

16.148 A sphere of mass m and radius r is dropped with no initial velocity on a belt which moves with a constant velocity $\mathbf{v}_0$. At first the sphere will both rotate and slide on the belt. Denoting by μ the coefficient of friction between the sphere and the belt, determine the distance the sphere will move before it starts rolling without sliding.

16.149 A section of pipe, of mass 50 kg and radius 250 mm, rests on two corners as shown. Assuming that μ between the corners and the pipe is sufficient to prevent sliding, determine (*a*) the angular acceleration of the pipe just after corner B is removed, (*b*) the corresponding magnitude of the reaction at A.

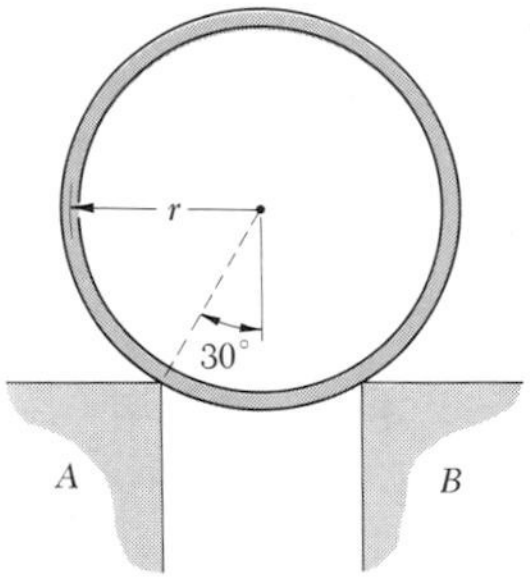

Fig. P16.149

16.150 A block B of mass m is attached to a cord wrapped around a cylinder of the same mass m and of radius r. The cylinder rolls without sliding on a horizontal surface. Determine the components of the accelerations of the center A of the cylinder and of the block B immediately after the system has been released from rest if (*a*) the block hangs freely, (*b*) the motion of the block is guided by a rigid member DAE, frictionless and of negligible mass, which is hinged to the cylinder at A.

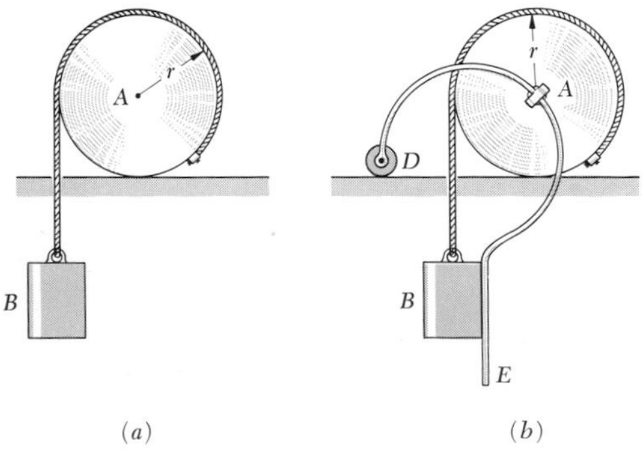

Fig. P16.150

Plane Motion of Rigid Bodies: Energy and Momentum Methods

CHAPTER

17

17.1 Principle of Work and Energy for a Rigid Body. In the first part of this chapter, the principle of work and energy will be used to analyze the plane motion of rigid bodies and of systems of rigid bodies. As was pointed out in Chap. 13, the method of work and energy is particularly well adapted to the solution of problems involving velocities and displacements. Its main advantage resides in the fact that the work of forces and the kinetic energy of particles are scalar quantities.

In order to apply the principle of work and energy to the analysis of the motion of a rigid body, we shall again assume that the rigid body is made of a large number n of particles of mass Δm_i. Recalling Eq. (14.30) of Sec. 14.7, we write

$$T_1 + U_{1\to 2} = T_2 \qquad (17.1)$$

where T_1, T_2 = initial and final values of total kinetic energy of the particles forming the rigid body

$U_{1\to 2}$ = work of all forces acting on the various particles of the body

The total kinetic energy

$$T = \frac{1}{2}\sum_{i=1}^{n} (\Delta m_i)v_i^2 \qquad (17.2)$$

is obtained by adding positive scalar quantities and is itself a positive scalar quantity. We shall see later how T may be determined for various types of motion of a rigid body.

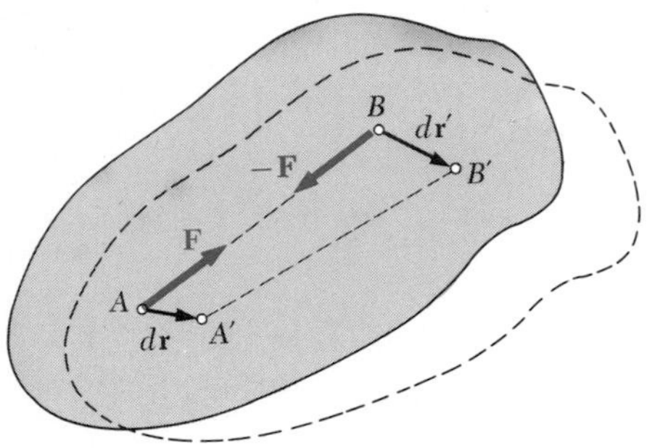

Fig. 17.1

The expression $U_{1\to2}$ in (17.1) represents the work of all the forces acting on the various particles of the body, whether these forces are internal or external. However, as we shall see presently, the total work of the internal forces holding together the particles of a rigid body is zero. Consider two particles A and B of a rigid body and the two equal and opposite forces $\mathbf{F}$ and $-\mathbf{F}$ they exert on each other (Fig. 17.1). While, in general, small displacements $d\mathbf{r}$ and $d\mathbf{r}'$ of the two particles are different, the components of these displacements along AB must be equal; otherwise, the particles would not remain at the same distance from each other, and the body would not be rigid. Therefore, the work of $\mathbf{F}$ is equal in magnitude and opposite in sign to the work of $-\mathbf{F}$, and their sum is zero. Thus, the total work of the internal forces acting on the particles of a rigid body is zero, and *the expression $U_{1\to2}$ in Eq.* (17.1) *reduces to the work of the external forces* acting on the body during the displacement considered.

17.2 Work of Forces Acting on a Rigid Body. We saw in Sec. 13.2 that the work of a force $\mathbf{F}$ during a displacement of its point of application from A_1 to A_2 is

$$U_{1\to2} = \int_{A_1}^{A_2} \mathbf{F}\cdot d\mathbf{r} \qquad (17.3)$$

or

$$U_{1\to2} = \int_{s_1}^{s_2} (F\cos\alpha)ds \qquad (17.3')$$

where F is the magnitude of the force, α the angle it forms with the direction of motion of its point of application A, and s the variable of integration which measures the distance traveled by A along its path.

In computing the work of the external forces acting on a rigid body, it is often convenient to determine the work of a couple without considering separately the work of each of the two forces forming the couple. Consider the two forces $\mathbf{F}$ and $-\mathbf{F}$

forming a couple of moment **M** and acting on a rigid body (Fig. 17.2). Any small displacement of the rigid body bringing A and B, respectively, into A' and B'' may be divided into two parts, one in which points A and B undergo equal displacements $d\mathbf{r}_1$, the other in which A' remains fixed while B' moves into B'' through a displacement $d\mathbf{r}_2$ of magnitude $ds_2 = r\,d\theta$. In the first part of the motion, the work of **F** is equal in magnitude and opposite in sign to the work of $-\mathbf{F}$ and their sum is zero. In the second part of the motion, only force **F** works, and its work is $dU = F\,ds_2 = Fr\,d\theta$. But the product Fr is equal to the magnitude M of the moment of the couple. Thus, the work of a couple of moment **M** acting on a rigid body is

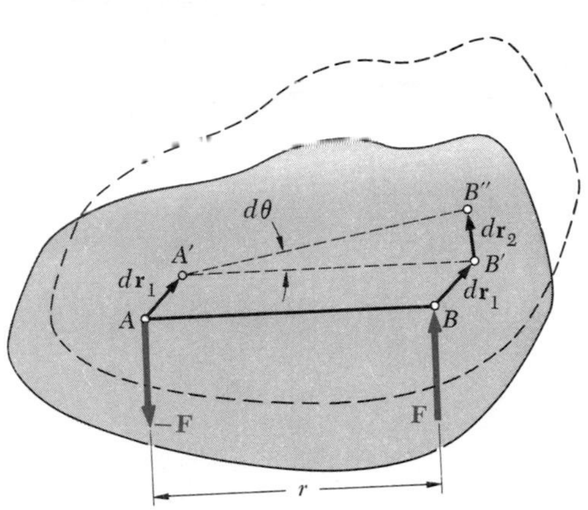

Fig. 17.2

$$dU = M\,d\theta \tag{17.4}$$

where $d\theta$ is the small angle expressed in radians through which the body rotates. We again note that work should be expressed in units obtained by multiplying units of force by units of length. The work of the couple during a finite rotation of the rigid body is obtained by integrating both members of (17.4) from the initial value θ_1 of the angle θ to its final value θ_2. We write

$$U_{1\rightarrow 2} = \int_{\theta_1}^{\theta_2} M\,d\theta \tag{17.5}$$

When the moment **M** *of the couple is constant,* formula (17.5) reduces to

$$U_{1\rightarrow 2} = M(\theta_2 - \theta_1) \tag{17.6}$$

It was pointed out in Sec. 13.2 that a number of forces encountered in problems of kinetics *do no work.* They are forces applied to fixed points or acting in a direction perpendicular to the displacement of their point of application. Among the forces which do no work the following have been listed: the reaction at a frictionless pin when the body supported rotates about the pin, the reaction at a frictionless surface when the body in contact moves along the surface, the weight of a body when its center of gravity moves horizontally. We should also indicate now that, *when a rigid body rolls without sliding on a fixed surface, the friction force* **F** *at the point of contact C does no work.* The velocity $\mathbf{v}_C$ of the point of contact C is zero, and the work of the friction force **F** during a small displacement of the rigid body is $dU = F\,ds_C = F(v_C\,dt) = 0$.

17.3 Kinetic Energy of a Rigid Body in Plane Motion. Consider a rigid body of mass m in plane motion. We recall from Sec. 14.6 that, if the absolute velocity $\mathbf{v}_i$ of each particle P_i of the body is expressed as the sum of the velocity $\overline{\mathbf{v}}$ of the mass center G of the body and of the velocity $\mathbf{v}'_i$ of the particle relative to a frame $Gx'y'$ attached to G and of fixed

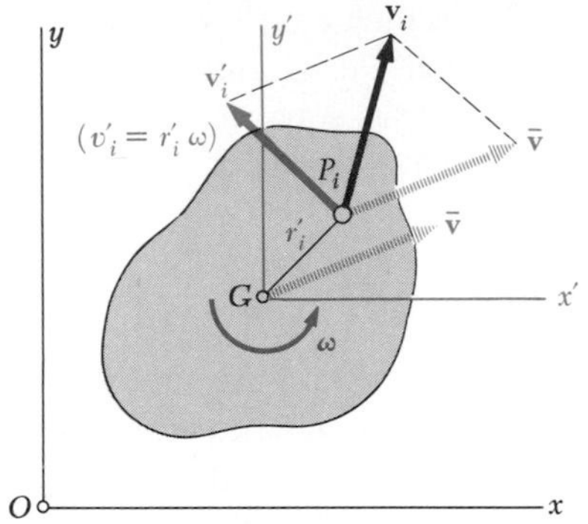

Fig. 17.3

orientation (Fig. 17.3), the kinetic energy of the system of particles forming the rigid body may be written in the form

$$T = \tfrac{1}{2}m\overline{v}^2 + \frac{1}{2}\sum_{i=1}^{n}(\Delta m_i)v_i'^2 \qquad (17.7)$$

But the magnitude v'_i of the relative velocity of P_i is equal to the product $r'_i\omega$ of the distance r'_i of P_i from the axis through G perpendicular to the plane of motion and of the magnitude ω of the angular velocity of the body at the instant considered. Substituting into (17.7), we have

$$T = \tfrac{1}{2}m\overline{v}^2 + \frac{1}{2}\left(\sum_{i=1}^{n} r_i'^2\,\Delta m_i\right)\omega^2 \qquad (17.8)$$

or, since the sum represents the moment of inertia $\overline{I}$ of the body about the axis through G,

$$T = \tfrac{1}{2}m\overline{v}^2 + \tfrac{1}{2}\overline{I}\omega^2 \qquad (17.9)$$

We note that, in the particular case of a body in translation ($\omega = 0$), the expression obtained reduces to $\frac{1}{2}m\overline{v}^2$, while, in the case of a centroidal rotation ($\overline{v} = 0$), it reduces to $\frac{1}{2}\overline{I}\omega^2$. We conclude that the kinetic energy of a rigid body in plane motion may be separated into two parts: (1) the kinetic energy $\frac{1}{2}m\overline{v}^2$ associated with the motion of the mass center G of the body, and (2) the kinetic energy $\frac{1}{2}\overline{I}\omega^2$ associated with the rotation of the body about G.

Noncentroidal Rotation. The relation (17.9) is valid for any type of plane motion and may, therefore, be used to express the kinetic energy of a rigid body rotating with an angular velocity $\boldsymbol{\omega}$ about a fixed axis through O (Fig. 17.4). In that case, however, the kinetic energy of the body may be expressed more directly by noting that the speed v_i of the particle P_i is equal to the product $r_i\omega$ of the distance r_i of P_i from the fixed axis and of the magnitude ω of the angular velocity of the body at the instant considered. Substituting into (17.2), we write

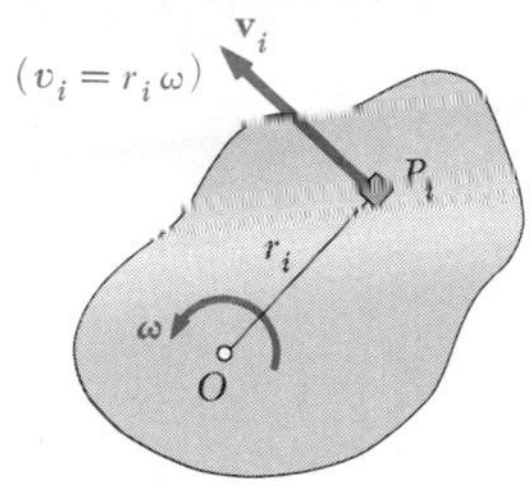

Fig. 17.4

$$T = \frac{1}{2}\sum_{i=1}^{n}(\Delta m_i)(r_i\omega)^2 = \frac{1}{2}\left(\sum_{i=1}^{n} r_i^2\,\Delta m_i\right)\omega^2$$

or, since the last sum represents the moment of inertia I_O of the body about the fixed axis through O,

$$T = \tfrac{1}{2}I_O\omega^2 \qquad (17.10)$$

We note that the results obtained are not limited to the motion of plane slabs or to the motion of bodies which are symmetrical with respect to the reference plane. They may be applied to the study of the plane motion of any rigid body, regardless of its shape.

17.4 Systems of Rigid Bodies. When a problem involves several rigid bodies, each rigid body may be considered separately, and the principle of work and energy may be applied to each body. Adding the kinetic energies of all the particles and considering the work of all the forces involved, we may also write the equation of work and energy for the entire system. We have

$$T_1 + U_{1\to 2} = T_2 \qquad (17.11)$$

where T represents the arithmetic sum of the kinetic energies of the rigid bodies forming the system (all terms are positive) and $U_{1\to 2}$ the work of all the forces acting on the various bodies, whether these forces are *internal* or *external* from the point of view of the system as a whole.

The method of work and energy is particularly useful in solving problems involving pin-connected members, or blocks and pulleys connected by inextensible cords, or meshed gears. In all these cases, the internal forces occur by pairs of equal and opposite forces, and the points of application of the forces in each pair *move through equal distances* during a small displacement of the system. As a result, the work of the internal forces is zero, and $U_{1\to 2}$ reduces to the work of the *forces external to the system.*

17.5 Conservation of Energy. We saw in Sec. 13.6 that the work of conservative forces, such as the weight of a body or the force exerted by a spring, may be expressed as a change in potential energy. When a rigid body, or a system of rigid bodies, moves under the action of conservative forces, the principle of work and energy stated in Sec. 17.1 may be expressed in a modified form. Substituting for $U_{1\to2}$ from (13.19′) into (17.1), we write

$$T_1 + V_1 = T_2 + V_2 \tag{17.12}$$

Formula (17.12) indicates that, when a rigid body, or a system of rigid bodies, moves under the action of conservative forces, *the sum of the kinetic energy and of the potential energy of the system remains constant.* It should be noted that, in the case of the plane motion of a rigid body, the kinetic energy of the body should include both the *translational* term $\frac{1}{2}m\bar{v}^2$ and the *rotational* term $\frac{1}{2}\bar{I}\omega^2$.

As an example of application of the principle of conservation of energy, we shall consider a slender rod AB, of length l and mass m, whose extremities are connected to blocks of negligible mass sliding along horizontal and vertical tracks. We assume that the rod is released with no initial velocity from a horizontal position (Fig. 17.5*a*), and we wish to determine its angular velocity after it has rotated through an angle θ (Fig. 17.5*b*).

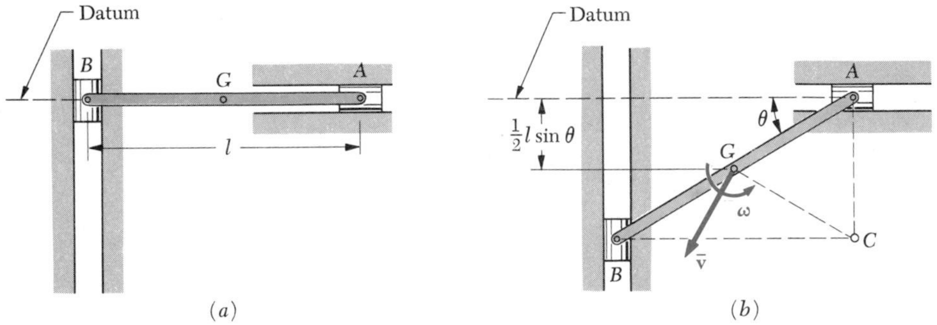

Fig. 17.5

Since the initial velocity is zero, we have $T_1 = 0$. Measuring the potential energy from the level of the horizontal track, we write $V_1 = 0$. After the rod has rotated through θ, the center of gravity G of the rod is at a distance $\frac{1}{2}l \sin\theta$ below the reference level and we have

$$V_2 = -\tfrac{1}{2}Wl \sin\theta = -\tfrac{1}{2}mgl \sin\theta$$

Observing that, in this position, the instantaneous center of the rod is located at C, and that $CG = \frac{1}{2}l$, we write $\bar{v}_2 = \frac{1}{2}l\omega$ and obtain

$$T_2 = \tfrac{1}{2}m\bar{v}_2^2 + \tfrac{1}{2}\bar{I}\omega_2^2 = \tfrac{1}{2}m(\tfrac{1}{2}l\omega)^2 + \tfrac{1}{2}(\tfrac{1}{12}ml^2)\omega^2$$
$$= \frac{1}{2}\frac{ml^2}{3}\omega^2$$

Applying the principle of conservation of energy, we write

$$T_1 + V_1 = T_2 + V_2$$
$$0 = \frac{1}{2}\frac{ml^2}{3}\omega^2 - \tfrac{1}{2}mgl\sin\theta$$
$$\omega = \left(\frac{3g}{l}\sin\theta\right)^{1/2}$$

We recall that the advantages of the method of work and energy, as well as its shortcomings, were indicated in Sec. 13.4. In this connection, we wish to mention that the method of work and energy must be supplemented by the application of D'Alembert's principle when reactions at fixed axles, at rollers, or at sliding blocks are to be determined. For example, in order to compute the reactions at the extremities A and B of the rod of Fig. 17.5*b*, a diagram should be drawn to express that the system of the external forces applied to the rod is equivalent to the vector $m\bar{\mathbf{a}}$ and the couple $\bar{I}\boldsymbol{\alpha}$. The angular velocity $\boldsymbol{\omega}$ of the rod, however, is determined by the method of work and energy before the equations of motion are solved for the reactions. The complete analysis of the motion of the rod and of the forces exerted on the rod requires, therefore, the combined use of the method of work and energy and of the principle of equivalence of the external and effective forces.

17.6. Power. *Power* was defined in Sec. 13.5 as the time rate at which work is done. In the case of a body acted upon by a force **F**, and moving with a velocity **v**, the power was expressed as follows:

$$\text{Power} = \frac{dU}{dt} = \mathbf{F}\cdot\mathbf{v} \qquad (13.13)$$

In the case of a rigid body rotating with an angular velocity $\boldsymbol{\omega}$ and acted upon by a couple of moment **M** parallel to the axis of rotation, we have, by (17.4),

$$\text{Power} = \frac{dU}{dt} = \frac{M\,d\theta}{dt} = M\omega \qquad (17.13)$$

The various units used to measure power, such as the watt and the horsepower, were defined in Sec. 13.5.

SAMPLE PROBLEM 17.1

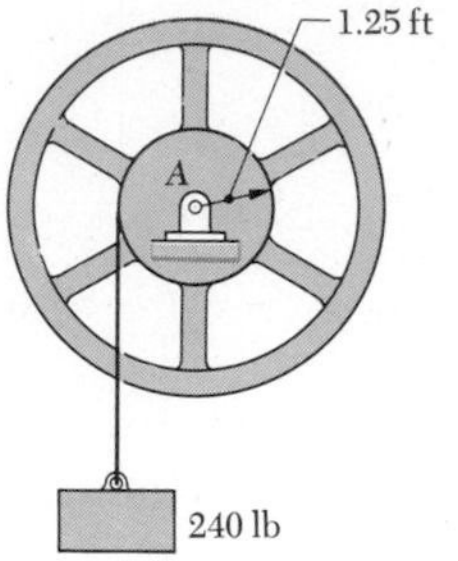

A 240-lb block is suspended from an inextensible cable which is wrapped around a drum of 1.25-ft radius rigidly attached to a flywheel. The drum and flywheel have a combined centroidal moment of inertia $\bar{I} = 10.5 \text{ lb} \cdot \text{ft} \cdot \text{s}^2$. At the instant shown, the velocity of the block is 6 ft/s directed downward. Knowing that the bearing at A is poorly lubricated and that the bearing friction is equivalent to a couple **M** of magnitude 60 lb·ft, determine the velocity of the block after it has moved 4 ft downward.

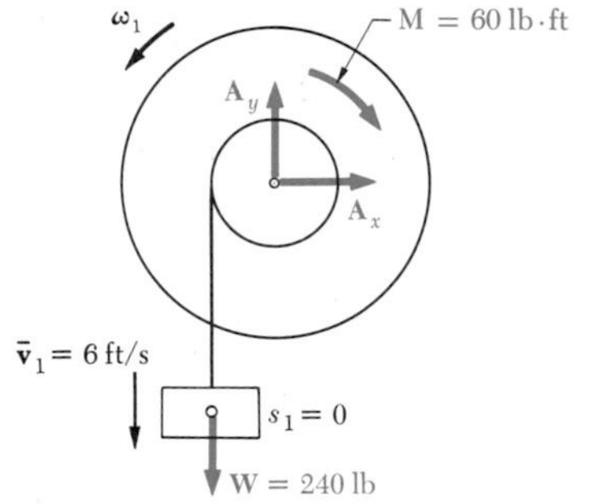

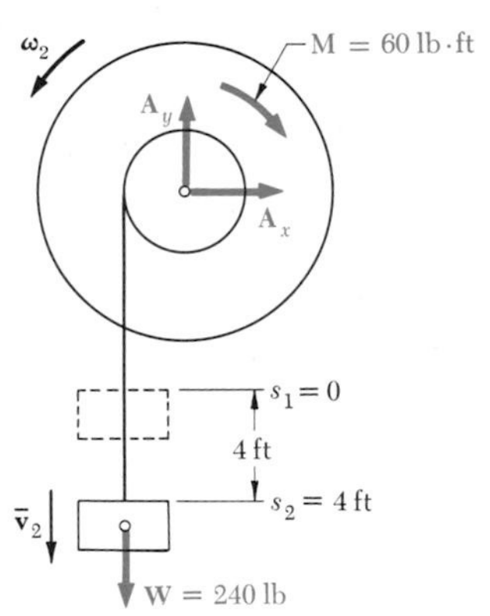

Solution. We consider the system formed by the flywheel and the block. Since the cable is inextensible, the work done by the internal forces exerted by the cable cancels. The initial and final positions of the system and the external forces acting on the system are as shown.

Kinetic Energy. *Position 1.* We have

$$\bar{v}_1 = 6 \text{ ft/s} \qquad \omega_1 = \frac{\bar{v}_1}{r} = \frac{6 \text{ ft/s}}{1.25 \text{ ft}} = 4.80 \text{ rad/s}$$

$$\begin{aligned} T_1 &= \tfrac{1}{2}m\bar{v}_1^2 + \tfrac{1}{2}\bar{I}\omega_1^2 \\ &= \frac{1}{2}\frac{240 \text{ lb}}{32.2 \text{ ft/s}^2}(6 \text{ ft/s})^2 + \tfrac{1}{2}(10.5 \text{ lb} \cdot \text{ft} \cdot \text{s}^2)(4.80 \text{ rad/s})^2 \\ &= 255 \text{ ft} \cdot \text{lb} \end{aligned}$$

Position 2. Noting that $\omega_2 = \bar{v}_2/1.25$, we write

$$\begin{aligned} T_2 &= \tfrac{1}{2}m\bar{v}_2^2 + \tfrac{1}{2}\bar{I}\omega_2^2 \\ &= \frac{1}{2}\frac{240}{32.2}(\bar{v}_2)^2 + (\tfrac{1}{2})(10.5)\left(\frac{\bar{v}_2}{1.25}\right)^2 = 7.09\bar{v}_2^2 \end{aligned}$$

Work. During the motion, only the weight **W** of the block and the friction couple **M** do work. Noting that **W** does positive work and that the friction couple **M** does negative work, we write

$$s_1 = 0 \qquad s_2 = 4 \text{ ft}$$

$$\theta_1 = 0 \qquad \theta_2 = \frac{s_2}{r} = \frac{4 \text{ ft}}{1.25 \text{ ft}} = 3.20 \text{ rad}$$

$$\begin{aligned} U_{1\to2} &= W(s_2 - s_1) - M(\theta_2 - \theta_1) \\ &= (240 \text{ lb})(4 \text{ ft}) - (60 \text{ lb} \cdot \text{ft})(3.20 \text{ rad}) \\ &= 768 \text{ ft} \cdot \text{lb} \end{aligned}$$

Principle of Work and Energy

$$T_1 + U_{1\to2} = T_2$$

$$255 \text{ ft} \cdot \text{lb} + 768 \text{ ft} \cdot \text{lb} = 7.09\bar{v}_2^2$$

$$\bar{v}_2 = 12.01 \text{ ft/s} \qquad \bar{\mathbf{v}}_2 = 12.01 \text{ ft/s} \downarrow \quad ◄$$

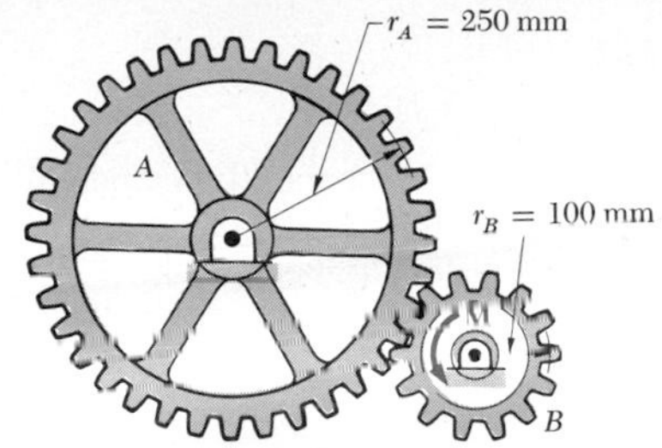

SAMPLE PROBLEM 17.2

Gear A has a mass of 10 kg and a radius of gyration of 200 mm, while gear B has a mass of 3 kg and a radius of gyration of 80 mm. The system is at rest when a couple $\mathbf{M}$ of magnitude 6 N·m is applied to gear B. Neglecting friction, determine (*a*) the number of revolutions executed by gear B before its angular velocity reaches 600 rpm, (*b*) the tangential force which gear B exerts on gear A.

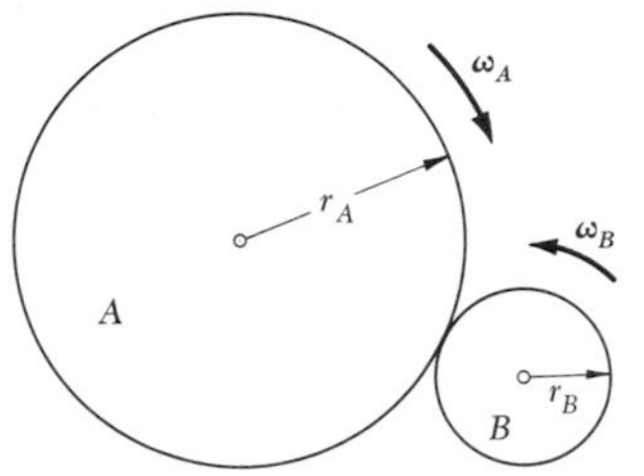

Motion of Entire System. Noting that the peripheral speeds of the gears are equal, we write

$$r_A\omega_A = r_B\omega_B \qquad \omega_A = \omega_B\frac{r_B}{r_A} = \omega_B\frac{100\text{ mm}}{250\text{ mm}} = 0.40\omega_B$$

For $\omega_B = 600$ rpm, we have

$$\omega_B = 62.8\text{ rad/s} \qquad \omega_A = 0.40\omega_B = 25.1\text{ rad/s}$$
$$\bar{I}_A = m_A\bar{k}_A^2 = (10\text{ kg})(0.200\text{ m})^2 = 0.400\text{ kg}\cdot\text{m}^2$$
$$\bar{I}_B = m_B\bar{k}_B^2 = (3\text{ kg})(0.080\text{ m})^2 = 0.0192\text{ kg}\cdot\text{m}^2$$

Kinetic Energy. Since the system is initially at rest, $T_1 = 0$. Adding the kinetic energies of the two gears when $\omega_B = 600$ rpm, we obtain

$$T_2 = \tfrac{1}{2}\bar{I}_A\omega_A^2 + \tfrac{1}{2}\bar{I}_B\omega_B^2$$
$$= \tfrac{1}{2}(0.400\text{ kg}\cdot\text{m}^2)(25.1\text{ rad/s})^2 + \tfrac{1}{2}(0.0192\text{ kg}\cdot\text{m}^2)(62.8\text{ rad/s})^2$$
$$= 163.9\text{ J}$$

Work. Denoting by θ_B the angular displacement of gear B, we have

$$U_{1\to2} = M\theta_B = (6\text{ N}\cdot\text{m})(\theta_B\text{ rad}) = (6\,\theta_B)\text{ J}$$

Principle of Work and Energy

$$T_1 + U_{1\to2} = T_2: \qquad 0 + (6\,\theta_B)\text{ J} = 163.9\text{ J}$$
$$\theta_B = 27.32\text{ rad} \qquad \theta_B = 4.35\text{ rev} \blacktriangleleft$$

Motion of Gear A. ***Kinetic Energy.*** Initially, gear A is at rest, $T_1 = 0$. When $\omega_B = 600$ rpm, the kinetic energy of gear A is

$$T_2 = \tfrac{1}{2}\bar{I}_A\omega_A^2 = \tfrac{1}{2}(0.400\text{ kg}\cdot\text{m}^2)(25.1\text{ rad/s})^2 = 126.0\text{ J}$$

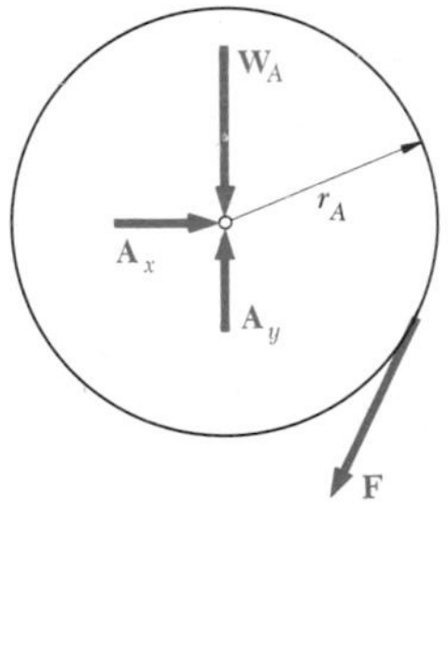

Work. The forces acting on gear A are as shown. The tangential force $\mathbf{F}$ does work equal to the product of its magnitude and of the length $\theta_A r_A$ of the arc described by the point of contact. Since $\theta_A r_A = \theta_B r_B$, we have

$$U_{1\to2} = F(\theta_B r_B) = F(27.3\text{ rad})(0.100\text{ m}) = F(2.73\text{ m})$$

Principle of Work and Energy

$$T_1 + U_{1\to2} = T_2: \qquad 0 + F(2.73\text{ m}) = 126.0\text{ J}$$
$$F = +46.1\text{ N} \qquad \mathbf{F} = 46.1\text{ N} \swarrow \blacktriangleleft$$

SAMPLE PROBLEM 17.3

A sphere, a cylinder, and a hoop, each having the same mass and the same radius, are released from rest on an incline. Determine the velocity of each body after it has rolled through a distance corresponding to a change in elevation h.

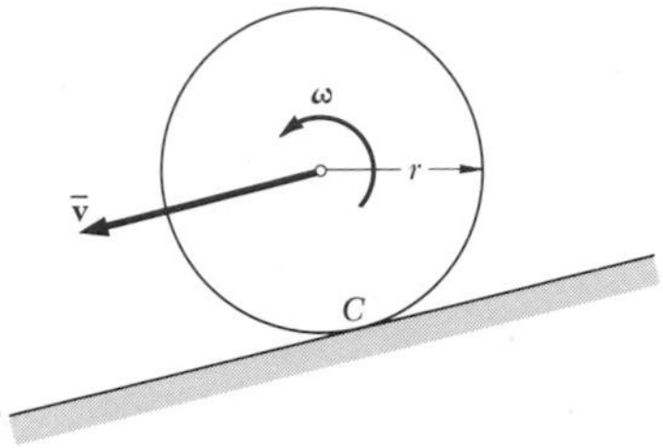

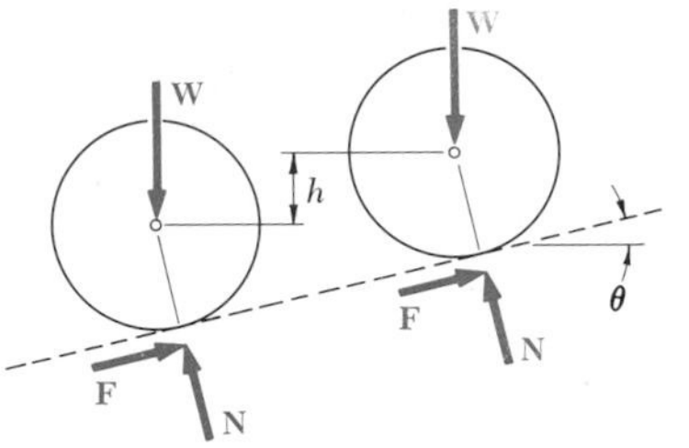

Solution. We shall first solve the problem in general terms and then find particular results for each body. We denote the mass by m, the centroidal moment of inertia by $\bar{I}$, the weight by W, and the radius by r.

Since each body rolls, the instantaneous center of rotation is located at C and we write

$$\omega = \frac{\bar{v}}{r}$$

Kinetic Energy

$$T_1 = 0$$

$$T_2 = \tfrac{1}{2}m\bar{v}^2 + \tfrac{1}{2}\bar{I}\omega^2$$

$$= \tfrac{1}{2}m\bar{v}^2 + \tfrac{1}{2}\bar{I}\left(\frac{\bar{v}}{r}\right)^2 = \tfrac{1}{2}\left(m + \frac{\bar{I}}{r^2}\right)\bar{v}^2$$

Work. Since the friction force $\mathbf{F}$ in rolling motion does no work,

$$U_{1\to2} = Wh$$

Principle of Work and Energy

$$T_1 + U_{1\to2} = T_2$$

$$0 + Wh = \tfrac{1}{2}\left(m + \frac{\bar{I}}{r^2}\right)\bar{v}^2 \qquad \bar{v}^2 = \frac{2Wh}{m + \bar{I}/r^2}$$

Noting that $W = mg$, we rearrange the result and obtain

$$\bar{v}^2 = \frac{2gh}{1 + \bar{I}/mr^2}$$

Velocities of Sphere, Cylinder, and Hoop. Introducing successively the particular expressions for $\bar{I}$, we obtain

Sphere:	$\bar{I} = \frac{2}{5}mr^2$	$\bar{v} = 0.845\sqrt{2gh}$ ◄
Cylinder:	$\bar{I} = \frac{1}{2}mr^2$	$\bar{v} = 0.816\sqrt{2gh}$ ◄
Hoop:	$\bar{I} = mr^2$	$\bar{v} = 0.707\sqrt{2gh}$ ◄

Remark. We may compare the results with the velocity attained by a frictionless block sliding through the same distance. The solution is identical to the above solution except that $\omega = 0$; we find $\bar{v} = \sqrt{2gh}$.

Comparing the results, we note that the velocity of the body is independent of both its mass and radius. However, the velocity does depend upon the quotient $\bar{I}/mr^2 = \bar{k}^2/r^2$, which measures the ratio of the rotational kinetic energy to the translational kinetic energy. Thus the hoop, which has the largest $\bar{k}$ for a given radius r, attains the smallest velocity, while the sliding block, which does not rotate, attains the largest velocity.

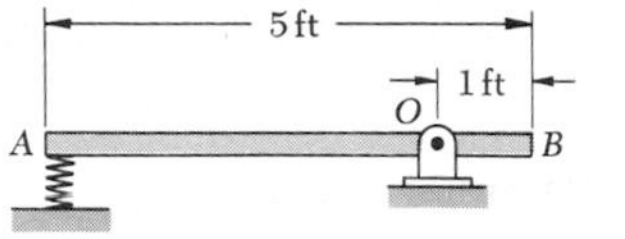

SAMPLE PROBLEM 17.4

A 30-lb slender rod AB is 5 ft long and is pivoted about a point O which is 1 ft from end B. The other end is pressed against a spring of constant $k = 1800$ lb/in. until the spring is compressed 1 in. The rod is then in a horizontal position. If the rod is released from this position, determine its angular velocity and the reaction at the pivot O as the rod passes through a vertical position.

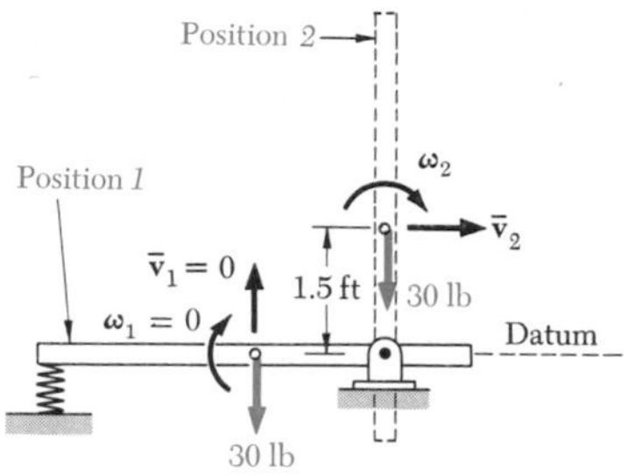

Position 1. *Potential Energy.* Since the spring is compressed 1 in., we have $x_1 = 1$ in.

$$V_e = \tfrac{1}{2}kx_1^2 = \tfrac{1}{2}(1800 \text{ lb/in.})(1 \text{ in.})^2 = 900 \text{ in} \cdot \text{lb}$$

Choosing the datum as shown, we have $V_g = 0$; therefore,

$$V_1 = V_e + V_g = 900 \text{ in} \cdot \text{lb} = 75 \text{ ft} \cdot \text{lb}$$

Kinetic Energy. Since the velocity in position *1* is zero, we have $T_1 = 0$.

Position 2. *Potential Energy.* The elongation of the spring is zero, and we have $V_e = 0$. Since the center of gravity of the rod is now 1.5 ft above the datum,

$$V_g = (30 \text{ lb})(+1.5 \text{ ft}) = 45 \text{ ft} \cdot \text{lb}$$
$$V_2 = V_e + V_g = 45 \text{ ft} \cdot \text{lb}$$

Kinetic Energy. Denoting by $\boldsymbol{\omega}_2$ the angular velocity of the rod in position *2*, we note that the rod rotates about O and write $\bar{v}_2 = \bar{r}\omega_2 = 1.5\omega_2$.

$$\bar{I} = \tfrac{1}{12}ml^2 = \frac{1}{12}\frac{30 \text{ lb}}{32.2 \text{ ft/s}^2}(5 \text{ ft})^2 = 1.941 \text{ lb} \cdot \text{ft} \cdot \text{s}^2$$

$$T_2 = \tfrac{1}{2}m\bar{v}_2^2 + \tfrac{1}{2}\bar{I}\omega_2^2 = \frac{1}{2}\frac{30}{32.2}(1.5\omega_2)^2 + \tfrac{1}{2}(1.941)\omega_2^2 = 2.019\omega_2^2$$

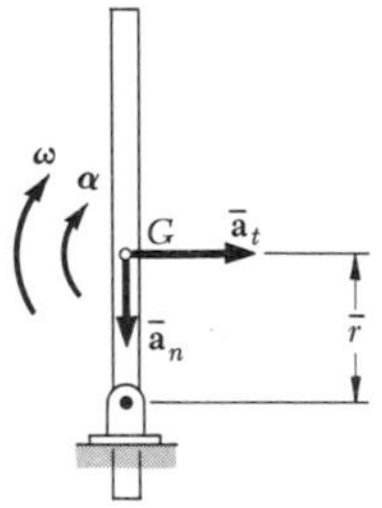

Conservation of Energy

$$T_1 + V_1 = T_2 + V_2: \qquad 0 + 75 \text{ ft} \cdot \text{lb} = 2.019\omega_2^2 + 45 \text{ ft} \cdot \text{lb}$$

$$\omega_2 = 3.86 \text{ rad/s} \circlearrowright \blacktriangleleft$$

Reaction in Position 2. Since $\omega_2 = 3.86$ rad/s, the components of the acceleration of G as the rod passes through position *2* are

$$\bar{a}_n = \bar{r}\omega_2^2 = (1.5 \text{ ft})(3.86 \text{ rad/s})^2 = 22.3 \text{ ft/s}^2 \qquad \bar{\mathbf{a}}_n = 22.3 \text{ ft/s}^2 \downarrow$$
$$\bar{a}_t = \bar{r}\alpha \qquad \bar{\mathbf{a}}_t = \bar{r}\alpha \rightarrow$$

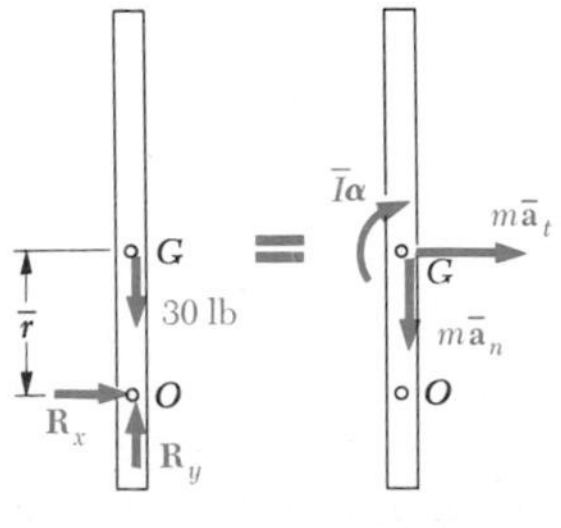

We express that the system of external forces is equivalent to the system of effective forces represented by the vector of components $m\bar{\mathbf{a}}_t$ and $m\bar{\mathbf{a}}_n$ attached at G and the couple $\bar{I}\boldsymbol{\alpha}$.

$$+\circlearrowright \Sigma M_O = \Sigma(M_O)_{\text{eff}}: \qquad 0 = \bar{I}\alpha + m(\bar{r}\alpha)\bar{r} \qquad \alpha = 0$$
$$\overset{+}{\rightarrow} \Sigma F_x = \Sigma(F_x)_{\text{eff}}: \qquad R_x = m(\bar{r}\alpha) \qquad R_x = 0$$
$$+\uparrow \Sigma F_y = \Sigma(F_y)_{\text{eff}}: \qquad R_y - 30 \text{ lb} = -m\bar{a}_n$$

$$R_y - 30 \text{ lb} = -\frac{30 \text{ lb}}{32.2 \text{ ft/s}^2}(22.3 \text{ ft/s}^2)$$

$$R_y = +9.22 \text{ lb} \qquad \mathbf{R} = 9.22 \text{ lb} \uparrow \blacktriangleleft$$

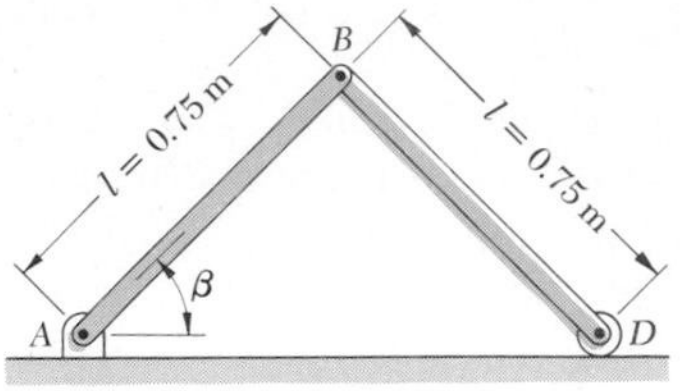

SAMPLE PROBLEM 17.5

Each of the two slender rods shown is 0.75 m long and has a mass of 6 kg. If the system is released from rest when $\beta = 60°$, determine (*a*) the angular velocity of rod *AB* when $\beta = 20°$, (*b*) the velocity of point *D* at the same instant.

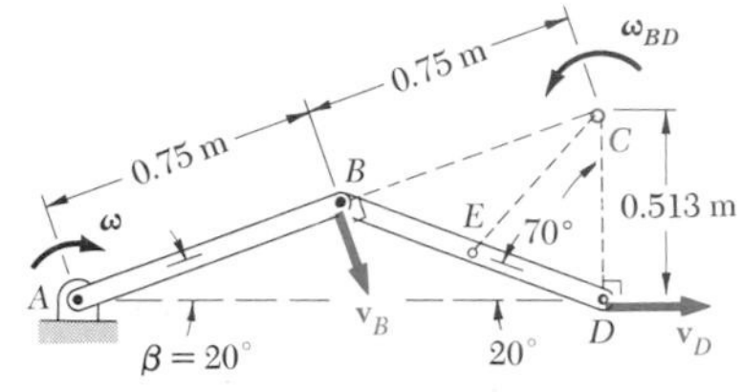

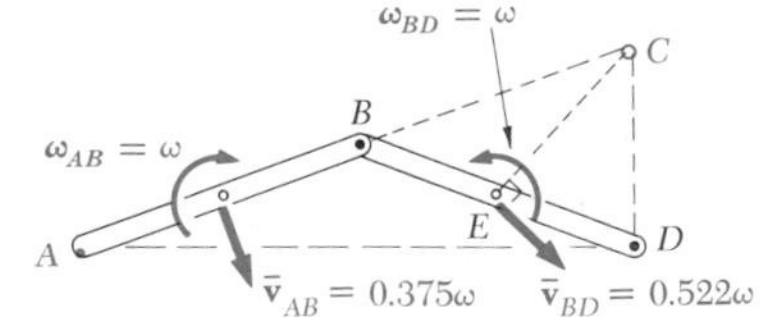

Kinematics of Motion When $\beta = 20°$. Since $\mathbf{v}_B$ is perpendicular to the rod *AB* and $\mathbf{v}_D$ is horizontal, the instantaneous center of rotation of rod *BD* is located at *C*. Considering the geometry of the figure, we obtain

$$BC = 0.75 \text{ m} \qquad CD = 2(0.75 \text{ m}) \sin 20° = 0.513 \text{ m}$$

Applying the law of cosines to triangle *CDE*, where *E* is located at the mass center of rod *BD*, we find $EC = 0.522$ m. Denoting by ω the angular velocity of rod *AB*, we have

$$\bar{v}_{AB} = (0.375 \text{ m})\omega \qquad \bar{\mathbf{v}}_{AB} = 0.375\omega \searrow$$
$$v_B = (0.75 \text{ m})\omega \qquad \mathbf{v}_B = 0.75\omega \searrow$$

Since rod *BD* seems to rotate about point *C*, we may write

$$v_B = (BC)\omega_{BD} \qquad (0.75 \text{ m})\omega = (0.75 \text{ m})\omega_{BD} \qquad \omega_{BD} = \omega \circlearrowleft$$
$$\bar{v}_{BD} = (EC)\omega_{BD} = (0.522 \text{ m})\omega \qquad \bar{\mathbf{v}}_{BD} = 0.522\omega \searrow$$

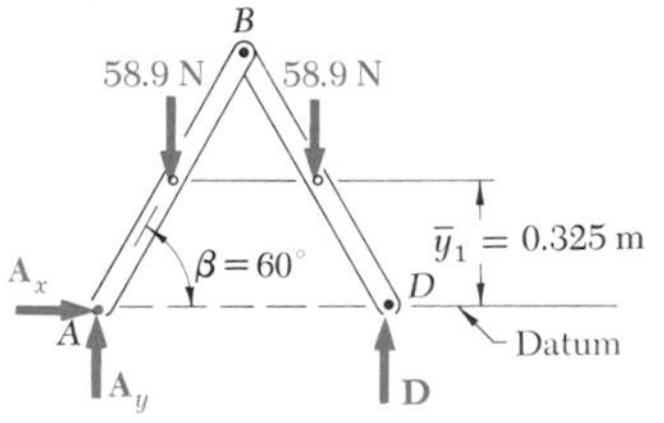

Position *1*

Position *1*. ***Potential Energy.*** Choosing the datum as shown, and observing that $W = (6 \text{ kg})(9.81 \text{ m/s}^2) = 58.9$ N, we have

$$V_1 = 2W\bar{y}_1 = 2(58.9 \text{ N})(0.325 \text{ m}) = 38.3 \text{ J}$$

Kinetic Energy. Since the system is at rest, $T_1 = 0$.

Position 2. ***Potential Energy***

$$V_2 = 2W\bar{y}_2 = 2(58.9 \text{ N})(0.128 \text{ m}) = 15.1 \text{ J}$$

Kinetic Energy

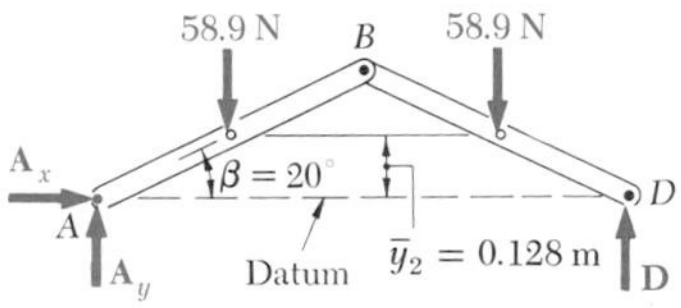

Position 2

$$\bar{I}_{AB} = \bar{I}_{BD} = \tfrac{1}{12}ml^2 = \tfrac{1}{12}(6 \text{ kg})(0.75 \text{ m})^2 = 0.281 \text{ kg} \cdot \text{m}^2$$
$$T_2 = \tfrac{1}{2}m\bar{v}_{AB}^2 + \tfrac{1}{2}\bar{I}_{AB}\omega_{AB}^2 + \tfrac{1}{2}m\bar{v}_{BD}^2 + \tfrac{1}{2}\bar{I}_{BD}\omega_{BD}^2$$
$$= \tfrac{1}{2}(6)(0.375\omega)^2 + \tfrac{1}{2}(0.281)\omega^2 + \tfrac{1}{2}(6)(0.522\omega)^2 + \tfrac{1}{2}(0.281)\omega^2$$
$$= 1.520\omega^2$$

Conservation of Energy

$$T_1 + V_1 = T_2 + V_2: \qquad 0 + 38.3 \text{ J} = 1.520\omega^2 + 15.1 \text{ J}$$
$$\omega = 3.91 \text{ rad/s} \qquad \omega_{AB} = 3.91 \text{ rad/s} \circlearrowright \blacktriangleleft$$

Velocity of Point *D*

$$v_D = (CD)\omega = (0.513 \text{ m})(3.91 \text{ rad/s}) = 2.01 \text{ m/s}$$
$$\mathbf{v}_D = 2.01 \text{ m/s} \rightarrow \blacktriangleleft$$

PROBLEMS

17.1 The rotor of a generator has an angular velocity of 3600 rpm when the generator is taken off line. The 150-kg rotor, which has a centroidal radius of gyration of 250 mm, then coasts to rest. Knowing that the kinetic friction of the rotor produces a couple of magnitude 2 N · m, determine the number of revolutions that the rotor executes before coming to rest.

17.2 A large flywheel of mass 1800 kg has a radius of gyration of 0.75 m. It is observed that 2500 revolutions are required for the flywheel to coast from an angular velocity of 450 rpm to rest. Determine the average magnitude of the couple due to kinetic friction in the bearings.

17.3 Two disks of the same material are attached to a shaft as shown. Disk A is of radius r and has a thickness $3b$, while disk B is of radius nr and thickness b. A couple $\mathbf{M}$ of constant magnitude is applied when the system is at rest and is removed after the system has executed two revolutions. Determine the value of n which results in the largest final speed for a point on the rim of disk B.

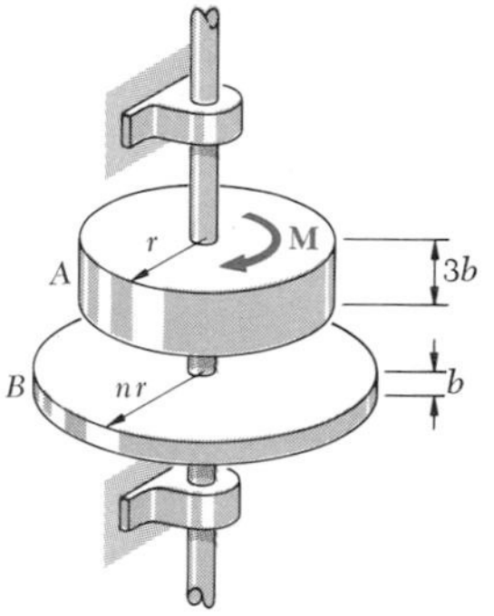

Fig. P17.3 and P17.4

17.4 Two disks of the same material are attached to a shaft as shown. Disk A weighs 30 lb and has a radius $r = 5$ in. Disk B is one-third as thick as disk A. A couple $\mathbf{M}$ of magnitude 10 lb·ft is applied to disk A when the system is at rest. Determine the radius nr of disk B if the angular velocity of the system is to be 450 rpm after 5 revolutions.

17.5 The flywheel of a small punch rotates at 240 rpm. It is known that 1500 ft·lb of work must be done each time a hole is punched. It is desired that the speed of the flywheel after one punching be not less than 90 percent of the original speed of 240 rpm. (*a*) Determine the required moment of inertia of the flywheel. (*b*) If a constant 20-lb·ft couple is applied to the shaft of the flywheel, determine the number of revolutions which must occur between each punching, knowing that the initial velocity is to be 240 rpm at the start of each punching.

17.6 The flywheel of a punching machine weighs 600 lb and has a radius of gyration of 24 in. Each punching operation requires 1500 ft·lb of work. (*a*) Knowing that the speed of the flywheel is 300 rpm just before a punching, determine the speed immediately after the punching. (*b*) If a constant 15-lb·ft couple is applied to the shaft of the flywheel, determine the number of revolutions executed before the speed is again 300 rpm.

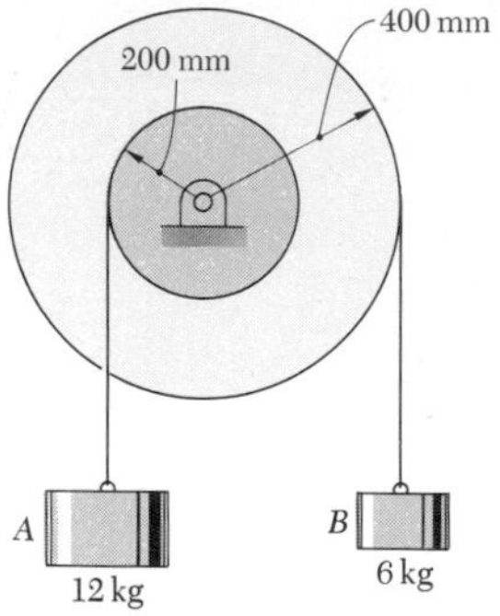

Fig. P17.7

17.7 Two cylinders are attached by cords to an 18-kg double pulley which has a radius of gyration of 300 mm. When the system is at rest and in equilibrium, a 3-kg collar is added to the 12-kg cylinder. Neglecting friction, determine the velocity of each cylinder after the pulley has completed one revolution.

17.8 Solve Prob. 17.7, assuming that the 3-kg collar is added to the 6-kg cylinder.

17.9 Using the principle of work and energy, solve Prob. 16.36*b*.

17.10 Using the principle of work and energy, solve Prob. 16.34*c*.

17.11 A disk of constant thickness and initially at rest is placed in contact with the belt, which moves with a constant velocity **v**. Denoting by μ the coefficient of friction between the disk and the belt, derive an expression for the number of revolutions executed by the disk before it reaches a constant angular velocity.

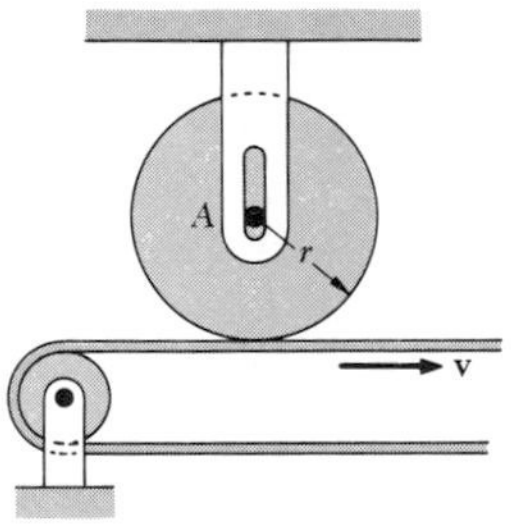

Fig. P17.11 and P17.12

17.12 Disk *A*, of weight 5 lb and radius $r = 3$ in., is at rest when it is placed in contact with the belt, which moves with a constant speed $v = 50$ ft/s. Knowing that $\mu = 0.20$ between the disk and the belt, determine the number of revolutions executed by the disk before it reaches a constant angular velocity.

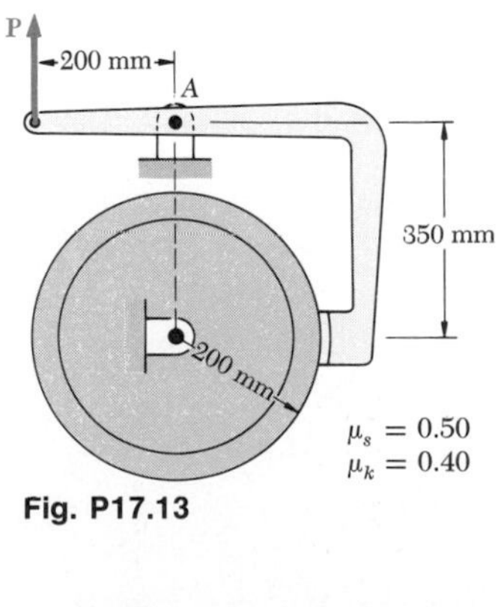

Fig. P17.13

17.13 The 200-mm-radius brake drum is attached to a larger flywheel which is not shown. The total mass moment of inertia of the flywheel and drum is 8 kg·m². Knowing that the initial angular velocity is 120 rpm clockwise, determine the force **P** which must be applied if the system is to come to rest in 8 revolutions.

17.14 Solve Prob. 17.13, assuming that the initial angular velocity of the flywheel is 120 rpm counterclockwise.

17.15 Each of the gears A and B has a mass of 2 kg and a radius of gyration of 70 mm, while gear C has a mass of 10 kg and a radius of gyration of 175 mm. A couple **M** of constant magnitude 12 N · m is applied to gear C. Determine (a) the number of revolutions required for the angular velocity of gear C to increase from 100 to 450 rpm, (b) the corresponding tangential force acting on gear A.

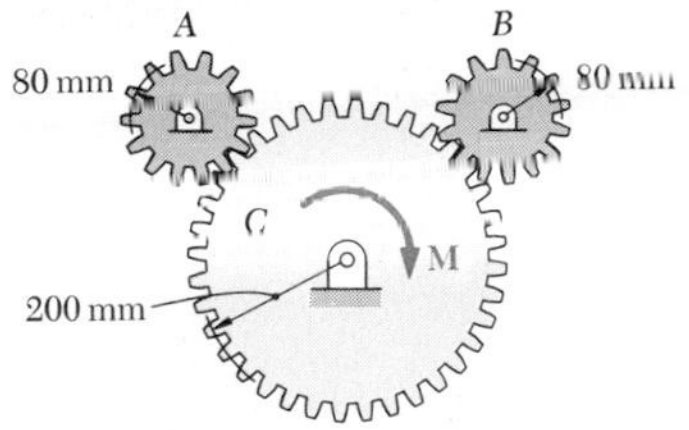

Fig. P17.15

17.16 Solve Prob. 17.15, assuming that the 12-N·m couple is applied to gear B.

17.17 A cord is wrapped around a cylinder of radius r and mass m as shown. If the cylinder is released from rest, determine the velocity of the center of the cylinder after it has moved through a distance h.

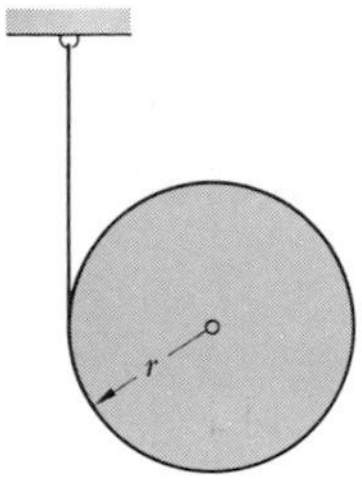

Fig. P17.17

17.18 Two 10-kg disks, each of radius $r = 0.3$ m, are connected by a cord. At the instant shown, the angular velocity of disk B is 20 rad/s clockwise. Determine how far disk A will rise before the angular velocity of disk B is reduced to 4 rad/s.

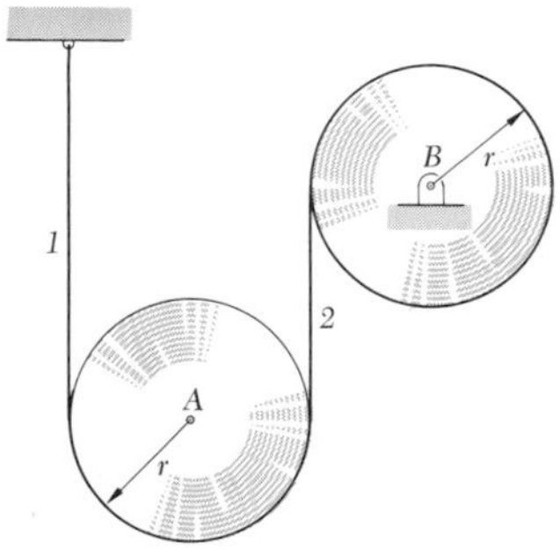

Fig. P17.18

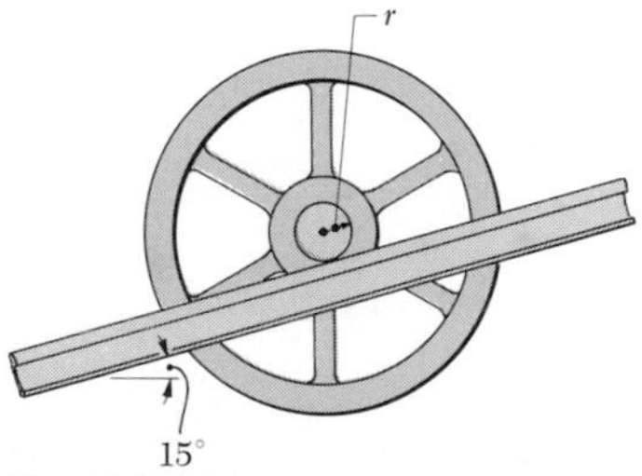

Fig. P17.19

17.19 A flywheel is rigidly attached to a $1\frac{1}{2}$-in.-radius shaft which rolls without sliding along parallel rails. The system is released from rest and attains a speed of 6 in./s after moving 75 in. along the rails. Determine the centroidal radius of gyration of the system.

17.20 A hemisphere of mass m and radius r is released from rest in the position shown. Assuming that the hemisphere rolls without sliding, determine (a) its angular velocity after it has rolled through 90°, (b) the normal reaction at the surface at the same instant. [*Hint.* Note that $GO = 3r/8$ and that, by the parallel-axis theorem, $\bar{I} = \frac{2}{5}mr^2 - m(GO)^2$.]

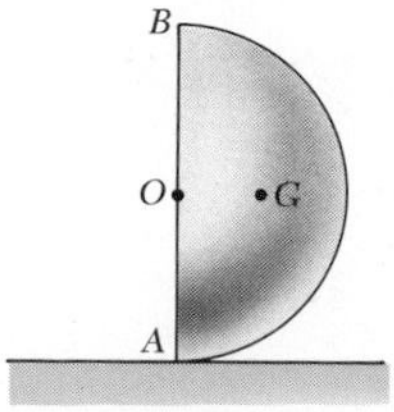

Fig. P17.20

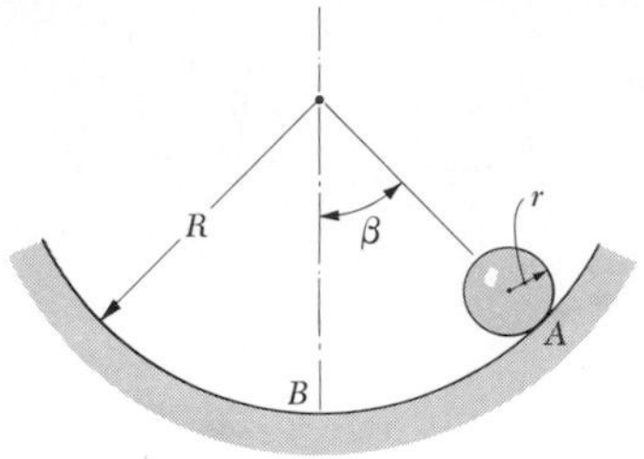

Fig. P17.21

17.21 A sphere of mass m and radius r rolls without slipping inside a curved surface of radius R. Knowing that the sphere is released from rest in the position shown, derive an expression (*a*) for the linear velocity of the sphere as it passes through B, (*b*) for the magnitude of the vertical reaction at that instant.

17.22 Solve Prob. 17.21, assuming that the sphere is replaced by a uniform cylinder of mass m and radius r.

17.23 A slender rod of length l and mass m is pivoted at one end as shown. It is released from rest in a horizontal position and swings freely. (*a*) Determine the angular velocity of the rod as it passes through a vertical position and the corresponding reaction at the pivot. (*b*) Solve part *a* for $m = 1.5$ kg and $l = 0.9$ m.

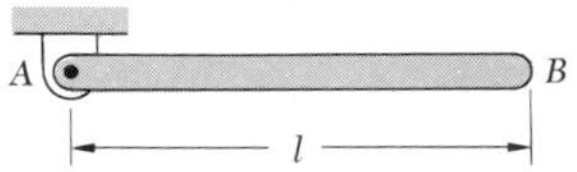

Fig. P17.23

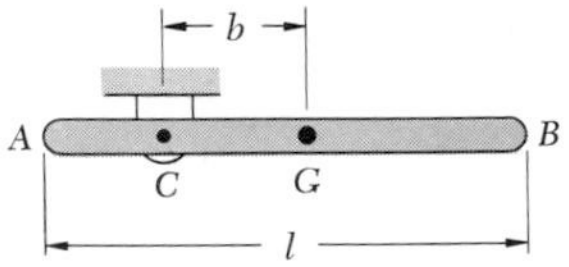

Fig. P17.24

17.24 A uniform rod of length l is pivoted about a point C located at a distance b from its center G. The rod is released from rest in a horizontal position. Determine (*a*) the distance b so that the angular velocity of the rod as it passes through a vertical position is maximum, (*b*) the value of the maximum angular velocity.

17.25 A 6- by 8-in. rectangular plate is suspended by two pins at A and B. The pin at B is removed and the plate swings about point A. Determine (*a*) the angular velocity of the plate after it has rotated through 90°, (*b*) the maximum angular velocity attained by the plate as it swings freely.

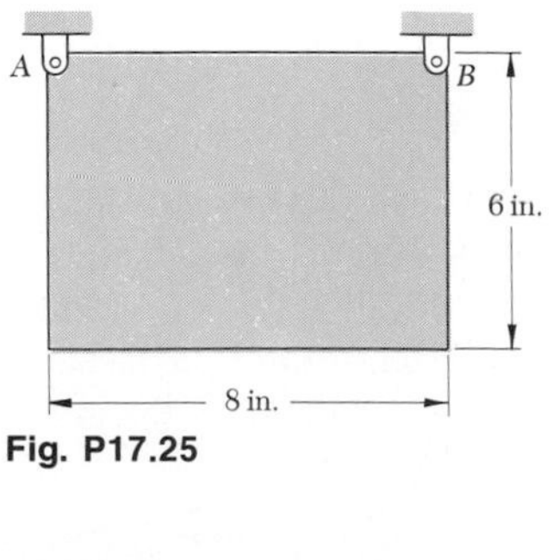

Fig. P17.25

17.26 and 17.27 Gear C weighs 6 lb and has a centroidal radius of gyration of 3 in. The uniform bar AB weighs 5 lb, and gear D is stationary. If the system is released from rest in the position shown, determine the velocity of point B after bar AB has rotated through 90°.

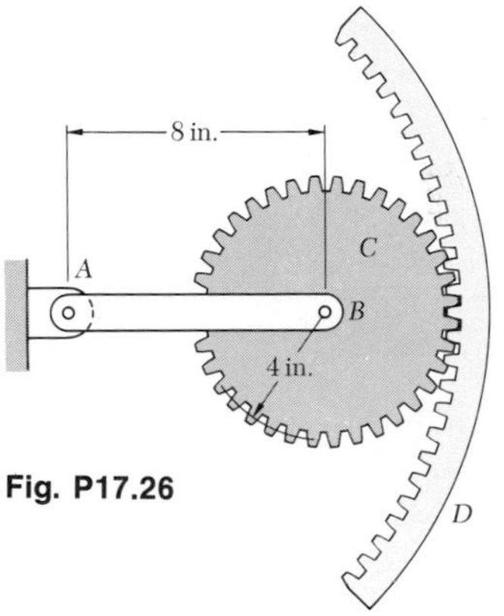

Fig. P17.26

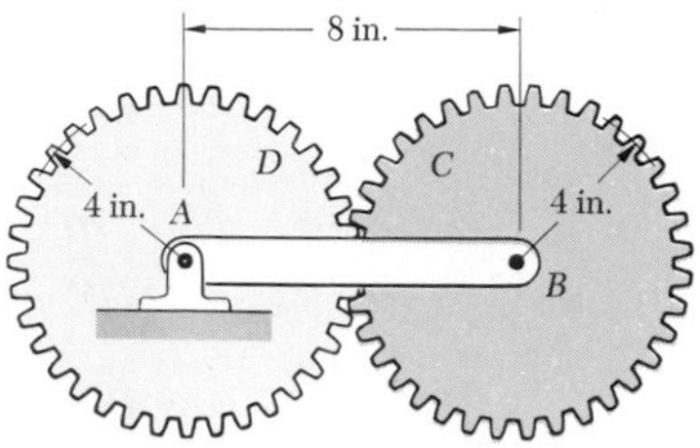

Fig. P17.27

17.28 The mass center G of a 1.5-kg wheel of radius $R = 150$ mm is located at a distance $r = 50$ mm from its geometric center C. The centroidal radius of gyration of the wheel is $\bar{k} = 75$ mm. As the wheel rolls without sliding, its angular velocity is observed to vary. Knowing that in position *1* the angular velocity is 10 rad/s, determine the angular velocity of the wheel (*a*) in position *2*, (*b*) in position *3*.

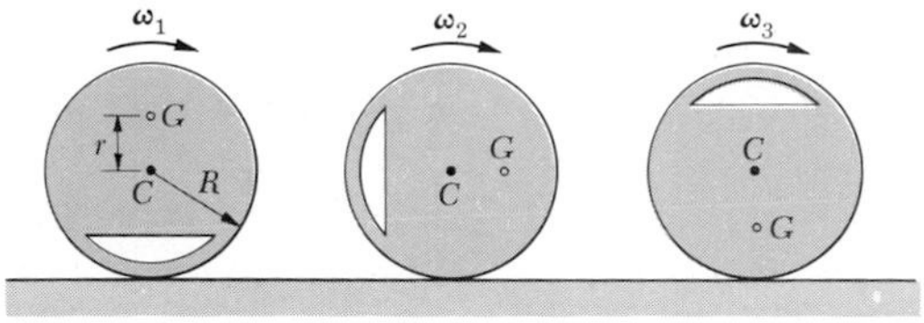

Fig. P17.28 and P17.29

17.29 The mass center G of a wheel of radius R is located at a distance r from its geometric center C. The centroidal radius of gyration of the wheel is denoted by $\bar{k}$. As the wheel rolls freely and without sliding on a horizontal plane, its angular velocity is observed to vary. Denoting by ω_1, ω_2, and ω_3, respectively, the angular velocity of the wheel when G is directly above C, level with C, and directly below C, show that ω_1, ω_2, and ω_3 satisfy the relation

$$\frac{\omega_2^2 - \omega_1^2}{\omega_3^2 - \omega_2^2} = \frac{g/R + \omega_1^2}{g/R + \omega_3^2}$$

17.30 and 17.31 The 12-lb carriage is supported as shown by two uniform disks, each of weight 8 lb and radius 3 in. Knowing that the system is initially at rest, determine the velocity of the carriage after it has moved 3 ft. Assume that the disks roll without sliding.

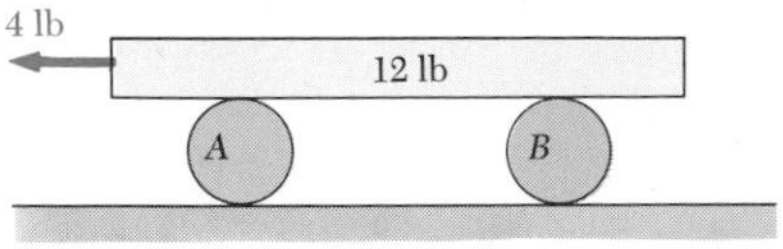

Fig. P17.30

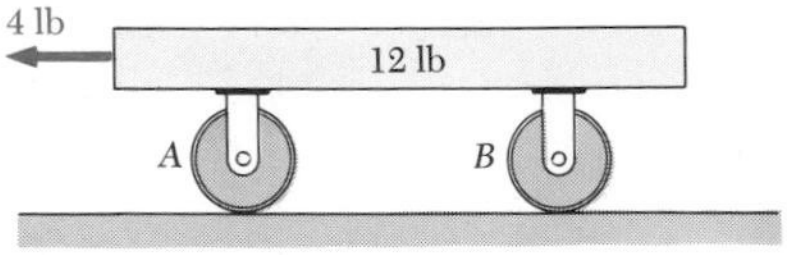

Fig. P17.31

17.32 The motion of the 240-mm rod AB is guided by pins at A and B which slide freely in the slots shown. If the rod is released from rest in position *1*, determine the velocity of A and B when the rod is (*a*) in position *2*, (*b*) in position *3*.

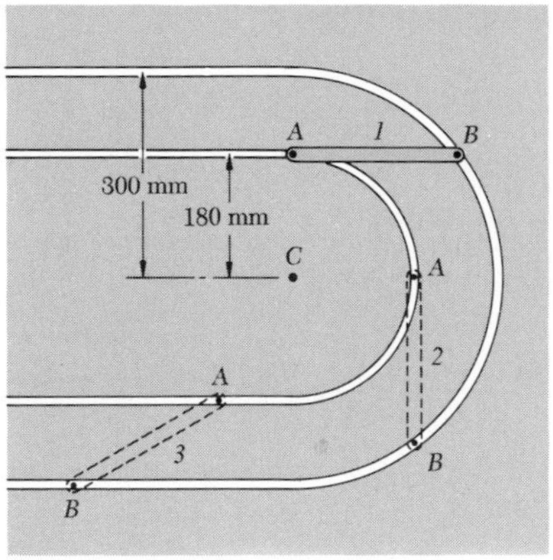

Fig. P17.32

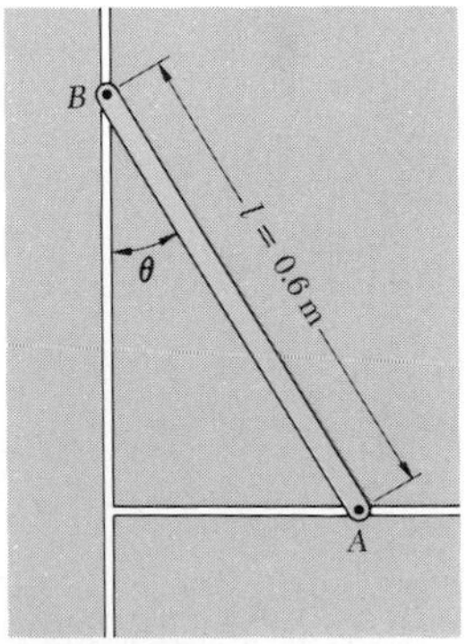

Fig. P17.33

17.33 The motion of a 0.6-m slender rod is guided by pins at A and B which slide freely in the slots shown. Knowing that the rod is released from rest when $\theta = 0$ and that end A is given a slight push to the right, determine (*a*) the angle θ for which the speed of end A is maximum, (*b*) the corresponding maximum speed of A.

17.34 In Prob. 17.33, determine the velocity of ends A and B (*a*) when $\theta = 30°$, (*b*) when $\theta = 90°$.

17.35 The ends of a 25-lb rod AB are constrained to move along the slots shown. A spring of constant 3 lb/in. is attached to end A. Knowing that the rod is released from rest when $\theta = 0$ and that the initial tension in the spring is zero, determine the maximum distance through which end A will move.

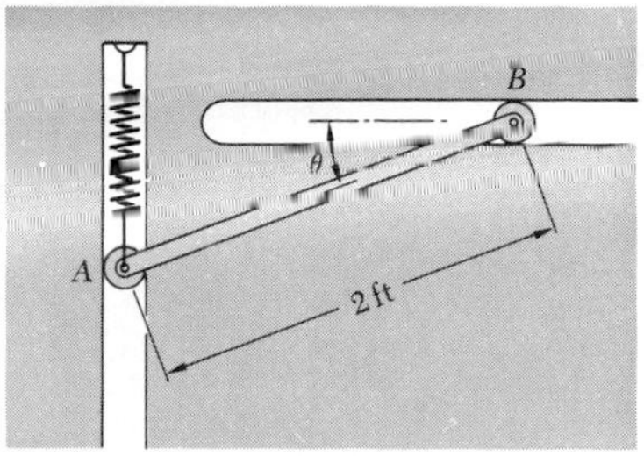

Fig. P17.35 and P17.36

17.36 The ends of a 25-lb rod AB are constrained to move along the slots shown. A spring of constant 10 lb/in. is attached to end A in such a way that its tension is zero when $\theta = 0$. If the rod is released from rest when $\theta = 60°$, determine the angular velocity of the rod when $\theta = 30°$.

17.37 Determine the velocity of pin B as the rods of Sample Prob. 17.5 strike the horizontal surface.

17.38 The uniform rods AB and BC are of mass 4.5 and 1.5 kg respectively. If the system is released from rest in the position shown, determine the angular velocity of rod BC as it passes through a vertical position.

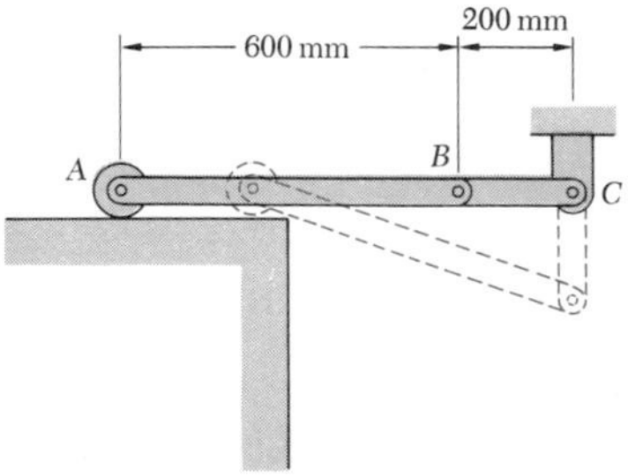

Fig. P17.38

17.39 In Prob. 17.38, determine the angular velocity of rod BC after it has rotated 45°.

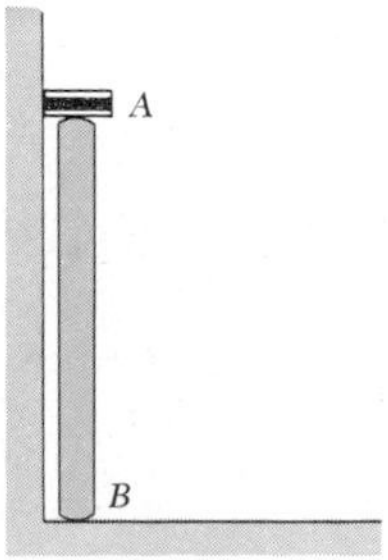

Fig. P17.40

***17.40** A small matchbox is placed on top of the rod AB. End B of the rod is given a slight horizontal push, causing it to slide on the horizontal floor. Assuming no friction and neglecting the weight of the matchbox, determine the angle θ through which the rod will have rotated when the matchbox loses contact with the rod.

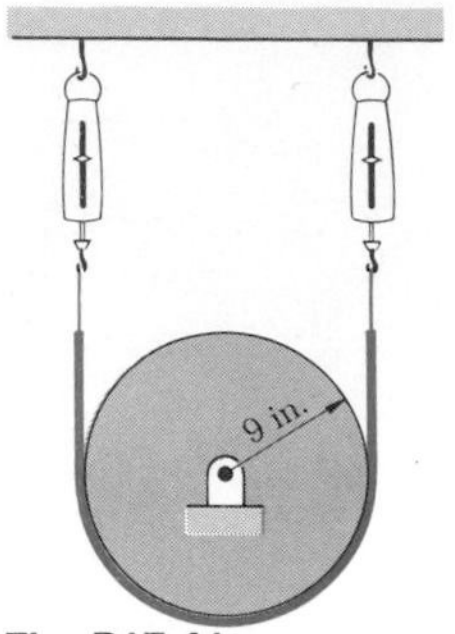

Fig. P17.41

17.41 The experimental setup shown is used to measure the power output of a small turbine. When the turbine is operating at 200 rpm, the readings of the two spring scales are 10 and 22 lb, respectively. Determine the power being developed by the turbine.

17.42 In Sample Prob. 17.2 determine the power being delivered to gear B at the instant when (*a*) the gear starts rotating, (*b*) the gear attains an angular velocity $\omega_B = 300$ rpm.

17.43 Knowing that the maximum allowable couple which can be applied to a shaft is 12 kN · m, determine the maximum power which can be transmitted by the shaft (*a*) at 100 rpm, (*b*) at 1000 rpm.

17.44 Determine the moment of the couple which must be exerted by a motor to develop $\frac{1}{4}$ hp at a speed of (*a*) 3600 rpm, (*b*) 720 rpm.

17.7. Principle of Impulse and Momentum for the Plane Motion of a Rigid Body. We shall now apply the principle of impulse and momentum to the analysis of the plane motion of rigid bodies and of systems of rigid bodies. As was pointed out in Chap. 13, the method of impulse and momentum is particularly well adapted to the solution of problems involving time and velocities. Moreover, the principle of impulse and momentum provides the only practicable method for the solution of problems involving impulsive motion or impact (Secs. 17.10 and 17.11).

Considering again a rigid body as made of a large number of particles P_i, we recall from Sec. 14.8 that the system formed by the momenta of the particles at time t_1 and the system of the impulses of the external forces applied from t_1 to t_2 are together equipollent to the system formed by the momenta of the particles at time t_2. Since the vectors associated with a rigid body may be considered as sliding vectors, it follows (Sec. 3.18) that the systems of vectors shown in Fig. 17.6 are not only equipollent but

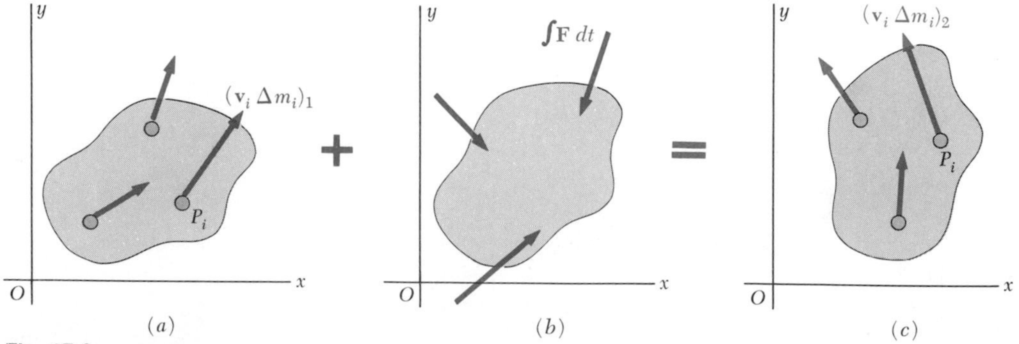

Fig. 17.6

truly *equivalent* in the sense that the vectors on the left-hand side of the equals sign may be transformed into the vectors on the right-hand side through the use of the fundamental operations listed in Sec. 3.12. We therefore write

$$\textbf{Syst Momenta}_1 + \textbf{Syst Ext Imp}_{1\to 2} = \textbf{Syst Momenta}_2 \qquad (17.14)$$

But the momenta $\mathbf{v}_i\,\Delta m_i$ of the particles may be reduced to a vector attached at G, equal to their sum

$$\mathbf{L} = \sum_{i=1}^{n} \mathbf{v}_i\,\Delta m_i$$

and to a couple of moment equal to the sum of their moments about G

$$\mathbf{H}_G = \sum_{i=1}^{n} \mathbf{r}'_i \times \mathbf{v}_i\,\Delta m_i$$

We recall from Sec. 14.2 that $\mathbf{L}$ and $\mathbf{H}_G$ define, respectively, the linear momentum and the angular momentum about G of the system of particles forming the rigid body. We also note from Eq. (14.14) that $\mathbf{L} = m\overline{\mathbf{v}}$. On the other hand, restricting the present analysis to the plane motion of a rigid slab or of a rigid body symmetrical with respect to the reference plane, we recall from Eq. (16.4) that $\mathbf{H}_G = \overline{I}\boldsymbol{\omega}$. We thus conclude that the system of the momenta $\mathbf{v}_i\,\Delta m_i$ is equivalent to the *linear momentum vector* $m\overline{\mathbf{v}}$ attached at G and to the *angular momentum couple* $\overline{I}\boldsymbol{\omega}$ (Fig. 17.7). Observing that the system of momenta reduces to the vector $m\overline{\mathbf{v}}$ in the particular case of a translation ($\boldsymbol{\omega} = 0$) and to the couple $\overline{I}\boldsymbol{\omega}$ in the particular case of a centroidal rotation ($\overline{\mathbf{v}} = 0$), we verify once more that the plane motion of a rigid body symmetrical with respect to the reference plane may be resolved into a translation with the mass center G and a rotation about G.

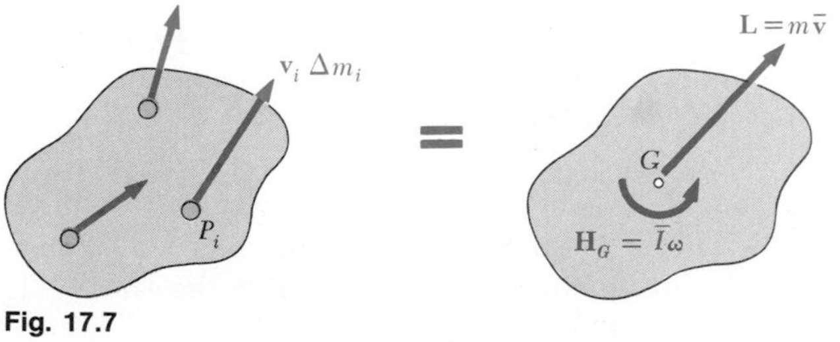

Fig. 17.7

Replacing the system of momenta in parts *a* amd *c* of Fig. 17.6 by the equivalent linear momentum vector and angular momentum couple, we obtain the three diagrams shown in Fig. 17.8.

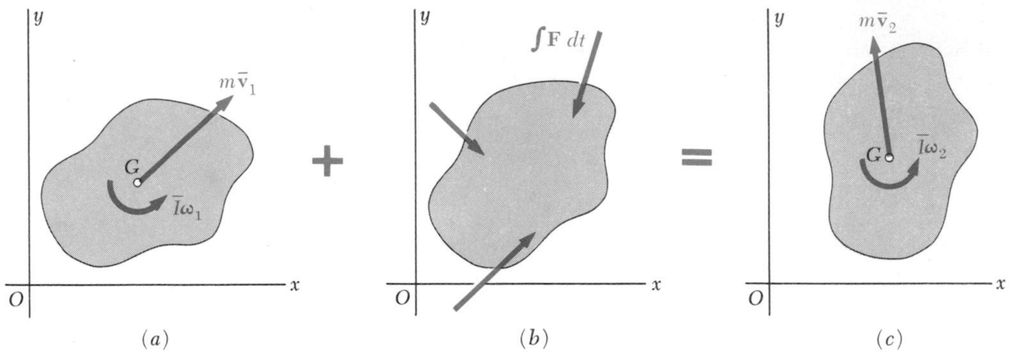

Fig. 17.8

This figure expresses graphically the fundamental relation (17.14) in the case of the plane motion of a rigid slab or of a rigid body symmetrical with respect to the reference plane.

Three equations of motion may be derived from Fig. 17.8. Two equations are obtained by summing and equating the x and y *components* of the momenta and impulses, and the third by summing and equating the *moments* of these vectors *about any given point.* The coordinate axes may be chosen fixed in space, or they may be allowed to move with the mass center of the body while maintaining a fixed direction. In either case, the point about which moments are taken should keep the same position relative to the coordinate axes during the interval of time considered.

In deriving the three equations of motion for a rigid body, care should be taken not to add indiscriminately linear and angular momenta. Confusion will be avoided if it is kept in mind that $m\bar{v}_x$ and $m\bar{v}_y$ represent the *components of a vector,* namely, the linear momentum vector $m\bar{\mathbf{v}}$, while $\bar{I}\omega$ represents the *magnitude of a couple,* namely, the angular momentum couple $\bar{I}\boldsymbol{\omega}$. Thus the quantity $\bar{I}\omega$ should be added only to the *moment* of the linear momentum $m\bar{\mathbf{v}}$, never to this vector itself nor to its components. All quantities involved will then be expressed in the same units, namely N·m·s or lb·ft·s.

Noncentroidal Rotation. In this particular case of plane motion, the magnitude of the velocity of the mass center of the body is $\bar{v} = \bar{r}\omega$, where $\bar{r}$ represents the distance from the mass center to the fixed axis of rotation and ω the angular velocity of the body at the instant considered; the magnitude of the momentum vector attached at G is thus $m\bar{v} = m\bar{r}\omega$. Summing the moments about O of the momentum vector and momentum couple (Fig. 17.9) and using the parallel-axis theorem for moments of inertia, we find that the angular momentum $\mathbf{H}_O$ of the body about O has the magnitude†

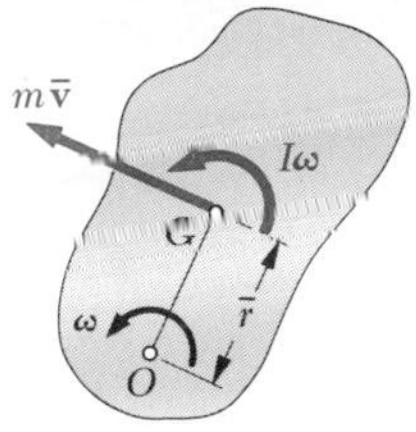

Fig. 17.9

$$\bar{I}\omega + (m\bar{r}\omega)\bar{r} = (\bar{I} + m\bar{r}^2)\omega = I_O\omega \tag{17.15}$$

Equating the moments about O of the momenta and impulses in (17.14), we write

$$I_O\omega_1 + \sum\int_{t_1}^{t_2} M_O\,dt = I_O\omega_2 \tag{17.16}$$

In the general case of plane motion of a rigid body symmetrical with respect to the reference plane, Eq. (17.16) may be used with respect to the instantaneous axis of rotation under certain conditions. It is recommended, however, that all problems of plane motion be solved by the general method described earlier in this section.

17.8. Systems of Rigid Bodies. The motion of several rigid bodies may be analyzed by applying the principle of impulse and momentum to each body separately (Sample Prob. 17.6).

However, in solving problems involving no more than three unknowns (including the impulses of unknown reactions), it is often found convenient to apply the principle of impulse and momentum to the system as a whole. The momentum and impulse diagrams are drawn for the entire system of bodies. The diagrams of momenta should include a momentum vector, a momentum couple, or both, for each moving part of the system.

†Note that the sum $\mathbf{H}_A$ of the moments about an arbitrary point A of the momenta of the particles of a rigid slab is, in general, *not* equal to $I_A\boldsymbol{\omega}$. (See Prob. 17.59.)

Impulses of forces internal to the system may be omitted from the impulse diagram since they occur in pairs of equal and opposite vectors. Summing and equating successively the x components, y components, and moments of all vectors involved, one obtains three relations which express that the momenta at time t_1 and the impulses of the external forces form a system equipollent to the system of the momenta at time t_2.† Again, care should be taken not to add indiscriminately linear and angular momenta; each equation should be checked to make sure that consistent units have been used. This approach has been used in Sample Prob. 17.8 and, further on, in Sample Probs. 17.9 and 17.10.

17.9. Conservation of Angular Momentum. When no external force acts on a rigid body or a system of rigid bodies, the impulses of the external forces are zero and the system of the momenta at time t_1 is equipollent to the system of the momenta at time t_2. Summing and equating successively the x components, y components, and moments of the momenta at times t_1 and t_2, we conclude that the total linear momentum of the system is conserved in any direction and that its total angular momentum is conserved about any point.

There are many engineering applications, however, in which *the linear momentum is not conserved*, yet in which *the angular momentum* $\mathbf{H}_O$ of the system about a given point O is *conserved:*

$$(\mathbf{H}_O)_1 = (\mathbf{H}_O)_2 \tag{17.17}$$

Such cases occur when the lines of action of all external forces pass through O or, more generally, when the sum of the angular impulses of the external forces about O is zero.

Problems involving *conservation of angular momentum* about a point O may be solved by the general method of impulse and momentum, i.e., by drawing momentum and impulse diagrams as described in Secs. 17.7 and 17.8. Equation (17.17) is then obtained by summing and equating moments about O (Sample Prob. 17.8). As we shall see later in Sample Prob. 17.9, two additional equations may be written by summing and equating x and y components; these equations may be used to determine two unknown linear impulses, such as the impulses of the reaction components at a fixed point.

† Note that, as in Sec. 16.7, we cannot speak of *equivalent* systems since we are not dealing with a single rigid body.

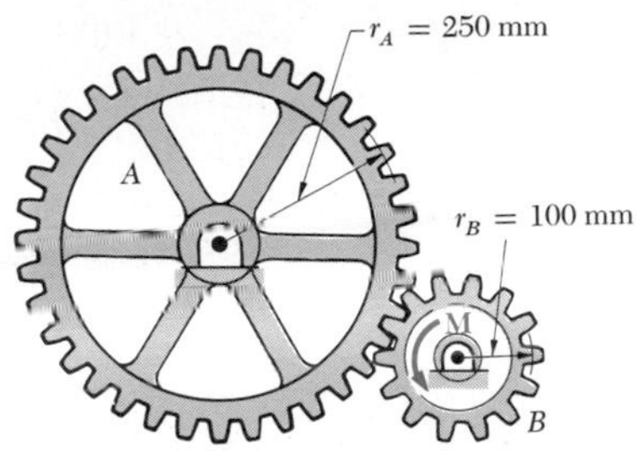

SAMPLE PROBLEM 17.6

Gear A has a mass of 10 kg and a radius of gyration of 200 mm, while gear B has a mass of 3 kg and a radius of gyration of 80 mm. The system is at rest when a couple $\mathbf{M}$ of magnitude 6 N·m is applied to gear B. Neglecting friction, determine (*a*) the time required for the angular velocity of gear B to reach 600 rpm, (*b*) the tangential force which gear B exerts on gear A. These gears have been previously considered in Sample Prob. 17.2.

Solution. We apply the principle of impulse and momentum to each gear separately. Since all forces and the couple are constant, their impulses are obtained by multiplying them by the unknown time t. We recall from Sample Prob. 17.2 that the centroidal moments of inertia and the final angular velocities are

$$\bar{I}_A = 0.400 \text{ kg}\cdot\text{m}^2 \qquad \bar{I}_B = 0.0192 \text{ kg}\cdot\text{m}^2$$
$$(\omega_A)_2 = 25.1 \text{ rad/s} \qquad (\omega_B)_2 = 62.8 \text{ rad/s}$$

Principle of Impulse and Momentum for Gear A. The systems of initial momenta, impulses, and final momenta are shown in three separate sketches.

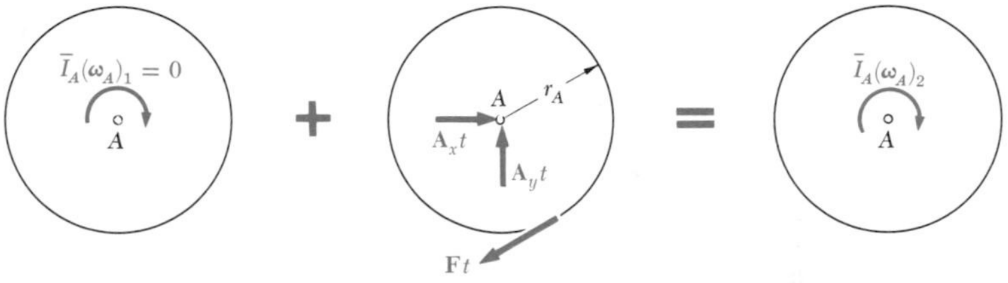

Syst Momenta$_1$ + Syst Ext Imp$_{1\to2}$ = Syst Momenta$_2$

+↺ moments about A: $\qquad 0 - Ftr_A = -\bar{I}_A(\omega_A)_2$

$$Ft(0.250 \text{ m}) = (0.400 \text{ kg}\cdot\text{m}^2)(25.1 \text{ rad/s})$$
$$Ft = 40.2 \text{ N}\cdot\text{s}$$

Principle of Impulse and Momentum for Gear B.

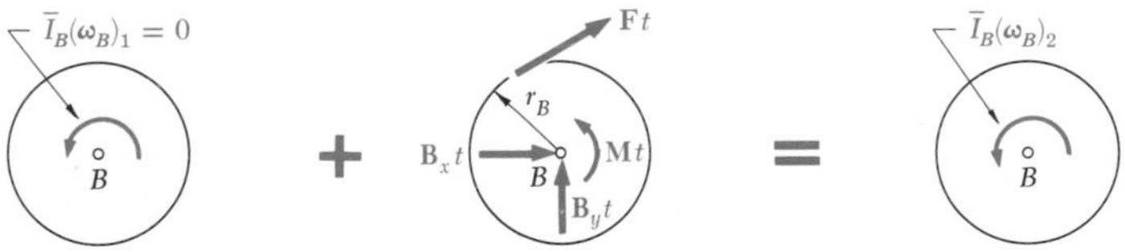

Syst Momenta$_1$ + Syst Ext Imp$_{1\to2}$ = Syst Momenta$_2$

+↺ moments about B: $\qquad 0 + Mt - Ftr_B = \bar{I}_B(\omega_B)_2$

$$+(6 \text{ N}\cdot\text{m})t - (40.2 \text{ N}\cdot\text{s})(0.100 \text{ m}) = (0.0192 \text{ kg}\cdot\text{m}^2)(62.8 \text{ rad/s})$$

$t = 0.871$ s ◄

Recalling that $Ft = 40.2$ N·s, we write

$$F(0.871 \text{ s}) = 40.2 \text{ N}\cdot\text{s} \qquad F = +46.2 \text{ N}$$

Thus, the force exerted by gear B on gear A is $\quad \mathbf{F} = 46.2 \text{ N} \swarrow$ ◄

SAMPLE PROBLEM 17.7

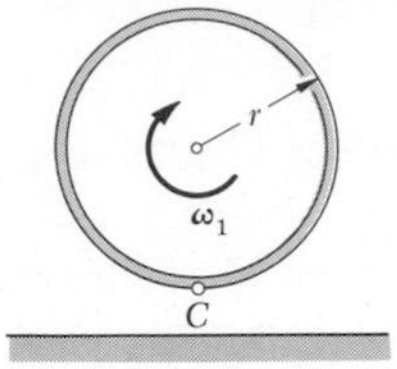

A hoop of radius r and mass m is placed on a horizontal surface with no linear velocity but with a clockwise angular velocity $\boldsymbol{\omega}_1$. Denoting by μ the coefficient of friction between the hoop and the surface, determine (*a*) the time t_2 at which the hoop will start rolling without sliding, (*b*) the linear and angular velocities of the hoop at time t_2.

Solution. Since the entire mass m is located at a distance r from the center of the hoop, we have $\bar{I} = mr^2$. While the hoop is sliding relative to the surface, it is acted upon by the normal force **N**, the friction force **F**, and its weight **W** of magnitude $W = mg$.

Principle of Impulse and Momentum. We apply the principle of impulse and momentum to the hoop from the time t_1 when it is placed on the surface until the time t_2 when it starts rolling without sliding.

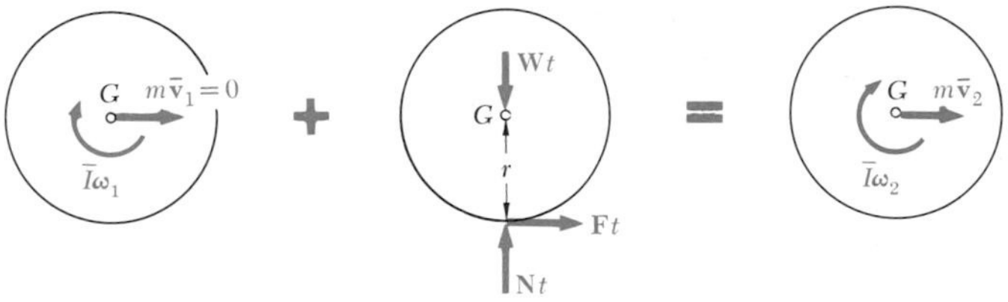

$$\textbf{Syst Momenta}_1 + \textbf{Syst Ext Imp}_{1\to 2} = \textbf{Syst Momenta}_2$$

$+\uparrow y$ components: $\quad Nt - Wt = 0 \quad (1)$

$\xrightarrow{+} x$ components: $\quad Ft = m\bar{v}_2 \quad (2)$

$+\circlearrowright$ moments about G: $\quad -\bar{I}\omega_1 + Ftr = -\bar{I}\omega_2 \quad (3)$

From (1) we obtain $N = W = mg$. For $t < t_2$, sliding occurs at point C and we have $F = \mu N = \mu mg$. Substituting for F into (2), we write

$$\mu mgt = m\bar{v}_2 \qquad \bar{v}_2 = \mu gt \qquad (4)$$

Substituting $F = \mu mg$ and $\bar{I} = mr^2$ into (3),

$$-mr^2\omega_1 + \mu mgtr = -mr^2\omega_2 \qquad \omega_2 = \omega_1 - \frac{\mu g}{r}t \qquad (5)$$

The hoop will start rolling without sliding when the velocity $\mathbf{v}_C$ of the point of contact is zero. At that time, $t = t_2$, point C becomes the instantaneous center of rotation, and we have $\bar{v}_2 = r\omega_2$. Substituting from (4) and (5), we write

$$\bar{v}_2 = r\omega_2 \qquad \mu g t_2 = r\left(\omega_1 - \frac{\mu g}{r}t_2\right) \qquad t_2 = \frac{r\omega_1}{2\mu g} \quad ◄$$

Substituting this expression for t_2 into (4),

$$\bar{v}_2 = \mu g t_2 = \mu g\frac{r\omega_1}{2\mu g} \qquad \bar{v}_2 = \tfrac{1}{2}r\omega_1 \qquad \bar{\mathbf{v}}_2 = \tfrac{1}{2}r\omega_1 \rightarrow \quad ◄$$

$$\omega_2 = \frac{\bar{v}_2}{r} \qquad \omega_2 = \tfrac{1}{2}\omega_1 \qquad \boldsymbol{\omega}_2 = \tfrac{1}{2}\omega_1 \circlearrowright \quad ◄$$

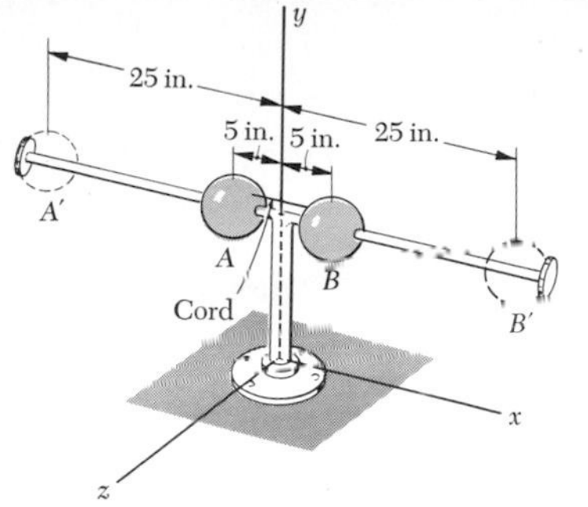

SAMPLE PROBLEM 17.8

Two solid spheres of radius 3 in., weighing 2 lb each, are mounted at A and B on the horizontal rod $A'B'$, which rotates freely about the vertical with a counterclockwise angular velocity of 6 rad/s. The spheres are held in position by a cord which is suddenly cut. Knowing that the centroidal moment of inertia of the rod and pivot is $\bar{I}_R = 0.25 \text{ lb} \cdot \text{ft} \cdot \text{s}^2$, determine (*a*) the angular velocity of the rod after the spheres have moved to positions A' and B', (*b*) the energy lost due to the plastic impact of the spheres and the stops at A' and B'.

***a.* Principle of Impulse and Momentum.** In order to determine the final angular velocity of the rod, we shall express that the initial momenta of the various parts of the system and the impulses of the external forces are together equipollent to the final momenta of the system.

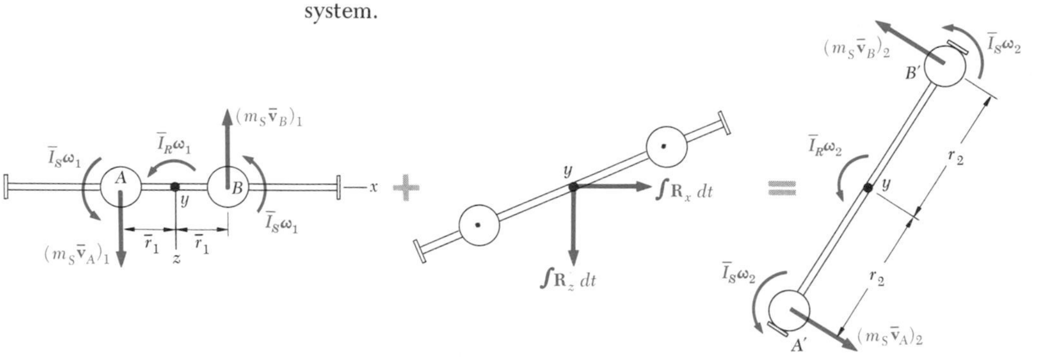

Observing that the external forces consist of the weights and the reaction at the pivot, which have no moment about the y axis, and noting that $\bar{v}_A = \bar{v}_B = \bar{r}\omega$, we write

$+\circlearrowleft$ moments about y axis:

$$2(m_S\bar{r}_1\omega_1)\bar{r}_1 + 2\bar{I}_S\omega_1 + \bar{I}_R\omega_1 = 2(m_S\bar{r}_2\omega_2)\bar{r}_2 + 2\bar{I}_S\omega_2 + \bar{I}_R\omega_2$$

$$(2m_S\bar{r}_1^2 + 2\bar{I}_S + \bar{I}_R)\omega_1 = (2m_S\bar{r}_2^2 + 2\bar{I}_S + \bar{I}_R)\omega_2 \qquad (1)$$

which expresses that *the angular momentum of the system about the y axis is conserved.* We now compute

$$\bar{I}_S = \tfrac{2}{5}m_S a^2 = \frac{2}{5}\left(\frac{2 \text{ lb}}{32.2 \text{ ft/s}^2}\right)(\tfrac{3}{12}\text{ ft})^2 = 0.00155 \text{ lb} \cdot \text{ft} \cdot \text{s}^2$$

$$m_S\bar{r}_1^2 = \frac{2}{32.2}\left(\frac{5}{12}\right)^2 = 0.0108 \qquad m_S\bar{r}_2^2 = \frac{2}{32.2}\left(\frac{25}{12}\right)^2 = 0.2696$$

Substituting these values and $\bar{I}_R = 0.25$, $\omega_1 = 6$ rad/s into (1):

$$0.275(6 \text{ rad/s}) = 0.792\omega_2 \qquad \omega_2 = 2.08 \text{ rad/s} \circlearrowleft \blacktriangleleft$$

***b.* Energy Lost.** The kinetic energy of the system at any instant is

$$T = 2(\tfrac{1}{2}m_S\bar{v}^2 + \tfrac{1}{2}\bar{I}_S\omega^2) + \tfrac{1}{2}\bar{I}_R\omega^2 = \tfrac{1}{2}(2m_S\bar{r}^2 + 2\bar{I}_S + \bar{I}_R)\omega^2$$

Recalling the numerical values found above, we have

$$T_1 = \tfrac{1}{2}(0.275)(6)^2 = 4.95 \text{ ft} \cdot \text{lb} \qquad T_2 = \tfrac{1}{2}(0.792)(2.08)^2 = 1.713 \text{ ft} \cdot \text{lb}$$

$$\Delta T = T_2 - T_1 = 1.71 - 4.95 \qquad \Delta T = -3.24 \text{ ft} \cdot \text{lb} \blacktriangleleft$$

PROBLEMS

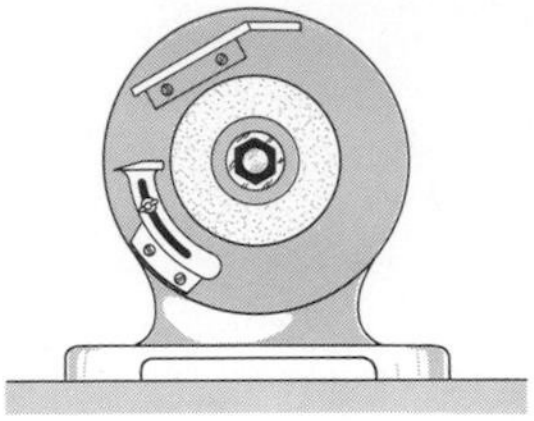

Fig. P17.45

17.45 A small grinding wheel is attached to the shaft of an electric motor which has a rated speed of 3600 rpm. When the power is turned off, the unit coasts to rest in 70 s. The grinding wheel and rotor have a combined weight of 6 lb and a combined radius of gyration of 2 in. Determine the average magnitude of the couple due to kinetic friction in the bearings of the motor.

17.46 A turbine-generator unit is shut off when its rotor is rotating at 3600 rpm; it is observed that the rotor coasts to rest in 7.10 min. Knowing that the 1850-kg rotor has a radius of gyration of 234 mm, determine the average magnitude of the couple due to bearing friction.

17.47 A bolt located 50 mm from the center of an automobile wheel is tightened by applying the couple shown for 0.1 s. Assuming that the wheel is free to rotate and is initially at rest, determine the resulting angular velocity of the wheel. The 20-kg wheel has a radius of gyration of 250 mm.

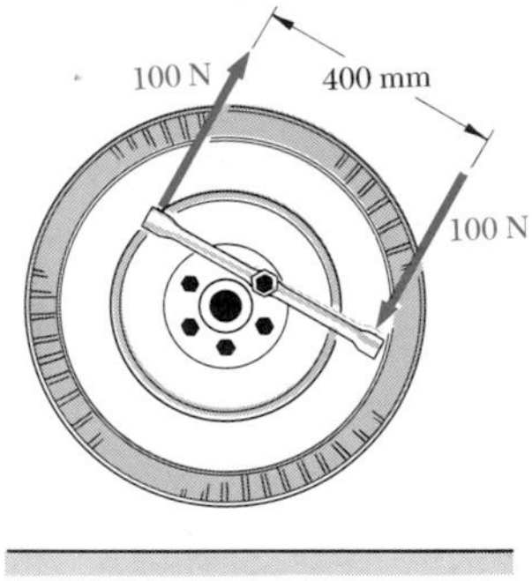

Fig. P17.47

17.48 Solve Prob. 17.3, assuming that the couple **M** is applied for a time t_0 and then removed.

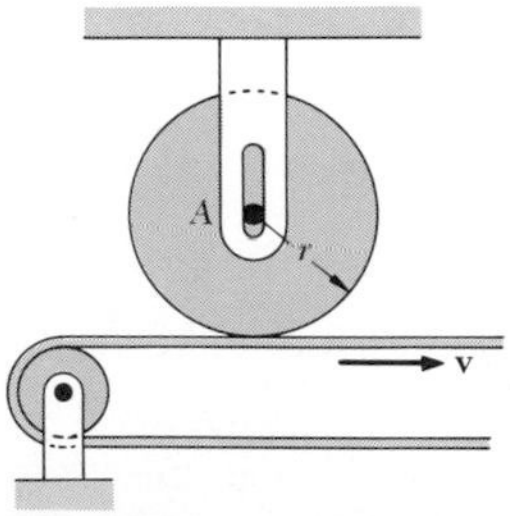

Fig. P17.49 and P17.50

17.49 A disk of constant thickness, initially at rest, is placed in contact with a belt which moves with a constant velocity **v**. Denoting by μ the coefficient of friction between the disk and the belt, derive an expression for the time required for the disk to reach a constant angular velocity.

17.50 Disk A, of weight 5 lb and radius $r = 3$ in., is at rest when it is placed in contact with the belt, which moves with a constant speed $v = 50$ ft/s. Knowing that $\mu = 0.20$ between the disk and the belt, determine the time required for the disk to reach a constant angular velocity.

17.51 Solve Prob. 12.17*b*, assuming that each pulley is of 8-in. radius and has a centroidal moment of inertia of 0.25 lb·ft·s^2.

17.52 Using the principle of impulse and momentum, solve Prob. 10.34*b*.

17.53 Disks *A* and *B* are of mass 5 and 1.8 kg, respectively. The disks are initially at rest and the coefficient of friction between them is 0.20. A couple **M** of magnitude 4 N · m is applied to disk *A* for 1.50 s and then removed. Determine (*a*) whether slipping occurs between the disks, (*b*) the final angular velocity of each disk.

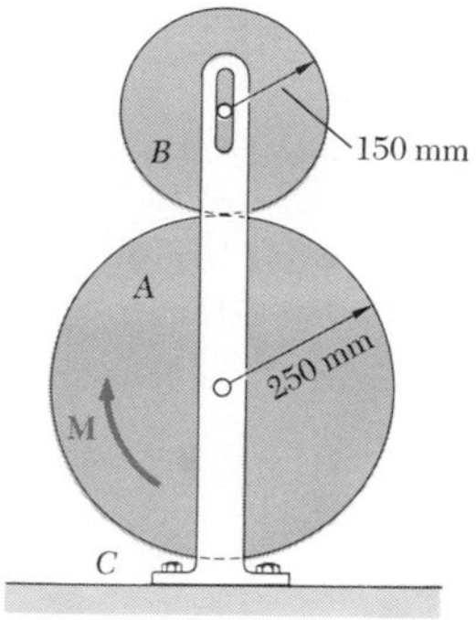

Fig. P17.53

17.54 In Prob. 17.53, determine (*a*) the largest couple **M** for which no slipping occurs, (*b*) the corresponding final angular velocity of each disk.

17.55 Two disks *A* and *B* are connected by a belt as shown. Each disk weighs 30 lb and has a radius of 1.5 ft. The shaft of disk *B* rests in a slotted bearing and is held by a spring which exerts a constant force of 15 lb. If a 20-lb·ft couple is applied to disk *A*, determine (*a*) the time required for the disks to attain a speed of 600 rpm, (*b*) the tension in both portions of the belt, (*c*) the minimum coefficient of friction if no slipping is to occur.

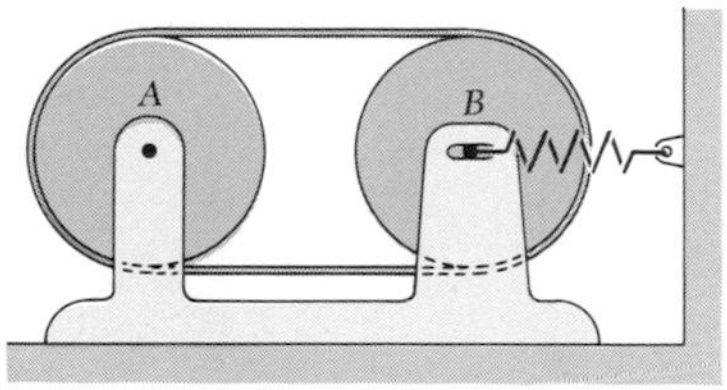

Fig. P17.55

17.56 Solve Prob. 17.55, assuming that disk *A* weighs 10 lb and disk *B* weighs 50 lb.

17.57 Show that the system of momenta for a rigid slab in plane motion reduces to a single vector, and express the distance from the mass center *G* to the line of action of this vector in terms of the centroidal radius of gyration $\bar{k}$ of the slab, the magnitude $\bar{v}$ of the velocity of *G*, and the angular velocity ω.

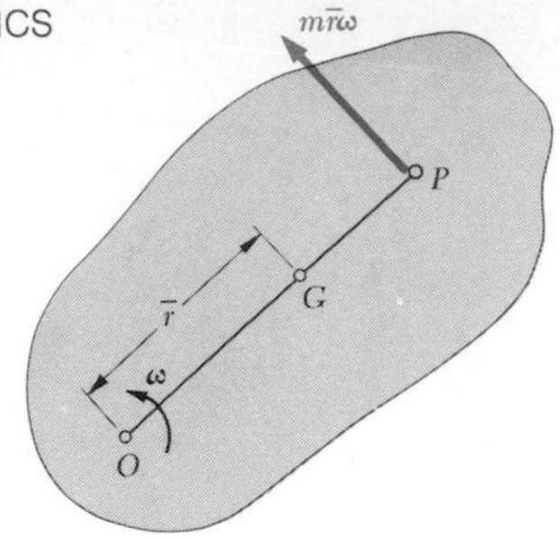

Fig. P17.58

17.58 Show that, when a rigid slab rotates about a fixed axis through O perpendicular to the slab, the system of momenta of its particles is equivalent to a single vector of magnitude $m\bar{r}\omega$, perpendicular to the line OG, and applied to a point P on this line, called the *center of percussion*, at a distance $GP = \bar{k}^2/\bar{r}$ from the mass center of the slab.

17.59 Show that the sum $\mathbf{H}_A$ of the moments about a point A of the momenta of the particles of a rigid slab in plane motion is equal to $I_A\boldsymbol{\omega}$, where $\boldsymbol{\omega}$ is the angular velocity of the slab at the instant considered and I_A the moment of inertia of the slab about A, *if and only if* one of the following conditions is satisfied: (*a*) A is the mass center of the slab, (*b*) A is the instantaneous center of rotation, (*c*) the velocity of A is directed along a line joining point A and the mass center G.

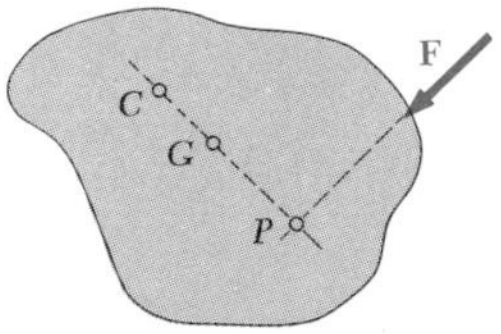

Fig. P17.60

17.60 Consider a rigid slab initially at rest and subjected to an impulsive force **F** contained in the plane of the slab. We define the *center of percussion* P as the point of intersection of the line of action of **F** with the perpendicular drawn from G. (*a*) Show that the instantaneous center of rotation C of the slab is located on line GP at a distance $GC = \bar{k}^2/GP$ on the opposite side of G. (*b*) Show that, if the center of percussion were located at C, the instantaneous center of rotation would be located at P.

17.61 A cord is wrapped around a solid cylinder of radius r and mass m as shown. If the cylinder is released from rest at time $t = 0$, determine the velocity of the center of the cylinder at a time t.

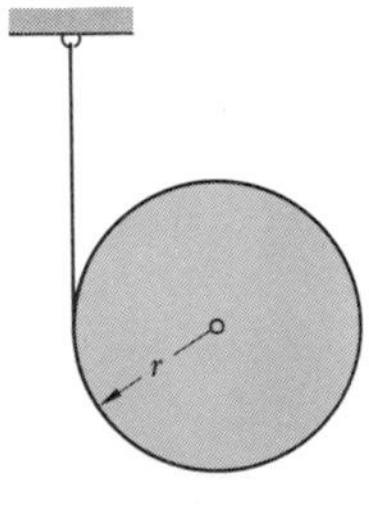

Fig. P17.61

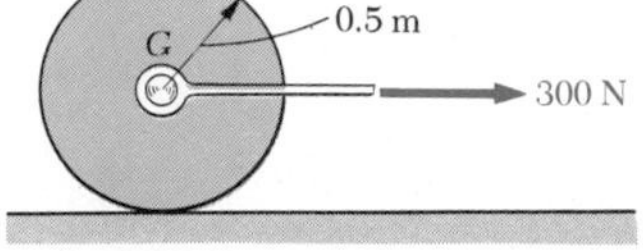

Fig. P17.62

17.62 A 100-kg cylindrical roller is initially at rest and is acted upon by a 300-N force as shown. Assuming that the body rolls without slipping, determine (*a*) the velocity of the center G after 6 s, (*b*) the friction force required to prevent slipping.

17.63 A section of thin-walled pipe of radius r is released from rest at time $t = 0$. Assuming that the pipe rolls without slipping, determine (*a*) the velocity of the center at time t, (*b*) the coefficient of friction required to prevent slipping.

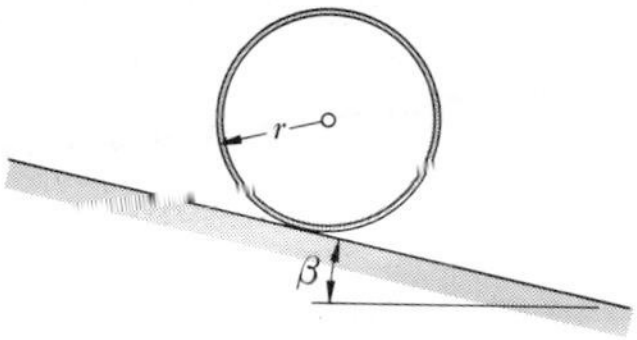

Fig. P17.63

17.64 Two disks, each of weight 12 lb and radius 6 in., which roll without slipping, are connected by a drum of radius r and of negligible weight. A rope is wrapped around the drum and is pulled horizontally with a force **P** of magnitude 8 lb. Knowing that $r = 3$ in. and that the disks are initially at rest, determine (*a*) the velocity of the center G after 3 s, (*b*) the friction force required to prevent slipping.

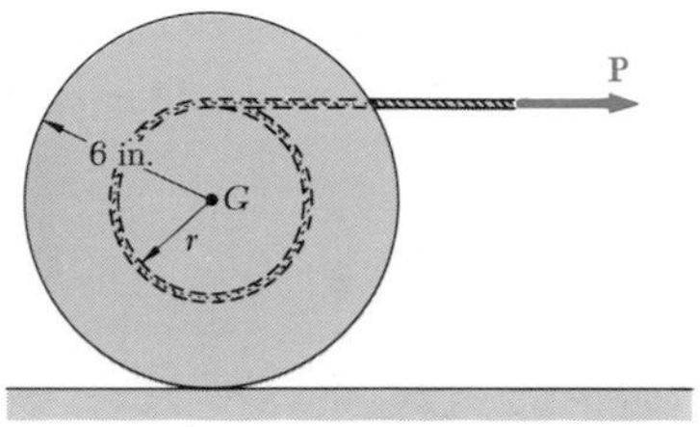

Fig. P17.64

17.65 In Prob. 17.64, determine the required value of r and the corresponding velocity after 3 s if the friction force is to be zero.

17.66 and 17.67 The 12-lb carriage is supported as shown by two uniform disks, each of weight 8 lb and radius 3 in. Knowing that the carriage is initially at rest, determine the velocity of the carriage 3 s after the 4-lb force is applied. Assume that the disks roll without sliding.

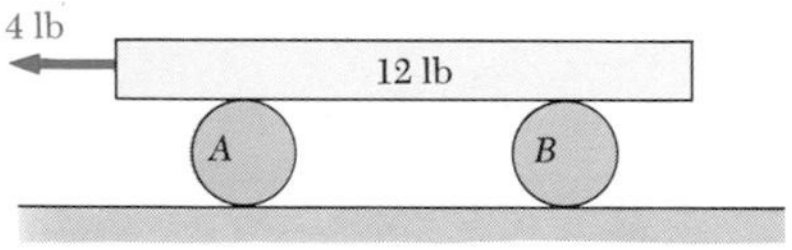

Fig. P17.66

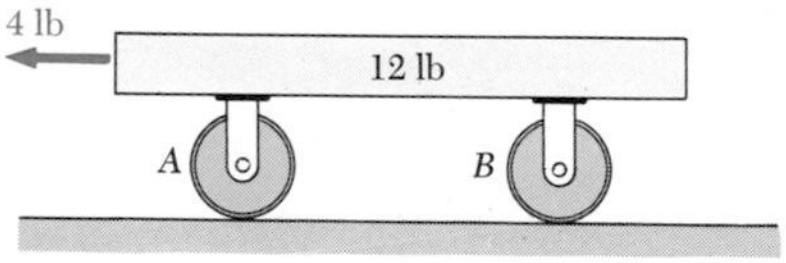

Fig. P17.67

17.68 A sphere of mass m and radius r is projected along a rough horizontal surface with a linear velocity $\bar{\mathbf{v}}_0$ but with no angular velocity ($\boldsymbol{\omega}_0 = 0$). Determine (*a*) the final velocity of the sphere, (*b*) the time at which the velocity of the sphere becomes constant in terms of $\bar{v}_0$ and μ.

17.69 A sphere of mass m and radius r is projected along a rough horizontal surface with the initial velocities indicated. If the final velocity of the sphere is to be zero, express (*a*) the required ω_0 in terms of $\bar{v}_0$ and r, (*b*) the time required for the sphere to come to rest in terms of $\bar{v}_0$ and μ.

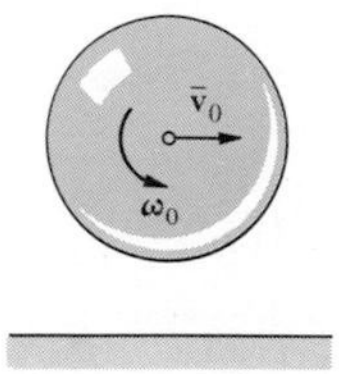

Fig. P17.68 and P17.69

17.70 Solve Sample Prob. 17.7, assuming that the hoop is replaced by a uniform sphere of radius r and mass m.

17.71 Solve Sample Prob. 17.8, assuming that, after the cord is cut, sphere B moves to position B' but that an obstruction prevents sphere A from moving.

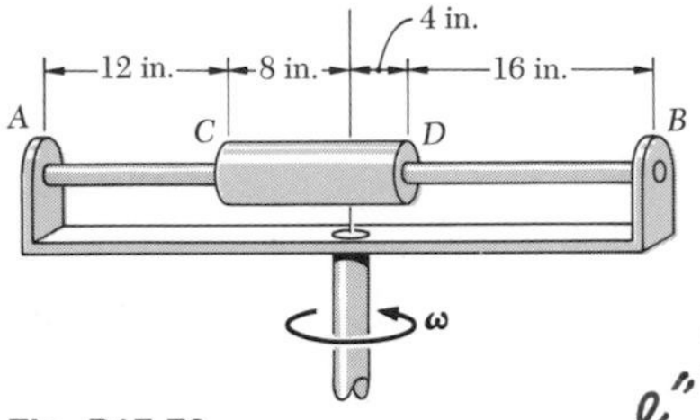

Fig. P17.72

17.72 An 8-lb tube CD may slide freely on rod AB, which in turn may rotate freely in a horizontal plane. At the instant shown, the assembly is rotating with an angular velocity of magnitude $\omega = 8$ rad/s and the tube is moving toward A with a speed of 5 ft/s relative to the rod. Knowing that the centroidal moment of inertia about a vertical axis is 0.022 lb·ft·s^2 for the tube and 0.400 lb·ft·s^2 for the rod and bracket, determine (*a*) the angular velocity of the assembly after the tube has moved to end A, (*b*) the energy lost due to the plastic impact at A.

17.73 Four rectangular panels, each of length b and height $\frac{1}{2}b$, are attached with hinges to a circular plate of diameter $\sqrt{2}b$ and held by a wire loop in the position shown. The plate and the panels are made of the same material and have the same thickness. The entire assembly is rotating with an angular velocity $\boldsymbol{\omega}_0$ when the wire breaks. Determine the angular velocity of the assembly after the panels have come to rest in a horizontal position.

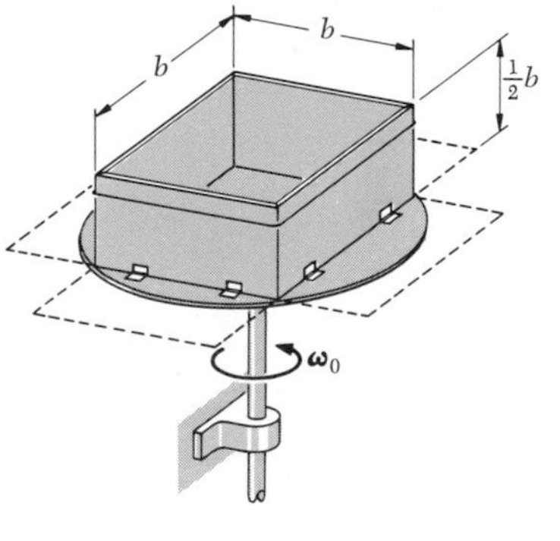

Fig. P17.73

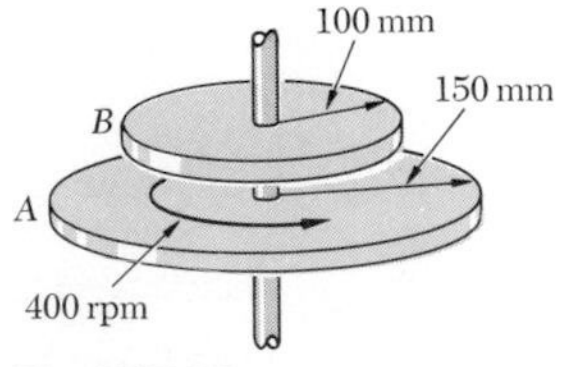

Fig. P17.74

17.74 Disks A and B are made of the same material and are of the same thickness; they may rotate freely about the vertical shaft. Disk B is at rest when it is dropped onto disk A which is rotating with an angular velocity of 400 rpm. Knowing that the mass of disk A is 4 kg, determine (*a*) the final angular velocity of the disks, (*b*) the change in kinetic energy of the system.

17.75 In Prob. 17.74, show that if both disks are initially rotating, the change in kinetic energy ΔT of the system depends only upon the initial relative velocity $\omega_{B/A}$ of the disks, and derive an expression for ΔT in terms of $\omega_{B/A}$.

17.76 A small 250-g ball may slide in a slender tube of length 1 m and of mass 1 kg which rotates freely about a vertical axis passing through its center C. If the angular velocity of the tube is 10 rad/s as the ball passes through C, determine the angular velocity of the tube (a) just before the ball leaves the tube, (b) just after the ball has left the tube.

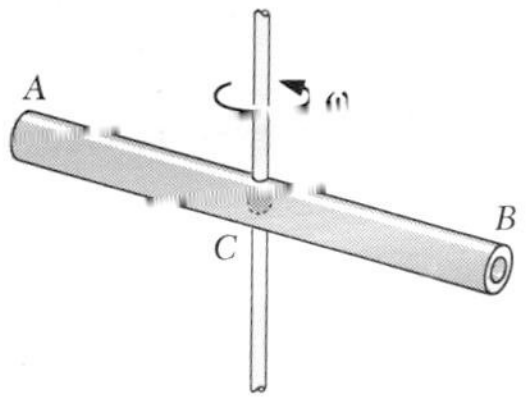

Fig. P17.76

17.77 The rod AB is of mass m and slides freely inside the tube CD which is also of mass m. The angular velocity of the assembly was ω_1 when the rod was entirely inside the tube ($x = 0$). Neglecting the effect of friction, determine the angular velocity of the assembly when $x = \frac{1}{2}L$.

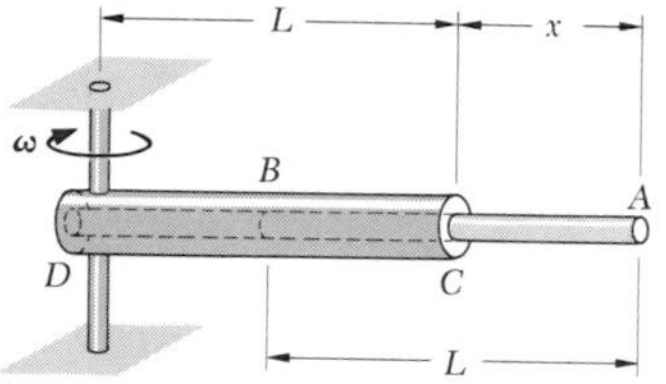

Fig. P17.77

17.78 In the helicopter shown, a vertical tail propeller is used to prevent rotation of the cab as the speed of the main blades is changed. Assuming that the tail propeller is not operating, determine the final angular velocity of the cab after the speed of the main blades has been changed from 180 to 240 rpm. The speed of the main blades is measured relative to the cab, which has a centroidal moment of inertia of 650 lb·ft·s^2. Each of the four main blades is assumed to be a 14-ft slender rod weighing 55 lb.

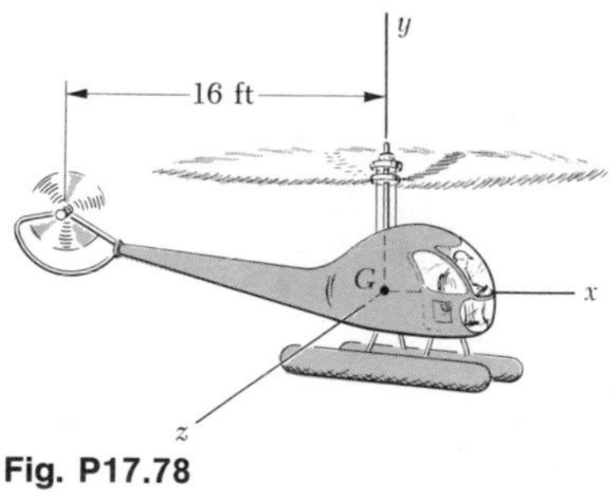

Fig. P17.78

17.79 Assuming that the tail propeller in Prob. 17.78 is operating and that the angular velocity of the cab remains zero, determine the final horizontal velocity of the cab when the speed of the main blades is changed from 180 to 240 rpm. The cab weighs 1250 lb and is initially at rest. Also determine the force exerted by the tail propeller if this change in speed takes places uniformly in 12 s.

17.80 The 5-kg disk is attached to the arm AB which is free to rotate about the vertical axle CD. The arm and motor unit has a moment of inertia of $0.03\ \text{kg}\cdot\text{m}^2$ with respect to the axle CD, and the normal operating speed of the motor is 360 rpm. Knowing that the system is initially at rest, determine the angular velocities of the arm and of the disk when the motor reaches a speed of 360 rpm.

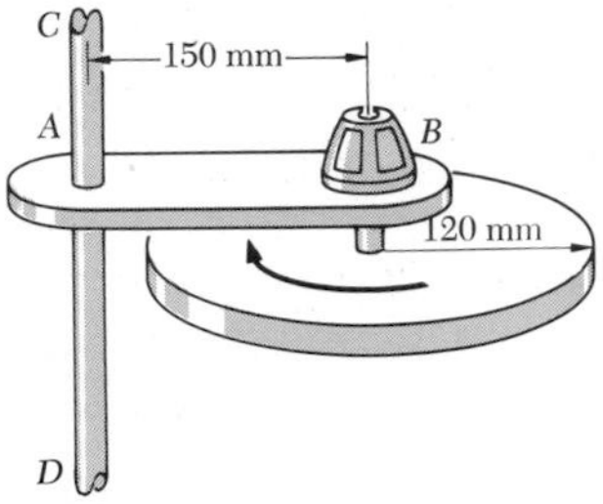

Fig. P17.80

17.81 In Prob. 17.77, determine the velocity of the rod relative to the tube when $x = \frac{1}{2}L$.

17.82 Knowing that in Prob. 17.76 the speed of the ball is 1.2 m/s as it passes through C, determine the radial and transverse components of the velocity of the ball as it leaves the tube at B.

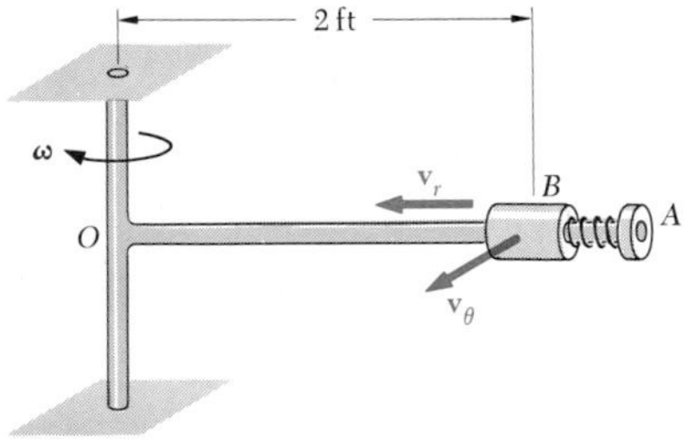

Fig. P17.83

17.83 Collar B weighs 3 lb and may slide freely on rod OA which in turn may rotate freely in the horizontal plane. The assembly is rotating with an angular velocity $\omega = 1.5$ rad/s when a spring located between A and B is released, projecting the collar along the rod with an initial relative speed $v_r = 5$ ft/s. Knowing that the moment of inertia about O of the rod and spring is $0.15\ \text{lb}\cdot\text{ft}\cdot\text{s}^2$, determine (*a*) the minimum distance between the collar and point O in the ensuing motion, (*b*) the corresponding angular velocity of the assembly.

17.84 In Prob. 17.83, determine the required magnitude of the initial relative velocity $\mathbf{v}_r$ if during the ensuing motion the minimum distance between collar B and point O is to be 1 ft.

17.85 Solve Prob. 17.83, assuming that the initial relative speed of the collar is $v_r = 10$ ft/s.

17.10. Impulsive Motion. We saw in Chap. 13 that the method of impulse and momentum is the only practicable method for the solution of problems involving impulsive motion. Now we shall also find that, compared with the various problems considered in the preceding sections, problems involving impulsive motion are particularly well adapted to a solution by the method of impulse and momentum. The computation of linear impulses and angular impulses is quite simple, since, the time interval considered being very short, the bodies involved may be assumed to occupy the same position during that time interval.

17.11. Eccentric Impact. In Secs. 13.13 and 13.14, we learned to solve problems of *central impact*, i.e., problems in which the mass centers of the two colliding bodies are located on the line of impact. We shall now analyze the *eccentric impact* of two rigid bodies. Consider two bodies which collide, and denote by $\mathbf{v}_A$ and $\mathbf{v}_B$ the velocities before impact *of the two points of contact* A and B (Fig. 17.10*a*). Under the impact, the two bodies will *deform* and, at the end of the period of deformation, the velocities $\mathbf{u}_A$ and $\mathbf{u}_B$ of A and B will have equal components along the line of impact nn (Fig. 17.10*b*). A period of *restitution* will then take place, at the end of which A and B will have velocities $\mathbf{v}'_A$ and $\mathbf{v}'_B$ (Fig. 17.10*c*). Assuming the bodies frictionless, we find that the forces they exert on each other are directed along the line of impact. Denoting, respectively, by $\int P\,dt$ and $\int R\,dt$ the magnitude of the impulse of one of these forces during the period of deformation and during the period of restitution, we recall that the coefficient of restitution e is defined as the ratio

$$e = \frac{\int R\,dt}{\int P\,dt} \qquad (17.18)$$

We propose to show that the relation established in Sec. 13.13 between the relative velocities of two particles before and after impact also holds between the components along the line of impact of the relative velocities of the two points of contact A and B. We propose to show, therefore, that

$$(v'_B)_n - (v'_A)_n = e[(v_A)_n - (v_B)_n] \qquad (17.19)$$

We shall first assume that the motion of each of the two colliding bodies of Fig. 17.10 is unconstrained. Thus the only impulsive forces exerted on the bodies during the impact are applied at A and B respectively. Consider the body to which point A belongs and draw the three momentum and impulse

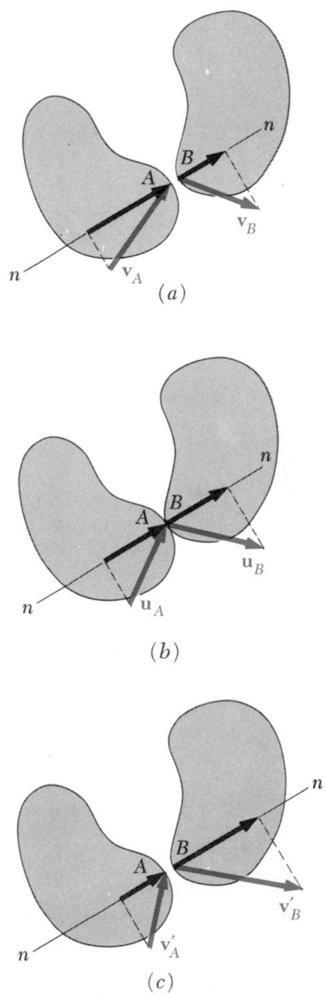

Fig. 17.10

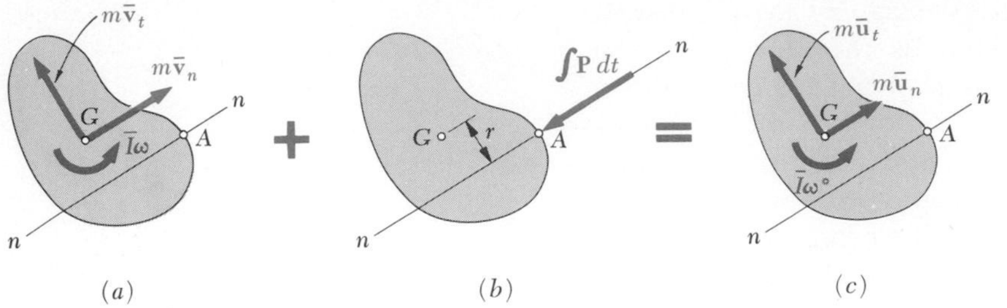

Fig. 17.11

diagrams corresponding to the period of deformation (Fig. 17.11). We denote by $\bar{\mathbf{v}}$ and $\bar{\mathbf{u}}$, respectively, the velocity of the mass center at the beginning and at the end of the period of deformation, and by $\boldsymbol{\omega}$ and $\boldsymbol{\omega}^*$ the angular velocity of the body at the same instants. Summing and equating the components of the momenta and impulses along the line of impact nn, we write

$$m\bar{v}_n - \int P\, dt = m\bar{u}_n \qquad (17.20)$$

Summing and equating the moments about G of the momenta and impulses, we also write

$$\bar{I}\omega - r\int P\, dt = \bar{I}\omega^* \qquad (17.21)$$

where r represents the perpendicular distance from G to the line of impact. Considering now the period of restitution, we obtain in a similar way

$$m\bar{u}_n - \int R\, dt = m\bar{v}'_n \qquad (17.22)$$

$$\bar{I}\omega^* - r\int R\, dt = \bar{I}\omega' \qquad (17.23)$$

where $\bar{\mathbf{v}}'$ and $\boldsymbol{\omega}'$ represent, respectively, the velocity of the mass center and the angular velocity of the body after impact. Solving (17.20) and (17.22) for the two impulses and substituting into (17.18), and then solving (17.21) and (17.23) for the same two impulses and substituting again into (17.18), we obtain the following two alternate expressions for the coefficient of restitution:

$$e = \frac{\bar{u}_n - \bar{v}'_n}{\bar{v}_n - \bar{u}_n} \qquad e = \frac{\omega^* - \omega'}{\omega - \omega^*} \qquad (17.24)$$

Multiplying by r the numerator and denominator of the second expression obtained for e, and adding respectively to the numerator and denominator of the first expression, we have

$$e = \frac{\bar{u}_n + r\omega^* - (\bar{v}'_n + r\omega')}{\bar{v}_n + r\omega - (\bar{u}_n + r\omega^*)} \qquad (17.25)$$

Observing that $\bar{v}_n + r\omega$ represents the component $(v_A)_n$ along nn of the velocity of the point of contact A and that, similarly,

$\bar{u}_n + r\omega^*$ and $\bar{v}'_n + r\omega'$ represent, respectively, the components $(u_A)_n$ and $(v'_A)_n$, we write

$$e = \frac{(u_A)_n - (v'_A)_n}{(v_A)_n - (u_A)_n} \tag{17.26}$$

The analysis of the motion of the second body leads to a similar expression for e in terms of the components along nn of the successive velocities of point B. Recalling that $(u_A)_n = (u_B)_n$, and eliminating these two velocity components by a manipulation similar to the one used in Sec. 13.13, we obtain relation (17.19).

If one or both of the colliding bodies is constrained to rotate about a fixed point O, as in the case of a compound pendulum (Fig. 17.12*a*), an impulsive reaction will be exerted at O (Fig. 17.12*b*). We shall verify that, while their derivation must be

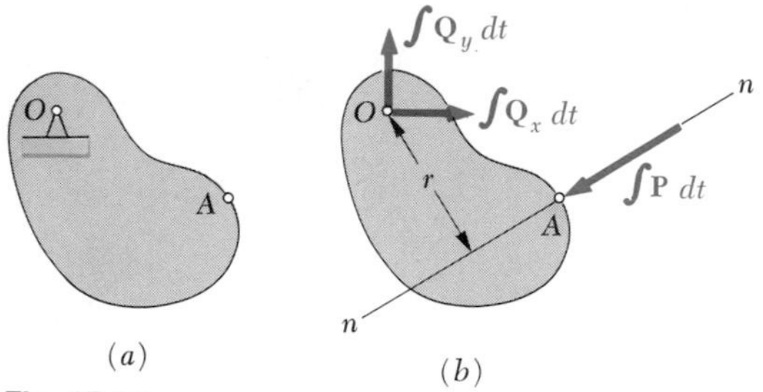

Fig. 17.12

modified, Eqs. (17.26) and (17.19) remain valid. Applying formula (17.16) to the period of deformation and to the period of restitution, we write

$$I_O\omega - r\int P\,dt = I_O\omega^* \tag{17.27}$$
$$I_O\omega^* - r\int R\,dt = I_O\omega' \tag{17.28}$$

where r represents the perpendicular distance from the fixed point O to the line of impact. Solving (17.27) and (17.28) for the two impulses and substituting into (17.18), and then observing that $r\omega$, $r\omega^*$, and $r\omega'$ represent the components along nn of the successive velocities of point A, we write

$$e = \frac{\omega^* - \omega'}{\omega - \omega^*} = \frac{r\omega^* - r\omega'}{r\omega - r\omega^*} = \frac{(u_A)_n - (v'_A)_n}{(v_A)_n - (u_A)_n}$$

and check that Eq. (17.26) still holds. Thus Eq. (17.19) remains valid when one or both of the colliding bodies is constrained to rotate about a fixed point O.

In order to determine the velocities of the two colliding bodies after impact, relation (17.19) should be used in conjunction with one or several other equations obtained by applying the principle of impulse and momentum (Sample Prob. 17.10).

SAMPLE PROBLEM 17.9

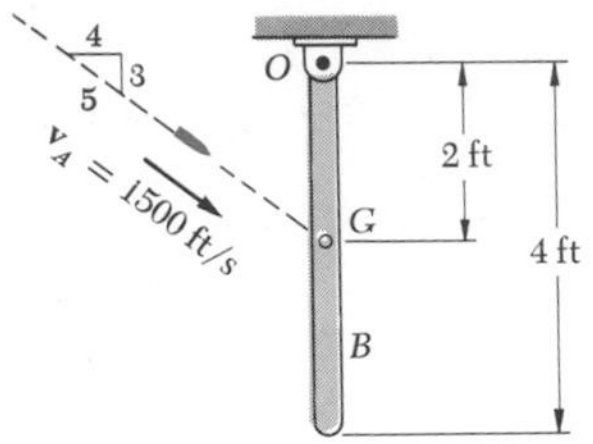

A 0.125-lb bullet A is fired with an initial velocity of 1500 ft/s into a 50-lb wooden beam B which is suspended from a hinge at O. Knowing that the beam is initially at rest, determine (*a*) the angular velocity of the beam immediately after the bullet becomes embedded in the beam, (*b*) the impulsive reaction at the hinge, assuming that the bullet becomes embedded in 0.0002 s.

Solution. ***Principle of Impulse and Momentum.*** We consider the bullet and the beam as a single system and express that the initial momenta of the bullet and beam and the impulses of the external forces are together equipollent to the final momenta of the system. Since the time interval $\Delta t = 0.0002$ s is very short, we neglect all nonimpulsive forces and consider only the external impulses $\mathbf{R}_x \Delta t$ and $\mathbf{R}_y \Delta t$.

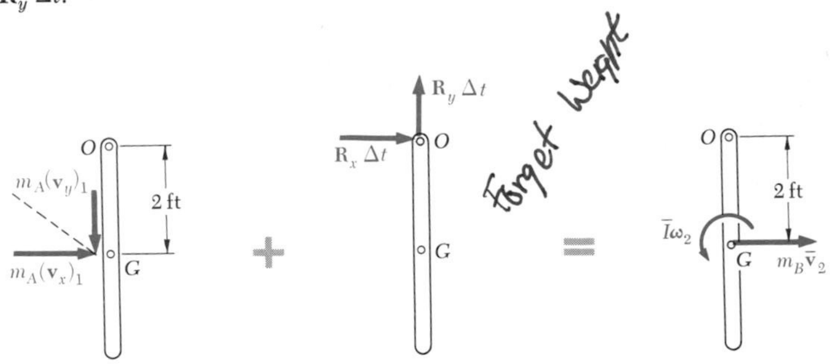

$+\circlearrowleft$ moments about O:	$m_A(v_x)_1(2\text{ ft}) + 0 = \bar{I}\omega_2 + m_B\bar{v}_2(2\text{ ft})$	(1)
$\overset{+}{\rightarrow}$ x components:	$m_A(v_x)_1 + R_x \Delta t = m_B\bar{v}_2$	(2)
$+\uparrow$ y components:	$-m_A(v_y)_1 + R_y \Delta t = 0$	(3)

The components of the velocity of the bullet and the centroidal moment of inertia of the beam are

$$(v_x)_1 = \tfrac{4}{5}(1500\text{ ft/s}) = 1200\text{ ft/s} \qquad (v_y)_1 = \tfrac{3}{5}(1500\text{ ft/s}) = 900\text{ ft/s}$$

$$\bar{I} = \tfrac{1}{12}ml^2 = \frac{1}{12}\frac{50\text{ lb}}{32.2\text{ ft/s}^2}(4\text{ ft})^2 = 2.07\text{ lb}\cdot\text{ft}\cdot\text{s}^2$$

Substituting these values into (1) and noting that $\bar{v}_2 = (2\text{ ft})\omega_2$:

$$(0.125/32.2)(1200)(2) = 2.07\omega_2 + (50/32.2)(2\omega_2)(2)$$

$$\omega_2 = 1.125\text{ rad/s} \qquad \omega_2 = 1.125\text{ rad/s} \circlearrowleft \blacktriangleleft$$

Substituting $\bar{v}_2 = (2\text{ ft})(1.125\text{ rad/s}) = 2.25\text{ ft/s}$ into (2), we solve Eqs. (2) and (3) for R_x and R_y, respectively.

$$(0.125/32.2)(1200) + R_x(0.0002) = (50/32.2)(2.25)$$

$$R_x = -5820\text{ lb} \qquad \mathbf{R}_x = 5820\text{ lb} \leftarrow \blacktriangleleft$$

$$-(0.125/32.2)(900) + R_y(0.0002) = 0$$

$$R_y = +17{,}470\text{ lb} \qquad \mathbf{R}_y = 17{,}470\text{ lb} \uparrow \blacktriangleleft$$

SAMPLE PROBLEM 17.10

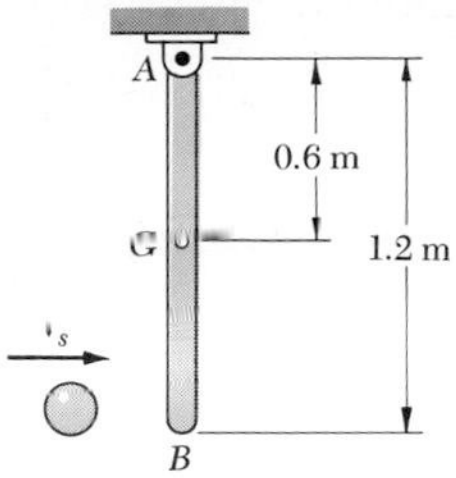

A 2-kg sphere moving horizontally to the right with an initial velocity of 5 m/s strikes the lower end of an 8-kg rigid rod AB. The rod is suspended from a hinge at A and is initially at rest. Knowing that the coefficient of restitution between the rod and sphere is 0.80, determine the angular velocity of the rod and the velocity of the sphere immediately after the impact.

Principle of Impulse and Momentum. We consider the rod and sphere as a single system and express that the initial momenta of the rod and sphere and the impulses of the external forces are together equipollent to the final momenta of the system. We note that the only impulsive force external to the system is the impulsive reaction at A.

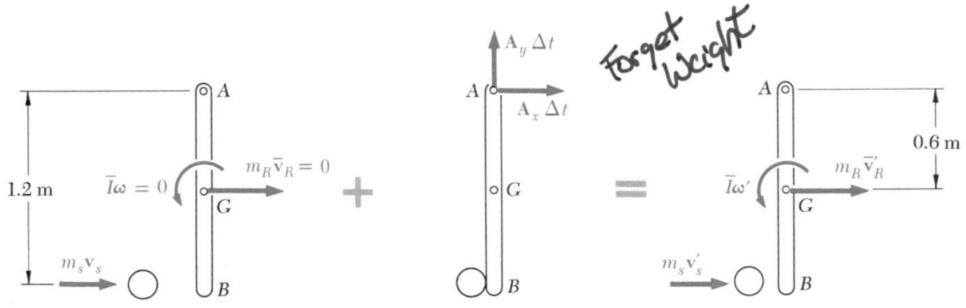

$+\circlearrowleft$ moments about A:

$$m_s v_s(1.2\text{ m}) = m_s v'_s(1.2\text{ m}) + m_R\bar{v}'_R(0.6\text{ m}) + \bar{I}\omega' \quad (1)$$

Since the rod rotates about A, we have $\bar{v}'_R = \bar{r}\omega' = (0.6\text{ m})\omega'$. Also,

$$\bar{I} = \tfrac{1}{12}mL^2 = \tfrac{1}{12}(8\text{ kg})(1.2\text{ m})^2 = 0.96\text{ kg}\cdot\text{m}^2$$

Substituting these values and the given data into Eq. (1), we have

$$(2\text{ kg})(5\text{ m/s})(1.2\text{ m})$$
$$= (2\text{ kg})v'_s(1.2\text{ m}) + (8\text{ kg})(0.6\text{ m})\omega'(0.6\text{ m}) + (0.96\text{ kg}\cdot\text{m}^2)\omega'$$
$$12 = 2.4v'_s + 3.84\omega' \quad (2)$$

Relative Velocities. Choosing positive to the right, we write

$$v'_B - v'_s = e(v_s - v_B)$$

Substituting $v_s = 5$ m/s, $v_B = 0$, and $e = 0.80$, we obtain

$$v'_B - v'_s = 0.80(5\text{ m/s}) \quad (3)$$

Again noting that the rod rotates about A, we write

$$v'_B = (1.2\text{ m})\omega' \quad (4)$$

Solving Eqs. (2) to (4) simultaneously, we obtain

$\omega' = +3.21$ rad/s $\qquad \omega' = 3.21$ rad/s $\circlearrowleft$ ◀

$v'_s = -0.143$ m/s $\qquad \mathbf{v}'_s = 0.143$ m/s $\leftarrow$ ◀

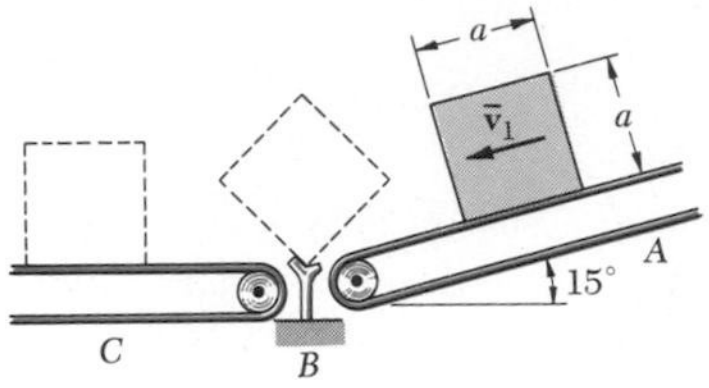

SAMPLE PROBLEM 17.11

A square package of side a and mass m moves down a conveyor belt A with a constant velocity $\bar{\mathbf{v}}_1$. At the end of the conveyor belt, the corner of the package strikes a rigid support at B. Assuming that the impact at B is perfectly plastic, derive an expression for the smallest magnitude of the velocity $\bar{\mathbf{v}}_1$ for which the package will rotate about B and reach conveyor belt C.

Principle of Impulse and Momentum. Since the impact between the package and the support is perfectly plastic, the package rotates about B during the impact. We apply the principle of impulse and momentum to the package and note that the only impulsive force external to the package is the impulsive reaction at B.

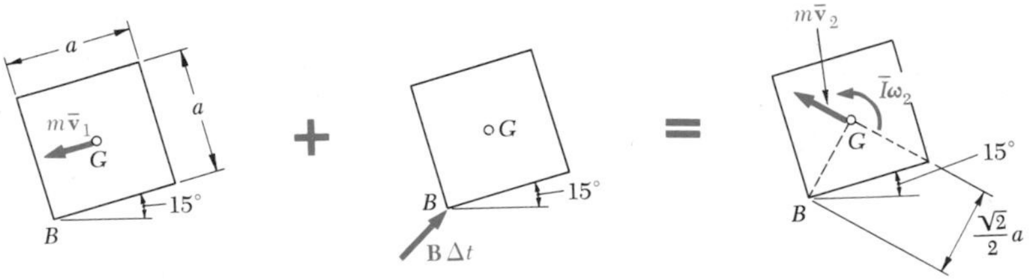

Syst Momenta$_1$ + Syst Ext Imp$_{1\to2}$ = Syst Momenta$_2$

+↺ moments about B: $\quad (m\bar{v}_1)(\tfrac{1}{2}a) + 0 = (m\bar{v}_2)(\tfrac{1}{2}\sqrt{2}a) + \bar{I}\omega_2 \qquad (1)$

Since the package rotates about B, we have $\bar{v}_2 = (GB)\omega_2 = \tfrac{1}{2}\sqrt{2}a\omega_2$. We substitute this expression, together with $\bar{I} = \tfrac{1}{6}ma^2$, into Eq. (1):

$$(m\bar{v}_1)(\tfrac{1}{2}a) = m(\tfrac{1}{2}\sqrt{2}a\omega_2)(\tfrac{1}{2}\sqrt{2}a) + \tfrac{1}{6}ma^2\omega_2 \qquad \bar{v}_1 = \tfrac{4}{3}a\omega_2 \quad (2)$$

Position 2

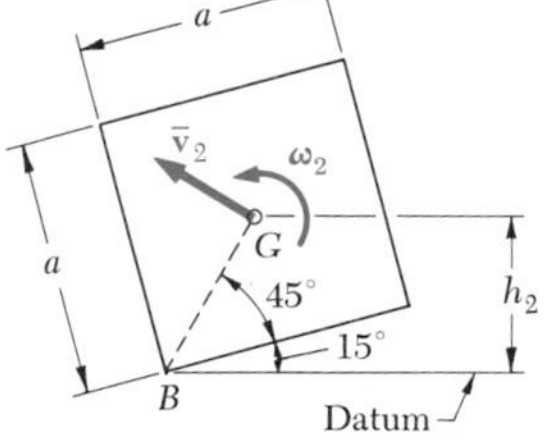

$GB = \tfrac{1}{2}\sqrt{2}a = 0.707a$
$h_2 = GB\sin(45° + 15°)$
$= 0.612a$

Principle of Conservation of Energy. We apply the principle of conservation of energy between position 2 and position 3.

Position 2. $V_2 = Wh_2$. Recalling that $\bar{v}_2 = \tfrac{1}{2}\sqrt{2}a\omega_2$, we write

$$T_2 = \tfrac{1}{2}m\bar{v}_2^2 + \tfrac{1}{2}\bar{I}\omega_2^2 = \tfrac{1}{2}m(\tfrac{1}{2}\sqrt{2}a\omega_2)^2 + \tfrac{1}{2}(\tfrac{1}{6}ma^2)\omega_2^2 = \tfrac{1}{3}ma^2\omega_2^2$$

Position 3. Since the package must reach conveyor belt B, it must pass through position 3 where G is directly above B. Also, since we wish to determine the smallest velocity for which the package will reach this position, we choose $\bar{v}_3 = \omega_3 = 0$. Therefore $T_3 = 0$ and $V_3 = Wh_3$.

Position 3

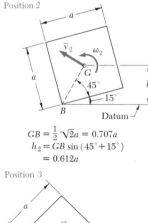

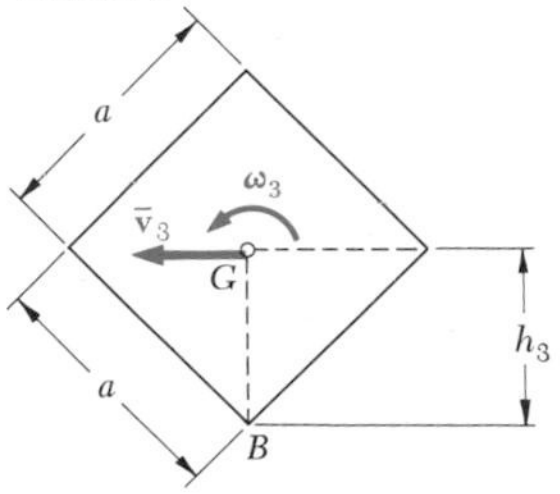

$h_3 = GB = 0.707a$

Conservation of Energy

$$T_2 + V_2 = T_3 + V_3$$
$$\tfrac{1}{3}ma^2\omega_2^2 + Wh_2 = 0 + Wh_3$$
$$\omega_2^2 = \frac{3W}{ma^2}(h_3 - h_2) = \frac{3g}{a^2}(h_3 - h_2) \qquad (3)$$

Substituting the computed values of h_2 and h_3 into Eq. (3), we obtain

$$\omega_2^2 = \frac{3g}{a^2}(0.707a - 0.612a) = \frac{3g}{a^2}(0.095a) \qquad \omega_2 = \sqrt{0.285g/a}$$

$$\bar{v}_1 = \tfrac{4}{3}a\omega_2 = \tfrac{4}{3}a\sqrt{0.285g/a} \qquad v_1 = 0.712\sqrt{ga} \quad ◄$$

PROBLEMS

17.86 A 45-g bullet is fired with a horizontal velocity of 400 m/s into a 9-kg square panel of side $b = 200$ mm. Knowing that $h = 200$ mm and that the panel is initially at rest, determine (*a*) the velocity of the center of the panel immediately after the bullet becomes embedded, (*b*) the impulsive reaction at *A*, assuming that the bullet becomes embedded in 1 ms.

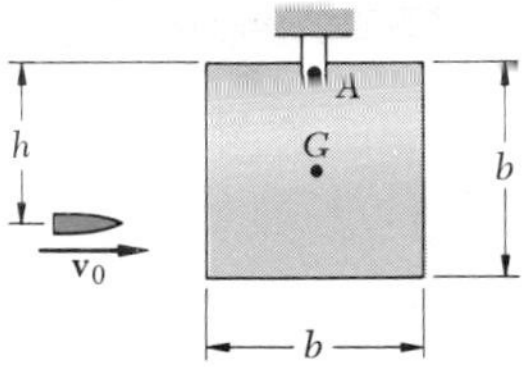

Fig. P17.86

17.87 In Prob. 17.86, determine (*a*) the required distance *h* if the impulsive reaction at *A* is to be zero, (*b*) the corresponding velocity of the center of the panel after the bullet becomes embedded.

17.88 A bullet weighing 0.08 lb is fired with a horizontal velocity of 1800 ft/s into the 15-lb wooden rod *AB* of length $L = 30$ in. The rod, which is initially at rest, is suspended by a cord of length $L = 30$ in. Knowing that $h = 6$ in., determine the velocity of each end of the rod immediately after the bullet becomes embedded.

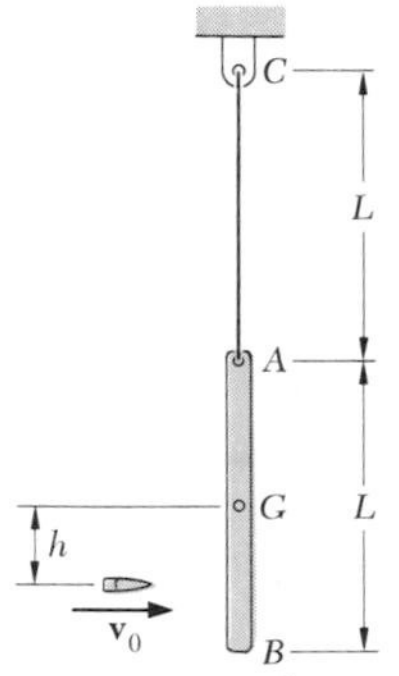

Fig. P17.88

17.89 In Prob. 17.88, determine the distance *h* for which, immediately after the bullet becomes embedded, the instantaneous center of rotation of the rod is point *C*.

17.90 A bullet of mass *m* is fired with a horizontal velocity $\mathbf{v}_0$ and at a height $h = \frac{1}{2}R$ into a wooden disk of much larger mass *M* and radius *R*. The disk rests on a horizontal plane and the coefficient of friction between the disk and the plane is finite. (*a*) Determine the linear velocity $\overline{\mathbf{v}}_1$ and the angular velocity $\boldsymbol{\omega}_1$ of the disk immediately after the bullet has penetrated the disk. (*b*) Describe the ensuing motion of the disk and determine its linear velocity after the motion has become uniform.

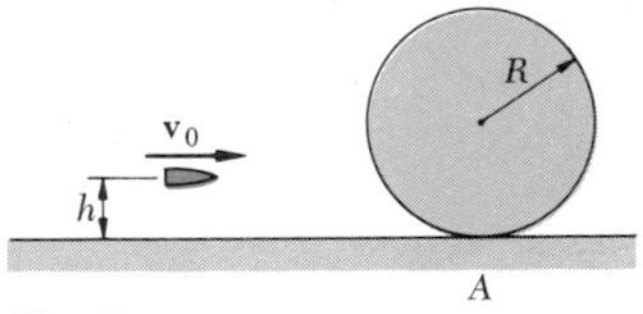

Fig. P17.90

17.91 Determine the height *h* at which the bullet of Prob. 17.90 should be fired (*a*) if the disk is to roll without sliding immediately after impact, (*b*) if the disk is to slide without rolling immediately after impact.

17.92 A uniform slender rod *AB* is equipped at both ends with the hooks shown and is supported by a frictionless horizontal table. Initially the rod is hooked at *A* to a fixed pin *C* about which it rotates with the constant angular velocity $\boldsymbol{\omega}_1$. Suddenly end *B* of the rod hits and gets hooked to the pin *D*, causing end *A* to be released. Determine the magnitude of the angular velocity $\boldsymbol{\omega}_2$ of the rod in its subsequent rotation about *D*.

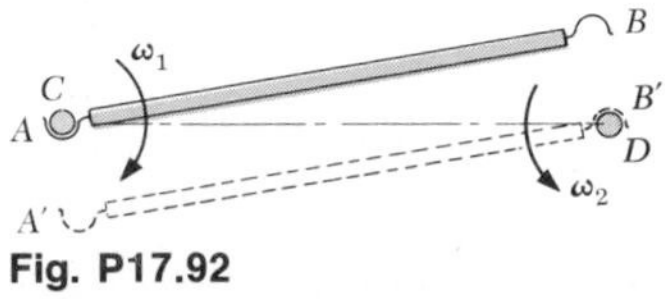

Fig. P17.92

17.93 A uniform disk of radius r and mass m is supported by a frictionless horizontal table. Initially the disk is spinning freely about its mass center G with a constant angular velocity ω_1. Suddenly a latch B is moved to the right and is struck by a small stop A welded to the edge of the disk. Assuming that the impact of A and B is perfectly plastic, determine the angular velocity of the disk and the velocity of its mass center immediately after impact.

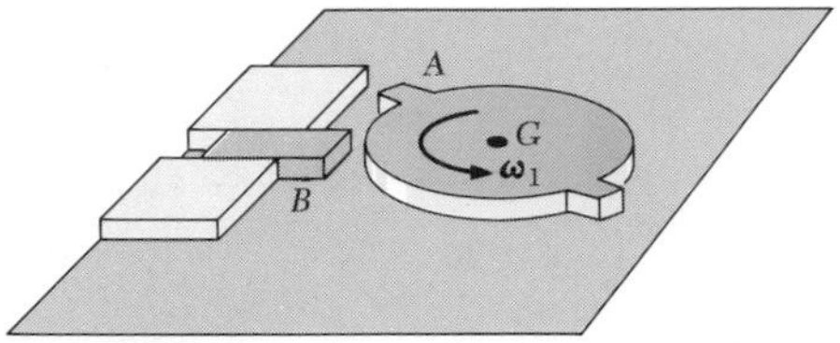

Fig. P17.93

17.94 Solve Prob. 17.93, assuming that the impact of A and B is perfectly elastic.

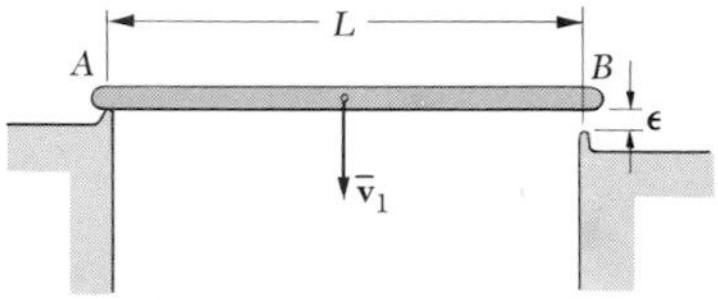

Fig. P17.95

17.95 A uniform slender rod of length L is dropped onto rigid supports at A and B. Immediately before striking A the velocity of the rod is $\bar{\mathbf{v}}_1$. Since support B is slightly lower than support A, the rod strikes A before it strikes B. Assuming perfectly elastic impact at both A and B, determine the angular velocity of the rod and the velocity of its mass center immediately after the rod (*a*) strikes support A, (*b*) strikes support B, (*c*) again strikes support A.

17.96 A square block of mass m moves along a frictionless horizontal surface and strikes a small obstruction at B. Assuming that the impact between corner A and the obstruction B is perfectly plastic, determine the angular velocity of the block and the velocity of its mass center G immediately after the impact.

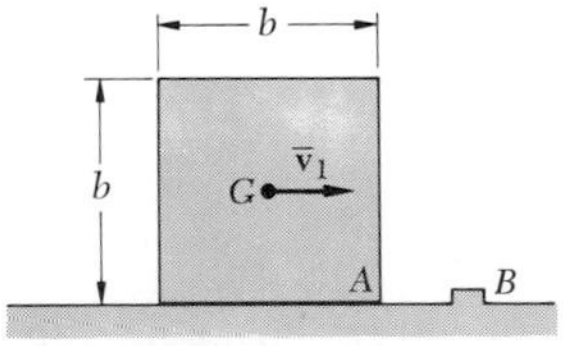

Fig. P17.96

17.97 Solve Prob. 17.96, assuming that the impact between corner A and the obstruction B is perfectly elastic.

17.98 A uniformly loaded square crate is released from rest with its corner D directly above A; it rotates about A until its corner B strikes the floor, and then rotates about B. The floor is sufficiently rough to prevent slipping and the impact at B is perfectly plastic. Denoting by ω_0 the angular velocity of the crate immediately before B strikes the floor, determine (*a*) the angular velocity of the crate immediately after B strikes the floor, (*b*) the fraction of the kinetic energy of the crate lost during the impact, (*c*) the angle θ through which the crate will rotate after B strikes the floor.

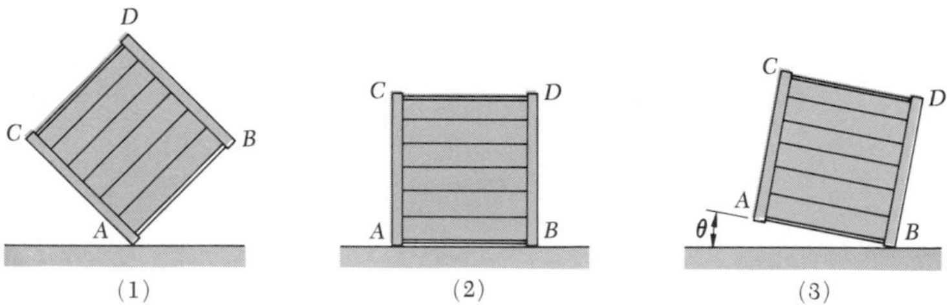

Fig. P17.98

17.99 A uniform sphere of radius r rolls without slipping down the incline shown. It hits the horizontal surface and, after slipping for a while, starts rolling again. Assuming that the sphere does not bounce as it hits the horizontal surface, determine its angular velocity and the velocity of its mass center after it has resumed rolling.

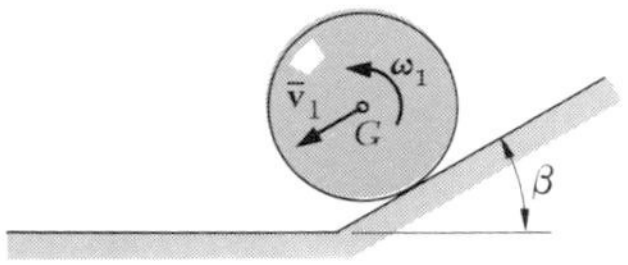

Fig. P17.99

17.100 A sphere A of mass m and radius r rolls without slipping with a velocity $\mathbf{v}_0$ on a horizontal plane. It hits squarely an identical sphere B which is at rest. Denoting by μ the coefficient of friction between the spheres and the plane, neglecting the friction between the spheres, and assuming perfectly elastic impact ($e = 1$), determine (*a*) the linear and angular velocity of each sphere immediately after impact, (*b*) the velocity of each sphere after it has started rolling uniformly. (*c*) Discuss the special case when $\mu = 0$.

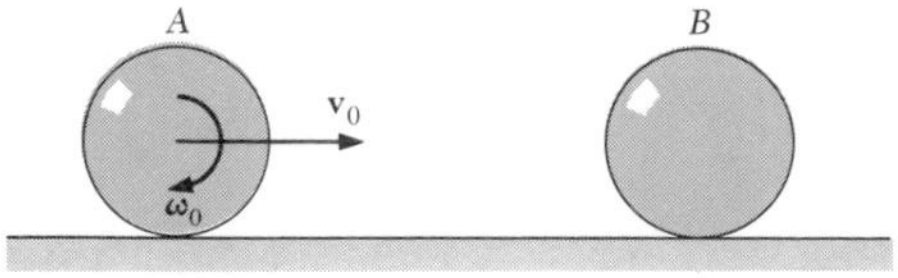

Fig. P17.100

17.101 A slender rod of length l strikes a frictionless floor at A with a vertical velocity $\overline{\mathbf{v}}_1$ and no angular velocity. Assuming that the impact at A is perfectly elastic, derive an expression for the angular velocity of the rod immediately after impact.

17.102 Solve Prob. 17.101, assuming that the impact at A is perfectly plastic.

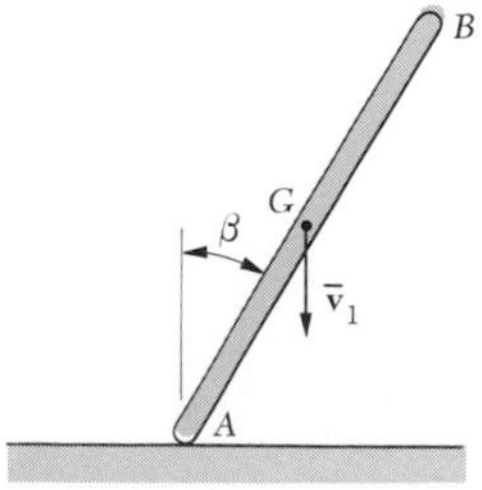

Fig. P17.101

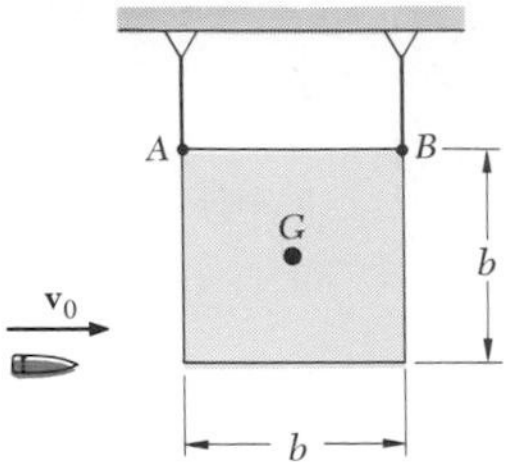

Fig. P17.103

17.103 A bullet of mass m is fired with a horizontal velocity $\mathbf{v}_0$ into the lower corner of a square panel of much larger mass M. The panel is held by two vertical wires as shown. Determine the velocity of the center G of the panel immediately after the bullet becomes embedded.

17.104 Two uniform rods, each of mass m, form the L-shaped rigid body ABC which is initially at rest on the frictionless horizontal surface when hook D of the carriage E engages a small pin at C. Knowing that the carriage is pulled to the right with a constant velocity $\mathbf{v}_0$, determine immediately after the impact (a) the angular velocity of the body, (b) the velocity of corner B. Assume that the velocity of the carriage is unchanged and that the impact is perfectly plastic.

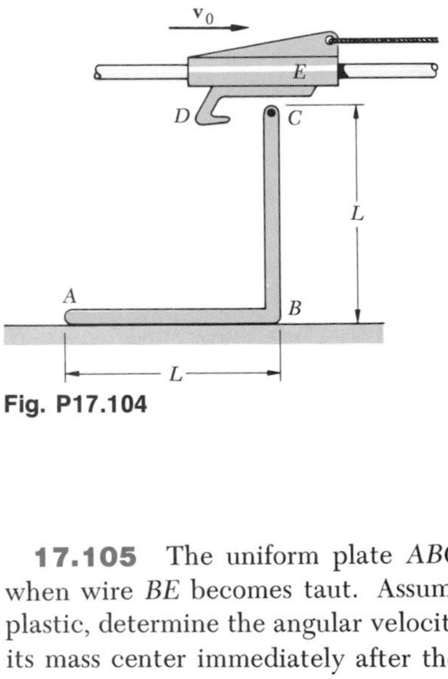

Fig. P17.104

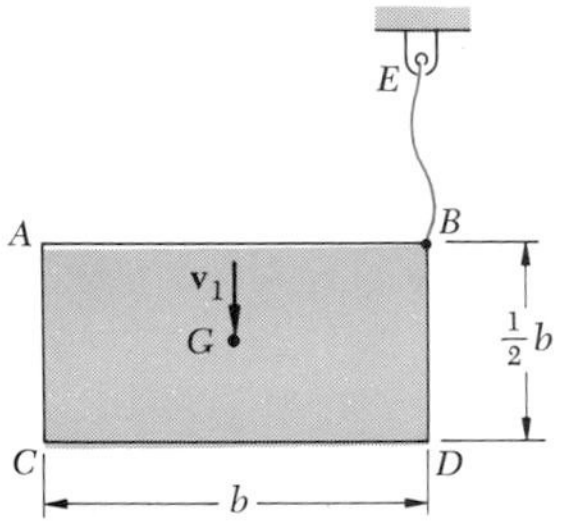

Fig. P17.105

17.105 The uniform plate $ABCD$ is falling with a velocity $\mathbf{v}_1$ when wire BE becomes taut. Assuming that the impact is perfectly plastic, determine the angular velocity of the plate and the velocity of its mass center immediately after the impact.

17.106 In Prob. 17.96, determine the line of action of the impulsive force exerted on the block by the obstruction at B.

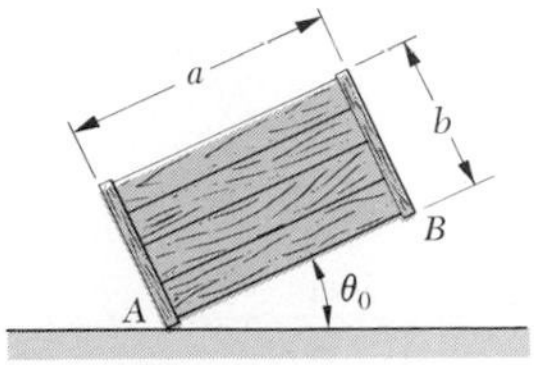

Fig. P17.107

17.107 A uniformly loaded rectangular crate is released from rest in the position shown. Assuming that the floor is sufficiently rough to prevent slipping and that the impact at B is perfectly plastic, determine the largest value of the ratio b/a for which corner A will remain in contact with the floor.

17.108 A slender rod of mass m and length l is held in the position shown. Roller B is given a slight push to the right and moves along the horizontal plane, while roller A is constrained to move vertically. Determine the magnitudes of the impulses exerted on the rollers A and B as roller A strikes the ground. Assume perfectly plastic impact.

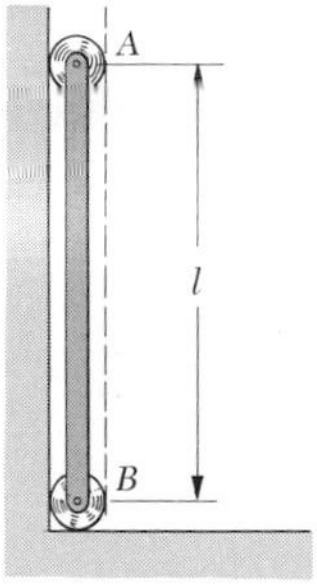

Fig. P17.108

17.109 In a game of billiards, ball A is rolling without slipping with a velocity $\mathbf{v}_0$ as it hits obliquely ball B which is at rest. Denoting by r the radius of each ball, by μ the coefficient of friction between the balls and the table, neglecting friction between the balls, and assuming perfectly elastic impact ($e = 1$), determine (a) the linear and angular velocity of each ball immediately after impact, (b) the velocity of B after it has started rolling uniformly.

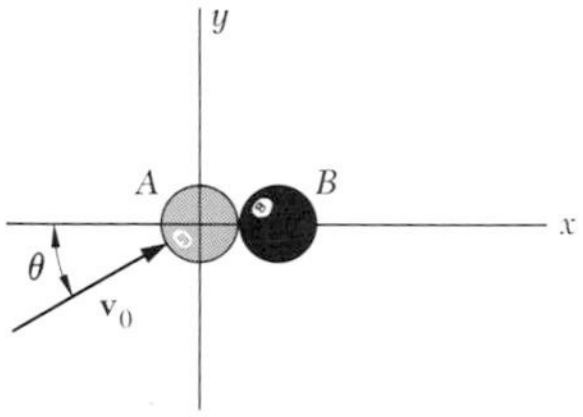

Fig. P17.109

17.110 In Prob. 17.109, determine the equation of the path described by the center of ball A while the ball is slipping.

17.111 For the billiard balls of Prob. 17.109, determine (a) the velocity of ball A after it has started rolling again without slipping, (b) the angle ϕ formed by the velocities of balls A and B after they have finished slipping. (Compare the result obtained here with the one obtained for the pucks of Prob. 13.139 when $e = 1$.)

17.112 A small rubber ball of radius r is thrown against a rough floor with a velocity $\mathbf{v}_A$ of magnitude v_0 and a "backspin" $\boldsymbol{\omega}_A$ of magnitude ω_0. It is observed that the ball bounces from A to B, then from B to A, then from A to B, etc. Assuming perfectly elastic impact, determine (a) the required magnitude ω_0 of the "backspin" in terms of v_0 and r, (b) the minimum required value of the coefficient of friction.

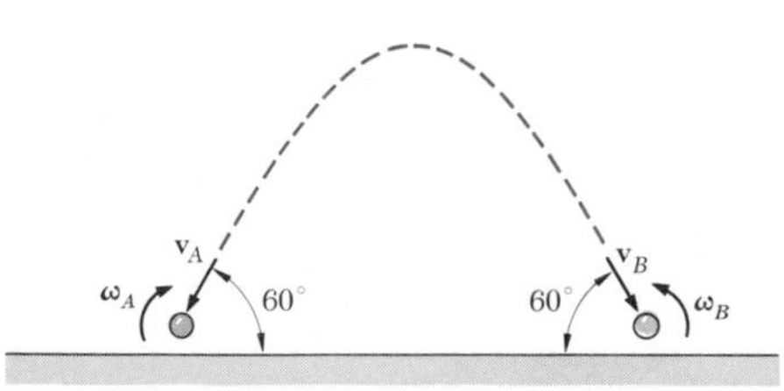

Fig. P17.112

17.113 Two identical rods AB and CD, each of length L, may move freely on a frictionless horizontal surface. Rod AB is rotating about its mass center with an angular velocity $\boldsymbol{\omega}_0$ when end B strikes end C of rod CD, which is at rest. Knowing that at the instant of impact the rods are parallel and assuming perfectly elastic impact ($e = 1$), determine the angular velocity of each rod and the velocity of its mass center immediately after impact.

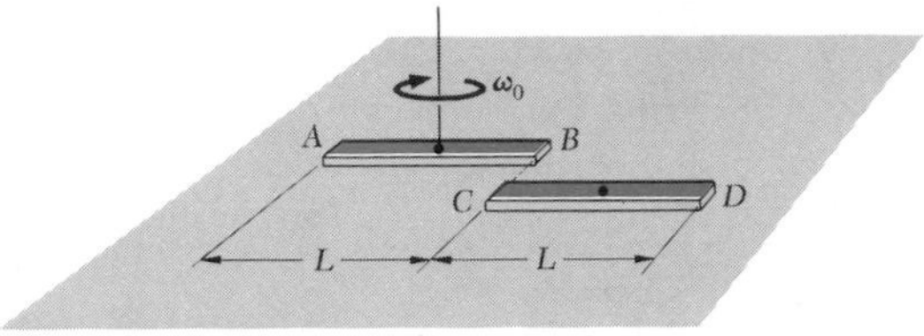

Fig. P17.113

17.114 Solve Prob. 17.113, assuming that the impact is perfectly plastic ($e = 0$).

REVIEW PROBLEMS

17.115 A small disk A is driven at a constant angular velocity of 1200 rpm and is pressed against disk B, which is initially at rest. The normal force between disks is 10 lb, and $\mu_k = 0.20$. Knowing that disk B weighs 50 lb, determine the number of revolutions executed by disk B before its speed reaches 120 rpm.

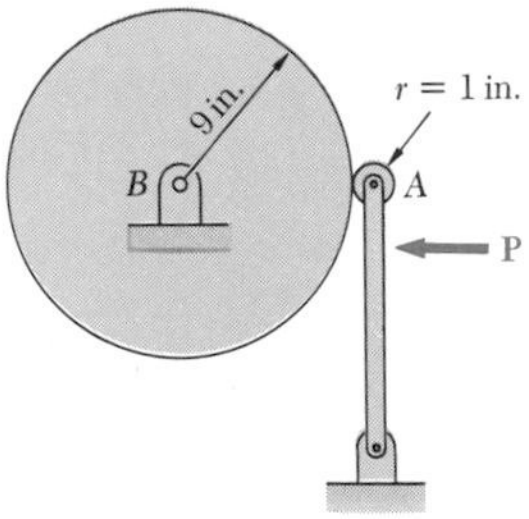

Fig. P17.115

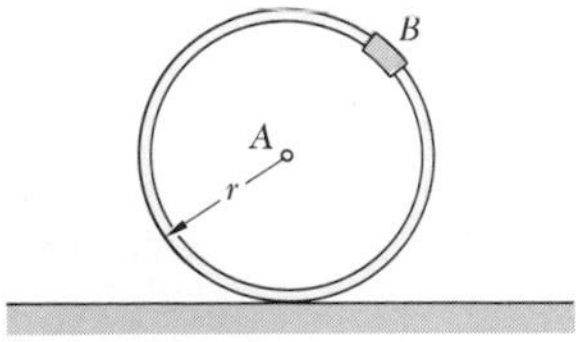

Fig. P17.116

17.116 A small collar of mass m is attached at B to the rim of a hoop of mass m and radius r. The hoop rolls without sliding on a horizontal plane. Find the angular velocity $\boldsymbol{\omega}_1$ of the hoop when B is directly above the center A in terms of g and r, knowing that the angular velocity of the hoop is $3\boldsymbol{\omega}_1$ when B is directly below A.

17.117 The gear train shown consists of four gears of the same thickness and of the same material; two gears are of radius r, and the other two are of radius nr. The system is at rest when the couple $\mathbf{M}_0$ is applied to shaft C. Denoting by I_0 the moment of inertia of a gear of radius r, determine the angular velocity of shaft A if the couple $\mathbf{M}_0$ is applied (a) for one revolution of shaft C, (b) for t seconds.

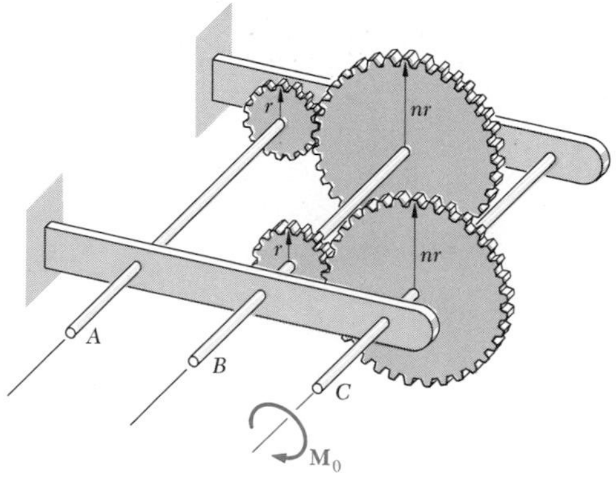

Fig. P17.117

17.118 The motion of a 16-kg sliding panel is guided by rollers at B and C. The counterweight A has a mass of 12 kg and is attached to a cable as shown. If the system is released from rest, determine for each case shown the velocity of the counterweight as it strikes the ground. Neglect the effect of friction.

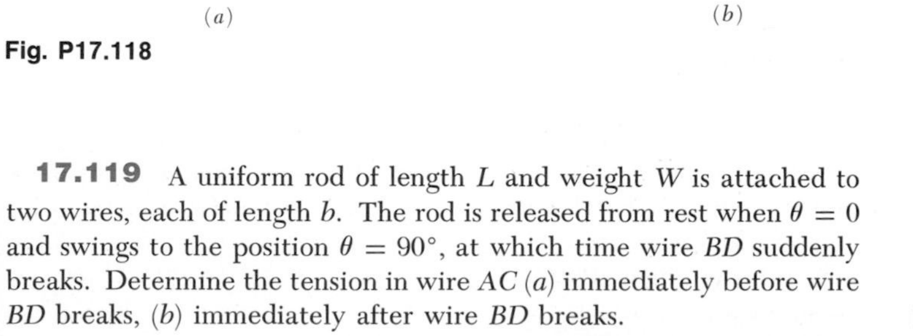

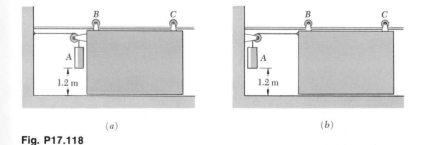

Fig. P17.118

17.119 A uniform rod of length L and weight W is attached to two wires, each of length b. The rod is released from rest when $\theta = 0$ and swings to the position $\theta = 90°$, at which time wire BD suddenly breaks. Determine the tension in wire AC (a) immediately before wire BD breaks, (b) immediately after wire BD breaks.

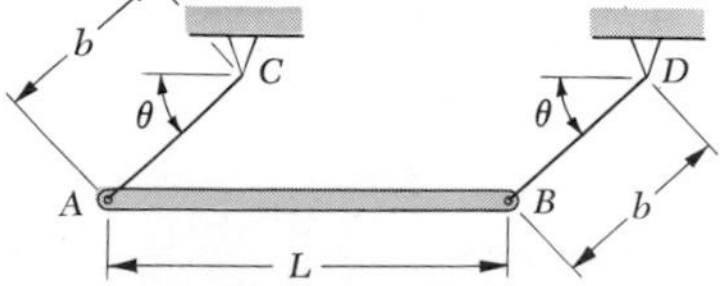

Fig. P17.119

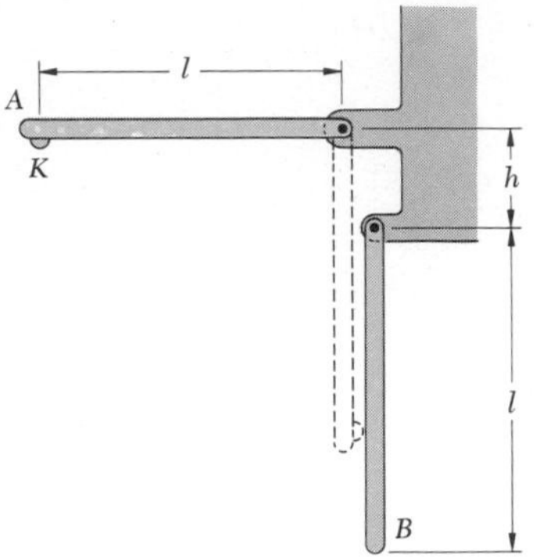

Fig. P17.120

17.120 Two identical slender rods may swing freely from the pivots shown. Rod A is released from rest in a horizontal position and swings to a vertical position, at which time the small knob K strikes rod B which was at rest. If $h = \frac{1}{2}l$ and $e = \frac{1}{2}$, determine (*a*) the angle through which rod B will swing, (*b*) the angle through which rod A will rebound.

17.121 Solve Prob. 17.120, assuming $e = 1$.

17.122 The motor shown runs a machine attached to the shaft at A. The motor develops 4 hp and runs at a constant speed of 300 rpm. Determine the magnitude of the couple exerted (*a*) by the shaft on pulley A, (*b*) by the motor on pulley B.

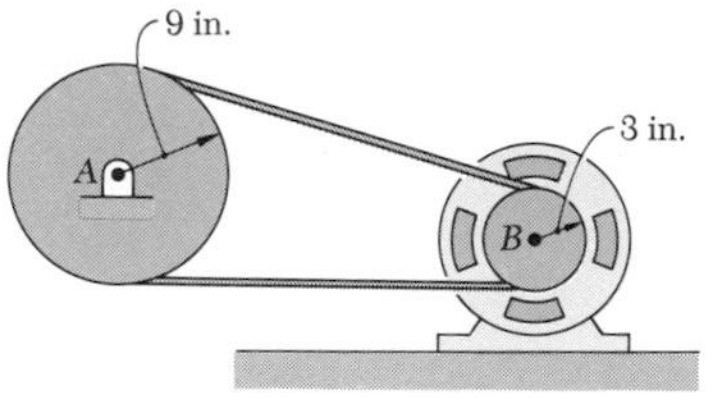

Fig. P17.122

17.123 The plank CDE of mass m_P rests on top of a small pivot at D. A gymnast A of mass m stands on the plank at end C; a second gymnast B of the same mass m jumps from a height h and strikes the plank at E. Assuming perfectly plastic impact, determine the height to which gymnast A will rise. (Assume that gymnast A stands completely rigid.)

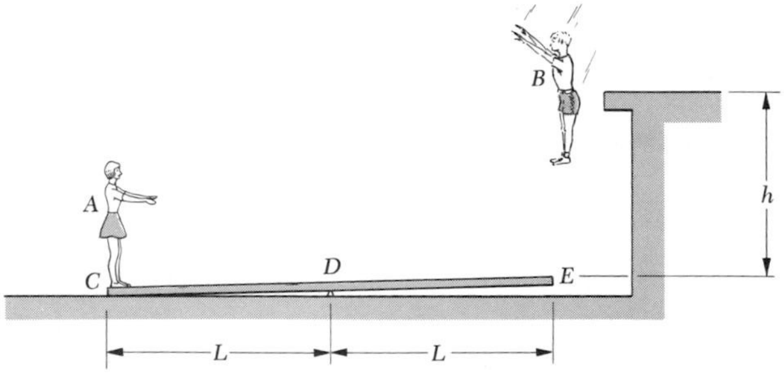

Fig. P17.123

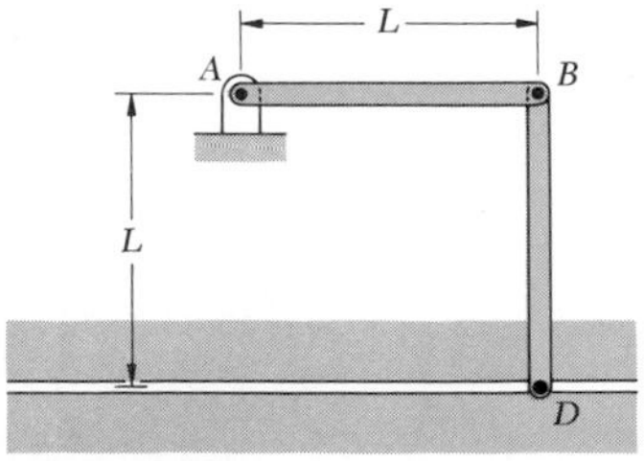

Fig. P17.124

17.124 Two uniform rods, each of mass m and length L, are connected to form the linkage shown. End D of rod BD may slide freely in the horizontal slot, while end A of rod AB is attached to a fixed pin support. If end D is moved slightly to the left and then released, determine its velocity (*a*) when D is directly below A, (*b*) when rod AB is vertical.

17.125 A rectangular slab of mass m_S moves across a series of rollers, each of which is equivalent to a uniform disk of mass m_R and is initially at rest. Since the length of the slab is slightly less than three times the distance b between two adjacent rollers, the slab leaves a roller just before it reaches another one. Each time a new roller enters into contact with the slab, slipping occurs between the roller and the slab for a short period of time (less than the time needed for the slab to move through the distance b). Denoting by $\mathbf{v}_0$ the velocity of the slab in the position shown, determine the velocity of the slab after it has moved (*a*) a distance b, (*b*) a distance nb.

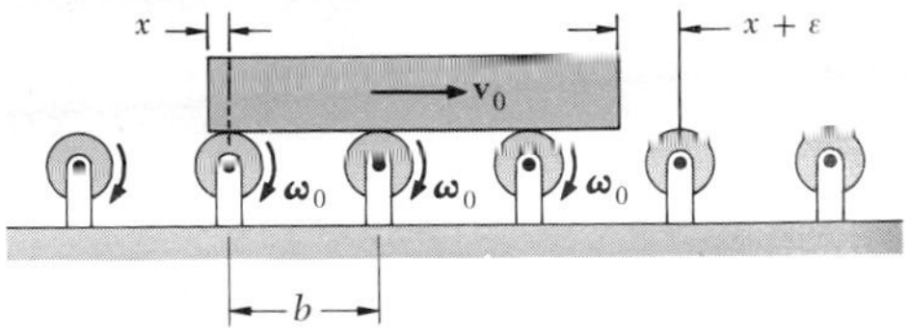

Fig. P17.125

17.126 Solve Prob. 17.125, assuming $v_0 = 5$ m/s, $m_S = 17$ kg, $m_R = 2$ kg, and $n = 5$.

CHAPTER

18 Kinetics of Rigid Bodies in Three Dimensions

***18.1. Introduction.** In Chaps. 16 and 17 we were concerned with the plane motion of rigid bodies and of systems of rigid bodies. In Chap. 16 and in the second half of Chap. 17 (momentum method), our study was further restricted to that of plane slabs and of bodies symmetrical with respect to the reference plane. However, many of the fundamental results obtained in these two chapters remain valid in the case of the motion of a rigid body in three dimensions.

For example, the two fundamental equations

$$\Sigma\mathbf{F} = m\bar{\mathbf{a}} \tag{18.1}$$

$$\Sigma\mathbf{M}_G = \dot{\mathbf{H}}_G \tag{18.2}$$

on which the analysis of the plane motion of a rigid body was based, remain valid in the most general case of motion of a rigid body. As it was indicated in Sec. 16.2, these equations express that the system of the external forces is equipollent to the system consisting of the vector $m\bar{\mathbf{a}}$ attached at G and the couple of moment $\dot{\mathbf{H}}_G$ (Fig. 18.1). However, the relation $\mathbf{H}_G = \bar{I}\omega$, which enabled us to determine the angular momentum of a rigid slab and which played an important part in the solution of problems involving the plane motion of slabs and bodies symmetrical with respect to the reference plane, ceases to be valid in the case of nonsymmetrical bodies or three-dimensional motion. It will thus be necessary for us to develop in Sec. 18.2 a more general

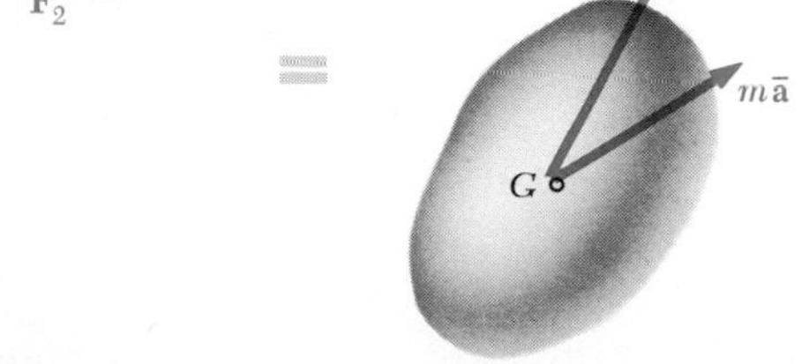

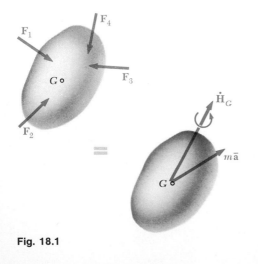

Fig. 18.1

method for the computation of the angular momentum $\mathbf{H}_G$ of a rigid body in three dimensions.

Similarly, the main feature of the impulse-momentum method discussed in Sec. 17.7, namely the reduction of the momenta of the particles of a rigid body to a linear momentum vector $m\bar{\mathbf{v}}$ attached at the mass center G of the body and an angular momentum couple $\mathbf{H}_G$, remains valid. Here again, however, the relation $\mathbf{H}_G = \bar{I}\boldsymbol{\omega}$ will have to be discarded and replaced by the more general relation to be developed in Sec. 18.2.

Finally, we note that the work-energy principle (Sec. 17.1) and the principle of conservation of energy (Sec. 17.5) still apply in the case of the motion of a rigid body in three dimensions. However, the expression obtained in Sec. 17.3 for the kinetic energy of a rigid body in plane motion will be replaced by a new expression to be developed in Sec. 18.4 for a rigid body in three-dimensional motion.

***18.2. Angular Momentum of a Rigid Body in Three Dimensions.** We shall see in this section how the angular momentum $\mathbf{H}_G$ of the body about its mass center G may be determined from the angular velocity $\boldsymbol{\omega}$ of the body in the case of three-dimensional motion.

According to Eq. (14.24), the angular momentum of the body about G may be expressed as

$$\mathbf{H}_G = \sum_{i=1}^{n} (\mathbf{r}'_i \times \mathbf{v}'_i \, \Delta m_i) \tag{18.3}$$

where $\mathbf{r}'_i$ and $\mathbf{v}'_i$ denote, respectively, the position vector and the velocity of the particle P_i, of mass Δm_i, relative to the centroidal frame $Gxyz$ (Fig. 18.2). But $\mathbf{v}'_i = \boldsymbol{\omega} \times \mathbf{r}'_i$, where $\boldsymbol{\omega}$ is the angular velocity of the body at the instant considered. Substituting into (18.3), we have

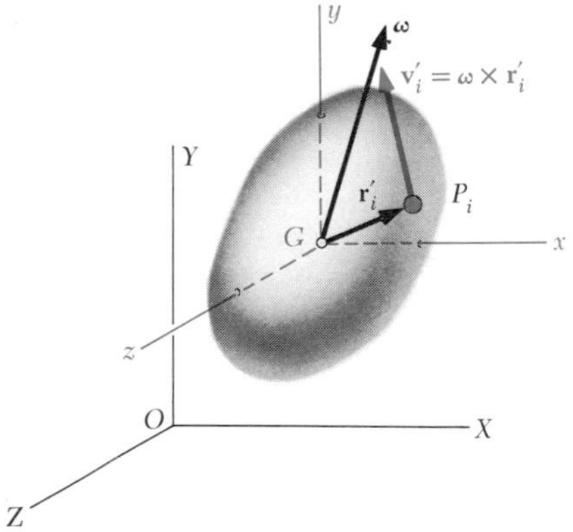

Fig. 18.2

$$\mathbf{H}_G = \sum_{i=1}^{n} [\mathbf{r}'_i \times (\boldsymbol{\omega} \times \mathbf{r}'_i) \, \Delta m_i]$$

Recalling the rule for determining the rectangular components of a vector product (Sec. 3.4), we obtain the following expression for the x component of the angular momentum:

$$\begin{aligned} H_x &= \sum_{i=1}^{n} [y_i(\boldsymbol{\omega} \times \mathbf{r}'_i)_z - z_i(\boldsymbol{\omega} \times \mathbf{r}'_i)_y] \, \Delta m_i \\ &= \sum_{i=1}^{n} [y_i(\omega_x y_i - \omega_y x_i) - z_i(\omega_z x_i - \omega_x z_i)] \, \Delta m_i \\ &= \omega_x \sum_i (y_i^2 + z_i^2) \, \Delta m_i - \omega_y \sum_i x_i y_i \, \Delta m_i - \omega_z \sum_i z_i x_i \, \Delta m_i \end{aligned}$$

Replacing the sums by integrals in this expression and in the two similar expressions which are obtained for H_y and H_z, we have

$$\begin{aligned} H_x &= \omega_x \int (y^2 + z^2)\, dm - \omega_y \int xy\, dm - \omega_z \int zx\, dm \\ H_y &= -\omega_x \int xy\, dm + \omega_y \int (z^2 + x^2)\, dm - \omega_z \int yz\, dm \\ H_z &= -\omega_x \int zx\, dm - \omega_y \int yz\, dm + \omega_z \int (x^2 + y^2)\, dm \end{aligned} \tag{18.4}$$

We note that the integrals containing squares represent the *centroidal mass moments of inertia* of the body about the x, y, and z axes, respectively (Sec. 9.10); we have

$$\bar{I}_x = \int (y^2 + z^2)\, dm \qquad \bar{I}_y = \int (z^2 + x^2)\, dm \qquad \bar{I}_z = \int (x^2 + y^2)\, dm \tag{18.5}$$

Similarly, the integrals containing products of coordinates represent the *centroidal mass products of inertia* of the body (Sec. 9.15); we have

$$\bar{P}_{xy} = \int xy\, dm \qquad \bar{P}_{yz} = \int yz\, dm \qquad \bar{P}_{zx} = \int zx\, dm \tag{18.6}$$

Substituting from (18.5) and (18.6) into (18.4), we obtain the components of the angular momentum $\mathbf{H}_G$ of the body about its mass center G:

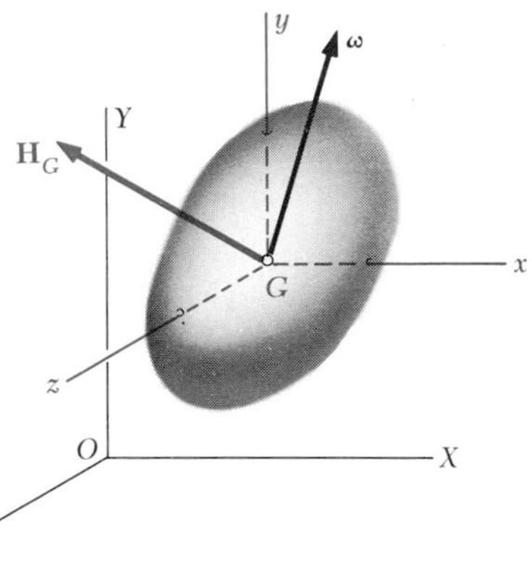

Fig. 18.3

$$\begin{aligned} H_x &= +\bar{I}_x\omega_x - \bar{P}_{xy}\omega_y - \bar{P}_{xz}\omega_z \\ H_y &= -\bar{P}_{yx}\omega_x + \bar{I}_y\omega_y - \bar{P}_{yz}\omega_z \\ H_z &= -\bar{P}_{zx}\omega_x - \bar{P}_{zy}\omega_y + \bar{I}_z\omega_z \end{aligned} \tag{18.7}$$

The relations (18.7) show that the operation which transforms the vector $\boldsymbol{\omega}$ into the vector $\mathbf{H}_G$ (Fig. 18.3) is characterized by the array of moments and products of inertia

$$\begin{pmatrix} \bar{I}_x & -\bar{P}_{xy} & -\bar{P}_{xz} \\ -\bar{P}_{yx} & \bar{I}_y & -\bar{P}_{yz} \\ -\bar{P}_{zx} & -\bar{P}_{zy} & \bar{I}_z \end{pmatrix} \tag{18.8}$$

The array (18.8) defines the *inertia tensor* of the body at its mass center G.† A new array of moments and products of inertia

† Setting $\bar{I}_x = I_{11}$, $\bar{I}_y = I_{22}$, $\bar{I}_z = I_{33}$, and $-\bar{P}_{xy} = I_{12}$, $-\bar{P}_{xz} = I_{13}$, etc., we may write the inertia tensor in the standard form

$$\begin{pmatrix} I_{11} & I_{12} & I_{13} \\ I_{21} & I_{22} & I_{23} \\ I_{31} & I_{32} & I_{33} \end{pmatrix}$$

Denoting by H_1, H_2, H_3 the components of the angular momentum $\mathbf{H}_G$ and by ω_1, ω_2, ω_3 the components of the angular velocity $\boldsymbol{\omega}$, we may write the relations (18.7) in the form

$$H_i = \sum_j I_{ij}\omega_j$$

where i and j take the values 1, 2, 3. The quantities I_{ij} are said to be the *components* of the inertia tensor. Since $I_{ij} = I_{ji}$, the inertia tensor is a *symmetric tensor of the second order*.

would be obtained if a different system of axes were used. The transformation characterized by this new array, however, would still be the same. Clearly, the angular momentum $\mathbf{H}_G$ corresponding to a given angular velocity $\boldsymbol{\omega}$ is independent of the choice of the coordinate axes. As it was shown in Sec. 9.16, it is always possible to select a system of axes $Gx'y'z'$, called *principal axes of inertia*, with respect to which all the products of inertia of a given body are zero. The array (18.8) takes then the diagonalized form

$$\begin{pmatrix} \bar{I}_{x'} & 0 & 0 \\ 0 & \bar{I}_{y'} & 0 \\ 0 & 0 & \bar{I}_{z'} \end{pmatrix} \tag{18.9}$$

where $\bar{I}_{x'}, \bar{I}_{y'}, \bar{I}_{z'}$ represent the *principal centroidal moments of inertia* of the body, and the relations (18.7) reduce to

$$H_{x'} = \bar{I}_{x'}\omega_{x'} \qquad H_{y'} = \bar{I}_{y'}\omega_{y'} \qquad H_{z'} = \bar{I}_{z'}\omega_{z'} \tag{18.10}$$

We note that, if the three principal centroidal moments of inertia $\bar{I}_{x'}, \bar{I}_{y'}, \bar{I}_{z'}$ are equal, the components $H_{x'}, H_{y'}, H_{z'}$ of the angular momentum about G are proportional to the components $\omega_{x'}, \omega_{y'}, \omega_{z'}$ of the angular velocity, and the vectors $\mathbf{H}_G$ and $\boldsymbol{\omega}$ are collinear. In general, however, the principal moments of inertia will be different, and the vectors $\mathbf{H}_G$ and $\boldsymbol{\omega}$ *will have different directions,* except when two of the three components of $\boldsymbol{\omega}$ happen to be zero, i.e., when $\boldsymbol{\omega}$ is directed along one of the coordinate axes. Thus, *the angular momentum* $\mathbf{H}_G$ *of a rigid body and its angular velocity* $\boldsymbol{\omega}$ *have the same direction if, and only if,* $\boldsymbol{\omega}$ *is directed along a principal axis of inertia.*† Since this condition is satisfied in the case of the plane motion of a rigid body symmetrical with respect to the reference plane, we were able in Secs. 16.3 and 17.7 to represent the angular momentum $\mathbf{H}_G$ of such a body by the vector $\bar{I}\boldsymbol{\omega}$. We must realize, however, that this result cannot be extended to the case of the plane motion of a nonsymmetrical body, or to the case of the three-dimensional motion of a rigid body. Except when $\boldsymbol{\omega}$ happens to be directed along a principal axis of inertia, the angular momentum and angular velocity of a rigid body have different directions, and the relation (18.7) or (18.10) must be used to determine $\mathbf{H}_G$ from $\boldsymbol{\omega}$.

† In the particular case when $\bar{I}_{x'} = \bar{I}_{y'} = \bar{I}_{z'}$, any line through G may be considered as a principal axis of inertia, and the vectors $\mathbf{H}_G$ and $\boldsymbol{\omega}$ are always collinear.

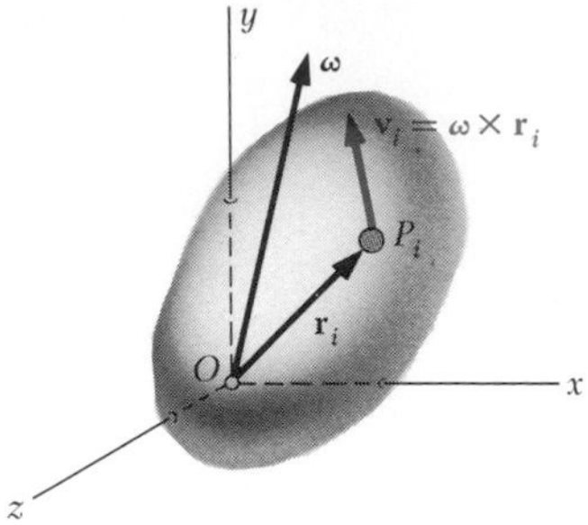

Fig. 18.4

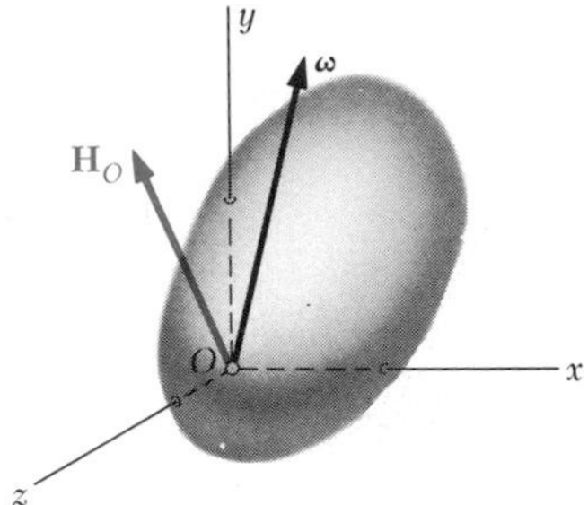

Fig. 18.5

Angular Momentum of a Rigid Body About a Fixed Point. In the particular case of a rigid body rotating in three-dimensional space about a fixed point O (Fig. 18.4), it is sometimes useful to determine the angular momentum $\mathbf{H}_O$ of the body about the fixed point O. Recalling Eq. (14.7), we write

$$\mathbf{H}_O = \sum_{i=1}^{n} (\mathbf{r}_i \times \mathbf{v}_i \, \Delta m_i) \tag{18.11}$$

where $\mathbf{r}_i$ and $\mathbf{v}_i$ denote, respectively, the position vector and the velocity of the particle P_i with respect to the fixed frame $Oxyz$. Substituting $\mathbf{v}_i = \boldsymbol{\omega} \times \mathbf{r}_i$, and after manipulations similar to the ones used above, we find that the components of the angular momentum $\mathbf{H}_O$ (Fig. 18.5) are given by the relations

$$\begin{aligned} H_x &= +\, I_x\omega_x - P_{xy}\omega_y - P_{xz}\omega_z \\ H_y &= -P_{yx}\omega_x + I_y\omega_y - P_{yz}\omega_z \\ H_z &= -P_{zx}\omega_x - P_{zy}\omega_y + I_z\omega_z \end{aligned} \tag{18.12}$$

where the moments of inertia I_x, I_y, I_z and the products of inertia P_{xy}, P_{yz}, P_{zx} are computed with respect to the frame $Oxyz$ centered at the fixed point O.

∗18.3. Application of the Principle of Impulse and Momentum to the Three-dimensional Motion of a Rigid Body. Before we can apply the fundamental equation (18.2) to the solution of problems involving the three-dimensional motion of a rigid body, we shall have to learn to compute the derivative of the vector $\mathbf{H}_G$. This will be done in Sec. 18.5. We may, however, immediately use the results obtained in the preceding section to solve problems by the impulse-momentum method.

Recalling from Sec. 17.7 that the system formed by the momenta of the particles of a rigid body reduces to a linear momentum vector $m\bar{\mathbf{v}}$ attached at the mass center G of the body and an angular momentum couple $\mathbf{H}_G$, we represent graphically the fundamental relation

$$\textbf{Syst Momenta}_1 + \textbf{Syst Ext Imp}_{1\to 2} = \textbf{Syst Momenta}_2 \tag{17.14}$$

by means of the three sketches shown in Fig. 18.6. To solve a given problem, we may use these sketches to write appropriate component and moment equations, keeping in mind that the components of the angular momentum $\mathbf{H}_G$ are related to the components of the angular velocity $\boldsymbol{\omega}$ by Eqs. (18.7) of the preceding section.

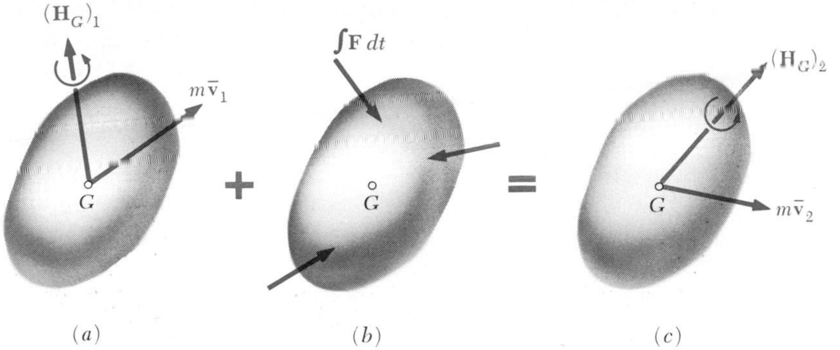

Fig. 18.6

In solving problems dealing with the motion of a body rotating about a fixed point O, it will be convenient to eliminate the impulse of the reaction at O by writing an equation involving the moments of the momenta and impulses about O. We note in this connection that the angular momentum $\mathbf{H}_O$ of the body about the fixed point O may be obtained directly from Eqs. (18.12).

***18.4. Kinetic Energy of a Rigid Body in Three Dimensions.** Consider a rigid body of mass m in three-dimensional motion. We recall from Sec. 14.6 that, if the absolute velocity $\mathbf{v}_i$ of each particle P_i of the body is expressed as the sum of the velocity $\overline{\mathbf{v}}$ of the mass center G of the body and of the velocity $\mathbf{v}'_i$ of the particle relative to a frame $Gxyz$ attached to G and of fixed orientation (Fig. 18.7), the kinetic energy of the system of particles forming the rigid body may be written in the form

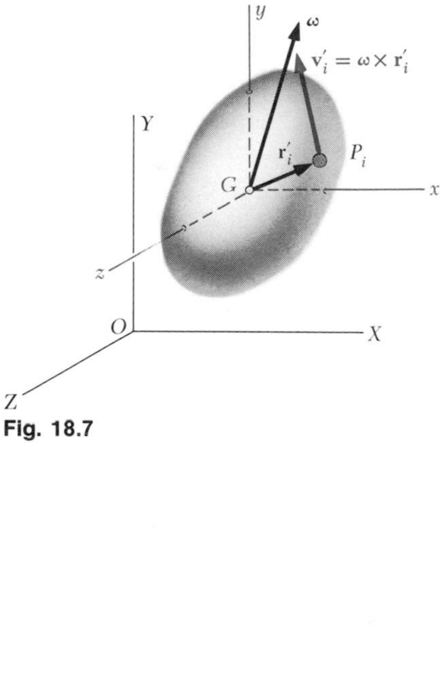

Fig. 18.7

$$T = \tfrac{1}{2}m\bar{v}^2 + \frac{1}{2}\sum_{i=1}^{n}(\Delta m_i)v_i'^2 \tag{18.13}$$

where the last term represents the kinetic energy T' of the body relative to the centroidal frame $Gxyz$. Since $\mathbf{v}'_i = \boldsymbol{\omega} \times \mathbf{r}'_i$, we write

$$T' = \frac{1}{2}\sum_{i=1}^{n}(\Delta m_i)v_i'^2 = \frac{1}{2}\sum_{i=1}^{n}(\boldsymbol{\omega} \times \mathbf{r}'_i)^2\,\Delta m_i \tag{18.14}$$

Expressing the square of the vector product in terms of its rectangular components, and replacing the sums by integrals, we have

$$\begin{aligned} T' &= \int[(\omega_x y - \omega_y x)^2 + (\omega_y z - \omega_z y)^2 + (\omega_z x - \omega_x z)^2]\,dm \\ &= \omega_x^2\int(y^2 + z^2)\,dm + \omega_y^2\int(z^2 + x^2)\,dm + \omega_z^2\int(x^2 + y^2)\,dm \\ &\qquad - 2\omega_x\omega_y\int xy\,dm - 2\omega_y\omega_z\int yz\,dm - 2\omega_z\omega_x\int zx\,dm \end{aligned}$$

or, recalling the relations (18.5) and (18.6),

$$T' = \tfrac{1}{2}(\bar{I}_x\omega_x^2 + \bar{I}_y\omega_y^2 + \bar{I}_z\omega_z^2 - 2\bar{P}_{xy}\omega_x\omega_y - 2\bar{P}_{yz}\omega_y\omega_z - 2\bar{P}_{zx}\omega_z\omega_x) \tag{18.15}$$

Substituting into (18.13) the expression (18.15) we have just obtained for the kinetic energy of the body relative to centroidal axes, we write

$$T = \tfrac{1}{2}m\bar{v}^2 + \tfrac{1}{2}(\bar{I}_x\omega_x^2 + \bar{I}_y\omega_y^2 + \bar{I}_z\omega_z^2 - 2\bar{P}_{xy}\omega_x\omega_y - 2\bar{P}_{yz}\omega_y\omega_z - 2\bar{P}_{zx}\omega_z\omega_x) \tag{18.16}$$

If the axes of coordinates are chosen so that they coincide at the instant considered with the principal axes x', y', z' of the body, the relation obtained reduces to

$$T = \tfrac{1}{2}m\bar{v}^2 + \tfrac{1}{2}(\bar{I}_{x'}\omega_{x'}^2 + \bar{I}_{y'}\omega_{y'}^2 + \bar{I}_{z'}\omega_{z'}^2) \tag{18.17}$$

where $\bar{\mathbf{v}}$ = velocity of mass center
$\boldsymbol{\omega}$ = angular velocity
m = mass of rigid body
$\bar{I}_{x'}, \bar{I}_{y'}, \bar{I}_{z'}$ = principal centroidal moments of inertia

The results we have obtained enable us to extend to the three-dimensional motion of a rigid body the application of the principle of work and energy (Sec. 17.1) and of the principle of conservation of energy (Sec. 17.5).

Kinetic Energy of a Rigid Body With a Fixed Point. In the particular case of a rigid body rotating in three-dimensional space about a fixed point O, the kinetic energy of the body may be expressed in terms of its moments and products of inertia with respect to axes attached at O (Fig. 18.8). Recalling the definition of the kinetic energy of a system of particles, and substituting $\mathbf{v}_i = \boldsymbol{\omega} \times \mathbf{r}_i$, we write

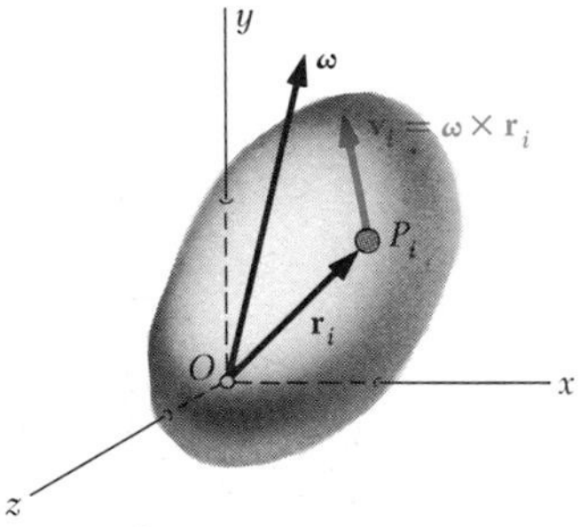

Fig. 18.8

$$T = \frac{1}{2}\sum_{i=1}^{n}(\Delta m_i)v_i^2 = \frac{1}{2}\sum_{i=1}^{n}(\boldsymbol{\omega} \times \mathbf{r}_i)^2\,\Delta m_i \tag{18.18}$$

Manipulations similar to those used to derive (18.15) from (18.14) yield

$$T = \tfrac{1}{2}(I_x\omega_x^2 + I_y\omega_y^2 + I_z\omega_z^2 - 2P_{xy}\omega_x\omega_y - 2P_{yz}\omega_y\omega_z - 2P_{zx}\omega_z\omega_x) \tag{18.19}$$

or, if the principal axes x', y', z' of the body at the origin O are chosen as coordinate axes,

$$T = \tfrac{1}{2}(I_{x'}\omega_{x'}^2 + I_{y'}\omega_{y'}^2 + I_{z'}\omega_{z'}^2) \tag{18.20}$$

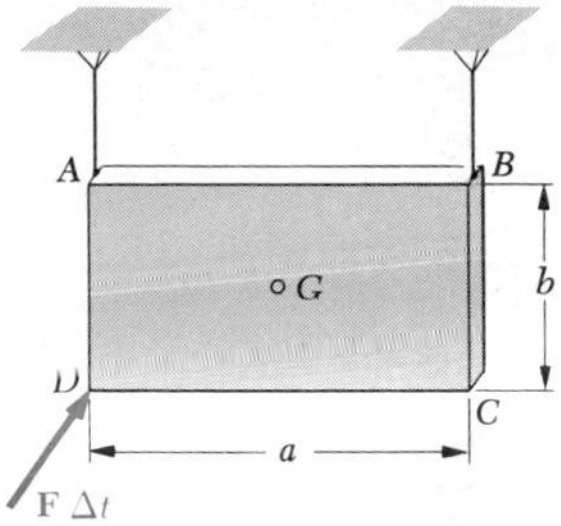

SAMPLE PROBLEM 18.1

A rectangular plate of mass m suspended from two wires at A and B is hit at D in a direction perpendicular to the plate. Denoting by $\mathbf{F}\,\Delta t$ the impulse applied at D, determine immediately after impact (a) the velocity of the mass center G, (b) the angular velocity of the plate.

Solution. We shall assume that the wires remain taut and, thus, that the components $\bar{v}_y$ of $\bar{\mathbf{v}}$ and ω_z of $\boldsymbol{\omega}$ are zero after impact. We have therefore

$$\bar{\mathbf{v}} = \bar{v}_x\mathbf{i} + \bar{v}_z\mathbf{k} \qquad \boldsymbol{\omega} = \omega_x\mathbf{i} + \omega_y\mathbf{j}$$

and, since the x, y, z axes are principal axes of inertia,

$$\mathbf{H}_G = \bar{I}_x\omega_x\mathbf{i} + \bar{I}_y\omega_y\mathbf{j} \qquad \mathbf{H}_G = \tfrac{1}{12}mb^2\omega_x\mathbf{i} + \tfrac{1}{12}ma^2\omega_y\mathbf{j} \qquad (1)$$

Principle of Impulse and Momentum. Since the initial momenta are zero, the system of the impulses must be equivalent to the system of the final momenta:

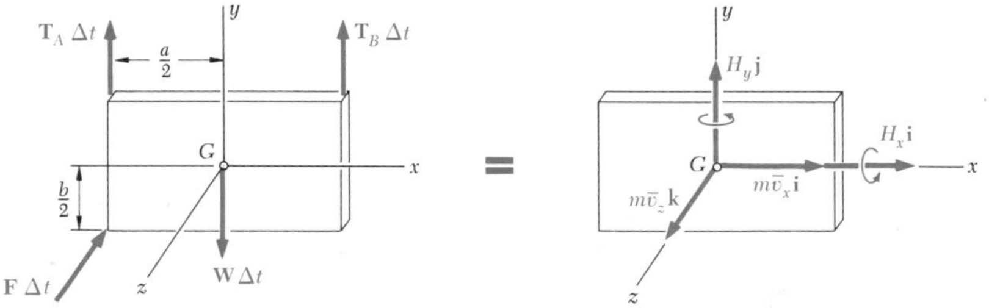

a. Velocity of Mass Center. Equating the components of the impulses and momenta in the x and z directions:

x comp.: $\quad 0 = m\bar{v}_x \qquad \bar{v}_x = 0$

z comp.: $\quad -F\,\Delta t = m\bar{v}_z \qquad \bar{v}_z = -F\,\Delta t/m$

$$\bar{\mathbf{v}} = \bar{v}_x\mathbf{i} + \bar{v}_z\mathbf{k} \qquad \bar{\mathbf{v}} = -(F\,\Delta t/m)\mathbf{k} \blacktriangleleft$$

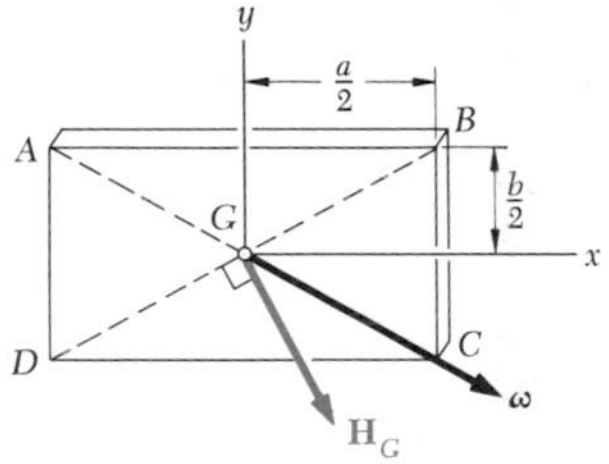

b. Angular Velocity. Equating the moments of the impulses and momenta about the x and y axes:

About x axis: $\quad \tfrac{1}{2}bF\,\Delta t = H_x$

About y axis: $\quad -\tfrac{1}{2}aF\,\Delta t = H_y$

$$\mathbf{H}_G = H_x\mathbf{i} + H_y\mathbf{j} \qquad \mathbf{H}_G = \tfrac{1}{2}bF\,\Delta t\,\mathbf{i} - \tfrac{1}{2}aF\,\Delta t\,\mathbf{j} \qquad (2)$$

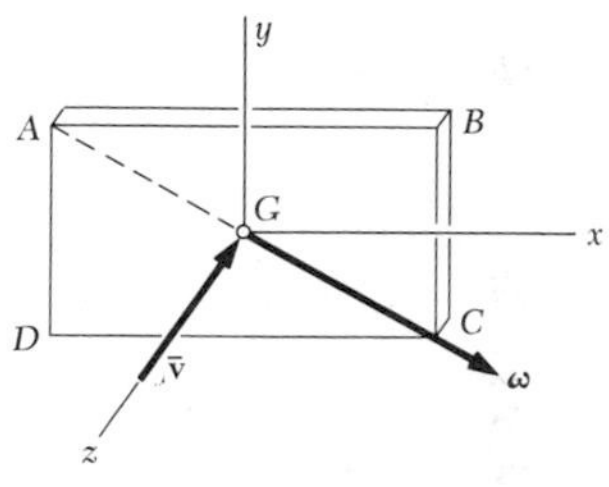

Comparing Eqs. (1) and (2), we conclude that

$$\omega_x = 6F\,\Delta t/mb \qquad \omega_y = -6F\,\Delta t/ma$$

$$\boldsymbol{\omega} = \omega_x\mathbf{i} + \omega_y\mathbf{j} \qquad \boldsymbol{\omega} = (6F\,\Delta t/mab)(a\mathbf{i} - b\mathbf{j}) \blacktriangleleft$$

We note that $\boldsymbol{\omega}$ is directed along the diagonal AC.

Remark: Equating the y components of the impulses and momenta, and their moments about the z axis, we obtain two additional equations which yield $T_A = T_B = \tfrac{1}{2}W$. We thus verify that the wires remain taut and that our assumption was correct.

SAMPLE PROBLEM 18.2

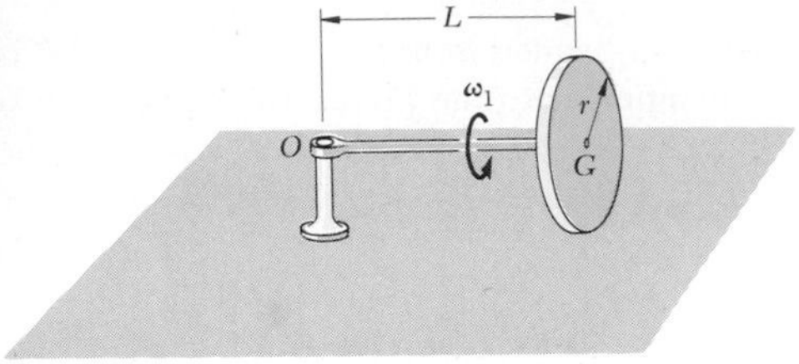

A homogeneous disk of radius r and mass m is mounted on an axle OG of length L and negligible mass. The axle is pivoted at the fixed point O, and the disk is constrained to roll on a horizontal floor. Knowing that the disk rotates counterclockwise at the rate ω_1 about the axle OG, determine (*a*) the angular velocity of the disk, (*b*) its angular momentum about O, (*c*) its kinetic energy, (*d*) the vector and couple at G equivalent to the momenta of the particles of the disk.

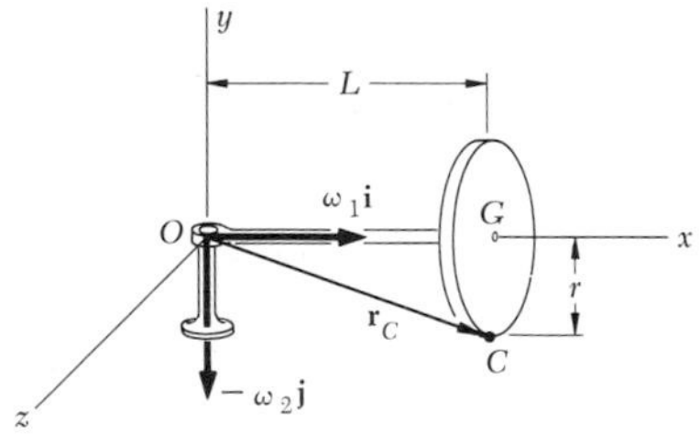

***a.* Angular Velocity.** As the disk rotates about the axle OG it also rotates with the axle about the y axis at a rate ω_2 clockwise. The total angular velocity of the disk is therefore

$$\boldsymbol{\omega} = \omega_1\mathbf{i} - \omega_2\mathbf{j} \qquad (1)$$

To determine ω_2 we write that the velocity of C is zero:

$$\mathbf{v}_C = \boldsymbol{\omega} \times \mathbf{r}_C = 0$$
$$(\omega_1\mathbf{i} - \omega_2\mathbf{j}) \times (L\mathbf{i} - r\mathbf{j}) = 0$$
$$(L\omega_2 - r\omega_1)\mathbf{k} = 0 \qquad \omega_2 = r\omega_1/L$$

Substituting into (1) for ω_2: $\qquad \boldsymbol{\omega} = \omega_1\mathbf{i} - (r\omega_1/L)\mathbf{j}$ ◄

***b.* Angular Momentum About *O*.** Assuming the axle to be part of the disk, we may consider the disk to have a fixed point at O. Since the x, y, and z axes are principal axes of inertia for the disk,

$$H_x = I_x\omega_x = (\tfrac{1}{2}mr^2)\omega_1$$
$$H_y = I_y\omega_y = (mL^2 + \tfrac{1}{4}mr^2)(-r\omega_1/L)$$
$$H_z = I_z\omega_z = (mL^2 + \tfrac{1}{4}mr^2)\,0 = 0$$
$$\mathbf{H}_O = \tfrac{1}{2}mr^2\omega_1\mathbf{i} - m(L^2 + \tfrac{1}{4}r^2)(r\omega_1/L)\mathbf{j} \quad ◄$$

***c.* Kinetic Energy.** Using the values obtained for the moments of inertia and the components of $\boldsymbol{\omega}$, we have

$$T = \tfrac{1}{2}(I_x\omega_x^2 + I_y\omega_y^2 + I_z\omega_z^2) = \tfrac{1}{2}[\tfrac{1}{2}mr^2\omega_1^2 + m(L^2 + \tfrac{1}{4}r^2)(-r\omega_1/L)^2]$$

$$T = \tfrac{1}{8}mr^2\left(6 + \frac{r^2}{L^2}\right)\omega_1^2 \quad ◄$$

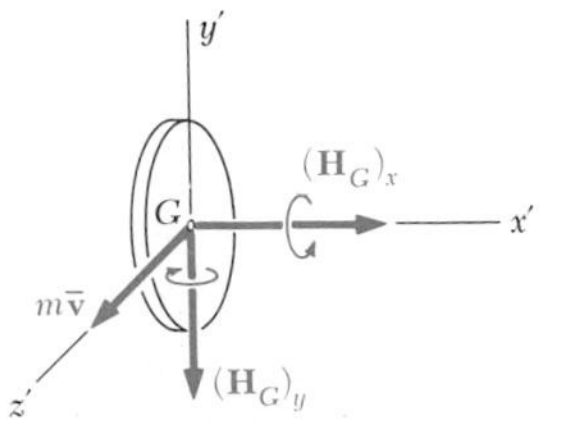

***d.* Momentum Vector and Couple at *G*.** The linear momentum vector $m\bar{\mathbf{v}}$ and the angular momentum couple $\mathbf{H}_G$ are

$$m\bar{\mathbf{v}} = mr\omega_1\mathbf{k} \quad ◄$$

and

$$\mathbf{H}_G = \bar{I}_{x'}\omega_x\mathbf{i} + \bar{I}_{y'}\omega_y\mathbf{j} + \bar{I}_{z'}\omega_z\mathbf{k} = \tfrac{1}{2}mr^2\omega_1\mathbf{i} + \tfrac{1}{4}mr^2(-r\omega_1/L)\mathbf{j}$$

$$\mathbf{H}_G = \tfrac{1}{2}mr^2\omega_1\left(\mathbf{i} - \frac{r}{2L}\mathbf{j}\right) \quad ◄$$

PROBLEMS

18.1 A thin homogeneous rod of mass m and length L rotates with a constant angular velocity $\boldsymbol{\omega}$ about a vertical axis through its mass center G. Determine the magnitude and direction of the angular momentum $\mathbf{H}_G$ of the rod about its mass center.

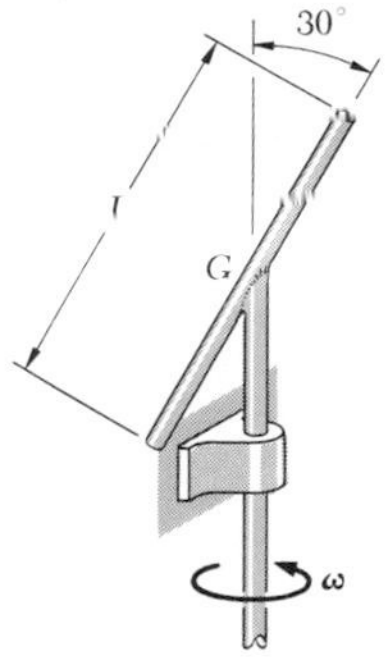

Fig. P18.1

18.2 A thin homogeneous disk of mass m and radius r spins at the constant rate ω_2 about an axle held by a fork-ended horizontal rod which rotates at the constant rate ω_1. Determine the angular momentum of the disk about its mass center.

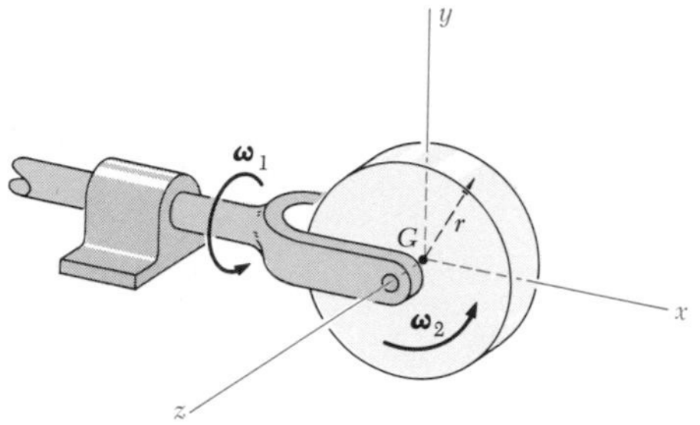

Fig. P18.2

18.3 A thin homogeneous disk of mass m and radius r is mounted on the vertical axle AB. The plane of the disk forms an angle $\beta = 30°$ with the horizontal. Knowing that the axle rotates with an angular velocity $\boldsymbol{\omega}$, determine the angle θ formed by the axle and the angular momentum of the disk about G.

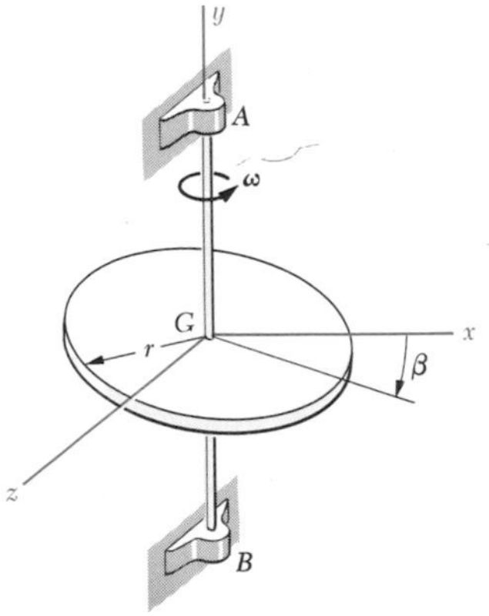

Fig. P18.3

18.4 A thin rectangular plate of mass 9 kg is attached to a shaft as shown. If the angular velocity $\boldsymbol{\omega}$ of the plate is 4 rad/s at the instant shown, determine its angular momentum about its mass center G.

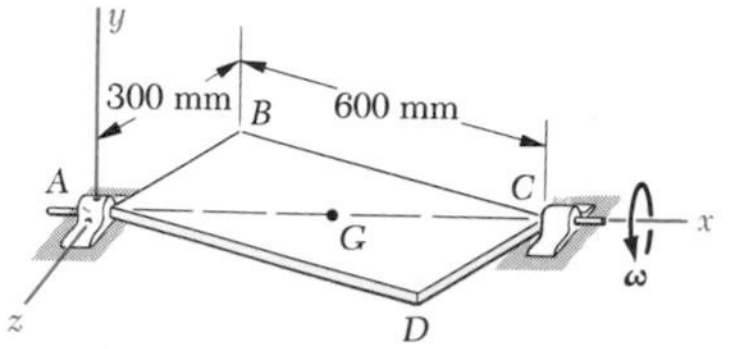

Fig. P18.4

18.5 In Prob. 18.3, determine the value of β for which the angle θ formed by the axle and the angular momentum $\mathbf{H}_G$ is maximum.

18.6 A thin homogeneous disk of mass 800 g and radius 100 mm rotates at a constant rate $\omega_2 = 20$ rad/s with respect to the arm ABC, which itself rotates at a constant rate $\omega_1 = 10$ rad/s about the x axis. Determine the angular momentum of the disk about point C.

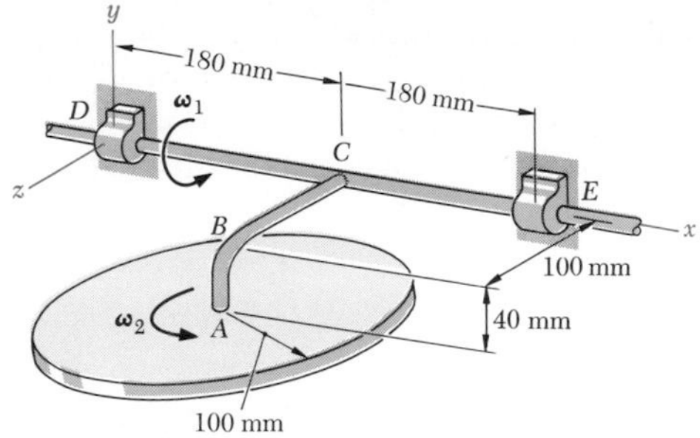

Fig. P18.6

18.7 Determine the angular momentum of the disk of Prob. 18.6 about point D.

18.8 A thin homogeneous triangular plate weighing 12 lb is welded to a light axle which can rotate freely in bearings at A and B. Knowing that the plate rotates at a constant rate $\omega = 5$ rad/s, determine its angular momentum about A.

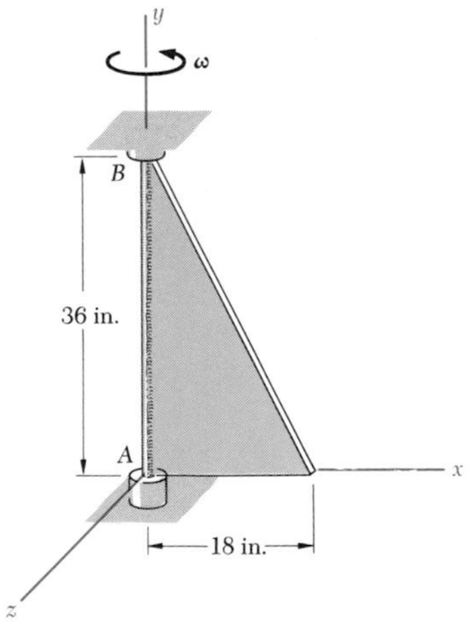

Fig. P18.8

18.9 Determine the angular momentum of the plate of Prob. 18.8 about its mass center.

18.10 Each element of the crankshaft shown is a homogeneous rod of mass m per unit length. Knowing that the crankshaft rotates with a constant angular velocity $\boldsymbol{\omega}$, determine (a) the angular momentum of the crankshaft about G, (b) the angle formed by the angular momentum and the axis AB.

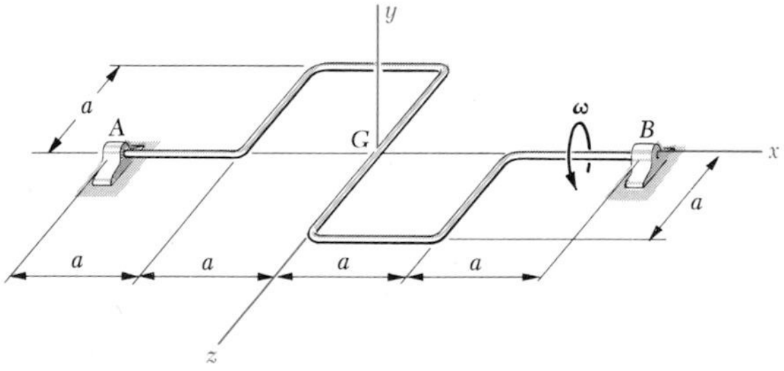

Fig. P18.10

18.11 Determine the angular momentum of the crankshaft of Prob. 18.10 about point A.

18.12 Show that, when a rigid body rotates about a fixed axis, its angular momentum is the same about any two points A and B on the fixed axis ($\mathbf{H}_A = \mathbf{H}_B$) if, and only if, the mass center G of the body is located on the fixed axis.

18.13 Two L-shaped arms, each weighing 6 lb, are welded at the third points of the 3-ft shaft AB. Knowing that shaft AB rotates at a constant rate $\omega = 300$ rpm, determine (*a*) the angular momentum of the body about A, (*b*) the angle formed by the angular momentum and the shaft AB.

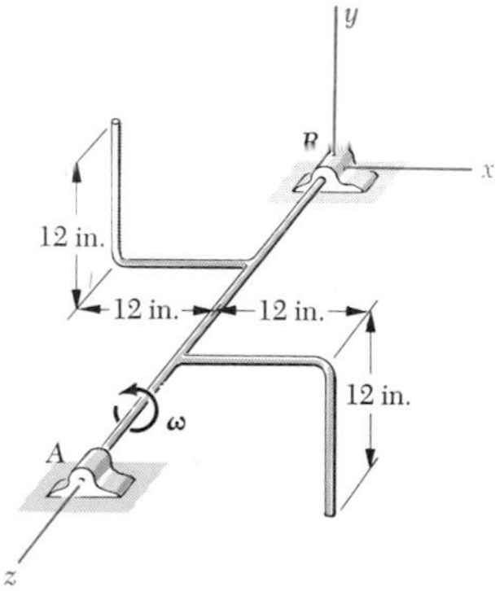

Fig. P18.13

18.14 At a given instant during its flight, a launch vehicle has an angular velocity $\boldsymbol{\omega} = (0.3 \text{ rad/s})\mathbf{j} + (2 \text{ rad/s})\mathbf{k}$ and its mass center G has a velocity $\mathbf{v} = (6 \text{ m/s})\mathbf{i} + (9 \text{ m/s})\mathbf{j} + (1800 \text{ m/s})\mathbf{k}$, where $\mathbf{i}$, $\mathbf{j}$, and $\mathbf{k}$ are the unit vectors corresponding to the principal centroidal axes of inertia of the vehicle. Knowing that the vehicle has a mass of 40 Mg and that its centroidal radii of gyration are $\bar{k}_x = \bar{k}_y = 6$ m and $\bar{k}_z = 1.5$ m, determine (*a*) the linear momentum $m\bar{\mathbf{v}}$ and the angular momentum $\mathbf{H}_G$, (*b*) the angle between the vectors representing $m\bar{\mathbf{v}}$ and $\mathbf{H}_G$.

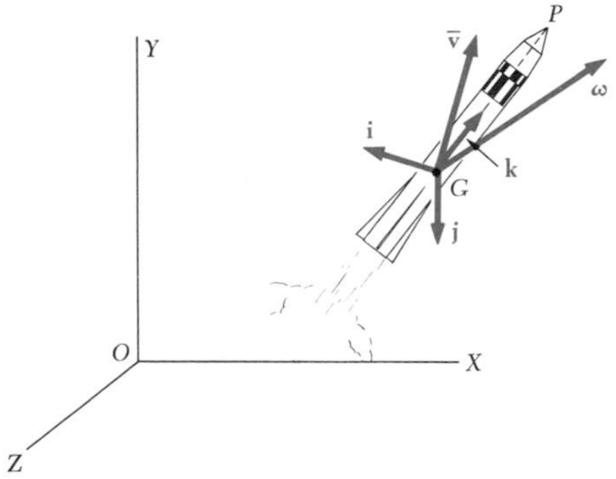

Fig. P18.14

18.15 For the launch vehicle of Prob. 18.14, determine the sum $\mathbf{H}_P$ of the moments about P of the momenta of the particles of the vehicle, knowing that the distance from G to P is 10 m.

18.16 A homogeneous wire, of weight 2 lb/ft, is used to form the wire figure shown, which is suspended from point A. If an impulse $\mathbf{F}\Delta t = -(10 \text{ lb}\cdot\text{s})\mathbf{k}$ is applied at point D of coordinates $x = 3$ ft, $y = 2$ ft, $z = 3$ ft, determine (*a*) the velocity of the mass center of the wire figure, (*b*) the angular velocity of the figure.

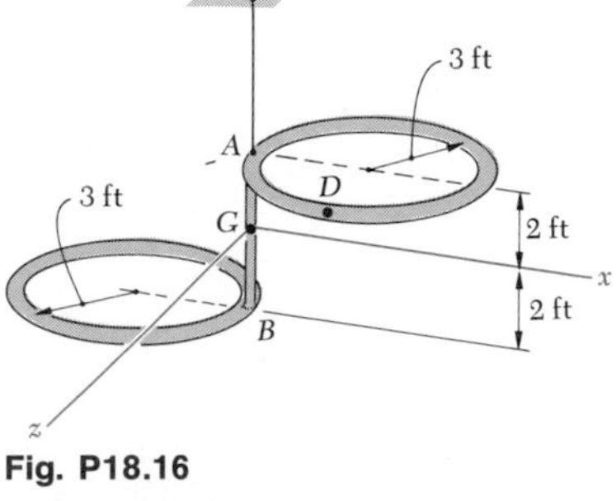

Fig. P18.16

18.17 Solve Prob. 18.16, assuming that the impulse applied at point D is $\mathbf{F}\Delta t = (10 \text{ lb}\cdot\text{s})\mathbf{j}$.

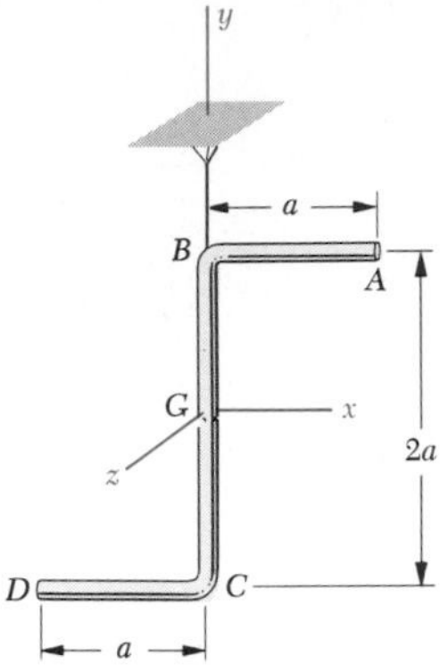

Fig. P18.18

18.18 A uniform rod of total mass m is bent into the shape shown and is suspended by a wire attached at B. The bent rod is hit at D in a direction perpendicular to the plane containing the rod (in the negative z direction). Denoting the corresponding impulse by $\mathbf{F}\Delta t$, determine (a) the velocity of the mass center of the rod, (b) the angular velocity of the rod.

18.19 Solve Prob. 18.18, assuming that the bent rod is hit at C.

18.20 Three slender homogeneous rods, each of mass m and length d, are welded together to form the assembly shown, which hangs from a wire at G. The assembly is hit at A in a vertical downward direction. Denoting the corresponding impulse by $\mathbf{F}\,\Delta t$, determine immediately after impact (a) the velocity of the mass center G, (b) the angular velocity of the assembly.

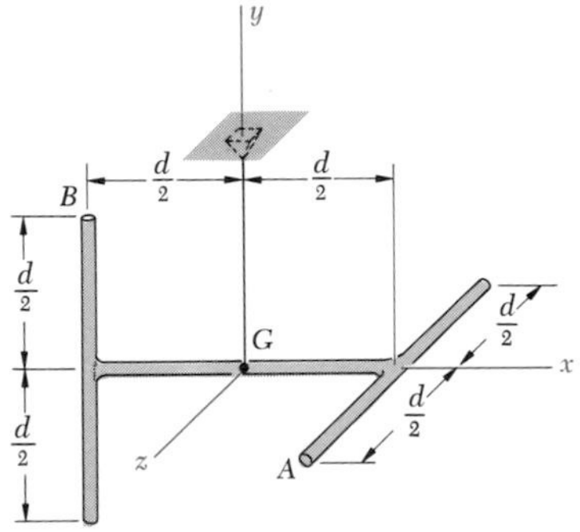

Fig. P18.20

18.21 Solve Prob. 18.20, assuming that the assembly is hit at B in a direction opposite to that of the z axis.

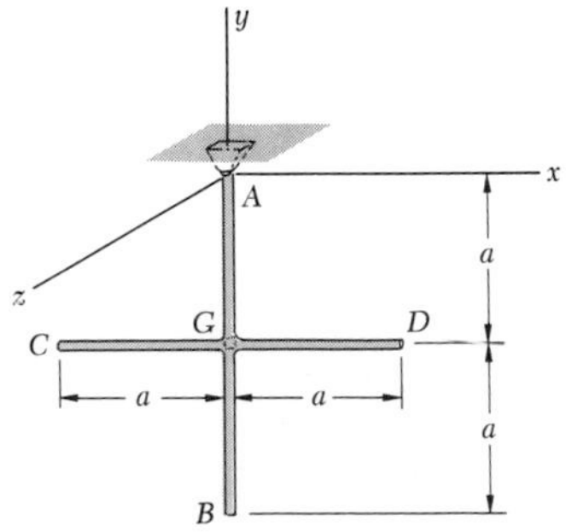

Fig. P18.22

18.22 A cross of total mass m, made of two rods AB and CD, each of length $2a$ and welded together at G, is suspended from a ball-and-socket joint at A. The cross is hit at C in a direction perpendicular to its plane (in the negative z direction). Denoting the corresponding impulse by $\mathbf{F}\,\Delta t$, determine immediately after impact (a) the angular velocity of the cross, (b) its instantaneous axis of rotation.

18.23 A bullet of mass m_0 is fired with an initial velocity $\mathbf{v}_0$ into a heavy circular plate of mass m which is suspended from a ball-and-socket joint at O. Knowing that the bullet strikes point A and becomes embedded in the plate, determine immediately after impact (*a*) the angular velocity of the plate, (*b*) its instantaneous axis of rotation.

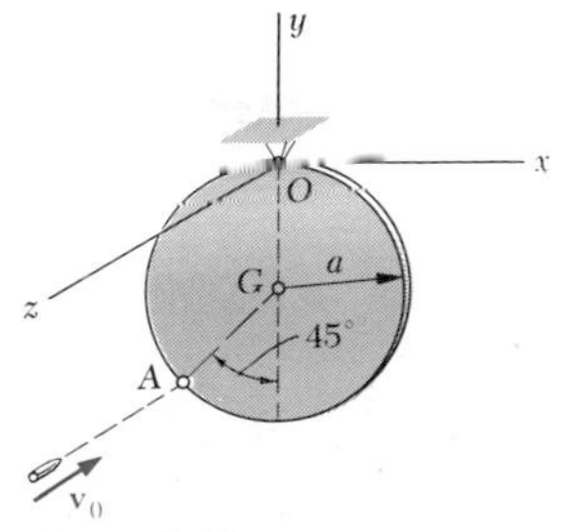

Fig. P18.23

18.24 A circular plate of radius a and mass m supported by a ball-and-socket joint at O was rotating about the y axis with a constant angular velocity $\boldsymbol{\omega} = \omega_0\mathbf{j}$ when an obstruction was suddenly introduced at A. Assuming that the impact at A is perfectly plastic, determine immediately after impact (*a*) the angular velocity of the plate, (*b*) the velocity of the mass center G.

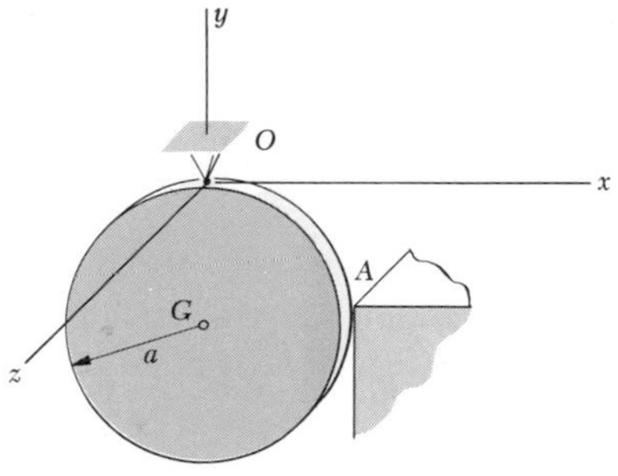

Fig. P18.24

18.25 Solve Prob. 18.24, assuming that, before the obstruction was introduced, the plate was rotating about the x axis with a constant angular velocity $\boldsymbol{\omega} = \omega_0\mathbf{i}$.

18.26 The angular velocity of a 1000-kg space capsule is $\boldsymbol{\omega} = (0.02 \text{ rad/s})\mathbf{i} + (0.10 \text{ rad/s})\mathbf{j}$ when two small jets are activated at A and B, each in a direction parallel to the z axis. Knowing that the radii of gyration of the capsule are $\bar{k}_x = \bar{k}_z = 1.00$ m and $\bar{k}_y = 1.25$ m, and that each jet produces a thrust of 50 N, determine (*a*) the required operating time of each jet if the angular velocity of the capsule is to be reduced to zero, (*b*) the resulting change in the velocity of the mass center G.

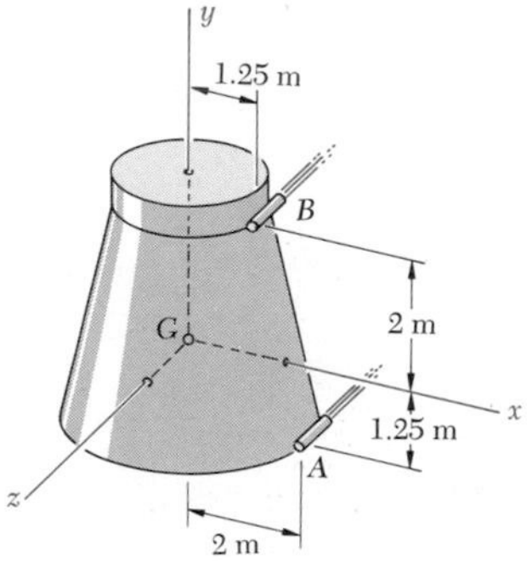

Fig. P18.26

18.27 If jet B in Prob. 18.26 is inoperative, determine (*a*) the required operating time of jet A to reduce the x component of the angular velocity $\boldsymbol{\omega}$ of the capsule to zero, (*b*) the resulting final angular velocity $\boldsymbol{\omega}$, (*c*) the resulting change in the velocity of the mass center G.

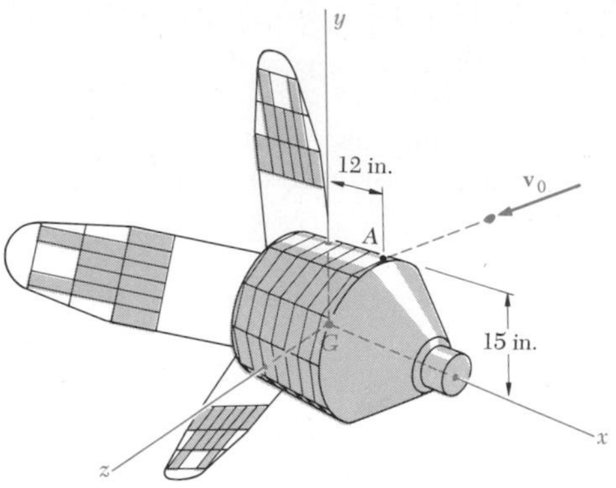

Fig. P18.28

18.28 A satellite weighing 320 lb has no angular velocity when it is struck at A by a 0.04-lb meteorite traveling with a velocity $\mathbf{v}_0 = -(2400 \text{ ft/s})\mathbf{i} - (1800 \text{ ft/s})\mathbf{j} + (4000 \text{ ft/s})\mathbf{k}$ relative to the satellite. Knowing that the radii of gyration of the satellite are $\bar{k}_x = 12$ in. and $\bar{k}_y = \bar{k}_z = 16$ in., determine the angular velocity of the satellite in rpm immediately after the meteorite has become imbedded.

18.29 Solve Prob. 18.28, assuming that, initially, the satellite was spinning about its axis of symmetry with an angular velocity of 12 rpm clockwise as viewed from the positive x axis.

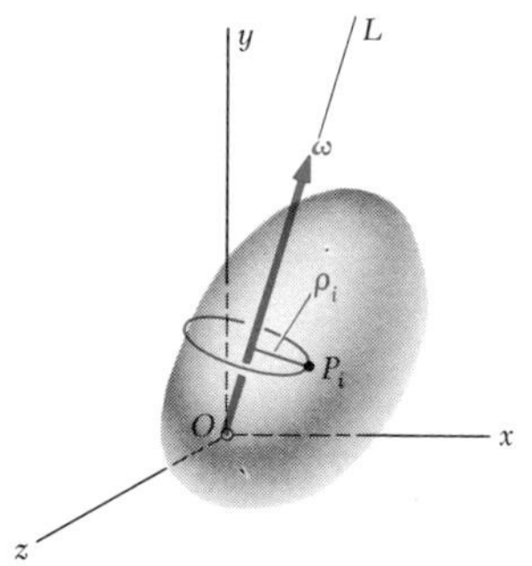

Fig. P18.30

18.30 Show that the kinetic energy of a rigid body with a fixed point O may be expressed as

$$T = \tfrac{1}{2} I_{OL} \omega^2$$

where $\boldsymbol{\omega}$ is the instantaneous angular velocity of the body and I_{OL} its moment of inertia about the line of action OL of $\boldsymbol{\omega}$. Derive this expression (a) from Eqs. (9.46) and (18.19), (b) by considering T as the sum of the kinetic energies of particles P_i describing circles of radius ρ_i about line OL.

18.31 Denoting respectively by $\boldsymbol{\omega}$, $\mathbf{H}_O$, and T the angular velocity, the angular momentum, and the kinetic energy of a rigid body with a fixed point O, (a) prove that

$$\mathbf{H}_O \cdot \boldsymbol{\omega} = 2T$$

(b) show that the angle θ between $\boldsymbol{\omega}$ and $\mathbf{H}_O$ will always be acute.

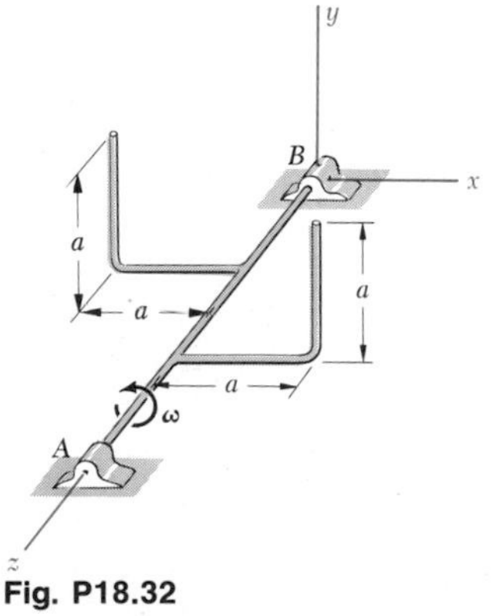

Fig. P18.32

18.32 The body shown is made of slender, homogeneous rods and may rotate freely in bearings at A and B. If the body is at rest when it is given a slight push, determine its angular velocity after it has rotated through 180°.

18.33 Determine the angular velocity of the body of Prob. 18.32 after it has rotated through 90°.

18.34 Determine the kinetic energy of the plate of Prob. 18.4.

18.35 Determine the kinetic energy of the disk of Prob. 18.3.

18.36 Determine the change in kinetic energy of the plate of Prob. 18.24 due to its impact with the obstruction.

18.37 Determine the change in kinetic energy of the plate of Prob. 18.25 due to its impact with the obstruction.

18.38 Determine the change in the kinetic energy of the satellite of Prob. 18.28 in its motion about its mass center due to the impact of the meteorite, knowing that before the impact the satellite was spinning about its axis of symmetry with an angular velocity of 12 rpm clockwise as viewed from the positive x axis.

18.39 Gear A rolls on the fixed gear B and rotates about the axle AD of length $L = 500$ mm which is rigidly attached at D to the vertical shaft DE. The shaft DE is made to rotate with a constant angular velocity $\boldsymbol{\omega}_1$ of magnitude 4 rad/s. Assuming that gear A can be approximated by a thin disk of mass 2 kg and radius $a = 100$ mm, and that $\beta = 30°$, determine (*a*) the angular momentum of gear A about point D, (*b*) the kinetic energy of gear A.

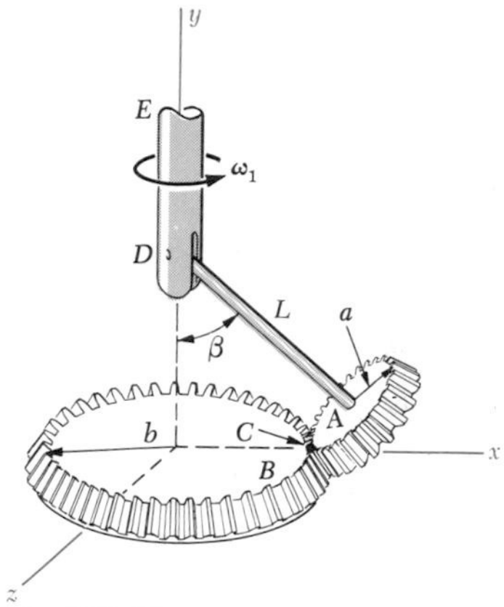

Fig. P18.39

*18.5. Motion of a Rigid Body in Three Dimensions.

As was indicated in Sec. 18.2, the fundamental equations

$$\Sigma \mathbf{F} = m\bar{\mathbf{a}} \quad (18.1)$$
$$\Sigma \mathbf{M}_G = \dot{\mathbf{H}}_G \quad (18.2)$$

remain valid in the most general case of the motion of a rigid body. Before Eq. (18.2) could be applied to the three-dimensional motion of a rigid body, however, it was necessary to derive Eqs. (18.7), which relate the components of the angular momentum $\mathbf{H}_G$ and of the angular velocity $\boldsymbol{\omega}$. It still remains for us to find an effective and convenient way for computing the components of the derivative $\dot{\mathbf{H}}_G$ of the angular momentum.

Since $\mathbf{H}_G$ represents the angular momentum of the body in its motion relative to centroidal axes $GX'Y'Z'$ of fixed orientation (Fig. 18.9), and since $\dot{\mathbf{H}}_G$ represents the rate of change of $\mathbf{H}_G$ with respect to the same axes, it would seem natural to use components of $\boldsymbol{\omega}$ and $\mathbf{H}_G$ along the axes X', Y', Z' in writing the relations (18.7). But, since the body rotates, its moments and products of inertia would change continuously, and it would be necessary to determine their values as functions of the time. It is therefore more convenient to use axes x, y, z attached to the body, thus making sure that its moments and products of inertia will maintain the same values during the motion. This is permis-

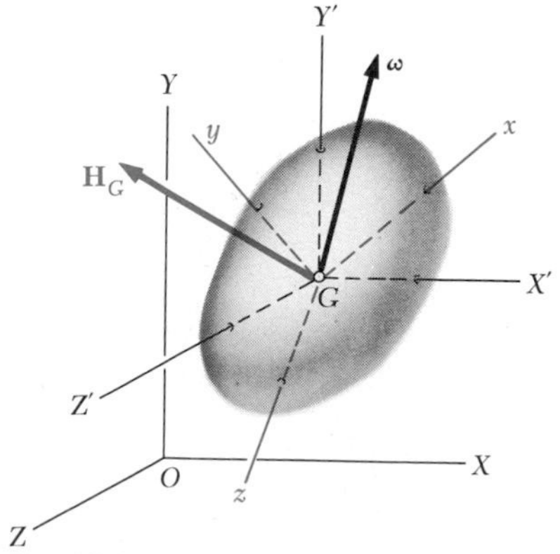

Fig. 18.9

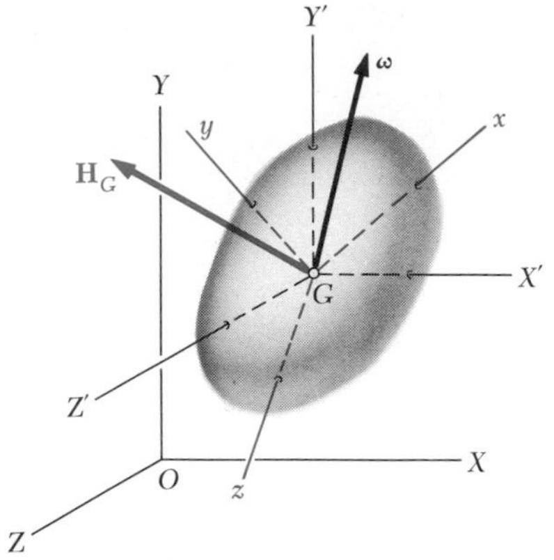

Fig. 18.9 (*repeated*)

sible since, as indicated earlier, the transformation of $\boldsymbol{\omega}$ into $\mathbf{H}_G$ is independent of the system of coordinate axes which has been selected. The angular velocity $\boldsymbol{\omega}$, however, should still be *defined* with respect to the frame $GX'Y'Z'$ of fixed orientation. The vector $\boldsymbol{\omega}$ may then be *resolved* into components along the rotating x, y, and z axes. Applying the relations (18.7), we obtain the *components* of the vector $\mathbf{H}_G$ along the rotating axes. The vector $\mathbf{H}_G$, however, represents the angular momentum about G of the body *in its motion relative to the frame* $GX'Y'Z'$.

Differentiating with respect to t the components of the angular momentum in (18.7), we define the rate of change of the vector $\mathbf{H}_G$ with respect to the rotating frame $Gxyz$:

$$(\dot{\mathbf{H}}_G)_{Gxyz} = \dot{H}_x\mathbf{i} + \dot{H}_y\mathbf{j} + \dot{H}_z\mathbf{k} \qquad (18.21)$$

where $\mathbf{i}$, $\mathbf{j}$, $\mathbf{k}$ are the unit vectors along the rotating axes. Recalling from Sec. 15.10 that the rate of change $\dot{\mathbf{H}}_G$ of the vector $\mathbf{H}_G$ with respect to the frame $GX'Y'Z'$ may be obtained by adding to $(\dot{\mathbf{H}}_G)_{Gxyz}$ the vector product $\boldsymbol{\Omega} \times \mathbf{H}_G$, where $\boldsymbol{\Omega}$ denotes the angular velocity of the rotating frame, we write

$$\dot{\mathbf{H}}_G = (\dot{\mathbf{H}}_G)_{Gxyz} + \boldsymbol{\Omega} \times \mathbf{H}_G \qquad (18.22)$$

where $\mathbf{H}_G$ = angular momentum of the body with respect to the frame $GX'Y'Z'$ of fixed orientation

$(\dot{\mathbf{H}}_G)_{Gxyz}$ = rate of change of $\mathbf{H}_G$ with respect to the rotating frame $Gxyz$, to be computed from the relations (18.7) and (18.21)

$\boldsymbol{\Omega}$ = angular velocity of the rotating frame $Gxyz$

Substituting for $\dot{\mathbf{H}}_G$ from (18.22) into (18.2), we have

$$\Sigma\mathbf{M}_G = (\dot{\mathbf{H}}_G)_{Gxyz} + \boldsymbol{\Omega} \times \mathbf{H}_G \qquad (18.23)$$

If, as it has been assumed in this discussion, the rotating frame is attached to the body, its angular velocity $\boldsymbol{\Omega}$ is identically equal to the angular velocity $\boldsymbol{\omega}$ of the body. There are many applications, however, where it is advantageous to use a frame of reference which is not actually attached to the body, but rotates in an independent manner. For example, if the body considered is axisymmetrical, as in Sample Prob. 18.5 or Sec. 18.9, it is possible to select a frame of reference with respect to which the moments and products of inertia of the body remain constant, but which rotates less than the body itself.† As a result, simpler expressions

† More specifically, the frame of reference will have no spin (see Sec. 18.9).

may be obtained for the angular velocity $\boldsymbol{\omega}$ and the angular momentum $\mathbf{H}_G$ of the body than would have been possible if the frame of reference had actually been attached to the body. It is clear that in such cases the angular velocity $\boldsymbol{\Omega}$ of the rotating frame and the angular velocity $\boldsymbol{\omega}$ of the body are different.

***18.6. Euler's Equations of Motion. Extension of D'Alembert's Principle to the Motion of a Rigid Body in Three Dimensions.** If the x, y, and z axes are chosen to coincide with the principal axes of inertia of the body, the simplified relations (18.10) may be used to determine the components of the angular momentum $\mathbf{H}_G$. Omitting the primes from the subscripts, we write

$$\mathbf{H}_G = \bar{I}_x\omega_x\mathbf{i} + \bar{I}_y\omega_y\mathbf{j} + \bar{I}_z\omega_z\mathbf{k} \tag{18.24}$$

where $\bar{I}_x$, $\bar{I}_y$, and $\bar{I}_z$ denote the principal centroidal moments of inertia of the body. Substituting for $\mathbf{H}_G$ from (18.24) into (18.23) and setting $\boldsymbol{\Omega} = \boldsymbol{\omega}$, we obtain the three scalar equations

$$\begin{aligned}
\Sigma M_x &= \bar{I}_x\dot{\omega}_x - (\bar{I}_y - \bar{I}_z)\omega_y\omega_z \\
\Sigma M_y &= \bar{I}_y\dot{\omega}_y - (\bar{I}_z - \bar{I}_x)\omega_z\omega_x \\
\Sigma M_z &= \bar{I}_z\dot{\omega}_z - (\bar{I}_x - \bar{I}_y)\omega_x\omega_y
\end{aligned} \tag{18.25}$$

These equations, called *Euler's equations of motion* after the Swiss mathematician Leonhard Euler (1707–1783), may be used to analyze the motion of a rigid body about its mass center. In the following sections, however, we shall use Eq. (18.23) in preference to Eqs. (18.25), since the former is more general, and the compact vectorial form in which it is expressed is easier to remember.

Writing Eq. (18.1) in scalar form, we obtain the three additional equations

$$\Sigma F_x = m\bar{a}_x \qquad \Sigma F_y = m\bar{a}_y \qquad \Sigma F_z = m\bar{a}_z \tag{18.26}$$

which, together with Euler's equations, form a system of six differential equations. Given appropriate initial conditions, these differential equations have a unique solution. Thus, the motion of a rigid body in three dimensions is completely defined by the resultant and the moment resultant of the external forces acting on it. This result will be recognized as a generalization of a similar result obtained in Sec. 16.4 in the case of the plane motion of a rigid slab. It follows that, in three as well as in two dimensions, two systems of forces which are equipollent are also equivalent; i.e., they have the same effect on a given rigid body.

Considering in particular the system of the external forces acting on a rigid body (Fig. 18.10*a*) and the system of the effective forces associated with the particles forming the rigid body (Fig. 18.10*b*), we may state that the two systems—which were shown in Sec. 14.1 to be equipollent—are also equivalent. This is the extension of D'Alembert's principle to the three-dimensional motion of a rigid body. Replacing the effective forces in Fig. (18.10*b*) by an equivalent force-couple system, we verify that the system of the external forces acting on a rigid body in three-dimensional motion is equivalent to the system consisting of the vector $m\bar{\mathbf{a}}$ attached at the mass center G of the body and the couple of moment $\dot{\mathbf{H}}_G$ (Fig. 18.11), where $\dot{\mathbf{H}}_G$ is obtained from the relations (18.7) and (18.22). Problems involving the three-dimensional motion of a rigid body may be solved by drawing the two sketches shown in Fig. 18.11 and writing appropriate equations relating the components or moments of the external and effective forces (see Sample Prob. 18.3).

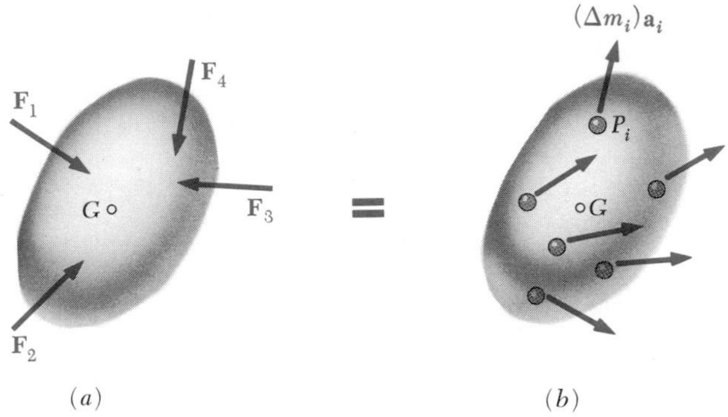

Fig. 18.10

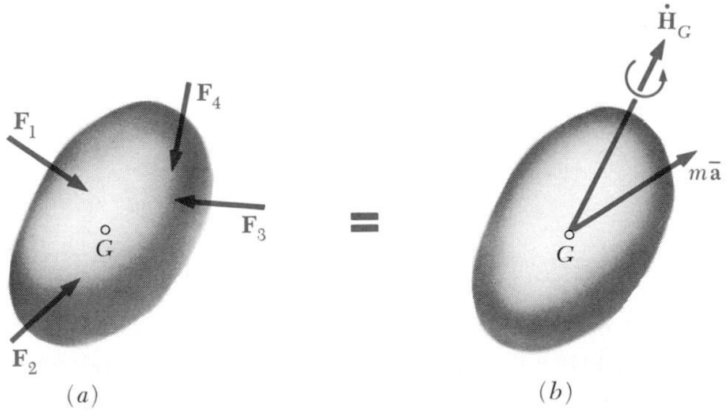

Fig. 18.11

***18.7. Motion of a Rigid Body about a Fixed Point.** When a rigid body is constrained to rotate about a fixed point O, it is desirable to write an equation involving the moments about O of the external and effective forces, since this equation will not contain the unknown reaction at O. While such an equation may be obtained from Fig. 18.11, it may be more convenient to write it by considering the rate of change of the angular momentum $\mathbf{H}_O$ of the body about the fixed point O

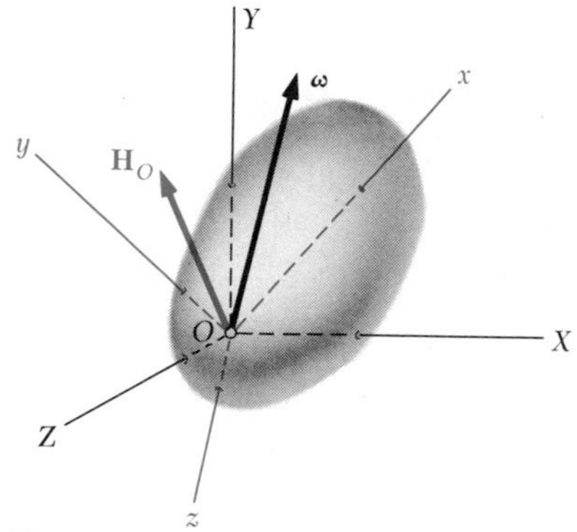

Fig. 18.12

(Fig. 18.12). Recalling Eq. (14.11), we write

$$\Sigma \mathbf{M}_O = \dot{\mathbf{H}}_O \qquad (18.27)$$

where $\dot{\mathbf{H}}_O$ denotes the rate of change of the vector $\mathbf{H}_O$ with respect to the fixed frame $OXYZ$. A derivation similar to that used in Sec. 18.5 enables us to relate $\dot{\mathbf{H}}_O$ to the rate of change $(\dot{\mathbf{H}}_O)_{Oxyz}$ of $\mathbf{H}_O$ with respect to the rotating frame $Oxyz$. Substitution into (18.27) leads to the equation

$$\Sigma \mathbf{M}_O = (\dot{\mathbf{H}}_O)_{Oxyz} + \mathbf{\Omega} \times \mathbf{H}_O \qquad (18.28)$$

where $\Sigma \mathbf{M}_O$ = sum of the moments about O of the forces applied to the rigid body
$\mathbf{H}_O$ = angular momentum of the body with respect to the fixed frame $OXYZ$
$(\dot{\mathbf{H}}_O)_{Oxyz}$ = rate of change of $\mathbf{H}_O$ with respect to the rotating frame $Oxyz$, to be computed from the relations (18.12)
$\mathbf{\Omega}$ = angular velocity of the rotating frame $Oxyz$†

† Read last paragraph of Sec. 18.5, replacing $\mathbf{H}_G$ by $\mathbf{H}_O$.

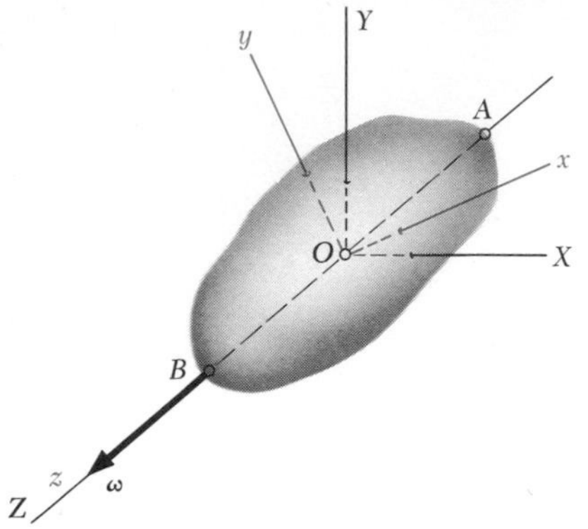

Fig. 18.13

∗18.8 Rotation of a Rigid Body about a Fixed Axis. We shall use Eq. (18.28), which was derived in the preceding section, to analyze the motion of a rigid body constrained to rotate about a fixed axis AB (Fig. 18.13). First, we note that the angular velocity of the body with respect to the fixed frame $OXYZ$ is represented by the vector $\boldsymbol{\omega}$ directed along the axis of rotation. Attaching the moving frame of reference $Oxyz$ to the body, with the z axis along AB, we have $\boldsymbol{\omega} = \omega\mathbf{k}$. Substituting $\omega_x = 0$, $\omega_y = 0$, $\omega_z = \omega$ into the relations (18.12), we obtain the components along the rotating axes of the angular momentum $\mathbf{H}_O$ of the body about O:

$$H_x = -P_{xz}\omega \qquad H_y = -P_{yz}\omega \qquad H_z = I_z\omega$$

Since the frame $Oxyz$ is attached to the body, we have $\boldsymbol{\Omega} = \boldsymbol{\omega}$ and Eq. (18.28) yields

$$\begin{aligned}\Sigma\mathbf{M}_O &= (\dot{\mathbf{H}}_O)_{Oxyz} + \boldsymbol{\omega} \times \mathbf{H}_O \\ &= (-P_{xz}\mathbf{i} - P_{yz}\mathbf{j} + I_z\mathbf{k})\dot{\omega} + \omega\mathbf{k} \times (-P_{xz}\mathbf{i} - P_{yz}\mathbf{j} + I_z\mathbf{k})\omega \\ &= (-P_{xz}\mathbf{i} - P_{yz}\mathbf{j} + I_z\mathbf{k})\alpha + (-P_{xz}\mathbf{j} + P_{yz}\mathbf{i})\omega^2\end{aligned}$$

The result obtained may be expressed by the three scalar equations

$$\begin{aligned}\Sigma M_x &= -P_{xz}\alpha + P_{yz}\omega^2 \\ \Sigma M_y &= -P_{yz}\alpha - P_{xz}\omega^2 \\ \Sigma M_z &= I_z\alpha\end{aligned} \qquad (18.29)$$

When the forces applied to the body are known, the angular acceleration α may be obtained from the last of Eqs. (18.29). The angular velocity ω is then determined by integration and the values obtained for α and ω may be substitued into the first two equations (18.29). These equations, plus the three equations (18.26), which define the motion of the mass center of the body, may then be used to determine the reactions at the bearings A and B.

It should be noted that axes other than the ones shown in Fig. 18.12 may be selected to analyze the rotation of a rigid body about a fixed axis. In many cases, the principal axes of inertia of the body will be found more advantageous. It is wise, therefore, to revert to Eq. (18.28) and to select the system of axes which best fits the problem under consideration.

If the rotating body is symmetrical with respect to the xy plane, the products of inertia P_{xz} and P_{yz} are equal to zero and Eqs. (18.29) reduce to

$$\Sigma M_x = 0 \qquad \Sigma M_y = 0 \qquad \Sigma M_z = I_z\alpha \qquad (18.30)$$

which is in accord with the results obtained in Chap. 16. If, on the other hand, the products of inertia P_{xz} and P_{yz} are different from zero, the sum of the moments of the external forces about the x and y axes will also be different from zero, even when the body rotates at a constant rate ω. Indeed, in the latter case, Eqs. (18.29) yield

$$\Sigma M_x = P_{yz}\omega^2 \qquad \Sigma M_y = -P_{xz}\omega^2 \qquad \Sigma M_z = 0 \quad (18.31)$$

This last observation leads us to discuss the *balancing of rotating shafts*. Consider, for instance, the crankshaft shown in Fig. 18.14*a*, which is symmetrical about its mass center G. We first observe that, when the crankshaft is at rest, it exerts no lateral thrust on its supports, since its center of gravity G is located directly above A. The shaft is said to be *statically balanced*. The reaction at A, often referred to as a *static reaction*, is vertical and its magnitude is equal to the weight W of the shaft. Let us now assume that the shaft rotates with a constant angular velocity $\boldsymbol{\omega}$. Attaching our frame of reference to the shaft, with its origin at G, the z axis along AB, and the y axis in the plane of symmetry of the shaft (Fig. 18.14*b*), we note that P_{xz} is zero and that P_{yz} is positive. According to Eqs. (18.31), the external forces must include a couple of moment $P_{yz}\omega^2\mathbf{i}$. Since this couple is formed by the reaction at B and the horizontal component of the reaction at A, we have

$$\mathbf{A}_y = \frac{P_{yz}\omega^2}{l}\mathbf{j} \qquad \mathbf{B} = -\frac{P_{yz}\omega^2}{l}\mathbf{j} \qquad (18.32)$$

Since the bearing reactions are proportional to ω^2, the shaft will have a tendency to tear away from its bearings when rotating at high speeds. Moreover, since the bearing reactions $\mathbf{A}_y$ and $\mathbf{B}$, called *dynamic reactions*, are contained in the yz plane, they rotate with the shaft and cause the structure supporting it to vibrate. These undesirable effects will be avoided if, by rearranging the distribution of mass around the shaft, or by adding corrective masses, we let P_{yz} become equal to zero. The dynamic reactions $\mathbf{A}_y$ and $\mathbf{B}$ will vanish and the reactions at the bearings will reduce to the static reaction $\mathbf{A}_z$, the direction of which is fixed. The shaft will then be *dynamically as well as statically balanced*.

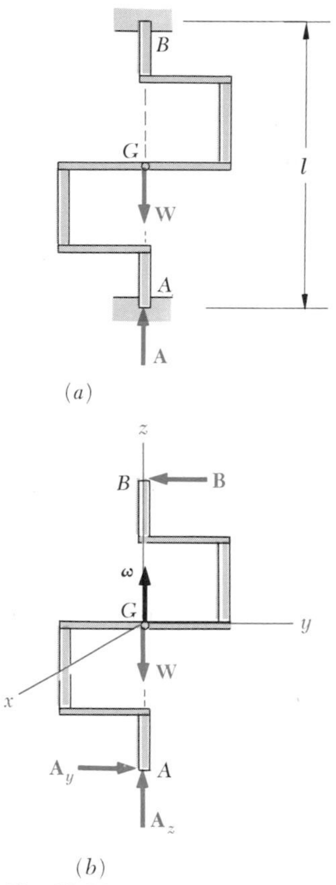

Fig. 18.14

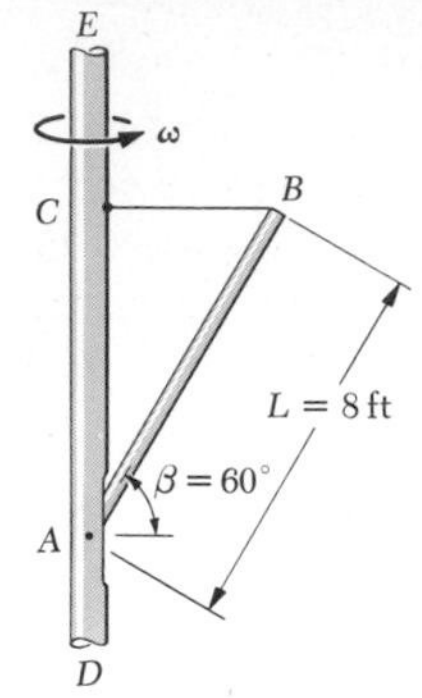

SAMPLE PROBLEM 18.3

A slender rod AB of length $L = 8$ ft and weight $W = 40$ lb is pinned at A to a vertical axle DE which rotates with a constant angular velocity $\boldsymbol{\omega}$ of 15 rad/s. The rod is maintained in position by means of a horizontal wire BC attached to the axle and to the end B of the rod. Determine the tension in the wire and the reaction at A.

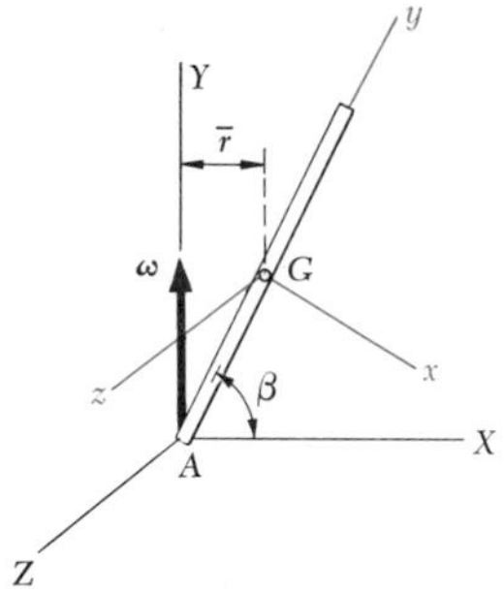

Solution. The effective forces reduce to the vector $m\bar{\mathbf{a}}$ attached at G and the couple $\dot{\mathbf{H}}_G$. Since G describes a horizontal circle of radius $\bar{r} = \frac{1}{2}L \cos \beta$ at the constant rate ω, we have

$$\bar{\mathbf{a}} = \mathbf{a}_n = -\bar{r}\omega^2 \mathbf{I} = -(\tfrac{1}{2}L \cos \beta)\omega^2 \mathbf{I} = -(450 \text{ ft/s}^2)\mathbf{I}$$
$$m\bar{\mathbf{a}} = \frac{40}{g}(-450\mathbf{I}) = -(559 \text{ lb})\mathbf{I}$$

Determination of $\dot{\mathbf{H}}_G$. We first compute the angular momentum $\mathbf{H}_G$. Using the principal centroidal axes of inertia x, y, z, we write

$$\bar{I}_x = \tfrac{1}{12}mL^2 \qquad \bar{I}_y = 0 \qquad \bar{I}_z = \tfrac{1}{12}mL^2$$
$$\omega_x = -\omega \cos \beta \qquad \omega_y = \omega \sin \beta \qquad \omega_z = 0$$
$$\mathbf{H}_G = \bar{I}_x\omega_x \mathbf{i} + \bar{I}_y\omega_y \mathbf{j} + \bar{I}_z\omega_z \mathbf{k}$$
$$\mathbf{H}_G = -\tfrac{1}{12}mL^2\omega \cos \beta \, \mathbf{i}$$

The rate of change $\dot{\mathbf{H}}_G$ of $\mathbf{H}_G$ with respect to axes of fixed orientation is obtained from Eq. (18.22). Observing that the rate of change $(\dot{\mathbf{H}}_G)_{Gxyz}$ of $\mathbf{H}_G$ with respect to the rotating frame $Gxyz$ is zero, and that the angular velocity $\boldsymbol{\Omega}$ of that frame is equal to the angular velocity $\boldsymbol{\omega}$ of the rod, we have

$$\dot{\mathbf{H}}_G = (\dot{\mathbf{H}}_G)_{Gxyz} + \boldsymbol{\omega} \times \mathbf{H}_G$$
$$\dot{\mathbf{H}}_G = 0 + (-\omega \cos \beta \, \mathbf{i} + \omega \sin \beta \, \mathbf{j}) \times (-\tfrac{1}{12}mL^2\omega \cos \beta \, \mathbf{i})$$
$$\dot{\mathbf{H}}_G = \tfrac{1}{12}mL^2\omega^2 \sin \beta \cos \beta \, \mathbf{k} = (645 \text{ lb}\cdot\text{ft})\mathbf{k}$$

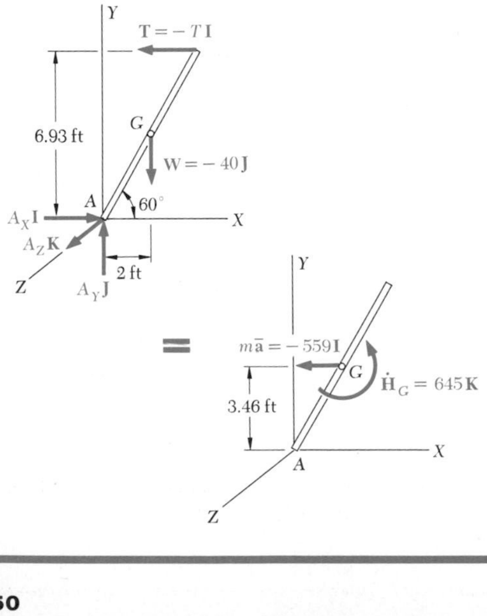

Equations of Motion. Expressing that the system of the external forces is equivalent to the system of the effective forces, we write

$\Sigma \mathbf{M}_A = \Sigma(\mathbf{M}_A)_{\text{eff}}$:
$$6.93\mathbf{J} \times (-T\mathbf{I}) + 2\mathbf{I} \times (-40\mathbf{J}) = 3.46\mathbf{J} \times (-559\mathbf{I}) + 645\mathbf{K}$$
$$(6.93T - 80)\mathbf{K} = (1934 + 645)\mathbf{K} \qquad T = 384 \text{ lb} \blacktriangleleft$$

$\Sigma \mathbf{F} = \Sigma \mathbf{F}_{\text{eff}}$: $\quad A_X\mathbf{I} + A_Y\mathbf{J} + A_Z\mathbf{K} - 384\mathbf{I} - 40\mathbf{J} = -559\mathbf{I}$
$$\mathbf{A} = -(175 \text{ lb})\mathbf{I} + (40 \text{ lb})\mathbf{J} \blacktriangleleft$$

Remark. The value of T could have been obtained from $\mathbf{H}_A$ and Eq. (18.28). However, the method used here also yields the reaction at A. Moreover, it draws attention to the effect of the asymmetry of the rod on the solution of the problem by clearly showing that both the vector $m\bar{\mathbf{a}}$ and the couple $\dot{\mathbf{H}}_G$ must be used to represent the effective forces.

SAMPLE PROBLEM 18.4

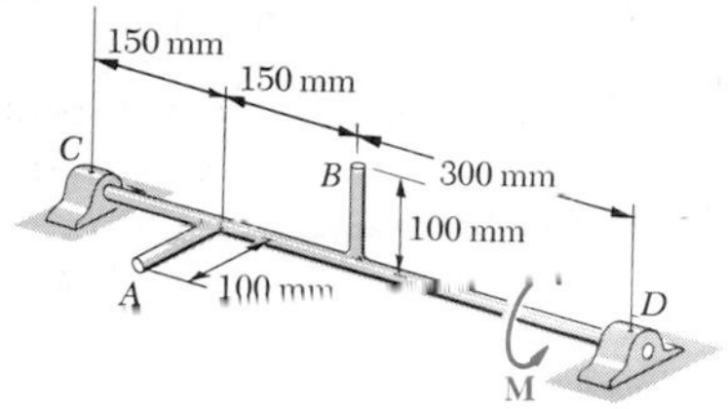

Two 100-mm rods A and B, each of mass 300 g, are welded to the shaft CD which is supported by bearings at C and D. If a couple $\mathbf{M}$ of magnitude equal to 6 N·m is applied to the shaft, determine the components of the dynamic reactions at C and D at the instant when the shaft has reached an angular velocity of 1200 rpm. Neglect the moment of inertia of the shaft itself.

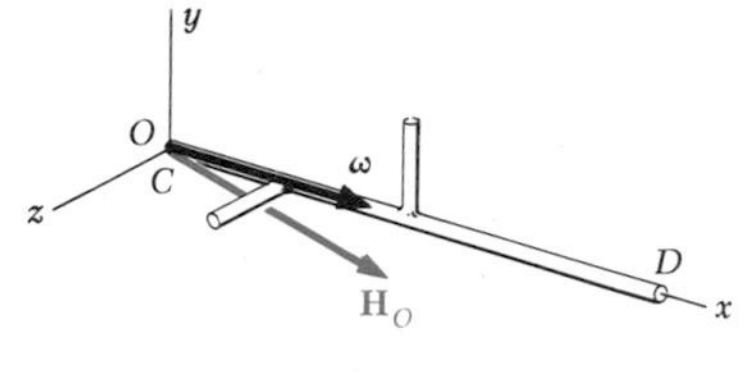

Angular Momentum about O. We attach to the body the frame of reference $Oxyz$ and note that the axes chosen are not principal axes of inertia for the body. Since the body rotates about the x axis, we have $\omega_x = \omega$ and $\omega_y = \omega_z = 0$. Substituting into Eqs. (18.12),

$$H_x = I_x\omega \qquad H_y = -P_{xy}\omega \qquad H_z = -P_{xz}\omega$$
$$\mathbf{H}_O = (I_x\mathbf{i} - P_{xy}\mathbf{j} - P_{xz}\mathbf{k})\omega$$

Moments of the External Forces about O. Since the frame of reference rotates with the angular velocity $\boldsymbol{\omega}$, Eq. (18.28) yields

$$\begin{aligned}\Sigma\mathbf{M}_O &= (\dot{\mathbf{H}}_O)_{Oxyz} + \boldsymbol{\omega} \times \mathbf{H}_O \\ &= (I_x\mathbf{i} - P_{xy}\mathbf{j} - P_{xz}\mathbf{k})\alpha + \omega\mathbf{i} \times (I_x\mathbf{i} - P_{xy}\mathbf{j} - P_{xz}\mathbf{k})\omega \\ &= I_x\alpha\mathbf{i} - (P_{xy}\alpha - P_{xz}\omega^2)\mathbf{j} - (P_{xz}\alpha + P_{xy}\omega^2)\mathbf{k} \qquad (1)\end{aligned}$$

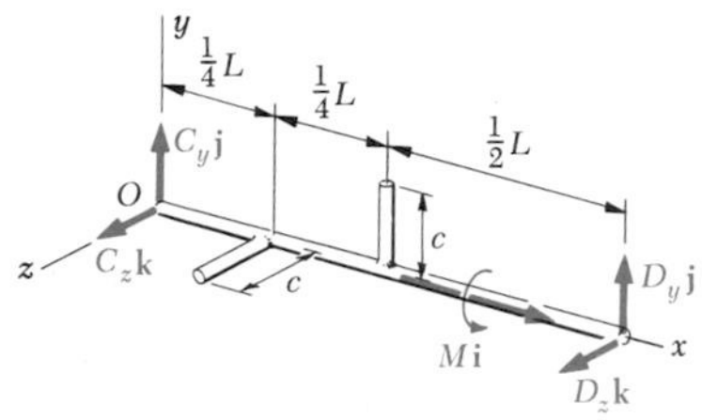

Dynamic Reaction at D. The external forces consist of the weights of the shaft and rods, the couple $\mathbf{M}$, the static reactions at C and D, and the dynamic reactions at C and D. Since the weights and static reactions are balanced, the external forces reduce to the couple $\mathbf{M}$ and the dynamic reactions $\mathbf{C}$ and $\mathbf{D}$ as shown in the figure. Taking moments about O, we have

$$\Sigma\mathbf{M}_O = L\mathbf{i} \times (D_y\mathbf{j} + D_z\mathbf{k}) + M\mathbf{i} = M\mathbf{i} - D_zL\mathbf{j} + D_yL\mathbf{k} \qquad (2)$$

Equating the coefficients of the unit vector $\mathbf{i}$ in (1) and (2):

$$M = I_x\alpha \qquad M = 2(\tfrac{1}{3}mc^2)\alpha \qquad \alpha = 3M/2mc^2$$

Equating the coefficients of $\mathbf{k}$ and $\mathbf{j}$ in (1) and (2):

$$D_y = -(P_{xz}\alpha + P_{xy}\omega^2)/L \qquad D_z = (P_{xy}\alpha - P_{xz}\omega^2)/L \qquad (3)$$

Using the parallel-axis theorem, and noting that the product of inertia of each rod is zero with respect to centroidal axes, we have

$$P_{xy} = \Sigma m\bar{x}\bar{y} = m(\tfrac{1}{2}L)(\tfrac{1}{2}c) = \tfrac{1}{4}mLc$$
$$P_{xz} = \Sigma m\bar{x}\bar{z} = m(\tfrac{1}{4}L)(\tfrac{1}{2}c) = \tfrac{1}{8}mLc$$

Substituting into (3) the values found for P_{xy}, P_{xz}, and α:

$$D_y = -\tfrac{3}{16}(M/c) - \tfrac{1}{4}mc\omega^2 \qquad D_z = \tfrac{3}{8}(M/c) - \tfrac{1}{8}mc\omega^2$$

Substituting $\omega = 1200$ rpm $= 125.7$ rad/s, $c = 0.100$ m, $M = 6$ N·m, and $m = 0.300$ kg, we have

$$D_y = -129.8 \text{ N} \qquad D_z = -36.8 \text{ N} \blacktriangleleft$$

Dynamic Reaction at C. Using a frame of reference attached at D, we obtain equations similar to Eqs. (3), which yield

$$C_y = -152.2 \text{ N} \qquad C_z = -155.2 \text{ N} \blacktriangleleft$$

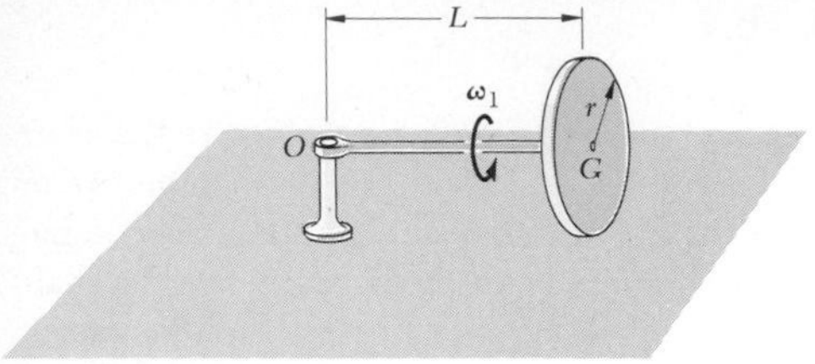

SAMPLE PROBLEM 18.5

A homogeneous disk of radius r and mass m is mounted on an axle OG of length L and negligible mass. The axle is pivoted at the fixed point O and the disk is constrained to roll on a horizontal floor. Knowing that the disk rotates counterclockwise at the constant rate ω_1 about the axle, determine (*a*) the force (assumed vertical) exerted by the floor on the disk, (*b*) the reaction at the pivot O.

Solution. The effective forces reduce to the vector $m\bar{\mathbf{a}}$ attached at G and the couple $\dot{\mathbf{H}}_G$. Recalling from Sample Prob. 18.2 that the axle rotates about the y axis at the rate $\omega_2 = r\omega_1/L$, we write

$$m\bar{\mathbf{a}} = -mL\omega_2^2\mathbf{i} = -mL(r\omega_1/L)^2\mathbf{i} = -(mr^2\omega_1^2/L)\mathbf{i} \qquad (1)$$

Determination of $\dot{\mathbf{H}}_G$. We recall from Sample Prob. 18.2 that the angular momentum of the disk about G is

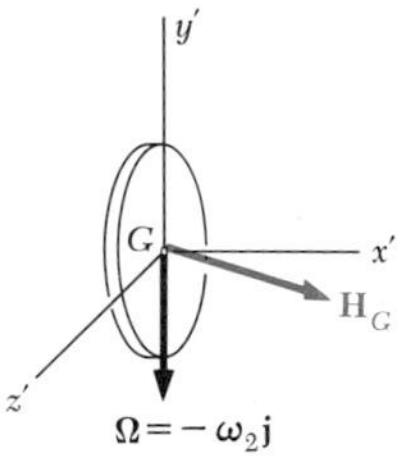

$$\mathbf{H}_G = \tfrac{1}{2}mr^2\omega_1\left(\mathbf{i} - \frac{r}{2L}\mathbf{j}\right)$$

where $\mathbf{H}_G$ is resolved into components along the rotating axes x', y', z', with x' along OG and y' vertical. The rate of change $\dot{\mathbf{H}}_G$ of $\mathbf{H}_G$ with respect to axes of fixed orientation is obtained from Eq. (18.22). Noting that the rate of change $(\dot{\mathbf{H}}_G)_{Gx'y'z'}$ of $\mathbf{H}_G$ with respect to the rotating frame is zero, and that the angular velocity $\boldsymbol{\Omega}$ of that frame is

$$\boldsymbol{\Omega} = -\omega_2\mathbf{j} = -(r\omega_1/L)\mathbf{j}$$

we have

$$\dot{\mathbf{H}}_G = (\dot{\mathbf{H}}_G)_{Gx'y'z'} + \boldsymbol{\Omega} \times \mathbf{H}_G$$

$$= 0 - \frac{r\omega_1}{L}\mathbf{j} \times \tfrac{1}{2}mr^2\omega_1\left(\mathbf{i} - \frac{r}{2L}\mathbf{j}\right)$$

$$= \tfrac{1}{2}mr^2(r/L)\omega_1^2\mathbf{k} \qquad (2)$$

Equations of Motion. Expressing that the system of the external forces is equivalent to the system of the effective forces, we write

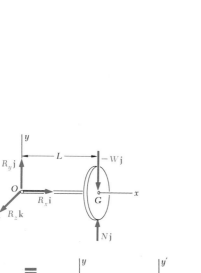

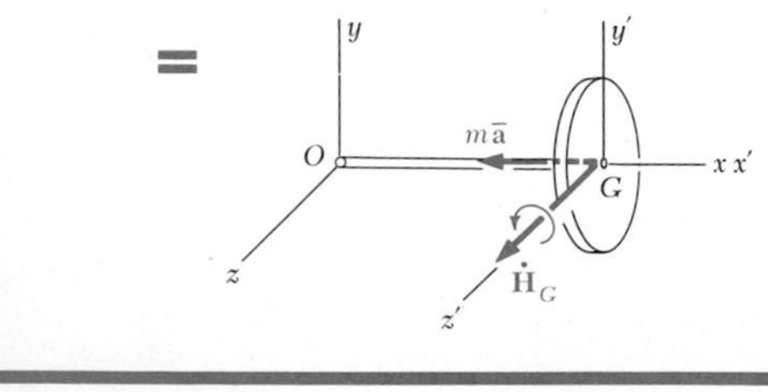

$\Sigma\mathbf{M}_O = \Sigma(\mathbf{M}_O)_{\text{eff}}$: $\qquad L\mathbf{i} \times (N\mathbf{j} - W\mathbf{j}) = \dot{\mathbf{H}}_G$

$$(N - W)L\mathbf{k} = \tfrac{1}{2}mr^2(r/L)\omega_1^2\mathbf{k}$$

$$N = W + \tfrac{1}{2}mr(r/L)^2\omega_1^2 \qquad \mathbf{N} = [W + \tfrac{1}{2}mr(r/L)^2\omega_1^2]\mathbf{j} \quad (3) \blacktriangleleft$$

$\Sigma\mathbf{F} = \Sigma\mathbf{F}_{\text{eff}}$: $\qquad \mathbf{R} + N\mathbf{j} - W\mathbf{j} = m\bar{\mathbf{a}}$

Substituting for N from (3), for $m\bar{\mathbf{a}}$ from (1), and solving for $\mathbf{R}$:

$$\mathbf{R} = -(mr^2\omega_1^2/L)\mathbf{i} - \tfrac{1}{2}mr(r/L)^2\omega_1^2\mathbf{j}$$

$$\mathbf{R} = -\frac{mr^2\omega_1^2}{L}\left(\mathbf{i} + \frac{r}{2L}\mathbf{j}\right) \blacktriangleleft$$

PROBLEMS

18.40 Determine the rate of change $\dot{\mathbf{H}}_G$ of the angular momentum $\mathbf{H}_G$ of the disk of Prob. 18.2.

18.41 Determine the rate of change $\dot{\mathbf{H}}_G$ of the angular momentum $\mathbf{H}_G$ of the disk of Prob. 18.3, assuming that the angular velocity $\boldsymbol{\omega}$ of axle *AB* remains constant.

18.42 Determine the rate of change $\dot{\mathbf{H}}_G$ of the angular momentum $\mathbf{H}_G$ of the plate of Prob. 18.4, assuming that its angular velocity $\boldsymbol{\omega}$ remains constant.

18.43 Determine the rate of change $\dot{\mathbf{H}}_A$ of the angular momentum $\mathbf{H}_A$ of the disk of Prob. 18.6.

18.44 Determine the rate of change $\dot{\mathbf{H}}_G$ of the angular momentum $\mathbf{H}_G$ of the plate of Prob. 18.4 if, at the instant considered, the angular velocity $\boldsymbol{\omega}$ of the plate is 4 rad/s and is increasing at the rate of 8 rad/s^2.

18.45 Determine the rate of change $\dot{\mathbf{H}}_G$ of the angular momentum $\mathbf{H}_G$ of the disk of Prob. 18.3 if axle *AB* has an angular acceleration $\boldsymbol{\alpha}$.

18.46 Two 600-mm rods *BE* and *CF*, each of mass 4 kg, are attached to the shaft *AD* which rotates at a constant speed of 20 rad/s. Knowing that the two rods and the shaft lie in the same plane, determine the dynamic reactions at *A* and *D*.

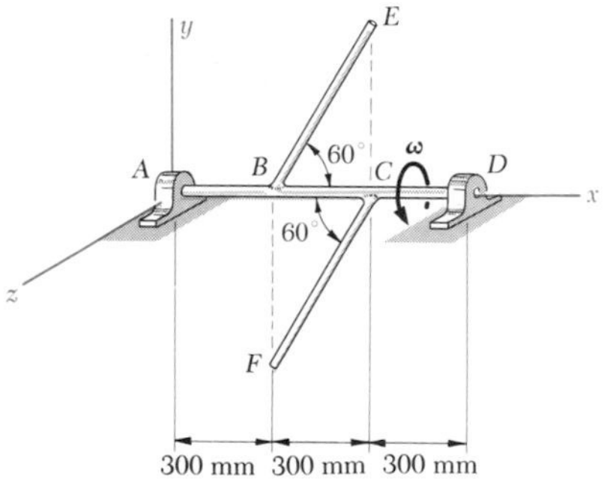

Fig. P18.46

18.47 Two triangular plates weighing 10 lb each are welded to a vertical shaft *AB*. Knowing that the system rotates at the constant rate $\omega = 6$ rad/s, determine the dynamic reactions at *A* and *B*.

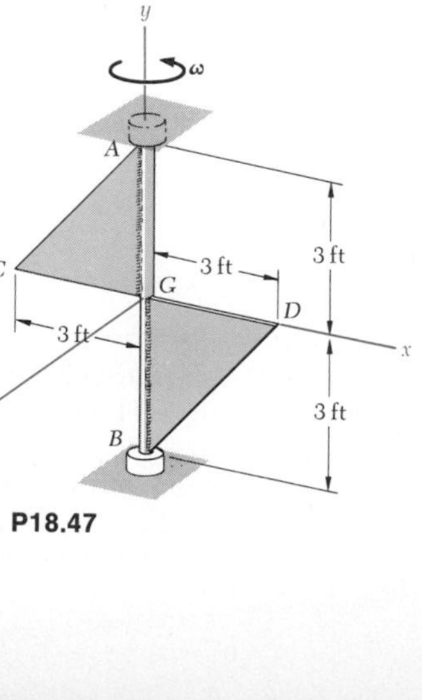

Fig. P18.47

18.48 Each element of the crankshaft shown is a homogeneous rod of weight w per unit length. Knowing that the crankshaft rotates with a constant angular velocity $\boldsymbol{\omega}$, determine the dynamic reactions at A and B.

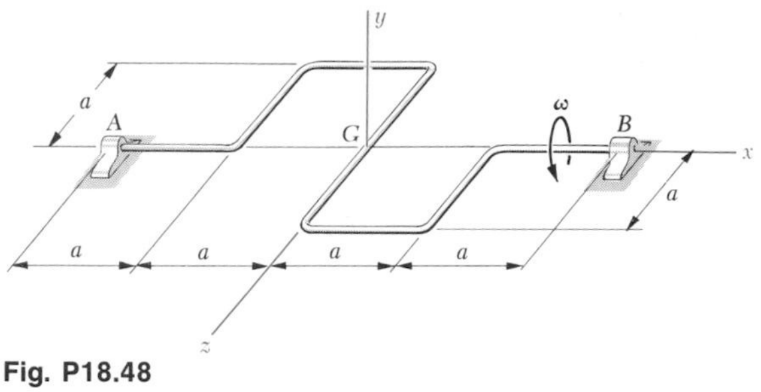

Fig. P18.48

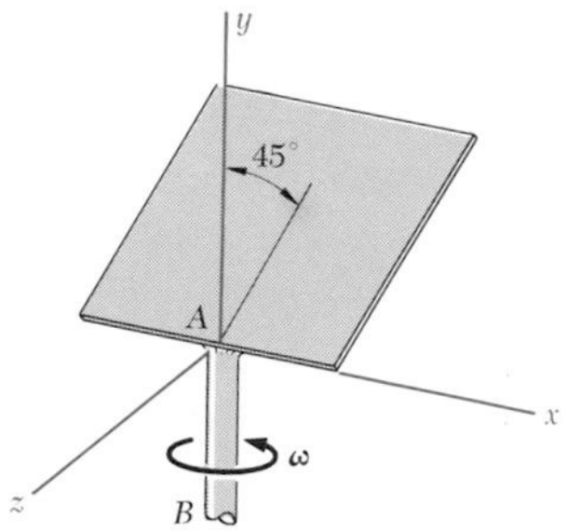

Fig. P18.49

18.49 A thin homogeneous square plate of mass m and side a is welded to a vertical shaft AB with which it forms an angle of 45°. Knowing that the shaft rotates with a constant angular velocity $\boldsymbol{\omega}$, determine the force-couple system representing the dynamic reaction at A.

18.50 The shaft of Prob. 18.48 is initially at rest ($\boldsymbol{\omega} = 0$) and is accelerated at the rate $\alpha = \dot{\omega} = 100$ rad/s^2. Knowing that $w = 4$ lb/ft and $a = 3$ in., determine (*a*) the couple $\mathbf{M}$ required to cause the acceleration, (*b*) the corresponding dynamic reactions at A and B.

18.51 The system of Prob. 18.47 is initially at rest ($\boldsymbol{\omega} = 0$) and has an angular acceleration $\boldsymbol{\alpha} = (30 \text{ rad/s}^2)\mathbf{j}$. Determine (*a*) the couple $\mathbf{M}$ required to cause the acceleration, (*b*) the corresponding dynamic reactions at A and B.

18.52 The square plate of Prob. 18.49 is at rest ($\boldsymbol{\omega} = 0$) when a couple of moment $M_0\mathbf{j}$ is applied to the shaft. Determine (*a*) the angular acceleration of the plate, (*b*) the force-couple system representing the dynamic reaction at A at that instant.

18.53 Two uniform rods CD and DE, each of mass 2 kg, are welded to the shaft AB, which is at rest. If a couple $\mathbf{M}$ of magnitude 10 N·m is applied to the shaft, determine the dynamic reactions at A and B.

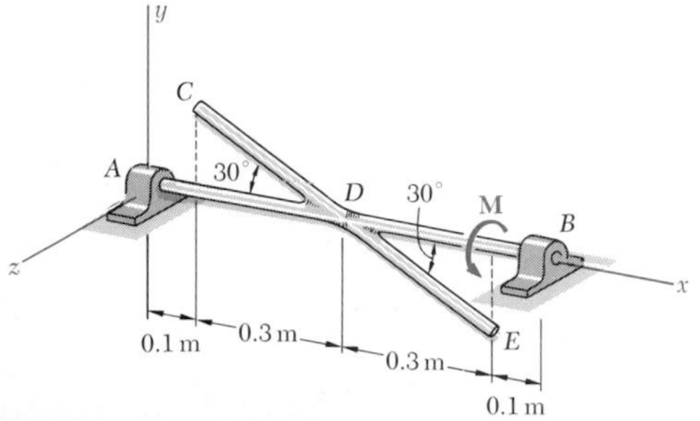

Fig. P18.53 and P18.54

18.54 Two uniform rods CD and DE, each of mass 2 kg, are welded to the shaft AB. At the instant shown the angular velocity of the shaft is 15 rad/s and the angular acceleration is 100 rad/s^2, both counterclockwise when viewed from the positive x axis. Determine (*a*) the couple $\mathbf{M}$ which must be applied to the shaft, (*b*) the corresponding dynamic reactions at A and B.

18.55 Two L-shaped arms, each weighing 6 lb, are welded at the third points of the 3-ft shaft AB. A couple $\mathbf{M} = (15 \text{ lb}\cdot\text{ft})\mathbf{k}$ is applied to the shaft, which is initially at rest. Determine (*a*) the angular acceleration of the shaft, (*b*) the dynamic reactions at A and B as the shaft reaches an angular velocity of 10 rad/s.

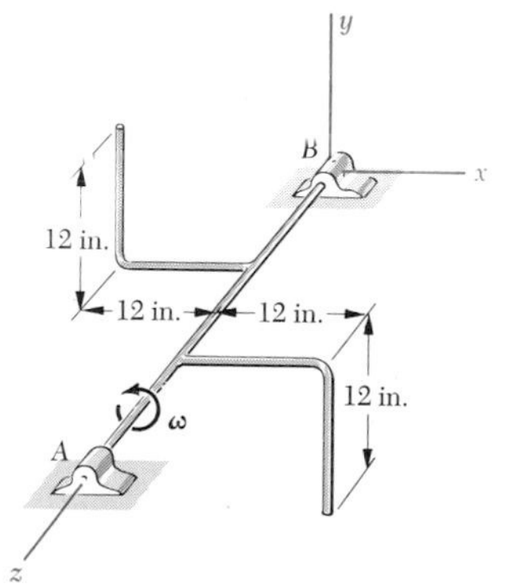

Fig. P18.55

18.56 The blade of a portable saw and the rotor of its motor have a combined mass of 1.2 kg and a radius of gyration of 35 mm. Determine the couple that a man must exert on the handle to rotate the saw about the y axis with a constant angular velocity of 3 rad/s clockwise, as viewed from above, when the blade rotates at the rate $\omega = 1800$ rpm as shown.

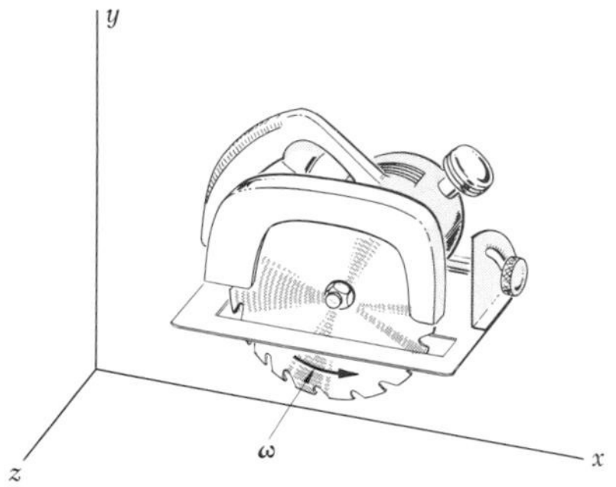

Fig. P18.56

18.57 A three-bladed airplane propeller has a mass of 120 kg and a radius of gyration of 900 mm. Knowing that the propeller rotates at 1500 rpm, determine the magnitude of the couple applied by the propeller to its shaft when the airplane travels in a circular path of 360-m radius at 600 km/h.

18.58 The flywheel of an automobile engine, which is mounted on the crankshaft, is equivalent to a 16-in.-diameter steel plate of $\frac{15}{16}$-in. thickness. At a time when the flywheel is rotating at 4000 rpm the automobile is traveling around a curve of 600-ft radius at a speed of 60 mi/h. Determine, at that time, the magnitude of the couple exerted by the flywheel on the horizontal crankshaft. (Specific weight of steel = 490 lb/ft^3.)

18.59 The essential structure of a certain type of aircraft turn indicator is shown. Springs AC and BD are initially stretched and exert equal vertical forces at A and B when the airplane is traveling in a straight path. Knowing that the disk weighs $\frac{1}{2}$ lb and spins at the rate of 10,000 rpm, determine the angle through which the yoke will rotate when the airplane executes a horizontal turn of radius 2500 ft at a speed of 500 mi/h. The constant of each spring is 2 lb/in.

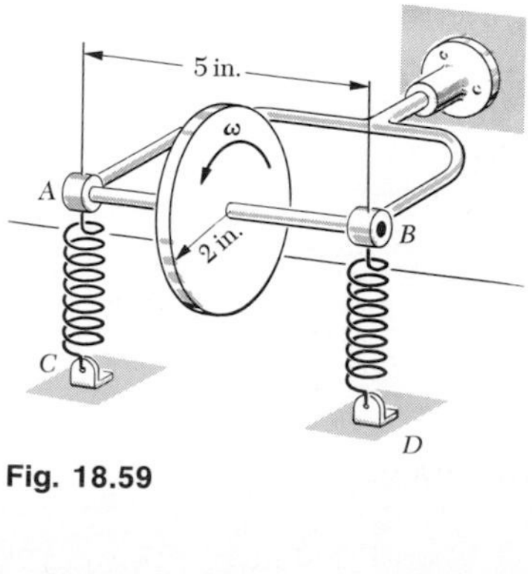

Fig. 18.59

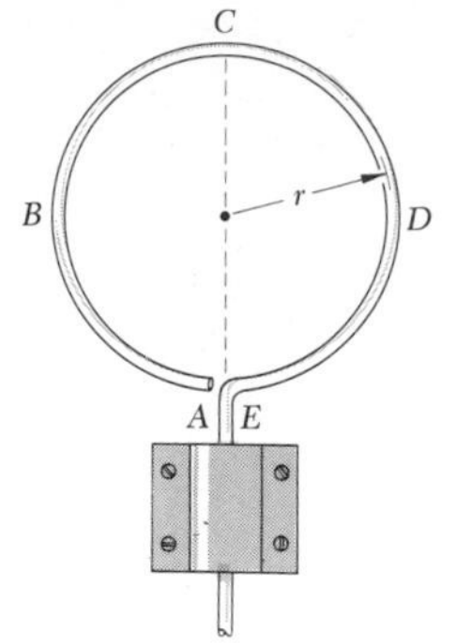

Fig. P18.60

18.60 A thin homogeneous wire, of mass m per unit length and in the shape of a circle of radius r, is made to rotate about a vertical shaft with a constant angular velocity $\boldsymbol{\omega}$. Determine the bending moment in the wire (*a*) at point C, (*b*) at point E, (*c*) at point B. (Neglect the effect of gravity.)

18.61 A thin homogeneous disk of mass m and radius r spins at the constant rate ω_2 about a horizontal axle held by a fork-ended vertical rod which rotates at the constant rate ω_1. Determine the couple $\mathbf{M}$ exerted by the rod on the disk.

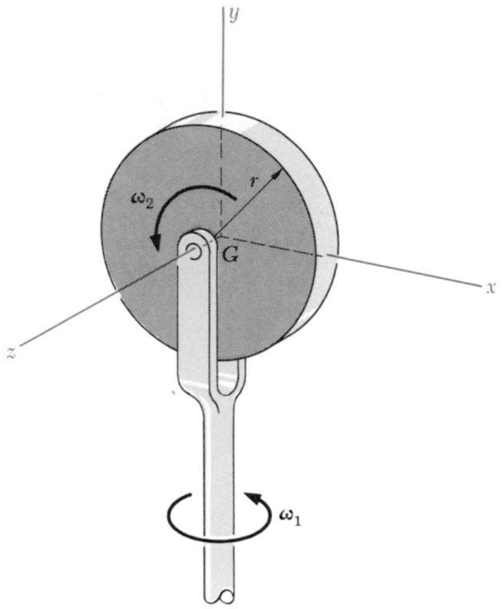

Fig. P18.61

18.62 A thin ring of radius a is attached by a collar at A to a vertical shaft which rotates with a constant angular velocity $\boldsymbol{\omega}$. Derive an expression (*a*) for the constant angle β that the plane of the ring forms with the vertical, (*b*) for the maximum value of ω for which the ring will remain vertical ($\beta = 0$).

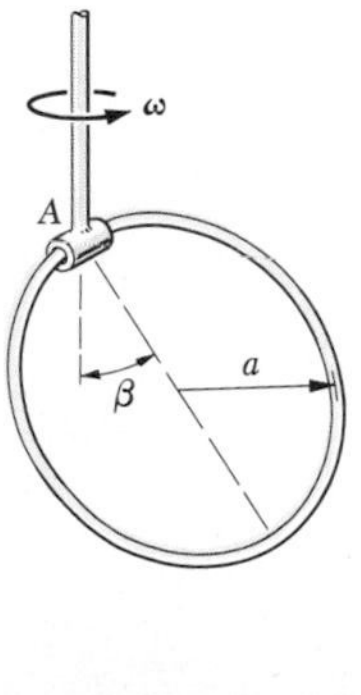

Fig. 18.62

18.63 A uniform disk of radius r is welded to a rod AB of negligible weight, which is attached to the pin of a clevis which rotates with a constant angular velocity $\boldsymbol{\omega}$. Derive an expression (a) for the constant angle β that the rod forms with the vertical, (b) for the maximum value of ω for which the rod will remain vertical ($\beta = 0$).

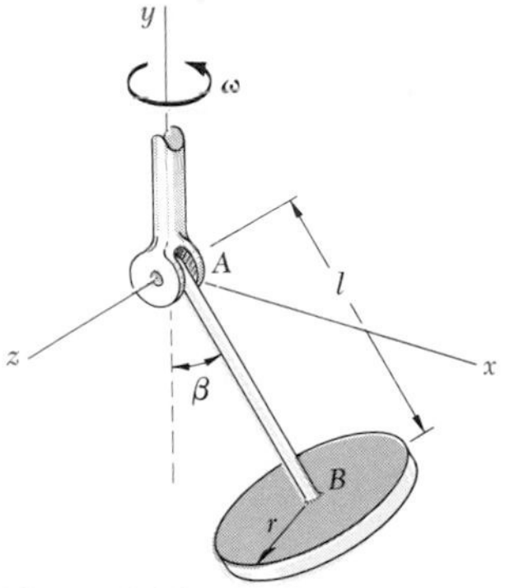

Fig. P18.63

18.64 A disk of mass m and radius r rotates at a constant rate ω_2 with respect to the arm OA, which itself rotates at a constant rate ω_1 about the y axis. Determine the force-couple system representing the dynamic reaction at O.

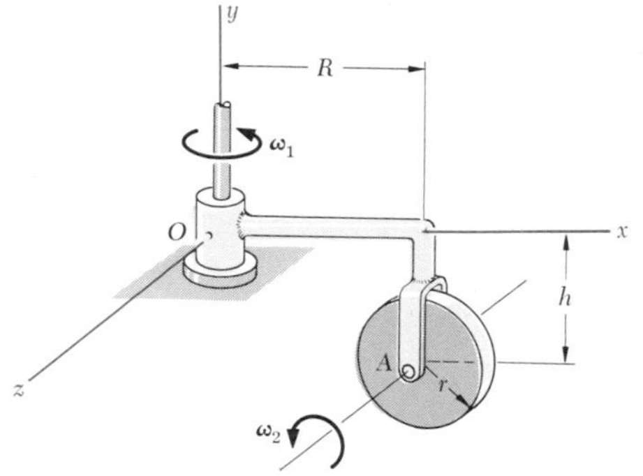

Fig. P18.64

18.65 Two disks, each of mass 5 kg and radius 300 mm, spin as shown at 1200 rpm about the rod AB, which is attached to shaft CD. The entire system is made to rotate about the z axis with an angular velocity $\boldsymbol{\Omega}$ of 60 rpm. (a) Determine the dynamic reactions at C and D as the system passes through the position shown. (b) Solve part a assuming that the direction of spin of disk B is reversed.

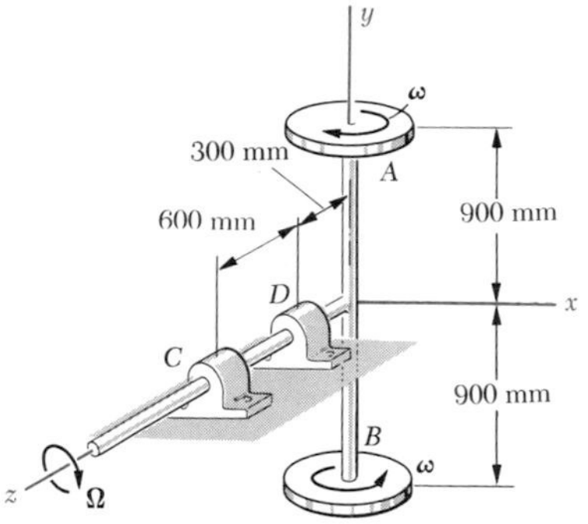

Fig. P18.65

18.66 A stationary horizontal plate is attached to the ceiling by means of a fixed vertical tube. A wheel of radius a and mass m is mounted on a light axle AC which is attached by means of a clevis at A to a rod AB fitted inside the vertical tube. The rod AB is made to rotate with a constant angular velocity $\boldsymbol{\Omega}$ causing the wheel to roll on the lower face of the stationary plate. Determine the minimum angular velocity Ω for which contact is maintained between the wheel and the plate. Consider the particular cases (a) when the mass of the wheel is concentrated in the rim, (b) when the wheel is equivalent to a thin disk of radius a.

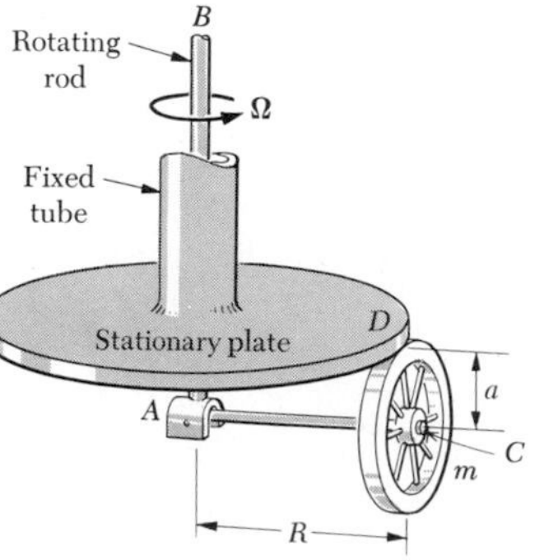

Fig. P18.66

18.67 Assuming that the wheel of Prob. 18.66 weighs 8 lb, has a radius $a = 4$ in. and a radius of gyration of 3 in., and that $R = 20$ in., determine the force exerted by the plate on the wheel when $\Omega = 25$ rad/s.

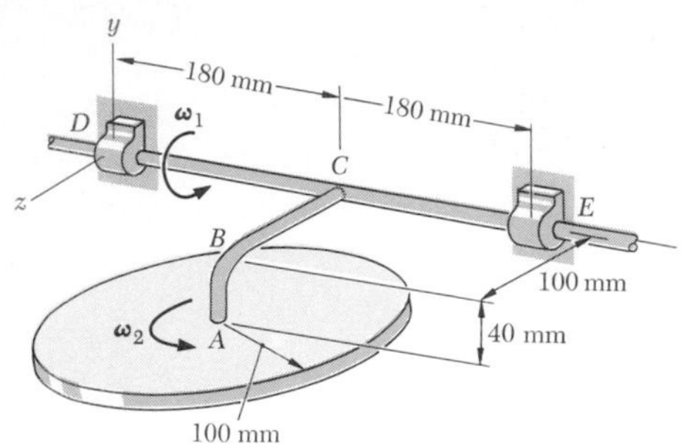

Fig. P18.68

18.68 A thin homogeneous disk of mass 800 g and radius 100 mm rotates at a constant rate $\omega_2 = 20$ rad/s with respect to the arm ABC, which itself rotates at a constant rate $\omega_1 = 10$ rad/s about the x axis. For the position shown, determine the dynamic reactions at the bearings D and E.

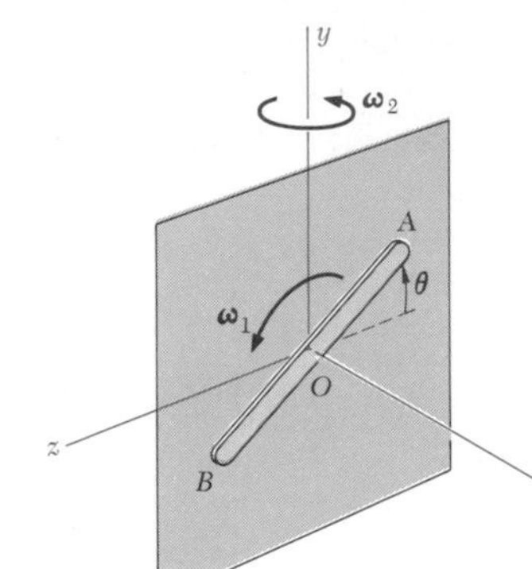

Fig. P18.69

18.69 A slender homogeneous rod AB of mass m and length L is made to rotate at the constant rate ω_1 about the horizontal x axis, while the vertical plane in which it rotates is made to rotate at the constant rate ω_2 about the vertical y axis. Express as a function of the angle θ (*a*) the couple $M_1\mathbf{i}$ required to maintain the rotation of the rod in the vertical plane, (*b*) the couple $M_2\,\mathbf{j}$ required to maintain the rotation of that plane.

*18.9. Motion of a Gyroscope. Eulerian Angles.

A *gyroscope* consists essentially of a rotor which may spin freely about its geometric axis. When mounted in a Cardan's suspension (Fig. 18.15), a gyroscope may assume any orientation, but its mass center must remain fixed in space. In order to define the position of a gyroscope at a given instant, we shall select a fixed frame of reference $OXYZ$, with the origin O located at the mass center of the gyroscope and the Z axis directed along the line defined by the bearings A and A' of the outer gimbal, and we shall consider a reference position of the gyroscope in which the two gimbals and a given diameter DD' of the rotor are located in the fixed YZ plane (Fig. 18.15*a*). The gyroscope may be brought from this reference position into any arbitrary position (Fig. 18.15*b*) by means of the following steps: (1) a rotation of the outer gimbal through an angle ϕ about the axis AA', (2) a rotation of the inner gimbal through θ about BB', (3) a rotation of the rotor through ψ about CC'. The angles ϕ, θ, and ψ are called the *Eulerian angles;* they completely characterize the position of the gyroscope at any given instant. Their derivatives $\dot{\phi}$, $\dot{\theta}$, and $\dot{\psi}$ define, respectively, the rate of *precession,* the rate of *nutation,* and the rate of *spin* of the gyroscope at the instant considered.

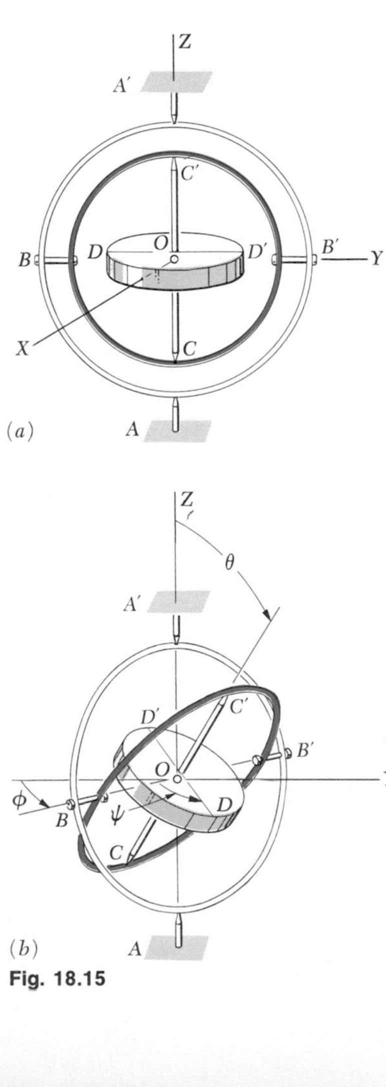

Fig. 18.15

In order to compute the components of the angular velocity and of the angular momentum of the gyroscope, we shall use a rotating system of axes $Oxyz$ *attached to the inner gimbal*, with the y axis along BB' and the z axis along CC' (Fig. 18.16). These axes are principal axes of inertia for the gyroscope but, while they follow it in its precession and nutation, they do not spin. For that reason, they are more convenient to use than axes actually attached to the gyroscope. We shall now express the angular velocity $\boldsymbol{\omega}$ of the gyroscope with respect to the fixed frame of reference $OXYZ$ as the sum of three partial angular velocities corresponding respectively to the precession, the nutation, and the spin of the gyroscope. Denoting by **i**, **j**, **k** the unit vectors along the rotating axes, and by **K** the unit vector along the fixed Z axis, we have

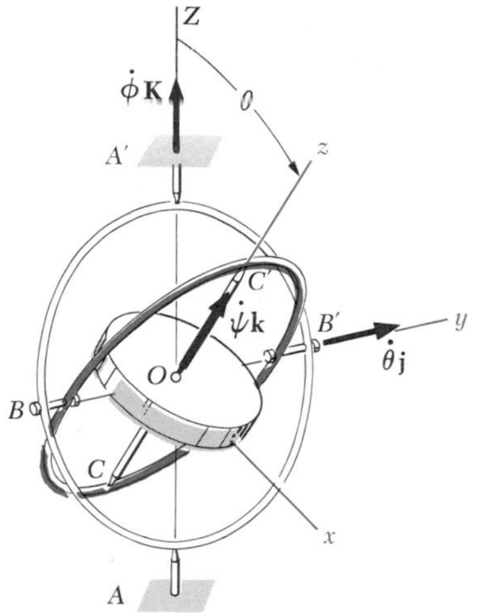

Fig. 18.16

$$\boldsymbol{\omega} = \dot{\phi}\mathbf{K} + \dot{\theta}\mathbf{j} + \dot{\psi}\mathbf{k} \tag{18.33}$$

Since the vector components obtained for $\boldsymbol{\omega}$ in (18.33) are not orthogonal (Fig. 18.16), we shall resolve the unit vector **K** into components along the x and z axes; we write

$$\mathbf{K} = -\sin\theta\,\mathbf{i} + \cos\theta\,\mathbf{k} \tag{18.34}$$

and, substituting for **K** into (18.33),

$$\boldsymbol{\omega} = -\dot{\phi}\sin\theta\,\mathbf{i} + \dot{\theta}\mathbf{j} + (\dot{\psi} + \dot{\phi}\cos\theta)\mathbf{k} \tag{18.35}$$

Since the coordinate axes are principal axes of inertia, the components of the angular momentum $\mathbf{H}_O$ may be obtained by multiplying the components of $\boldsymbol{\omega}$ by the moments of inertia of the rotor about the x, y, and z axes, respectively. Denoting by I the moment of inertia of the rotor about its spin axis, by I' its moment of inertia about a transverse axis through O, and neglecting the mass of the gimbals, we write

$$\mathbf{H}_O = -I'\dot{\phi}\sin\theta\,\mathbf{i} + I'\dot{\theta}\mathbf{j} + I(\dot{\psi} + \dot{\phi}\cos\theta)\mathbf{k} \tag{18.36}$$

Recalling that the rotating axes are attached to the inner gimbal, and thus do not spin, we express their angular velocity as the sum

$$\boldsymbol{\Omega} = \dot{\phi}\mathbf{K} + \dot{\theta}\mathbf{j} \tag{18.37}$$

or, substituting for **K** from (18.34),

$$\boldsymbol{\Omega} = -\dot{\phi}\sin\theta\,\mathbf{i} + \dot{\theta}\mathbf{j} + \dot{\phi}\cos\theta\,\mathbf{k} \tag{18.38}$$

Substituting for $\mathbf{H}_O$ and $\boldsymbol{\Omega}$ from (18.36) and (18.38) into the equation

$$\Sigma \mathbf{M}_O = (\dot{\mathbf{H}}_O)_{Oxyz} + \boldsymbol{\Omega} \times \mathbf{H}_O \tag{18.28}$$

we obtain the three differential equations

$$\begin{aligned} \Sigma M_x &= -I'(\ddot{\phi} \sin\theta + 2\dot{\theta}\dot{\phi}\cos\theta) + I\dot{\theta}(\dot{\psi} + \dot{\phi}\cos\theta) \\ \Sigma M_y &= I'(\ddot{\theta} - \dot{\phi}^2 \sin\theta\cos\theta) + I\dot{\phi}\sin\theta(\dot{\psi} + \dot{\phi}\cos\theta) \\ \Sigma M_z &= I\frac{d}{dt}(\dot{\psi} + \dot{\phi}\cos\theta) \end{aligned} \tag{18.39}$$

The equations (18.39) define the motion of a gyroscope subjected to a given system of forces when the mass of its gimbals is neglected. They may also be used to define the motion of an *axisymmetrical body* (or body of revolution) attached at a point on its axis of symmetry, or the motion of an axisymmetrical body about its mass center. While the gimbals of the gyroscope helped us visualize the Eulerian angles, it is clear that these angles may be used to define the position of any rigid body with respect to axes centered at a point of the body, regardless of the way in which the body is actually supported.

Since the equations (18.39) are nonlinear, it will not be possible, in general, to express the Eulerian angles ϕ, θ, and ψ as analytical functions of the time *t*, and numerical methods of solution may have to be used. However, as we shall see in the following sections, there are several particular cases of interest which may be analyzed easily.

***18.10. Steady Precession of a Gyroscope.** We shall consider in this section the particular case of gyroscopic motion in which the angle θ, the rate of precession $\dot{\phi}$, and the rate of spin $\dot{\psi}$ remain constant. We propose to determine the forces which must be applied to the gyroscope to maintain this motion, known as the *steady precession* of a gyroscope.

Instead of applying the general equations (18.39), we shall determine the sum of the moments of the required forces by computing the rate of change of the angular momentum of the gyroscope in the particular case considered. We first note that the angular velocity $\boldsymbol{\omega}$ of the gyroscope, its angular momentum $\mathbf{H}_O$, and the angular velocity $\boldsymbol{\Omega}$ of the rotating frame of reference (Fig. 18.17) reduce, respectively, to

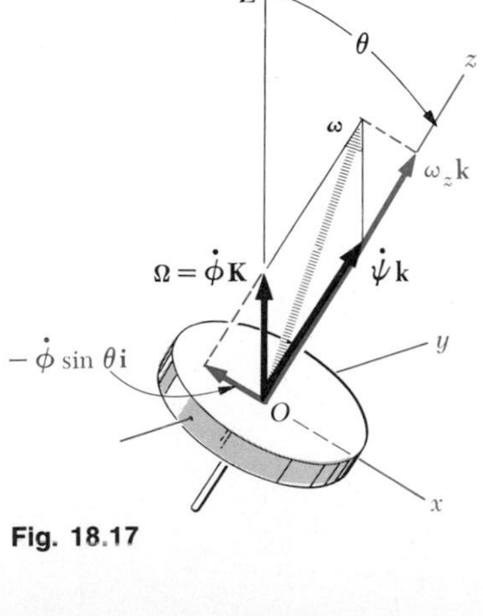

Fig. 18.17

$$\boldsymbol{\omega} = -\dot{\phi}\sin\theta\,\mathbf{i} + \omega_z\,\mathbf{k} \tag{18.40}$$

$$\mathbf{H}_O = -I'\dot{\phi}\sin\theta\,\mathbf{i} + I\omega_z\,\mathbf{k} \tag{18.41}$$

$$\boldsymbol{\Omega} = -\dot{\phi}\sin\theta\,\mathbf{i} + \dot{\phi}\cos\theta\,\mathbf{k} \tag{18.42}$$

where $\omega_z = \dot{\psi} + \dot{\phi}\cos\theta$ = component along the spin axis of the total angular velocity of the gyroscope

Since θ, $\dot{\phi}$, and $\dot{\psi}$ are constant, the vector $\mathbf{H}_O$ is constant in magnitude and direction with respect to the rotating frame of reference, and its rate of change $(\dot{\mathbf{H}}_O)_{Oxyz}$ with respect to that frame is zero. Thus Eq. (18.28) reduces to

$$\Sigma \mathbf{M}_O = \boldsymbol{\Omega} \times \mathbf{H}_O \tag{18.43}$$

which yields, after substitutions from (18.41) and (18.42),

$$\Sigma \mathbf{M}_O = (I\omega_z - I'\dot{\phi}\cos\theta)\dot{\phi}\sin\theta\,\mathbf{j} \tag{18.44}$$

Since the mass center of the gyroscope is fixed in space, we have, by (18.1), $\Sigma\mathbf{F} = 0$; thus, the forces which must be applied to the gyroscope to maintain its steady precession reduce to a couple of moment equal to the right-hand member of Eq. (18.44). We note that *this couple should be applied about an axis perpendicular to the precession axis and to the spin axis of the gyroscope* (Fig. 18.18).

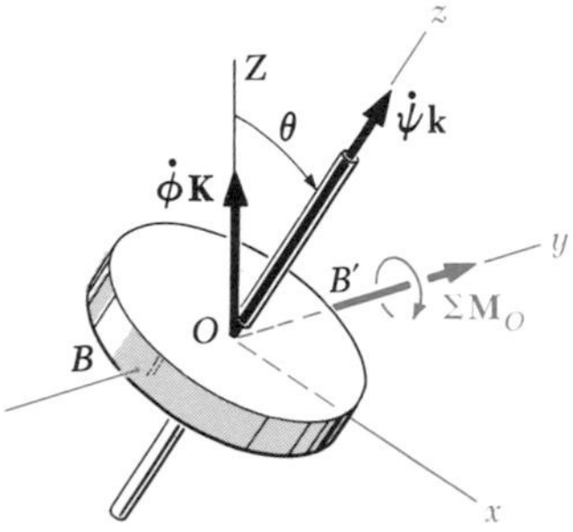

Fig. 18.18

In the particular case when the precession axis and the spin axis are at a right angle to each other, we have $\theta = 90°$ and Eq. (18.44) reduces to

$$\Sigma \mathbf{M}_O = I\dot{\psi}\dot{\phi}\mathbf{j} \tag{18.45}$$

Thus, if we apply to the gyroscope a couple $\mathbf{M}_O$ about an axis perpendicular to its axis of spin, the gyroscope will precess about an axis perpendicular to both the spin axis and the couple axis, in a sense such that the vectors representing respectively the spin, the couple, and the precession form a right-handed triad (Fig. 18.19).

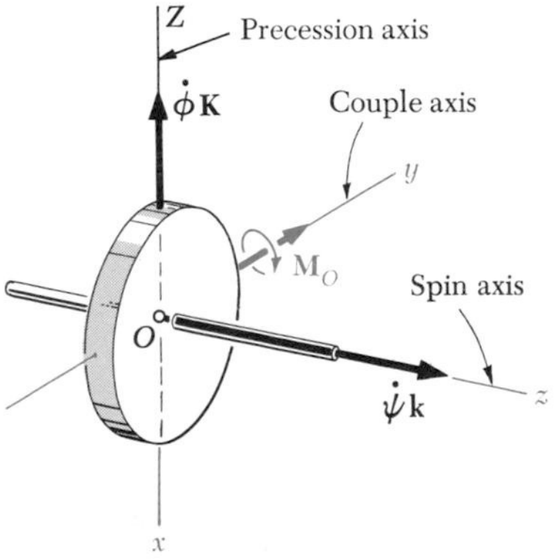

Fig. 18.19

Because of the relatively large couples required to change the orientation of their axles, gyroscopes are used as stabilizers in torpedoes and ships. Spinning bullets and shells remain tangent to their trajectory because of gyroscopic action. And a bicycle is easier to keep balanced at high speeds because of the stabilizing effect of its spinning wheels. However, gyroscopic action is not always welcome and must be taken into account in the design of bearings supporting rotating shafts subjected to forced precession. The reactions exerted by its propellers on an airplane which changes its direction of flight must also be taken into consideration and compensated for whenever possible.

***18.11. Motion of an Axisymmetrical Body under No Force.** We shall consider in this section the motion about its mass center of an axisymmetrical body under no force, except its own weight. Examples of such a motion are furnished by projectiles, if air resistance is neglected, and by artificial satellites and space vehicles after burnout of their launching rockets.

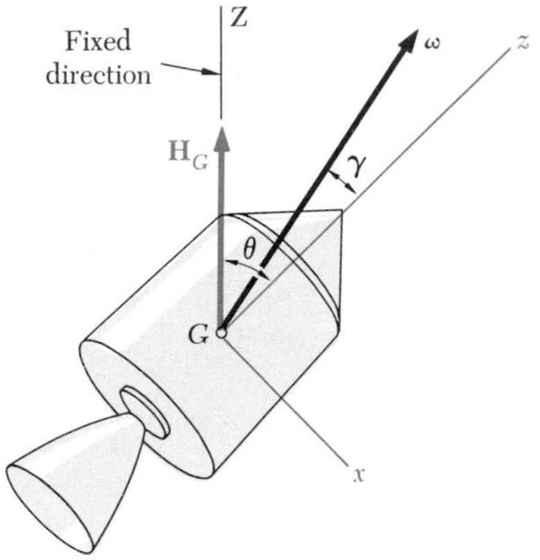

Fig. 18.20

Since the sum of the moments of the external forces about the mass center G of the body is zero, Eq. (18.2) yields $\dot{\mathbf{H}}_G = 0$. It follows that the angular momentum $\mathbf{H}_G$ of the body about G is constant. Thus, the direction of $\mathbf{H}_G$ is fixed in space and may be used to define the Z axis, or axis of precession (Fig. 18.20). Selecting a rotating system of axes $Gxyz$ with the z axis along the axis of symmetry of the body and the x axis in the plane defined by the Z and z axes, we have

$$H_x = -H_G \sin\theta \qquad H_y = 0 \qquad H_z = H_G \cos\theta \qquad (18.46)$$

where θ represents the angle formed by the Z and z axes, and H_G denotes the constant magnitude of the angular momentum of the body about G. Since the x, y, and z axes are principal axes of inertia for the body considered, we may write

$$H_x = I'\omega_x \qquad H_y = I'\omega_y \qquad H_z = I\omega_z \qquad (18.47)$$

where I denotes the moment of inertia of the body about its axis of symmetry, and I' its moment of inertia about a transverse axis through G. It follows from Eqs. (18.46) and (18.47) that

$$\omega_x = -\frac{H_G \sin\theta}{I'} \qquad \omega_y = 0 \qquad \omega_z = \frac{H_G \cos\theta}{I} \qquad (18.48)$$

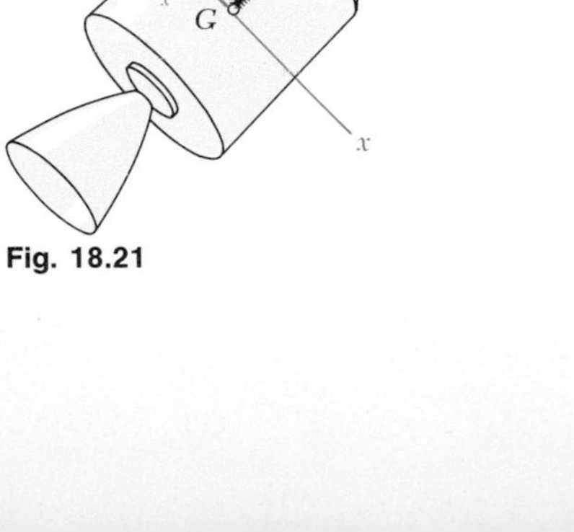

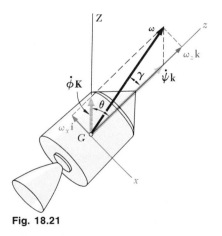

Fig. 18.21

The second of the relations obtained shows that the angular velocity $\boldsymbol{\omega}$ has no component along the y axis, i.e., along an axis perpendicular to the Zz plane. Thus, the angle θ formed by the Z and z axes remains constant and *the body is in steady precession about the Z axis.*

Dividing the first and third of the relations (18.48) member by member, and observing from Fig. 18.21 that $-\omega_x/\omega_z = \tan\gamma$, we obtain the following relation between the angles γ and θ that the vectors $\boldsymbol{\omega}$ and $\mathbf{H}_G$ respectively form with the axis of symmetry of the body:

$$\tan\gamma = \frac{I}{I'}\tan\theta \qquad (18.49)$$

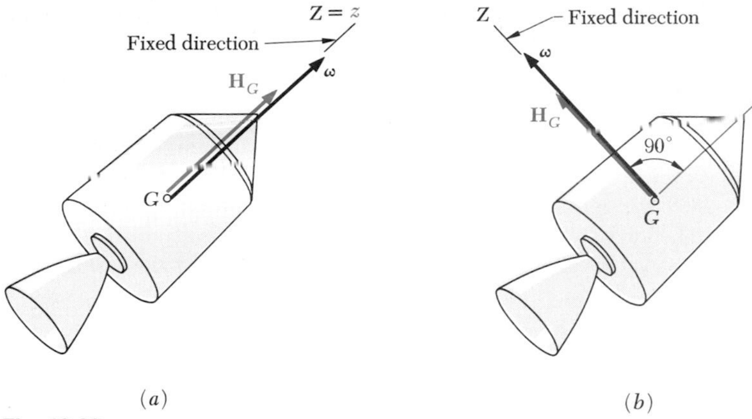

Fig. 18.22

There are two particular cases of motion of an axisymmetrical body under no force which involve no precession: (1) If the body is set to spin about its axis of symmetry, we have $\omega_x = 0$ and, by (18.47), $H_x = 0$; the vectors $\boldsymbol{\omega}$ and $\mathbf{H}_G$ have the same orientation and the body keeps spinning about its axis of symmetry (Fig. 18.22*a*). (2) If the body is set to spin about a transverse axis, we have $\omega_z = 0$ and, by (18.47), $H_z = 0$; again $\boldsymbol{\omega}$ and $\mathbf{H}_G$ have the same orientation and the body keeps spinning about the given transverse axis (Fig. 18.22*b*).

Considering now the general case represented in Fig. 18.21, we recall from Sec. 15.12 that the motion of a body about a fixed point—or about its mass center—may be represented by the motion of a body cone rolling on a space cone. In the case of steady precession, the two cones are circular, since the angles γ and $\theta - \gamma$ that the angular velocity $\boldsymbol{\omega}$ forms, respectively, with the axis of symmetry of the body and with the precession axis are constant. Two cases should be distinguished:

1. $I < I'$. This is the case of an elongated body, such as the space vehicle of Fig. 18.23. By (18.49) we have $\gamma < \theta$; the vector $\boldsymbol{\omega}$ lies inside the angle ZGz; the space cone and the body cone are tangent externally; the spin and the precession are both observed as counterclockwise from the positive z axis. The precession is said to be *direct*.
2. $I > I'$. This is the case of a flattened body, such as the satellite of Fig. 18.24. By (18.49) we have $\gamma > \theta$; since the vector $\boldsymbol{\omega}$ must lie outside the angle ZGz, the vector $\dot{\psi}\mathbf{k}$ has a sense opposite to that of the z axis; the space cone is inside the body cone; the precession and the spin have opposite senses; the precession is said to be *retrograde*.

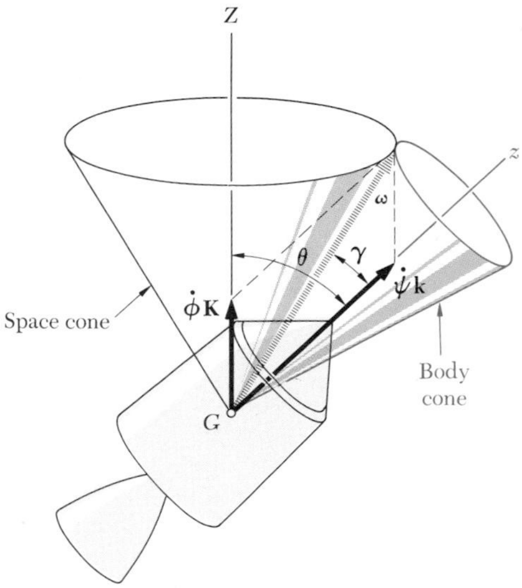

Fig. 18.23

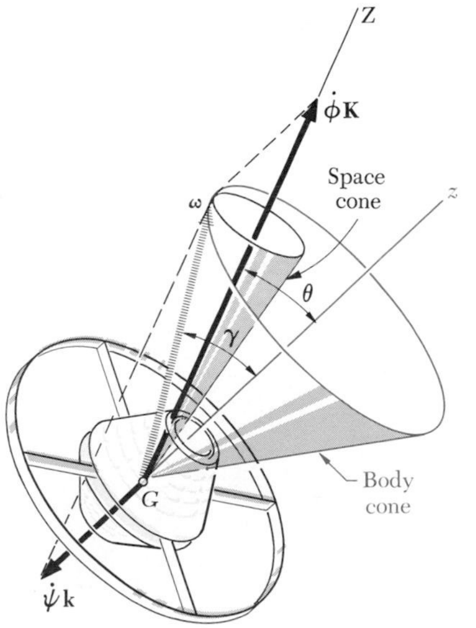

Fig. 18.24

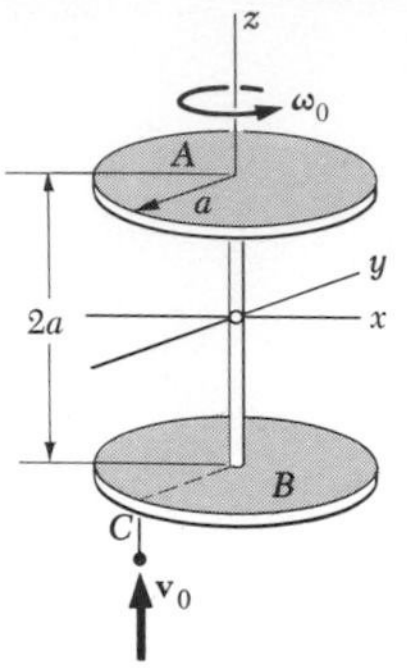

SAMPLE PROBLEM 18.6

A space satellite of mass m is known to be dynamically equivalent to two thin disks of equal mass. The disks are of radius $a = 800$ mm and are rigidly connected by a light rod of length $2a$. Initially the satellite is spinning freely about its axis of symmetry at the rate $\omega_0 = 60$ rpm. A meteorite, of mass $m_0 = m/1000$ and traveling with a velocity $\mathbf{v}_0$ of 2000 m/s relative to the satellite, strikes the satellite and becomes embedded at C. Determine (*a*) the angular velocity of the satellite immediately after impact, (*b*) the precession axis of the ensuing motion, (*c*) the rates of precession and spin of the ensuing motion.

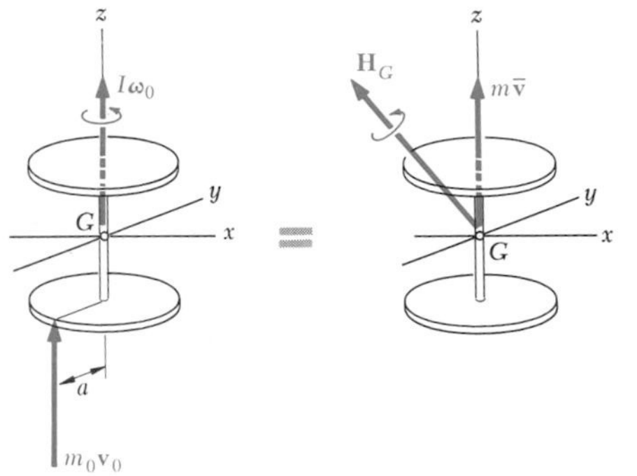

Solution. *Moments of Inertia.* We note that the axes shown are principal axes of inertia for the satellite and write

$$I = I_z = \tfrac{1}{2}ma^2 \qquad I' = I_x = I_y = 2[\tfrac{1}{4}(\tfrac{1}{2}m)a^2 + (\tfrac{1}{2}m)a^2] = \tfrac{5}{4}ma^2$$

Principle of Impulse and Momentum. We consider the satellite and the meteorite as a single system. Since no external force acts on this system, the momenta before and after impact are equipollent. Taking moments about G we write

$$-a\mathbf{j} \times m_0 v_0 \mathbf{k} + I\omega_0 \mathbf{k} = \mathbf{H}_G$$
$$\mathbf{H}_G = -m_0 v_0 a\mathbf{i} + I\omega_0\mathbf{k} \qquad (1)$$

Angular Velocity after Impact. Substituting the values obtained for the components of $\mathbf{H}_G$ and for the moments of inertia into

$$H_x = I_x\omega_x \qquad H_y = I_y\omega_y \qquad H_z = I_z\omega_z$$

we write

$$-m_0 v_0 a = I'\omega_x = \tfrac{5}{4}ma^2\omega_x \qquad 0 = I'\omega_y \qquad I\omega_0 = I\omega_z$$

$$\omega_x = -\frac{4}{5}\frac{m_0 v_0}{ma} \qquad \omega_y = 0 \qquad \omega_z = \omega_0 \qquad (2)$$

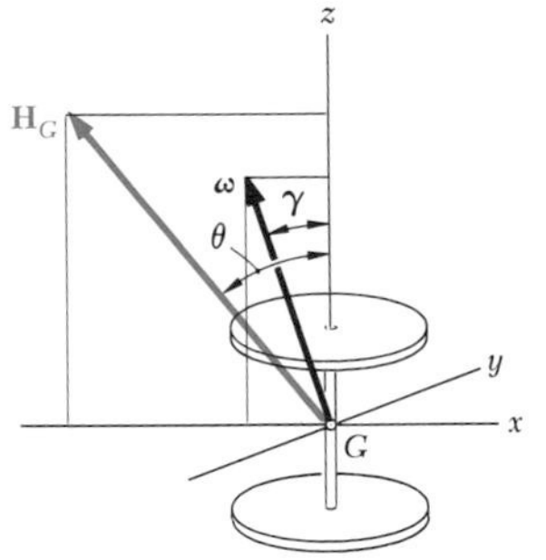

For the satellite considered we have $\omega_0 = 60$ rpm $= 6.28$ rad/s, $m_0/m = 1/1000$, $a = 0.800$ m, and $v_0 = 2000$ m/s; we find

$$\omega_x = -2 \text{ rad/s} \qquad \omega_y = 0 \qquad \omega_z = 6.28 \text{ rad/s}$$

$$\omega = \sqrt{\omega_x^2 + \omega_z^2} = 6.59 \text{ rad/s} \qquad \tan\gamma = \frac{-\omega_x}{\omega_z} = +0.3185$$

$\omega = 63.0$ rpm $\qquad \gamma = 17.7°$ ◄

Precession Axis. Since, in free motion, the direction of the angular momentum $\mathbf{H}_G$ is fixed in space, the satellite will precess about this direction. The angle θ formed by the precession axis and the z axis is

$$\tan\theta = \frac{-h_x}{h_z} = \frac{m_0 v_0 a}{I\omega_0} = \frac{2m_0 v_0}{ma\omega_0} = 0.796 \qquad \theta = 38.5°$$ ◄

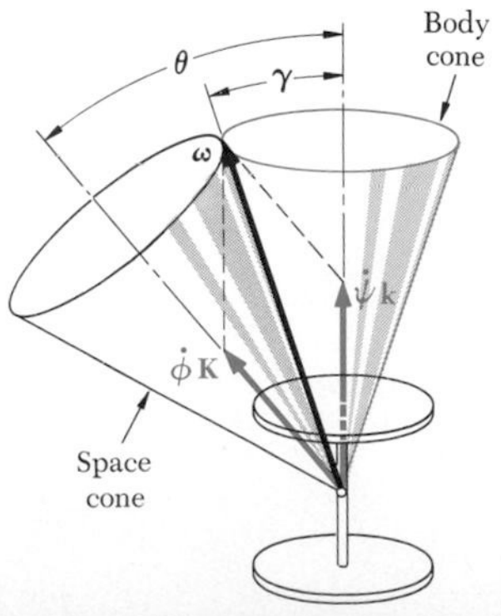

Rates of Precession and Spin. We sketch the space and body cones for the free motion of the satellite. Using the law of sines, we compute the rates of precession and spin.

$$\frac{\omega}{\sin\theta} = \frac{\dot{\phi}}{\sin\gamma} = \frac{\dot{\psi}}{\sin(\theta - \gamma)}$$

$\dot{\phi} = 30.7$ rpm $\qquad \dot{\psi} = 36.0$ rpm ◄

PROBLEMS

18.70 The rate of steady precession $\dot{\phi}$ of the cone shown about the vertical is observed to be 30 rpm. Knowing that $r = 75$ mm and $h = 300$ mm, determine the rate of spin $\dot{\psi}$ of the cone about its axis of symmetry if $\beta = 120°$.

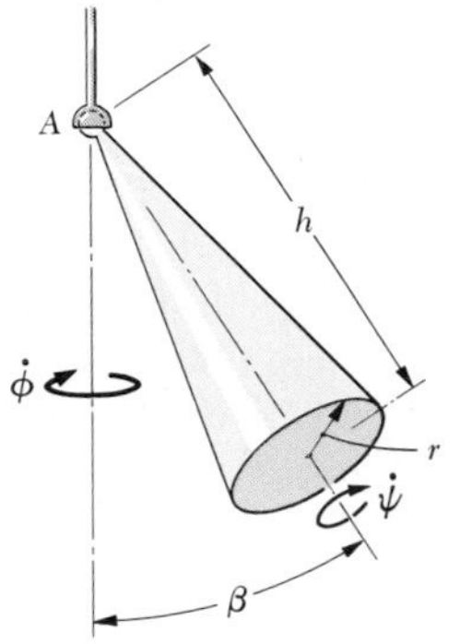

Fig. P18.70

18.71 Solve Prob. 18.70, assuming the same rate of steady precession and $\beta = 60°$.

18.72 A 5-lb disk of 9-in. diameter is attached to the end of a rod AB of negligible weight which is supported by a ball-and-socket joint at A. If the rate of steady precession $\dot{\phi}$ of the disk about the vertical is observed to be 24 rpm, determine the rate of spin $\dot{\psi}$ of the disk about AB when $\beta = 60°$.

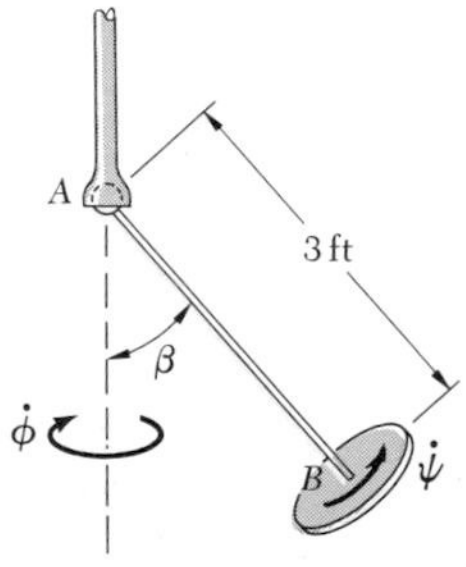

Fig. P18.72

18.73 Solve Prob. 18.72, assuming the same rate of steady precession and $\beta = 30°$.

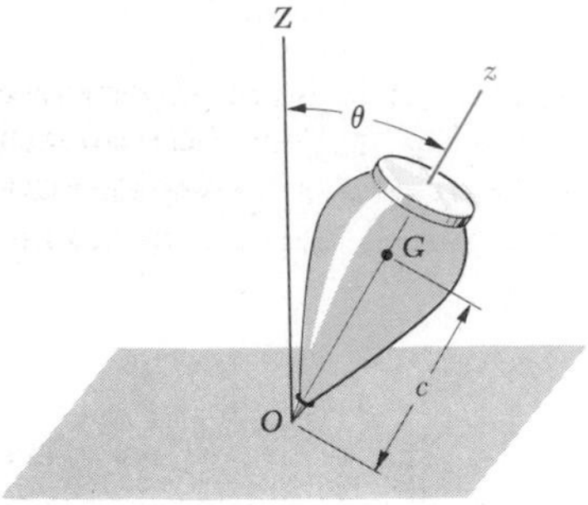

Fig. P18.74, P18.75, and 18.76

18.74 The top shown is supported at the fixed point O. Denoting by I and I', respectively, the moments of inertia of the top about its axis of symmetry and about a transverse axis through O, show that the condition for steady precession is

$$(I\omega_z - I'\dot{\phi}\cos\theta)\dot{\phi} = Wc$$

where $\dot{\phi}$ is the rate of precession and ω_z the component of the angular velocity along the axis of symmetry of the top.

18.75 Show that, if the rate of spin $\dot{\psi}$ of a top is very large compared to its rate of precession $\dot{\phi}$, the condition for steady precession is $I\dot{\psi}\dot{\phi} \approx Wc$.

18.76 The top shown weighs 0.2 lb and is supported at the fixed point O. The radii of gyration of the top with respect to its axis of symmetry and with respect to a transverse axis through O are 0.75 in. and 1.75 in., respectively. It is known that $c = 1.50$ in. and that the rate of spin of the top with respect to its axis of symmetry is 1600 rpm. (*a*) Using the relation of Prob. 18.74, determine the two possible rates of steady precession corresponding to $\theta = 30°$. (*b*) Determine the relative error introduced when the slower of the two rates obtained in part *a* is approximated by the relation of Prob. 18.75.

18.77 If the earth were a sphere, the gravitational attraction of the sun, moon, and planets would at all times be equivalent to a single force **R** acting at the mass center of the earth. However, the earth is actually an oblate spheroid and the gravitational system acting on the earth is equivalent to a force **R** and a couple **M**. Knowing that the effect of the couple **M** is to cause the axis of the earth to precess about the axis GA at the rate of one revolution in 25,800 years, determine the average magnitude of the couple **M** applied to the earth. Assume that the average density of the earth is 5.51, that the average radius of the earth is 3960 mi, and that $\bar{I} = \frac{2}{5}mR^2$. (*Note.* This forced precession is known as the precession of the equinoxes and is not to be confused with the free precession discussed in Prob. 18.85.)

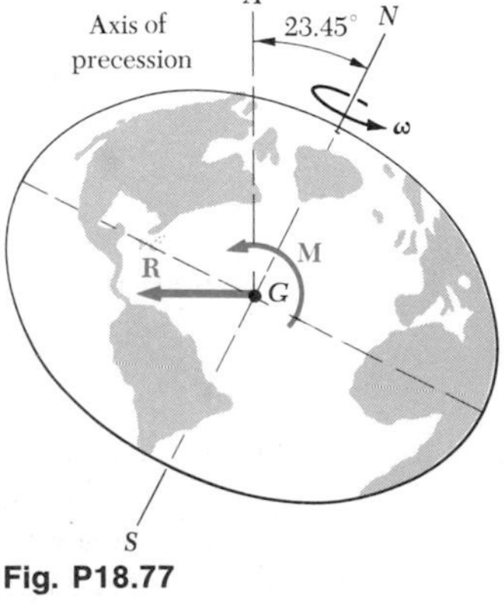

Fig. P18.77

18.78 A high-speed photographic record shows that a certain projectile was fired with a horizontal velocity $\bar{\mathbf{v}}$ of 2000 ft/s and with its axis of symmetry forming an angle $\beta = 3°$ with the horizontal. The rate of spin $\dot{\psi}$ of the projectile was 6000 rpm, and the atmospheric drag was equivalent to a force **D** of 25 lb acting at the center of pressure C_P located at a distance $c = 3$ in. from G. (*a*) Knowing that the projectile weighs 40 lb and has a radius of gyration of 2 in. with respect to its axis of symmetry, determine its approximate rate of steady precession. (*b*) If it is further known that the radius of gyration of the projectile with respect to a transverse axis through G is 8 in., determine the exact values of the two possible rates of precession.

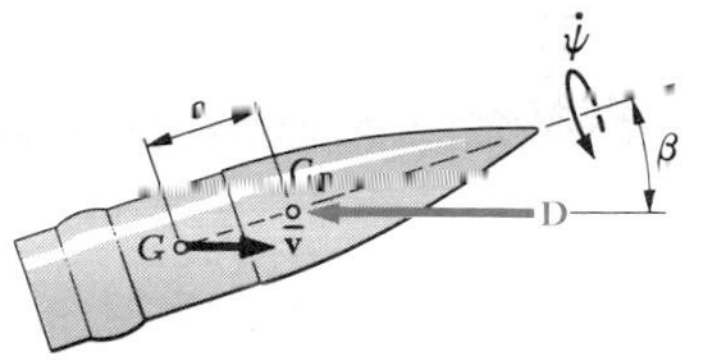

Fig. P18.78

18.79 The essential features of the gyrocompass are shown. The rotor spins at the rate $\dot{\psi}$ about an axis mounted in a single gimbal, which may rotate freely about the vertical axis AB. The angle formed by the axis of the rotor and the plane of the meridian is denoted by θ and the latitude of the position on the earth is denoted by λ. We note that the line OC is parallel to the axis of the earth and we denote by $\boldsymbol{\omega}_e$ the angular velocity of the earth about its axis.

(*a*) Show that the equations of motion of the gyrocompass are

$$I'\ddot{\theta} + I\omega_z\omega_e \cos\lambda \sin\theta - I'\omega_e^2 \cos^2\lambda \sin\theta \cos\theta = 0$$
$$I\dot{\omega}_z = 0$$

where ω_z is the component of the total angular velocity along the axis of the rotor, and I and I' are the moments of inertia of the rotor with respect to its axis of symmetry and a transverse axis through O, respectively.

(*b*) Neglecting the term containing ω_e^2, show that, for small values of θ, we have

$$\ddot{\theta} + \frac{I\omega_z\omega_e \cos\lambda}{I'}\theta = 0$$

and that the axis of the gyrocompass oscillates about the north-south direction.

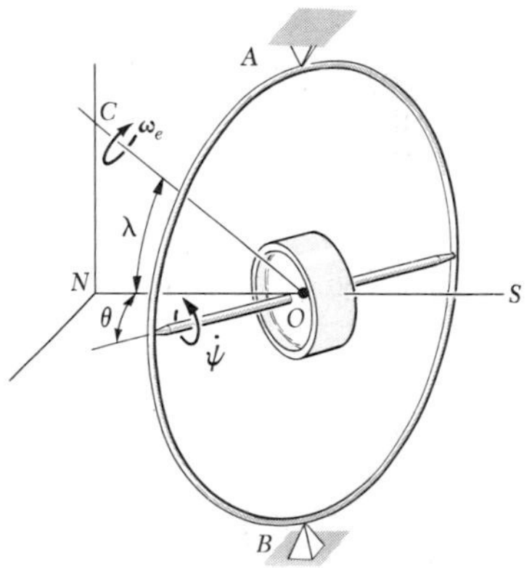

Fig. P18.79

18.80 Show that, for an axisymmetrical body under no force, the rates of precession and spin may be expressed, respectively, as

$$\dot{\phi} = \frac{H_G}{I'}$$

and

$$\dot{\psi} = \frac{H_G \cos\theta\,(I' - I)}{II'}$$

where H_G is the constant value of the angular momentum of the body.

18.81 (*a*) Show that, for an axisymmetrical body under no force, the rate of precession may be expressed as

$$\dot{\phi} = \frac{I\omega_z}{I' \cos\theta}$$

where ω_z is the component of $\boldsymbol{\omega}$ along the axis of symmetry of the body. (*b*) Use this result to check that the condition (18.44) for steady precession is satisfied by an axisymmetrical body under no force.

18.82 Show that the angular velocity vector $\boldsymbol{\omega}$ of an axisymmetrical body under no force is observed from the body itself to rotate about the axis of symmetry at the constant rate

$$n = \frac{I' - I}{I'}\omega_z$$

where ω_z is the component of $\boldsymbol{\omega}$ along the axis of symmetry of the body.

18.83 For an axisymmetrical body under no force, prove (*a*) that the rate of retrograde precession can never be less than twice the rate of spin of the body about its axis of symmetry, (*b*) that in Fig. 18.24 the axis of symmetry of the body can never lie within the space cone.

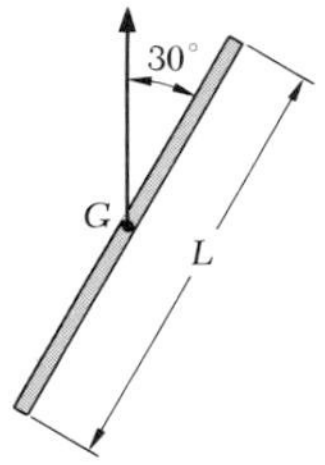

Fig. P18.84

18.84 Determine the precession axis and the rates of precession and spin of a rod which is given an initial angular velocity $\boldsymbol{\omega}$ of 12 rad/s in the direction shown.

18.85 Using the relation given in Prob. 18.82, determine the period of precession of the north pole of the earth about the axis of symmetry of the earth. The earth may be approximated by an oblate spheroid of axial moment of inertia I and of transverse moment of inertia $I' = 0.9967I$. (*Note.* Actual observations show a period of precession of the north pole of about 432.5 mean solar days; the difference between the observed and computed periods is due to the fact that the earth is not a perfectly rigid body. The free precession considered here should not be confused with the much slower precession of the equinoxes, which is a forced precession. See Prob. 18.77.)

18.86 Determine the precession axis and the rates of precession and spin of the satellite of Prob. 18.28 after the impact.

18.87 Determine the precession axis and the rates of precession and spin of the satellite of Prob. 18.28 knowing that, before impact, the angular velocity of the satellite was $\boldsymbol{\omega}_0 = -(12 \text{ rpm})\mathbf{i}$.

18.88 The space capsule has no angular velocity when the jet at A is activated for 1 s in a direction parallel to the x axis. Knowing that the capsule has a mass of 1000 kg, that its radii of gyration are $\bar{k}_x = \bar{k}_y = 1.00$ m and $\bar{k}_z = 1.25$ m, and that the jet at A produces a thrust of 50 N, determine the axis of precession and the rates of precession and spin after the jet has stopped.

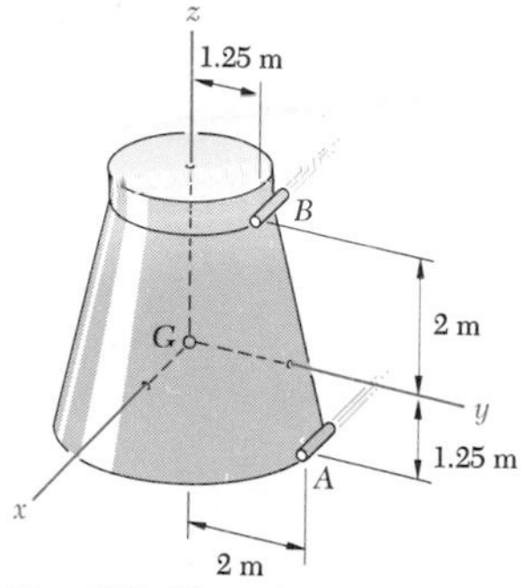

Fig. P18.88 and P18.89

18.89 The space capsule has an angular velocity $\boldsymbol{\omega} = (0.02 \text{ rad/s})\mathbf{j} + (0.10 \text{ rad/s})\mathbf{k}$ when the jet at B is activated for 1 s in a direction parallel to the x axis. Knowing that the capsule has a mass of 1000 kg, that its radii of gyration are $\bar{k}_x = \bar{k}_y = 1.00$ m and $\bar{k}_z = 1.25$ m, and that the jet at B produces a thrust of 50 N, determine the axis of precession and the rates of precession and spin after the jet has stopped.

18.90 The space station shown is known to precess about the fixed direction OC at the rate of one revolution per hour. Assuming that the station is dynamically equivalent to a homogeneous cylinder of length 100 ft and radius 10 ft, determine the rate of spin of the station about its axis of symmetry.

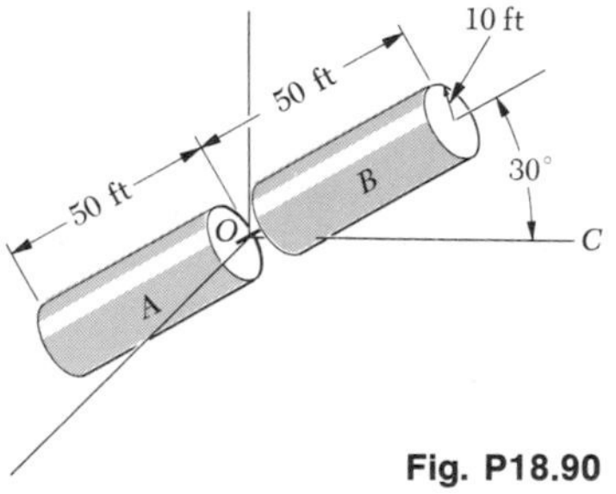

Fig. P18.90

18.91 The link connecting portions A and B of the space station of Prob. 18.90 may be severed to allow each portion to move freely. Each portion of the station is dynamically equivalent to a cylinder of length 50 ft and radius 10 ft. Knowing that, when the link is severed, the station is oriented as shown, determine for portion B the axis of precession, the rate of precession, and the rate of spin about the axis of symmetry.

18.92 Solve Sample Prob. 18.6, assuming that the meteorite strikes the satellite at C with a velocity $\mathbf{v}_0 = -(2000 \text{ m/s})\mathbf{i}$.

18.93 After the motion determined in Sample Prob. 18.6 has been established, the rod connecting disks A and B of the satellite breaks, and disk A moves freely as a separate body. Knowing that the rod and the z axis coincide when the rod breaks, determine the precession axis, the rate of precession, and the rate of spin for the ensuing motion of disk A.

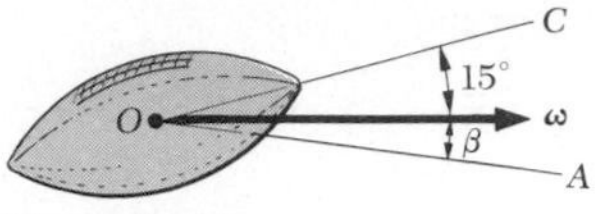

Fig. P18.94

18.94 The angular velocity vector of a football which has just been kicked is horizontal, and its axis of symmetry OC is oriented as shown. Knowing that the magnitude of the angular velocity is 180 rpm and that the ratio of the axial and transverse moments of inertia is $I/I' = 1/3$, determine (*a*) the orientation of the axis of precession OA, (*b*) the rates of precession and spin.

18.95 A slender homogeneous rod OA of mass m and length L is supported by a ball-and-socket joint at O and may swing freely under its own weight. If the rod is held in a horizontal position ($\theta = 90°$) and given an initial angular velocity $\dot{\phi}_0 = \sqrt{8g/L}$ about the vertical OB, determine (*a*) the smallest value of θ in the ensuing motion, (*b*) the corresponding value of the angular velocity $\dot{\phi}$ of the rod about OB. (*Hint.* Apply the principle of conservation of energy and the principle of impulse and momentum, observing that, since $\Sigma M_{OB} = 0$, the component of $\mathbf{H}_O$ along OB must be constant.)

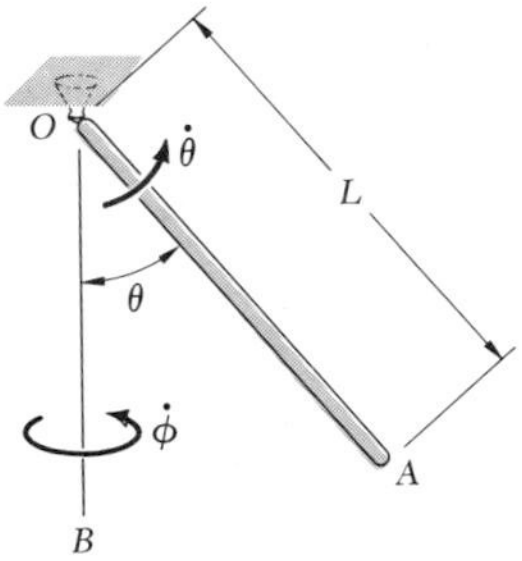

Fig. P18.95 and P18.96

18.96 A slender homogeneous rod OA of mass m and length L is supported by a ball-and-socket joint at O and may swing freely under its own weight. If the rod is held in a horizontal position ($\theta = 90°$), what initial angular velocity $\dot{\phi}_0$ should be given to the rod about the vertical OB if the smallest value of θ in the ensuing motion is to be 60°? (See hint of Prob. 18.95.)

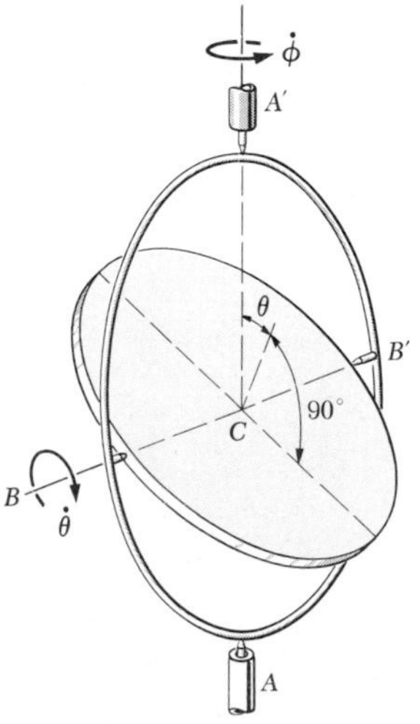

Fig. P18.97

18.97 The gimbal $ABA'B'$ is of negligible mass and may rotate freely about the vertical AA'. The uniform disk of radius a and mass m may rotate freely about its diameter BB', which is also the horizontal diameter of the gimbal. (*a*) Applying the principle of conservation of energy, and observing that, since $\Sigma M_{AA'} = 0$, the component of the angular momentum of the disk along the fixed axis AA' must be constant, write two first-order differential equations defining the motion of the disk. (*b*) Given the initial conditions $\theta_0 \neq 0$, $\dot{\phi}_0 \neq 0$, and $\dot{\theta}_0 = 0$, express the rate of nutation $\dot{\theta}$ as a function of θ. (*c*) Show that the angle θ will never be larger than θ_0 during the ensuing motion.

***18.98** The top shown is supported at the fixed point O. We denote by ϕ, θ, and ψ the Eulerian angles defining the position of the top with respect to a fixed frame of reference. We shall consider the general motion of the top in which all Eulerian angles vary.

(*a*) Observing that $\Sigma M_Z = 0$ and $\Sigma M_z = 0$, and denoting by I and I', respectively, the moments of inertia of the top about its axis of symmetry and about a transverse axis through O, derive the two first-order differential equations of motion

$$I'\dot{\phi}\sin^2\theta + I(\dot{\psi} + \dot{\phi}\cos\theta)\cos\theta = \alpha \qquad (1)$$
$$I(\dot{\psi} + \dot{\phi}\cos\theta) = \beta \qquad (2)$$

where α and β are constants depending upon the initial conditions. These equations express that the angular momentum of the top is conserved about both the Z and z axes, i.e., that the rectangular component of $\mathbf{H}_O$ along each of these axes is constant.

(*b*) Use Eqs. (1) and (2) to show that the component ω_z of the angular velocity of the top is constant and that the rate of precession $\dot{\phi}$ depends upon the value of the angle of nutation θ.

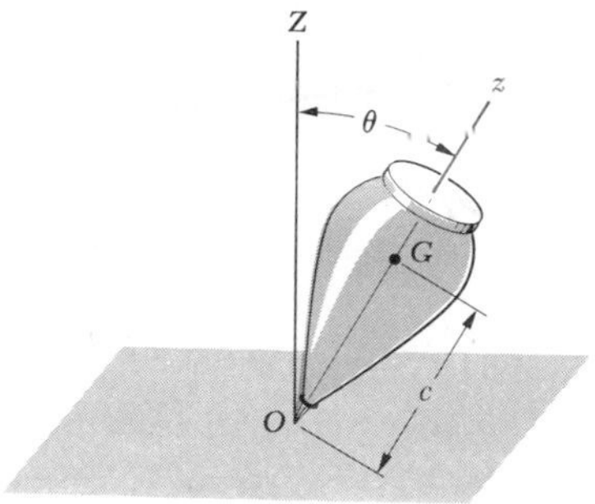

Fig. P18.98

***18.99** (*a*) Applying the principle of conservation of energy, derive a third differential equation for the general motion of the top of Prob. 18.98. (*b*) Eliminating the derivatives $\dot{\phi}$ and $\dot{\psi}$ from the equation obtained and from the two equations of Prob. 18.98, express the rate of nutation $\dot{\theta}$ as a function of the angle θ.

***18.100** A thin homogeneous disk of radius a and mass m is mounted on a light axle OA of length a which is held by a ball-and-socket support at O. The disk is released in the position $\beta = 0$ with a rate of spin $\dot{\psi}_0$, clockwise as viewed from O, and with no precession or nutation. Knowing that the largest value of β in the ensuing motion is 30°, determine in terms of $\dot{\psi}_0$ the rates of precession and spin of the disk when $\beta = 30°$. (*Hint.* The angular momentum of the disk is conserved about both the Z and z axes; see Prob. 18.98, part *a*.)

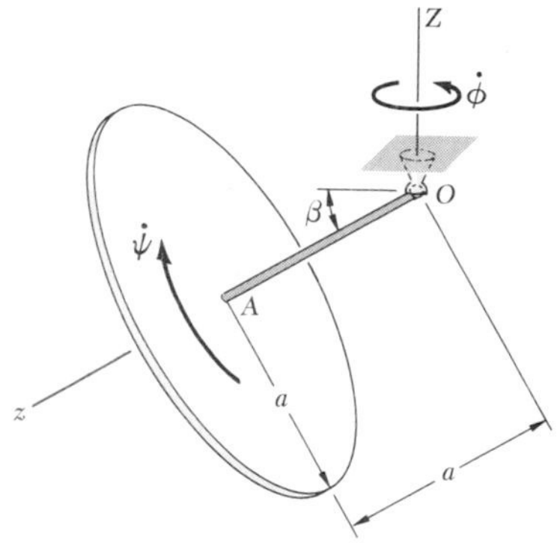

Fig. P18.100

***18.101** For the disk of Prob. 18.100, determine the initial value $\dot{\psi}_0$ of the spin, knowing that the largest value of β in the ensuing motion is 30°. (*Hint.* Use the principle of conservation of energy and the answers obtained for Prob. 18.100.)

***18.102** A solid homogeneous cone of mass m, radius a, and height $h = \frac{3}{2}a$, is held by a ball-and-socket support O. Initially the axis of symmetry of the cone is vertical ($\theta = 0$) with the cone spinning about it at the constant rate $\dot{\psi}_0$, counterclockwise as viewed from above. However, after being slightly disturbed, the cone starts falling and precessing. If the largest value of θ in the ensuing motion is 90°, determine (*a*) the rate of spin $\dot{\psi}_0$ of the cone in its initial vertical position, (*b*) the rates of precession and spin as the cone passes through its lowest position ($\theta = 90°$). (*Hint.* Use the principle of conservation of energy and the fact that the angular momentum of the cone is conserved about both the Z and z axes; see Prob. 18.98, part *a*.)

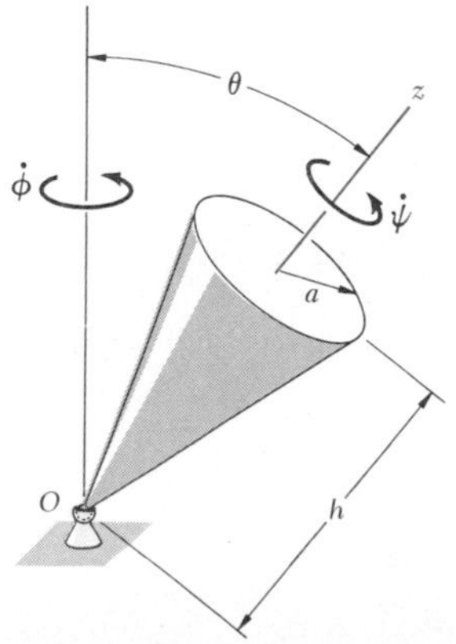

Fig. P18.102

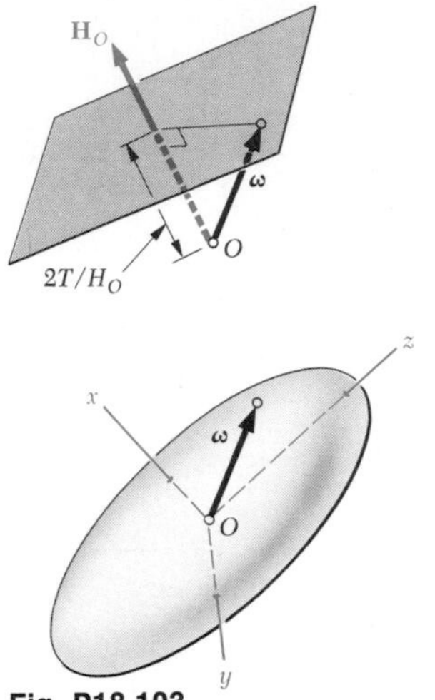

Fig. P18.103

∗18.103 Consider a rigid body of arbitrary shape which is attached at its mass center O and subjected to no force other than its weight and the reaction of the support at O. (*a*) Prove that the angular momentum $\mathbf{H}_O$ of the body about the fixed point O is constant in magnitude and direction, that the kinetic energy T of the body is constant, and that the projection along $\mathbf{H}_O$ of the angular velocity $\boldsymbol{\omega}$ of the body is constant. (*b*) Show that the tip of the vector $\boldsymbol{\omega}$ describes a curve on a fixed plane in space (called the *invariable plane*), which is perpendicular to $\mathbf{H}_O$ and at a distance $2T/H_O$ from O. (*c*) Show that, with respect to a frame of reference attached to the body and coinciding with its principal axes of inertia, the tip of the vector $\boldsymbol{\omega}$ appears to describe a curve on an ellipsoid of equation

$$I_x\omega_x^2 + I_y\omega_y^2 + I_z\omega_z^2 = 2T = \text{constant}$$

This ellipsoid (called the *Poinsot ellipsoid*) is rigidly attached to the body and is of the same shape as the ellipsoid of inertia, but of a different size.

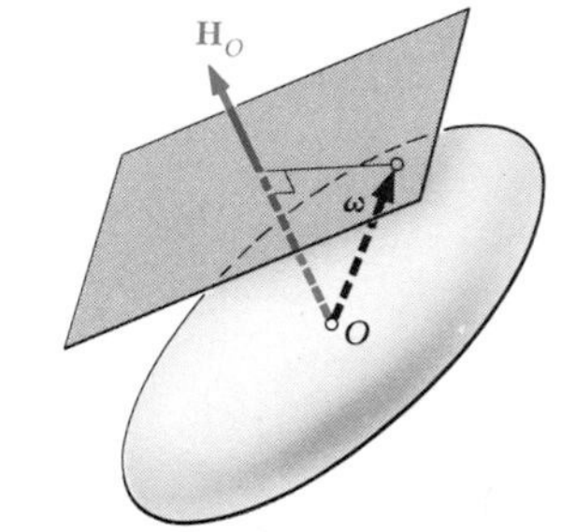

Fig. P18.104

∗18.104 Referring to Prob. 18.103, (*a*) prove that the Poinsot ellipsoid is tangent to the invariable plane, (*b*) show that the motion of the rigid body must be such that the Poinsot ellipsoid appears to roll on the invariable plane. (*Hint.* In part *a*, show that the normal to the Poinsot ellipsoid at the tip of $\boldsymbol{\omega}$ is parallel to $\mathbf{H}_O$. It is recalled that the direction of the normal to a surface of equation $F(x,y,z) = \text{constant}$ at a point P is the same as that of **grad** F at point P.)

∗18.105 Using the results obtained in Probs. 18.103 and 18.104, show that, for an axisymmetrical body attached at its mass center O and under no force other than its weight and the reaction at O, the Poinsot ellipsoid is an ellipsoid of revolution and the space and body cones are both circular and are tangent to each other. Further show that (*a*) the two cones are tangent externally, and the precession is direct, when $I < I'$, where I and I' denote, respectively, the axial and transverse moment of inertia of the body, (*b*) the space cone is inside the body cone, and the precession is retrograde, when $I > I'$.

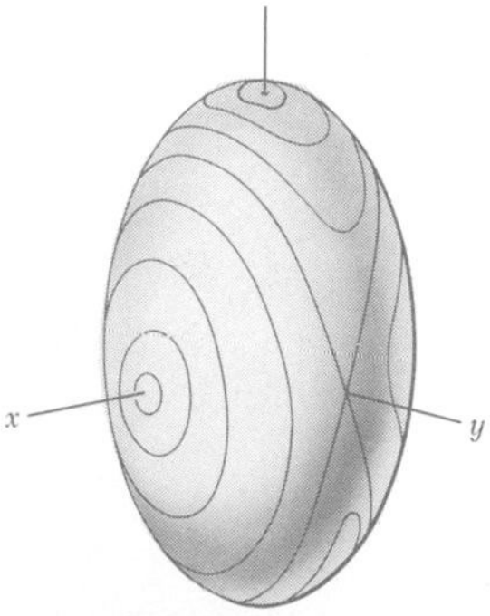

Fig. P18.106

∗18.106 Referring to Probs. 18.103 and 18.104, (*a*) show that the curve (called *polhode*) described by the tip of the vector $\boldsymbol{\omega}$ with respect to a frame of reference coinciding with the principal axes of inertia of the rigid body is defined by the equations

$$I_x\omega_x^2 + I_y\omega_y^2 + I_z\omega_z^2 = 2T = \text{constant} \qquad (1)$$

$$I_x^2\omega_x^2 + I_y^2\omega_y^2 + I_z^2\omega_z^2 = H_O^2 = \text{constant} \qquad (2)$$

and that this curve may, therefore, be obtained by intersecting the Poinsot ellipsoid with the ellipsoid defined by Eq. (2). (*b*) Further show, assuming $I_x > I_y > I_z$, that the polhodes obtained for various values of H_O have the shapes indicated in the figure. (*c*) Using the result obtained in part *b*, show that a rigid body under no force can rotate about a fixed centroidal axis if, and only if, that axis coincides with one of the principal axes of inertia of the body, and that the motion will be stable if the axis of rotation coincides with the major or minor axis of the Poinsot ellipsoid (z or x axis in the figure) and unstable if it coincides with the intermediate axis (y axis).

REVIEW PROBLEMS

18.107 A thin rectangular plate of mass 9 kg is attached to a shaft as shown. If at the instant shown the angular velocity $\boldsymbol{\omega}$ of the plate is 4 rad/s and is increasing at the rate of 8 rad/s^2, determine (*a*) the couple **M** which must be applied to the shaft, (*b*) the corresponding dynamic reactions at *A* and *C*.

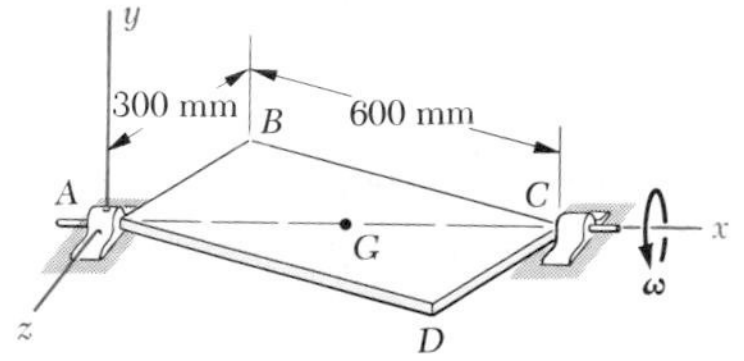

Fig. P18.107 and P18.108

18.108 A thin rectangular plate of mass 9 kg is attached to a shaft as shown. A couple of moment (3N · m)**i** is applied to the plate which is initially at rest. Determine (*a*) the angular acceleration of the plate, (*b*) the dynamic reactions at *A* and *C* as the plate reaches an angular velocity of 10 rad/s.

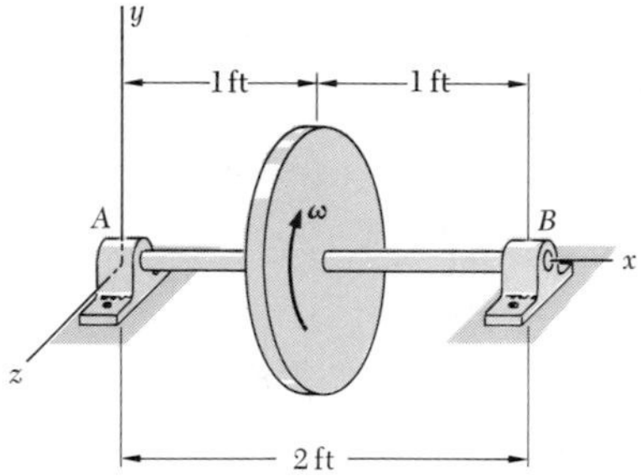

Fig. P18.109

18.109 The rotor of a given turbine may be approximated by a 50-lb disk of 12-in. radius. Knowing that the turbine rotates clockwise at 10,000 rpm as viewed from the positive x axis, determine the components due to gyroscopic action of the forces exerted by the bearings on axle *AB* if the instantaneous angular velocity of the turbine housing is 2 rad/s clockwise as viewed from (*a*) the positive y axis, (*b*) the positive x axis.

18.110 The rectangular plate shown is falling with a velocity $\overline{\mathbf{v}}_0$ and no angular velocity when its corner *A* strikes an obstruction. Assuming the impact at *A* is perfectly plastic, determine immediately after impact (*a*) the angular velocity of the plate, (*b*) the velocity of the mass center *G* of the plate.

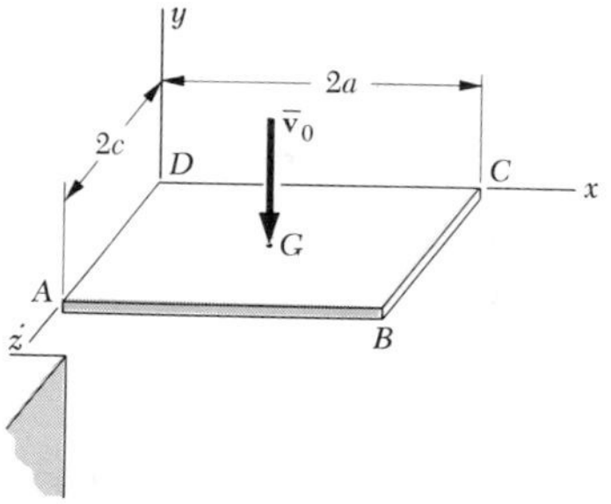

Fig. P18.110

***18.111** Solve Prob. 18.110, assuming that the impact at *A* is perfectly elastic.

18.112 A rigid square frame *ABCD* consisting of four slender uniform bars, each 1.2 m long, is suspended by a wire attached at *A*. Bars *AB* and *CD* have each a mass of 25 kg, while bars *AD* and *BC* have each a mass of 5 kg. The frame is hit at *B* in a direction perpendicular to, and into the plane of, the frame. Knowing that the corresponding impulse applied to the frame is 75 N·s, determine immediately after the impact (*a*) the velocity of the mass center of the frame, (*b*) the angular velocity of the frame.

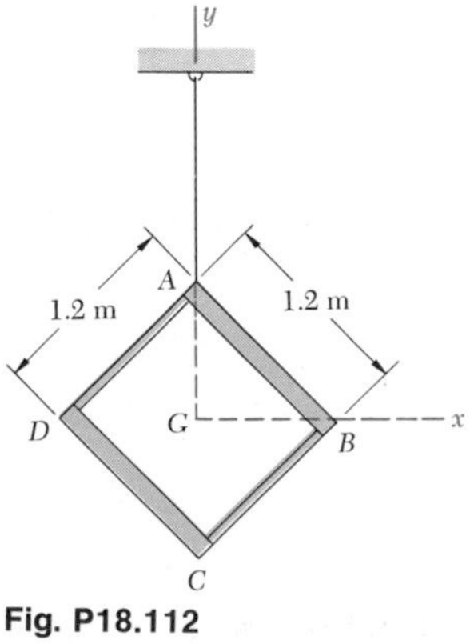

Fig. P18.112

18.113 Solve Prob. 18.112, assuming that the frame is hit at corner *C*.

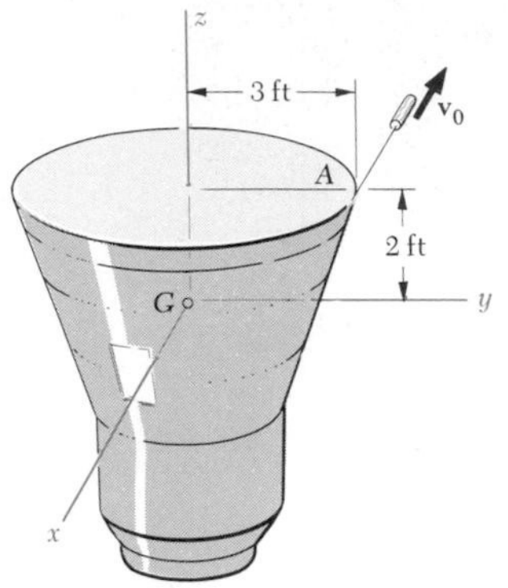

Fig. P18.114

18.114 The 800-lb space capsule is spinning with an angular velocity $\boldsymbol{\omega}_0 = (100 \text{ rpm})\mathbf{k}$ when a 1-lb projectile is fired from A in a direction parallel to the x axis and with a velocity $\mathbf{v}_0$ of 4000 ft/s. Knowing that the radii of gyration of the capsule are $\bar{k}_x = \bar{k}_y = 1.50$ ft and $\bar{k}_z = 2.00$ ft, determine immediately after the projectile has been fired (*a*) the angular velocity of the capsule, (*b*) the kinetic energy of the capsule.

18.115 Determine the precession axis and the rates of precession and spin of the capsule of Prob. 18.114 after the projectile has been fired.

18.116 A coin is tossed into the air. During the free motion the angle β between the plane of the coin and the horizontal is observed to be constant. (*a*) Derive an expression for the angle formed by the angular velocity of the coin and the vertical. (*b*) Denoting by $\dot{\psi}$ the rate of spin of the coin about its axis of symmetry, derive an expression for the rate of precession. (*c*) Solve parts *a* and *b* for the case $\beta = 10°$.

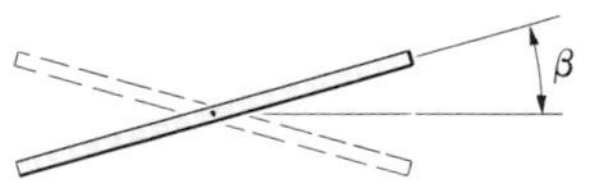

Fig. P18.116

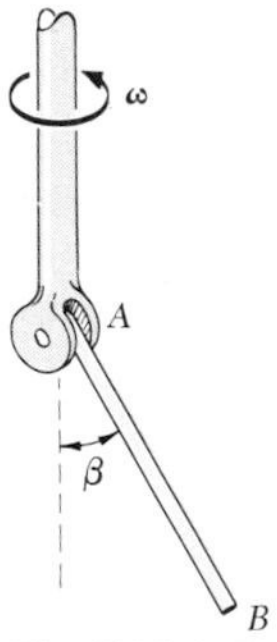

Fig. P18.117

18.117 A uniform rod AB of length l and mass m is attached to the pin of a clevis which rotates with a constant angular velocity $\boldsymbol{\omega}$. Derive an expression (*a*) for the constant angle β that the rod forms with the vertical, (*b*) for the maximum value of ω for which the rod will remain vertical ($\beta = 0$).

18.118 A homogeneous sphere of radius a and mass m is attached to a light rod of length $4a$. The rod forms an angle of 30° with the vertical and rotates about AC at the constant rate $\Omega = \sqrt{g/a}$. (*a*) Assuming that the sphere does not spin about the rod ($\dot{\psi} = 0$), determine the tension in the cord BC and the kinetic energy of the sphere. (*b*) Determine the spin $\dot{\psi}$ (magnitude and sense) which should be given to the sphere if the tension in the cord BC is to be zero. What is the corresponding kinetic energy of the sphere?

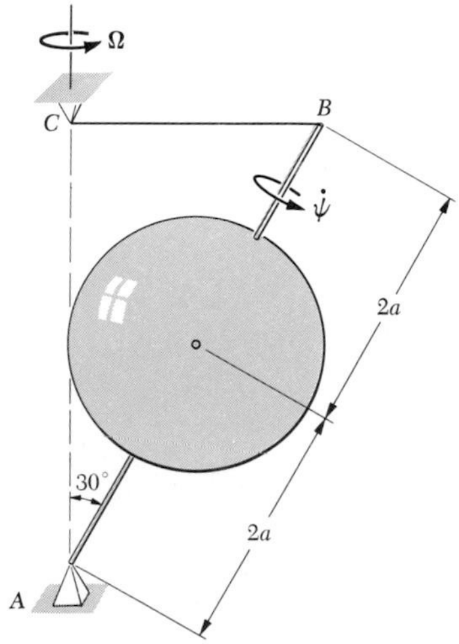

Fig. P18.118

Mechanical Vibrations

CHAPTER 19

19.1. Introduction. A *mechanical vibration* is the motion of a particle or a body which oscillates about a position of equilibrium. Most vibrations in machines and structures are undesirable because of the increased stresses and energy losses which accompany them. They should therefore be eliminated or reduced as much as possible by appropriate design. The analysis of vibrations has become increasingly important in recent years owing to the current trend toward higher-speed machines and lighter structures. There is every reason to expect that this trend will continue and that an even greater need for vibration analysis will develop in the future.

The analysis of vibrations is a very extensive subject to which entire texts have been devoted. We shall therefore limit our present study to the simpler types of vibrations, namely, the vibrations of a body or a system of bodies with one degree of freedom.

A mechanical vibration generally results when a system is displaced from a position of stable equilibrium. The system tends to return to this position under the action of restoring forces (either elastic forces, as in the case of a mass attached to a spring, or gravitational forces, as in the case of a pendulum). But the system generally reaches its original position with a certain acquired velocity which carries it beyond that position. Since the process can be repeated indefinitely, the system keeps moving back and forth across its position of equilibrium. The time interval required for the system to complete a full cycle of motion is called the *period* of the vibration. The number of cycles per unit time defines the *frequency*, and the maximum displacement of the system from its position of equilibrium is called the *amplitude* of the vibration.

When the motion is maintained by the restoring forces only, the vibration is said to be a *free vibration* (Secs. 19.2 to 19.6). When a periodic force is applied to the system, the resulting motion is described as a *forced vibration* (Sec. 19.7). When the effects of friction may be neglected, the vibrations are said to be *undamped*. However, all vibrations are actually *damped* to some degree. If a free vibration is only slightly damped, its amplitude slowly decreases until, after a certain time, the motion comes to a stop. But damping may be large enough to prevent any true vibration; the system then slowly regains its original position (Sec. 19.8). A damped forced vibration is maintained as long as the periodic force which produces the vibration is applied. The amplitude of the vibration, however, is affected by the magnitude of the damping forces (Sec. 19.9).

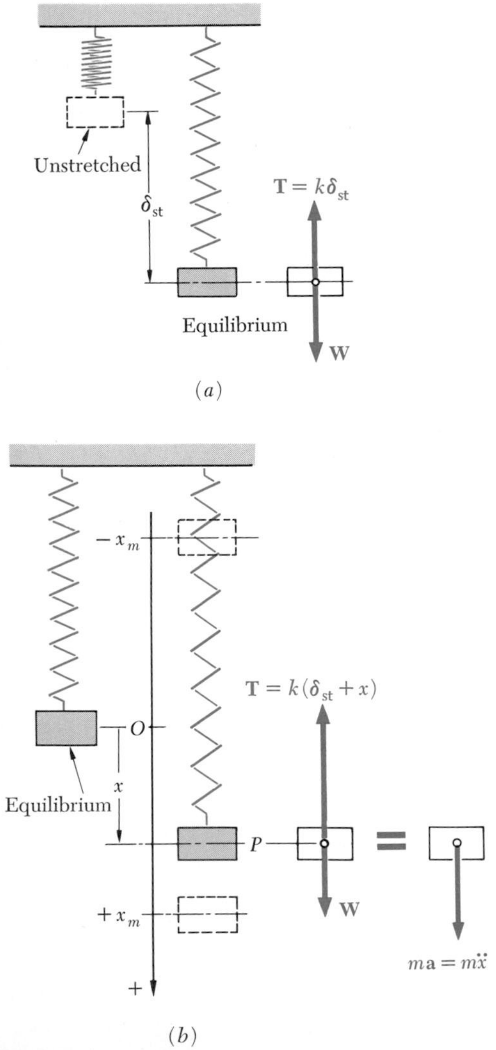

Fig. 19.1

VIBRATIONS WITHOUT DAMPING

19.2. Free Vibrations of Particles. Simple Harmonic Motion. Consider a body of mass m attached to a spring of constant k (Fig. 19.1*a*). Since, at the present time, we are concerned only with the motion of its mass center, we shall refer to this body as a particle. When the particle is in static equilibrium, the forces acting on it are its weight $\mathbf{W}$ and the force $\mathbf{T}$ exerted by the spring, of magnitude $T = k\delta_{st}$, where δ_{st} denotes the elongation of the spring. We have, therefore,

$$W = k\delta_{st} \tag{19.1}$$

Suppose now that the particle is displaced through a distance x_m from its equilibrium position and released with no initial velocity. If x_m has been chosen smaller than δ_{st}, the particle will move back and forth through its equilibrium position; a vibration of amplitude x_m has been generated. Note that the vibration may also be produced by imparting a certain initial velocity to the particle when it is in its equilibrium position $x = 0$ or, more generally, by starting the particle from any given position $x = x_0$ with a given initial velocity $\mathbf{v}_0$.

To analyze the vibration, we shall consider the particle in a position P at some arbitrary time t (Fig. 19.1*b*). Denoting by x the displacement OP measured from the equilibrium position

O (positive downward), we note that the forces acting on the particle are its weight **W** and the force **T** exerted by the spring which, in this position, has a magnitude $T = k(\delta_{st} + x)$. Recalling (19.1), we find that the magnitude of the resultant **F** of the two forces (positive downward) is

$$F = W - k(\delta_{st} + x) = -kx \tag{19.2}$$

Thus the *resultant* of the forces exerted on the particle is proportional to the displacement *OP measured from the equilibrium position.* Recalling the sign convention, we note that **F** is always directed *toward* the equilibrium position O. Substituting for F into the fundamental equation $F = ma$ and recalling that a is the second derivative $\ddot{x}$ of x with respect to t, we write

$$m\ddot{x} + kx = 0 \tag{19.3}$$

Note that the same sign convention should be used for the acceleration $\ddot{x}$ and for the displacement x, namely, positive downward.

Equation (19.3) is a linear differential equation of the second order. Setting

$$p^2 = \frac{k}{m} \tag{19.4}$$

we write (19.3) in the form

$$\ddot{x} + p^2x = 0 \tag{19.5}$$

The motion defined by Eq. (19.5) is called *simple harmonic motion.* It is characterized by the fact that *the acceleration is proportional to the displacement and of opposite direction.* We note that each of the functions $x_1 = \sin pt$ and $x_2 = \cos pt$ satisfies (19.5). These functions, therefore, constitute two *particular solutions* of the differential equation (19.5). As we shall see presently, the *general solution* of (19.5) may be obtained by multiplying the two particular solutions by arbitrary constants A and B and adding. We write

$$x = Ax_1 + Bx_2 = A \sin pt + B \cos pt \tag{19.6}$$

Differentiating, we obtain successively the velocity and acceleration at time t,

$$v = \dot{x} = Ap \cos pt - Bp \sin pt \tag{19.7}$$

$$a = \ddot{x} = -Ap^2 \sin pt - Bp^2 \cos pt \tag{19.8}$$

Substituting from (19.6) and (19.8) into (19.5), we verify that the expression (19.6) provides a solution of the differential equation (19.5). Since this expression contains two arbitrary constants A

and B, the solution obtained is the general solution of the differential equation. The values of the constants A and B depend upon the *initial conditions* of the motion. For example, we have $A = 0$ if the particle is displaced from its equilibrium position and released at $t = 0$ with no initial velocity, and we have $B = 0$ if P is started from O at $t = 0$ with a certain initial velocity. In general, substituting $t = 0$ and the initial values x_0 and v_0 of the displacement and velocity into (19.6) and (19.7), we find $A = v_0/p$ and $B = x_0$.

The expressions obtained for the displacement, velocity, and acceleration of a particle may be written in a more compact form if we observe that (19.6) expresses that the displacement $x = OP$ is the sum of the x components of two vectors $\mathbf{A}$ and $\mathbf{B}$, respectively of magnitude A and B, directed as shown in Fig. 19.2*a*. As t varies, both vectors rotate clockwise; we also note that the magnitude of their resultant $\overrightarrow{OQ}$ is equal to the maximum displacement x_m. The simple harmonic motion of P along the x axis may thus be obtained by projecting on this axis the motion of a point Q describing an *auxiliary circle* of radius x_m *with a constant angular velocity* p. Denoting by ϕ the angle formed by the vectors $\overrightarrow{OQ}$ and $\mathbf{A}$, we write

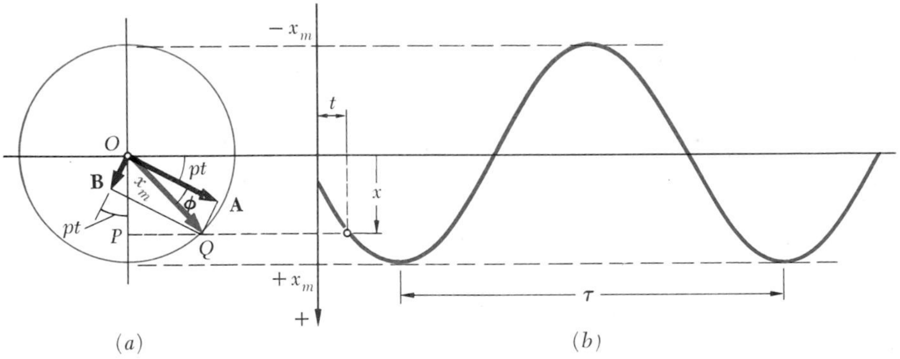

Fig. 19.2

$$OP = OQ \sin(pt + \phi) \tag{19.9}$$

which leads to new expressions for the displacement, velocity, and acceleration of P,

$$x = x_m \sin(pt + \phi) \tag{19.10}$$

$$v = \dot{x} = x_m p \cos(pt + \phi) \tag{19.11}$$

$$a = \ddot{x} = -x_m p^2 \sin(pt + \phi) \tag{19.12}$$

The displacement-time curve is represented by a sine curve (Fig. 19.2*b*), and the maximum value x_m of the displacement is called the *amplitude* of the vibration. The angular velocity p of the point Q which describes the auxiliary circle is known as the *circular frequency* of the vibration and is measured in rad/s, while the angle ϕ which defines the initial position of Q on the circle is called the *phase angle*. We note from Fig. 19.2 that a full *cycle* has been described after the angle pt has increased by 2π rad. The corresponding value of t, denoted by τ, is called the *period* of the vibration and is measured in seconds. We have

$$\text{Period} = \tau = \frac{2\pi}{p} \tag{19.13}$$

The number of cycles described per unit of time is denoted by f and is known as the *frequency* of the vibration. We write

$$\text{Frequency} = f = \frac{1}{\tau} = \frac{p}{2\pi} \tag{19.14}$$

The unit of frequency is a frequency of 1 cycle per second, corresponding to a period of 1 s. In terms of base units the unit of frequency is thus 1/s or s^{-1}. It is called a *hertz* (Hz) in the SI system of units. It also follows from Eq. (19.14) that a frequency of $1\ \text{s}^{-1}$ or 1 Hz corresponds to a circular frequency of 2π rad/s. In problems involving angular velocities expressed in revolutions per minute (rpm), we have $1\ \text{rpm} = \frac{1}{60}\ \text{s}^{-1} = \frac{1}{60}\ \text{Hz}$, or $1\ \text{rpm} = (2\pi/60)\ \text{rad/s}$.

Recalling that p was defined in (19.4) in terms of the constant k of the spring and the mass m of the particle, we observe that the period and the frequency are independent of the initial conditions and of the amplitude of the vibration. Note that τ and f depend on the *mass* rather than on the *weight* of the particle and thus are independent of the value of g.

The velocity-time and acceleration-time curves may be represented by sine curves of the same period as the displacement-time curve, but with different phase angles. From (19.11) and (19.12), we note that the maximum values of the magnitudes of the velocity and acceleration are

$$v_m = x_m p \qquad a_m = x_m p^2 \tag{19.15}$$

Since the point Q describes the auxiliary circle, of radius x_m, at the constant angular velocity p, its velocity and acceleration are equal, respectively, to the expressions (19.15). Recalling Eqs. (19.11) and (19.12), we find, therefore, that the velocity and acceleration of P may be obtained at any instant by projecting on the x axis vectors of magnitudes $v_m = x_m p$ and $a_m = x_m p^2$ representing respectively the velocity and acceleration of Q at the same instant (Fig. 19.3).

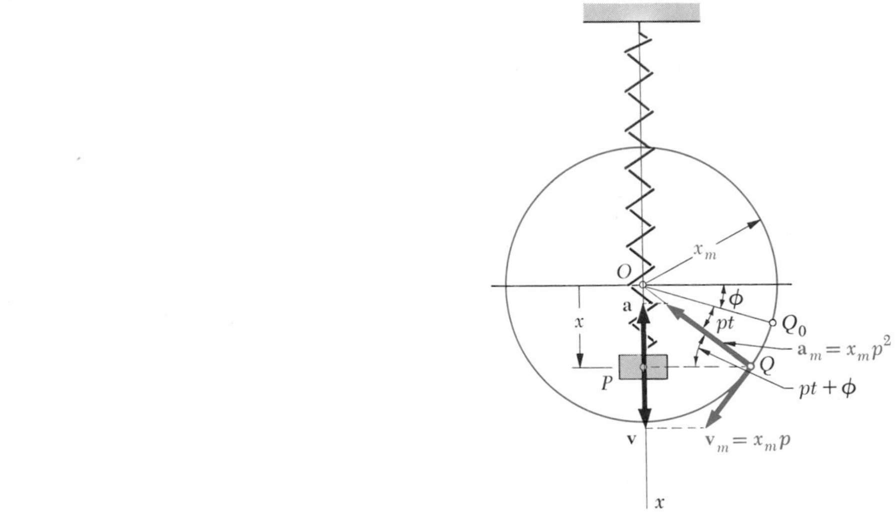

Fig. 19.3

The results obtained are not limited to the solution of the problem of a mass attached to a spring. They may be used to analyze the rectilinear motion of a particle *whenever the resultant* **F** *of the forces acting on the particle is proportional to the displacement x and directed toward O.* The fundamental equation of motion $F = ma$ may then be written in the form (19.5), which characterizes simple harmonic motion. Observing that the coefficient of x in (19.5) represents the square of the circular frequency p of the vibration, we easily obtain p and, after substitution into (19.13) and (19.14), the period τ and the frequency f of the vibration.

19.3. Simple Pendulum (Approximate Solution). Most of the vibrations encountered in engineering applications may be represented by a simple harmonic motion. Many others, although of a different type, may be *approximated* by a simple harmonic motion, provided that their amplitude remains small. Consider for example a *simple pendulum,* consisting of a bob of mass m attached to a cord of length l, which may oscillate in a vertical plane (Fig. 19.4*a*). At a given time t, the cord forms an angle θ with the vertical. The forces acting on the bob are its weight **W** and the force **T** exerted by the cord (Fig. 19.4*b*). Resolving the vector $m\mathbf{a}$ into tangential and normal components,

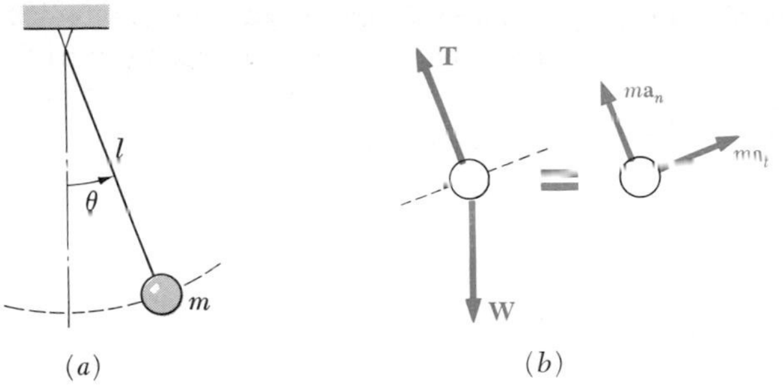

Fig. 19.4

with $m\mathbf{a}_t$ directed to the right, i.e., in the direction corresponding to increasing values of θ, and observing that $a_t = l\alpha = l\ddot{\theta}$, we write

$$\Sigma F_t = ma_t: \qquad -W \sin \theta = ml\ddot{\theta}$$

Noting that $W = mg$ and dividing through by ml, we obtain

$$\ddot{\theta} + \frac{g}{l} \sin \theta = 0 \tag{19.16}$$

For oscillations of small amplitude, we may replace $\sin \theta$ by θ, expressed in radians, and write

$$\ddot{\theta} + \frac{g}{l} \theta = 0 \tag{19.17}$$

Comparison with (19.5) shows that the equation obtained is that of a simple harmonic motion and that the circular frequency p of the oscillations is equal to $(g/l)^{1/2}$. Substitution into (19.13) yields the period of the small oscillations of a pendulum of length l,

$$\tau = \frac{2\pi}{p} = 2\pi \sqrt{\frac{l}{g}} \tag{19.18}$$

***19.4. Simple Pendulum (Exact Solution).** Formula (19.18) is only approximate. To obtain an exact expression for the period of the oscillations of a simple pendulum, we must return to (19.16). Multiplying both terms by $2\dot{\theta}$ and integrating from an initial position corresponding to the maximum deflection, that is, $\theta = \theta_m$ and $\dot{\theta} = 0$, we write

$$\dot{\theta}^2 = \frac{2g}{l} (\cos \theta - \cos \theta_m)$$

or

$$\left(\frac{d\theta}{dt}\right)^2 = \frac{2g}{l} (\cos \theta - \cos \theta_m)$$

Replacing $\cos\theta$ by $1 - 2\sin^2(\theta/2)$ and $\cos\theta_m$ by a similar expression, solving for dt, and integrating over a quarter period from $t = 0$, $\theta = 0$ to $t = \tau/4$, $\theta = \theta_m$, we have

$$\tau = 2\sqrt{\frac{l}{g}}\int_0^{\theta_m} \frac{d\theta}{\sqrt{\sin^2(\theta_m/2) - \sin^2(\theta/2)}}$$

The integral in the right-hand member is known as an *elliptic integral;* it cannot be expressed in terms of the usual algebraic or trigonometric functions. However, setting

$$\sin(\theta/2) = \sin(\theta_m/2)\sin\phi$$

we may write

$$\tau = 4\sqrt{\frac{l}{g}}\int_0^{\pi/2} \frac{d\phi}{\sqrt{1 - \sin^2(\theta_m/2)\sin^2\phi}} \qquad (19.19)$$

where the integral obtained, commonly denoted by K, may be found in *tables of elliptic integrals* for various values of $\theta_m/2$.† In order to compare the result just obtained with that of the preceding section, we write (19.19) in the form

$$\tau = \frac{2K}{\pi}\left(2\pi\sqrt{\frac{l}{g}}\right) \qquad (19.20)$$

Formula (19.20) shows that the actual value of the period of a simple pendulum may be obtained by multiplying the approximate value (19.18) by the correction factor $2K/\pi$. Values of the correction factor are given in Table 19.1 for various values of

Table 19.1 Correction Factor for the Period of a Simple Pendulum

θ_m	0°	10°	20°	30°	60°	90°	120°	150°	180°
K	1.571	1.574	1.583	1.598	1.686	1.854	2.157	2.768	∞
$2K/\pi$	1.000	1.002	1.008	1.017	1.073	1.180	1.373	1.762	∞

the amplitude θ_m. We note that for ordinary engineering computations the correction factor may be omitted as long as the amplitude does not exceed 10°.

† See, for example, Dwight, "Table of Integrals and Other Mathematical Data," The Macmillan Company, or Peirce, "A Short Table of Integrals," Ginn and Company.

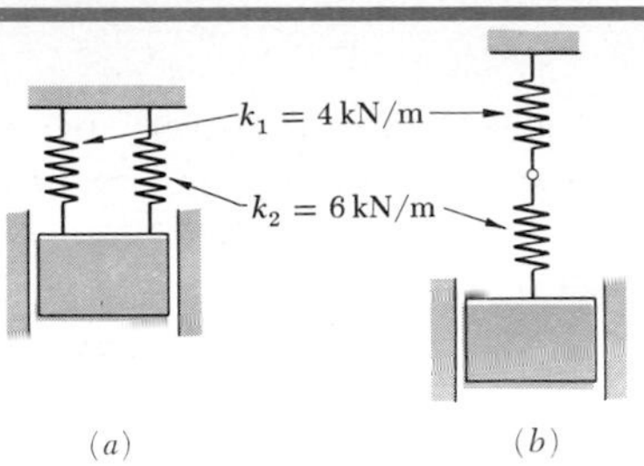

SAMPLE PROBLEM 19.1

A 50-kg block moves between vertical guides as shown. The block is pulled 40 mm down from its equilibrium position and released. For each spring arrangement, determine the period of the vibration, the maximum velocity of the block, and the maximum acceleration of the block.

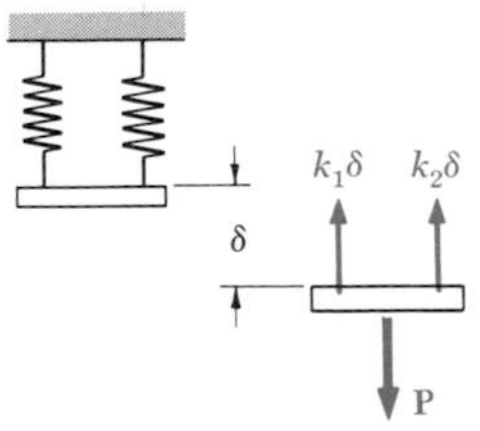

a. Springs Attached in Parallel. We first determine the constant k of a single spring equivalent to the two springs *by finding the magnitude of the force* **P** required to cause a given deflection δ. Since for a deflection δ the magnitudes of the forces exerted by the springs are, respectively, $k_1\delta$ and $k_2\delta$, we have

$$P = k_1\delta + k_2\delta = (k_1 + k_2)\delta$$

The constant k of the single equivalent spring is

$$k = \frac{P}{\delta} = k_1 + k_2 = 4 \text{ kN/m} + 6 \text{ kN/m} = 10 \text{ kN/m} = 10^4 \text{ N/m}$$

Period of Vibration: Since $m = 50$ kg, Eq. (19.4) yields

$$p^2 = \frac{k}{m} = \frac{10^4 \text{ N/m}}{50 \text{ kg}} \qquad p = 14.14 \text{ rad/s}$$

$$\tau = 2\pi/p \qquad\qquad \tau = 0.444 \text{ s} \blacktriangleleft$$

Maximum Velocity: $v_m = x_m p = (0.040 \text{ m})(14.14 \text{ rad/s})$

$$v_m = 0.566 \text{ m/s} \qquad \mathbf{v}_m = 0.566 \text{ m/s} \updownarrow \blacktriangleleft$$

Maximum Acceleration: $a_m = x_m p^2 = (0.040 \text{ m})(14.14 \text{ rad/s})^2$

$$a_m = 8.00 \text{ m/s}^2 \qquad \mathbf{a}_m = 8.00 \text{ m/s}^2 \updownarrow \blacktriangleleft$$

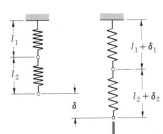

b. Springs Attached in Series. We first determine the constant k of a single spring equivalent to the two springs *by finding the total elongation* δ of the springs under a given static load **P**. To facilitate the computation, a static load of magnitude $P = 12$ kN is used.

$$\delta = \delta_1 + \delta_2 = \frac{P}{k_1} + \frac{P}{k_2} = \frac{12 \text{ kN}}{4 \text{ kN/m}} + \frac{12 \text{ kN}}{6 \text{ kN/m}} = 5 \text{ m}$$

$$k = \frac{P}{\delta} = \frac{12 \text{ kN}}{5 \text{ m}} = 2.4 \text{ kN/m} = 2400 \text{ N/m}$$

Period of Vibration: $p^2 = \frac{k}{m} = \frac{2400 \text{ N/m}}{50 \text{ kg}} \qquad p = 6.93 \text{ rad/s}$

$$\tau = \frac{2\pi}{p} \qquad\qquad \tau = 0.907 \text{ s} \blacktriangleleft$$

Maximum Velocity: $v_m = x_m p = (0.040 \text{ m})(6.93 \text{ rad/s})$

$$v_m = 0.277 \text{ m/s} \qquad \mathbf{v}_m = 0.277 \text{ m/s} \updownarrow \blacktriangleleft$$

Maximum Acceleration: $a_m = x_m p^2 = (0.040 \text{ m})(6.93 \text{ rad/s})^2$

$$a_m = 1.920 \text{ m/s}^2 \qquad \mathbf{a}_m = 1.920 \text{ m/s}^2 \updownarrow \blacktriangleleft$$

PROBLEMS

19.1 A particle moves in simple harmonic motion with an amplitude of 4 in. and a period of 0.60 s. Find the maximum velocity and the maximum acceleration.

19.2 The analysis of the motion of a particle shows a maximum acceleration of 30 m/s^2 and a frequency of 120 cycles per minute. Assuming that the motion is simple harmonic, determine (*a*) the amplitude, (*b*) the maximum velocity.

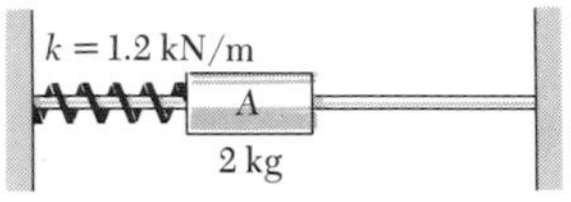

Fig. P19.3

19.3 Collar A is attached to the spring shown and may slide without friction on the horizontal rod. If the collar is moved 75 mm from its equilibrium position and released, determine the period, the maximum velocity, and the maximum acceleration of the resulting motion.

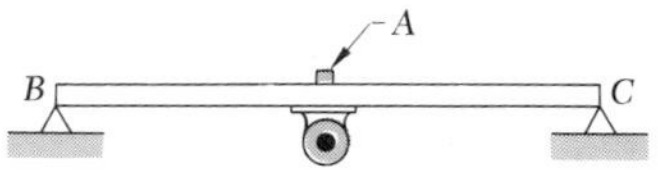

Fig. P19.4

19.4 A variable-speed motor is rigidly attached to the beam BC. The rotor is slightly unbalanced and causes the beam to vibrate with a circular frequency equal to the motor speed. When the speed of the motor is less than 600 rpm or more than 1200 rpm, a small object placed at A is observed to remain in contact with the beam. For speeds between 600 and 1200 rpm the object is observed to "dance" and actually to lose contact with the beam. Determine the amplitude of the motion of A when the speed of the motor is (*a*) 600 rpm, (*b*) 1200 rpm. Give answers in both SI and U.S. customary units.

19.5 The 6-lb collar rests on, but is not attached to, the spring shown. The collar is depressed 2 in. and released. If the ensuing motion is to be simple harmonic, determine (*a*) the largest permissible value of the spring constant k, (*b*) the corresponding frequency of the motion.

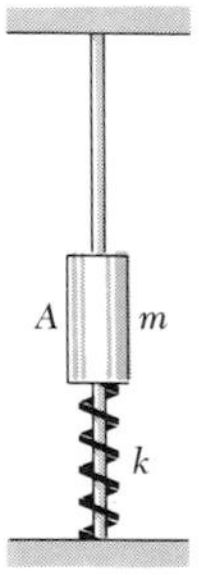

Fig. P19.5, P19.6, and P19.8

19.6 The 5-kg collar is attached to a spring of constant $k = 800$ N/m as shown. If the collar is given a displacement of 50 mm from its equilibrium position and released, determine for the ensuing motion (*a*) the period, (*b*) the maximum velocity of the collar, (*c*) the maximum acceleration of the collar.

19.7 In Prob. 19.6, determine the position, velocity, and acceleration of the collar 0.20 s after it has been released.

19.8 An 8-lb collar is attached to a spring of constant $k = 5$ lb/in. as shown. If the collar is given a displacement of 2 in. downward from its equilibrium position and released, determine (*a*) the time required for the collar to move 3 in. upward, (*b*) the corresponding velocity and acceleration of the collar.

19.9 and 19.10 A 35-kg block is supported by the spring arrangement shown. If the block is moved vertically downward from its equilibrium position and released, determine (*a*) the period and frequency of the resulting motion, (*b*) the maximum velocity and acceleration of the block if the amplitude of the motion is 20 mm.

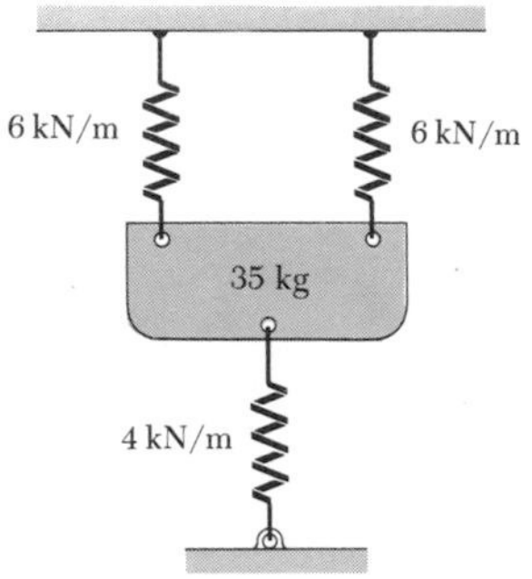

Fig. P19.9

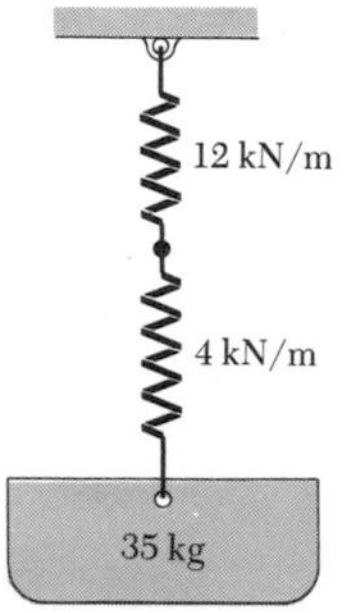

Fig. P19.10

19.11 Denoting by δ_{st} the static deflection of a beam under a given load, show that the frequency of vibration of the load is

$$f = \frac{1}{2\pi}\sqrt{\frac{g}{\delta_{st}}}$$

Neglect the mass of the beam, and assume that the load remains in contact with the beam.

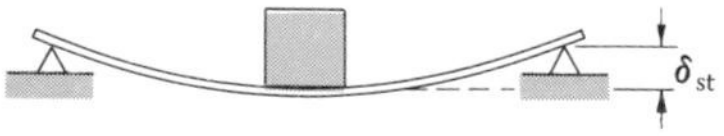

Fig. P19.11

19.12 The period of vibration of the system shown is observed to be 0.8 s. If block *A* is removed, the period is observed to be 0.7 s. Determine (*a*) the weight of block *C*, (*b*) the period of vibration when both blocks *A* and *B* have been removed.

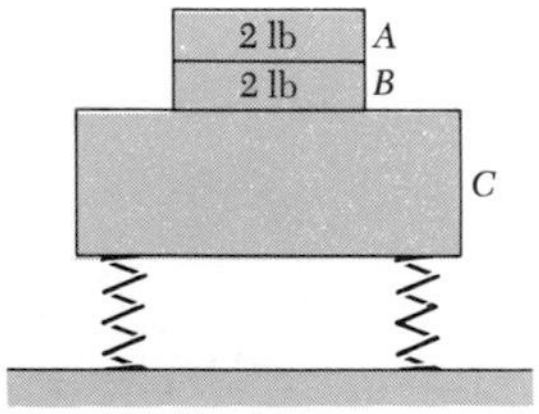

Fig. P19.12

19.13 A simple pendulum of length l is suspended in an elevator. A mass m is attached to a spring of constant k and is carried in the same elevator. Determine the period of vibration of both the pendulum and the mass if the elevator has an upward acceleration **a**.

19.14 Determine (*a*) the required length l of a simple pendulum if the period of small oscillations is to be 2 s, (*b*) the required amplitude of this pendulum if the maximum velocity of the bob is to be 200 mm/s.

19.15 A small bob is attached to a cord of length 4 ft and is released from rest when $\theta_A = 5°$. Knowing that $d = 2$ ft, determine (*a*) the time required for the bob to return to point *A*, (*b*) the amplitude θ_C.

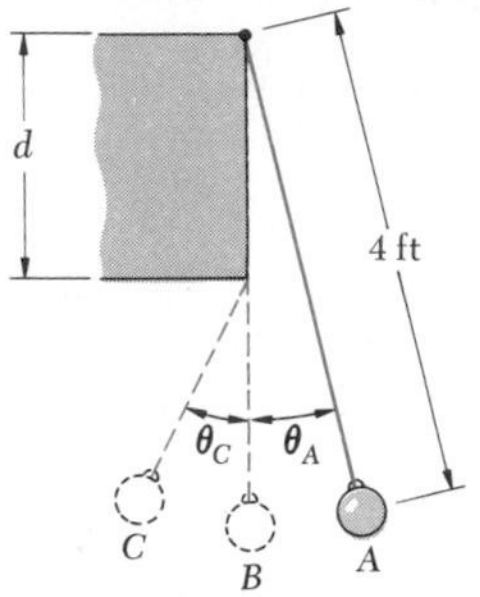

Fig. P19.15 and P19.16

19.16 A small bob is attached to a cord of length 4 ft and released from rest at *A* when $\theta_A = 4°$. Determine the distance *d* for which the bob will return to point *A* in 2 s.

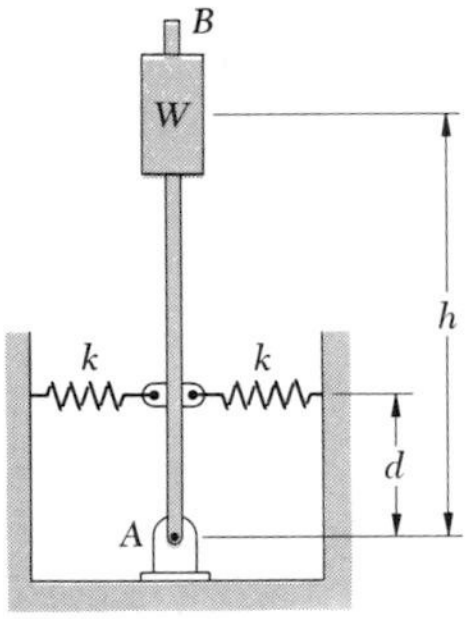

Fig. P19.17 and P19.18

19.17 The rod *AB* is attached to a hinge at *A* and to two springs each of constant *k*. When $h = 24$ in., $d = 10$ in., and $W = 50$ lb, determine the value of *k* for which the period of small oscillations is (*a*) 1 s, (*b*) infinite. Neglect the weight of the rod and assume that each spring can act in either tension or compression.

19.18 If $d = 16$ in., $h = 24$ in., and each spring has a constant $k = 4$ lb/in., determine the load *W* for which the period of small oscillations is (*a*) 0.50 s, (*b*) infinite. Neglect the weight of the rod and assume that each spring can act in either tension or compression.

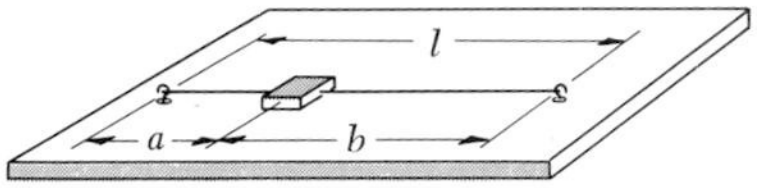

Fig. P19.19 and P19.20

19.19 A small block of mass *m* rests on a frictionless horizontal surface and is attached to a taut string. Denoting by *T* the tension in the string, determine the period and frequency of small oscillations of the block in a direction perpendicular to the string. Show that the longest period occurs when $a = b = \frac{1}{2}l$.

19.20 A 3-lb block rests on a frictionless horizontal surface and is attached to a taut string. Knowing that the tension in the string is 8 lb, determine the frequency of small oscillations of the block when $a = 12$ in. and $b = 18$ in.

***19.21** A particle is placed with no initial velocity on a frictionless *plane* tangent to the surface of the earth. (*a*) Show that the particle will theoretically execute simple harmonic motion with a period of oscillation equal to that of a simple pendulum of length equal to the radius of the earth. (*b*) Compute the theoretical period of oscillation and show that it is equal to the periodic time of an earth satellite describing a low-altitude circular orbit. [*Hint.* See Eq. (12.44).]

***19.22** Expanding the integrand in (19.19) into a series of even powers of $\sin\phi$ and integrating, show that the period of a simple pendulum of length l may be approximated by the formula

$$\tau = 2\pi\sqrt{\frac{l}{g}}\left(1 + \tfrac{1}{4}\sin^2\frac{\theta_m}{2}\right)$$

where θ_m is the amplitude of the oscillations.

***19.23** Using the data of Table 19.1, determine the period of a simple pendulum of length 800 mm (*a*) for small oscillations, (*b*) for oscillations of amplitude $\theta_m = 30°$, (*c*) for oscillations of amplitude $\theta_m = 90°$.

***19.24** Using the formula given in Prob. 19.22, determine the amplitude θ_m for which the period of a simple pendulum is $\frac{1}{2}$ percent longer than the period of the same pendulum for small oscillations.

***19.25** Using a table of elliptic integrals, determine the period of a simple pendulum of length $l = 800$ mm if the amplitude of the oscillations is $\theta_m = 40°$.

19.5. Free Vibrations of Rigid Bodies. The analysis of the vibrations of a rigid body or of a system of rigid bodies possessing a single degree of freedom is similar to the analysis of the vibrations of a particle. An appropriate variable, such as a distance x or an angle θ, is chosen to define the position of the body or system of bodies, and an equation relating this variable and its second derivative with respect to t is written. If the equation obtained is of the same form as (19.5), i.e., if we have

$$\ddot{x} + p^2x = 0 \qquad \text{or} \qquad \ddot{\theta} + p^2\theta = 0 \tag{19.21}$$

the vibration considered is a simple harmonic motion. The period and frequency of the vibration may then be obtained by identifying p and substituting into (19.13) and (19.14).

In general, a simple way to obtain one of Eqs. (19.21) is to express that the system of the external forces is equivalent to the system of the effective forces by drawing a diagram of the body for an arbitrary value of the variable and writing the appropriate equation of motion. We recall that our goal should be *the determination of the coefficient* of the variable x or θ, *not* the determination of the variable itself or of the derivatives $\ddot{x}$ or $\ddot{\theta}$. Setting this coefficient equal to p^2, we obtain the circular frequency p, from which τ and f may be determined.

The method we have outlined may be used to analyze vibrations which are truly represented by a simple harmonic motion, or vibrations of small amplitude which can be *approximated* by a simple harmonic motion. As an example, we shall determine the period of the small oscillations of a square plate of side $2b$

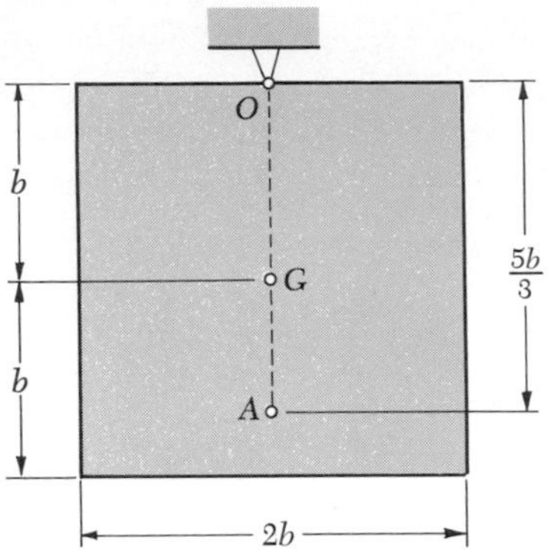

(a)

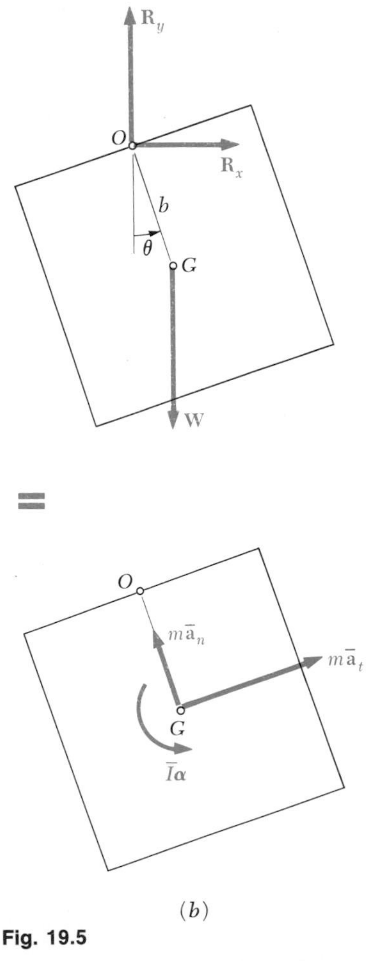

(b)

Fig. 19.5

which is suspended from the midpoint O of one side (Fig. 19.5*a*). We consider the plate in an arbitrary position defined by the angle θ that the line OG forms with the vertical and draw a diagram to express that the weight $\mathbf{W}$ of the plate and the components $\mathbf{R}_x$ and $\mathbf{R}_y$ of the reaction at O are equivalent to the vectors $m\bar{\mathbf{a}}_t$ and $m\bar{\mathbf{a}}_n$ and to the couple $\bar{I}\boldsymbol{\alpha}$ (Fig. 19.5*b*). Since the angular velocity and angular acceleration of the plate are equal, respectively, to $\dot{\theta}$ and $\ddot{\theta}$, the magnitudes of the two vectors are, respectively, $mb\ddot{\theta}$ and $mb\dot{\theta}^2$, while the moment of the couple is $\bar{I}\ddot{\theta}$. In previous applications of this method (Chap. 16), we tried whenever possible to assume the correct sense for the acceleration. Here, however, we must assume the same positive sense for θ and $\ddot{\theta}$ in order to obtain an equation of the form (19.21). Consequently, the angular acceleration $\ddot{\theta}$ will be assumed positive counterclockwise, even though this assumption is obviously unrealistic. Equating moments about O, we write

$$+\circlearrowleft \qquad -W(b\sin\theta) = (mb\ddot{\theta})b + \bar{I}\ddot{\theta}$$

Noting that $\bar{I} = \frac{1}{12}m[(2b)^2 + (2b)^2] = \frac{2}{3}mb^2$ and $W = mg$, we obtain

$$\ddot{\theta} + \frac{3}{5}\frac{g}{b}\sin\theta = 0 \tag{19.22}$$

For oscillations of small amplitude, we may replace $\sin\theta$ by θ, expressed in radians, and write

$$\ddot{\theta} + \frac{3}{5}\frac{g}{b}\theta = 0 \tag{19.23}$$

Comparison with (19.21) shows that the equation obtained is that of a simple harmonic motion and that the circular frequency p of the oscillations is equal to $(3g/5b)^{1/2}$. Substituting into (19.13), we find that the period of the oscillations is

$$\tau = \frac{2\pi}{p} = 2\pi\sqrt{\frac{5b}{3g}} \tag{19.24}$$

The result obtained is valid only for oscillations of small amplitude. A more accurate description of the motion of the plate is obtained by comparing Eqs. (19.16) and (19.22). We note that the two equations are identical if we choose l equal to $5b/3$. This means that the plate will oscillate as a simple pendulum of length $l = 5b/3$, and the results of Sec. 19.4 may be used to correct the value of the period given in (19.24). The point A of the plate located on line OG at a distance $l = 5b/3$ from O is defined as the *center of oscillation* corresponding to O (Fig. 19.5*a*).

SAMPLE PROBLEM 19.2

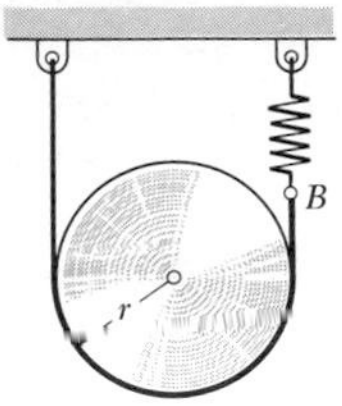

A cylinder of weight W and radius r is suspended from a looped cord as shown. One end of the cord is attached directly to a rigid support, while the other end is attached to a spring of constant k. Determine the period and frequency of vibration of the cylinder.

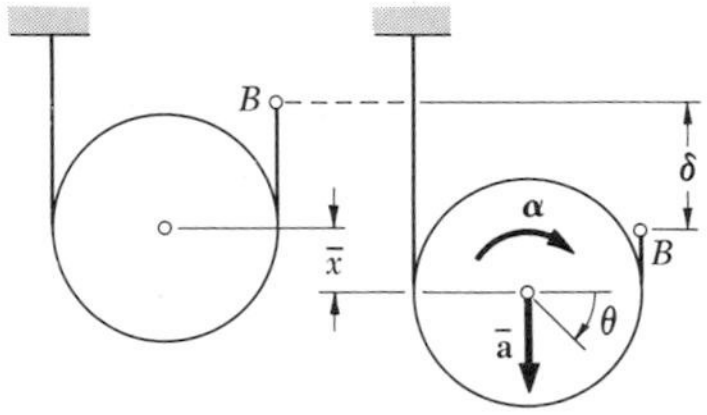

Kinematics of Motion. We express the linear displacement and the acceleration of the cylinder in terms of the angular displacement θ. Choosing the positive sense clockwise and measuring the displacements from the equilibrium position, we write

$$\bar{x} = r\theta \qquad \delta = 2\bar{x} = 2r\theta$$

$$\boldsymbol{\alpha} = \ddot{\theta} \downarrow \qquad \bar{a} = r\alpha = r\ddot{\theta} \qquad \bar{\mathbf{a}} = r\ddot{\theta} \downarrow \tag{1}$$

Equations of Motion. The system of external forces acting on the cylinder consists of the weight $\mathbf{W}$ and of the forces $\mathbf{T}_1$ and $\mathbf{T}_2$ exerted by the cord. We express that this system is equivalent to the system of effective forces represented by the vector $m\bar{\mathbf{a}}$ attached at G and the couple $\bar{I}\boldsymbol{\alpha}$.

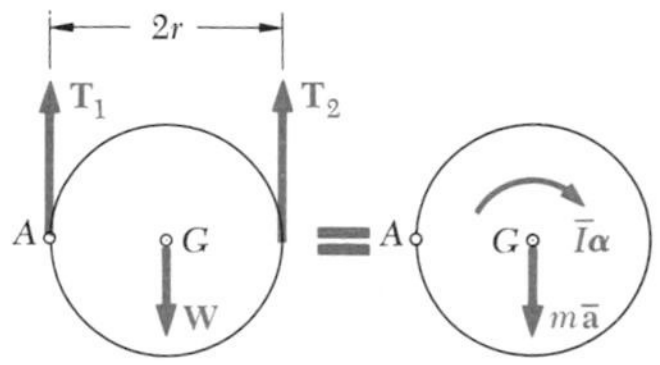

$$+\downarrow \Sigma M_A = \Sigma(M_A)_{\text{eff}}: \qquad Wr - T_2(2r) = m\bar{a}r + \bar{I}\alpha \tag{2}$$

When the cylinder is in its position of equilibrium, the tension in the cord is $T_0 = \frac{1}{2}W$. We note that, for an angular displacement θ, the magnitude of $\mathbf{T}_2$ is

$$T_2 = T_0 + k\delta = \tfrac{1}{2}W + k\delta = \tfrac{1}{2}W + k(2r\theta) \tag{3}$$

Substituting from (1) and (3) into (2), and recalling that $\bar{I} = \frac{1}{2}mr^2$, we write

$$Wr - (\tfrac{1}{2}W + 2kr\theta)(2r) = m(r\ddot{\theta})r + \tfrac{1}{2}mr^2\ddot{\theta}$$

$$\ddot{\theta} + \frac{8}{3}\frac{k}{m}\theta = 0$$

The motion is seen to be simple harmonic, and we have

$$p^2 = \frac{8}{3}\frac{k}{m} \qquad p = \sqrt{\frac{8}{3}\frac{k}{m}}$$

$$\tau = \frac{2\pi}{p} \qquad \tau = 2\pi\sqrt{\frac{3}{8}\frac{m}{k}} \blacktriangleleft$$

$$f = \frac{p}{2\pi} \qquad f = \frac{1}{2\pi}\sqrt{\frac{8}{3}\frac{k}{m}} \blacktriangleleft$$

SAMPLE PROBLEM 19.3

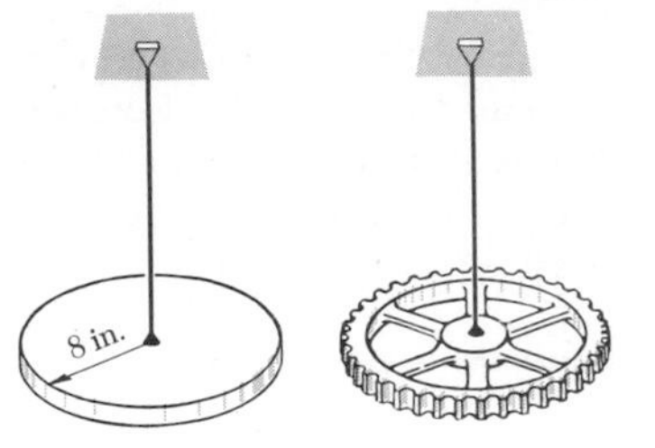

A circular disk, weighing 20 lb and of radius 8 in., is suspended from a wire as shown. The disk is rotated (thus twisting the wire) and then released; the period of the torsional vibration is observed to be 1.13 s. A gear is then suspended from the same wire, and the period of torsional vibration for the gear is observed to be 1.93 s. Assuming that the moment of the couple exerted by the wire is proportional to the angle of twist, determine (*a*) the torsional spring constant of the wire, (*b*) the centroidal moment of inertia of the gear, (*c*) the maximum angular velocity reached by the gear if it is rotated through 90° and released.

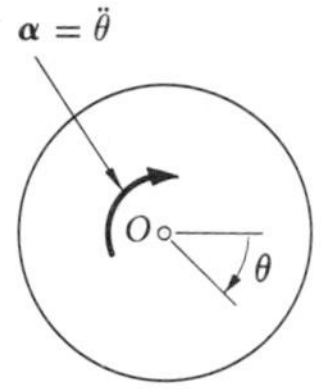

***a.* Vibration of Disk.** Denoting by θ the angular displacement of the disk, we express that the magnitude of the couple exerted by the wire is $M = K\theta$, where K is the torsional spring constant of the wire. Since this couple must be equivalent to the couple $\bar{I}\boldsymbol{\alpha}$ representing the effective forces of the disk, we write

$$+\circlearrowleft \Sigma M_O = \Sigma(M_O)_{\text{eff}}: \qquad +K\theta = -\bar{I}\ddot{\theta}$$

$$\ddot{\theta} + \frac{K}{\bar{I}}\theta = 0$$

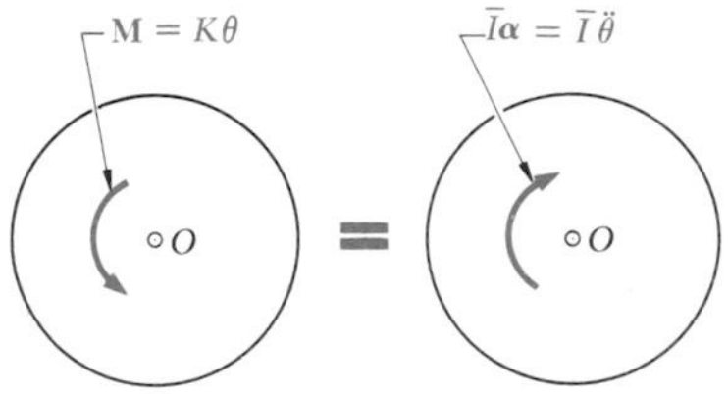

The motion is seen to be simple harmonic, and we have

$$p^2 = \frac{K}{\bar{I}} \qquad \tau = \frac{2\pi}{p} \qquad \tau = 2\pi\sqrt{\frac{\bar{I}}{K}} \qquad (1)$$

For the disk, we have

$$\tau = 1.13\text{ s} \qquad \bar{I} = \tfrac{1}{2}mr^2 = \frac{1}{2}\left(\frac{20\text{ lb}}{32.2\text{ ft/s}^2}\right)(\tfrac{8}{12}\text{ ft})^2 = 0.138\text{ lb}\cdot\text{ft}\cdot\text{s}^2$$

Substituting into (1), we obtain

$$1.13 = 2\pi\sqrt{\frac{0.138}{K}} \qquad K = 4.27\text{ lb}\cdot\text{ft/rad} \quad ◀$$

***b.* Vibration of Gear.** Since the period of vibration of the gear is 1.93 s and $K = 4.27$ lb · ft/rad, Eq. (1) yields

$$1.93 = 2\pi\sqrt{\frac{\bar{I}}{4.27}} \qquad \bar{I}_{\text{gear}} = 0.403\text{ lb}\cdot\text{ft}\cdot\text{s}^2 \quad ◀$$

***c.* Maximum Angular Velocity of the Gear.** Since the motion is simple harmonic, we have

$$\theta = \theta_m \sin pt \qquad \omega = \theta_m p \cos pt \qquad \omega_m = \theta_m p$$

Recalling that $\theta_m = 90° = 1.571$ rad and $\tau = 1.93$ s, we write

$$\omega_m = \theta_m p = \theta_m(2\pi/\tau) = (1.571\text{ rad})(2\pi/1.93\text{ s})$$

$$\omega_m = 5.11\text{ rad/s} \quad ◀$$

PROBLEMS

19.26 and 19.27 The uniform rod shown weighs 8 lb and is attached to a spring of constant $k = 2.5$ lb/in. If end A of the rod is depressed 2 in. and released, determine (a) the period of vibration, (b) the maximum velocity of end A.

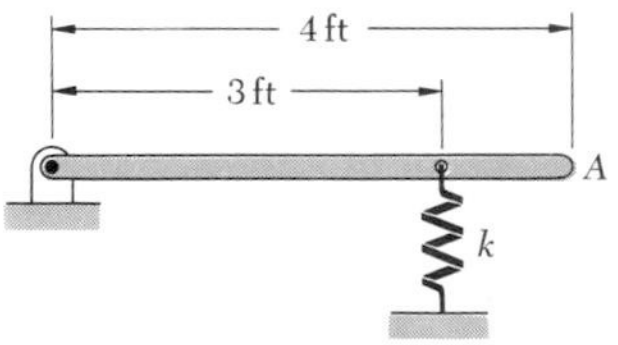

Fig. P19.26

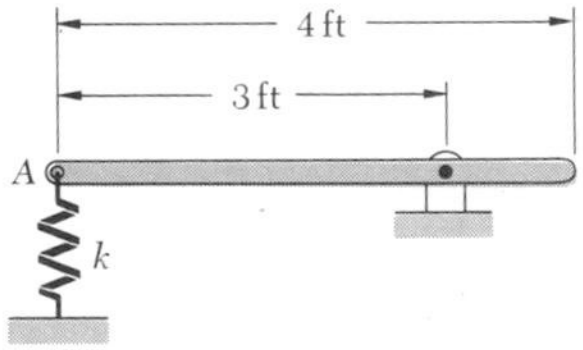

Fig. P19.27

19.28 A belt is placed over the rim of a 15-kg disk as shown and then attached to a 5-kg cylinder and to a spring of constant $k = 600$ N/m. If the cylinder is moved 50 mm down from its equilibrium position and released, determine (a) the period of vibration, (b) the maximum velocity of the cylinder. Assume friction is sufficient to prevent the belt from slipping on the rim.

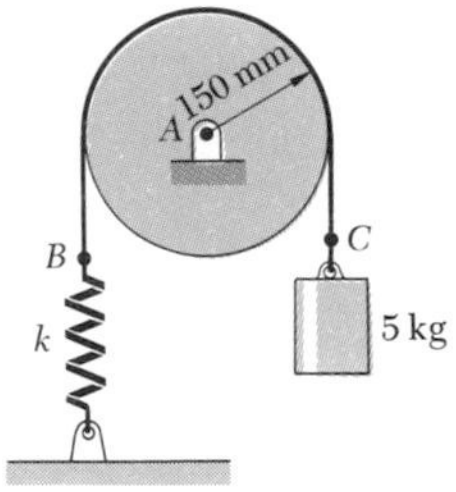

Fig. P19.28

19.29 In Prob. 19.28, determine (a) the frequency of vibration, (b) the maximum tension which occurs in the belt at B and at C.

19.30 A 600-lb flywheel has a diameter of 4 ft and a radius of gyration of 20 in. A belt is placed around the rim and attached to two springs, each of constant $k = 75$ lb/in. The initial tension in the belt is sufficient to prevent slipping. If the end C of the belt is pulled 1 in. down and released, determine (a) the period of vibration, (b) the maximum angular velocity of the flywheel.

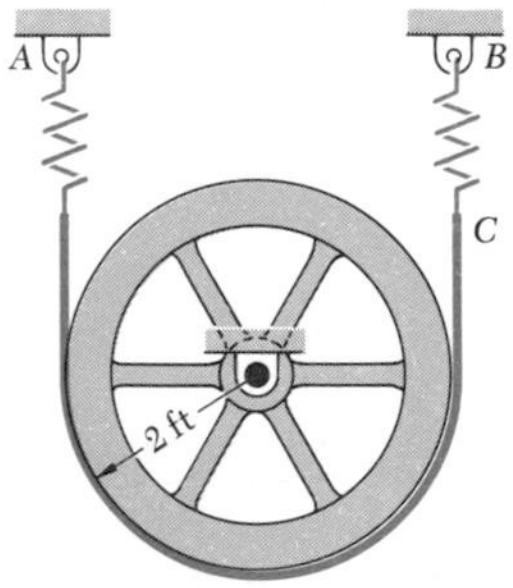

Fig. P19.30

19.31 A uniform square plate of mass m is supported in a horizontal plane by a vertical pin at B and is attached at A to a spring of constant k. If corner A is given a small displacement and released, determine the period of the resulting motion.

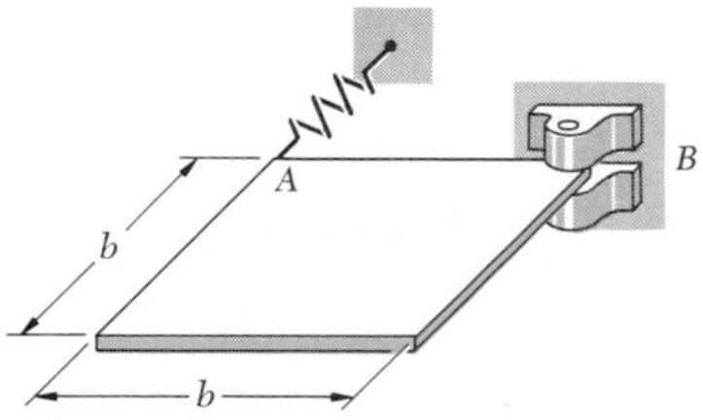

Fig. P19.31

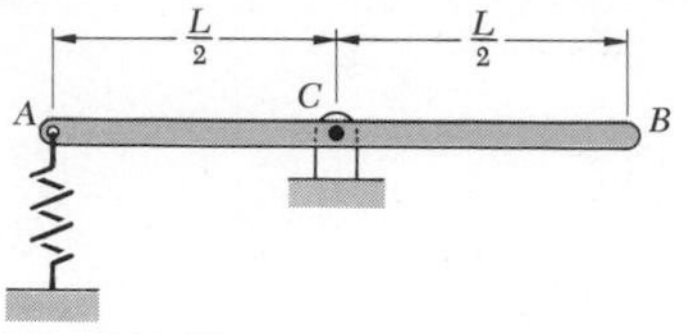

Fig. P19.32

19.32 A uniform rod of mass m is supported by a pin at its midpoint C and is attached to a spring of constant k. If end A is given a small displacement and released, determine the frequency of the resulting motion.

19.33 A uniform circular plate of mass m and radius r is held by four springs, each of constant k. Determine the frequency of the resulting vibration if the plate is (a) given a small vertical displacement and released, (b) rotated through a small angle about diameter AC and released, (c) rotated through a small angle about any other diameter and released.

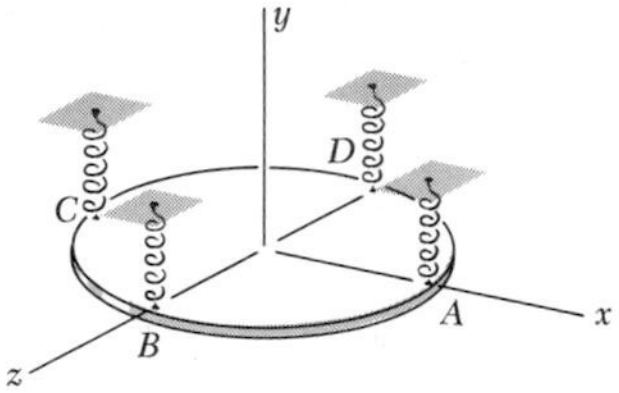

Fig. P19.33

19.34 A *compound pendulum* is defined as a rigid slab which oscillates about a fixed point O, called the center of suspension. Show that the period of oscillation of a compound pendulum is equal to the period of a simple pendulum of length OA, where the distance from A to the mass center G is $GA = \bar{k}^2/\bar{r}$. Point A is defined as the center of oscillation and coincides with the center of percussion defined in Prob. 17.58.

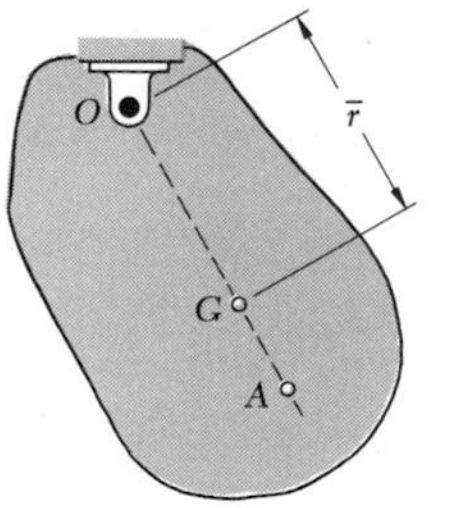

Fig. P19.34 and P19.36

19.35 Show that, if the compound pendulum of Prob. 19.34 is suspended from A instead of O, the period of oscillation is the same as before and the new center of oscillation is located at O.

19.36 A rigid slab oscillates about a fixed point O. Show that the smallest period of oscillation occurs when the distance $\bar{r}$ from point O to the mass center G is equal to $\bar{k}$.

19.37 A uniform bar of length l may oscillate about a hinge at A located a distance c from its mass center G. (*a*) Determine the frequency of small oscillations if $c = \frac{1}{2}l$. (*b*) Determine a second value of c for which the frequency of small oscillations is the same as that found in part *a*.

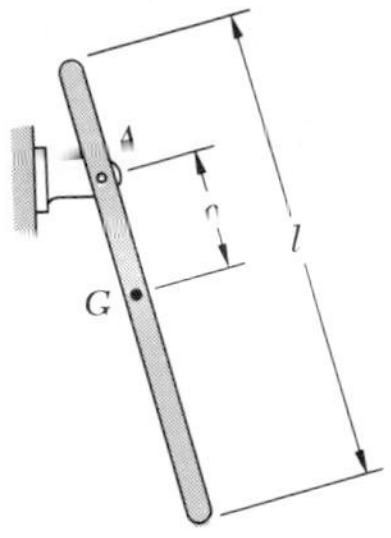

Fig. P19.37

19.38 For the rod of Prob. 19.37, determine (*a*) the distance c for which the frequency of oscillation is maximum, (*b*) the corresponding minimum period.

19.39 A thin hoop of radius r and mass m is suspended from a rough rod as shown. Determine the frequency of small oscillations of the hoop (*a*) in the plane of the hoop, (*b*) in a direction perpendicular to the plane of the hoop. Assume that μ is sufficiently large to prevent slipping at A.

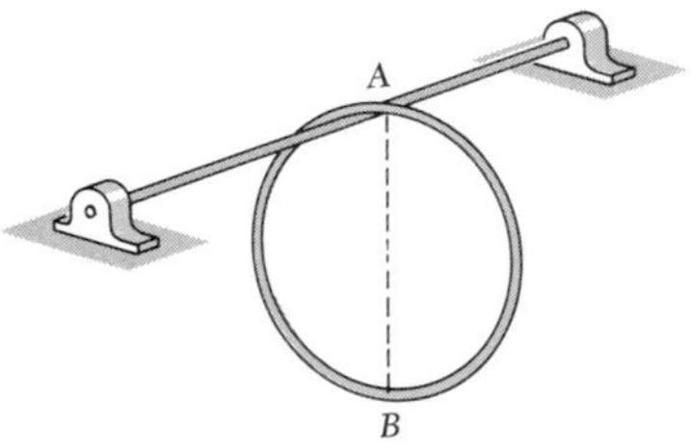

Fig. P19.39

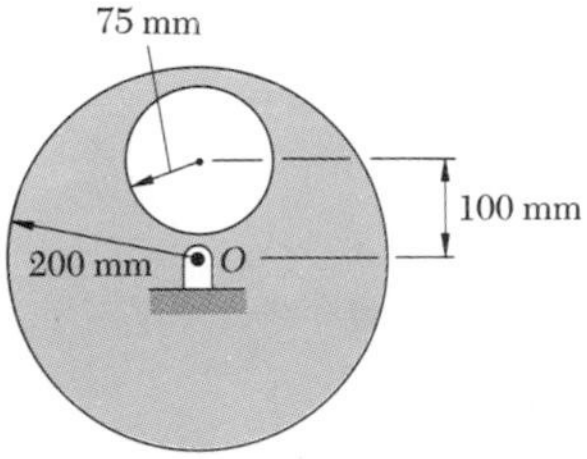

Fig. P19.40

19.40 A 75-mm-radius hole is cut in a 200-mm-radius uniform disk which is attached to a frictionless pin at its geometric center O. Determine (*a*) the period of small oscillations of the disk, (*b*) the length of a simple pendulum which has the same period.

19.41 A uniform rectangular plate is suspended from a pin located at the midpoint of one edge as shown. Considering the dimension b constant, determine (*a*) the ratio c/b for which the period of oscillation of the plate is minimum, (*b*) the ratio c/b for which the period of oscillation of the plate is the same as the period of a simple pendulum of length c.

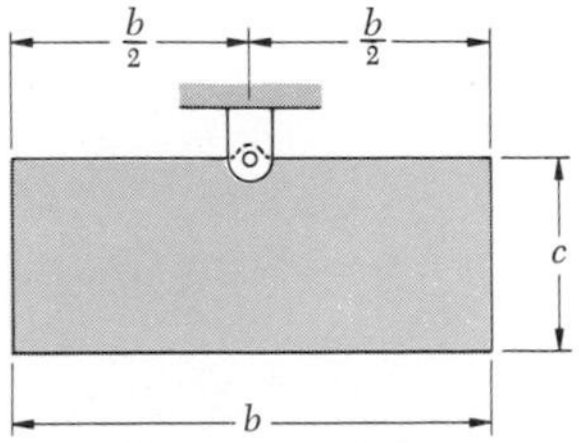

Fig. P19.41 and P19.42

19.42 A uniform rectangular plate is suspended from a pin located at the midpoint of one edge as shown. (*a*) Determine the period of small oscillations if $c = b$. (*b*) Considering the dimension b constant, determine a second value of c for which the period of oscillations is the same as that found in part *a*.

19.43 The period of small oscillations about A of a connecting rod is observed to be 1.12 s. Knowing that the distance r_a is 7.50 in., determine the centroidal radius of gyration of the connecting rod.

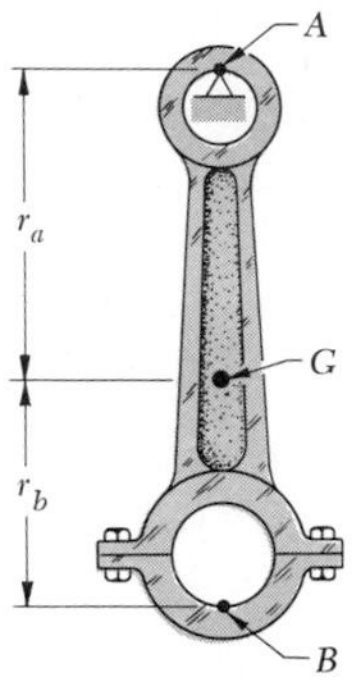

Fig. P19.43 and 19.44

19.44 A connecting rod is supported by a knife-edge at point A; the period of small oscillations is observed to be 0.945 s. The rod is then inverted and supported by a knife-edge at point B and the period of small oscillations is observed to be 0.850 s. Knowing that $r_a + r_b = 11.50$ in., determine (*a*) the location of the mass center G, (*b*) the centroidal radius of gyration $\bar{k}$.

19.45 A slender rod of length l is suspended from two vertical wires of length h, each located a distance $\frac{1}{2}b$ from the mass center G. Determine the period of oscillation when (*a*) the rod is rotated through a small angle about a vertical axis passing through G and released, (*b*) the rod is given a small horizontal translation along AB and released.

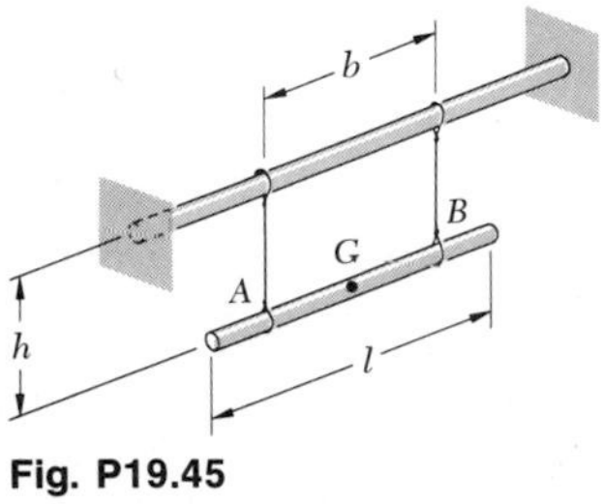

Fig. P19.45

19.46 A uniform disk of weight 5 lb is suspended from a steel wire which is known to have a torsional spring constant $K = 0.30$ lb·in./rad. If the disk is rotated through 360° about the vertical and then released, determine (*a*) the period of oscillation, (*b*) the maximum velocity of a point on the rim of the disk.

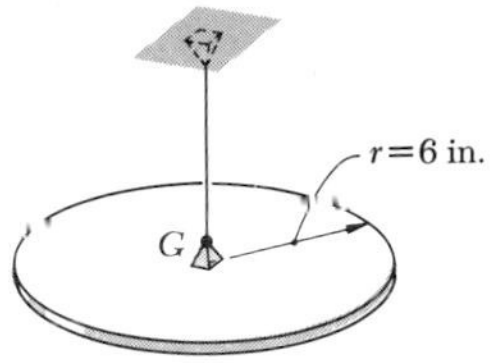

Fig. P19.46

19.47 A uniform disk of radius 200 mm and of mass 8 kg is attached to a vertical shaft which is rigidly held at *B*. It is known that the disk rotates through 3° when a 4 N·m static couple is applied to the disk. If the disk is rotated through 6° and then released, determine (*a*) the period of the resulting vibration, (*b*) the maximum velocity of a point on the rim of the disk.

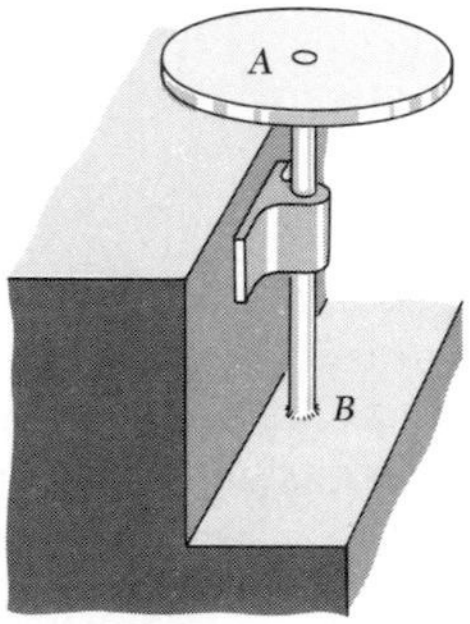

Fig. P19.47

19.48 A steel casting is rigidly bolted to the disk of Prob. 19.47. Knowing that the period of torsional vibration of the disk and casting is 0.650 s, determine the moment of inertia of the casting with respect to the shaft *AB*.

19.49 A torsion pendulum may be used to determine experimentally the moment of inertia of a given object. The horizontal platform *P* is held by several rigid bars which are connected to a vertical wire. The period of oscillation of the platform is found equal to τ_0 when the platform is empty and to τ_A when an object of known moment of inertia I_A is placed on the platform so that its mass center is directly above the center of the plate. (*a*) Show that the moment of inertia I_0 of the platform and its supports may be expressed as $I_0 = I_A\tau_0^2/(\tau_A^2 - \tau_0^2)$. (*b*) If a period of oscillation τ_B is observed when an object *B* of unknown moment of inertia I_B is placed on the platform, show that $I_B = I_A(\tau_B^2 - \tau_0^2)/(\tau_A^2 - \tau_0^2)$.

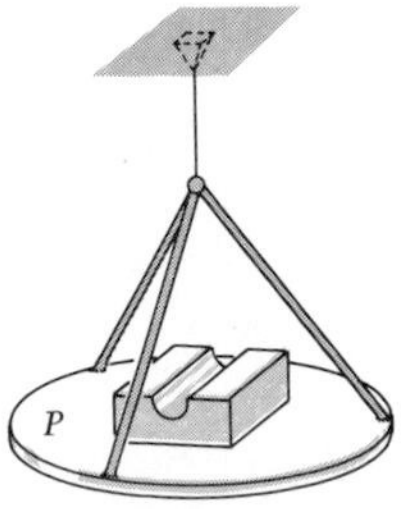

Fig P19.49

19.6. Application of the Principle of Conservation of Energy. We saw in Sec. 19.2 that, when a particle of mass m is in simple harmonic motion, the resultant $\mathbf{F}$ of the forces exerted on the particle has a magnitude proportional to the displacement x measured from the position of equilibrium O and is directed toward O; we write $F = -kx$. Referring to Sec. 13.6, we note that $\mathbf{F}$ is a *conservative force* and that the corresponding potential energy is $V = \frac{1}{2}kx^2$, where V is assumed equal to zero in the equilibrium position $x = 0$. Since the velocity of the particle is equal to $\dot{x}$, its kinetic energy is $T = \frac{1}{2}m\dot{x}^2$ and we may express that the total energy of the particle is conserved by writing

$$T + V = \text{constant} \qquad \tfrac{1}{2}m\dot{x}^2 + \tfrac{1}{2}kx^2 = \text{constant}$$

Setting $p^2 = k/m$ as in (19.4), where p is the circular frequency of the vibration, we have

$$\dot{x}^2 + p^2x^2 = \text{constant} \tag{19.25}$$

Equation (19.25) is characteristic of simple harmonic motion; it may be obtained directly from (19.5) by multiplying both terms by $2\dot{x}$ and integrating.

The principle of conservation of energy provides a convenient way for determining the period of vibration of a rigid body or of a system of rigid bodies possessing a single degree of freedom, once it has been established that the motion of the system is a simple harmonic motion, or that it may be approximated by a simple harmonic motion. Choosing an appropriate variable, such as a distance x or an angle θ, we consider two particular positions of the system:

1. *The displacement of the system is maximum;* we have $T_1 = 0$, and V_1 may be expressed in terms of the amplitude x_m or θ_m (choosing $V = 0$ in the equilibrium position).
2. *The system passes through its equilibrium position;* we have $V_2 = 0$, and T_2 may be expressed in terms of the maximum velocity $\dot{x}_m$ or $\dot{\theta}_m$.

We then express that the total energy of the system is conserved and write $T_1 + V_1 = T_2 + V_2$. Recalling from (19.15) that for simple harmonic motion the maximum velocity is equal to the product of the amplitude and of the circular frequency p, we find that the equation obtained may be solved for p.

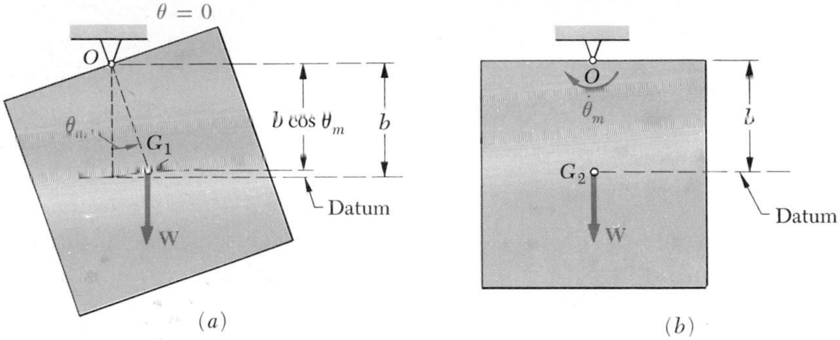

Fig. 19.6

As an example, we shall consider again the square plate of Sec. 19.5. In the position of maximum displacement (Fig. 19.6*a*), we have

$$T_1 = 0 \qquad V_1 = W(b - b\cos\theta_m) = Wb(1 - \cos\theta_m)$$

or, since $1 - \cos\theta_m = 2\sin^2(\theta_m/2) \approx 2(\theta_m/2)^2 = \theta_m^2/2$ for oscillations of small amplitude,

$$T_1 = 0 \qquad V_1 = \tfrac{1}{2}Wb\theta_m^2 \tag{19.26}$$

As the plate passes through its position of equilibrium (Fig. 19.6*b*), its velocity is maximum and we have

$$T_2 = \tfrac{1}{2}m\bar{v}_m^2 + \tfrac{1}{2}\bar{I}\omega_m^2 = \tfrac{1}{2}mb^2\dot{\theta}_m^2 + \tfrac{1}{2}\bar{I}\dot{\theta}_m^2 \qquad V_2 = 0$$

or, recalling from Sec. 19.5 that $\bar{I} = \frac{2}{3}mb^2$,

$$T_2 = \tfrac{1}{2}(\tfrac{5}{3}mb^2)\dot{\theta}_m^2 \qquad V_2 = 0 \tag{19.27}$$

Substituting from (19.26) and (19.27) into $T_1 + V_1 = T_2 + V_2$, and noting that the maximum velocity $\dot{\theta}_m$ is equal to the product $\theta_m p$, we write

$$\tfrac{1}{2}Wb\theta_m^2 = \tfrac{1}{2}(\tfrac{5}{3}mb^2)\theta_m^2 p^2 \tag{19.28}$$

which yields $p^2 = 3g/5b$ and

$$\tau = \frac{2\pi}{p} = 2\pi\sqrt{\frac{5b}{3g}} \tag{19.29}$$

as previously obtained.

SAMPLE PROBLEM 19.4

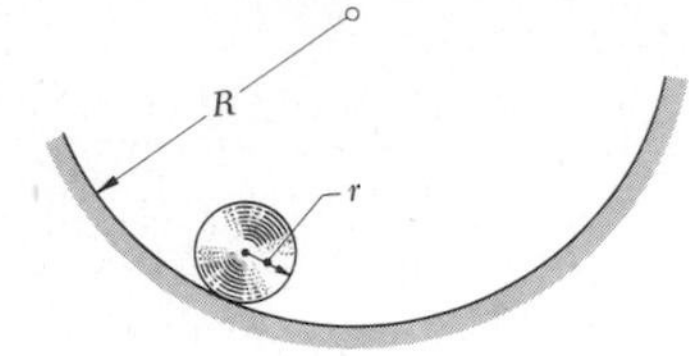

Determine the period of small oscillations of a cylinder of radius r which rolls without slipping inside a curved surface of radius R.

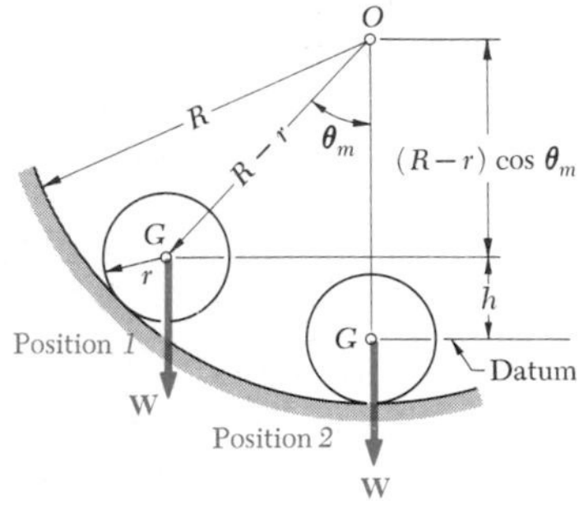

Solution. We denote by θ the angle which line OG forms with the vertical. Since the cylinder rolls without slipping, we may apply the principle of conservation of energy between position *1*, where $\theta = \theta_m$, and position 2, where $\theta = 0$.

Position *1*. ***Kinetic Energy.*** Since the velocity of the cylinder is zero, we have $T_1 = 0$.

Potential Energy. Choosing a datum as shown and denoting by W the weight of the cylinder, we have

$$V_1 = Wh = W(R - r)(1 - \cos\theta)$$

Noting that for small oscillations $(1 - \cos\theta) = 2\sin^2(\theta/2) \approx \theta^2/2$, we have

$$V_1 = W(R - r)\frac{\theta_m^2}{2}$$

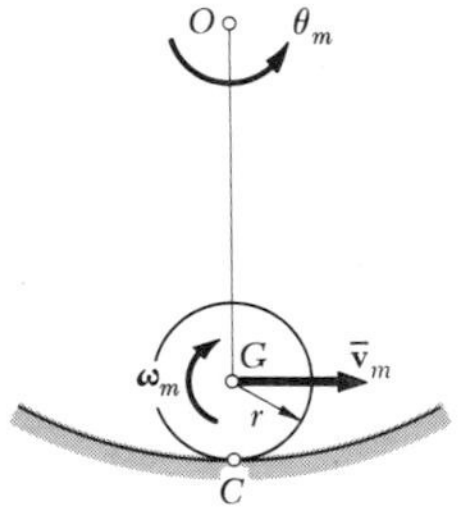

Position 2. Denoting by $\dot{\theta}_m$ the angular velocity of line OG as the cylinder passes through position 2, and observing that point C is the instantaneous center of rotation of the cylinder, we write

$$\bar{v}_m = (R - r)\dot{\theta}_m \qquad \omega_m = \frac{\bar{v}_m}{r} = \frac{R - r}{r}\dot{\theta}_m$$

Kinetic Energy:

$$\begin{aligned} T_2 &= \tfrac{1}{2}m\bar{v}_m^2 + \tfrac{1}{2}\bar{I}\omega_m^2 \\ &= \tfrac{1}{2}m(R - r)^2\dot{\theta}_m^2 + \tfrac{1}{2}(\tfrac{1}{2}mr^2)\left(\frac{R - r}{r}\right)^2\dot{\theta}_m^2 \\ &= \tfrac{3}{4}m(R - r)^2\dot{\theta}_m^2 \end{aligned}$$

Potential Energy: $V_2 = 0$

Conservation of Energy

$$T_1 + V_1 = T_2 + V_2$$

$$0 + W(R - r)\frac{\theta_m^2}{2} = \tfrac{3}{4}m(R - r)^2\dot{\theta}_m^2 + 0$$

Since $\dot{\theta}_m = p\theta_m$ and $W = mg$, we write

$$mg(R - r)\frac{\theta_m^2}{2} = \tfrac{3}{4}m(R - r)^2(p\theta_m)^2 \qquad p^2 = \frac{2}{3}\frac{g}{R - r}$$

$$\tau = \frac{2\pi}{p} \qquad \tau = 2\pi\sqrt{\frac{3}{2}\frac{R - r}{g}} \quad ◀$$

PROBLEMS

19.50 Using the method of Sec. 19.6, determine the period of a simple pendulum of length l.

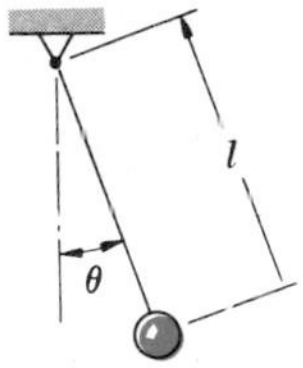

Fig. P19.50

19.51 The springs of an automobile are observed to expand 8 in. to an undeformed position as the body is lifted by several jacks. Assuming that each spring carries an equal portion of the weight of the automobile, determine the frequency of the free vertical vibrations of the body.

19.52 Using the method of Sec. 19.6, solve Prob. 19.6.

19.53 Using the method of Sec. 19.6, solve Prob. 19.9.

19.54 Using the method of Sec. 19.6, solve Prob. 19.10.

19.55 Neglecting fluid friction, determine the frequency of oscillation of the liquid in the U-tube manometer shown. Show that this frequency is independent of the density of the liquid and of the amplitude of the oscillation.

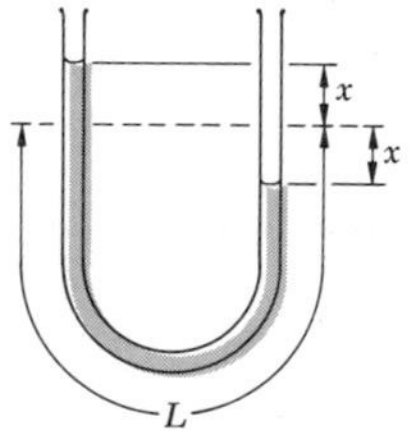

Fig. P19.55

19.56 Two collars, each of weight W, are attached as shown to a hoop of radius r and of negligible weight. (*a*) Show that for any value of β the period is $\tau = 2\pi\sqrt{2r/g}$. (*b*) Show that the same result is obtained if the weight of the hoop is not neglected.

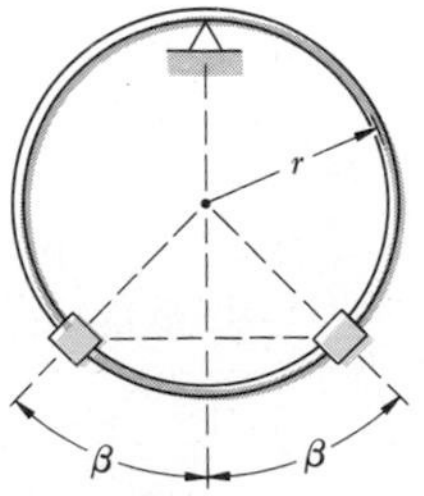

Fig. P19.56

19.57 A thin homogeneous wire is bent into the shape of a square of side l and suspended as shown. Determine the period of oscillation when the wire figure is given a small displacement to the right and released.

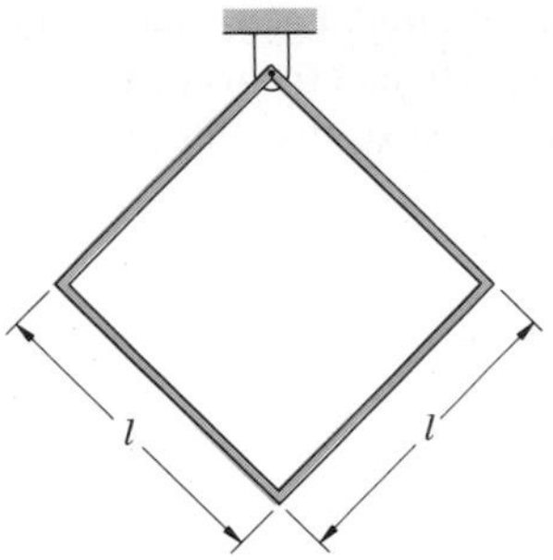

Fig. P19.57

19.58 Solve Prob. 19.57, assuming that the wire square is suspended from a pin located at the midpoint of one side.

19.59 Using the method of Sec. 19.6, solve Prob. 19.45.

19.60 Using the method of Sec. 19.6, solve Prob. 19.40.

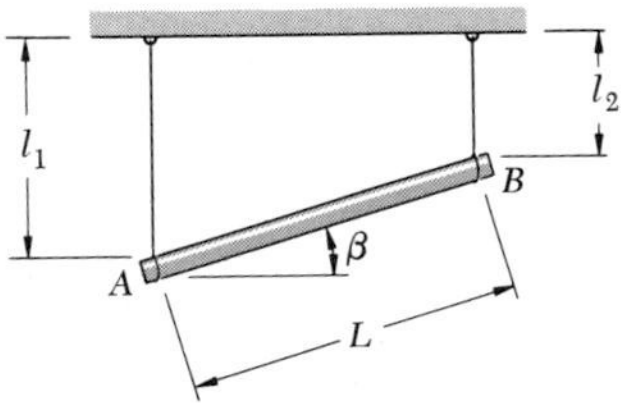

Fig. P19.61

19.61 A section of uniform pipe is suspended from two vertical cables attached at A and B. Determine the period of oscillation in terms of l_1 and l_2 when point B is given a small horizontal displacement to the right and released.

19.62 The motion of the uniform rod AB is guided by the cord BC and by the small roller at A. Determine the frequency of oscillation when the end B of the rod is given a small horizontal displacement and released.

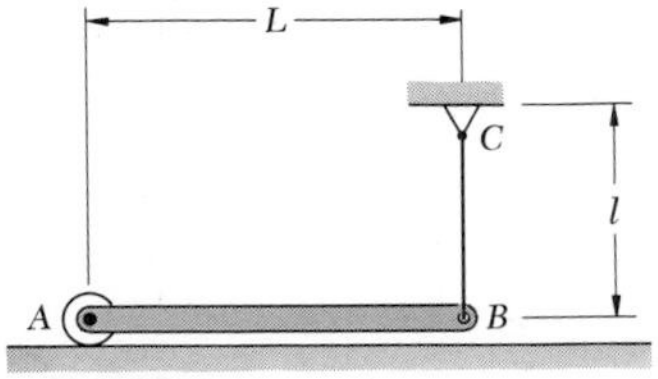

Fig. P19.62

19.63 The 20-lb rod AB is attached to 8-lb disks as shown. Knowing that the disks roll without sliding, determine the frequency of small oscillations of the system.

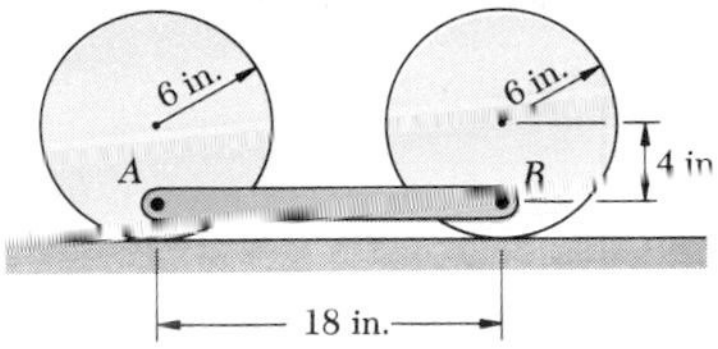

Fig. P19.63

19.64 Blade AB of the experimental wind-turbine generator shown is to be temporarily removed. Motion of the turbine generator about the y axis is prevented, but the remaining three blades may oscillate as a unit about the x axis. Assuming that each blade is equivalent to a 40-ft slender rod, determine the period of small oscillations of the blades.

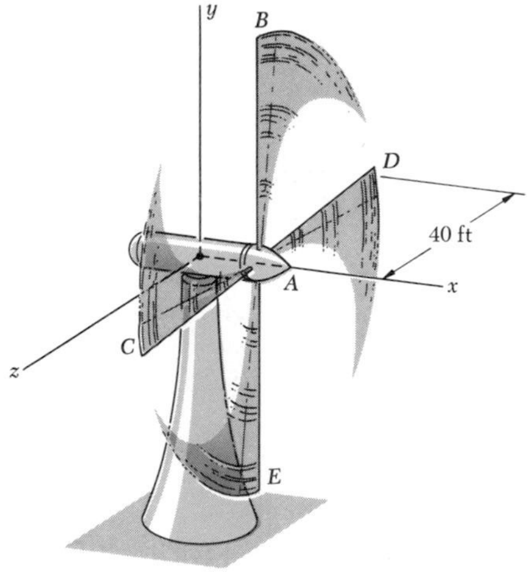

Fig. P19.64

19.65 The 8-kg rod AB is bolted to the 12-kg disk. Knowing that the disk rolls without sliding, determine the period of small oscillations of the system.

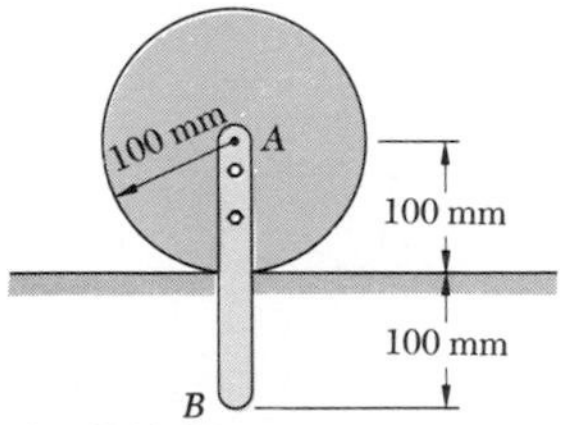

Fig. P19.65

19.66 Using the method of Sec. 19.6, solve Prob. 19.32.

19.67 Using the method of Sec. 19.6, solve Prob. 19.31.

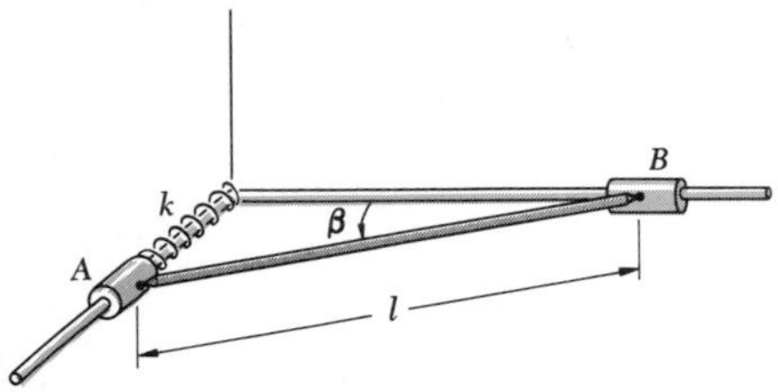

Fig. P19.68 and P19.70

19.68 The slender rod AB of mass m is attached to two collars of negligible mass. Knowing that the system lies in a horizontal plane and is in equilibrium in the position shown, determine the period of vibration if the collar A is given a small displacement and released.

19.69 Solve Prob. 19.68, assuming that rod AB is of mass m and that each collar is of mass m_c.

19.70 A slender rod AB of negligible mass connects two collars, each of mass m_c. Knowing that the system lies in a horizontal plane and is in equilibrium in the position shown, determine the period of vibration if the collar at A is given a small displacement and released.

19.71 Two uniform rods AB and CD, each of length l and weight W, are attached to two gears as shown. Neglecting the mass of the gears, determine for each of the positions shown the period of small oscillations of the system.

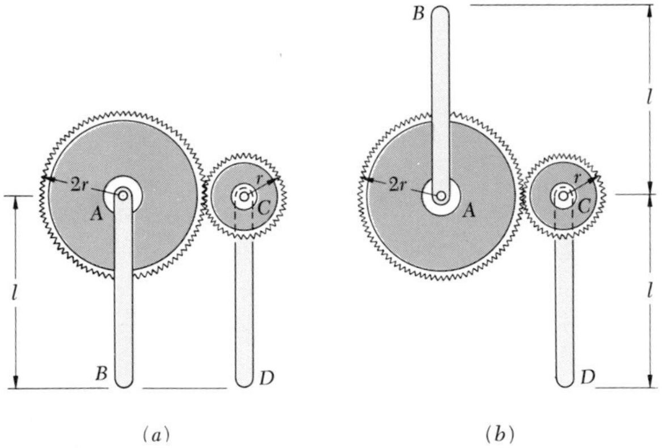

Fig. P19.71

19.72 Solve Prob. 19.71, assuming that $l = 18$ in., $W = 4$ lb, and $r = 3$ in.

19.73 A thin circular plate of radius r is suspended from three vertical wires of length h equally spaced around the perimeter of the plate. Determine the period of oscillation when (*a*) the plate is rotated through a small angle about a vertical axis passing through its mass center and released, (*b*) the plate is given a small horizontal translation and released.

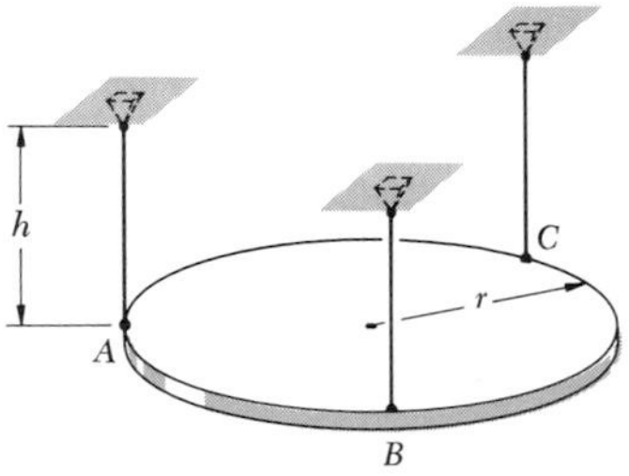

Fig. P19.73

19.74 Solve Prob. 19.73, assuming that $r = 500$ mm and $h = 300$ mm.

*__19.75__ As a submerged body moves through a fluid, the particles of the fluid flow around the body and thus acquire kinetic energy. In the case of a sphere moving in an ideal fluid, the total kinetic energy acquired by the fluid is $\frac{1}{4}\rho V v^2$, where ρ is the mass density of the fluid, V the volume of the sphere, and v the velocity of the sphere. Consider a 1-lb hollow spherical shell of radius 3 in. which is held submerged in a tank of water by a spring of constant 3 lb/in. (*a*) Neglecting fluid friction, determine the period of vibration of the shell when it is displaced vertically and then released. (*b*) Solve part *a*, assuming that the tank is accelerated upward at the constant rate of 10 ft/s^2.

Fig. P19.75

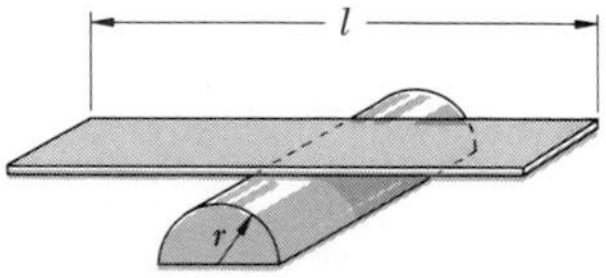

Fig. P19.76

*__19.76__ A thin plate of length l rests on a half cylinder of radius r. Derive an expression for the period of small oscillations of the plate.

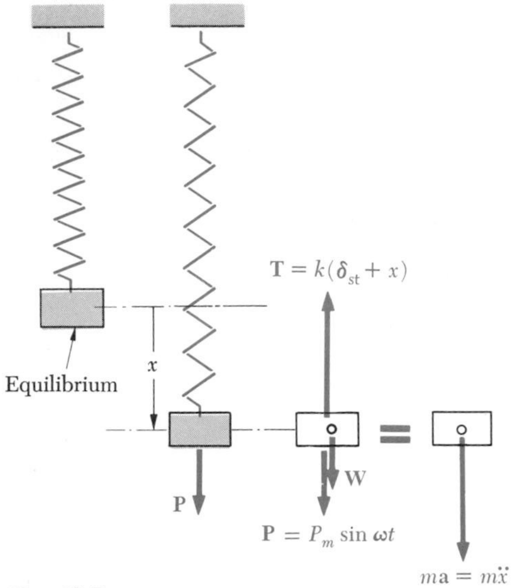

Fig. 19.7

19.7. Forced Vibrations. The most important vibrations from the point of view of engineering applications are the *forced vibrations* of a system. These vibrations occur when a system is subjected to a periodic force or when it is elastically connected to a support which has an alternating motion.

Consider first the case of a body of mass m suspended from a spring and subjected to a periodic force $\mathbf{P}$ of magnitude $P = P_m \sin \omega t$ (Fig. 19.7). This force may be an actual external force applied to the body, or it may be a centrifugal force produced by the rotation of some unbalanced part of the body. (See Sample Prob. 19.5.) Denoting by x the displacement of the body measured from its equilibrium position, we write the equation of motion,

$$+\downarrow \Sigma F = ma: \qquad P_m \sin \omega t + W - k(\delta_{st} + x) = m\ddot{x}$$

Recalling that $W = k\delta_{st}$, we have

$$m\ddot{x} + kx = P_m \sin \omega t \tag{19.30}$$

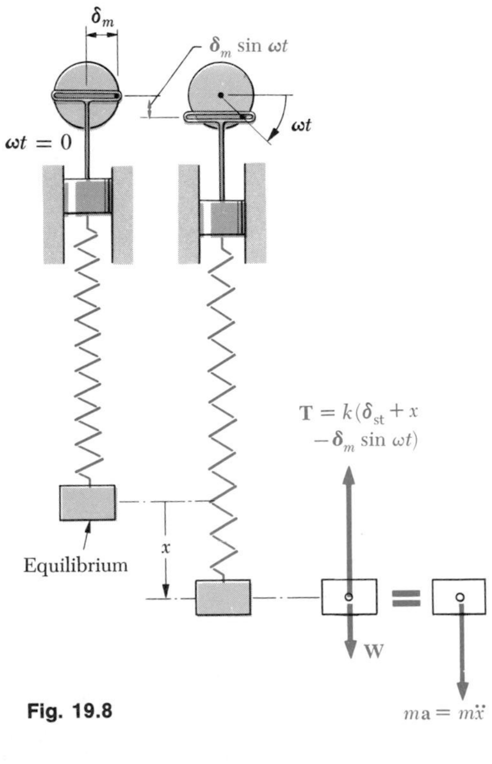

Fig. 19.8

Next we consider the case of a body of mass m suspended from a spring attached to a moving support whose displacement δ is equal to $\delta_m \sin \omega t$ (Fig. 19.8). Measuring the displacement x of the body from the position of static equilibrium corresponding to $\omega t = 0$, we find that the total elongation of the spring at time t is $\delta_{st} + x - \delta_m \sin \omega t$. The equation of motion is thus

$$+\downarrow \Sigma F = ma: \qquad W - k(\delta_{st} + x - \delta_m \sin \omega t) = m\ddot{x}$$

Recalling that $W = k\delta_{st}$, we have

$$m\ddot{x} + kx = k\delta_m \sin \omega t \tag{19.31}$$

We note that Eqs. (19.30) and (19.31) are of the same form and that a solution of the first equation will satisfy the second if we set $P_m = k\delta_m$.

A differential equation like (19.30) or (19.31), possessing a right-hand member different from zero, is said to be *nonhomogeneous*. Its general solution is obtained by adding a particular solution of the given equation to the general solution of the corresponding *homogeneous* equation (with right-hand member equal to zero). A *particular solution* of (19.30) or (19.31) may be obtained by trying a solution of the form

$$x_{part} = x_m \sin \omega t \tag{19.32}$$

Substituting x_{part} for x into (19.30), we find

$$-m\omega^2 x_m \sin \omega t + kx_m \sin \omega t = P_m \sin \omega t$$

which may be solved for the amplitude,

$$x_m = \frac{P_m}{k - m\omega^2}$$

Recalling from (19.4) that $k/m = p^2$, where p is the circular frequency of the free vibration of the body, we write

$$x_m = \frac{P_m/k}{1 - (\omega/p)^2} \qquad (19.33)$$

Substituting from (19.32) into (19.31), we obtain in a similar way

$$x_m = \frac{\delta_m}{1 - (\omega/p)^2} \qquad (19.33')$$

The homogeneous equation corresponding to (19.30) or (19.31) is Eq. (19.3), defining the free vibration of the body. Its general solution, called the *complementary function*, was found in Sec. 19.2,

$$x_{comp} = A \sin pt + B \cos pt \qquad (19.34)$$

Adding the particular solution (19.32) and the complementary function (19.34), we obtain the *general solution* of Eqs. (19.30) and (19.31),

$$x = A \sin pt + B \cos pt + x_m \sin \omega t \qquad (19.35)$$

We note that the vibration obtained consists of two superposed vibrations. The first two terms in (19.35) represent a free vibration of the system. The frequency of this vibration, called the *natural frequency* of the system, depends only upon the constant k of the spring and the mass m of the body, and the constants A and B may be determined from the initial conditions. This free vibration is also called a *transient* vibration since, in actual practice, it will soon be damped out by friction forces (Sec. 19.9).

The last term in (19.35) represents the *steady-state* vibration produced and maintained by the impressed force or impressed support movement. Its frequency is the *forced frequency* imposed by this force or movement, and its amplitude x_m, defined by (19.33) or (19.33′), depends upon the *frequency ratio* ω/p. The ratio of the amplitude x_m of the steady-state vibration to the static deflection P_m/k caused by a force P_m, or to the amplitude δ_m of the support movement, is called the *magnification factor*. From (19.33) and (19.33′), we obtain

$$\text{Magnification factor} = \frac{x_m}{P_m/k} = \frac{x_m}{\delta_m} = \frac{1}{1-(\omega/p)^2} \tag{19.36}$$

The magnification factor has been plotted in Fig. 19.9 against the frequency ratio ω/p. We note that, when $\omega = p$, the amplitude of the forced vibration becomes infinite. The impressed force or impressed support movement is said to be in *resonance* with the given system. Actually, the amplitude of the vibration remains finite because of damping forces (Sec. 19.9); nevertheless, such a situation should be avoided, and the forced frequency should not be chosen too close to the natural frequency of the system. We also note that for $\omega < p$ the coefficient of $\sin \omega t$ in (19.35) is positive, while for $\omega > p$ this coefficient is negative. In the first case the forced vibration is *in phase* with the impressed force or impressed support movement, while in the second case it is 180° *out of phase*.

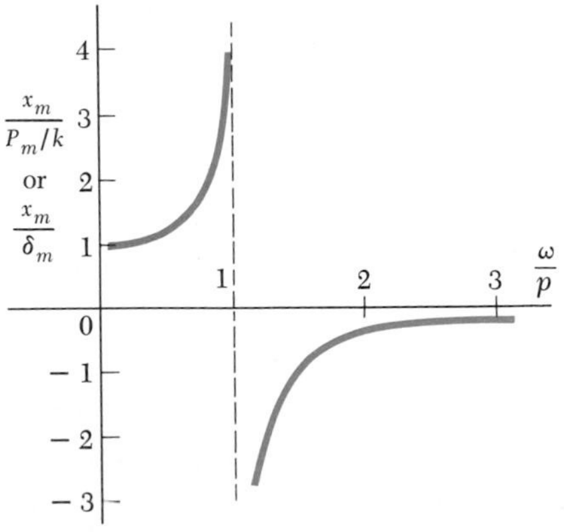

Fig. 19.9

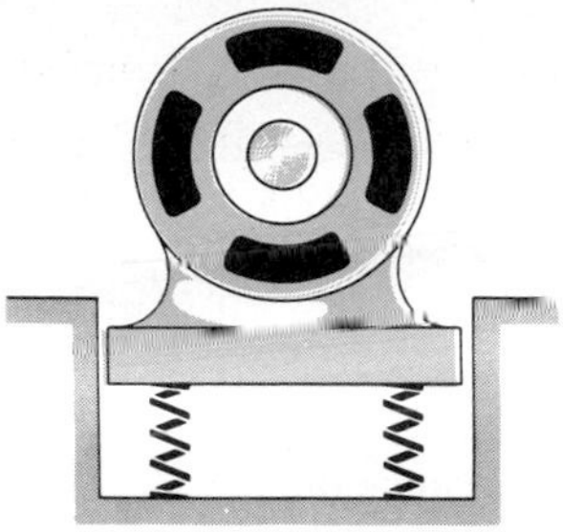

SAMPLE PROBLEM 19.5

A motor weighing 350 lb is supported by four springs, each having a constant of 750 lb/in. The unbalance of the rotor is equivalent to a weight of 1 oz located 6 in. from the axis of rotation. Knowing that the motor is constrained to move vertically, determine (*a*) the speed in rpm at which resonance will occur, (*b*) the amplitude of the vibration of the motor at a speed of 1200 rpm.

***a.* Resonance Speed.** The resonance speed is equal to the circular frequency (in rpm) of the free vibration of the motor. The mass of the motor and the equivalent constant of the supporting springs are

$$m = \frac{350 \text{ lb}}{32.2 \text{ ft/s}^2} = 10.87 \text{ lb} \cdot \text{s}^2/\text{ft}$$

$$k = 4(750 \text{ lb/in.}) = 3000 \text{ lb/in.} = 36{,}000 \text{ lb/ft}$$

$$p = \sqrt{\frac{k}{m}} = \sqrt{\frac{36{,}000}{10.87}} = 57.5 \text{ rad/s} = 549 \text{ rpm}$$

Resonance speed = 549 rpm ◄

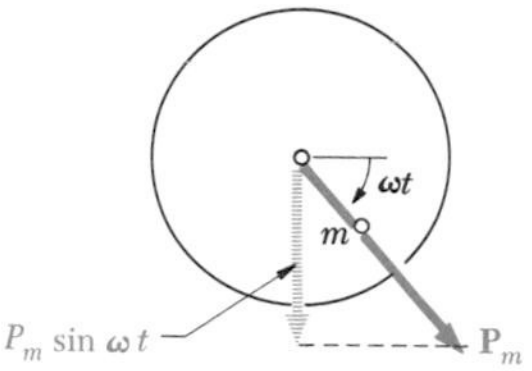

***b.* Amplitude of Vibration at 1200 rpm.** The angular velocity of the motor and the mass of the equivalent 1-oz weight are

$$\omega = 1200 \text{ rpm} = 125.7 \text{ rad/s}$$

$$m = (1 \text{ oz})\frac{1 \text{ lb}}{16 \text{ oz}}\frac{1}{32.2 \text{ ft/s}^2} = 0.00194 \text{ lb} \cdot \text{s}^2/\text{ft}$$

The magnitude of the centrifugal force due to the unbalance of the rotor is

$$P_m = ma_n = mr\omega^2 = (0.00194 \text{ lb} \cdot \text{s}^2/\text{ft})(\tfrac{6}{12} \text{ ft})(125.7 \text{ rad/s})^2 = 15.3 \text{ lb}$$

The static deflection that would be caused by a constant load P_m is

$$\frac{P_m}{k} = \frac{15.3 \text{ lb}}{3000 \text{ lb/in.}} = 0.00510 \text{ in.}$$

Substituting the value of P_m/k together with the known values of ω and p into Eq. (19.33), we obtain

$$x_m = \frac{P_m/k}{1 - (\omega/p)^2} = \frac{0.00510 \text{ in.}}{1 - (125.7/57.5)^2}$$

$x_m = 0.00135$ in. ◄

Note. Since $\omega > p$, the vibration is 180° out of phase with the centrifugal force due to the unbalance of the rotor. For example, when the unbalanced mass is directly below the axis of rotation, the position of the motor is $x_m = 0.00135$ in. above the position of equilibrium.

PROBLEMS

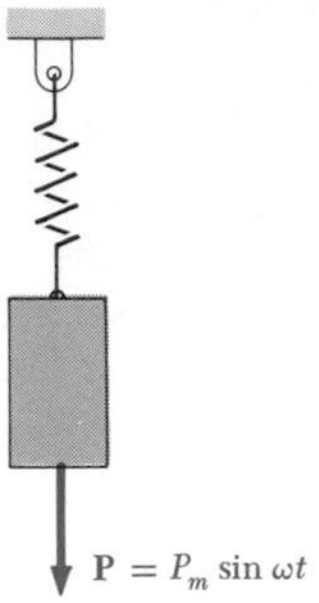

Fig. P19.77

19.77 A block of mass m is suspended from a spring of constant k and is acted upon by a vertical periodic force of magnitude $P = P_m \sin \omega t$. Determine the range of values of ω for which the amplitude of the vibration exceeds twice the static deflection caused by a constant force of magnitude P_m.

19.78 In Prob. 19.77, determine the range of values of ω for which the amplitude of the vibration is less than the static deflection caused by a constant force of magnitude P_m.

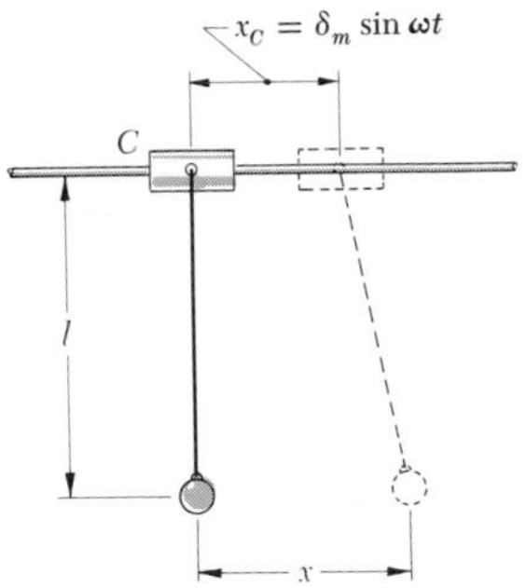

Fig. P19.79

19.79 A simple pendulum of length l is suspended from a collar C which is forced to move horizontally according to the relation $x_C = \delta_m \sin \omega t$. Determine the range of values of ω for which the amplitude of the motion of the bob exceeds $2\delta_m$. (Assume δ_m is small compared to the length l of the pendulum.)

19.80 In Prob. 19.79, determine the range of values of ω for which the amplitude of the motion of the bob is less than δ_m.

19.81 A 500-lb motor is supported by a light horizontal beam. The unbalance of the rotor is equivalent to a weight of 1 oz located 10 in. from the axis of rotation. Knowing that the static deflection of the beam due to the weight of the motor is 0.220 in., determine (*a*) the speed (in rpm) at which resonance will occur, (*b*) the amplitude of the steady-state vibration of the motor at a speed of 800 rpm.

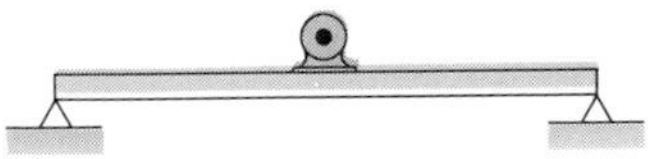

Fig. P19.81

19.82 Solve Prob. 19.81, assuming that the 500-lb motor is supported by a nest of springs having a total constant of 400 lb/in.

19.83 A motor of mass 45 kg is supported by four springs, each of constant 100 kN/m. The motor is constrained to move vertically, and the amplitude of its movement is observed to be 0.5 mm at a speed of 1200 rpm. Knowing that the mass of the rotor is 14 kg, determine the distance between the mass center of the rotor and the axis of the shaft.

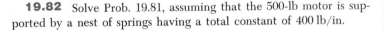

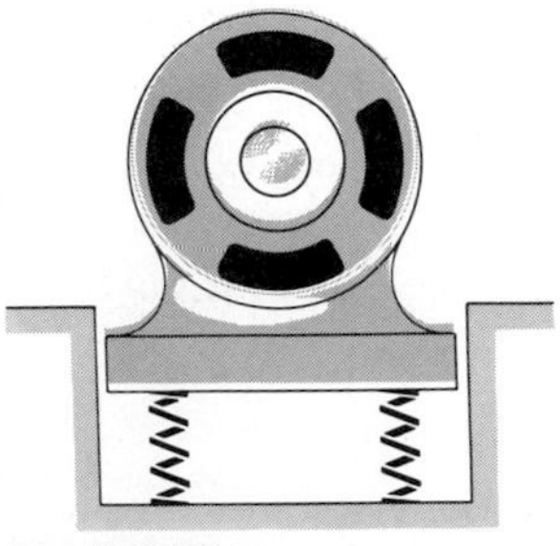

Fig. P19.83

19.84 In Prob. 19.83, determine the amplitude of the vertical movement of the motor at a speed of (*a*) 200 rpm, (*b*) 1600 rpm, (*c*) 900 rpm.

19.85 Rod AB is rigidly attached to the frame of a motor running at a constant speed. When a collar of mass m is placed on the spring, it is observed to vibrate with an amplitude of 0.5 in. When two collars, each of mass m, are placed on the spring, the amplitude is observed to be 0.6 in. What amplitude of vibration should be expected when three collars, each of mass m, are placed on the spring? (Obtain two answers.)

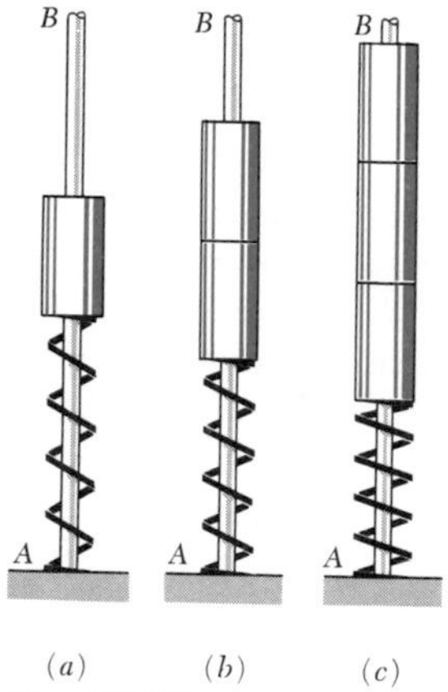

Fig. P19.85

19.86 Solve Prob. 19.85, assuming that the speed of the motor is changed and that one collar has an amplitude of 0.60 in. and two collars have an amplitude of 0.20 in.

19.87 A disk of mass m is attached to the midpoint of a vertical shaft which revolves at an angular velocity ω. Denoting by k the spring constant of the system for a horizontal movement of the disk and by e the eccentricity of the disk with respect to the shaft, show that the deflection of the center of the shaft may be written in the form

$$r = \frac{e(\omega/p)^2}{1 - (\omega/p)^2}$$

where $p = \sqrt{k/m}$.

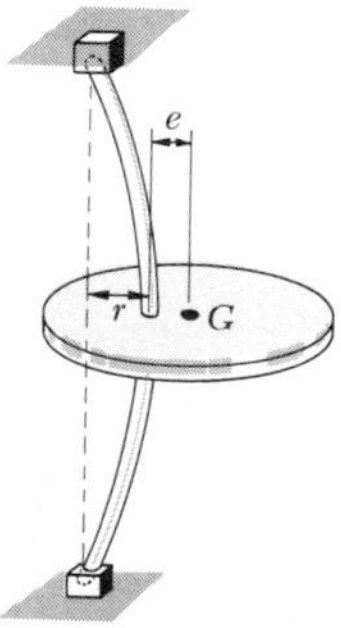

Fig. P19.87 and P19.88

19.88 A disk of mass 30 kg is attached to the midpoint of a shaft. Knowing that a static force of 200 N will deflect the shaft 0.6 mm, determine the speed of the shaft in rpm at which resonance will occur.

19.89 Knowing that the disk of Prob. 19.88 is attached to the shaft with an eccentricity $e = 0.2$ mm, determine the deflection r of the shaft at a speed of 900 rpm.

19.90 A variable-speed motor is rigidly attached to the beam BC. When the speed of the motor is less than 1000 rpm or more than 2000 rpm, a small object placed at A is observed to remain in contact with the beam. For speeds between 1000 and 2000 rpm the object is observed to "dance" and actually to lose contact with the beam. Determine the speed at which resonance will occur.

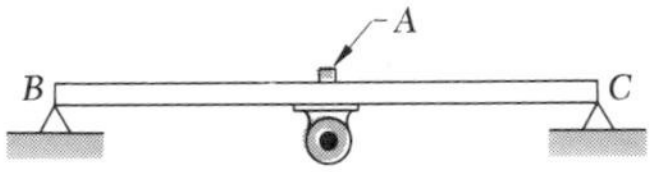

Fig. P19.90

19.91 As the speed of a spring-supported motor is slowly increased from 150 to 200 rpm, the amplitude of the vibration due to the unbalance of the rotor is observed to decrease continuously from 0.150 to 0.080 in. Determine the speed at which resonance will occur.

19.92 In Prob. 19.91, determine the speed for which the amplitude of the vibration is 0.200 in.

19.93 The amplitude of the motion of the pendulum bob shown is observed to be 3 in. when the amplitude of the motion of collar C is $\frac{3}{4}$ in. Knowing that the length of the pendulum is $l = 36$ in., determine the two possible values of the frequency of the horizontal movement of the collar C.

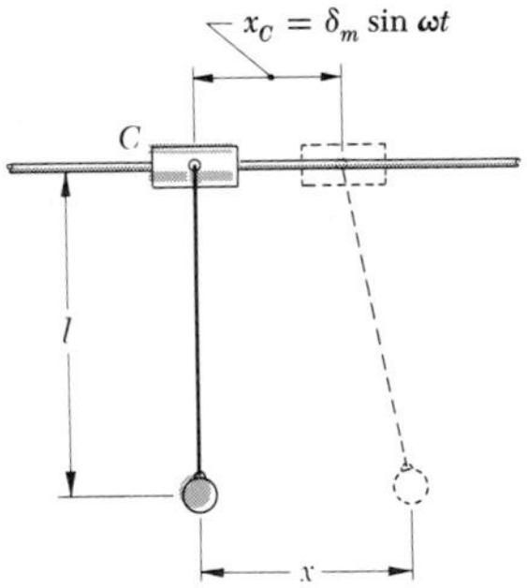

Fig. P19.93

19.94 A certain vibrometer used to measure vibration amplitudes consists essentially of a box containing a slender rod to which a mass m is attached; the natural frequency of the mass-rod system is known to be 5 Hz. When the box is rigidly attached to the casing of a motor rotating at 600 rpm, the mass is observed to vibrate with an amplitude of 1.6 mm relative to the box. Determine the amplitude of the vertical motion of the motor.

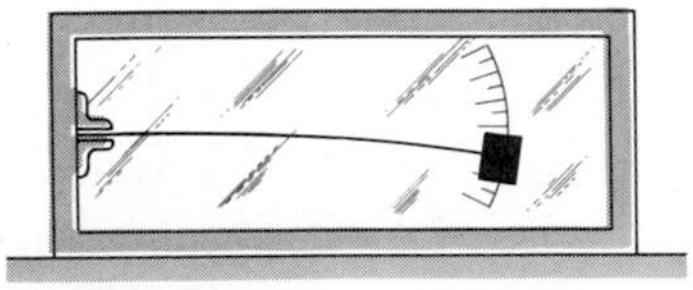

Fig. P19.94

19.95 A small trailer of mass 200 kg with its load is supported by two springs, each of constant 20 kN/m. The trailer is pulled over a road, the surface of which may be approximated by a sine curve of amplitude 30 mm and of period 5 m (i.e., the distance between two successive crests is 5 m, and the vertical distance from a crest to a trough is 60 mm). Determine (*a*) the speed at which resonance will occur, (*b*) the amplitude of the vibration of the trailer at a speed of 60 km/h.

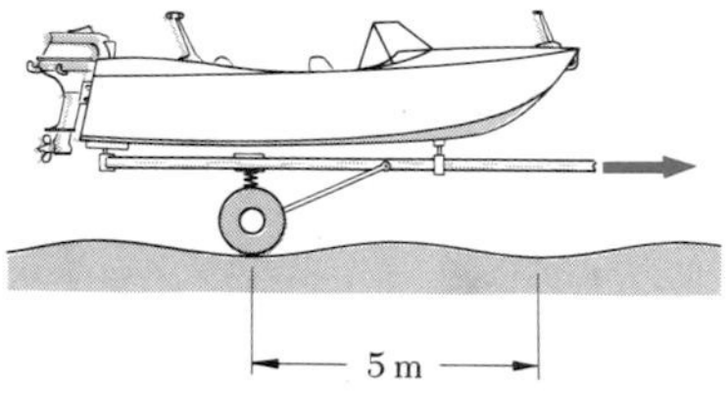

Fig. P19.95

19.96 Knowing that the amplitude of the vibration of the trailer of Prob. 19.95 is not to exceed 15 mm, determine the smallest speed at which the trailer can be pulled over the road.

DAMPED VIBRATIONS

***19.8. Damped Free Vibrations.** The vibrating systems considered in the first part of this chapter were assumed free of damping. Actually all vibrations are damped to some degree by friction forces. These forces may be caused by *dry friction*, or *Coulomb friction*, between rigid bodies, by *fluid friction* when a rigid body moves in a fluid, or by *internal friction* between the molecules of a seemingly elastic body.

A type of damping of special interest is the *viscous damping* caused by fluid friction at low and moderate speeds. Viscous damping is characterized by the fact that the friction force is *directly proportional to the speed* of the moving body. As an example, we shall consider again a body of mass m suspended from a spring of constant k, and we shall assume that the body is attached to the plunger of a dashpot (Fig. 19.10). The magnitude of the friction force exerted on the plunger by the surrounding fluid is equal to $c\dot{x}$, where the constant c, expressed in N · s/m or lb · s/ft and known as the *coefficient of viscous damping*, depends upon the physical properties of the fluid and the construction of the dashpot. The equation of motion is

$$+\downarrow \Sigma F = ma: \qquad W - k(\delta_{st} + x) - c\dot{x} = m\ddot{x}$$

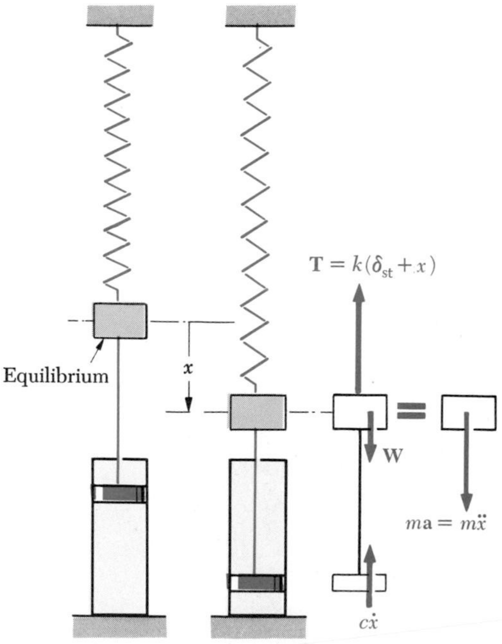

Fig. 19.10

Recalling that $W = k\delta_{st}$, we write

$$m\ddot{x} + c\dot{x} + kx = 0 \tag{19.37}$$

Substituting $x = e^{\lambda t}$ into (19.37) and dividing through by $e^{\lambda t}$, we write the *characteristic equation*

$$m\lambda^2 + c\lambda + k = 0 \tag{19.38}$$

and obtain the roots

$$\lambda = -\frac{c}{2m} \pm \sqrt{\left(\frac{c}{2m}\right)^2 - \frac{k}{m}} \tag{19.39}$$

Defining the *critical damping coefficient* c_c as the value of c which makes the radical in (19.39) equal to zero, we write

$$\left(\frac{c_c}{2m}\right)^2 - \frac{k}{m} = 0 \qquad c_c = 2m\sqrt{\frac{k}{m}} = 2mp \tag{19.40}$$

where p is the circular frequency of the system in the absence of damping. We may distinguish three different cases of damping, depending upon the value of the coefficient c.

1. *Heavy damping:* $c > c_c$. The roots λ_1 and λ_2 of the characteristic equation (19.38) are real and distinct, and the general solution of the differential equation (19.37) is

   $$x = Ae^{\lambda_1 t} + Be^{\lambda_2 t} \tag{19.41}$$

   This solution corresponds to a nonvibratory motion. Since λ_1 and λ_2 are both negative, x approaches zero as t increases indefinitely. However, the system actually regains its equilibrium position after a finite time.
2. *Critical damping:* $c = c_c$. The characteristic equation has a double root $\lambda = -c_c/2m = -p$, and the general solution of (19.37) is

   $$x = (A + Bt)e^{-pt} \tag{19.42}$$

   The motion obtained is again nonvibratory. Critically damped systems are of special interest in engineering applications since they regain their equilibrium position in the shortest possible time without oscillation.
3. *Light damping:* $c < c_c$. The roots of (19.38) are complex and conjugate, and the general solution of (19.37) is of the form

   $$x = e^{-(c/2m)t}(A \sin qt + B \cos qt) \tag{19.43}$$

where q is defined by the relation

$$q^2 = \frac{k}{m} - \left(\frac{c}{2m}\right)^2$$

Substituting $k/m = p^2$ and recalling (19.40), we write

$$q = p\sqrt{1 - \left(\frac{c}{c_c}\right)^2} \qquad (19.44)$$

where the constant c/c_c is known as the *damping factor*. A substitution similar to the one used in Sec. 19.2 enables us to write the general solution of (19.37) in the form

$$x = x_m e^{-(c/2m)t} \sin(qt + \phi) \qquad (19.45)$$

The motion defined by (19.45) is vibratory with diminishing amplitude (Fig. 19.11). Although this motion does not actually repeat itself, the time interval $\tau = 2\pi/q$, corresponding to two successive points where the curve (19.45) touches one of the limiting curves shown in Fig. 19.11, is commonly referred to as the period of the damped vibration. Recalling (19.44), we observe that τ is larger than the period of vibration of the corresponding undamped system.

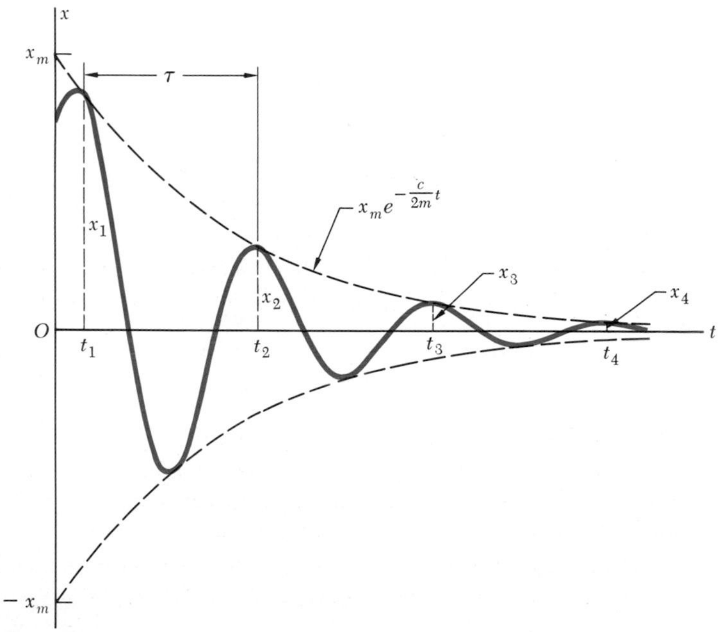

Fig. 19.11

***19.9. Damped Forced Vibrations.** If the system considered in the preceding section is subjected to a periodic force **P** of magnitude $P = P_m \sin \omega t$, the equation of motion becomes

$$m\ddot{x} + c\dot{x} + kx = P_m \sin \omega t \tag{19.46}$$

The general solution of (19.46) is obtained by adding a particular solution of (19.46) to the complementary function or general solution of the homogeneous equation (19.37). The complementary function is given by (19.41), (19.42), or (19.43), depending upon the type of damping considered. It represents a *transient* motion which is eventually damped out.

Our interest in this section is centered on the steady-state vibration represented by a particular solution of (19.46) of the form

$$x_{\text{part}} = x_m \sin (\omega t - \varphi) \tag{19.47}$$

Substituting x_{part} for x into (19.46), we obtain

$$-m\omega^2 x_m \sin (\omega t - \varphi) + c\omega x_m \cos (\omega t - \varphi) + kx_m \sin (\omega t - \varphi) = P_m \sin \omega t$$

Making $\omega t - \varphi$ successively equal to 0 and to $\pi/2$, we write

$$c\omega x_m = P_m \sin \varphi \tag{19.48}$$

$$(k - m\omega^2)x_m = P_m \cos \varphi \tag{19.49}$$

Squaring both members of (19.48) and (19.49) and adding, we have

$$[(k - m\omega^2)^2 + (c\omega)^2]x_m^2 = P_m^2 \tag{19.50}$$

Solving (19.50) for x_m and dividing (19.48) and (19.49) member by member, we obtain, respectively,

$$x_m = \frac{P_m}{\sqrt{(k - m\omega^2)^2 + (c\omega)^2}} \qquad \tan \varphi = \frac{c\omega}{k - m\omega^2} \tag{19.51}$$

Recalling from (19.4) that $k/m = p^2$, where p is the circular frequency of the undamped free vibration, and from (19.40) that $2mp = c_c$, where c_c is the critical damping coefficient of the system, we write

$$\frac{x_m}{P_m/k} = \frac{x_m}{\delta_m} = \frac{1}{\sqrt{[1 - (\omega/p)^2]^2 + [2(c/c_c)(\omega/p)]^2}} \tag{19.52}$$

$$\tan \varphi = \frac{2(c/c_c)(\omega/p)}{1 - (\omega/p)^2} \tag{19.53}$$

Formula (19.52) expresses the magnification factor in terms of the frequency ratio ω/p and damping factor c/c_c. It may be used to determine the amplitude of the steady-state vibration produced by an impressed force of magnitude $P = P_m \sin \omega t$ or by an impressed support movement $\delta = \delta_m \sin \omega t$. Formula (19.53) defines in terms of the same parameters the *phase difference* φ between the impressed force or impressed support movement and the resulting steady-state vibration of the damped system. The magnification factor has been plotted against the frequency ratio in Fig. 19.12 for various values of the damping factor. We observe that the amplitude of a forced vibration may be kept small by choosing a large coefficient of viscous damping c or by keeping the natural and forced frequencies far apart.

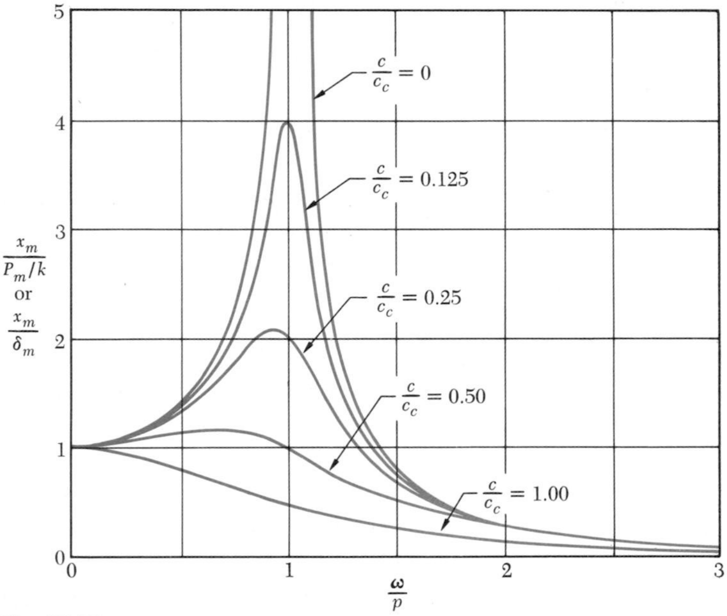

Fig. 19.12

***19.10. Electrical Analogues.** Oscillating electrical circuits are characterized by differential equations of the same type as those obtained in the preceding sections. Their analysis is therefore similar to that of a mechanical system, and the results obtained for a given vibrating system may be readily extended to the equivalent circuit. Conversely, any result obtained for an electrical circuit will also apply to the corresponding mechanical system.

Consider an electrical circuit consisting of an inductor of inductance L, a resistor of resistance R, and a capacitor of

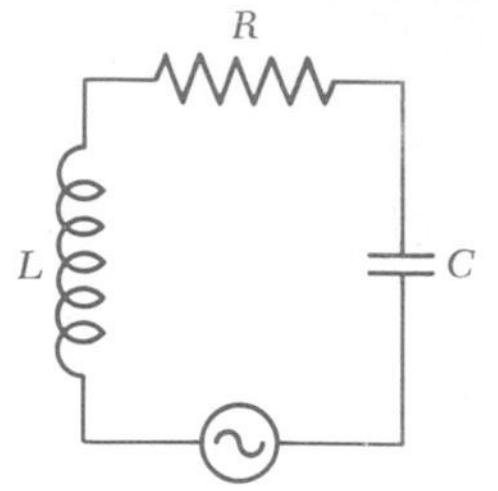

Fig. 19.13

capacitance C, connected in series with a source of alternating voltage $E = E_m \sin \omega t$ (Fig. 19.13). It is recalled from elementary electromagnetic theory† that, if i denotes the current in the circuit and q the electric charge on the capacitor, the drop in potential is $L(di/dt)$ across the inductor, Ri across the resistor, and q/C across the capacitor. Expressing that the algebraic sum of the applied voltage and of the drops in potential around the circuit loop is zero, we write

$$E_m \sin \omega t - L\frac{di}{dt} - Ri - \frac{q}{C} = 0 \qquad (19.54)$$

Rearranging the terms and recalling that, at any instant, the current i is equal to the rate of change $\dot{q}$ of the charge q, we have

$$L\ddot{q} + R\dot{q} + \frac{1}{C}q = E_m \sin \omega t \qquad (19.55)$$

We verify that Eq. (19.55), which defines the oscillations of the electrical circuit of Fig. 19.13, is of the same type as Eq. (19.46), which characterizes the damped forced vibrations of the mechanical system of Fig. 19.10. By comparing the two equations, we may construct a table of the analogous mechanical and electrical expressions.

Table 19.2 may be used to extend to their electrical analogues the results obtained in the preceding sections for various mechanical systems. For instance, the amplitude i_m of the current in the circuit of Fig. 19.13 may be obtained by noting that it corresponds to the maximum value v_m of the velocity in the analogous mechanical system. Recalling that $v_m = \omega x_m$, substituting for x_m from Eq. (19.51), and replacing the constants of

Table 19.2 Characteristics of a Mechanical System and of Its Electrical Analogue

Mechanical System		Electrical Circuit	
m	mass	L	Inductance
c	Coefficient of viscous damping	R	Resistance
k	Spring constant	$1/C$	Reciprocal of capacitance
x	Displacement	q	Charge
v	Velocity	i	Current
P	Applied force	E	Applied voltage

† See Hammond, "Electrical Engineering," McGraw-Hill Book Company, or Smith, "Circuits, Devices, and Systems," John Wiley & Sons.

the mechanical system by the corresponding electrical expressions, we have

$$i_m = \frac{\omega E_m}{\sqrt{\left(\frac{1}{C} - L\omega^2\right)^2 + (R\omega)^2}}$$

$$i_m = \frac{E_m}{\sqrt{R^2 + \left(L\omega - \frac{1}{C\omega}\right)^2}} \qquad (19.56)$$

The radical in the expression obtained is known as the *impedance* of the electrical circuit.

The analogy between mechanical systems and electrical circuits holds for transient as well as steady-state oscillations. The oscillations of the circuit shown in Fig. 19.14, for instance, are analogous to the damped free vibrations of the system of Fig. 19.10. As far as the initial conditions are concerned, we may note that closing the switch S when the charge on the capacitor is $q = q_0$ is equivalent to releasing the mass of the mechanical system with no initial velocity from the position $x = x_0$. We should also observe that, if a battery of constant voltage E is introduced in the electrical circuit of Fig. 19.14, closing the switch S will be equivalent to suddenly applying a force of constant magnitude P to the mass of the mechanical system of Fig. 19.10.

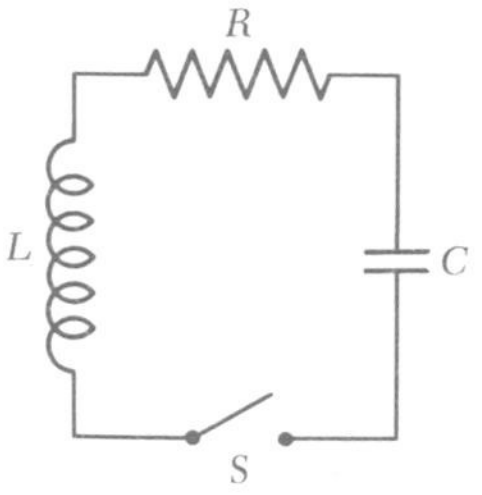

Fig. 19.14

The above discussion would be of questionable value if its only result were to make it possible for mechanics students to analyze electrical circuits without learning the elements of electromagnetism. It is hoped, rather, that this discussion will encourage the students to apply to the solution of problems in mechanical vibrations the mathematical techniques they may learn in later courses in electrical circuits theory. The chief value of the concept of electrical analogue, however, resides in its application to *experimental methods* for the determination of the characteristics of a given mechanical system. Indeed, an electrical circuit is much more easily constructed than a mechanical model, and the fact that its characteristics may be modified by varying the inductance, resistance, or capacitance of its various components makes the use of the electrical analogue particularly convenient.

To determine the electrical analogue of a given mechanical system, we shall focus our attention on each moving mass in the system and observe which springs, dashpots, or external forces are applied directly to it. An equivalent electrical loop may then be constructed to match each of the mechanical units

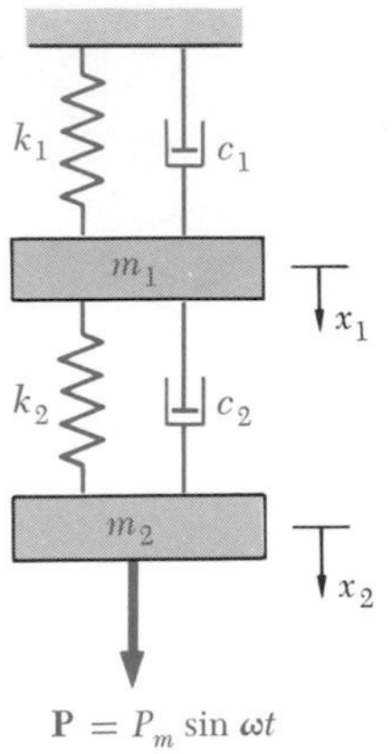

Fig. 19.15

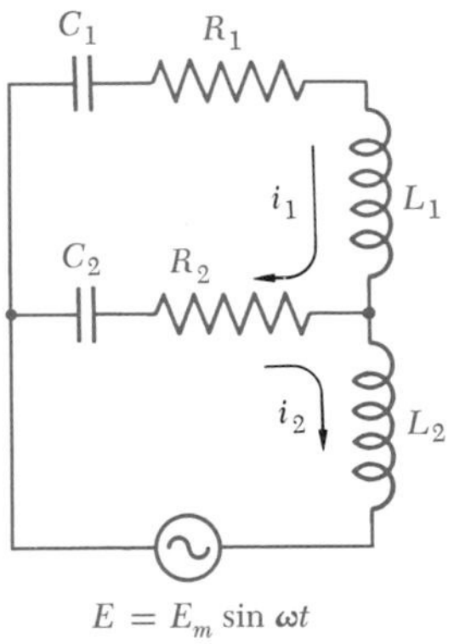

Fig. 19.16

thus defined; the various loops obtained in that way will form together the desired circuit. Consider, for instance, the mechanical system of Fig. 19.15. We observe that the mass m_1 is acted upon by two springs of constants k_1 and k_2 and by two dashpots characterized by the coefficients of viscous damping c_1 and c_2. The electrical circuit should therefore include a loop consisting of an inductor of inductance L_1 proportional to m_1, of two capacitors of capacitance C_1 and C_2 inversely proportional to k_1 and k_2, respectively, and of two resistors of resistance R_1 and R_2, proportional to c_1 and c_2, respectively. Since the mass m_2 is acted upon by the spring k_2 and the dashpot c_2, as well as by the force $P = P_m \sin \omega t$, the circuit should also include a loop containing the capacitor C_2, the resistor R_2, the new inductor L_2, and the voltage source $E = E_m \sin \omega t$ (Fig. 19.16).

To check that the mechanical system of Fig. 19.15 and the electrical circuit of Fig. 19.16 actually satisfy the same differential equations, we shall first derive the equations of motion for m_1 and m_2. Denoting respectively by x_1 and x_2 the displacements of m_1 and m_2 from their equilibrium positions, we observe that the elongation of the spring k_1 (measured from the equilibrium position) is equal to x_1, while the elongation of the spring k_2 is equal to the relative displacement $x_2 - x_1$ of m_2 with respect to m_1. The equations of motion for m_1 and m_2 are therefore

$$m_1\ddot{x}_1 + c_1\dot{x}_1 + c_2(\dot{x}_1 - \dot{x}_2) + k_1x_1 + k_2(x_1 - x_2) = 0 \quad (19.57)$$

$$m_2\ddot{x}_2 + c_2(\dot{x}_2 - \dot{x}_1) + k_2(x_2 - x_1) = P_m \sin \omega t \quad (19.58)$$

Consider now the electrical circuit of Fig. 19.16; we denote respectively by i_1 and i_2 the current in the first and second loops, and by q_1 and q_2 the integrals $\int i_1\,dt$ and $\int i_2\,dt$. Noting that the charge on the capacitor C_1 is q_1, while the charge on C_2 is $q_1 - q_2$, we express that the sum of the potential differences in each loop is zero:

$$L_1\ddot{q}_1 + R_1\dot{q}_1 + R_2(\dot{q}_1 - \dot{q}_2) + \frac{q_1}{C_1} + \frac{q_1 - q_2}{C_2} = 0 \quad (19.59)$$

$$L_2\ddot{q}_2 + R_2(\dot{q}_2 - \dot{q}_1) + \frac{q_2 - q_1}{C_2} = E_m \sin \omega t \quad (19.60)$$

We easily check that Eqs. (19.59) and (19.60) reduce to (19.57) and (19.58), respectively, when the substitutions indicated in Table 19.2 are performed.

PROBLEMS

19.97 Show that, in the case of heavy damping ($c > c_c$), a body never passes through its position of equilibrium O (*a*) if it is released with no initial velocity from an arbitrary position or (*b*) if it is started from O with an arbitrary initial velocity.

19.98 Show that, in the case of heavy damping ($c > c_c$), a body released from an arbitrary position with an arbitrary initial velocity cannot pass more than once through its equilibrium position.

19.99 In the case of light damping, the displacements x_1, x_2, x_3, etc., shown in Fig. 19.11 may be assumed equal to the maximum displacements. Show that the ratio of any two successive maximum displacements x_n and x_{n+1} is a constant and that the natural logarithm of this ratio, called the *logarithmic decrement,* is

$$\ln \frac{x_n}{x_{n+1}} = \frac{2\pi(c/c_c)}{\sqrt{1-(c/c_c)^2}}$$

19.100 In practice, it is often difficult to determine the logarithmic decrement defined in Prob. 19.99 by measuring two successive maximum displacements. Show that the logarithmic decrement may also be expressed as $(1/n)\ln(x_1/x_{n+1})$, where n is the number of cycles between readings of the maximum displacement.

19.101 Successive maximum displacements of a spring-mass-dashpot system similar to that shown in Fig. 19.10 are 75, 60, 48, and 38.4 mm. Knowing that $m = 20$ kg and $k = 800$ N/m, determine (*a*) the damping factor c/c_c, (*b*) the value of the coefficient of viscous damping c. (*Hint.* See Probs. 19.99 and 19.100.)

19.102 In a system with light damping ($c < c_c$), the period of vibration is commonly defined as the time interval $\tau = 2\pi/q$ corresponding to two successive points where the displacement-time curve touches one of the limiting curves shown in Fig. 19.11. Show that the interval of time (*a*) between a maximum positive displacement and the following maximum negative displacement is $\frac{1}{2}\tau$, (*b*) between two successive zero displacements is $\frac{1}{2}\tau$, (*c*) between a maximum positive displacement and the following zero displacement is greater than $\frac{1}{4}\tau$.

19.103 The barrel of a field gun weighs 1200 lb and is returned into firing position after recoil by a recuperator of constant $k = 8000$ lb/ft. Determine the value of the coefficient of damping of the recoil mechanism which causes the barrel to return into firing position in the shortest possible time without oscillation.

19.104 A critically damped system is released from rest at an arbitrary position x_0 when $t = 0$. (*a*) Determine the position of the system at any time *t*. (*b*) Apply the result obtained in part *a* to the barrel of the gun of Prob. 19.103, and determine the time at which the barrel is halfway back to its firing position.

19.105 Assuming that the barrel of the gun of Prob. 19.103 is modified, with a resulting increase in weight of 300 lb, determine the constant *k* of the recuperator which should be used if the recoil mechanism is to remain critically damped.

19.106 In the case of the forced vibration of a system with a given damping factor c/c_c, determine the frequency ratio ω/p for which the amplitude of the vibration is maximum.

19.107 Show that for a small value of the damping factor c/c_c, (*a*) the maximum amplitude of a forced vibration occurs when $\omega \approx p$, (*b*) the corresponding value of the magnification factor is approximately $\frac{1}{2}(c_c/c)$.

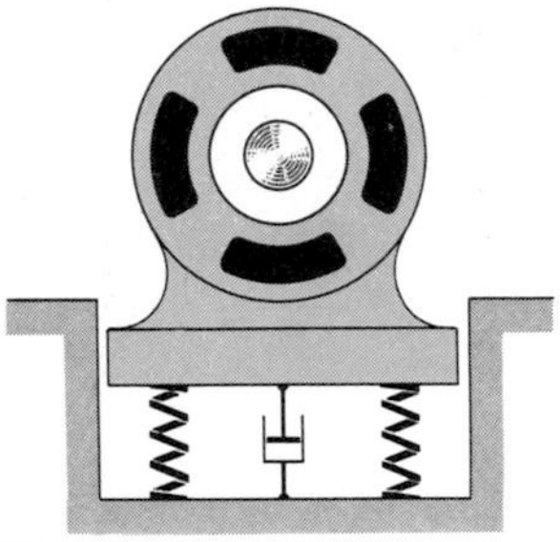

Fig. P19.108

19.108 A motor of mass 25 kg is supported by four springs, each having a constant of 200 kN/m. The unbalance of the rotor is equivalent to a mass of 30 g located 125 mm from the axis of rotation. Knowing that the motor is constrained to move vertically, determine the amplitude of the steady-state vibration of the motor at a speed of 1800 rpm, assuming (*a*) that no damping is present, (*b*) that the damping factor c/c_c is equal to 0.125.

19.109 Assume that the 25-kg motor of Prob. 19.108 is directly supported by a light horizontal beam. The static deflection of the beam due to the weight of the motor is observed to be 5.75 mm, and the amplitude of the vibration of the motor is 0.5 mm at a speed of 400 rpm. Determine (*a*) the damping factor c/c_c, (*b*) the coefficient of damping *c*.

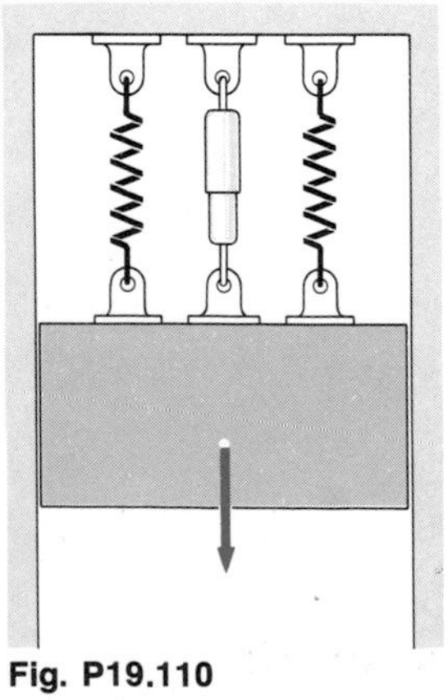

Fig. P19.110

19.110 A machine element weighing 800 lb is supported by two springs, each having a constant of 200 lb/in. A periodic force of maximum value 30 lb is applied to the element with a frequency of 2.5 cycles per second. Knowing that the coefficient of damping is 8 lb · s/in., determine the amplitude of the steady-state vibration of the element.

19.111 In Prob. 19.110, determine the required value of the coefficient of damping if the amplitude of the steady-state vibration of the element is to be 0.15 in.

19.112 A platform of mass 100 kg, supported by a set of springs equivalent to a single spring of constant $k = 80$ kN/m, is subjected to a periodic force of maximum magnitude 500 N. Knowing that the coefficient of damping is 2 kN · s/m, determine (*a*) the natural frequency in rpm of the platform *if* there were no damping, (*b*) the frequency in rpm of the periodic force corresponding to the maximum value of the magnification factor, assuming damping, (*c*) the amplitude of the *actual* motion of the platform for each of the frequencies found in parts *a* and *b*.

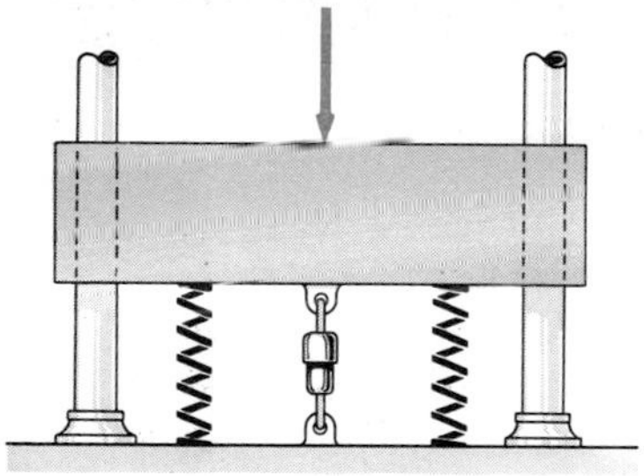

Fig. P19.112

***19.113** The suspension of an automobile may be approximated by the simplified spring-and-dashpot system shown. (*a*) Write the differential equation defining the absolute motion of the mass m when the system moves at a speed v over a road of sinusoidal cross section as shown. (*b*) Derive an expression for the amplitude of the absolute motion of m.

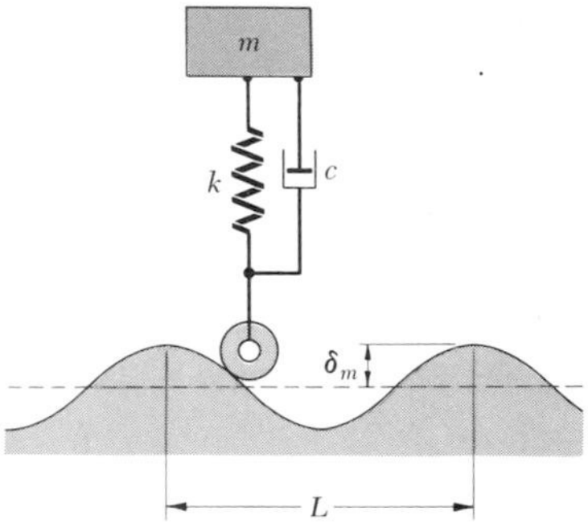

Fig. P19.113

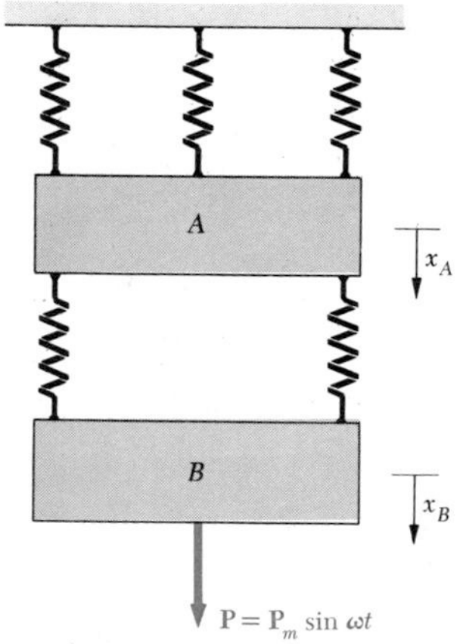

Fig. P19.114

***19.114** Two loads A and B, each of mass m, are suspended as shown by means of five springs of the same constant k. Load B is subjected to a force of magnitude $P = P_m \sin \omega t$. Write the differential equations defining the displacements x_A and x_B of the two loads from their equilibrium positions.

19.115 Determine the range of values of the resistance R for which oscillations will take place in the circuit shown when the switch S is closed.

19.116 Consider the circuit of Prob. 19.115 when the capacitance C is equal to zero. If the switch S is closed at time $t = 0$, determine (*a*) the final value of the current in the circuit, (*b*) the time t at which the current will have reached $(1 - 1/e)$ times its final value. (The desired value of t is known as the *time constant* of the circuit.)

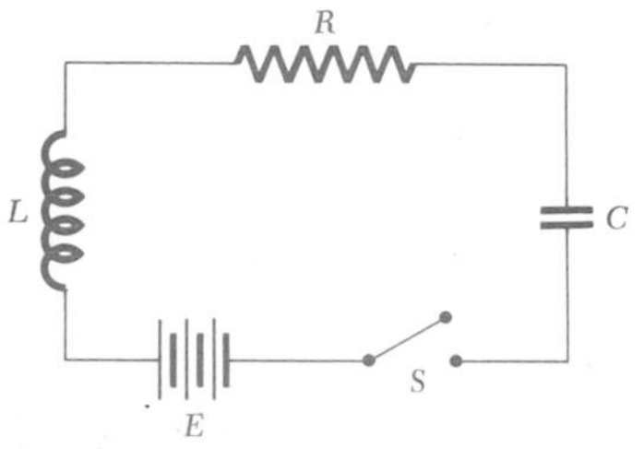

Fig. P19.115

19.117 through 19.120 Draw the electrical analogue of the mechanical system shown. (*Hint.* In Probs. 19.117 and 19.118, draw the loops corresponding to the free bodies m and A.)

19.121 and 19.122 Write the differential equations defining (*a*) the displacements of mass m and point A, (*b*) the currents in the corresponding loops of the electrical analogue.

19.123 and 19.124 Write the differential equations defining (*a*) the displacements of the masses m_1 and m_2, (*b*) the currents in the corresponding loops of the electrical analogue.

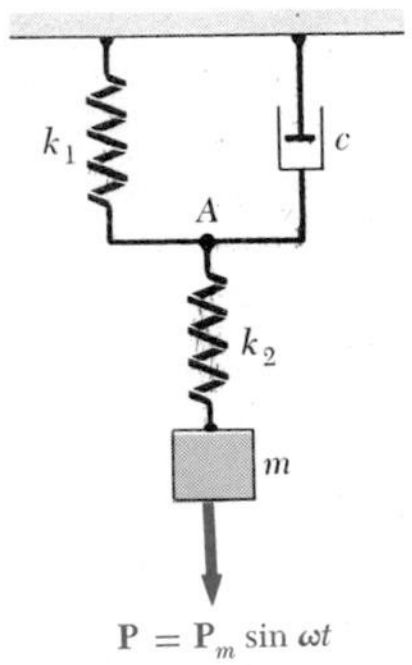

Fig. P19.117 and P19.121

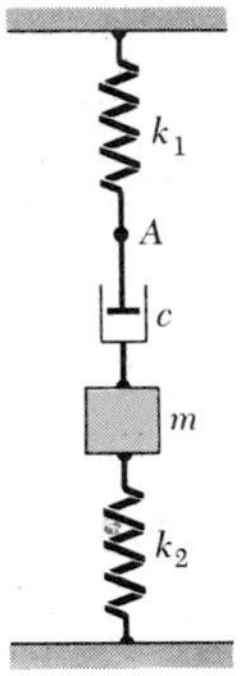

Fig. P19.118 and P19.122

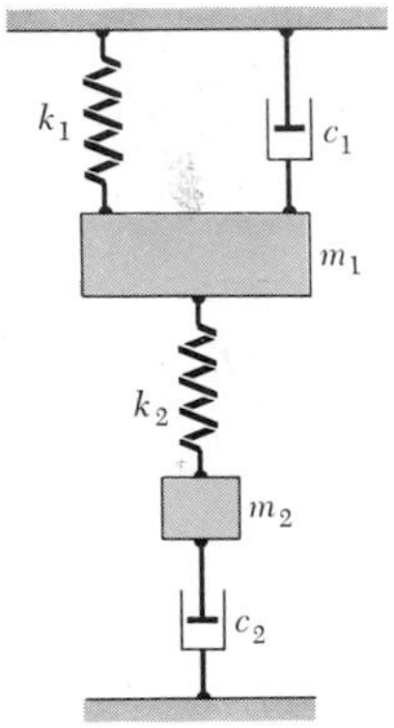

Fig. P19.119 and P19.123

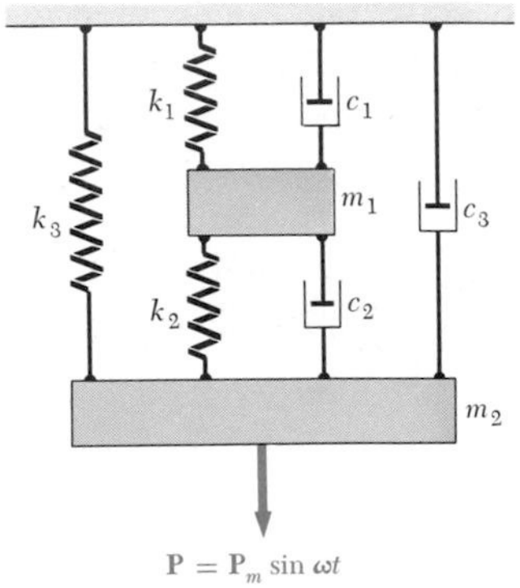

Fig. P19.120 and P19.124

REVIEW PROBLEMS

19.125 A homogeneous wire of length $2l$ is bent as shown and allowed to oscillate about a frictionless pin at B. Denoting by τ_0 the period of small oscillations when $\beta = 0$, determine the angle β for which the period of small oscillations is $2\tau_0$.

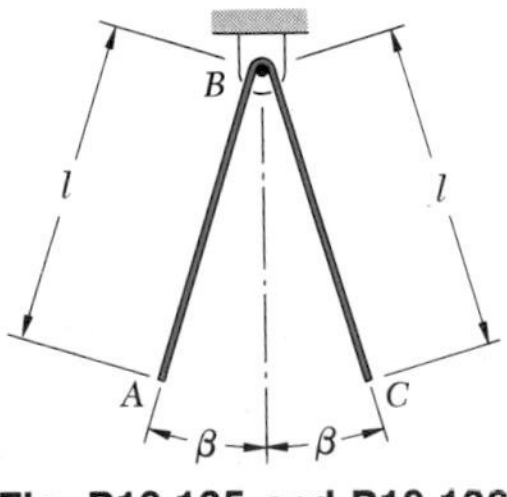

Fig. P19.125 and P19.126

19.126 Knowing that $l = 0.6$ m and $\beta = 30°$, determine the period of oscillation of the bent homogeneous wire shown.

19.127 A period of 4.10 s is observed for the angular oscillations of a 1-lb gyroscope rotor suspended from a wire as shown. Knowing that a period of 6.20 s is obtained when a 2-in.-diameter steel sphere is suspended in the same fashion, determine the centroidal radius of gyration of the rotor. (Specific weight of steel = 490 lb/ft^3.)

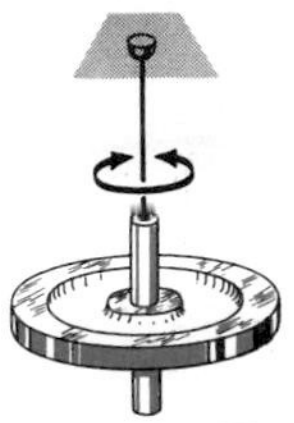

Fig. P19.127

19.128 An automobile wheel-and-tire assembly of total weight 47 lb is attached to a mounting plate of negligible weight which is suspended from a steel wire. The torsional spring constant of the wire is known to be $K = 0.40$ lb · in./rad. The wheel is rotated through 90° about the vertical and then released. Knowing that the period of oscillation is observed to be 30 s, determine the centroidal mass moment of inertia and the centroidal radius of gyration of the wheel-and-tire assembly.

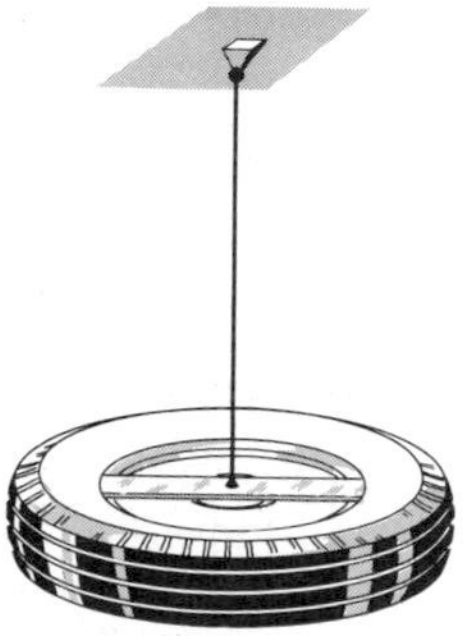

Fig. P19.128

19.129 A homogeneous wire is bent to form a square of side l which is supported by a ball-and-socket joint at A. Determine the period of small oscillations of the square (a) in the plane of the square, (b) in a direction perpendicular to the square.

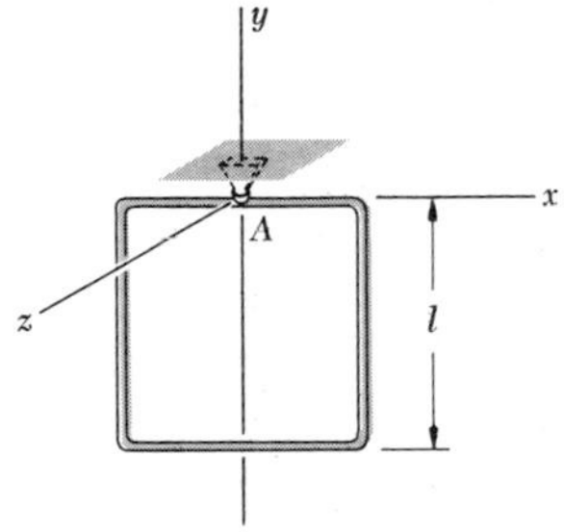

Fig. P19.129

19.130 A 150-kg electromagnet is at rest and is holding 100 kg of scrap steel when the current is turned off and the steel is dropped. Knowing that the cable and the supporting crane have a total stiffness equivalent to a spring of constant 200 kN/m, determine (a) the frequency, the amplitude, and the maximum velocity of the resulting motion, (b) the minimum tension which will occur in the cable during the motion, (c) the velocity of the magnet 0.03 s after the current is turned off.

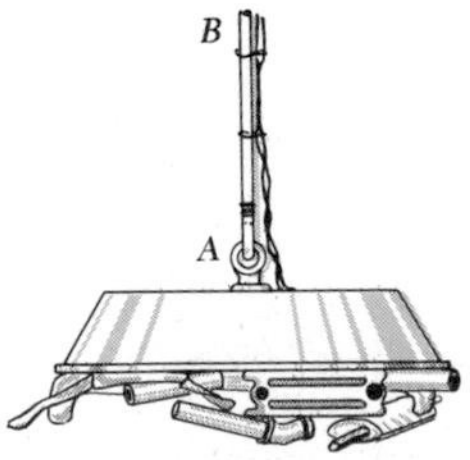

Fig. P19.130

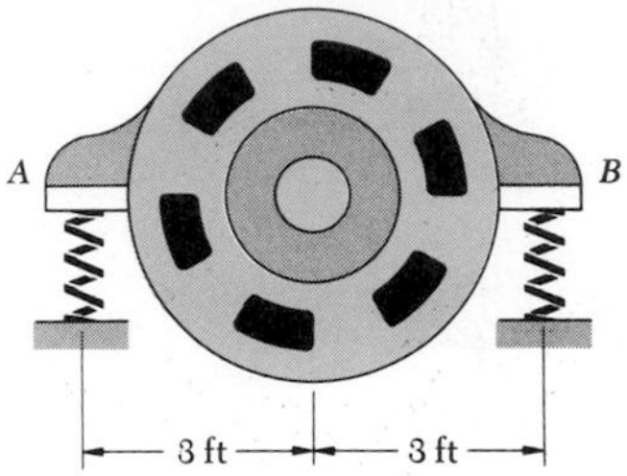

Fig. P19.131

19.131 During the normal operation of a single-phase generator, the transmission of undesirable vibrations is prevented by four springs mounted as shown (two springs at A and two springs at B). Knowing that the stator of the generator weighs 400 lb and has a centroidal radius of gyration of 24 in., determine the required constant of each spring if the frequency of the free *angular* vibration of the stator is to be 15 cycles per second.

19.132 The *rotor* of the generator of Prob. 19.131 weighs 300 lb and has a centroidal radius of gyration of 18 in. Knowing that the springs have been chosen so that the angular frequency of the *stator* alone is 15 cycles per second, determine the frequency of the *angular* vibration of the generator if the bearings are frozen so that the rotor and stator move as a single rigid body.

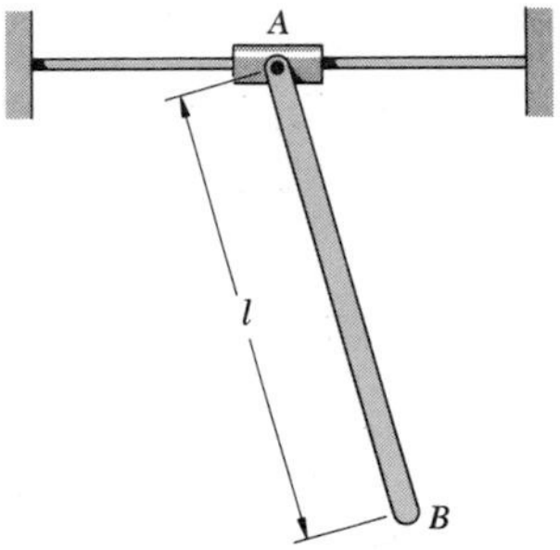

Fig. P19.133

19.133 A slender bar of length l is attached by a smooth pin at A to a collar of negligible mass. Determine the period of small oscillations of the bar, assuming that the coefficient of friction between the collar and the horizontal rod (a) is sufficient to prevent any movement of the collar, (b) is zero.

19.134 A 2-kg instrument is spring-mounted on the casing of a motor rotating at 1800 rpm. The motor is unbalanced and the amplitude of the motion of its casing is 0.5 mm. Knowing that $k = 9000$ N/m, determine the amplitude of the motion of the instrument.

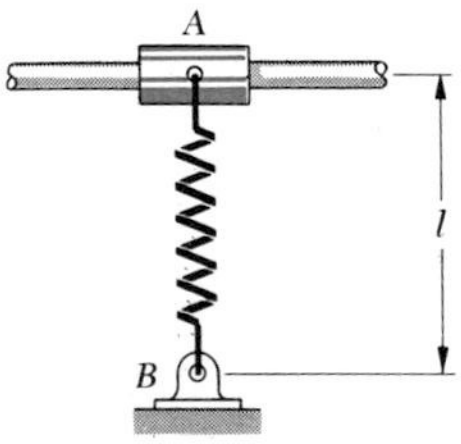

Fig. P19.135

19.135 A collar of mass m slides without friction on a horizontal rod and is attached to a spring AB of constant k. (a) If the unstretched length of the spring is just equal to l, show that the collar does not execute simple harmonic motion even when the amplitude of the oscillations is small. (b) If the unstretched length of the spring is less than l, show that the motion may be approximated by a simple harmonic motion for small oscillations.

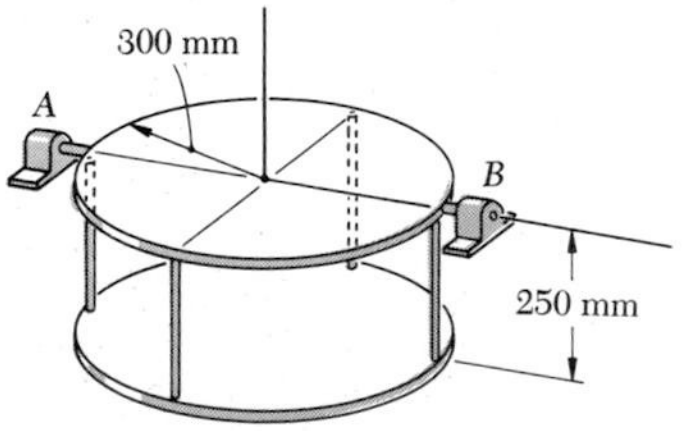

Fig. P19.136

19.136 The assembly shown consists of two thin disks, each of mass 3 kg, and four slender rods, each of mass 0.4 kg. Determine the period of oscillation of the assembly.

APPENDIX

A

Some Useful Definitions and Properties of Vector Algebra

The following definitions and properties of vector algebra were discussed fully in Chaps. 2 and 3 of *Vector Mechanics for Engineers: Statics*. They are summarized here for the convenience of the reader, with references to the appropriate sections of the *Statics* volume. Equation and illustration numbers are those used in the original presentation.

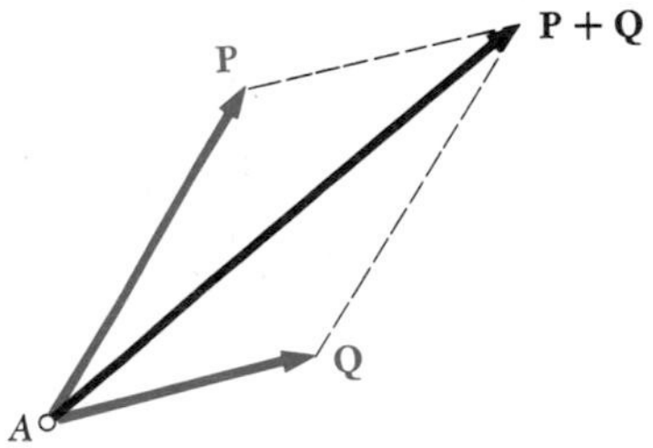

Fig. 2.6

A.1. Addition of Vectors (Secs. 2.2 and 2.3). Vectors are defined as *mathematical expressions possessing magnitude and direction, which add according to the parallelogram law*. Thus the sum of two vectors **P** and **Q** is obtained by attaching the two vectors to the same point A and constructing a parallelogram, using **P** and **Q** as two sides of the parallelogram (Fig. 2.6). The diagonal that passes through A represents the sum of the vectors **P** and **Q**, and this sum is denoted by **P** + **Q**. Vector addition is *associative* and *commutative*.

The *negative vector* of a given vector **P** is defined as a vector having the same magnitude P and a direction opposite to that of **P** (Fig. 2.5); the negative of the vector **P** is denoted by −**P**. Clearly, we have

$$\mathbf{P} + (-\mathbf{P}) = 0$$

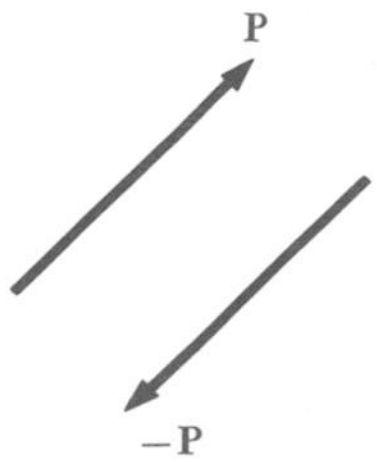

Fig. 2.5

A.2. Product of a Scalar and a Vector (Sec. 2.3). The product $k\mathbf{P}$ of a scalar k and a vector $\mathbf{P}$ is defined as a vector having the same direction as $\mathbf{P}$ (if k is positive), or a direction opposite to that of $\mathbf{P}$ (if k is negative), and a magnitude equal to the product of the magnitude P and of the absolute value of k (Fig. 2.13).

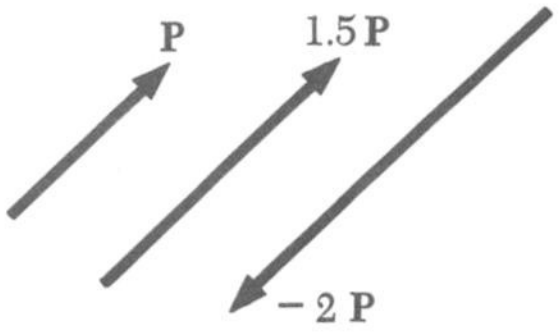

Fig. 2.13

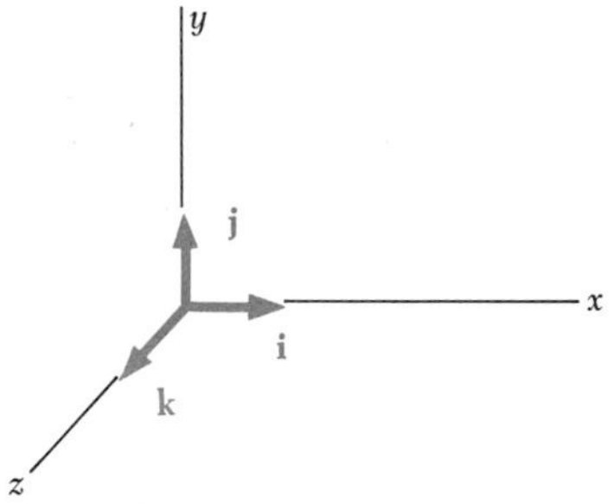

Fig. 2.32

A.3. Unit Vectors. Resolution of a Vector into Rectangular Components (Secs. 2.6 and 2.11). The vectors **i**, **j**, and **k**, called *unit vectors*, are defined as vectors of magnitude 1, directed respectively along the positive x, y, and z axes (Fig. 2.32).

Denoting by F_x, F_y, and F_z the scalar components of a vector **F**, we have (Fig. 2.33)

$$\mathbf{F} = F_x\mathbf{i} + F_y\mathbf{j} + F_z\mathbf{k} \tag{2.20}$$

In the particular case of a unit vector $\boldsymbol{\lambda}$ directed along a line forming angles θ_x, θ_y, and θ_z with the coordinate axes, we have

$$\boldsymbol{\lambda} = \cos\theta_x\mathbf{i} + \cos\theta_y\mathbf{j} + \cos\theta_z\mathbf{k} \tag{2.22}$$

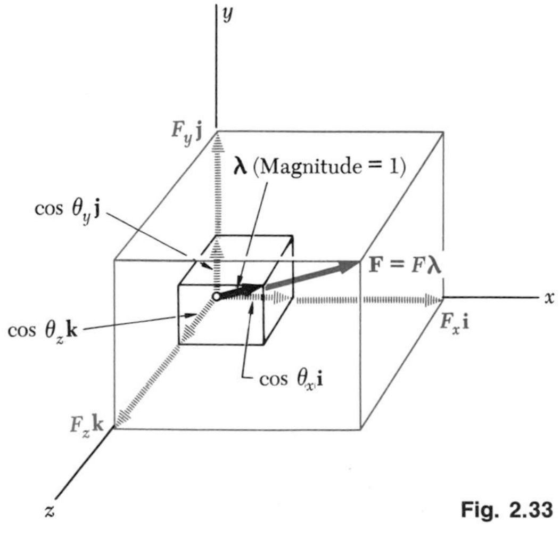

Fig. 2.33

A.4. Vector Product of Two Vectors (Secs. 3.3 and 3.4). The vector product, or *cross product,* of two vectors **P** and **Q** is defined as the vector

$$\mathbf{V} = \mathbf{P} \times \mathbf{Q}$$

which satisfies the following conditions:

1. The line of action of **V** is perpendicular to the plane containing **P** and **Q** (Fig. 3.6).
2. The magnitude of **V** is the product of the magnitudes of **P** and **Q** and of the sine of the angle θ formed by **P** and **Q** (the measure of which will always be 180° or less); we thus have

$$V = PQ \sin\theta \tag{3.1}$$

3. The sense of **V** is such that a man located at the tip of **V** will observe as counterclockwise the rotation through θ which brings the vector **P** in line with the vector **Q**; note that if **P** and **Q** do not have a common point of application, they should first be redrawn from the same point. The three vectors **P**, **Q**, and **V**—taken in that order—form a *right-handed triad.*

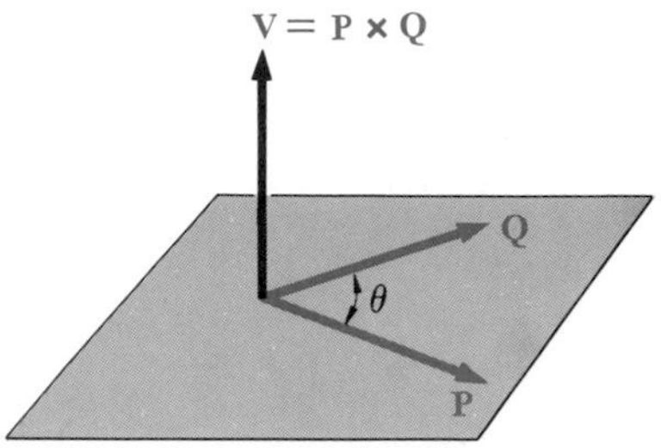

Fig. 3.6

Vector products are *distributive,* but they are *not commutative.* We have

$$\mathbf{Q} \times \mathbf{P} = -(\mathbf{P} \times \mathbf{Q}) \tag{3.4}$$

Vector Products of Unit Vectors. It follows from the definition of the vector product of two vectors that

$$\begin{array}{lll} \mathbf{i} \times \mathbf{i} = 0 & \mathbf{j} \times \mathbf{i} = -\mathbf{k} & \mathbf{k} \times \mathbf{i} = \mathbf{j} \\ \mathbf{i} \times \mathbf{j} = \mathbf{k} & \mathbf{j} \times \mathbf{j} = 0 & \mathbf{k} \times \mathbf{j} = -\mathbf{i} \\ \mathbf{i} \times \mathbf{k} = -\mathbf{j} & \mathbf{j} \times \mathbf{k} = \mathbf{i} & \mathbf{k} \times \mathbf{k} = 0 \end{array} \tag{3.7}$$

Rectangular Components of Vector Product. Resolving the vectors **P** and **Q** into rectangular components, we obtain the following expressions for the components of their vector product **V**:

$$\begin{aligned} V_x &= P_yQ_z - P_zQ_y \\ V_y &= P_zQ_x - P_xQ_z \\ V_z &= P_xQ_y - P_yQ_x \end{aligned} \tag{3.9}$$

In determinant form, we have

$$\mathbf{V} = \mathbf{P} \times \mathbf{Q} = \begin{vmatrix} \mathbf{i} & \mathbf{j} & \mathbf{k} \\ P_x & P_y & P_z \\ Q_x & Q_y & Q_z \end{vmatrix} \tag{3.10}$$

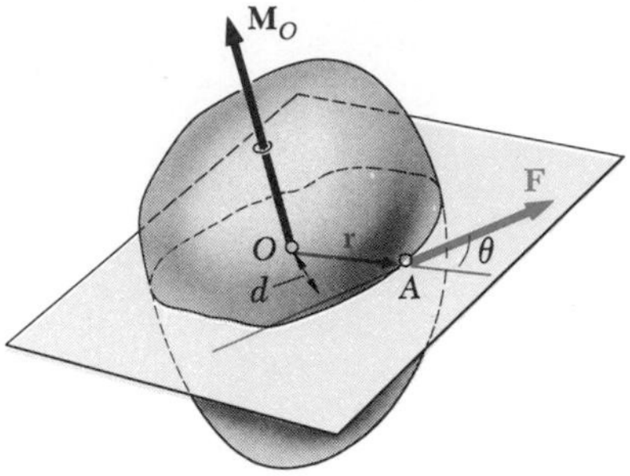

Fig. 3.12

A.5. Moment of a Force about a Point (Secs. 3.5 and 3.7). The moment of a force **F** (or, more generally, of a vector **F**) about a point O is defined as the vector product

$$\mathbf{M}_O = \mathbf{r} \times \mathbf{F} \tag{3.11}$$

where **r** denotes the *position vector* of the point of application A of **F** (Fig. 3.12).

Rectangular Components of the Moment of a Force. Denoting by x, y, and z the coordinates of the point of application A of **F**, we obtain the following expressions for the components of the moment $\mathbf{M}_O$ of **F**:

$$\begin{aligned} M_x &= yF_z - zF_y \\ M_y &= zF_x - xF_z \\ M_z &= xF_y - yF_x \end{aligned} \tag{3.18}$$

In determinant form, we have

$$\mathbf{M}_O = \mathbf{r} \times \mathbf{F} = \begin{vmatrix} \mathbf{i} & \mathbf{j} & \mathbf{k} \\ x & y & z \\ F_x & F_y & F_z \end{vmatrix} \tag{3.19}$$

To compute the moment $\mathbf{M}_B$ about an arbitrary point B of a force **F** applied at A, we must use the vector $\Delta\mathbf{r} = \mathbf{r}_A - \mathbf{r}_B$ instead of the vector **r**. We write

$$\mathbf{M}_B = \Delta\mathbf{r} \times \mathbf{F} = (\mathbf{r}_A - \mathbf{r}_B) \times \mathbf{F} \tag{3.20}$$

or, using the determinant form,

$$\mathbf{M}_B = \begin{vmatrix} \mathbf{i} & \mathbf{j} & \mathbf{k} \\ \Delta x & \Delta y & \Delta z \\ F_x & F_y & F_z \end{vmatrix} \tag{3.21}$$

where Δx, Δy, Δz are the components of the vector $\Delta\mathbf{r}$ joining A and B:

$$\Delta x = x_A - x_B \qquad \Delta y = y_A - y_B \qquad \Delta z = z_A - z_B$$

A.6. Scalar Product of Two Vectors (Sec. 3.8). The scalar product, or *dot product,* of two vectors **P** and **Q** is defined as the product of the magnitudes of **P** and **Q** and of the cosine of the angle θ formed by **P** and **Q** (Fig. 3.19). The scalar product of **P** and **Q** is denoted by $\mathbf{P} \cdot \mathbf{Q}$. We write

$$\mathbf{P} \cdot \mathbf{Q} = PQ \cos\theta \tag{3.24}$$

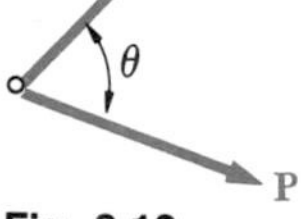

Fig. 3.19

Scalar products are *commutative* and *distributive.*

Scalar Products of Unit Vectors. It follows from the definition of the scalar product of two vectors that

$$\mathbf{i}\cdot\mathbf{i} = 1 \qquad \mathbf{j}\cdot\mathbf{j} = 1 \qquad \mathbf{k}\cdot\mathbf{k} = 1$$
$$\mathbf{i}\cdot\mathbf{j} = 0 \qquad \mathbf{j}\cdot\mathbf{k} = 0 \qquad \mathbf{k}\cdot\mathbf{i} = 0 \tag{3.29}$$

Scalar Product Expressed in Terms of Rectangular Components. Resolving the vectors **P** and **Q** into rectangular components, we obtain

$$\mathbf{P}\cdot\mathbf{Q} = P_xQ_x + P_yQ_y + P_zQ_z \tag{3.30}$$

Angle Formed by Two Vectors. It follows from (3.24) and (3.29) that

$$\cos\theta = \frac{\mathbf{P}\cdot\mathbf{Q}}{PQ} = \frac{P_xQ_x + P_yQ_y + P_zQ_z}{PQ} \tag{3.32}$$

Projection of a Vector on a Given Axis. The projection of a vector **P** on the axis OL defined by the unit vector $\boldsymbol{\lambda}$ (Fig. 3.23) is

$$P_{OL} = OA = \mathbf{P}\cdot\boldsymbol{\lambda} \tag{3.36}$$

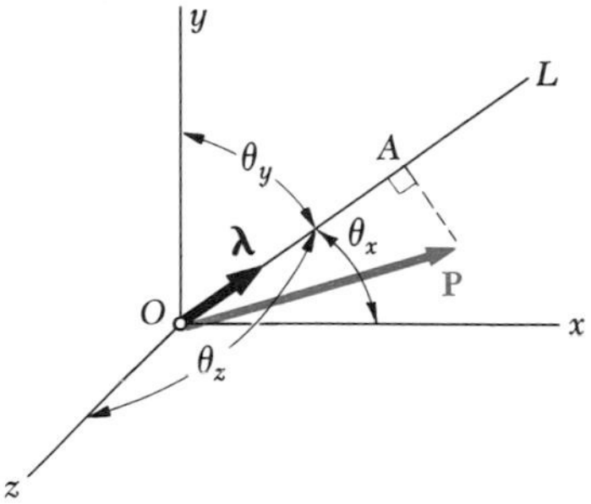

Fig. 3.23

A.7. Mixed Triple Product of Three Vectors (Sec. 3.9). The mixed triple product of the three vectors **S**, **P**, and **Q** is defined as the scalar expression

$$\mathbf{S}\cdot(\mathbf{P}\times\mathbf{Q}) \tag{3.38}$$

obtained by forming the scalar product of **S** with the vector product of **P** and **Q**. Mixed triple products are invariant under *circular permutations*, but change sign under any other permutation:

$$\begin{aligned}\mathbf{S}\cdot(\mathbf{P}\times\mathbf{Q}) &= \mathbf{P}\cdot(\mathbf{Q}\times\mathbf{S}) = \mathbf{Q}\cdot(\mathbf{S}\times\mathbf{P})\\ &= -\mathbf{S}\cdot(\mathbf{Q}\times\mathbf{P}) = -\mathbf{P}\cdot(\mathbf{S}\times\mathbf{Q}) = -\mathbf{Q}\cdot(\mathbf{P}\times\mathbf{S})\end{aligned} \tag{3.39}$$

Mixed Triple Product Expressed in Terms of Rectangular Components. The mixed triple product of **S**, **P**, and **Q** may be expressed in the form of a determinant:

$$\mathbf{S}\cdot(\mathbf{P}\times\mathbf{Q}) = \begin{vmatrix} S_x & S_y & S_z \\ P_x & P_y & P_z \\ Q_x & Q_y & Q_z \end{vmatrix} \tag{3.41}$$

The mixed triple product $\mathbf{S}\cdot(\mathbf{P}\times\mathbf{Q})$ measures the volume of the parallelepiped having the vectors **S**, **P**, and **Q** for sides (Fig. 3.25).

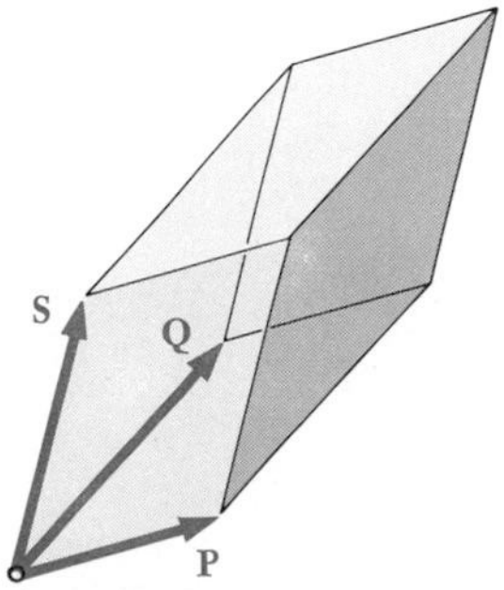

Fig. 3.25

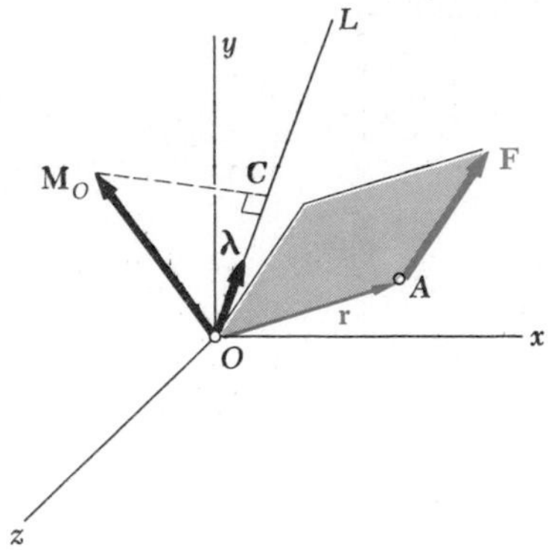

Fig. 3.27

A.8. Moment of a Force about a Given Axis (Sec. 3.10). The moment M_{OL} of a force $\mathbf{F}$ (or, more generally, of a vector $\mathbf{F}$) about an axis OL is defined as the projection OC on the axis OL of the moment $\mathbf{M}_O$ of $\mathbf{F}$ about O (Fig. 3.27). Denoting by $\boldsymbol{\lambda}$ the unit vector along OL, we have

$$M_{OL} = \boldsymbol{\lambda} \cdot \mathbf{M}_O = \boldsymbol{\lambda} \cdot (\mathbf{r} \times \mathbf{F}) \tag{3.42}$$

or, in determinant form,

$$M_{OL} = \begin{vmatrix} \lambda_x & \lambda_y & \lambda_z \\ x & y & z \\ F_x & F_y & F_z \end{vmatrix} \tag{3.43}$$

where $\lambda_x, \lambda_y, \lambda_z$ = direction cosines of axis OL
x, y, z = coordinates of point of application of $\mathbf{F}$
F_x, F_y, F_z = components of force $\mathbf{F}$

The moments of the force $\mathbf{F}$ about the three coordinate axes are given by the expressions (3.18) obtained earlier for the rectangular components of the moment $\mathbf{M}_O$ of $\mathbf{F}$ about O:

$$\begin{aligned} M_x &= yF_z - zF_y \\ M_y &= zF_x - xF_z \\ M_z &= xF_y - yF_x \end{aligned} \tag{3.18}$$

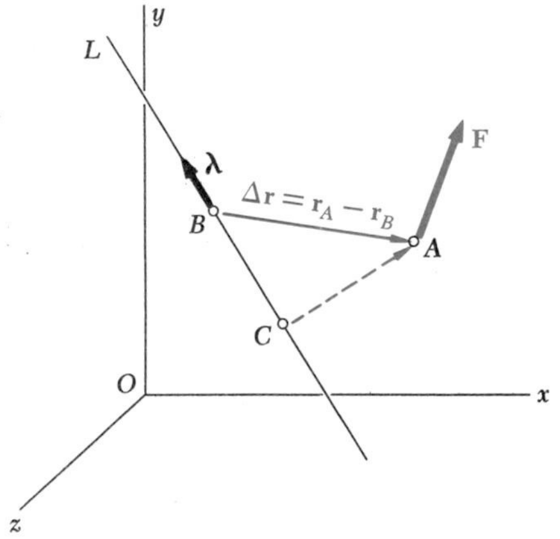

Fig. 3.29

More generally, the moment of a force $\mathbf{F}$ applied at A about an axis which does not pass through the origin is obtained by choosing an arbitrary point B on the axis (Fig. 3.29) and determining the projection on the axis BL of the moment $\mathbf{M}_B$ of $\mathbf{F}$ about B. We write

$$M_{BL} = \boldsymbol{\lambda} \cdot \mathbf{M}_B = \boldsymbol{\lambda} \cdot (\Delta\mathbf{r} \times \mathbf{F}) \tag{3.45}$$

where $\Delta\mathbf{r} = \mathbf{r}_A - \mathbf{r}_B$ represents the vector joining B and A. Expressing M_{BL} in the form of a determinant, we have

$$M_{BL} = \begin{vmatrix} \lambda_x & \lambda_y & \lambda_z \\ \Delta x & \Delta y & \Delta z \\ F_x & F_y & F_z \end{vmatrix} \tag{3.46}$$

where $\lambda_x, \lambda_y, \lambda_z$ = direction cosines of axis BL

$\Delta x = x_A - x_B,\ \Delta y = y_A - y_B,\ \Delta_z = z_A - z_B$
F_x, F_y, F_z = components of force $\mathbf{F}$

It should be noted that the result obtained is independent of the choice of the point B on the given axis; the same result would have been obtained if point C had been chosen instead of B.

APPENDIX

B

Moments of Inertia of Masses

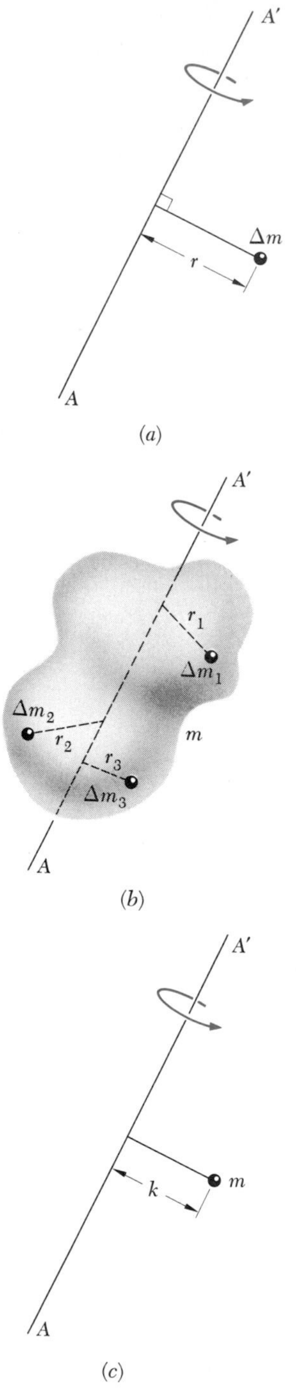

Fig. 9.20

MOMENTS OF INERTIA OF MASSES*

9.10. Moment of Inertia of a Mass. Consider a small mass Δm mounted on a rod of negligible mass which may rotate freely about an axis AA' (Fig. 9.20*a*). If a couple is applied to the system, the rod and mass, assumed initially at rest, will start rotating about AA'. The details of this motion will be studied later in dynamics. At present, we wish only to indicate that the time required for the system to reach a given speed of rotation is proportional to the mass Δm and to the square of the distance r. The product $r^2 \Delta m$ provides, therefore, a measure of the *inertia* of the system, i.e., of the resistance the system offers when we try to set it in motion. For this reason, the product $r^2 \Delta m$ is called the *moment of inertia* of the mass Δm with respect to the axis AA'.

Consider now a body of mass m which is to be rotated about an axis AA' (Fig. 9.20*b*). Dividing the body into elements of mass Δm_1, Δm_2, etc., we find that the resistance offered by the body is measured by the sum $r_1^2 \Delta m_1 + r_2^2 \Delta m_2 + \cdots$. This sum defines, therefore, the moment of inertia of the body with respect to the axis AA'. Increasing the number of elements, we find that the moment of inertia is equal, at the limit, to the integral

$$I = \int r^2\, dm \tag{9.28}$$

The *radius of gyration* k of the body with respect to the axis AA' is defined by the relation

$$I = k^2 m \qquad \text{or} \qquad k = \sqrt{\frac{I}{m}} \tag{9.29}$$

The radius of gyration k represents, therefore, the distance at which the entire mass of the body should be concentrated if its moment of inertia with respect to AA' is to remain unchanged (Fig. 9.20*c*). Whether it is kept in its original shape (Fig. 9.20*b*) or whether it is concentrated as shown in Fig. 9.20*c*, the mass m will react in the same way to a rotation, or *gyration*, about AA'.

If SI units are used, the radius of gyration k is expressed in meters and the mass m in kilograms. The moment of inertia of a mass, therefore, will be expressed in $\text{kg} \cdot \text{m}^2$. If U.S. customary units are used, the radius of gyration is expressed in feet

* This repeats Secs. 9.10 through 9.16 of the volume on statics.

and the mass in slugs, i.e., in lb · s²/ft. The moment of inertia of a mass, then, will be expressed in lb · ft · s².†

The moment of inertia of a body with respect to a coordinate axis may easily be expressed in terms of the coordinates x, y, z of the element of mass dm (Fig. 9.21). Noting, for example, that the square of the distance r from the element dm to the y axis is $z^2 + x^2$, we express the moment of inertia of the body with respect to the y axis as

$$I_y = \int r^2\, dm = \int (z^2 + x^2)\, dm$$

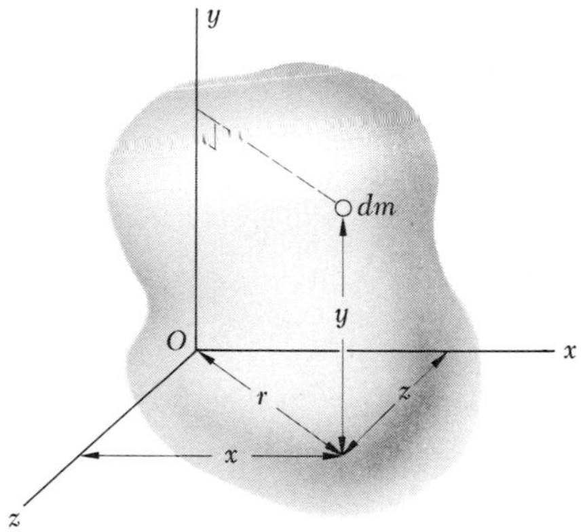

Fig. 9.21

Similar expressions may be obtained for the moments of inertia with respect to the x and z axes. We write

$$\begin{aligned} I_x &= \int (y^2 + z^2)\, dm \\ I_y &= \int (z^2 + x^2)\, dm \\ I_z &= \int (x^2 + y^2)\, dm \end{aligned} \qquad (9.30)$$

9.11. Parallel-Axis Theorem. Consider a body of mass m. Let $Oxyz$ be a system of rectangular coordinates with origin at an arbitrary point O, and $Gx'y'z'$ a system of parallel *centroidal axes,* i.e., a system with origin at the center of gravity G of the body‡ and with axes x', y', z', respectively parallel to x, y, z (Fig. 9.22). Denoting by $\bar{x}$, $\bar{y}$, $\bar{z}$ the coordinates of G with respect to $Oxyz$, we write the following relations between the coordinates x, y, z of the element dm with respect to $Oxyz$ and its coordinates x', y', z' with respect to the centroidal axes $Gx'y'z'$:

$$x = x' + \bar{x} \qquad y = y' + \bar{y} \qquad z = z' + \bar{z} \qquad (9.31)$$

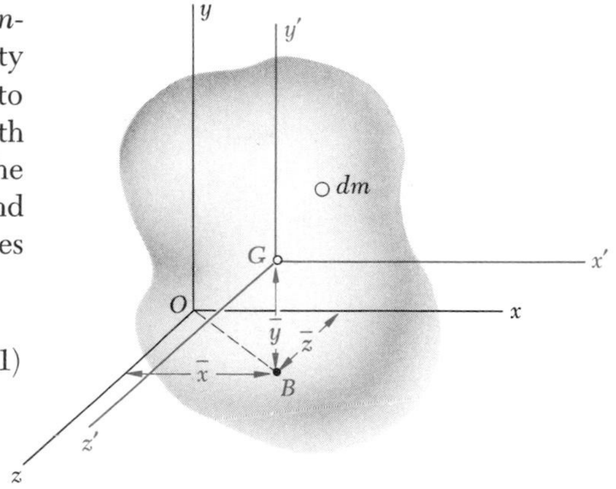

Fig. 9.22

†It should be kept in mind, when converting the moment of inertia of a mass from U.S. customary units to SI units, that the base unit pound used in the derived unit lb · ft · s² is a unit of force (*not* of mass) and, therefore, should be converted into newtons. We have

$$1 \text{ lb}\cdot\text{ft}\cdot\text{s}^2 = (4.45 \text{ N})(0.3048 \text{ m})(1 \text{ s})^2 = 1.356 \text{ N}\cdot\text{m}\cdot\text{s}^2$$

or, since N = kg · m/s²,

$$1 \text{ lb}\cdot\text{ft}\cdot\text{s}^2 = 1.356 \text{ kg}\cdot\text{m}^2$$

‡Note that the term centroidal is used to define an axis passing through the center of gravity G of the body, whether or not G coincides with the centroid of the volume of the body.

Referring to Eqs. (9.30), we may express the moment of inertia of the body with respect to the x axis as follows:

$$\begin{aligned} I_x &= \int(y^2 + z^2)\,dm = \int[(y' + \bar{y})^2 + (z' + \bar{z})^2]\,dm \\ &= \int(y'^2 + z'^2)\,dm + 2\bar{y}\int y'\,dm + 2\bar{z}\int z'\,dm + (\bar{y}^2 + \bar{z}^2)\int dm \end{aligned}$$

The first integral in the expression obtained represents the moment of inertia $\bar{I}_{x'}$ of the body with respect to the centroidal axis x'; the second and third integrals represent the first moment of the body with respect to the $z'x'$ and $x'y'$ planes, respectively, and, since both planes contain G, the two integrals are zero; the last integral is equal to the total mass m of the body. We write, therefore,

$$I_x = \bar{I}_{x'} + m(\bar{y}^2 + \bar{z}^2) \tag{9.32}$$

and, similarly,

$$I_y = \bar{I}_{y'} + m(\bar{z}^2 + \bar{x}^2) \qquad I_z = \bar{I}_{z'} + m(\bar{x}^2 + \bar{y}^2) \tag{9.32'}$$

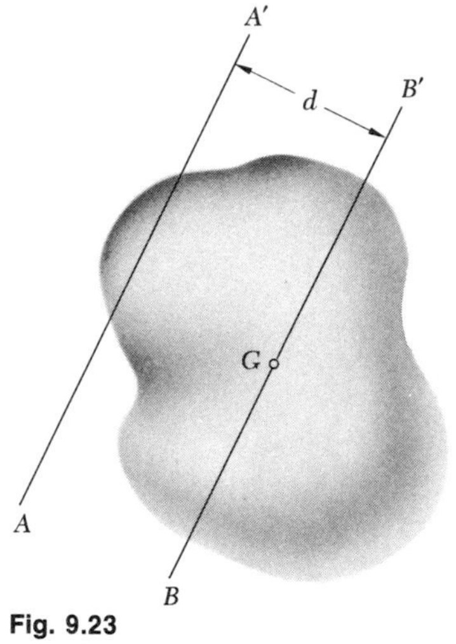

Fig. 9.23

We easily verify from Fig. 9.22 that the sum $\bar{z}^2 + \bar{x}^2$ represents the square of the distance OB between the y and y' axis. Similarly, $\bar{y}^2 + \bar{z}^2$ and $\bar{x}^2 + \bar{y}^2$ represent the squares of the distances between the x and x' axes, and the z and z' axes, respectively. Denoting by d the distance between an arbitrary axis AA' and a parallel centroidal axis BB' (Fig. 9.23), we may, therefore, write the following general relation between the moment of inertia I of the body with respect to AA' and its moment of inertia $\bar{I}$ with respect to BB':

$$I = \bar{I} + md^2 \tag{9.33}$$

Expressing the moments of inertia in terms of the corresponding radii of gyration, we may also write

$$k^2 = \bar{k}^2 + d^2 \tag{9.34}$$

where k and $\bar{k}$ represent the radii of gyration about AA' and BB', respectively.

9.12. Moments of Inertia of Thin Plates. Consider a thin plate of uniform thickness t, made of a homogeneous material of density ρ (density = mass per unit volume). The mass moment of inertia of the plate with respect to an axis AA' *contained in the plane* of the plate (Fig. 9.24*a*) is

$$I_{AA',\text{mass}} = \int r^2\, dm$$

Since $dm = \rho t\, dA$, we write

$$I_{AA',\text{mass}} = \rho t \int r^2\, dA$$

But r represents the distance of the element of area dA to the axis AA'; the integral is therefore equal to the moment of inertia of the area of the plate with respect to AA'. We have

$$I_{AA',\text{mass}} = \rho t I_{AA',\text{area}} \tag{9.35}$$

Similarly, we have with respect to an axis BB' perpendicular to AA' (Fig. 9.24*b*)

$$I_{BB',\text{mass}} = \rho t I_{BB',\text{area}} \tag{9.36}$$

Considering now the axis CC' *perpendicular* to the plate through the point of intersection C of AA' and BB' (Fig. 9.24*c*), we write

$$I_{CC',\text{ mass}} = \rho t J_{C,\text{ area}} \tag{9.37}$$

where J_C is the *polar* moment of inertia of the area of the plate with respect to point C.

Recalling the relation $J_C = I_{AA'} + I_{BB'}$ existing between polar and rectangular moments of inertia of an area, we write the following relation between the mass moments of inertia of a thin plate:

$$I_{CC'} = I_{AA'} + I_{BB'} \tag{9.38}$$

Rectangular Plate. In the case of a rectangular plate of sides a and b (Fig. 9.25), we obtain the following mass moments of inertia with respect to axes through the center of gravity of the plate:

$$I_{AA',\text{mass}} = \rho t I_{AA',\text{area}} = \rho t(\tfrac{1}{12}a^3 b)$$
$$I_{BB',\text{mass}} = \rho t I_{BB',\text{area}} = \rho t(\tfrac{1}{12}ab^3)$$

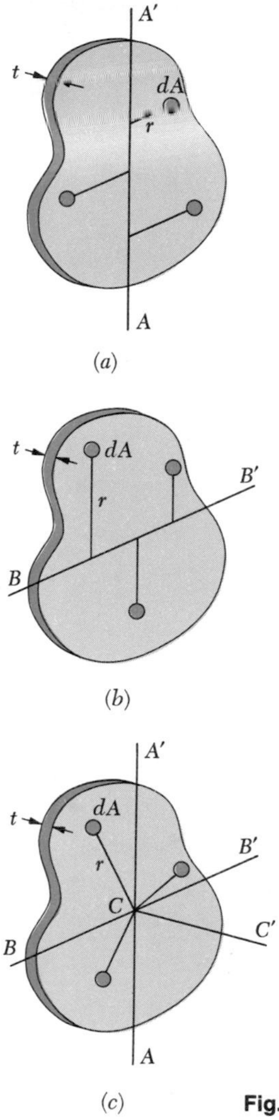

Fig. 9.24

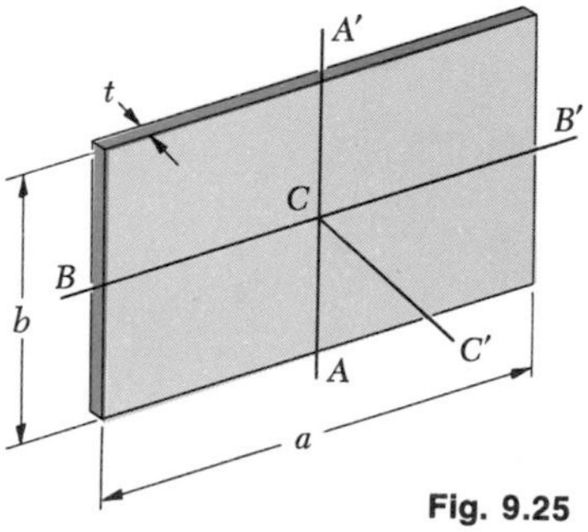

Fig. 9.25

Observing that the product ρabt is equal to the mass m of the plate, we write the mass moments of inertia of a thin rectangular plate as follows:

$$I_{AA'} = \tfrac{1}{12}ma^2 \qquad I_{BB'} = \tfrac{1}{12}mb^2 \tag{9.39}$$
$$I_{CC'} = I_{AA'} + I_{BB'} = \tfrac{1}{12}m(a^2 + b^2) \tag{9.40}$$

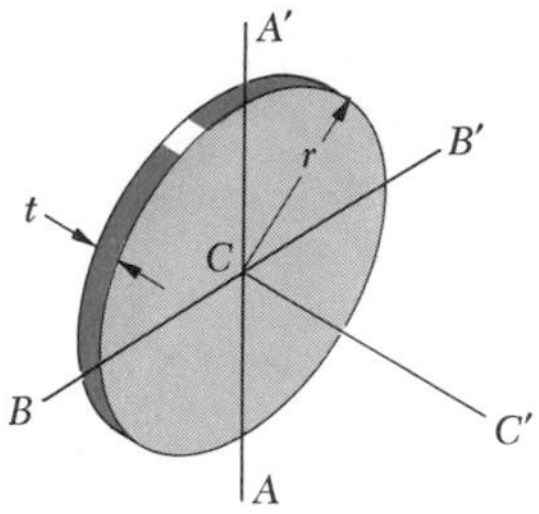

Fig. 9.26

Circular Plate. In the case of a circular plate, or disk, of radius r (Fig. 9.26), we write

$$I_{AA',\text{mass}} = \rho t I_{AA',\text{area}} = \rho t(\tfrac{1}{4}\pi r^4)$$

Observing that the product $\rho\pi r^2 t$ is equal to the mass m of the plate and that $I_{AA'} = I_{BB'}$, we write the mass moments of inertia of a circular plate as follows:

$$I_{AA'} = I_{BB'} = \tfrac{1}{4}mr^2 \tag{9.41}$$
$$I_{CC'} = I_{AA'} + I_{BB'} = \tfrac{1}{2}mr^2 \tag{9.42}$$

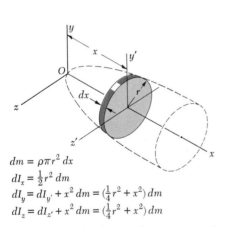

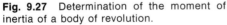

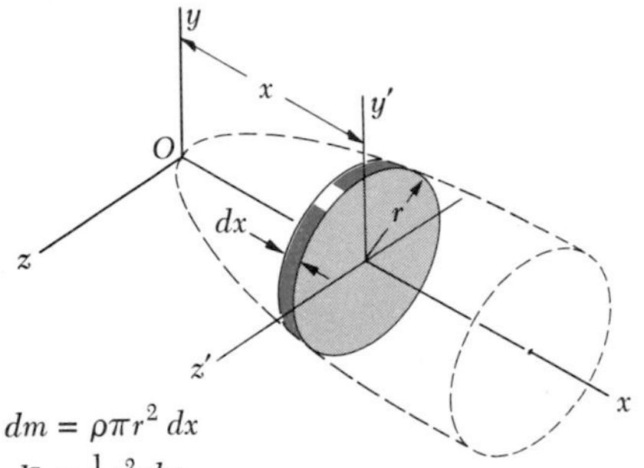

$dm = \rho\pi r^2\, dx$

$dI_x = \frac{1}{2}r^2\, dm$

$dI_y = dI_{y'} + x^2\, dm = (\frac{1}{4}r^2 + x^2)\, dm$

$dI_z = dI_{z'} + x^2\, dm = (\frac{1}{4}r^2 + x^2)\, dm$

Fig. 9.27 Determination of the moment of inertia of a body of revolution.

9.13. Determination of the Moment of Inertia of a Three-dimensional Body by Integration. The moment of inertia of a three-dimensional body is obtained by computing the integral $I = \int r^2\, dm$. If the body is made of a homogeneous material of density ρ, we have $dm = \rho\, dV$ and write $I = \rho\int r^2\, dV$. This integral depends only upon the shape of the body. In order to compute it, it will generally be necessary to perform a triple, or at least a double, integration.

However, if the body possesses two planes of symmetry, it is usually possible to determine its moment of inertia through a single integration by choosing as an element of mass dm the mass of a thin slab perpendicular to the planes of symmetry. In the case of bodies of revolution, for example, the element of mass should be a thin disk (Fig. 9.27). Using formula (9.42), the moment of inertia of the disk with respect to the axis of revolution may be readily expressed as indicated in Fig. 9.27. Its moment of inertia with respect to each of the other two axes of coordinates will be obtained by using formula (9.41) and the parallel-axis theorem. Integration of the expressions obtained will yield the desired moments of inertia of the body of revolution.

9.14. Moments of Inertia of Composite Bodies. The moments of inertia of a few common shapes are shown in Fig. 9.28. The moment of inertia with respect to a given axis of a body made of several of these simple shapes may be obtained by computing the moments of inertia of its component parts about the desired axis and adding them together. We should note, as we already have noted in the case of areas, that the radius of gyration of a composite body *cannot* be obtained by adding the radii of gyration of its component parts.

Slender rod	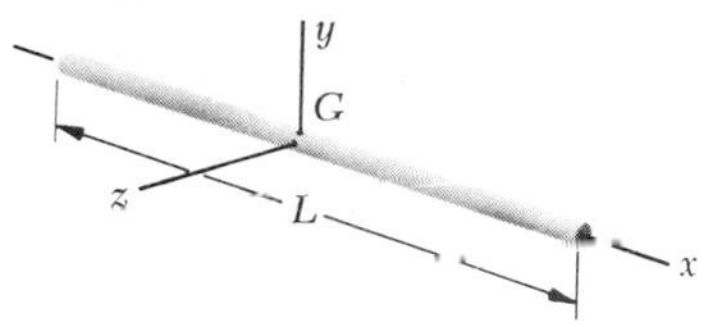	$I_y = I_z = \frac{1}{12}mL^2$
Thin rectangular plate	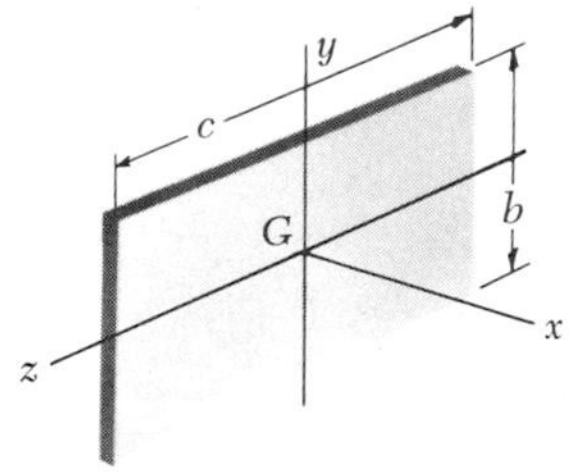	$I_x = \frac{1}{12}m(b^2 + c^2)$ $I_y = \frac{1}{12}mc^2$ $I_z = \frac{1}{12}mb^2$
Rectangular prism	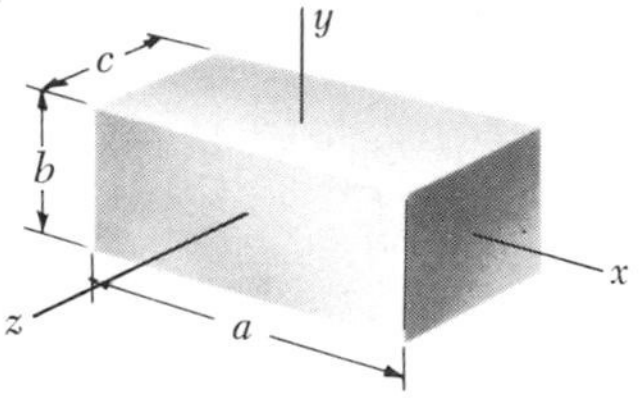	$I_x = \frac{1}{12}m(b^2 + c^2)$ $I_y = \frac{1}{12}m(c^2 + a^2)$ $I_z = \frac{1}{12}m(a^2 + b^2)$
Thin disk	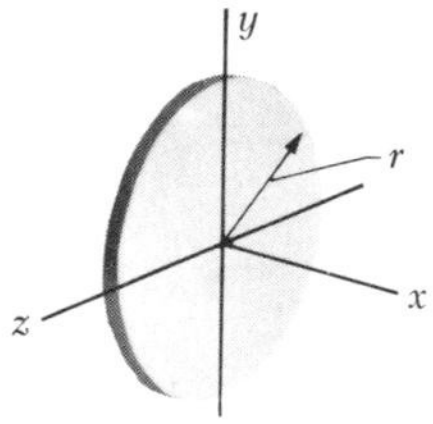	$I_x = \frac{1}{2}mr^2$ $I_y = I_z = \frac{1}{4}mr^2$
Circular cylinder	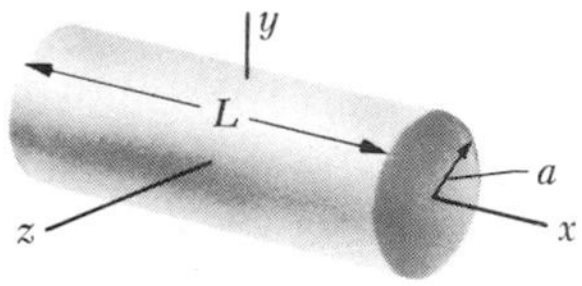	$I_x = \frac{1}{2}ma^2$ $I_y = I_z = \frac{1}{12}m(3a^2 + L^2)$
Circular cone	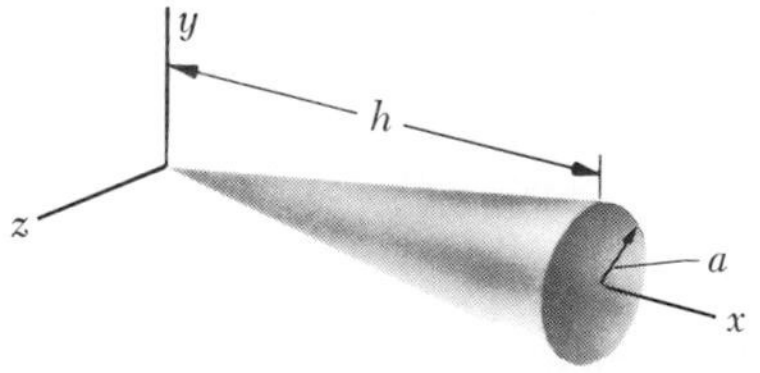	$I_x = \frac{3}{10}ma^2$ $I_y = I_z = \frac{3}{5}m(\frac{1}{4}a^2 + h^2)$
Sphere	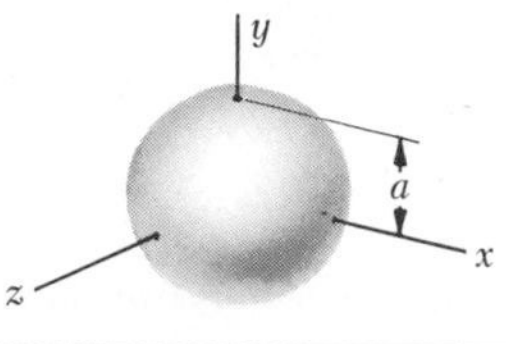	$I_x = I_y = I_z = \frac{2}{5}ma^2$

Fig. 9.28 Mass moments of inertia of common geometric shapes

SAMPLE PROBLEM 9.9

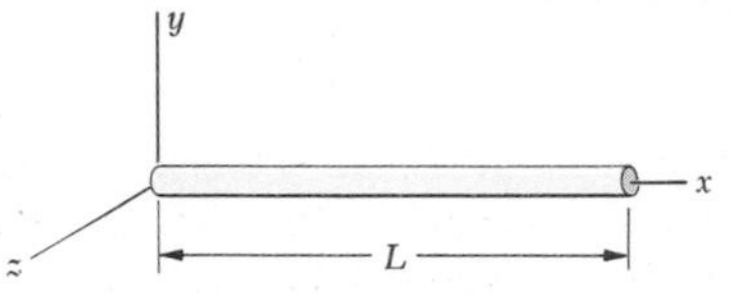

Determine the mass moment of inertia of a slender rod of length L and mass m with respect to an axis perpendicular to the rod and passing through one end of the rod.

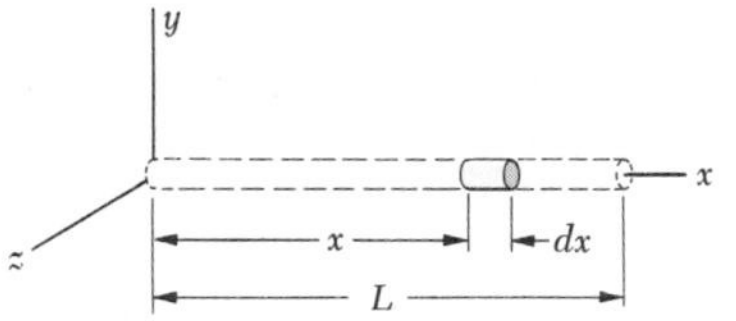

Solution. Choosing the differential element of mass shown, we write

$$dm = \frac{m}{L}\,dx$$

$$I_y = \int x^2\,dm = \int_0^L x^2 \frac{m}{L}\,dx = \left[\frac{m}{L}\frac{x^3}{3}\right]_0^L \qquad I_y = \frac{mL^2}{3} \blacktriangleleft$$

SAMPLE PROBLEM 9.10

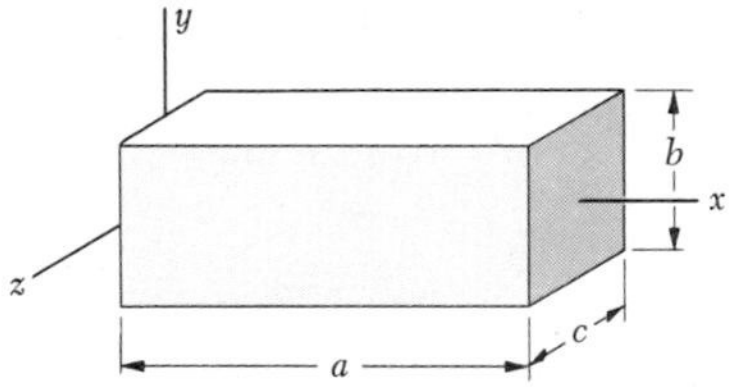

Determine the mass moment of inertia of the homogeneous rectangular prism shown with respect to the z axis.

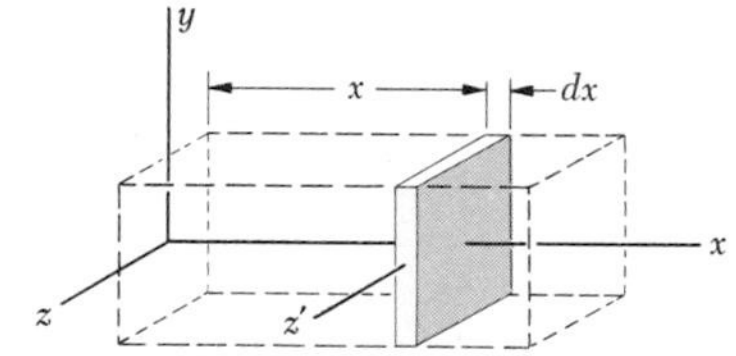

Solution. We choose as a differential element of mass the thin slab shown for which

$$dm = \rho bc\,dx$$

Referring to Sec. 9.12, we find that the moment of inertia of the element with respect to the z' axis is

$$dI_{z'} = \tfrac{1}{12}b^2\,dm$$

Applying the parallel-axis theorem, we obtain the mass moment of inertia of the slab with respect to the z axis.

$$dI_z = dI_{z'} + x^2\,dm = \tfrac{1}{12}b^2\,dm + x^2\,dm = (\tfrac{1}{12}b^2 + x^2)\,\rho bc\,dx$$

Integrating from $x = 0$ to $x = a$, we obtain

$$I_z = \int dI_z = \int_0^a (\tfrac{1}{12}b^2 + x^2)\,\rho bc\,dx = \rho abc\,(\tfrac{1}{12}b^2 + \tfrac{1}{3}a^2)$$

Since the total mass of the prism is $m = \rho abc$, we may write

$$I_z = m(\tfrac{1}{12}b^2 + \tfrac{1}{3}a^2) \qquad I_z = \tfrac{1}{12}m(4a^2 + b^2) \blacktriangleleft$$

We note that if the prism is slender, b is small compared to a and the expression for I_z reduces to $ma^2/3$, which is the result obtained in Sample Prob. 9.9 when $L = a$.

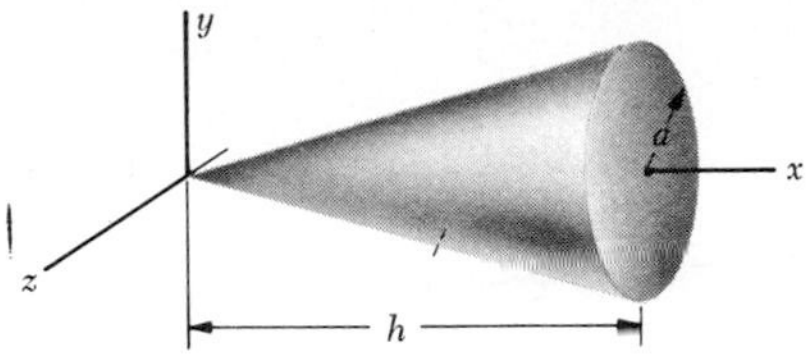

SAMPLE PROBLEM 9.11

Determine the mass moment of inertia of a right circular cone with respect to (*a*) its longitudinal axis, (*b*) an axis through the apex of the cone and perpendicular to its longitudinal axis, (*c*) an axis through the centroid of the cone and perpendicular to its longitudinal axis.

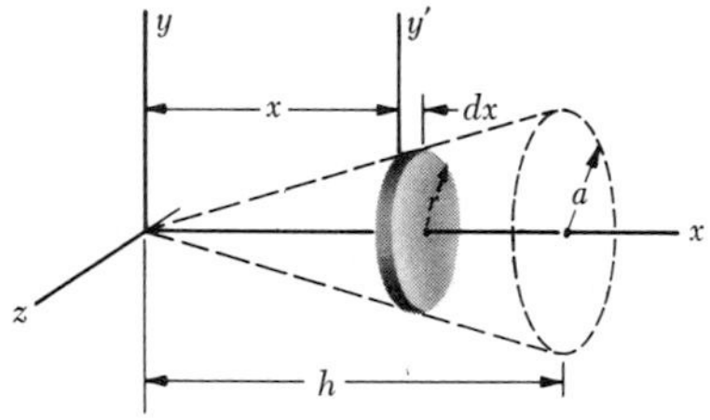

Solution. We choose the differential element of mass shown.

$$r = a\frac{x}{h} \qquad dm = \rho\pi r^2\,dx = \rho\pi\frac{a^2}{h^2}x^2\,dx$$

a. Moment of Inertia I_x. Using the expression derived in Sec. 9.12 for a thin disk, we compute the mass moment of inertia of the differential element with respect to the x axis.

$$dI_x = \tfrac{1}{2}r^2\,dm = \tfrac{1}{2}\left(a\frac{x}{h}\right)^2\left(\rho\pi\frac{a^2}{h^2}x^2\,dx\right) = \tfrac{1}{2}\rho\pi\frac{a^4}{h^4}x^4\,dx$$

Integrating from $x = 0$ to $x = h$, we obtain

$$I_x = \int dI_x = \int_0^h \tfrac{1}{2}\rho\pi\frac{a^4}{h^4}x^4\,dx = \tfrac{1}{2}\rho\pi\frac{a^4}{h^4}\frac{h^5}{5} = \tfrac{1}{10}\rho\pi a^4 h$$

Since the total mass of the cone is $m = \tfrac{1}{3}\rho\pi a^2 h$, we may write

$$I_x = \tfrac{1}{10}\rho\pi a^4 h = \tfrac{3}{10}a^2(\tfrac{1}{3}\rho\pi a^2 h) = \tfrac{3}{10}ma^2 \qquad I_x = \tfrac{3}{10}ma^2 \blacktriangleleft$$

b. Moment of Inertia I_y. The same differential element will be used. Applying the parallel-axis theorem and using the expression derived in Sec. 9.12 for a thin disk, we write

$$dI_y = dI_{y'} + x^2\,dm = \tfrac{1}{4}r^2\,dm + x^2\,dm = (\tfrac{1}{4}r^2 + x^2)\,dm$$

Substituting the expressions for r and dm, we obtain

$$dI_y = \left(\frac{1}{4}\frac{a^2}{h^2}x^2 + x^2\right)\left(\rho\pi\frac{a^2}{h^2}x^2\,dx\right) = \rho\pi\frac{a^2}{h^2}\left(\frac{a^2}{4h^2} + 1\right)x^4\,dx$$

$$I_y = \int dI_y = \int_0^h \rho\pi\frac{a^2}{h^2}\left(\frac{a^2}{4h^2} + 1\right)x^4\,dx = \rho\pi\frac{a^2}{h^2}\left(\frac{a^2}{4h^2} + 1\right)\frac{h^5}{5}$$

Introducing the total mass of the cone m, we rewrite I_y as follows:

$$I_y = \tfrac{3}{5}(\tfrac{1}{4}a^2 + h^2)\tfrac{1}{3}\rho\pi a^2 h \qquad I_y = \tfrac{3}{5}m(\tfrac{1}{4}a^2 + h^2) \blacktriangleleft$$

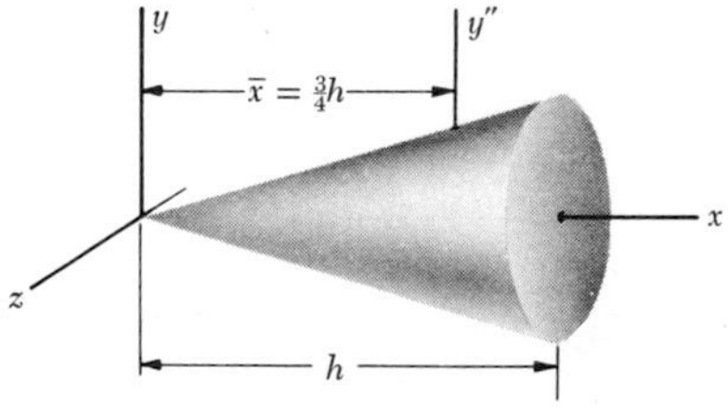

c. Moment of Inertia $\bar{I}_{y''}$. We apply the parallel-axis theorem and write

$$I_y = \bar{I}_{y''} + m\bar{x}^2$$

Solving for $\bar{I}_{y''}$ and recalling that $\bar{x} = \tfrac{3}{4}h$, we have

$$\bar{I}_{y''} = I_y - m\bar{x}^2 = \tfrac{3}{5}m(\tfrac{1}{4}a^2 + h^2) - m(\tfrac{3}{4}h)^2$$

$$\bar{I}_{y''} = \tfrac{3}{20}m(a^2 + \tfrac{1}{4}h^2) \blacktriangleleft$$

SAMPLE PROBLEM 9.12

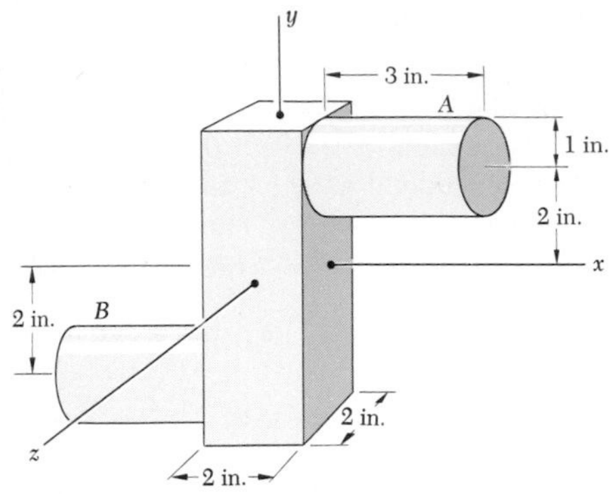

A steel forging consists of a rectangular prism 6 by 2 by 2 in. and of two cylinders of diameter 2 in. and length 3 in., as shown. Determine the mass moments of inertia with respect to the coordinate axes. (Specific weight of steel = 490 lb/ft³.)

Computation of Masses
Prism

$$V = 24 \text{ in}^3 \qquad W = \frac{(24 \text{ in}^3)(490 \text{ lb/ft}^3)}{1728 \text{ in}^3/\text{ft}^3} = 6.81 \text{ lb}$$

$$m = \frac{6.81 \text{ lb}}{32.2 \text{ ft/s}^2} = 0.211 \text{ lb} \cdot \text{s}^2/\text{ft}$$

Each Cylinder

$$V = \pi(1 \text{ in.})^2(3 \text{ in.}) = 9.42 \text{ in}^3$$

$$W = \frac{(9.42 \text{ in}^3)(490 \text{ lb/ft}^3)}{1728 \text{ in}^3/\text{ft}^3} = 2.67 \text{ lb}$$

$$m = \frac{2.67 \text{ lb}}{32.2 \text{ ft/s}^2} = 0.0829 \text{ lb} \cdot \text{s}^2/\text{ft}$$

Mass Moments of Inertia. The mass moments of inertia of each component are computed from Fig. 9.28, using the parallel-axis theorem when necessary. Note that all lengths should be expressed in feet.

Prism

$$I_x = I_z = \tfrac{1}{12}(0.211 \text{ lb} \cdot \text{s}^2/\text{ft})[(\tfrac{6}{12} \text{ ft})^2 + (\tfrac{2}{12} \text{ ft})^2] = 4.88 \times 10^{-3} \text{ lb} \cdot \text{ft} \cdot \text{s}^2$$

$$I_y = \tfrac{1}{12}(0.211 \text{ lb} \cdot \text{s}^2/\text{ft})[(\tfrac{2}{12} \text{ ft})^2 + (\tfrac{2}{12} \text{ ft})^2] = 0.977 \times 10^{-3} \text{ lb} \cdot \text{ft} \cdot \text{s}^2$$

Each cylinder

$$I_x = \tfrac{1}{2}ma^2 + m\bar{y}^2 = \tfrac{1}{2}(0.0829 \text{ lb} \cdot \text{s}^2/\text{ft})(\tfrac{1}{12} \text{ ft})^2 + (0.0829 \text{ lb} \cdot \text{s}^2/\text{ft})(\tfrac{2}{12} \text{ ft})^2 = 2.59 \times 10^{-3} \text{ lb} \cdot \text{ft} \cdot \text{s}^2$$

$$I_y = \tfrac{1}{12}m(3a^2 + L^2) + m\bar{x}^2 = \tfrac{1}{12}(0.0829 \text{ lb} \cdot \text{s}^2/\text{ft})[3(\tfrac{1}{12} \text{ ft})^2 + (\tfrac{3}{12} \text{ ft})^2] + (0.0829 \text{ lb} \cdot \text{s}^2/\text{ft})(\tfrac{2.5}{12} \text{ ft})^2 = 4.17 \times 10^{-3} \text{ lb} \cdot \text{ft} \cdot \text{s}^2$$

$$I_z = \tfrac{1}{12}m(3a^2 + L^2) + m(\bar{x}^2 + \bar{y}^2) = \tfrac{1}{12}(0.0829)[3(\tfrac{1}{12})^2 + (\tfrac{3}{12})^2] + (0.0829)[(\tfrac{2.5}{12})^2 + (\tfrac{2}{12})^2] = 6.48 \times 10^{-3} \text{ lb} \cdot \text{ft} \cdot \text{s}^2$$

Entire Body. Adding the values obtained:

$$I_x = 4.88 \times 10^{-3} + 2(2.59 \times 10^{-3})$$

$$I_x = 10.06 \times 10^{-3} \text{ lb} \cdot \text{ft} \cdot \text{s}^2 \quad ◄$$

$$I_y = 0.977 \times 10^{-3} + 2(4.17 \times 10^{-3})$$

$$I_y = 9.32 \times 10^{-3} \text{ lb} \cdot \text{ft} \cdot \text{s}^2 \quad ◄$$

$$I_z = 4.88 \times 10^{-3} + 2(6.48 \times 10^{-3})$$

$$I_z = 17.84 \times 10^{-3} \text{ lb} \cdot \text{ft} \cdot \text{s}^2 \quad ◄$$

SAMPLE PROBLEM 9.13

Solve Sample Prob. 9.12 using SI units.

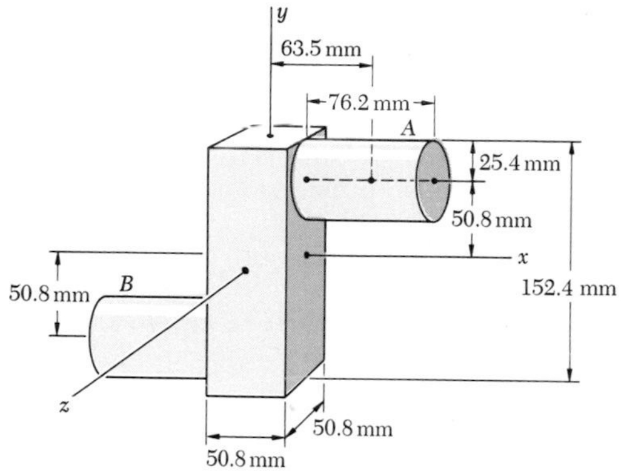

Solution. First, the dimensions are converted into millimeters (1 in. = 25.4 mm). Next, the density of steel ρ (mass per unit volume) is determined in SI units. Recalling that 1 ft = 0.3048 m and that the mass of a block weighing 1 lb is 0.454 kg, we have

$$\rho = (490 \text{ lb/ft}^3)\left(\frac{0.454 \text{ kg}}{1 \text{ lb}}\right)\left(\frac{1 \text{ ft}}{0.3048 \text{ m}}\right)^3 = 7850 \text{ kg/m}^3$$

Computation of Masses

Prism. $V = (50.8 \text{ mm})^2(152.4 \text{ mm}) = 0.393 \times 10^6 \text{ mm}^3$

or, since $1 \text{ mm}^3 = (10^{-3} \text{ m})^3 = 10^{-9} \text{ m}^3$,

$$V = 0.393 \times 10^6 \times 10^{-9} \text{ m}^3 = 0.393 \times 10^{-3} \text{ m}^3$$
$$m = \rho V = (7.85 \times 10^3 \text{ kg/m}^3)(0.393 \times 10^{-3} \text{ m}^3) = 3.09 \text{ kg}$$

Each Cylinder

$$V = \pi r^2 h = \pi(25.4 \text{ mm})^2(76.2 \text{ mm}) = 0.1544 \times 10^6 \text{ mm}^3 = 0.1544 \times 10^{-3} \text{ m}^3$$
$$m = \rho V = (7.85 \times 10^3 \text{ kg/m}^3)(0.1544 \times 10^{-3} \text{ m}^3) = 1.212 \text{ kg}$$

Mass Moments of Inertia. The mass moments of inertia of each component are computed from Fig. 9.28, using the parallel-axis theorem when necessary. Note that all lengths should be expressed in millimeters.

Prism

$$I_x = I_z = \tfrac{1}{12}(3.09 \text{ kg})[(152.4 \text{ mm})^2 + (50.8 \text{ mm})^2] = 6640 \text{ kg}\cdot\text{mm}^2$$
$$I_y = \tfrac{1}{12}(3.09 \text{ kg})[(50.8 \text{ mm})^2 + (50.8 \text{ mm})^2] = 1329 \text{ kg}\cdot\text{mm}^2$$

Each Cylinder

$$I_x = \tfrac{1}{2}ma^2 + m\bar{y}^2 = \tfrac{1}{2}(1.212 \text{ kg})(25.4 \text{ mm})^2 + (1.212 \text{ kg})(50.8 \text{ mm})^2 = 3520 \text{ kg}\cdot\text{mm}^2$$
$$I_y = \tfrac{1}{12}m(3a^2 + L^2) + m\bar{x}^2 = \tfrac{1}{12}(1.212 \text{ kg})[3(25.4 \text{ mm})^2 + (76.2 \text{ mm})^2] + (1.212 \text{ kg})(63.5 \text{ mm})^2 \doteq 5670 \text{ kg}\cdot\text{mm}^2$$
$$I_z = \tfrac{1}{12}m(3a^2 + L^2) + m(\bar{x}^2 + \bar{y}^2) = \tfrac{1}{12}(1.212 \text{ kg})[3(25.4 \text{ mm})^2 + (76.2 \text{ mm})^2] + (1.212 \text{ kg})[(63.5 \text{ mm})^2 + (50.8 \text{ mm})^2] = 8800 \text{ kg}\cdot\text{mm}^2$$

Entire Body. Adding the values obtained, and observing that $1 \text{ mm}^2 = (10^{-3} \text{ m})^2 = 10^{-6} \text{ m}^2$, we have

$$I_x = 6640 \text{ kg}\cdot\text{mm}^2 + 2(3520 \text{ kg}\cdot\text{mm}^2) = 13.68 \times 10^3 \text{ kg}\cdot\text{mm}^2$$
$$I_x = 13.68 \times 10^{-3} \text{ kg}\cdot\text{m}^2 \blacktriangleleft$$
$$I_y = 1329 \text{ kg}\cdot\text{mm}^2 + 2(5670 \text{ kg}\cdot\text{mm}^2) = 12.67 \times 10^3 \text{ kg}\cdot\text{mm}^2$$
$$I_y = 12.67 \times 10^{-3} \text{ kg}\cdot\text{m}^2 \blacktriangleleft$$
$$I_z = 6640 \text{ kg}\cdot\text{mm}^2 + 2(8800 \text{ kg}\cdot\text{mm}^2) = 24.2 \times 10^3 \text{ kg}\cdot\text{mm}^2$$
$$I_z = 24.2 \times 10^{-3} \text{ kg}\cdot\text{m}^2 \blacktriangleleft$$

Recalling that $1 \text{ lb}\cdot\text{ft}\cdot\text{s}^2 = 1.356 \text{ kg}\cdot\text{m}^2$ (see footnote, page 383) we may check these answers against the values obtained in Sample Prob. 9.12.

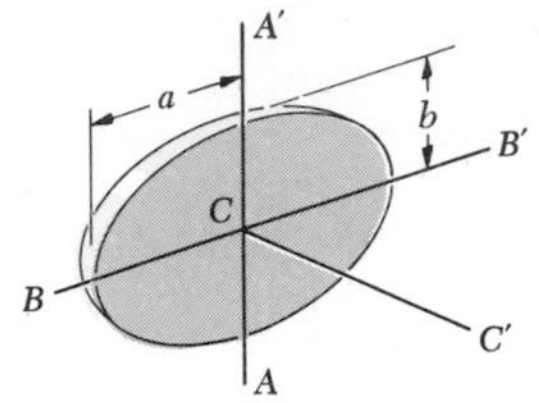

Fig. P9.72

PROBLEMS

9.72 Determine the mass moment of inertia of a thin elliptical plate of mass m with respect to (a) the axes AA' and BB' of the ellipse, (b) the axis CC' perpendicular to the plate.

9.73 Determine the mass moment of inertia of a ring of mass m, cut from a thin uniform plate, with respect to (a) the diameter AA' of the ring, (b) the axis CC' perpendicular to the plane of the ring.

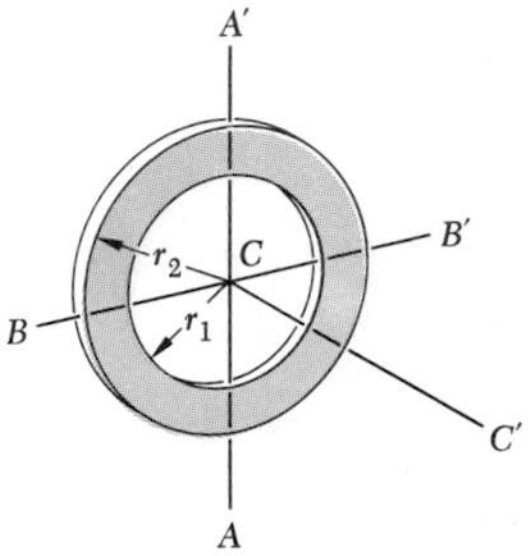

Fig. P9.73

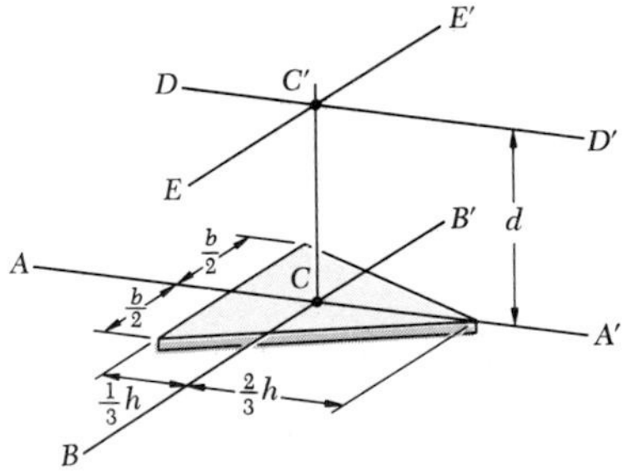

Fig. P9.74

9.74 A thin plate of mass m is cut in the shape of an isosceles triangle of base b and height h. Determine the mass moment of inertia of the plate with respect to (a) the centroidal axes AA' and BB' in the plane of the plate, (b) the centroidal axis CC' perpendicular to the plate.

9.75 Determine the mass moments of inertia of the plate of Prob. 9.74 with respect to the axes DD' and EE' parallel to the centroidal axes AA' and BB' respectively.

9.76 Determine by direct integration the mass moment of inertia with respect to the y axis of the right circular cylinder shown, assuming a uniform density and a mass m.

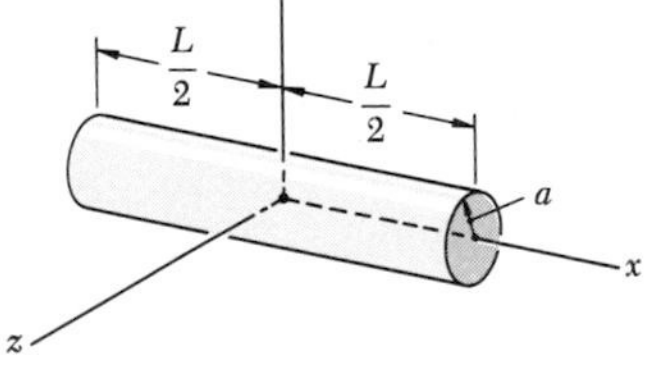

Fig. P9.76

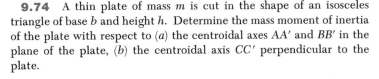

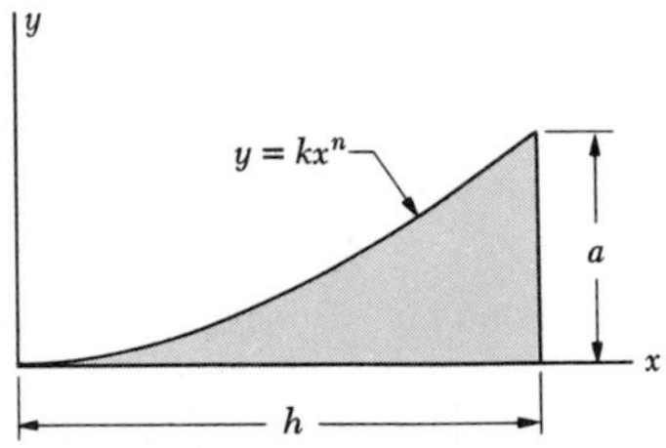

Fig. P9.77

9.77 The area shown is revolved about the x axis to form a homogeneous solid of revolution of mass m. Express the mass moment of inertia of the solid with respect to the x axis in terms of m, a, and n. The expression obtained may be used to verify (a) the value given in Fig. 9.28 for a cone (with $n = 1$), (b) the answer to Prob. 9.78 (with $n = \frac{1}{2}$), (c) the answer to Prob. 9.80 (with $n = 2$).

9.78 Determine by direct integration the mass moment of inertia and the radius of gyration with respect to the x axis of the paraboloid shown, assuming a uniform density and a mass m.

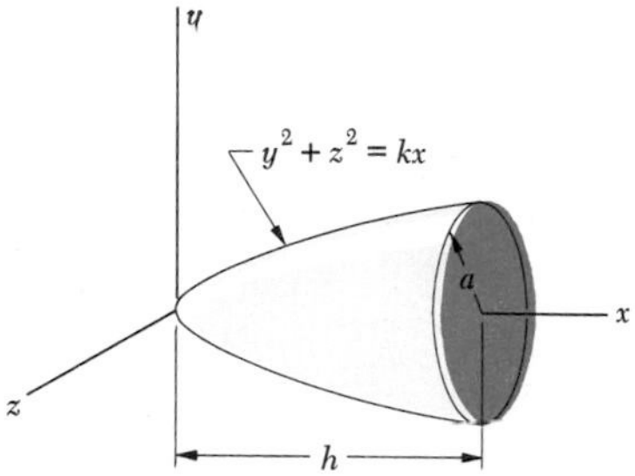

Fig. P9.78 and P9.79

9.79 Determine by direct integration the mass moment of inertia and the radius of gyration with respect to the y axis of the paraboloid shown, assuming a uniform density and a mass m.

9.80 The homogeneous solid shown was obtained by rotating the area of Prob. 9.77, with $n = 2$, through 360° about the x axis. Determine the mass moment of inertia $\bar{I}_x$ in terms of m and a.

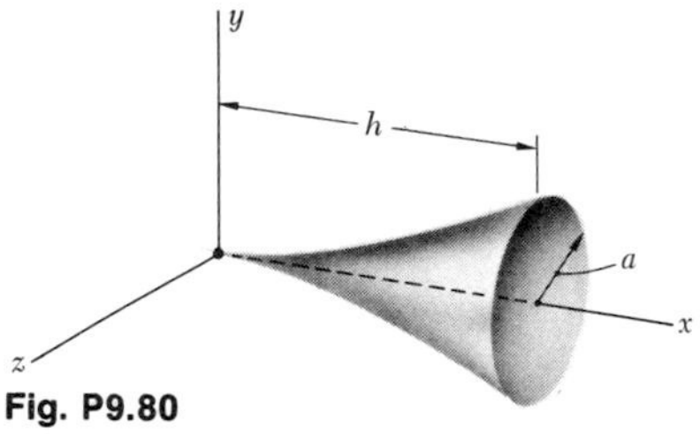

Fig. P9.80

9.81 Determine in terms of m and a the mass moment of inertia and the radius of gyration of the homogeneous solid of Prob. 9.80 with respect to the y axis.

9.82 Determine by direct integration the mass moment of inertia with respect to the x axis of the pyramid shown, assuming a uniform density and a mass m.

9.83 Determine by direct integration the mass moment of inertia with respect to the y axis of the pyramid shown, assuming a uniform density and a mass m.

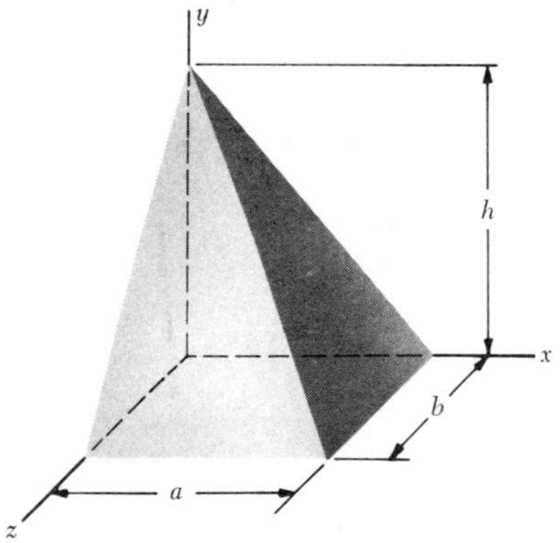

Fig. P9.82 and P9.83

9.84 Knowing that the thin hemispherical shell shown is of mass m and thickness t, determine the mass moment of inertia of the shell with respect to the x axis. (*Hint.* Consider the shell as formed by removing a hemisphere of radius r from a hemisphere of radius $r + t$; then neglect the terms containing t^2 and t^3, and keep those terms containing t.)

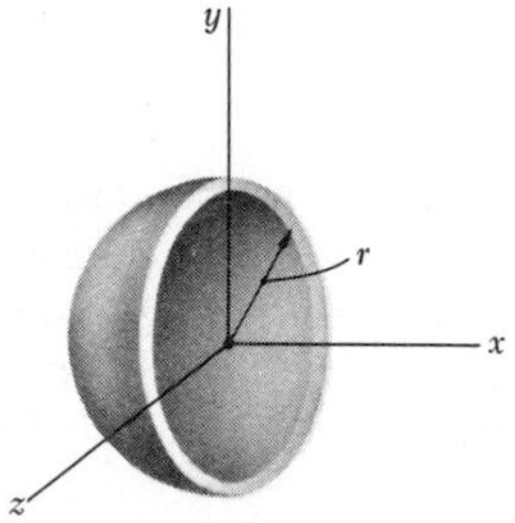

Fig. P9.84

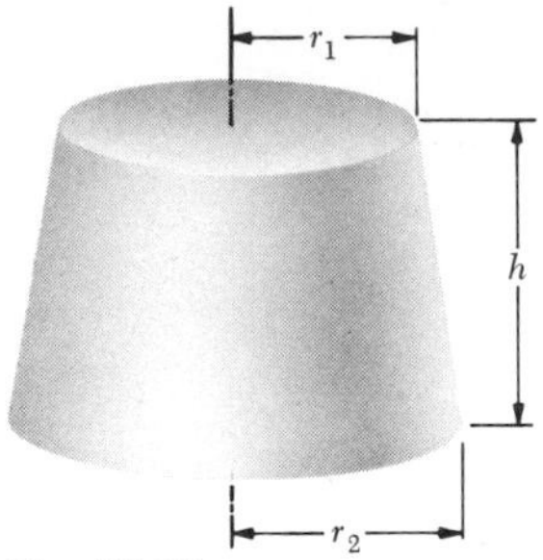

Fig. P9.85

9.85 Determine the mass moment of inertia of the frustum of a right circular cone of mass m with respect to its axis of symmetry.

9.86 Determine the mass moment of inertia and the radius of gyration of the steel flywheel shown with respect to the axis of rotation. The web of the flywheel consists of a solid plate 25 mm thick. (Density of steel = 7850 kg/m^3.)

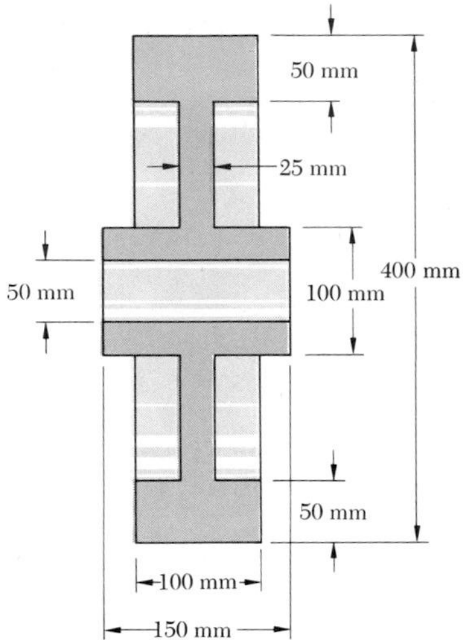

Fig. P9.86

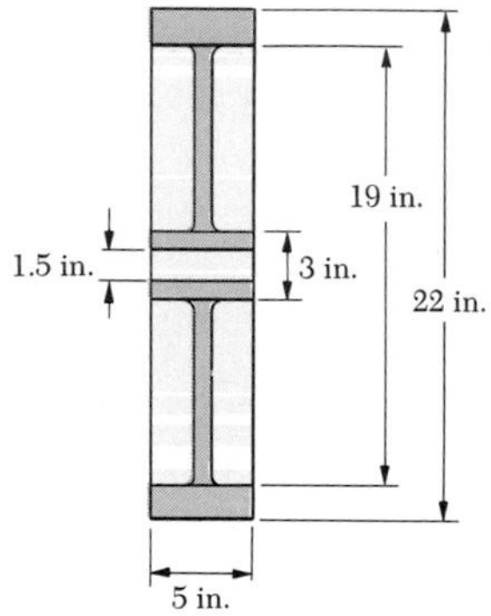

Fig. P9.87

9.87 The cross section of a small flywheel is shown. The rim and hub are connected by eight spokes (two of which are shown in the cross section). Each spoke has a cross-sectional area of 0.400 in^2. Determine the mass moment of inertia and radius of gyration of the flywheel with respect to the axis of rotation. (Specific weight of steel = 490 lb/ft^3.)

9.88 Three slender homogeneous rods are welded together as shown. Denoting the mass of each rod by m, determine the mass moment of inertia and the radius of gyration of the assembly with respect to (*a*) the x axis, (*b*) the y axis, (*c*) the z axis.

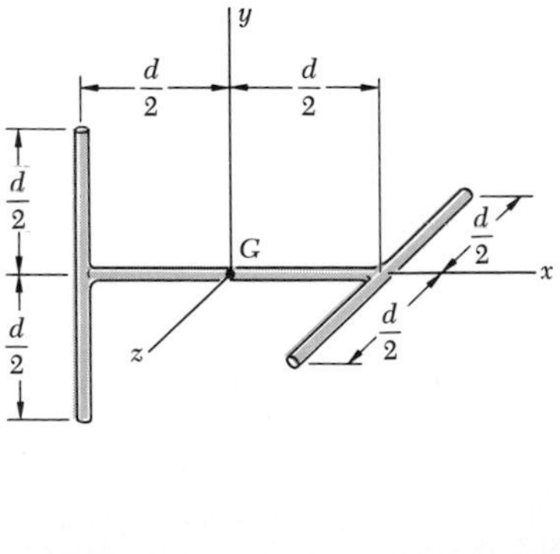

Fig. P9.88

9.89 In using the parallel-axis theorem, the error introduced by neglecting the centroidal moment of inertia is sometimes small. For a homogeneous sphere of radius a and mass m, (a) determine the mass moment of inertia with respect to an axis AA' at a distance R from the center of the sphere, (b) express as a function of a/R the relative error introduced by neglecting the centroidal moment of inertia, (c) determine the distance R in terms of a for which the relative error is 0.4 percent.

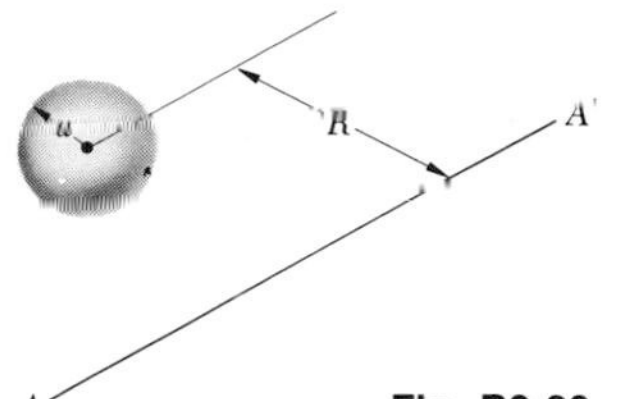

Fig. P9.89

9.90 A section of sheet steel, 2 mm thick, is cut and bent into the machine component shown. Knowing that the density of steel is 7850 kg/m³, determine the mass moment of inertia of the component with respect to (a) the x axis, (b) the y axis, (c) the z axis.

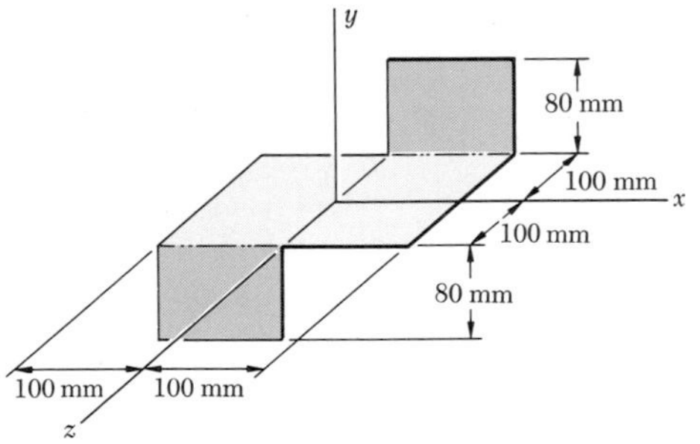

Fig. P9.90

9.91 Twelve uniform slender rods, each of length l, are welded together to form the cubical figure shown. Denoting by m the total mass of the twelve rods, determine the mass moment of inertia of the figure about the x axis.

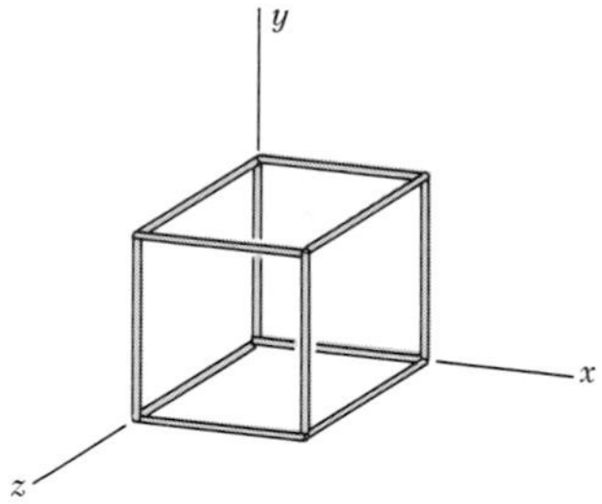

Fig. P9.91

9.92 and 9.93 Determine the mass moment of inertia and the radius of gyration of the steel machine element shown with respect to the x axis. (Specific weight of steel = 490 lb/ft³; density of steel = 7850 kg/m³.)

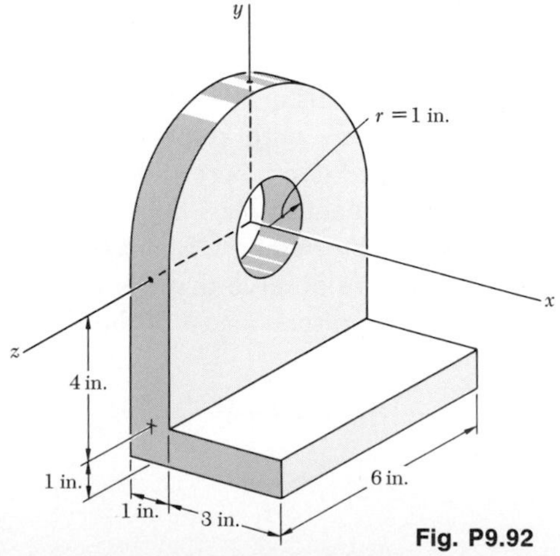

Fig. P9.92

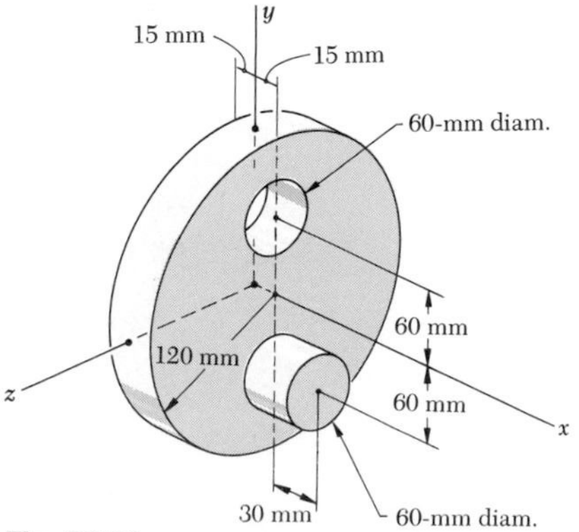

Fig. P9.93

9.94 A homogeneous wire, of weight 2 lb/ft, is used to form the figure shown. Determine the mass moment of inertia of the wire figure with respect to (*a*) the *x* axis, (*b*) the *y* axis, (*c*) the *z* axis.

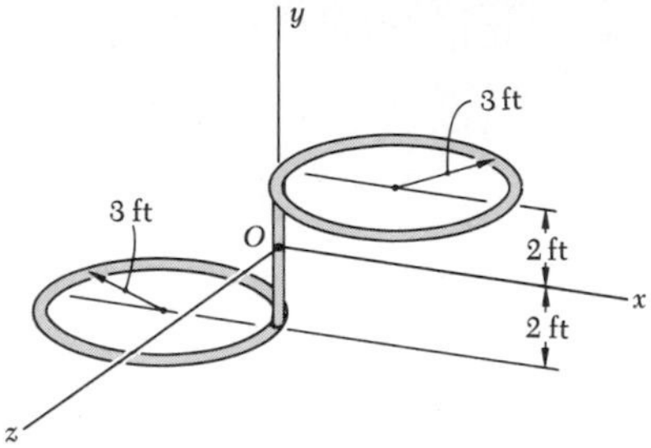

Fig. P9.94

9.95 Two holes, each of diameter 50 mm, are drilled through the steel block shown. Determine the mass moment of inertia of the body with respect to the axis of either of the holes. (Density of steel = 7850 kg/m³.)

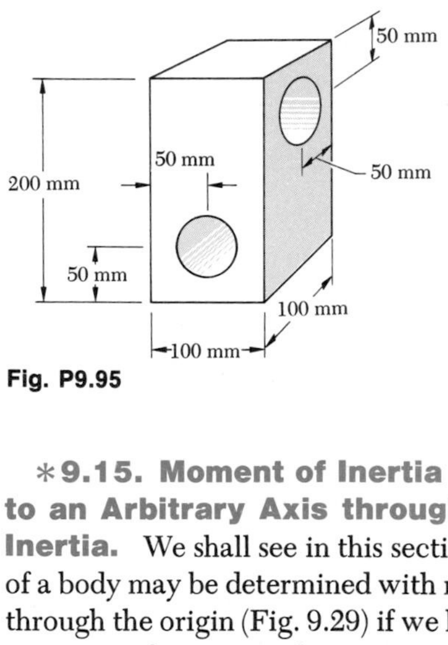

Fig. P9.95

***9.15. Moment of Inertia of a Body with Respect to an Arbitrary Axis through *O*. Mass Products of Inertia.** We shall see in this section how the moment of inertia of a body may be determined with respect to an arbitrary axis *OL* through the origin (Fig. 9.29) if we have computed beforehand its moments of inertia with respect to the three coordinate axes, as well as certain other quantities to be defined below.

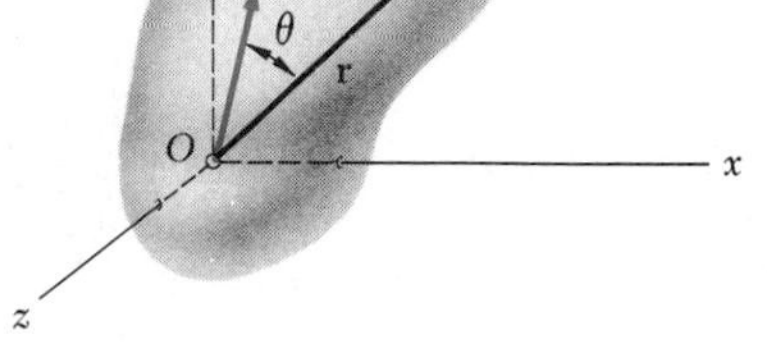

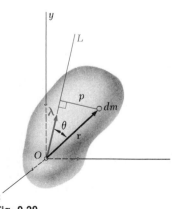

Fig. 9.29

The moment of inertia of the body with respect to *OL* is represented by the integral $I_{OL} = \int p^2\,dm$, where p denotes the perpendicular distance from the element of mass dm to the axis *OL*. But, denoting by $\boldsymbol{\lambda}$ the unit vector along *OL* and by $\mathbf{r}$ the position vector of the element dm, we observe that the perpendicular distance p is equal to the magnitude $r \sin\theta$ of the vector product $\boldsymbol{\lambda} \times \mathbf{r}$. We write therefore

$$I_{OL} = \int p^2\,dm = \int (\boldsymbol{\lambda} \times \mathbf{r})^2\,dm \qquad (9.43)$$

Expressing the square of the vector product in terms of its rectangular components, we have

$$I_{OL} = \int[(\lambda_x y - \lambda_y x)^2 + (\lambda_y z - \lambda_z y)^2 + (\lambda_z x - \lambda_x z)^2]\, dm$$

where the components λ_x, λ_y, λ_z of the unit vector $\boldsymbol{\lambda}$ represent the direction cosines of the axis OL, and the components x, y, z of $\mathbf{r}$ represent the coordinates of the element of mass dm. Expanding the squares in the expression obtained and rearranging the terms, we write

$$\begin{aligned} I_{OL} = {} & \lambda_x^2 \int (y^2 + z^2)\, dm + \lambda_y^2 \int (z^2 + x^2)\, dm + \lambda_z^2 \int (x^2 + y^2)\, dm \\ & - 2\lambda_x\lambda_y \int xy\, dm - 2\lambda_y\lambda_z \int yz\, dm - 2\lambda_z\lambda_x \int zx\, dm \end{aligned} \quad (9.44)$$

Referring to Eqs. (9.30), we note that the first three integrals in (9.44) represent, respectively, the moments of inertia I_x, I_y, and I_z of the body with respect to the coordinate axes. The last three integrals in (9.44), which involve products of coordinates, are called the *products of inertia* of the body with respect to the x and y axes, the y and z axes, and the z and x axes, respectively. We write

$$P_{xy} = \int xy\, dm \qquad P_{yz} = \int yz\, dm \qquad P_{zx} = \int zx\, dm \quad (9.45)$$

Substituting for the various integrals from (9.30) and (9.45) into (9.44), we have

$$I_{OL} = I_x\lambda_x^2 + I_y\lambda_y^2 + I_z\lambda_z^2 - 2P_{xy}\lambda_x\lambda_y - 2P_{yz}\lambda_y\lambda_z - 2P_{zx}\lambda_z\lambda_x \quad (9.46)$$

We note that the definition of the products of inertia of a mass given in Eqs. (9.45) is an extension of the definition of the product of inertia of an area (Sec. 9.7). Mass products of inertia reduce to zero under the same conditions of symmetry as products of inertia of areas, and the parallel-axis theorem for mass products of inertia is expressed by relations similar to the formula derived for the product of inertia of an area. Substituting for x, y, z from Eqs. (9.31) into Eqs. (9.45), we verify that

$$\begin{aligned} P_{xy} &= \overline{P}_{x'y'} + m\bar{x}\bar{y} \\ P_{yz} &= \overline{P}_{y'z'} + m\bar{y}\bar{z} \\ P_{zx} &= \overline{P}_{z'x'} + m\bar{z}\bar{x} \end{aligned} \quad (9.47)$$

where $\bar{x}$, $\bar{y}$, $\bar{z}$ are the coordinates of the center of gravity G of the body, and $\overline{P}_{x'y'}$, $\overline{P}_{y'z'}$, $\overline{P}_{z'x'}$ denote the products of inertia with respect to the centroidal axes x', y', z' (Fig. 9.22).

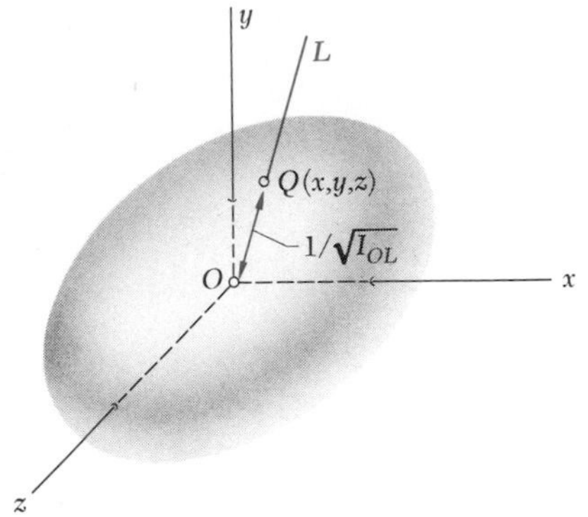

Fig. 9.30

***9.16. Ellipsoid of Inertia. Principal Axes of Inertia.** Let us assume that the moment of inertia of the body considered in the preceding section has been determined with respect to a large number of axes OL through the fixed point O, and that a point Q has been plotted on each axis OL at a distance $OQ = 1/\sqrt{I_{OL}}$ from O. The locus of the points Q thus obtained forms a surface (Fig. 9.30). The equation of that surface may be obtained by substituting $1/(OQ)^2$ for I_{OL} in (9.46) and multiplying both sides of the equation by $(OQ)^2$. Observing that

$$(OQ)\lambda_x = x \qquad (OQ)\lambda_y = y \qquad (OQ)\lambda_z = z$$

where x, y, z denote the rectangular coordinates of a point Q of the surface, we write

$$I_x x^2 + I_y y^2 + I_z z^2 - 2P_{xy}xy - 2P_{yz}yz - 2P_{zx}zx = 1 \quad (9.48)$$

The equation obtained is that of a *quadric*. Since the moment of inertia I_{OL} is different from zero for every axis OL, no point Q may be at an infinite distance from O. Thus, the quadric obtained is an *ellipsoid*. This ellipsoid, which defines the moment of inertia of the body with respect to any axis through O, is known as the *ellipsoid of inertia* of the body at O.

We observe that, if the axes in Fig. 9.30 are rotated, the coefficients of the equation defining the ellipsoid change, since these are equal to the moments and products of inertia of the body with respect to the rotated coordinate axes. However, the *ellipsoid itself remains unaffected*, since its shape depends only upon the distribution of mass in the body considered. Suppose now that we choose as coordinate axes the principal axes x', y', z' of the ellipsoid of inertia (Fig. 9.31). The equation of the ellipsoid with respect to these coordinate axes will be of the form

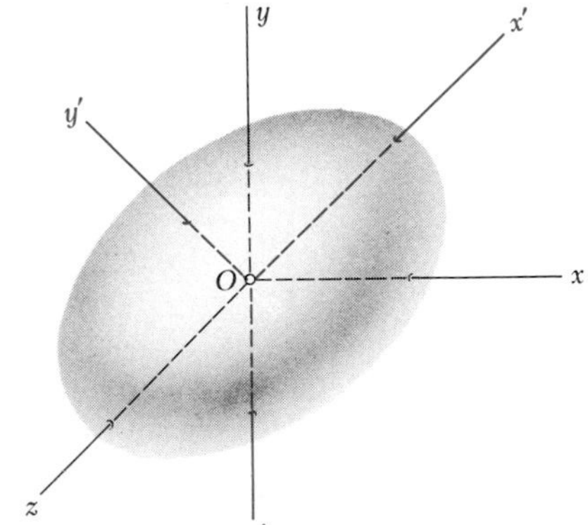

Fig. 9.31

$$I_{x'}x'^2 + I_{y'}y'^2 + I_{z'}z'^2 = 1 \quad (9.49)$$

which does not contain any product of coordinates. Thus, the products of inertia of the body with respect to the x', y', z' axes are zero. The x', y', z' axes are known as the *principal axes of inertia* of the body at O, and the coefficients $I_{x'}$, $I_{y'}$, $I_{z'}$ as the *principal moments of inertia* of the body at O. Note that, given a body of arbitrary shape and a point O, it is always possible to find axes which are the principal axes of inertia of the body at O, i.e., axes with respect to which the products of inertia of the body are zero. Indeed, no matter how odd or irregular the shape of the body may be, the moments of inertia of the body with respect to axes through O will define an ellipsoid, and this ellipsoid will have principal axes which, by definition, are the principal axes of the body at O.

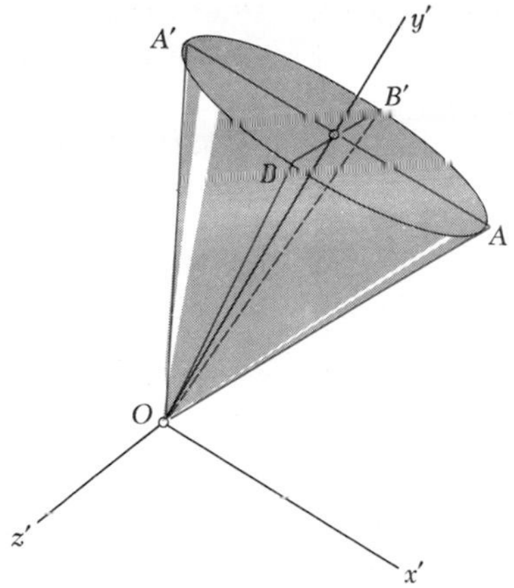

Fig. 9.32

If the principal axes of inertia x', y', z' are used as coordinate axes, the expression obtained in Eq. (9.46) for the moment of inertia of a body with respect to an arbitrary axis through O reduces to

$$I_{OL} = I_{x'}\lambda_{x'}^2 + I_{y'}\lambda_{y'}^2 + I_{z'}\lambda_{z'}^2 \qquad (9.50)$$

While the determination of the principal axes of inertia of a body of arbitrary shape is somewhat involved and requires solving a cubic equation,† there are many cases when these axes may be spotted immediately. Consider, for instance, the homogeneous cone of elliptical base shown in Fig. 9.32; this cone possesses two mutually perpendicular planes of symmetry OAA' and OBB'. We check from the definition (9.45) that, if the $x'y'$ and $y'z'$ planes are chosen to coincide with the two planes of symmetry, all the products of inertia are zero. The x', y', and z' axes thus selected are therefore the principal axes of inertia of the cone at O. In the case of the homogeneous regular tetrahedron $OABC$ shown in Fig. 9.33, the line joining the corner O to the center D of the opposite face is a principal axis of inertia at O and any line through O perpendicular to OD is also a principal axis of inertia at O. This property may be recognized if we observe that a rotation through 120° about OD leaves the shape and the mass distribution of the tetrahedron unchanged. It follows that the ellipsoid of inertia at O also remains unchanged under this rotation. The ellipsoid, therefore, is of revolution about OD, and the line OD, as well as any perpendicular line through O, must be a principal axis of the ellipsoid.

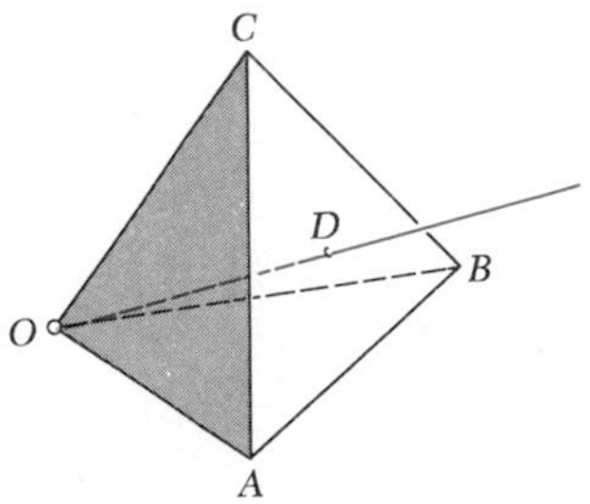

Fig. 9.33

† Cf. Synge and Griffith, *Principles of Mechanics*, McGraw-Hill Book Company, sec. 11.3.

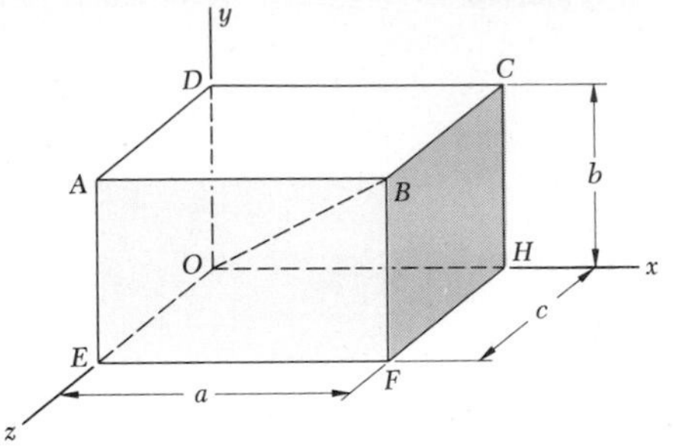

SAMPLE PROBLEM 9.14

Consider a rectangular prism of mass m and sides a, b, c. Determine (*a*) the mass moments and products of inertia of the prism with respect to the coordinate axes shown, (*b*) its moment of inertia with respect to the diagonal OB.

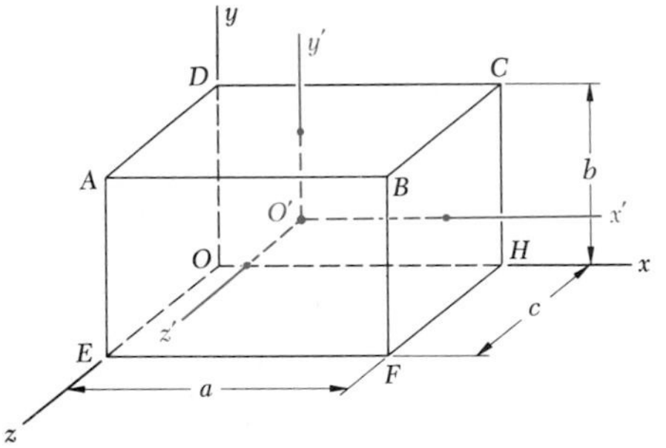

a.* Moments and Products of Inertia with Respect to the Coordinate Axes.** ***Moments of Inertia. Introducing the centroidal axes x', y', z', with respect to which the moments of inertia are given in Fig. 9.28, we apply the parallel-axis theorem:

$$I_x = \bar{I}_{x'} + m(\bar{y}^2 + \bar{z}^2) = \tfrac{1}{12}m(b^2 + c^2) + m(\tfrac{1}{4}b^2 + \tfrac{1}{4}c^2)$$

$$I_x = \tfrac{1}{3}m(b^2 + c^2) \quad ◀$$

Similarly: $\qquad I_y = \tfrac{1}{3}m(c^2 + a^2) \qquad I_z = \tfrac{1}{3}m(a^2 + b^2) \quad ◀$

Products of Inertia. Because of symmetry, the products of inertia with respect to the centroidal axes x', y', z' are zero and these axes are principal axes of inertia. Using the parallel-axis theorem, we have

$$P_{xy} = \bar{P}_{x'y'} + m\bar{x}\,\bar{y} = 0 + m(\tfrac{1}{2}a)(\tfrac{1}{2}b) \qquad P_{xy} = \tfrac{1}{4}mab \quad ◀$$

Similarly: $\qquad P_{yz} = \tfrac{1}{4}mbc \qquad P_{zx} = \tfrac{1}{4}mca \quad ◀$

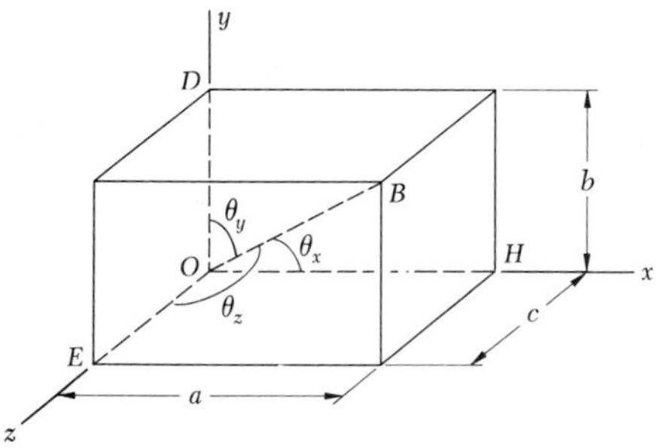

***b.* Moment of Inertia with Respect to *OB*.** We recall Eq. (9.46):

$$I_{OB} = I_x\lambda_x^2 + I_y\lambda_y^2 + I_z\lambda_z^2 - 2P_{xy}\lambda_x\lambda_y - 2P_{yz}\lambda_y\lambda_z - 2P_{zx}\lambda_z\lambda_x$$

where the direction cosines of OB are

$$\lambda_x = \cos\theta_x = (OH)/(OB) = a/(a^2 + b^2 + c^2)^{1/2}$$

$$\lambda_y = b/(a^2 + b^2 + c^2)^{1/2} \qquad \lambda_z = c/(a^2 + b^2 + c^2)^{1/2}$$

Substituting the values obtained for the moments and products of inertia and for the direction cosines:

$$I_{OB} = \frac{1}{a^2 + b^2 + c^2}[\tfrac{1}{3}m(b^2 + c^2)a^2 + \tfrac{1}{3}m(c^2 + a^2)b^2 + \tfrac{1}{3}m(a^2 + b^2)c^2$$

$$- \tfrac{1}{2}ma^2b^2 - \tfrac{1}{2}mb^2c^2 - \tfrac{1}{2}mc^2a^2]$$

$$I_{OB} = \frac{m}{6}\,\frac{a^2b^2 + b^2c^2 + c^2a^2}{a^2 + b^2 + c^2} \quad ◀$$

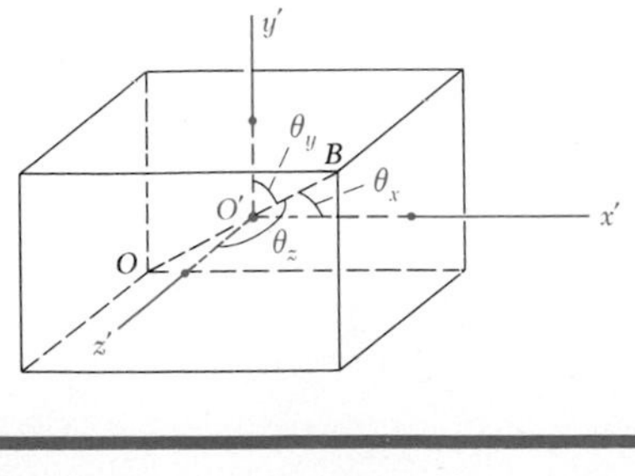

Alternate Solution. The moment of inertia I_{OB} may be obtained directly from the principal moments of inertia $\bar{I}_{x'}$, $\bar{I}_{y'}$, $\bar{I}_{z'}$, since the line OB passes through the centroid O'. Since the x', y', z' axes are principal axes of inertia, we use Eq. (9.50) and write

$$I_{OB} = \bar{I}_{x'}\lambda_x^2 + \bar{I}_{y'}\lambda_y^2 + \bar{I}_{z'}\lambda_z^2$$

$$= \frac{1}{a^2 + b^2 + c^2}\left[\frac{m}{12}(b^2 + c^2)a^2 + \frac{m}{12}(c^2 + a^2)b^2 + \frac{m}{12}(a^2 + b^2)c^2\right]$$

$$I_{OB} = \frac{m}{6}\,\frac{a^2b^2 + b^2c^2 + c^2a^2}{a^2 + b^2 + c^2} \quad ◀$$

PROBLEMS

9.96 and 9.97 Determine the mass products of inertia P_{xy}, P_{yz}, and P_{zx} of the steel machine element shown. (Specific weight of steel = 490 lb/ft^3; density of steel = 7850 kg/m^3.)

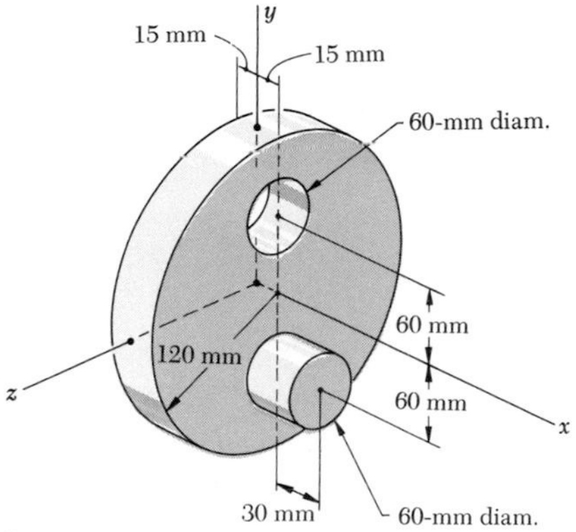

Fig. P9.96

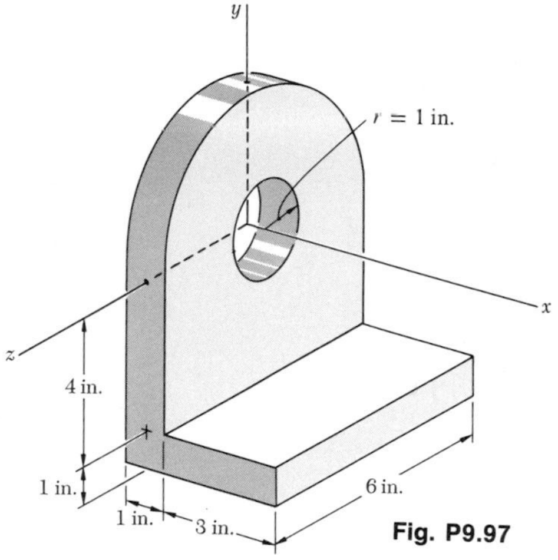

Fig. P9.97

9.98 A homogeneous wire, of weight 2 lb/ft, is used to form the figure shown. Determine the mass products of inertia P_{xy}, P_{yz}, and P_{zx} of the wire figure.

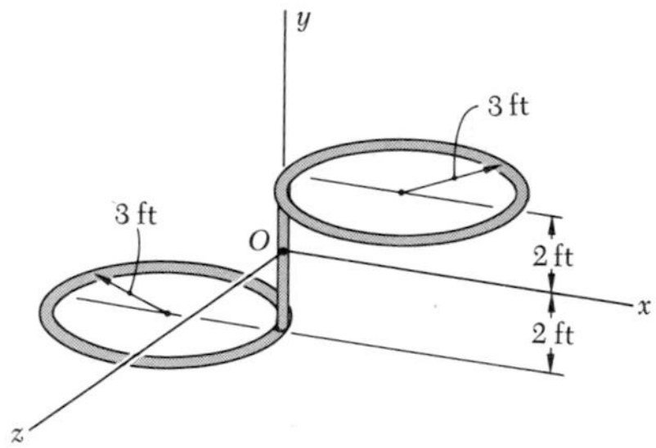

Fig. P9.98

9.99 A section of sheet steel, 2 mm thick, is cut and bent into the machine component shown. Knowing that the density of steel is 7850 kg/m^3, determine the mass products of inertia P_{xy}, P_{yz}, and P_{zx} of the component.

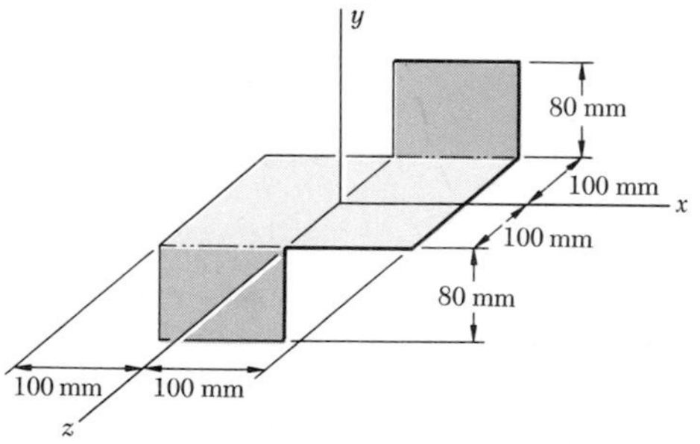

Fig. P9.99

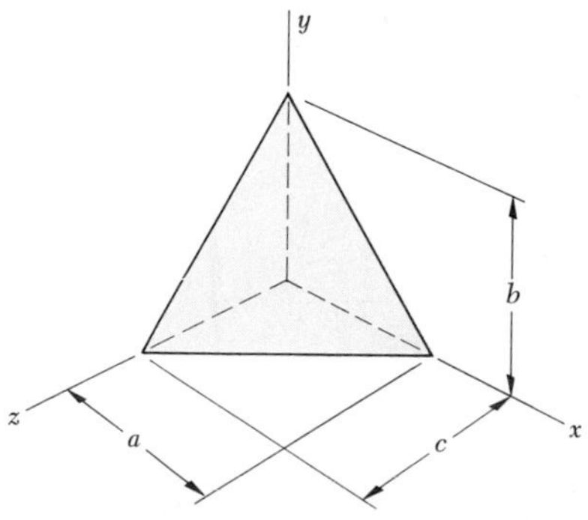

Fig. P9.100

9.100 For the homogeneous tetrahedron of mass m which is shown, (*a*) determine by direct integration the mass product of inertia P_{zx}, (*b*) deduce P_{yz} and P_{xy} from the result obtained in part *a*.

9.101 Complete the derivation of Eqs. (9.47), which express the parallel-axis theorem for mass products of inertia.

9.102 Determine the mass moment of inertia of the right circular cone of Sample Prob. 9.11 with respect to a generator of the cone.

9.103 Determine the mass moment of inertia of the rectangular prism of Sample Prob. 9.14 with respect to the diagonal *OF* of its base.

9.104 Determine the mass moment of inertia of the bent wire of Probs. 9.94 and 9.98 with respect to the axis through *O* which forms equal angles with the *x*, *y*, and *z* axes.

9.105 Determine the mass moment of inertia of the forging of Sample Prob. 9.12 with respect to an axis through *O* characterized by the unit vector $\boldsymbol{\lambda} = \frac{2}{3}\mathbf{i} + \frac{1}{3}\mathbf{j} + \frac{2}{3}\mathbf{k}$.

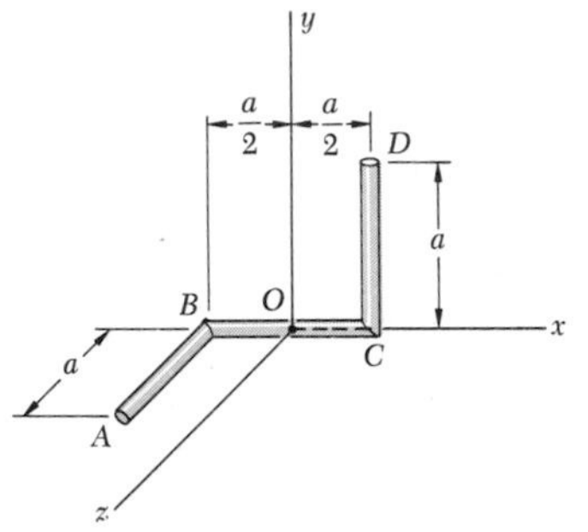

Fig. P9.106

9.106 Three uniform rods, each of mass *m*, are welded together as shown. Determine (*a*) the mass moments of inertia and the mass products of inertia with respect to the coordinate axes, (*b*) the mass moment of inertia with respect to a line joining the origin *O* and point *D*.

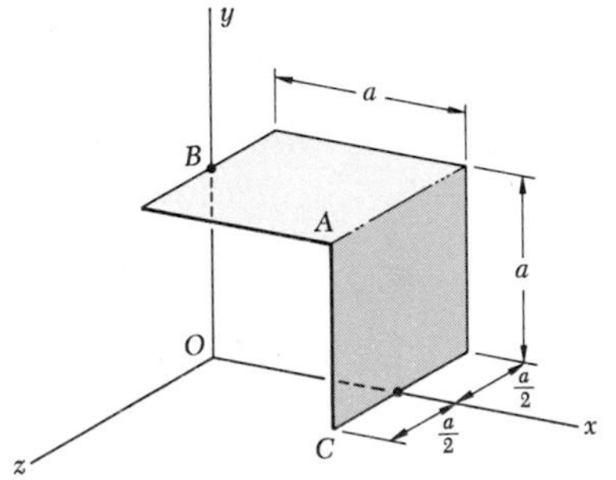

Fig. P9.107 and P9.108

9.107 The thin bent plate shown is of uniform density and mass *m*. Determine its mass moment of inertia with respect to a line joining the origin *O* and point *A*.

9.108 The thin bent plate shown is of uniform density and mass *m*. Determine its mass moment of inertia with respect to a line joining points *B* and *C*.

9.109 Consider a homogeneous circular cylinder of radius *a* and length *L*. Determine the value of the ratio a/L for which the ellipsoid of inertia of the cylinder is a sphere when computed (*a*) at the centroid of the cylinder, (*b*) at the center of one of its bases.

9.110 Determine the value of the ratio a/h for which the ellipsoid of inertia of the right circular cone of Sample Prob. 9.11 is a sphere when computed (*a*) at the apex of the cone, (*b*) at the centroid of the cone.

9.111 Given an arbitrary solid and three rectangular axes *x*, *y*, and *z*, prove that the mass moment of inertia of the solid with respect to any one of the three axes cannot be larger than the sum of the moments of inertia of the solid with respect to the other two axes; i.e., prove that the inequality $I_x \leq I_y + I_z$ is satisfied, as well as two similar inequalities. Further prove that, if the solid is homogeneous and of revolution, and if *x* is the axis of revolution and *y* a transverse axis, then $I_y \geq \frac{1}{2}I_x$.

9.112 Given a homogeneous solid of mass m and of arbitrary shape, and three rectangular axes x, y, and z of origin O, prove that the sum $I_x + I_y + I_z$ of the mass moments of inertia of the solid cannot be smaller than the similar sum computed for a sphere of the same mass and same material centered at O. Further prove, using the result of Prob. 9.111, that, if the solid is of revolution and if x is the axis of revolution, then its moment of inertia I_y about a transverse axis y must satisfy the inequality

$$I_y \geq \frac{3}{10} ma^2$$

where a is the radius of the sphere of the same mass and same material.

9.113 Consider a cube of mass m and side a. (*a*) Show that the ellipsoid of inertia at the center of the cube is a sphere, and use this property to determine the mass moment of inertia of the cube with respect to one of its diagonals. (*b*) Show that the ellipsoid of inertia at one of the corners of the cube is an ellipsoid of revolution, and determine the principal moments of inertia of the cube at that point.

Index

Index

Answers to Even-numbered Problems

CHAPTER 11

11.2 $t = 0$, $x = 12$ in., $a = -18$ in./s^2; $t = 3$ s, $x = -15$ in., $a = 18$ in./s^2. SI: $t = 0$, $x = 0.305$ m, $a = -0.457$ m/s^2; $t = 3$ s, $x = -0.381$ m, $a = 0.457$ m/s^2.

11.4 (a) 2 s, 4 s. (b) 8 m, 7.33 m.

11.6 -4 m/s; 12 m; 20 m.

11.8 (a) 3 s. (b) 116 in., -56 in./s. (c) 65 in. SI: (a) 3 s. (b) 2.95 m, -1.422 m/s. (c) 1.651 m.

11.10 25 s^{-2}.

11.12 (a) 384 in^3/s^2. (b) 13.86 in./s. SI: (a) 6.29×10^{-3} m^3/s^2. (b) 0.352 m/s.

11.14 (a) 55.5 m. (b) Infinite.

11.16 142.7 ft/s. SI: 43.5 m/s.

11.18 (a) $v = \frac{kT}{\pi}\left(1 - \cos\frac{\pi t}{T}\right)$; $x = \frac{kT^2}{\pi^2}\left(\frac{\pi t}{T} - \sin\frac{\pi t}{T}\right)$. (b) $2kT/\pi$. (c) $2kT^2/\pi$. (d) kT/π.

11.20 (a) 15,540 ft. (b) 318 mi. (c) Infinite. SI: (a) 4740 m. (b) 511 km. (c) Infinite.

11.22 (a) 6.90 m/s. (b) Infinite.

11.24 (a) 5 m/s. (b) 11 m/s. (c) 60 m.

11.26 (a) 30.7 ft/s. (b) 98.2 ft/s. SI: (a) 9.34 m/s. (b) 29.9 m/s.

11.28 $t = 15$ s; $x = 450$ ft. SI: $t = 15$ s; $x = 137.2$ m.

11.30 (a) 17.10 s; 171.0 m. (b) 81.5 km/h.

11.32 (a) 36 ft/s ↓. (b) 18 ft/s ↓. (c) 54 ft/s ↓. (d) 36 ft/s ↓. SI: (a) 10.97 m/s ↓. (b) 5.49 m/s ↓. (c) 16.46 m/s ↓. (d) 10.97 m/s ↓.

11.34 (a) 200 mm/s ←. (b) 200 mm/s ←; 400 mm/s ←. (c) 100 mm/s ←. (d) 200 mm/s →.

11.36 (a) 3 s. (b) 3.38 in. ↑. SI: (a) 3 s. (b) 85.7 mm ↑.

11.38 $v_A = 120$ mm/s ↓; $v_B = 40$ mm/s ↓; $v_C = 80$ mm/s ↑.

11.40 $v_A = 200$ mm/s ↓; $v_B = 40$ mm/s ↓; $v_C = 120$ mm/s ↑.

11.42 (a) 32 ft/s. (b) 192 ft. SI: (a) 9.75 m/s. (b) 58.5 m.

11.44 (a) 48 m. (b) 6 s, 13.75 s, 16.25 s.

11.46 19 s.

11.48 (a) 2.67 ft/s^2. (b) 23.2 mi/h. SI: (a) 0.813 m/s^2. (b) 37.3 km/h.

11.50 11 s; 70 m.

11.52 8.54 s; 58.3 mi/h. SI: 8.54 s; 93.8 km/h.

11.54 (*a*) 8.57 s. (*b*) 1.867 m/s^2; 1.400 m/s^2. (*c*) 68.6 m; 51.4 m.

11.56 (*a*) 150 in./s. (*b*) 800 in. (*c*) 100 in./s. SI: (*a*) 3.81 m/s. (*b*) 20.3 m. (*c*) 2.54 m/s.

11.58 (*a*) 10.0 m/s; 27.4 m. (*b*) 13.9 m/s; 51.5 m.

11.60 (*a*) −756 in./s^2. (*b*) −880 in./s^2. SI: (*a*) −19.20 m/s^2. (*b*) −22.4 m/s^2.

11.64 (*a*) 2.7 s. (*b*) 48.6 ft. SI: (*a*) 2.7 s. (*b*) 14.81 m.

11.66 (*a*) 12 m. (*b*) 48 m.

11.68 (*a*) 2 s. (*b*) 2 m/s ←; 2.24 m/s^2 $\measuredangle$ 26.6°.

11.70 $\mathbf{v}$ = 2.22 ft/s $\measuredangle$ 34.2°; $\mathbf{a}$ = 2.22 ft/s^2 $\measuredangle$ 34.2°. SI: $\mathbf{v}$ = 0.678 m/s $\measuredangle$ 34.2°; $\mathbf{a}$ = 0.678 m/s^2 $\measuredangle$ 34.2°.

11.74 $v = \sqrt{c^2 + R^2p^2}$; $a = Rp^2$.

11.76 4.20 m/s $\leqslant v_0 \leqslant$ 6.64 m/s.

11.78 44.0 ft/s; 38.1 ft/s. SI: 13.40 m/s; 11.61 m/s.

11.80 12.43 ft. SI: 3.79 m.

11.82 26.6° or 63.4°.

11.84 15° or 75°.

11.86 14.83 ft. SI: 4.52 m.

11.88 23.2 mi $\measuredangle$ 17.8°. SI: 37.3 km $\measuredangle$ 17.8°.

11.90 (*a*) 56.3° from rear of truck. (*b*) 16.63 m/s.

11.92 9.98 m/s $\measuredangle$ 81.2°.

11.94 22.4 mi/h from 63.4° east of north. SI: 36.0 km/h from 63.4° east of north.

11.96 10.18 m/s $\measuredangle$ 10.8°; 9.81 m/s^2 ↓.

11.98 (*a*) 2.08 m/s^2. (*b*) 63.6 km/h.

11.100 29.6 × 10^3 ft/s^2. SI: 9.02 × 10^3 m/s^2.

11.102 8.51 ft/s^2. SI: 2.59 m/s^2

11.104 1.2 m/s^2.

11.106 3810 m.

11.108 22,800 ft; 58,100 ft. SI: 6.96 km; 17.71 km.

11.110 $\rho = R + c^2/Rp^2$.

11.112 17,060 mi/h. SI: 27 400 km/h.

11.114 84.4 min.

11.116 (*a*) $\mathbf{v} = -(180\ \text{mm/s})\mathbf{i}_r$; $\mathbf{a} = -(240\ \text{mm/s}^2)\mathbf{i}_r - (4320\ \text{mm/s}^2)\mathbf{i}_\theta$.

11.118 (*a*) $\mathbf{v} = -4\pi b\mathbf{i}_r + 4\pi b\mathbf{i}_\theta$; $\mathbf{a} = -8\pi^2 b\mathbf{i}_r - 16\pi^2 b\mathbf{i}_\theta$. (*b*) $\mathbf{v} = 0$; $\mathbf{a} = 8\pi^2 b\mathbf{i}_r$.

11.120 $v = b \sec^2\theta\ \dot{\theta}$.

11.122 (*a*) $\mathbf{v} = bk\,\mathbf{i}_\theta$; $\mathbf{a} = -\frac{1}{2}bk^2\mathbf{i}_r$. (*b*) $\mathbf{v} = 2bk\,\mathbf{i}_r + 2bk\,\mathbf{i}_\theta$; $\mathbf{a} = 2bk^2\,\mathbf{i}_r + 4bk^2\,\mathbf{i}_\theta$.

11.124 $v = 2\pi\sqrt{A^2 + B^2n^2\cos^2 2\pi nt}$; $a = 4\pi^2\sqrt{A^2 + B^2n^4\sin^2 2\pi nt}$.

11.126 $v = h\tan\beta\sqrt{4\pi^2t^2 + \csc^2 2\pi t}$. $a = 4\pi h\tan\beta\sqrt{1 + \pi^2t^2}$.

11.128 $\tan^{-1}(Rp/c)$.

11.130 (*a*) $(\dot{x}\ddot{x} + \dot{y}\ddot{y} + \dot{z}\ddot{z})/(\dot{x}^2 + \dot{y}^2 + \dot{z}^2)^{1/2}$

$$(b)\ \left[\frac{(\dot{x}\ddot{y} - \dot{y}\ddot{x})^2 + (\dot{y}\ddot{z} - \dot{z}\ddot{y})^2 + (\dot{z}\ddot{x} - \dot{x}\ddot{z})^2}{\dot{x}^2 + \dot{y}^2 + \dot{z}^2}\right]^{1/2}$$

$$(c)\ \frac{[\dot{x}^2 + \dot{y}^2 + \dot{z}^2]^{3/2}}{[(\dot{x}\ddot{y} - \dot{y}\ddot{x})^2 + (\dot{y}\ddot{z} - \dot{z}\ddot{y})^2 + (\dot{z}\ddot{x} - \dot{x}\ddot{z})^2]^{1/2}}.$$

11.132 4.28 m/s; 0.188 m.

11.134 (*a*) 10 m. (*b*) 0.0693 s. (*c*) −1000 m/s^2.

11.136 398 m.

11.138 1609 ft. SI: 490 m.

11.140 (*a*) 60 s; 960 m. (*b*) 240 s; 5280 m.

11.142 0.816 s, later; 276 ft below ground. SI: 0.816 s, later; 84.0 m below ground.

CHAPTER 12

12.2 W = 16.49 lb; m = 100.00 lb; m = 3.11 lb · s^2/ft. SI: W = 73.4 N; m = 45.36 kg.

12.4 3.22 ft/s^2; 20 lb. SI: 0.981 m/s^2; 9.07 kg.

12.6 (*a*) 65.9 ft/s. (*b*) 2.73 s. SI: (*a*) 20.1 m/s. (*b*) 2.73 s.

12.8 (*a*) 3.37 m/s. (*b*) 10.28 m.

12.10 23.9 N.

12.12 (*a*) $\mathbf{a}_A = \mathbf{a}_B$ = 2.42 ft/s^2 ↙. (*b*) 1.160 lb ↗. SI: (*a*) $\mathbf{a}_A = \mathbf{a}_B$ = 0.739 m/s^2 ↙. (*b*) 5.16 N ↗.

12.14 (*a*) 180 N. (*b*) 26.4 kg.

12.16 (*a*) 0.956 m. (*b*) 1.064 m.

12.18 (*a*) 302 N. (*b*) 6.79 m/s ↑. (*c*) 1.346 m/s ↓.

12.20 (*a*) 24.9 lb →. (*b*) 7.96 lb. SI: (*a*) 110.6 N →. (*b*) 35.4 N.

12.22 (*a*) $\mathbf{a}_A$ = 8.92 ft/s^2 ←; $\mathbf{a}_B$ = 5.94 ft/s^2 ←. (*b*) 3.08 lb. SI: (*a*) $\mathbf{a}_A$ = 2.72 m/s^2 ←; $\mathbf{a}_B$ = 1.811 m/s^2 ←. (*b*) 13.69 N.

12.24 (*a*) 9.56 ft/s^2. (*b*) 19.71 ft/s^2. SI: (*a*) 2.91 m/s^2. (*b*) 6.00 m/s^2.

12.26 1.905 m/s.

12.28 (*a*) 10.73 ft/s^2 ←. (*b*) 18.67 lb. SI: (*a*) 3.27 m/s^2 ←. (*b*) 83.0 N.

12.30 (*a*) 4.56 m/s² ←. (*b*) 1.962 m/s² ←. (*c*) 2.60 m/s² →.
12.32 $a = (P/m)e^{-kt/m}$; $v = (P/k)[1 - e^{-kt/m}]$.
12.34 $a = -(kx/m)[1 - l/\sqrt{x^2 + l^2}]$.
12.36 $\mathbf{a}_A = 13.26$ ft/s² ↑; $\mathbf{a}_B = 1.894$ ft/s² ↓; $\mathbf{a}_C = 9.47$ ft/s² ↓. Block *C* strikes ground first. SI: $\mathbf{a}_A = 4.04$ m/s² ↑; $\mathbf{a}_B = 0.577$ m/s² ↓; $\mathbf{a}_C = 2.89$ m/s² ↓.
12.38 $\mathbf{a}_A = 0.577$ m/s² ↓; $\mathbf{a}_B = 2.89$ m/s² ↓; $\mathbf{a}_C = 4.04$ m/s² ↑.
12.40 $\mathbf{a}_A = 4.91$ m/s² ↑; $\mathbf{a}_B = 2.45$ m/s² ↓; $\mathbf{a}_C = 0$.
12.42 (*a*) 5.51 m/s. (*b*) 60.6°.
12.44 (*a*) 10.56 ft/s. (*b*) 7.32 lb. SI: (*a*) 3.22 m/s. (*b*) 32.6 N.
12.46 (*a*) $g \sin\theta$. (*b*) $\sqrt{2gl(\cos\theta - \cos\theta_0)}$. (*c*) $W(3 - 2\cos\theta_0)$. (*d*) 60°.
12.48 2.71 m/s.
12.50 *A*: 12.86 ft/s². *B*: 25.8 ft/s². *C*: 19.32 ft/s². SI: *A*: 3.92 m/s². *B*: 7.86 m/s². *C*: 5.89 m/s².
12.52 $v_{\min} = \sqrt{gr\tan(\theta - \phi)}$; $v_{\max} = \sqrt{gr\tan(\theta + \phi)}$.
12.54 22.5°.
12.56 10.36 ft/s. SI: 3.16 m/s.
12.58 $y = hx^2/b^2$.
12.60 $\delta = eVlL/mv_0^2 d$.
12.62 $\sqrt{eV/mv_0^2}$.
12.64 (*a*) $F_r = 4$ N, $F_\theta = 0$. (*b*) $F_r = -21.3$ N, $F_\theta = 21.3$ N.
12.66 (*a*) $F_r = -73.6$ lb, $F_\theta = 0$. (*b*) $F_r = -24.5$ lb, $F_\theta = -49.0$ lb. SI: (*a*) $F_r = -327$ N, $F_\theta = 0$. (*b*) $F_r = -109.1$ N, $F_\theta = -218$ N.
12.68 (*a*) 11.93 lb. (*b*) 2.98 lb. SI: (*a*) 53.0 N. (*b*) 13.26 N.
12.70 12.96 N.
12.72 $n = 0$: uniform circular motion; $n = 1$: uniform rectilinear motion.
12.74 (*a*) 24 in./s. (*b*) $\rho_A = \frac{2}{3}$ in., $\rho_B = 18$ in. SI: (*a*) 0.610 m/s. (*b*) $\rho_A = 16.93$ mm, $\rho_B = 457$ mm.
12.76 409×10^{21} lb · s²/ft or 13.17×10^{24} lb. SI: 5.97×10^{24} kg.
12.78 (*a*) 35 770 km or 22,230 mi. (*b*) 3070 m/s or 10,080 ft/s.
12.80 (*a*) 7.50 in./s. (*b*) Straight line. SI: (*a*) 0.1905 m/s.
12.82 2640 mi/h. SI: 4250 km/h.
12.84 (*a*) 6350 km/h. (*b*) 5940 km/h.
12.86 (*a*) $l_1^3 \sin^3\theta_1 \tan\theta_1 = l_2^3 \sin^3\theta_2 \tan\theta_2$. (*b*) 240 mm.
12.88 (*a*) 7910 ft/s. (*b*) 4800 ft/s. SI: (*a*) 2410 m/s. (*b*) 1462 m/s.
12.90 −30.4 m/s.
12.92 (*a*) 1537 km. (*b*) 4070 m/s. (*c*) 1.536.
12.94 (*a*) 5560 ft/s. (*b*) 61 ft/s. SI: (*a*) 1695 m/s. (*b*) 18.6 m/s.
12.96 45 h 30 min.
12.98 5 h 17 min.
12.100 79.7°.
12.102 197 ft/s. SI: 60 m/s.
12.106 (*a*) $\frac{1}{3}v_0$. (*b*) 75.9°.
12.108 (*a*) $v = R\sqrt{2g/r_0}\cos\frac{\theta}{2}$. (*b*) $\phi = \frac{1}{2}(\pi - \theta)$.
12.110 3.32 m.
12.112 (*a*) 9.91 ft/s² ↓. (*b*) 32.2 ft/s² ↓. SI: (*a*) 3.02 m/s² ↓. (*b*) 9.81 m/s² ↓.
12.114 3.39 m/s² ⦣ 60°.
12.116 $0.1438 v_t^2/g$.
12.118 (*a*) 35 200 km/h. (*b*) 5150 km/h.
12.120 (*a*) $\mathbf{a}_A = 0$; $\mathbf{a}_B = 1.591$ m/s² ↙. (*b*) $\mathbf{a}_A = \mathbf{a}_B = 0.643$ m/s² ↙.

CHAPTER 13

13.2 2.37 GJ.
13.4 (*a*) 3.37 m/s ↘. (*b*) 10.28 m.
13.6 8.72 ft/s. SI: 2.66 m/s.
13.8 14.40 N.
13.10 12.67 ft/s. SI: 3.86 m/s.
13.12 1.981 m/s.
13.14 (*a*) 9.27 ft/s. (*b*) 9.33 ft. SI: (*a*) 2.82 m/s. (*b*) 2.84 m.
13.16 10.99 ft/s. SI: 3.35 m/s.
13.22 34 in. SI: 0.86 m.
13.24 (*a*) 0.801 m/s. (*b*) 98.1 N.
13.26 19.67 in. SI: 0.500 m.
13.28 (*a*) 2.08 lb ⦣ 30°. (*b*) 2.83 lb ↑. SI: (*a*) 9.27 N ⦣ 30°. (*b*) 12.60 N ↑.
13.30 Loop 1: (*a*) $\sqrt{5gr}$ ←. (*b*) 3*W* →. Loop 2: (*a*) $\sqrt{4gr}$ ←. (*b*) 2*W* →.
13.32 1315 lb · in. SI: 148.5 J.
13.34 25,950 ft/s. SI: 7905 m/s.
13.36 14.13×10^3 km/h.
13.38 549 W; 628 W.
13.40 (*a*) 25.0 kW. (*b*) 6.13 kW.
13.42 (*a*) 8.18 hp. (*b*) 10.09 hp. SI: (*a*) 6.10 kW. (*b*) 7.52 kW.

13.44 (*a*) 55.2 kW. (*b*) 260 kW.
13.46 (*a*) 20.5 s; 701 ft. (*b*) 34.2 s; 1904 ft.
SI: (*a*) 20.5 s; 214 m. (*b*) 34.2 s; 580 m.
13.48 (*a*) 278 kW. (*b*) 6.43 km/h.
13.50 (*a*) $2\,kl^2\,(1 - \cos\theta)^2$. (*b*) $-mgl\sin\theta$.
13.54 (*b*) $V = (x^2 + y^2 + z^2)^{-1/2}$.
13.56 46.6 ft/s. SI: 14.21 m/s.
13.58 2.45 m/s.
13.60 (*a*) 4.71 m/s. (*b*) 4.03 m/s.
13.62 7.05 ft/s. SI: 2.15 m/s.
13.64 23.7 m/s.
13.66 104.9 N.
13.68 (*a*) 22.7 ft/s. (*b*) 7.75 ft.
SI: (*a*) 6.92 m/s. (*b*) 2.36 m.
13.70 6 *mg*.
13.72 1.600 in.; 24 lb. SI: 40.6 mm; 7.32 N.
13.74 36,700 ft/s. SI: 11.18 km/s.
13.76 (*a*) 0.943×10^6 ft · lb/lb.
(*b*) 0.447×10^6 ft · lb/lb.
SI: (*a*) 2.82 MJ/kg. (*b*) 1.336 MJ/kg.
13.80 (*a*) 1.155 m. (*b*) 5.20 m/s.
13.82 (*a*) 15.54 ft/s. (*b*) 5.18 ft/s. (*c*) 0.125 ft.
SI: (*a*) 4.74 m/s. (*b*) 1.579 m/s.
(*c*) 38.1 mm.
13.84 (*a*) 25.3 in. (*b*) 7.58 ft/s.
SI: (*a*) 0.643 m. (*b*) 2.31 m/s.
13.90 10,780 ft/s. SI: 3285 m/s.
13.92 8420 m/s; 74.4°.
13.94 5160 ft/s; 79.9°. SI: 1572 m/s; 79.9°.
13.96 $65.7° \leq \phi_0 \leq 114.3°$.
13.102 380 mi. SI: 610 km.
13.104 (*b*) $\frac{1}{3}\sqrt{6}\,v_{esc}$, $\frac{1}{2}\sqrt{2}\,v_{esc}$.
13.106 (*a*) and (*b*) 6 min 4 s.
13.108 (*a*) 11.42 s.
(*b*) $\mathbf{v} = -(125.5\text{ m/s})\mathbf{j} - (194.5\text{ m/s})\mathbf{k}$.
13.110 (*a*) 2.80 s. (*b*) 5.60 s.
13.114 (*a*) 38.9 s. (*b*) 10.71 kN *T*.
13.116 (*a*) 10.06 ft/s; 1.5 s. (*b*) 3 s.
SI: (*a*) 3.07 m/s; 1.5 s. (*b*) 3 s.
13.118 (*a*) 9.03 m/s. (*b*) 0.
13.120 (*a*) and (*b*) 111.1 kN.
13.122 48.4 lb ←, 188.0 lb ↓. SI: 215 N ←,
836 N ↓.
13.124 9.38 ft/s. SI: 2.86 m/s.
13.126 (*a*) 2 m/s ←. (*b*) $T_A = 3$J. $T_B = 9$J.
13.128 (*a*) 0.6 mi/h. (*b*) 4370 lb.
SI: (*a*) 0.966 km/h. (*b*) 19.45 kN.
13.130 0.742 m/s →.
13.132 (*a*) 135.6 N · s. (*b*) 108.5 N · s. (*c*) 368 J;
294 J.
13.134 (*a*) $\mathbf{v}_A = 1.125$ ft/s ←;
$\mathbf{v}_B = 13.875$ ft/s →. (*b*) 5.10 ft · lb.
SI: (*a*) $\mathbf{v}_A = 0.343$ m/s ←;
$\mathbf{v}_B = 4.23$ m/s →. (*b*) 6.91 J.
13.136 (*a*) $\mathbf{v}_A = 2.30$ m/s ←; $\mathbf{v}_B = 2.20$ m/s →.
(*b*) 2.84 J.
13.138 $\mathbf{v}_A = 3.50$ m/s ∠ 60°;
$\mathbf{v}_B = 4.03$ m/s ∠ 21.7°.
13.140 (*a*) $0.571\,\mathbf{v}_0$. (*b*) $1.333\,\mathbf{v}_0$.
13.144 (*a*) 0.943. (*b*) 28.4 in.; 15.08 in.
SI: (*a*) 0.943. (*b*) 0.722 m; 0.383 m.
13.146 (*a*) 0.883. (*b*) 11.30 in. SI: (*a*) 0.883.
(*b*) 0.287 m.
13.148 $\mathbf{v}_A = 0.721\,v_0$ ∠ 16.1°; $\mathbf{v}_B = 0.693\,v_0$ ←.
13.150 69.9°.
13.152 (*a*) 8.29 ft/s →. (*b*) 6.85 lb. (*c*) 1.068 ft.
SI: (*a*) 2.53 m/s →. (*b*) 30.5 N.
(*c*) 0.326 m.
13.154 2.57 in. SI: 65.3 mm.
13.156 (*a*) 34.7 mm. (*b*) 8.18 J.
13.158 (*a*) 8890 J. (*b*) 24 km/h.
13.160 (*a*) 9.32 ft · lb. (*b*) 8.10 ft · lb.
SI: (*a*) 12.63 J. (*b*) 10.98 J.
13.162 (*a*) Five. (*b*) 2 m/s →. (*c*) Same as original.
13.164 4.47 in. SI: 113.6 mm.
13.166 Impact at *A*: $\mathbf{v}_t = 1.333$ m/s →,
$\mathbf{v}_f = 0.333$ m/s →; impact at *B*: $\mathbf{v}_t = 0$,
$\mathbf{v}_f = 1$ m/s →.
13.168 5.29 m/s →.
13.170 4.89 ft. SI: 1.491 m.
13.172 317 N/m.

CHAPTER 14

14.2 (*a*) 5.20 km/h →. (*b*) 3.90 km/h →.
14.4 (*a*) 1670 ft/s →. (*b*) 1158 ft/s →.
SI: (*a*) 509 m/s →. (*b*) 353 m/s →.
14.6 (*a*) $v_x = 19.50$ ft/s, $v_y = 16.00$ ft/s.
(*b*) $-(1.118\text{ ft}\cdot\text{lb}\cdot\text{s})\mathbf{i}$.
SI: (*a*) $v_x = 5.94$ m/s, $v_y = 4.88$ m/s.
(*b*) $-(1.516\text{ kg}\cdot\text{m}^2/\text{s})\mathbf{i}$.
14.8 (*a*) $(3\text{ m})\mathbf{i} + (1.5\text{ m})\mathbf{j} + (1.5\text{ m})\mathbf{k}$.
(*b*) $(17\text{ kg}\cdot\text{m/s})\mathbf{i} + (19\text{ kg}\cdot\text{m/s})\mathbf{j} - (5\text{ kg}\cdot\text{m/s})\mathbf{k}$.
(*c*) $(2\text{ kg}\cdot\text{m}^2/\text{s})\mathbf{i} + (24.5\text{ kg}\cdot\text{m}^2/\text{s})\mathbf{j} + (25.5\text{ kg}\cdot\text{m}^2/\text{s})\mathbf{k}$.
14.10 $x = 780$ ft, $y = 17.55$ ft, $z = -21.0$ ft.
SI: $x = 238$ m, $y = 5.35$ m, $z = 6.40$ m.
14.12 $x = 100$ m, $y = -40.7$ m, $z = 16$ m.

14.14 $v_B = 4.57$ ft/s; $v_C = 5.78$ ft/s.
SI: $v_B = 1.393$ m/s; $v_C = 1.761$ m/s.
14.16 $v_A = 919$ m/s; $v_B = 717$ m/s;
$v_C = 919$ m/s.
14.22 9.55%
14.24 0.20%.
14.26 (*a*) $mv_0\mathbf{i}$; $\frac{3}{4}mlv_0\mathbf{k}$. (*b*) $\mathbf{v}_A = \frac{1}{4}v_0\mathbf{i} + \frac{3}{4}v_0\mathbf{j}$;
$\mathbf{v}_B = \frac{1}{4}v_0\mathbf{i} - \frac{1}{4}v_0\mathbf{j}$. (*c*) $\mathbf{v}_A = -\frac{1}{2}v_0\mathbf{i}$;
$\mathbf{v}_B = \frac{1}{2}v_0\mathbf{i}$.
14.28 $x = 181.7$ mm, $y = 0$, $z = 139.4$ mm.
14.30 $v_A = 1.500$ m/s; $v_B = 1.299$ m/s;
$v_C = 2.25$ m/s.
14.32 $\mathbf{v}_A = 34.3$ ft/s $\measuredangle$ 29.7°;
$\mathbf{v}_B = 17.59$ ft/s $\triangledown$ 40.1°.
SI: $\mathbf{v}_A = 10.46$ m/s $\measuredangle$ 29.7°;
$\mathbf{v}_B = 5.36$ m/s $\triangledown$ 40.1°.
14.34 (*a*) 13.00 ft/s →. (*b*) 10.82 ft/s $\triangledown$ 33.7°.
(*c*) $b = 8.33$ ft. SI: (*a*) 3.96 m/s →.
(*b*) 3.30 m/s $\triangledown$ 33.7°. (*c*) $b = 2.54$ m.
14.36 (*a*) 3 m/s $\measuredangle$ 36.9°.
(*b*) $\mathbf{H}_G = (4.80\text{ kg}\cdot\text{m}^2/\text{s})\mathbf{k}$; $T' = 48.0$ J.
(*c*) 0.600 m. (*d*) 20 rad/s.
14.38 $P_x = 800$ N; $P_y = 800$ N.
14.40 $\mathbf{B} = mv_0$ →; $\mathbf{C} = m\sqrt{2gh}$ $\measuredangle$ 30°.
14.42 $Q_1 = \frac{1}{2}Q(1 - \sin\theta)$; $Q_2 = \frac{1}{2}Q(1 + \sin\theta)$.
14.44 $\mathbf{C}_x = 83.3$ lb →, $\mathbf{C}_y = 30.3$ lb ↓,
$\mathbf{M}_C = 496$ lb · in. ↻. SI: $\mathbf{C}_x = 370$ N →,
$\mathbf{C}_y = 134.8$ N ↓, $\mathbf{M}_C = 56.1$ N · m ↻.
14.46 $\mathbf{C}_x = 475$ N ←, $\mathbf{C}_y = 675$ N ↑;
$\mathbf{D} = 865$ N →.
14.48 $\mathbf{C} = 321$ lb ↑; $\mathbf{D} = 479$ lb ↑.
SI: $\mathbf{C} = 1428$ N ↑; $\mathbf{D} = 2130$ N ↑.
14.50 (*a*) 10,250 lb. (*b*) 16,400 hp.
(*c*) 28,700 hp. SI: (*a*) 45.6 kN.
(*b*) 12.23 MW. (*c*) 21.4 MW.
14.52 (*a*) 26.4 kN. (*b*) 830 km/h.
14.54 (*a*) 1500 N. (*b*) 2500 N.
14.56 43.2 ft/s. SI: 13.17 m/s.
14.58 23.8 N.
14.60 24 rad/s ↺. (*b*) 0.400 N · m ↻.
14.62 216 rpm.
14.64 $\sin\theta = v^2/gL$.
14.66 $P = mv(v + gt)$.
14.68 qv.
14.70 $v = m_0v_0/(m_0 + qt)$;
$a = -m_0v_0q/(m_0 + qt)^2$.
14.72 (*a*) 40.3 lb/s. (*b*) 10.06 lb/s.
SI: (*a*) 18.26 kg/s. (*b*) 4.56 kg/s.
14.74 (*a*) 240 ft/s². (*b*) 960 ft/s².
SI: (*a*) 73.2 m/s². (*b*) 293 m/s².
14.76 (*a*) 6820 kg. (*b*) 341 s.
14.78 18,480 mi/h. SI: 29.7×10^3 km/h.
14.80 452,000 ft. SI: 137.9 km.
14.84 (*a*) 0.855 m/s². (*b*) 987 km/h.
14.86 (*a*) 5.54 ft/s. (*b*) 0.641 ft from *B*.
SI: (*a*) 1.687 m/s. (*b*) 0.1955 m from *B*.
14.88 (*a*) $\frac{2}{3}v$; (*BC*) 33.3%. (*b*) $\frac{2}{3}v$; (*BC*) 25%,
(*AB*) 8.33%.
14.90 (*a*) $\frac{1}{2}v_A$ →. (*b*) $\frac{1}{4}A\rho(1 - \cos\theta)v_A^3$.
(*c*) $2(V/v_A)[1 - (V/v_A)](1 - \cos\theta)$.
14.92 $\mathbf{C} = 89.3$ N ↓; $\mathbf{D} = 138.4$ N ↑.
14.94 (*a*) 10,560 lb; 1.922 ft below *B*.
(*b*) 8300 lb; 4.89 ft below *B*.
SI: (*a*) 47.0 kN; 0.586 m below *B*.
(*b*) 36.9 kN; 1.490 m below *B*.

CHAPTER 15

15.2 (*a*) −2.51 rad/s². (*b*) 18 000 rev.
15.4 (*a*) −2.42 rad/s². (*b*) 52 s.
15.6 $\mathbf{v}_H = -(32\text{ in./s})\mathbf{i} + (56\text{ in./s})\mathbf{k}$;
$\mathbf{a}_H = -(368\text{ in./s}^2)\mathbf{i} - (1040\text{ in./s}^2)\mathbf{j}$
$- (396\text{ in./s}^2)\mathbf{k}$.
SI: $\mathbf{v}_H = -(0.813\text{ m/s})\mathbf{i} + (1.422\text{ m/s})\mathbf{k}$;
$\mathbf{a}_H = -(9.35\text{ m/s}^2)\mathbf{i} - (26.4\text{ m/s}^2)\mathbf{j}$
$- (10.06\text{ m/s}^2)\mathbf{k}$.
15.8 $\mathbf{v}_E = (0.14\text{ m/s})\mathbf{i} - (0.48\text{ m/s})\mathbf{j}$
$- (0.96\text{ m/s})\mathbf{k}$; $\mathbf{a}_E = -(0.644\text{ m/s}^2)\mathbf{i}$
$+ (2.21\text{ m/s}^2)\mathbf{j} - (7.30\text{ m/s}^2)\mathbf{k}$.
15.10 $\mathbf{v}_C = -(1.8\text{ m/s})\mathbf{j} - (1.2\text{ m/s})\mathbf{k}$;
$\mathbf{a}_C = (7.8\text{ m/s}^2)\mathbf{i} - (12.6\text{ m/s}^2)\mathbf{j}$
$+ (7.2\text{ m/s}^2)\mathbf{k}$.
15.12 66,700 mi/h; 0.01947 ft/s².
SI: 107.3×10^3 km/h; 5.93 mm/s².
15.14 1.174 s; 7.05 rad/s.
15.16 (*a*) 2 rad/s ↻; 3 rad/s² ↺.
(*b*) 20 in./s² $\triangle$ 36.9°. SI: (*a*) 2 rad/s ↻;
3 rad/s² ↺. (*b*) 508 mm/s² $\triangle$ 36.9°.
15.18 (*a*) 10 rad/s. (*b*) $\mathbf{a}_B = 18$ m/s² ↓;
$\mathbf{a}_C = 6$ m/s² ↓.
15.20 3.49 s; 6.98 s; 13.96 s.
15.22 (*a*) $\boldsymbol{\alpha}_A = 4.19$ rad/s² ↺;
$\boldsymbol{\alpha}_B = 6.98$ rad/s² ↺. (*b*) 4.50 s.
15.24 $\alpha = bv^2/2\pi r^3$.
15.26 (*a*) 1.25 rad/s ↻. (*b*) 25 in./s $\triangle$ 60°.
SI: (*a*) 1.25 rad/s ↻.
(*b*) 635 mm/s $\triangle$ 60°.
15.28 (*a*) 2 rad/s² ↻.
(*b*) $\mathbf{v}_A = (80\text{ mm/s})\mathbf{i} + (440\text{ mm/s})\mathbf{j}$.

15.30 (*a*) $\mathbf{v}_B = -(160 \text{ mm/s})\mathbf{i} + (200 \text{ mm/s})\mathbf{j}$.
(*b*) $x = 220$ mm, $y = 80$ mm.

15.32 (*a*) $\boldsymbol{\omega}_A = \boldsymbol{\omega}_B = v/r$ ↻; $\boldsymbol{\omega}_C = v/2r$ ↺.
(*b*) $\mathbf{v}_D = 2v \rightarrow$; $\mathbf{v}_E = 0$;
$\mathbf{v}_F = \sqrt{2}v \measuredangle 45°$.

15.34 (*a*) 180 rpm ↻. (*b*) 2.83 m/s ↙.

15.36 (*a*) $\mathbf{v}_P = 0$; $\boldsymbol{\omega}_{BD} = 39.3$ rad/s ↺.
(*b*) $\mathbf{v}_P = 6.28$ m/s ↓; $\boldsymbol{\omega}_{BD} = 0$.
(*c*) $\mathbf{v}_P = 0$; $\boldsymbol{\omega}_{BD} = 39.3$ rad/s ↻.

15.38 $\boldsymbol{\omega}_{BD} = 2.94$ rad/s ↺; $\mathbf{v}_D = 31.8$ in./s ←.
SI: $\boldsymbol{\omega}_{BD} = 2.94$ rad/s ↺;
$\mathbf{v}_D = 0.807$ m/s ←.

15.40 $\boldsymbol{\omega}_{BD} = 1$ rad/s ↺; $\boldsymbol{\omega}_{DE} = 3$ rad/s ↺.

15.42 $\boldsymbol{\omega}_{BD} = 3.75$ rad/s ↻; $\boldsymbol{\omega}_{DE} = 2.25$ rad/s ↺.

15.44 (*a*) $\mathbf{v}_A = 10$ in./s →; $\boldsymbol{\omega}_{AC} = 0$.
(*b*) $\mathbf{v}_A = 45.4$ in./s →;
$\boldsymbol{\omega}_{AC} = 0.566$ rad/s ↻.
SI: (*a*) $\mathbf{v}_A = 0.254$ m/s →; $\boldsymbol{\omega}_{AC} = 0$.
(*b*) $\mathbf{v}_A = 1.152$ m/s →;
$\boldsymbol{\omega}_{AC} = 0.566$ rad/s ↻.

15.48 Vertical line intersecting zx plane at $x = 0$, $z = 9.34$ ft. SI: $x = 0$, $z = 2.85$ m.

15.50 (*a*) 2 rad/s ↺. (*b*) 12 in./s ←. (*c*) 9 in./s, wound.
SI: (*a*) 2 rad/s ↺. (*b*) 0.305 m/s ←.
(*c*) 0.229 m/s, wound.

15.52 (*a*) 0.6 rad/s ↻. (*b*) 24 mm/s →.

15.54 (*a*) 4 rad/s ↺. (*b*) 86.5 in./s $\measuredangle$16.1°.
SI: (*a*) 4 rad/s ↺. (*b*) 2.20 m/s $\measuredangle$16.1°.

15.56 (*a*) 6.67 rad/s ↻. (*b*) 2 m/s ←.
(*c*) 1.250 m/s $\measuredangle$36.9°.

15.58 $\cos^3\theta = b/l$.

15.60 (*a*) 2 rad/s ↺. (*b*) 18.33 in./s $\measuredangle$19.1°.
SI: (*a*) 2 rad/s ↺. (*b*) 0.466 m/s $\measuredangle$19.1°.

15.62 (*a*) 3 rad/s ↺. (*b*) On or inside a 2-in.-radius circle centered at a point 1.928 in. below *G*. SI: (*a*) 3 rad/s ↺.
(*b*) On or inside a 50.8-mm-radius circle centered at a point 49.0 mm below *G*.

15.64 (*a*) 0.9 rad/s ↻. (*b*) 144 mm/s ←.

15.66 Space centrode: Circle of 12-in. radius with center at intersection of tracks. Body centrode: Circle of 6-in. radius with center on rod at point equidistant from *A* and *B*.

15.76 (*a*) 0.5 rad/s² ↻. (*b*) 5.5 ft/s² ↑.
SI: (*a*) 0.5 rad/s² ↻. (*b*) 1.676 m/s² ↑.

15.78 (*a*) 0.4 m/s² ←. (*b*) 0.2 m/s² →.

15.80 $\mathbf{a}_C = 632$ m/s² ↑; $\mathbf{a}_D = 632$ m/s² $\measuredangle$60°.

15.82 (*a*) 10 in./s² ↑. (*b*) 18.36 in./s² $\measuredangle$60.6°.
(*c*) 42.2 in./s² $\measuredangle$22.3°.
SI: (*a*) 0.254 m/s² ↑.
(*b*) 0.466 m/s² $\measuredangle$60.6°.
(*c*) 1.071 m/s² $\measuredangle$22.3°.

15.84 $\boldsymbol{\omega}_{AB} = 0$; $\boldsymbol{\alpha}_{AB} = \frac{3}{2}\omega_0^2$ ↺; $\boldsymbol{\omega}_{BC} = \frac{1}{2}\omega_0$ ↺,
$\boldsymbol{\alpha}_{BC} = 0$.

15.86 (*a*) 157.0 m/s² ↑. (*b*) 592 m/s² ↓.

15.88 71.1 m/s² ↘.

15.90 (*a*) 0. (*b*) 2.67 rad/s² ↻.

15.92 (*a*) 3.46 rad/s² ↻.
(*b*) 15.59 in./s² $\measuredangle$30°.
SI: (*a*) 3.46 rad/s² ↻.
(*b*) 0.396 m/s² $\measuredangle$30°.

15.94 (*a*) 3.46 rad/s² ↻.
(*b*) 19.30 in./s² $\measuredangle$45.6°.
SI: (*a*) 3.46 rad/s² ↻.
(*b*) 0.490 m/s² $\measuredangle$45.6°.

15.96 1.814 m/s² $\measuredangle$60.3°.

15.98 $\omega = (v_B \sin\beta)/(l \cos\theta)$

15.100 (*a*) $v_B = r\omega\cos\theta$.
(*b*) $a_B = r\alpha\cos\theta - r\omega^2\sin\theta$.

15.102 $v_D = -2l\omega\sin\theta$;
$a_D = -2l\alpha\sin\theta - 2l\omega^2\cos\theta$.

15.104 $v_B = R\omega\sec^2\theta$;
$a_B = R\sec^2\theta(\alpha + 2\omega^2\tan\theta)$.

15.106 $\omega_{AB} = r\omega(a^2 + l^2 - 2al\cos\theta)^{1/2}/al\sin\theta$.

15.108 (*a*) $\omega = (v_A/b)\cos^2\theta$.
(*b*) $(\mathbf{v}_B)_x = (v_A L/b)\sin\theta\cos^2\theta$ ←,
$(\mathbf{v}_B)_y = v_A[(L/b)\cos^3\theta - 1]$ ↓.

15.110 (*a*) 2.58 rad/s ↻. (*b*) 19.75 in./s $\measuredangle$50°.
SI: (*a*) 2.58 rad/s ↻.
(*b*) 0.502 m/s $\measuredangle$50°.

15.112 $\boldsymbol{\omega}_{AP} = 1.958$ rad/s ↺; $\boldsymbol{\omega}_{BD} = 3.80$ rad/s ↺.

15.114 (*a*) $\boldsymbol{\omega}_{BD} = \omega$ ↺; $\mathbf{v}_{P/AH} = 0$;
$\mathbf{v}_{P/BD} = l\omega$ ↑. (*b*) $\boldsymbol{\omega}_{BD} = \omega$ ↺;
$\mathbf{v}_{P/AH} = 0.299 l\omega \measuredangle 15°$;
$\mathbf{v}_{P/BD} = 1.115\, l\omega \measuredangle 75°$.

15.116 $\mathbf{a}_1 = r\omega^2\mathbf{i} + 2u\omega\mathbf{j}$; $\mathbf{a}_2 = -2u\omega\mathbf{i} - r\omega^2\mathbf{j}$;
$\mathbf{a}_3 = (-r\omega^2 - u^2/r + 2u\omega)\mathbf{i}$;
$\mathbf{a}_4 = (r\omega^2 - 2u\omega)\mathbf{j}$.

15.118 (*a*) $\mathbf{v}_B = 735$ mm/s $\measuredangle$71.8°.
(*b*) $\mathbf{a}_B = 62.4$ mm/s² $\measuredangle$7.4°.

15.120 (*a*) $\mathbf{a}_B = (10.9 \text{ m/s}^2)\mathbf{j}$.
(*b*) $\mathbf{a}_D = -(0.1 \text{ m/s}^2)\mathbf{i} + (10.8 \text{ m/s}^2)\mathbf{j}$.
(*c*) $\mathbf{a}_E = (10.7 \text{ m/s}^2)\mathbf{j}$.

15.122 (*a*) 0.00582 ft/s² west.
(*b*) and (*c*) 0.00446 ft/s² west.
SI: (*a*) 1.773 mm/s² west.
(*b*) and (*c*) 1.358 mm/s² west.

15.124 $11.05\ \text{rad/s}^2 \circlearrowleft$.

15.126 (*a*) $476\ \text{ft/s}^2$.
(*b*) $307\ \text{ft/s}^2$.
SI: (*a*) $145.1\ \text{m/s}^2$.
(*b*) $93.7\ \text{m/s}^2$.

15.128 (*a*) $\boldsymbol{\omega}_{BD} = 2.4\ \text{rad/s} \circlearrowright$;
$\boldsymbol{\alpha}_{BD} = 34.6\ \text{rad/s}^2 \circlearrowright$.
(*b*) $\mathbf{v} = 1.342\ \text{m/s} \measuredangle 63.4°$;
$\mathbf{a} = 9.11\ \text{m/s}^2 \measuredangle 18.4°$.

15.130 $-120\ \text{mm/s}$.

15.132 (*a*) $\boldsymbol{\omega} = (2\ \text{rad/s})\mathbf{i} + (4\ \text{rad/s})\mathbf{j} + (3\ \text{rad/s})\mathbf{k}$.
(*b*) $\mathbf{v}_B = (5\ \text{in./s})\mathbf{i} - (10\ \text{in./s})\mathbf{j}$
$+ (10\ \text{in./s})\mathbf{k}$.
SI: (*b*) $\mathbf{v}_B = (127\ \text{mm/s})\mathbf{i}$
$- (254\ \text{mm/s})\mathbf{j} + (254\ \text{mm/s})\mathbf{k}$.

15.134 $\boldsymbol{\alpha} = (237\ \text{rad/s}^2)\mathbf{k}$.

15.136 $\boldsymbol{\alpha} = -(565\ \text{rad/s}^2)\mathbf{i} - (5\ \text{rad/s}^2)\mathbf{j}$.

15.138 (*a*) $\boldsymbol{\omega} = -(R\omega_1/r)\mathbf{i} + \omega_1\mathbf{j}$.
(*b*) $\boldsymbol{\alpha} = (R\omega_1^2/r)\mathbf{k}$.

15.140 $\omega_1 \cos 30°$.

15.142 (*a*) $\boldsymbol{\alpha} = (3\ \text{rad/s}^2)\mathbf{i} + (2.5\ \text{rad/s}^2)\mathbf{k}$.
(*b*) $\mathbf{a}_A = -(125\ \text{in./s}^2)\mathbf{i} + (50\ \text{in./s}^2)\mathbf{j}$
$+ (67.5\ \text{in./s}^2)\mathbf{k}$;
$\mathbf{a}_B = -(50\ \text{in./s}^2)\mathbf{i} + (170\ \text{in./s}^2)\mathbf{j}$
$- (180\ \text{in./s}^2)\mathbf{k}$.
SI: (*b*) $\mathbf{a}_A = -(3.18\ \text{m/s}^2)\mathbf{i}$
$+ (1.270\ \text{m/s}^2)\mathbf{j} + (1.715\ \text{m/s}^2)\mathbf{k}$;
$\mathbf{a}_B = -(1.270\ \text{m/s}^2)\mathbf{i} + (4.32\ \text{m/s}^2)\mathbf{j}$
$- (4.57\ \text{m/s}^2)\mathbf{k}$.

15.144 (*a*) $\boldsymbol{\omega} = -(4\ \text{rad/s})\mathbf{j} + (1.6\ \text{rad/s})\mathbf{k}$.
(*b*) $\boldsymbol{\alpha} = -(6.4\ \text{rad/s}^2)\mathbf{i}$.
(*c*) $\mathbf{v}_P = -(0.4\ \text{m/s})\mathbf{i} + (0.693\ \text{m/s})\mathbf{j}$
$+ (1.732\ \text{m/s})\mathbf{k}$.
$\mathbf{a}_P = -(8.04\ \text{m/s}^2)\mathbf{i} - (0.64\ \text{m/s}^2)\mathbf{j}$
$- (3.2\ \text{m/s}^2)\mathbf{k}$.

15.146 (*a*) $\boldsymbol{\alpha} = -\omega_1\omega_2\mathbf{j}$.
(*b*) $\mathbf{a}_P = -r\omega_2^2\mathbf{i} + 2r\omega_1\omega_2\mathbf{k}$.
(*c*) $\mathbf{a}_P = -r(\omega_1^2 + \omega_2^2)\mathbf{j}$.

15.148 (*a*) $\boldsymbol{\alpha} = -(150\ \text{rad/s}^2)\mathbf{k}$.
(*b*) $\mathbf{a} = -(225\ \text{in./s}^2)\mathbf{i} - (2400\ \text{in./s}^2)\mathbf{j}$.
SI: (*a*) $\boldsymbol{\alpha} = -(150\ \text{rad/s}^2)\mathbf{k}$.
(*b*) $\mathbf{a} = -(5.72\ \text{m/s}^2)\mathbf{i} - (61.0\ \text{m/s}^2)\mathbf{k}$.

15.150 (*a*) $\boldsymbol{\alpha} = -(8\ \text{rad/s}^2)\mathbf{k}$.
(*b*) $\mathbf{a}_C = (3.2\ \text{m/s}^2)\mathbf{i} - (0.8\ \text{m/s}^2)\mathbf{j}$.

15.152 $\mathbf{v}_B = (54\ \text{mm/s})\mathbf{i}$.

15.154 $\mathbf{v}_C = (32\ \text{in./s})\mathbf{j}$. SI: $\mathbf{v}_C = (0.813\ \text{m/s})\mathbf{j}$.

15.156 (*a*) $\boldsymbol{\omega} = (1.6\ \text{rad/s})\mathbf{i} + (15.2\ \text{rad/s})\mathbf{j}$
$- (3.2\ \text{rad/s})\mathbf{k}$.
(*b*) $\mathbf{v}_C = (32\ \text{in./s})\mathbf{j}$.
SI: (*b*) $\mathbf{v}_C = (0.813\ \text{m/s})\mathbf{j}$.

15.158 $\mathbf{v}_B = -(14.41\ \text{in./s})\mathbf{i} - (4.32\ \text{in./s})\mathbf{j}$.
SI: $\mathbf{v}_B = -(0.366\ \text{m/s})\mathbf{i} - (0.1098\ \text{m/s})\mathbf{j}$.

15.160 $\mathbf{a}_B = -(49.2\ \text{mm/s}^2)\mathbf{i}$

15.162 $\mathbf{a}_C = (1162\ \text{in./s}^2)\mathbf{j}$.
SI: $\mathbf{a}_C = (29.5\ \text{m/s}^2)\mathbf{j}$.

15.164 (*a*) $\mathbf{v}_D = (0.6\ \text{m/s})\mathbf{i} - (0.6\ \text{m/s})\mathbf{j}$
$+ (0.25\ \text{m/s})\mathbf{k}$.
(*b*) $\mathbf{a}_D = (3\ \text{m/s}^2)\mathbf{i} - (3.6\ \text{m/s}^2)\mathbf{k}$.

15.166 (*a*) $\mathbf{v}_D = -(20\ \text{in./s})\mathbf{i} - (34.6\ \text{in./s})\mathbf{j}$
$- (46.8\ \text{in./s})\mathbf{k}$. (*b*) $\mathbf{a}_D = -(652\ \text{in./s}^2)\mathbf{i}$
$+ (133.3\ \text{in./s}^2)\mathbf{j} + (360\ \text{in./s}^2)\mathbf{k}$.
SI: (*a*) $\mathbf{v}_D = -(0.508\ \text{m/s})\mathbf{i}$
$- (0.880\ \text{m/s})\mathbf{j} - (1.188\ \text{m/s})\mathbf{k}$.
(*b*) $\mathbf{a}_D = -(16.56\ \text{m/s}^2)\mathbf{i} + (3.39\ \text{m/s}^2)\mathbf{j}$
$+ (9.14\ \text{m/s}^2)\mathbf{k}$.

15.168 (*a*) $\mathbf{v}_D = (0.8\ \text{m/s})\mathbf{i} - (0.72\ \text{m/s})\mathbf{j}$
$+ (0.3\ \text{m/s})\mathbf{k}$. (*b*) $\mathbf{a}_D = (3\ \text{m/s}^2)\mathbf{i}$
$+ (2.4\ \text{m/s}^2)\mathbf{j} - (7.4\ \text{m/s}^2)\mathbf{k}$.

15.170 (*a*) $\mathbf{v}_P = -(1.701\ \text{m/s})\mathbf{i}$
$+ (5.95\ \text{m/s})\mathbf{j} - (3.12\ \text{m/s})\mathbf{k}$.
(*b*) $\mathbf{a}_P = -(4.29\ \text{m/s}^2)\mathbf{i}$
$- (0.201\ \text{m/s}^2)\mathbf{j} + (1.021\ \text{m/s}^2)\mathbf{k}$.

15.172 (*a*) $\boldsymbol{\omega} = \omega_1\mathbf{j} + \omega_2\mathbf{k}$; $\boldsymbol{\alpha} = \omega_1\omega_2\mathbf{i}$.
(*b*) $\mathbf{v}_B = r\omega_2\mathbf{j} - (R + r)\omega_1\mathbf{k}$;
$\mathbf{a}_B = -[(R + r)\omega_1^2 + r\omega_2^2]\mathbf{i}$.

15.174 (*a*) $\boldsymbol{\alpha} = -(0.314\ \text{rad/s}^2)\mathbf{k}$.
(*b*) $\mathbf{v}_B = (124.7\ \text{ft/s})\mathbf{k}$;
$\mathbf{a}_B = (25.0\ \text{ft/s}^2)\mathbf{i} - (395\ \text{ft/s}^2)\mathbf{j}$.
SI: (*b*) $\mathbf{v}_B = (38.0\ \text{m/s})\mathbf{k}$;
$\mathbf{a}_B = (7.63\ \text{m/s}^2)\mathbf{i} - (120.3\ \text{m/s}^2)\mathbf{j}$.

15.176 (*a*) $\boldsymbol{\alpha} = (200\ \text{rad/s}^2)\mathbf{k}$.
(*b*) $\mathbf{v}_D = -(1\ \text{m/s})\mathbf{j} - (2.4\ \text{m/s})\mathbf{k}$;
$\mathbf{a}_D = -(40\ \text{m/s}^2)\mathbf{i} + (44\ \text{m/s}^2)\mathbf{j}$
$- (10\ \text{m/s}^2)\mathbf{k}$.

15.178 $\mathbf{v}_A = -(18\ \text{in./s})\mathbf{j} + (160\ \text{in./s})\mathbf{k}$;
$\mathbf{v}_B = -(90\ \text{in./s})\mathbf{j} + (64\ \text{in./s})\mathbf{k}$;
$\mathbf{a}_A = -(1600\ \text{in./s}^2)\mathbf{j} - (360\ \text{in./s}^2)\mathbf{k}$;
$\mathbf{a}_B = -(880\ \text{in./s}^2)\mathbf{j} - (1000\ \text{in./s}^2)\mathbf{k}$.
SI: $\mathbf{v}_A = -(0.457\ \text{m/s})\mathbf{j} + (4.06\ \text{m/s})\mathbf{k}$;
$\mathbf{v}_B = -(2.29\ \text{m/s})\mathbf{j} + (1.626\ \text{m/s})\mathbf{k}$;
$\mathbf{a}_A = -(40.6\ \text{m/s}^2)\mathbf{j} - (9.14\ \text{m/s}^2)\mathbf{k}$;
$\mathbf{a}_B = -(22.4\ \text{m/s}^2)\mathbf{j} - (25.4\ \text{m/s}^2)\mathbf{k}$.

15.180 $\mathbf{v}_A = -(160\ \text{in./s})\mathbf{i} - (18\ \text{in./s})\mathbf{j}$;
$\mathbf{v}_B = -(40\ \text{in./s})\mathbf{i} + (24\ \text{in./s})\mathbf{k}$;
$\mathbf{a}_A = (360\ \text{in./s}^2)\mathbf{i} - (1600\ \text{in./s}^2)\mathbf{j}$;
$\mathbf{a}_B = -(400\ \text{in./s}^2)\mathbf{j} - (100\ \text{in./s}^2)\mathbf{k}$.
SI: $\mathbf{v}_A = -(4.06\ \text{m/s})\mathbf{i} - (0.457\ \text{m/s})\mathbf{j}$;
$\mathbf{v}_B = -(1.016\ \text{m/s})\mathbf{i} + (0.610\ \text{m/s})\mathbf{k}$;
$\mathbf{a}_A = (9.14\ \text{m/s}^2)\mathbf{i} - (40.6\ \text{m/s}^2)\mathbf{j}$;
$\mathbf{a}_B = -(10.16\ \text{m/s}^2)\mathbf{j} - (2.54\ \text{m/s}^2)\mathbf{k}$.

15.182 (*a*) $\mathbf{a}_B = (0.45 \text{ m/s}^2)\mathbf{j} - (1.979 \text{ m/s}^2)\mathbf{k}$.
(*b*) $\mathbf{a}_B = -(2.34 \text{ m/s}^2)\mathbf{i} + (0.346 \text{ m/s}^2)\mathbf{k}$.
(*c*) $\mathbf{a}_B = -(0.45 \text{ m/s}^2)\mathbf{j} + (2.67 \text{ m/s}^2)\mathbf{k}$.
15.184 $\mathbf{a}_B = -(3.03 \text{ m/s}^2)\mathbf{i} - (0.454 \text{ m/s}^2)\mathbf{k}$.
15.186 $\boldsymbol{\omega}_B = 40 \text{ rpm} \circlearrowright$; $\boldsymbol{\omega}_C = 20 \text{ rpm} \circlearrowleft$.
15.188 (*a*) $\boldsymbol{\alpha}_{AB} = \boldsymbol{\alpha}_{BC} = 0$;
$\boldsymbol{\alpha}_{DB} = 1.333 \text{ rad/s}^2 \circlearrowright$.
(*b*) $\mathbf{a}_A = 0.8 \text{ m/s}^2 \downarrow$; $\mathbf{a}_B = 0.4 \text{ m/s}^2 \downarrow$.
15.190 (*a*) $\mathbf{a}_1 = -(302 \text{ ft/s}^2)\mathbf{i} - (66.6 \text{ ft/s}^2)\mathbf{j}$.
(*b*) $\mathbf{a}_2 = -(59.2 \text{ ft/s}^2)\mathbf{i} + (190.6 \text{ ft/s}^2)\mathbf{j}$.
SI: (*a*) $\mathbf{a}_1 = -(91.9 \text{ m/s}^2)\mathbf{i} - (20.3 \text{ m/s}^2)\mathbf{j}$.
(*b*) $\mathbf{a}_2 = -(18.05 \text{ m/s}^2)\mathbf{i} + (58.1 \text{ m/s}^2)\mathbf{j}$.
15.192 (*a*) $\boldsymbol{\omega} = 1.996 \text{ rad/s} \circlearrowleft$;
$\boldsymbol{\alpha} = 1.068 \text{ rad/s}^2 \circlearrowright$.
(*b*) $\mathbf{v}_B = 5.63 \text{ m/s} \measuredangle 40°$;
$\mathbf{a}_B = 8.25 \text{ m/s}^2 \measuredangle 40°$.
15.194 $\boldsymbol{\omega} = 2.25 \text{ rad/s} \circlearrowleft$. $\boldsymbol{\alpha} = 23.3 \text{ rad/s}^2 \circlearrowright$.
15.196 $\mathbf{v}_B = 7.85 \text{ ft/s} \leftarrow$; $\mathbf{a}_B = 92.7 \text{ ft/s}^2 \rightarrow$.
SI: $\mathbf{v}_B = 2.39 \text{ m/s} \leftarrow$; $\mathbf{a}_B = 28.3 \text{ m/s}^2 \rightarrow$.

CHAPTER 16

16.2 (*a*) $5 \text{ m/s}^2 \leftarrow$. (*b*) $\mathbf{B} = 41.6 \text{ N} \uparrow$;
$\mathbf{C} = 36.9 \text{ N} \uparrow$.
16.4 (*a*) 3.75 lb. (*b*) $\mathbf{A} = 1.194 \text{ lb} \rightarrow$;
$\mathbf{B} = 1.194 \text{ lb} \leftarrow$.
SI: (*a*) 16.68 N. (*b*) $\mathbf{A} = 5.31 \text{ N} \rightarrow$;
$\mathbf{B} = 5.31 \text{ N} \leftarrow$.
16.6 (*a*) $0.297g$. (*b*) 5.
16.8 (*a*) $2.55 \text{ m/s}^2 \rightarrow$. (*b*) $h \leqslant 1.047$ m.
16.10 (*a*) $3710 \text{ N} \uparrow$. (*b*) $1411 \text{ N} \uparrow$.
16.12 (*a*) 25.8 ft/s^2. (*b*) 12.27 ft/s^2.
(*c*) 13.32 ft/s^2.
SI: (*a*) 7.85 m/s^2. (*b*) 3.74 m/s^2.
(*c*) 4.06 m/s^2.
16.14 (*a*) 43.2 kN. (*b*) $8.38 \text{ m/s}^2 \nwarrow$.
16.16 (*a*) 7.99 ft/s^2. (*b*) $N_B = 101.8$ lb,
$F_B = 10.18$ lb.
SI: (*a*) 2.43 m/s^2. (*b*) $N_B = 453$ N,
$F_B = 45.3$ N.
16.18 $F_{CE} = 8.72$ N *C*; $F_{DF} = 15.80$ N *C*.
16.20 $\mathbf{A} = 10.77 \text{ lb} \measuredangle 30°$;
$\mathbf{B} = 0.774 \text{ lb} \measuredangle 30°$.
SI: $\mathbf{A} = 47.9 \text{ N} \measuredangle 30°$;
$\mathbf{B} = 3.44 \text{ N} \measuredangle 30°$.
16.22 1381 N.
16.24 (*a*) $0.500g \measuredangle 30°$. (*b*) Platform:
$1.250g \measuredangle 30°$; block: $0.625g \downarrow$.
16.26 $V_B = -6.13$ N; $M_B = -3.07 \text{ N} \cdot \text{m}$.
16.30 $89.6 \text{ N} \cdot \text{m}$.
16.32 $45.1 \text{ rad/s}^2 \circlearrowright$.
16.34 (1) $19.62 \text{ rad/s}^2 \circlearrowleft$; $39.2 \text{ rad/s} \circlearrowleft$;
$19.81 \text{ rad/s} \circlearrowleft$. (2) $14.01 \text{ rad/s}^2 \circlearrowleft$;
$28.0 \text{ rad/s} \circlearrowleft$; $16.74 \text{ rad/s} \circlearrowleft$.
(3) $6.54 \text{ rad/s}^2 \circlearrowleft$; $13.08 \text{ rad/s} \circlearrowleft$;
$11.44 \text{ rad/s} \circlearrowleft$. (4) $10.90 \text{ rad/s}^2 \circlearrowleft$;
$21.8 \text{ rad/s} \circlearrowleft$; $10.44 \text{ rad/s} \circlearrowleft$.
16.36 (*a*) $5.66 \text{ ft/s}^2 \downarrow$. (*b*) $8.24 \text{ ft/s} \downarrow$.
SI: (*a*) $1.725 \text{ m/s}^2 \downarrow$. (*b*) $2.51 \text{ m/s} \downarrow$.
16.38 4.56 rad/s^2.
16.40 73.1 lb. SI: 325 N.
16.42 9.44 N.
16.44 (*a*) $\boldsymbol{\alpha}_A = 8.48 \text{ rad/s}^2 \circlearrowleft$;
$\boldsymbol{\alpha}_B = 39.2 \text{ rad/s}^2 \circlearrowleft$. (*b*) $\mathbf{C} = 66.7 \text{ N} \uparrow$,
$\mathbf{M}_C = 2.12 \text{ N} \cdot \text{m} \circlearrowleft$.
16.46 (*a*) $\boldsymbol{\alpha}_A = 51.5 \text{ rad/s}^2 \circlearrowleft$;
$\boldsymbol{\alpha}_B = 12.36 \text{ rad/s}^2 \circlearrowleft$.
(*b*) $\boldsymbol{\omega}_A = 343 \text{ rpm} \circlearrowleft$; $\boldsymbol{\omega}_B = 206 \text{ rpm} \circlearrowright$.
16.48 $I_R = \left(n + \dfrac{1}{n}\right)^2 I_0 + n^4 I_C$.
16.52 (*a*) $16.10 \text{ rad/s}^2 \circlearrowleft$. (*b*) $8.05 \text{ ft/s}^2 \rightarrow$.
(*c*) 12 in. from *B*.
SI: (*a*) $16.10 \text{ rad/s}^2 \circlearrowleft$. (*b*) $2.45 \text{ m/s}^2 \rightarrow$.
(*c*) 0.305 m from *B*.
16.54 (*a*) $-(2.37 \text{ rad/s}^2)\mathbf{j}$; 0.
(*b*) $-(1.778 \text{ rad/s}^2)\mathbf{j}$; $-(0.200 \text{ m/s}^2)\mathbf{i}$.
16.56 $T_A = 359$ lb; $T_B = 312$ lb.
SI: $T_A = 1595$ N; $T_B = 1388$ N.
16.58 $T_A = 1348$ N; $T_B = 1138$ N.
16.60 (*a*) $12 \text{ m/s}^2 \uparrow$. (*b*) $48 \text{ rad/s}^2 \circlearrowright$.
(*c*) $36 \text{ m/s}^2 \uparrow$.
16.62 (*a*) *W*. (*b*) $rg/\bar{k}^2 \circlearrowleft$.
16.64 (*a*) $3g/L \circlearrowright$. (*b*) $g \uparrow$. (*c*) $2g \downarrow$.
16.66 (*a*) $\bar{\mathbf{a}} = \mu g \leftarrow$; $\boldsymbol{\alpha} = 5\mu g/2r \circlearrowright$.
(*b*) $2v_0/7\mu g$. (*c*) $12v_0^2/49\mu g$.
(*d*) $\bar{\mathbf{v}} = 5v_0/7 \rightarrow$; $\boldsymbol{\omega} = 5v_0/7r \circlearrowright$.
16.68 $P = 4\mu W/\sqrt{58}$.
16.72 (*a*) 150 mm. (*b*) $125.0 \text{ rad/s}^2 \circlearrowright$.
16.74 (*a*) $3Pg/WL \circlearrowright$. (*b*) $\mathbf{A}_x = \frac{1}{2}P \leftarrow$,
$\mathbf{A}_y = W \uparrow$.
16.76 $\frac{1}{2}(m/l)\omega^2(l^2 - x^2)$.
16.78 (*a*) 1529 kg. (*b*) 2.90 mm.
16.80 (*a*) $4W/7 \uparrow$. (*b*) $3g/7 \uparrow$.
16.82 (*a*) $0.750g/l \circlearrowleft$. (*b*) $0.275g/l \circlearrowleft$.
16.84 (*a*) $20.6 \text{ rad/s}^2 \circlearrowright$. (*b*) $\mathbf{A}_x = 48.3 \text{ N} \leftarrow$,
$\mathbf{A}_y = 39.3 \text{ N} \uparrow$.
16.86 (*a*) $34.8 \text{ rad/s}^2 \circlearrowleft$.
(*b*) $\mathbf{A} = 66.6 \text{ lb} \measuredangle 60.9°$.
SI: (*b*) $\mathbf{A} = 296 \text{ N} \measuredangle 60.9°$.

16.88 (*a*) $3g/4L \circlearrowright$. (*b*) $\mathbf{N} = 13W/16 \uparrow$; $\mathbf{F} = 3\sqrt{3}W/16 \rightarrow$. (*c*) 0.400.
16.92 2.91 ft. SI: 0.887 m.
16.94 1.266 m.
16.96 (*a*) 24 rad/s$^2 \circlearrowright$; 3.84 m/s$^2 \rightarrow$. (*b*) 0.016.
16.98 (*a*) 8 rad/s$^2 \circlearrowleft$; 1.280 m/s$^2 \leftarrow$. (*b*) 0.220.
16.100 (*a*) Does not slide. (*b*) 23.2 rad/s$^2 \circlearrowright$; 15.46 ft/s$^2 \rightarrow$. SI: (*b*) 23.2 rad/s$^2 \circlearrowright$; 4.71 m/s$^2 \rightarrow$.
16.102 (*a*) Slides. (*b*) 12.88 rad/s$^2 \circlearrowleft$; 3.22 ft/s$^2 \leftarrow$. SI: (*b*) 12.88 rad/s$^2 \circlearrowleft$; 0.981 m/s$^2 \leftarrow$.
16.104 3.58 ft/s^2. SI: 1.091 m/s^2.
16.106 (*a*) $g/4r$. (*b*) $g\sqrt{2}/4 \measuredangle 45^\circ$.
16.108 23.6 rad/s$^2 \circlearrowleft$.
16.110 (*a*) 28.0 N. (*b*) $\mathbf{A} = 9.25$ N $\leftarrow$; $\mathbf{B} = 50.0$ N $\uparrow$.
16.112 (*a*) 8.18 rad/s$^2 \circlearrowright$. (*b*) $\mathbf{A} = 12.74$ N $\leftarrow$; $\mathbf{B} = 31.9$ N $\uparrow$.
16.114 (*a*) 13.23 rad/s$^2 \circlearrowright$. (*b*) $\mathbf{A} = 1.375$ lb $\uparrow$; $\mathbf{B} = 1.460$ lb $\measuredangle 30^\circ$. SI: (*b*) $\mathbf{A} = 6.12$ N $\uparrow$; $\mathbf{B} = 6.50$ N $\measuredangle 30^\circ$.
16.118 $\mathbf{A} = 105.9$ lb $\leftarrow$; $\mathbf{B} = 200$ lb $\rightarrow$. SI: $\mathbf{A} = 471$ N $\leftarrow$; $\mathbf{B} = 890$ N $\rightarrow$.
16.120 (*a*) $\boldsymbol{\alpha}_{AB} = 3.77$ rad/s$^2 \circlearrowright$; $\boldsymbol{\alpha}_{BC} = 3.77$ rad/s$^2 \circlearrowleft$. (*b*) $\mathbf{A}_x = 15.68$ N $\rightarrow$, $\mathbf{A}_y = 43.8$ N $\uparrow$; $\mathbf{C} = 30.2$ N $\uparrow$.
16.122 $\mathbf{A}_x = \frac{1}{4}mr^2\omega_0^2 \leftarrow$, $\mathbf{A}_y = 2mgr \uparrow$; $\mathbf{B}_x = 0$, $\mathbf{B}_y = mgr \downarrow$.
16.124 (1*a*) $a/3 \rightarrow$. (1*b*) $3d/2$. (2*a*) $2a/7 \rightarrow$. (2*b*) $7d/5$.
16.126 (*a*) 74.4 rad/s$^2 \circlearrowleft$. (*b*) 24.8 ft/s$^2 \downarrow$. SI: (*b*) 7.56 m/s$^2 \downarrow$.
16.128 13.82 N $\measuredangle 26.6^\circ$.
16.130 (*a*) 12.14 rad/s$^2 \circlearrowright$. (*b*) 11.21 m/s$^2 \measuredangle 30^\circ$. (*c*) 14.56 N $\measuredangle 60^\circ$.
16.132 $\boldsymbol{\alpha}_{AB} = 24.5$ rad/s$^2 \circlearrowright$; $\boldsymbol{\alpha}_{BC} = 122.7$ rad/s$^2 \circlearrowleft$.
16.134 (*a*) $\frac{1}{3}g \uparrow$. (*b*) $\frac{5}{3}g \downarrow$.
16.136 (On *AB*) 2.25 lb · ft $\circlearrowleft$. SI: (On *AB*) 3.05 N · m $\circlearrowleft$.
16.138 $V_{max} = \frac{1}{3}mg$ at *A*; $M_{max} = 4mgL/81$ at $\frac{1}{3}L$ to right of *A*.
16.140 (*a*) $\mathbf{a}_x = 0.3g \leftarrow$, $\mathbf{a}_y = 0.6g \downarrow$. (*b*) $\mathbf{a} = 0.630g \downarrow$.
16.142 (*a*) 21.5 ft/s$^2 \rightarrow$. (*b*) 15.95 ft/s$^2 \rightarrow$. SI: (*a*) 6.55 m/s$^2 \rightarrow$. (*b*) 4.86 m/s$^2 \rightarrow$.
16.144 (*a*) 1.634. (*b*) $0.1925m\omega^2r^3$.
16.146 6.33 in. SI: 160.8 mm.
16.148 $(v_0^2/2g)(\mu - \tan\theta)/\cos\theta\,(\frac{7}{2}\mu - \tan\theta)^2$.
16.150 (*a*) $\mathbf{a}_A = 2g/5 \leftarrow$; $\mathbf{a}_B = 2g/5 \downarrow$. (*b*) $\mathbf{a}_A = 2g/7 \leftarrow$; $(\mathbf{a}_B)_x = 2g/7 \leftarrow$, $(\mathbf{a}_B)_y = 2g/7 \downarrow$.

CHAPTER 17

17.2 71.6 N · m.
17.4 8.27 in. SI: 210 mm.
17.6 (*a*) 294 rpm. (*b*) 15.92 rev.
17.8 $\mathbf{v}_A = 1.293$ m/s $\uparrow$; $\mathbf{v}_B = 2.59$ m/s $\downarrow$.
17.12 61.8 rev.
17.14 338 N $\uparrow$.
17.16 (*a*) 2.40 rev. (*b*) 21.4 N $\swarrow$.
17.18 1.541 m.
17.20 (*a*) $1.074\sqrt{g/r}$. (*b*) $1.433mg \uparrow$.
17.22 (*a*) $\sqrt{\frac{4}{3}g(R - r)(1 - \cos\beta)}$. (*b*) $mg(7 - 4\cos\beta)/3$.
17.24 (*a*) $l/\sqrt{12}$. (*b*) $1.861\sqrt{g/l}$.
17.26 5.75 ft/s $\leftarrow$. SI: 1.752 m/s $\leftarrow$.
17.28 (*a*) 13.45 rad/s. (*b*) 20.4 rad/s.
17.30 6.55 ft/s $\leftarrow$. SI: 1.997 m/s $\leftarrow$.
17.32 (*a*) $\mathbf{v}_A = 1.922$ m/s $\downarrow$; $\mathbf{v}_B = 3.20$ m/s $\measuredangle 36.9^\circ$. (*b*) $\mathbf{v}_A = \mathbf{v}_B = 2.87$ m/s $\leftarrow$.
17.34 (*a*) $\mathbf{v}_A = 1.332$ m/s $\rightarrow$; $\mathbf{v}_B = 0.769$ m/s $\downarrow$. (*b*) $\mathbf{v}_A = 0$; $\mathbf{v}_B = 4.20$ m/s $\downarrow$.
17.36 14.63 rad/s $\circlearrowright$.
17.38 7.67 rad/s $\circlearrowleft$.
17.40 36.4°.
17.42 (*a*) Zero. (*b*) 188.5 W.
17.44 (*a*) 0.365 lb · ft. (*b*) 1.824 lb · ft. SI: (*a*) 0.495 N · m. (*b*) 2.47 N · m.
17.46 89.7 N · m.
17.48 1.000.
17.50 3.88 s.
17.54 (*a*) 3.33 N · m. (*b*) $\boldsymbol{\omega}_A = 23.5$ rad/s $\circlearrowright$; $\boldsymbol{\omega}_B = 39.2$ rad/s $\circlearrowleft$.
17.56 (*a*) 6.59 s. (*b*) 13.06 lb; 1.944 lb. (*c*) 0.61. SI: (*b*) 58.1 N; 8.65 N.
17.62 (*a*) 12 m/s $\rightarrow$. (*b*) 100 N $\leftarrow$.
17.64 (*a*) 32.2 ft/s $\rightarrow$. (*b*) Zero. SI: (*a*) 9.81 m/s $\rightarrow$. (*b*) Zero.
17.66 21.5 ft/s $\leftarrow$. SI: 6.54 m/s $\leftarrow$.
17.68 (*a*) $5\bar{v}_0/7 \rightarrow$. (*b*) $2\bar{v}_0/7\mu g$.
17.70 (*a*) $t_2 = 2r\omega_1/7\mu g$. (*b*) $\bar{\mathbf{v}}_2 = 2r\omega_1/7 \rightarrow$. $\boldsymbol{\omega}_2 = 2\omega_1/7 \circlearrowright$.
17.72 (*a*) 4.51 rad/s. (*b*) 9.09 ft · lb. SI: (*a*) 4.51 rad/s. (*b*) 12.32 J.

17.74 (a) 334 rpm. (b) −6.51 J.
17.76 (a) and (b) 5.71 rad/s.
17.78 24.4 rpm.
17.80 *Disk:* 287.4 rpm; *Arm:* 72.6 rpm.
17.82 $v_r = 3.97$ m/s; $v_\theta = 2.86$ m/s.
17.84 3.82 ft/s. SI: 1.164 m/s.
17.86 (a) 2.4 m/s →. (b) 3.6 kN →.
17.88 $\mathbf{v}_A = 1.920$ ft/s ←; $\mathbf{v}_B = 21.12$ ft/s →.
SI: $\mathbf{v}_A = 0.585$ m/s ←; $\mathbf{v}_B = 6.44$ m/s →.
17.90 (a) $\bar{\mathbf{v}}_1 = mv_0/M$ →; $\boldsymbol{\omega}_1 = mv_0/MR$ ↺. (b) $mv_0/3M$ →.
17.92 $\frac{1}{2}\omega_1$.
17.94 $\boldsymbol{\omega}_2 = \frac{1}{3}\omega_1$ ↻; $\bar{\mathbf{v}}_2 = \frac{2}{3}r\omega_1$ ↑.
17.96 $\boldsymbol{\omega} = \frac{3}{4}\bar{v}_1/b$ ↻; $\bar{\mathbf{v}} = \frac{3}{8}\sqrt{2}\,\bar{v}_1$ ∠45°.
17.98 (a) $\boldsymbol{\omega} = \frac{1}{4}\omega_0$ ↻. (b) $\frac{15}{16}$. (c) 1.5°.
17.100 (a) $\mathbf{v}_A = 0$, $\boldsymbol{\omega}_A = \omega_0$ ↻; $\mathbf{v}_B = v_0$ →, $\boldsymbol{\omega}_B = 0$. (b) $\mathbf{v}_A = 2v_0/7$ →, $\boldsymbol{\omega}_A = 2\omega_0/7$ ↻; $\mathbf{v}_B = 5v_0/7$ →, $\boldsymbol{\omega}_B = 5\omega_0/7$ ↻. (c) The motion of part *a* is the final motion.
17.102 $\boldsymbol{\omega}_2 = \dfrac{\bar{v}_1}{l}\dfrac{6\sin\beta}{3\sin^2\beta + 1}$ ↻.
17.104 (a) $0.9v_0/L$ ↻. (b) $0.1v_0$ →.
17.106 31.0° ⦣.
17.108 $\mathbf{A}\,\Delta t = m\sqrt{gl/3}$; $\mathbf{B}\,\Delta t = m\sqrt{gl/12}$.
17.110 $y^2 = (2v_0^2 \sin^2\theta/\mu g)x$.
17.112 (a) $\frac{5}{4}v_0/r$. (b) $1/\sqrt{3}$.
17.114 $\boldsymbol{\omega}_{AB} = \frac{5}{8}\omega_0$ ↻, $\bar{\mathbf{v}}_{AB} = \frac{1}{16}\omega_0 L$ ↑; $\boldsymbol{\omega}_{CD} = \frac{3}{8}\omega_0$ ↺, $\bar{\mathbf{v}}_{CD} = \frac{1}{16}\omega_0 L$ ↓.
17.116 $\sqrt{g/3r}$.
17.118 (a) 3.76 m/s ∠45°. (b) 3.18 m/s ↓.
17.120 (a) 50.2°. (b) 16.3°.
17.122 (a) 210 lb · ft. (b) 70.0 lb · ft.
SI: (a) 285 N · m. (b) 94.9 N · m.
17.124 (a) $0.926\sqrt{gL}$ ←. (b) $1.225\sqrt{gL}$ ←.
17.126 (a) 4.75 m/s →. (b) 3.87 m/s →.

CHAPTER 18

18.2 $\frac{1}{4}mr^2(\omega_1\mathbf{i} + 2\omega_2\mathbf{k})$.
18.4 $(0.432\ \text{kg}\cdot\text{m}^2/\text{s})\mathbf{i} - (0.324\ \text{kg}\cdot\text{m}^2/\text{s})\mathbf{k}$.
18.6 $(112.8\ \text{g}\cdot\text{m}^2/\text{s})\mathbf{i} + (80\ \text{g}\cdot\text{m}^2/\text{s})\mathbf{j}$.
18.8 $-(0.699\ \text{ft}\cdot\text{lb}\cdot\text{s})\mathbf{i} + (0.699\ \text{ft}\cdot\text{lb}\cdot\text{s})\mathbf{j}$.
SI: $-(0.947\ \text{kg}\cdot\text{m}^2/\text{s})\mathbf{i} + (0.947\ \text{kg}\cdot\text{m}^2/\text{s})\mathbf{j}$.
18.10 (a) $\frac{2}{3}ma^3\omega(5\mathbf{i} - 3\mathbf{k})$. (b) 31.0°.
18.14 (a) $m\mathbf{v} = (240\ \text{Mg}\cdot\text{m/s})\mathbf{i} + (360\ \text{Mg}\cdot\text{m/s})\mathbf{j} + (72 \times 10^3\ \text{Mg}\cdot\text{m/s})\mathbf{k}$;
$\mathbf{H}_G = (432\ \text{Mg}\cdot\text{m}^2/\text{s})\mathbf{j} + (180\ \text{Mg}\cdot\text{m}^2/\text{s})\mathbf{k}$.
(b) 67.1°.
18.16 (a) $-(3.86\ \text{ft/s})\mathbf{k}$; SI: $-(1.177\ \text{m/s})\mathbf{k}$.
(b) $-(0.643\ \text{rad/s})\mathbf{i} + (0.497\ \text{rad/s})\mathbf{j}$.
18.18 (a) $-(F\,\Delta t/m)\mathbf{k}$.
(b) $(12F\,\Delta t/7ma)(-\mathbf{i} - 5\mathbf{j})$.
18.20 (a) 0. (b) $(3F\,\Delta t/md)(\mathbf{i} - \frac{1}{4}\mathbf{k})$.
18.22 (a) $(6F\,\Delta t/7ma)(\mathbf{i} - 7\mathbf{j})$. (b) Axis through *A*, in *xy* plane, forming ∠81.9° with *x* axis.
18.24 (a) $\frac{1}{6}\omega_0(-\mathbf{i} + \mathbf{j})$. (b) $\frac{1}{6}\omega_0 a\mathbf{k}$.
18.26 (a) $\Delta t_A = 1.213$ s; $\Delta t_B = 0.558$ s.
(b) $\Delta\bar{\mathbf{v}} = (0.0886\ \text{m/s})\mathbf{k}$.
18.28 $(5.97\ \text{rpm})\mathbf{i} - (2.69\ \text{rpm})\mathbf{j} + (0.806\ \text{rpm})\mathbf{k}$.
18.32 $\sqrt{6g/5a}$.
18.34 0.864 J.
18.36 $-5ma^2\omega_0^2/48$.
18.38 −5.10 ft · lb. SI: −6.92 J.
18.40 $-\frac{1}{2}mr^2\omega_1\omega_2\mathbf{j}$.
18.42 $(1.296\ \text{N}\cdot\text{m})\mathbf{j}$.
18.44 $(0.864\ \text{N}\cdot\text{m})\mathbf{i} + (1.296\ \text{N}\cdot\text{m})\mathbf{j} - (0.648\ \text{N}\cdot\text{m})\mathbf{k}$.
18.46 $\mathbf{A} = (46.2\ \text{N})\mathbf{j}$; $\mathbf{D} = -(46.2\ \text{N})\mathbf{j}$.
18.48 $\mathbf{A} = \frac{1}{2}(w/g)a^2\omega^2\mathbf{k}$; $\mathbf{B} = -\mathbf{A}$.
18.50 (a) $\mathbf{M} = (0.647\ \text{lb}\cdot\text{ft})\mathbf{i}$.
(b) $\mathbf{A} = (0.388\ \text{lb})\mathbf{j}$; $\mathbf{B} = -(0.388\ \text{lb})\mathbf{j}$.
SI: (a) $\mathbf{M} = (0.877\ \text{N}\cdot\text{m})\mathbf{i}$.
(b) $\mathbf{A} = (1.727\ \text{N})\mathbf{j}$; $\mathbf{B} = -(1.727\ \text{N})\mathbf{j}$.
18.52 (a) $(4M_0/ma^2)\mathbf{j}$. (b) $\mathbf{R}_A = -(M_0\sqrt{2}/a)\mathbf{i}$; $\mathbf{M}_A = \frac{2}{3}M_0\mathbf{k}$.
18.54 (a) $\mathbf{M} = (4.00\ \text{N}\cdot\text{m})\mathbf{i}$.
(b) $\mathbf{A} = -(19.49\ \text{N})\mathbf{j} + (8.66\ \text{N})\mathbf{k}$; $\mathbf{B} = -\mathbf{A}$.
18.56 $-(0.831\ \text{N}\cdot\text{m})\mathbf{i}$.
18.58 22.7 lb · ft.
SI: 30.7 N · m.
18.60 (a) $2mr^3\omega^2$. (b) 0. (c) $\frac{1}{2}mr^3\omega^2$.
18.62 (a) $\cos\beta = 2g/3a\omega^2$. (b) $\sqrt{2g/3a}$.
18.64 $\mathbf{F} = -mR\omega_1^2\mathbf{i}$;
$\mathbf{M}_0 = \frac{1}{2}mr^2\omega_1\omega_2\mathbf{i} - mRh\omega_1^2\mathbf{k}$.
18.66 (a) $\sqrt{g/a}$. (b) $\sqrt{2g/a}$.
18.68 $\mathbf{D} = -(0.622\ \text{N})\mathbf{j} - (4.00\ \text{N})\mathbf{k}$.
$\mathbf{E} = (3.82\ \text{N})\mathbf{j} - (4.00\ \text{N})\mathbf{k}$.
18.70 4450 rpm.
18.72 3690 rpm.
18.76 (a) 36.1 rad/s; 7.40 rad/s. (b) −0.169.
18.78 (a) 2.75 rpm. (b) 2.77 rpm; 397 rpm.
18.84 Precession axis: ⦣30°; precession, 6.00 rad/s; spin, 10.39 rad/s.
18.86 Precession axis: $\theta_x = 39.9°$, $\theta_y = 127.9°$, $\theta_z = 79.4°$; precession, 4.38 rpm; spin, 2.61 rpm.

18.88 Precession axis: $\theta_x = 90°$, $\theta_y = 58.0°$, $\theta_z = 32.0°$; precession, 1.126 rpm (retrograde); spin, 0.343 rpm.
18.90 14.01 rev/h.
18.92 (*a*) $\omega = 22.7$ rpm, $\gamma = 57.3°$. (*b*) $\theta = 75.6°$. (*c*) Precession, 19.72 rpm; spin, 7.36 rpm.
18.94 (*a*) $\beta = 23.8°$. (*b*) Precession, 74.3 rpm; spin, 115.9 rpm.
18.96 $3\sqrt{g/2L}$.
18.100 $\dot{\phi} = 4\dot{\psi}_0/15$; $\dot{\psi} = 17\dot{\psi}_0/15$.
18.102 (*a*) $5\sqrt{3g/2a}$. (*b*) $\dot{\phi} = \sqrt{3g/2a}$; $\dot{\psi} = 5\sqrt{3g/2a}$.
18.108 (*a*) 27.8 rad/s². (*b*) $\mathbf{A} = (3.35\text{ N})\mathbf{j} + (12.08\text{ N})\mathbf{k}$; $\mathbf{C} = -\mathbf{A}$.
18.110 (*a*) $-\frac{3}{7}\bar{v}_0\left(\frac{1}{c}\mathbf{i} + \frac{1}{a}\mathbf{k}\right)$. (*b*) $-\frac{6}{7}\bar{v}_0\mathbf{j}$.
18.112 (*a*) $-(1.250\text{ m/s})\mathbf{k}$. (*b*) $(1.657\text{ rad/s})(\mathbf{i} + 3\mathbf{j})$.
18.114 (*a*) $(42.4\text{ rpm})\mathbf{j} + (64.2\text{ rpm})\mathbf{k}$. (*b*) 2800 ft · lb. SI: (*b*) 3790 J.
18.116 (*a*) Tangent of angle $= \dfrac{\tan\beta}{1 + 2\tan^2\beta}$. (*b*) $-2\dot{\psi}\sec\beta$. (*c*) 9.4°; $-2.03\dot{\psi}$.
18.118 (*a*) $F_{BC} = 0.789mg$; $T = 0.700mga$. (*b*) $\dot{\psi} = 13.66\sqrt{g/a}$; $T = 42.8mga$.

CHAPTER 19

19.2 (*a*) 0.1900 m. (*b*) 2.39 m/s.
19.4 (*a*) 2.49 mm; 0.0979 in. (*b*) 0.621 mm; 0.0245 in.
19.6 (*a*) 0.497 s. (*b*) 0.632 m/s. (*c*) 8.00 m/s².
19.8 (*a*) 0.1348 s. (*b*) 2.24 ft/s ↑; 20.1 ft/s² ↓. SI: (*b*) 0.683 m/s ↑; 6.13 m/s² ↓.
19.10 (*a*) 0.679 s; 1.473 Hz. (*b*) 0.1852 m/s; 1.714 m/s².
19.12 (*a*) 4.53 lb. (*b*) 0.583 s. SI: (*a*) 2.06 kg (mass). (*b*) 0.583 s.
19.14 (*a*) 0.994 m. (*b*) 3.67°.
19.16 1.400 ft. SI: 0.427 m.
19.18 (*a*) 7.90 lb. (*b*) 85.3 lb. SI: (*a*) 3.58 kg. (*b*) 38.7 kg.
19.20 1.904 Hz.
19.24 16.3°.
19.26 (*a*) 0.440 s. (*b*) 2.38 ft/s. SI: (*a*) 0.440 s. (*b*) 0.725 m/s.
19.28 (*a*) 0.907 s. (*b*) 0.346 m/s.
19.30 (*a*) 0.533 s. (*b*) 0.491 rad/s.
19.32 $f = (1/2\pi)\sqrt{3k/m}$.
19.38 (*a*) $l/\sqrt{12}$. (*b*) $4.77\sqrt{l/g}$.
19.40 (*a*) 2.28 s. (*b*) 1.294 m.
19.42 (*a*) $\tau = 2\pi\sqrt{5b/6g}$. (*b*) $c = \frac{1}{4}b$.
19.44 (*a*) $r_a = 7.09$ in. (*b*) 3.42 in. SI: (*a*) $r_a = 180.0$ mm. (*b*) 86.9 mm.
19.46 (*a*) 5.54 s. (*b*) 3.57 ft/s. SI: (*a*) 5.54 s. (*b*) 1.087 m/s.
19.48 0.658 kg · m².
19.50 $\tau = 2\pi\sqrt{l/g}$.
19.58 $\tau = 2\pi\sqrt{7l/6g}$.
19.62 $f = (1/2\pi)\sqrt{g/2l}$.
19.64 9.90 s.
19.68 $\tau = 2\pi\sqrt{m/3k\cos^2\beta}$.
19.70 $\tau = 2\pi\sqrt{m_c/k\cos^2\beta}$.
19.72 (*a*) 1.107 s. (*b*) 1.429 s.
19.74 (*a*) 0.777 s. (*b*) 1.099 s.
19.76 $\tau = \pi l/\sqrt{3gr}$.
19.78 $\omega > \sqrt{2k/m}$.
19.80 $\omega > \sqrt{2g/l}$.
19.82 (*a*) 168.0 rpm. (*b*) 0.00131 in. SI: (*a*) 168.0 rpm. (*b*) 33.3 μm.
19.84 (*a*) 11.38 μm. (*b*) 320 μm. (*c*) ∞.
19.86 0.0857 in. or 0.120 in. SI: 2.18 mm or 3.05 mm
19.88 1007 rpm.
19.90 1085 rpm.
19.92 109.5 rpm and 141.4 rpm.
19.94 1.200 mm.
19.96 70.1 km/h.
19.104 (*a*) $x = x_0 e^{-pt}(1 + pt)$. (*b*) 0.1146 s.
19.106 $\sqrt{1 - 2(c/c_c)^2}$.
19.108 (*a*) 1.509 mm. (*b*) 0.583 mm.
19.110 0.1791 in. SI: 4.55 mm.
19.112 (*a*) 270 rpm. (*b*) 234 rpm. (*c*) 8.84 mm; 9.45 mm.
19.114 $m\ddot{x}_A + 5kx_A - 2kx_B = 0$; $m\ddot{x}_B - 2kx_A + 2kx_B = P_m \sin\omega t$.
19.116 (*a*) E/R. (*b*) L/R.
19.122 (*a*) $m\ddot{x}_m + k_2 x_m + c(\dot{x}_m - \dot{x}_A) = 0$; $c(\dot{x}_A - \dot{x}_m) + k_1 x_A = 0$
(*b*) $L\ddot{q}_m + \dfrac{q_m}{C_2} + R(\dot{q}_m - \dot{q}_A) = 0$;
$R(\dot{q}_A - \dot{q}_m) + \dfrac{q_A}{C_1} = 0$.

19.124 (*a*) $m_1\ddot{x}_1 + c_1\dot{x}_1 + c_2(\dot{x}_1 - \dot{x}_2) + k_1x_1 + k_2(x_1 - x_2) = 0;$
$m_2\ddot{x}_2 + c_2(\dot{x}_2 - \dot{x}_1) + c_3\dot{x}_2 + k_2(x_2 - x_1) + k_3x_2 = P_m \sin \omega t.$
(*b*) $L_1\ddot{q}_1 + R_1\dot{q}_1 + R_2(\dot{q}_1 - \dot{q}_2) + \dfrac{q_1}{C_1} + \dfrac{q_1 - q_2}{C_2} = 0;$
$L_2\ddot{q}_2 + R_2(\dot{q}_2 - \dot{q}_1) + R_3\dot{q}_2 + \dfrac{(q_2 - q_1)}{C_2} + \dfrac{q_2}{C_3} = E_m \sin \omega t.$

19.126 1.363 s.

19.128 0.760 lb · ft · s^2; 8.66 in. SI: 1.030 kg · m^2; 0.220 m.

19.130 (*a*) 5.81 Hz; 4.91 mm; 179.2 mm/s. (*b*) 491 N. (*c*) 159.2 mm/s ↑ .

19.132 12.58 Hz.

19.134 72.5 μm.

19.136 1.346 s.

APPENDIX B

9.72 (*a*) $\frac{1}{4}ma^2$, $\frac{1}{4}mb^2$. (*b*) $\frac{1}{4}m(a^2 + b^2)$.

9.74 (*a*) $I_{AA'} = mb^2/24$; $I_{BB'} = mh^2/18$. (*b*) $I_{CC'} = m(3b^2 + 4h^2)/72$.

9.76 $m(3a^2 + L^2)/12$.

9.78 $\frac{1}{3}ma^2$; $a/\sqrt{3}$.

9.80 $5ma^2/18$

9.82 $m(2b^2 + h^2)/10$.

9.84 $2mr^2/3$.

9.86 1.514 kg · m^2; 155.7 mm.

9.88 (*a*) $md^2/6$. (*b*) $2md^2/3$. (*c*) $2md^2/3$.

9.90 (*a*) 5.14×10^{-3} kg · m^2. (*b*) 7.54×10^{-3} kg · m^2. (*c*) 3.47×10^{-3} kg · m^2.

9.92 0.0503 lb · ft · s^2; 3.73 in. SI: 0.0682 kg · m^2; 94.8 mm.

9.94 (*a*) 20.2 lb · ft · s^2. (*b*) 42.1 lb · ft · s^2. (*c*) 41.3 lb · ft · s^2. SI: (*a*) 27.4 kg · m^2. (*b*) 57.1 kg · m^2. (*c*) 56.0 kg · m^2.

9.96 $P_{xy} = -0.001199$ kg · m^2, $P_{yz} = P_{zx} = 0$.

9.98 $P_{xy} = 7.02$ lb · ft · s^2, $P_{yz} = P_{zx} = 0$. SI: $P_{xy} = 9.52$ kg · m^2, $P_{yz} = P_{zx} = 0$.

9.100 (*a*) $P_{zx} = mca/20$. (*b*) $P_{xy} = mab/20$; $P_{yz} = mbc/20$.

9.102 $3ma^2(a^2 + 6h^2)/20(a^2 + h^2)$.

9.104 29.9 lb · ft · s^2. SI: 40.5 kg · m^2.

9.106 (*a*) $I_x = 2ma^2/3$, $I_y = I_z = 11ma^2/12$, $P_{xy} = ma^2/4$, $P_{yz} = 0$, $P_{zx} = -ma^2/4$. (*b*) $2ma^2/3$.

9.108 $0.426\, ma^2$.

9.110 (*a*) 2. (*b*) 0.5.

Centroids of Common Shapes of Areas and Lines

Shape		$\bar{x}$	$\bar{y}$	Area
Triangular area	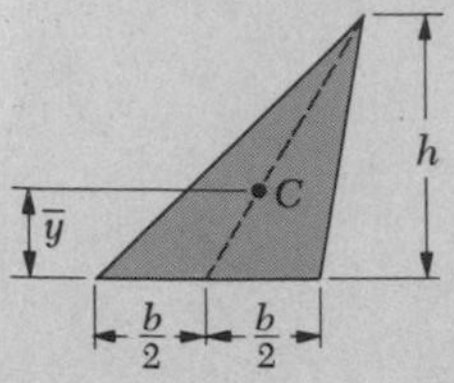		$\frac{h}{3}$	$\frac{bh}{2}$
Quarter-circular area	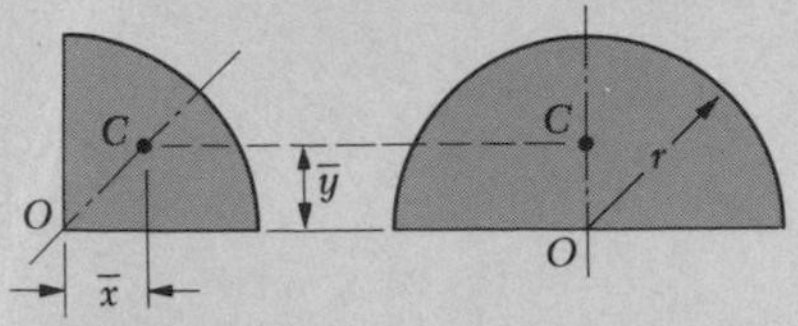	$\frac{4r}{3\pi}$	$\frac{4r}{3\pi}$	$\frac{\pi r^2}{4}$
Semicircular area		0	$\frac{4r}{3\pi}$	$\frac{\pi r^2}{2}$
Semiparabolic area	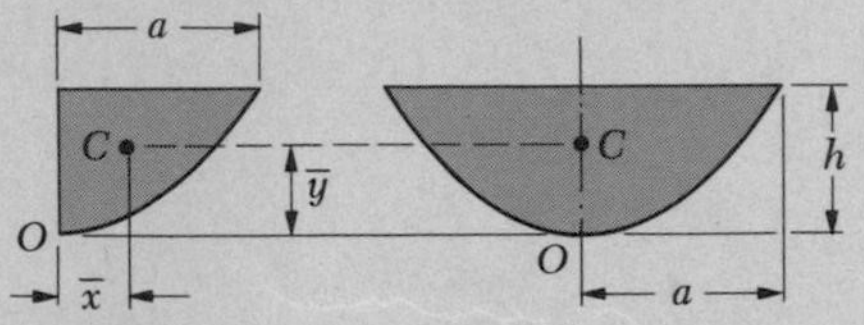	$\frac{3a}{8}$	$\frac{3h}{5}$	$\frac{2ah}{3}$
Parabolic area		0	$\frac{3h}{5}$	$\frac{4ah}{3}$
Parabolic spandrel	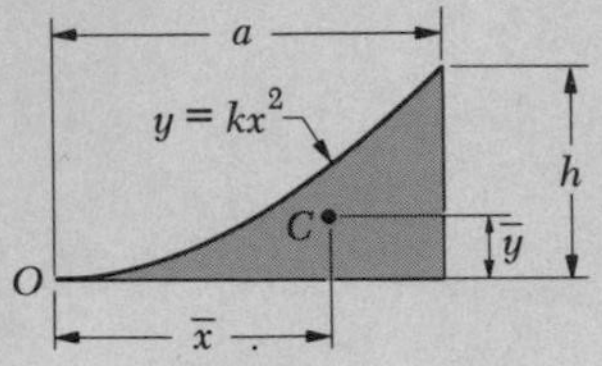	$\frac{3a}{4}$	$\frac{3h}{10}$	$\frac{ah}{3}$
Circular sector	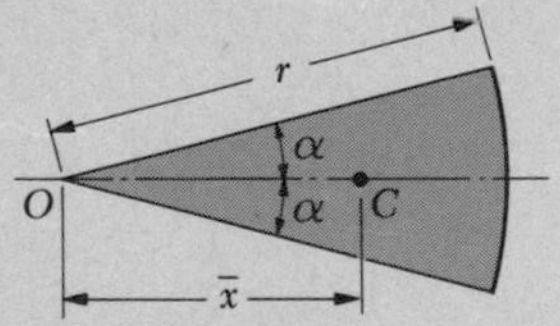	$\frac{2r \sin \alpha}{3\alpha}$	0	αr^2
Quarter-circular arc	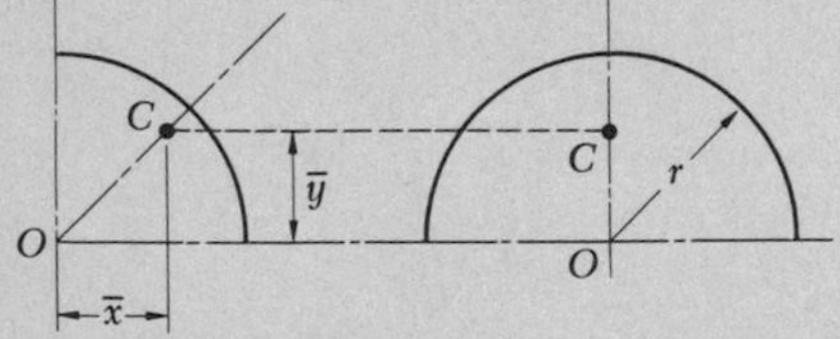	$\frac{2r}{\pi}$	$\frac{2r}{\pi}$	$\frac{\pi r}{2}$
Semicircular arc		0	$\frac{2r}{\pi}$	πr
Arc of circle	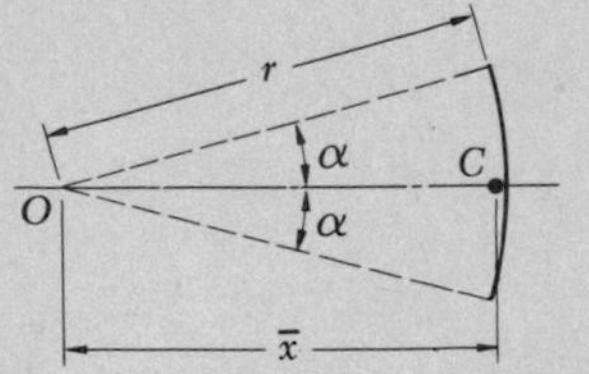	$\frac{r \sin \alpha}{\alpha}$	0	$2\alpha r$